GENES IN POPULATIONS

GENES IN POPULATIONS

Eliot B. Spiess

Professor of Biological Sciences
University of Illinois
at Chicago Circle

John Wiley & Sons
New York · Santa Barbara · London · Sydney · Toronto

Library of Congress Cataloging in Publication Data

Spiess, Eliot B
Genes in populations.

Bibliography: p.
Includes index.
1. Population genetics. I. Title.
QH455.S6779 575.1 77-3990
ISBN 0-471-81612-4

Printed in the United States of America

10 9 8 7 6 5 4 3 2 1

To the memory of
Theodosius Dobzhansky

PREFACE

From my experience at introducing students to exciting discoveries in experimental population genetics, it has become clear that they need a working knowledge of simple population dynamics and statics before they can grasp the full significance of observations made on populations. Study of current advances reported in the experimental literature or made in laboratory or field projects becomes meaningful only after fundamental principles are understood. This book, then, is intended as an introduction to the elementary dynamics of genes in populations. It is geared for those who have a knowledge of basic genetics and a facility in quantitative thinking. The coverage of topics may seem too "traditional" to those who are freshly trained as professional population geneticists; some of the exciting new experimental aspects may be omitted or too briefly discussed. For these shortcomings, I offer only this defense: students being introduced to this subject must become conversant in the "language" of the discipline. They will have no appreciation for arguments about forces at work on our populations unless they have a firm foundation in basic achievements already attained. For example, students may measure genetic polymorphism expressed as protein variation—a concept deceptively easy to grasp—but its causation and maintenance are of prime significance to us all. Without first exploring the principles of genetic equilibriums in populations, the consequences of high or low genetic variation cannot be understood.

Often, beginning students of genetics may fail to be excited by population genetics owing partly to their incomprehension and partly to the influence of the exciting, more publicized advances in molecular, physiological, and developmental genetics. This lack of interest generally has the following bases: first, an apprehension of mathematics and statistics; second, a lack of security with probability concepts compared with the more tangible phenomena of biology; third, a hearsay that mathematicians worked out the basic principles of population genetics long ago and that there is nothing further to discover in the field.

Although it is true that much of population theory has been developed by mathematicians, the basic principles and simple dynamics are comprehensible to genetics students with a rudimentary knowledge of algebra and statistics. However, to comprehend deeply, as with any discipline, special application to understanding the complexities is essential. In this book I introduce sufficient population genetics theory to disclose some of the main concepts. These concepts can be treated at an elementary level so that most biology students will find them useful in a populational analysis. Further probing by the student into either more mathematical or biological-experimental ramifications will thus be encouraged. The discovery of new population phenomena and new concepts from which to derive new principles cannot readily be appreciated by the student until an elementary understanding is achieved.

We are still a long way from being able to control genotypes of useful plants and animals, not to mention our ignorance of the human genotype. "Genetic surgery" is still very much in the future, although optimistic popularizers of molecular genetics often state the contrary. From a practical view, the matters of breeding improved plant and animal strains for human use, understanding medical genetic problems, comprehending the complexity of individual human genotypes, and living in a world using radioactivity, and thus increasing mutation rates in all organisms, behoove us to encourage more effort in experiment and understanding of the theory of genes in populations.

In modern times the trend toward practical application of principles is strong, but the need is clear for understanding the nature of this universe and the general principles associated with evolutionary mechanisms in the organic world. In the larger sense, those principles are associated with and derived from population genetic analysis. In an age when leisure time is becoming more available for the average person (although perhaps less available for teachers who must assimilate the explosion of knowledge), human curiosity could be more easily satisfied with understanding of such principles if a more elementary exposition were available.

Many books published in the last two decades have reduced the gap between the mathematical theoretician and the experimental geneticist. Consequently, population genetics has been brought closer to the comprehension of the average genetics student. Nonetheless, an elementary presentation aimed at understanding by the uninitiated is needed to help carry students into more sophisticated areas; only the most enlightened or energetic students will persist or struggle sufficiently with mathematical models to gain even a rudimentary knowledge of the field. Learning population genetics at such an advanced level does not serve those who need it most: the students in biological, medical, psychological, anthropological, sociological, and related fields.

The average sociologist's apprehension of (yet intuitive attraction to) population genetics is well illustrated in the following quotation from a popular book (*The Territorial Imperative*, Robert Ardrey, Atheneum, New York, 1966, p. 138):

> *Yet population genetics is a science so new, and so forbidding, in its theoretical complexities, that we tend to label it "Unfit for Human Consumption." It would be unwise, certainly, for you and me at this stage of inquiry to enter its mathematical labyrinth; we might never come out. Nevertheless if we are to grasp the profound link . . . , then we should knock around at the doors of population genetics even if we do not go inside.*

The material included in this textbook is based largely on lectures developed over several years at the University of Pittsburgh and at the University of Illinois at Chicago Circle. There is far more material than can be comfortably discussed in a short college term. Thus, students can, if they wish, pursue their study of the discipline beyond the limits of a formal course. The principles are illustrated wherever possible with either population data or numerical examples. Throughout illustrative examples with data from experiments are set off separately. The sequence is aimed at a logically increasing complexity of genetic mechanisms in natural and laboratory populations as well as their relevancy to human populations. Problems or exercises to help students become familiar with the principles illustrated are presented at the end of most chapters. Problem solving not only acquaints students with basic principles but also gives them facility in the "language" of population genetics and assists them to see relationships not apparent from words alone. Various tables and statistical supplements that are helpful for problem solving are presented in the Appendix. Examples and formulas are numbered within chapters according to the chapter number first.

On a more personal note, I credit C. C. Li and his remarkable facility for teaching biostatistics and population genetics for a considerable portion of the understanding I gained over the years 1952–1966. Few have his gift of clarity, and if this book achieves any success in elucidating the "mathematical labyrinth" of population genetics, it will be owing very much to the inspiration of Dr. Li, my colleague while at the University of Pittsburgh. I am indebted to him for his critical reading of Chapters 1 to 7.

Particularly, I am grateful to Jeffrey R. Powell and Max Levitan for their detailed comments on the entire book, and to three anonymous readers for constructive suggestions on Chapters 9 and 15. Many thanks are due to Paul D. Garst for computer programming in the simulations of stochastic changes in small populations (Chapters 12 to 14). Computing services were provided by the Computer Center of the University of Illinois at Chicago Circle, whose assistance is gratefully acknowledged. Richard C. Lewontin and his associates Jerry Coyne, Rama Singh, and Alex Felton kindly provided gel electrophoretic illustrations.

Finally, I owe my family—Luretta, Arthur, and Bruce—most gratitude for their persistent encouragement while I was writing this book as well as for their critical reading of many pages.

ELIOT B. SPIESS

CONTENTS

PAGE

GENES IN POPULATIONS

PART 1
INTRODUCTION AND HISTORY

One concept is of central importance to students beginning their study of genetics: the gene itself. The nature of the hereditary material, whether as a unit of segregation, recombination, mutation, or function, is the unifying idea basic to the field. Our knowledge of these self-reproducing particles has come mostly from detailed analysis of growth and reproduction of individual organisms; we know that genes are organized into larger units (chromosomes) and that they possess a remarkable ability to interact with specific environmental conditions in such a way as to direct biochemical pathways leading to individual phenotypes.

Many questions about heredity cannot be solved by studying individuals alone or by considering only the gene's organization and function, even though such knowledge is vital to our total understanding. Some legitimate questions that might be asked can be grouped according to three points of view: (1) the evolutionary (genetic change and its consequences), (2) the quantitative (breeding improved animals and plants for human needs), and (3) the medical-anthropological (organization of human genotypes).

EVOLUTIONARY QUESTIONS

Since the individual contains hundreds or thousands of genes acting together intricately, how did the total genotype evolve and at the same time maintain such remarkable integration that those genes act together "just right" in controlling growth and reproduction? How much change can the genotype stand? How can the genotype keep its balance and yet allow change to occur that may affect large numbers of individuals over many generations? To ask the question another way, what is the evolutionary significance of natural genetic variation? Answers to these questions cannot easily be obtained solely by examining detailed physiological and biochemical properties of genotypes. Of course, part of an answer does come from examination of primary gene products and their interaction. But genotypic changes taking place over many generations and affecting numerous individuals are changes we can discern only from a populational analysis. In the case of abnormal human hemoglobins, for example, we know many details of gene action: amino acid substitutions in the globin molecule, changes in the genetic code that bring about those substitutions, and the effect on the human individual when defective hemoglobin brings about collapse of red blood cells, as in sickle cell anemia. But why does the defective hemoglobin arise and then often persist in human populations? When a genic locus mutates, the question of the fate of its mutant allele in the future must be asked. Will it increase or decrease in frequency? Will it become incorporated into the entire species and thus bring about an evolutionary change? What are its chances of loss or fixation? What forces act either to utilize that mutation in transforming the genotype of the species or, in contrast, to prevent incorporation of the new heredity and thus help to preserve the genetic status quo?

These considerations lead us to ask another major question: how does the organism utilize its genetic variability in becoming adapted? Organisms must be able to carry on essential functions in specific environments. We also take note of the fact that every habitable environment contains organisms endowed with the ability to exploit that environment. Many environments would seem very adverse to us, but survival and exploitation can take place because of the remarkable ability of organisms to become adapted—to increase in their efficiency at living in their habitat. Many organisms have met "adversities" of deserts, polar and high mountain areas, deep oceans, underground areas in total darkness, salt lakes, boiling hot springs, antibiotic agents such as streptomycin, and even chronic radiation. How have organisms managed to adapt to such a great diversity of environments?

This adaptation, or being able to exploit an environment for successful living, is "the response of populations rather than of individuals who cannot react to the needs of a changing environment" (Li, 1955*b*, p. 251). The ability of a fish to live at great depths in the ocean or of a pine tree to live at timberline in high mountains takes a peculiar physiological system for each individual; this is not an ability that can be suddenly acquired merely by transporting any fish or any pine tree from the less to the more adverse environment. There are, of course, a few physiologically adaptable species with wide tolerances. Tolerance itself is an adaptive trait and is probably the result of selection in a cyclic and variable environment. But the acquisition of such a tolerance trait or any adaptive trait that allows a species to exploit a particular environment is not suddenly new for a single individual; it is a function of an entire population's adjustment. The individual is powerless to evolve. New genotypes that are adaptively superior to old ones usually become incorporated slowly over many generations. A populational genetic analysis is designed to describe the mechanics of that process and its predictability. More specifically, in such an analysis it becomes necessary to describe what genotypes exist in a population, how uniform or how variable they are, and then how the interbreeding group utilizes the genetic variation it has in becoming adapted.

QUANTITATIVE QUESTIONS

In an ever-growing world population, one of the chief jobs for the practical biologist is to answer the question: how can plants and animals useful to people be improved? Before the development of populational analysis, the answer was simply to "select and inbreed"; that is, to choose for propagation those parent phenotypes best suited for human needs, to select in one direction for a few generations, and to mate close relatives within lines hoping to establish uniform homozygous genotypes. It was soon discovered by breeders that inbreeding almost invariably leads to reduction in fertility, low survival, low yields, or depression of other traits related to reproductive potential ("Darwinian fitness"). With an understanding of genotypes of populations under natural conditions, the modern breeder finds it essential to maintain some genetic variability in stocks and herds instead of rigorously eliminating it by making all individuals homozygous (true breeding). Consequently, to improve a sizable portion of useful traits in a stock or herd, it has become indispensable to understand the dynamics of genes in populations, how genotypes arise and are maintained by a mating system, and how genotypes interact at a populational level, taking into account the evolutionary history of the species and how its total genotype is constantly being fashioned every generation.

A vast proportion of an organism's genome includes genes with minor, subtle, individually imperceptible phenotypic effects instead of genes with easily identified "major" effects, and special statistical methods have been devised to demonstrate their presence, magnitude, and predictability in terms of potential for improvement (Falconer, 1960). Artificial selection can thereby be made far more efficient at reaching useful goals. As a by-product, quantitative analysis of populations' genetic potential has led in turn to shedding more light on the problems of evolutionary and medical genetics.

MEDICAL-ANTHROPOLOGICAL QUESTIONS

Only a few years ago the human genotype was less well known than any of the classic genetic organisms, such as drosophila, neurospora, mouse, corn, bacteria, or virus, owing to the frequently stated limitations that "with people experimental crosses cannot be arranged and the length of a human generation is prohibitive" (Dobzhansky, 1965). What knowledge we did possess was due in no small measure to the contribution of population genetics as an indispensable supplement to pedigree analysis in establishing the modes of heredity for numerous traits (for example, the ability to taste phenylthiocarbamide as a simple Mendelian factor or the human blood antigens A and B as a series of multiple alleles). Answering questions of simple modality then was a classic function for populational analysis of human traits.

Recently, however, advances in biochemical genetics and current evolutionary population genetics have converged on a problem of major significance: how can we account for the origin and persistence of biochemical defects (the "inborn errors of metabolism") in human populations? A related question is: how many such defects can a population stand (the problem of genetic loads)? Advances in protein chemistry, deciphering the genetic code, and linking up protein structure with genetic architecture have opened a door to our comprehension of the human genotype to the point where at least the number of genetic entities being recognized and investigated eclipses the number in organisms such as drosophila. The reciprocal relationship between this extensive knowledge of human gene variation and the central problems and techniques of population genetics is now bringing the beginnings of answers to those problems. As important supporting information, our knowledge of mating systems, population structure, and the extent of chromosomal aberrations is becoming more extensive and refined.

From the biochemical refinement of gene structure, answers may be brought to bear on the question of human affinities, the retracement of evolutionary paths, and, perhaps, an entirely new physical anthropology. It appears that racial differences are mostly quantitative, not qualitative; that is, races differ in frequencies of particular alleles, not in the sharp presence or absence of particular genes. The question of how we should conceive of races, their mutual relationships, and their significance in human history and evolution is being resolved through the convergence of populational and biochemical genetics.

Finally, by gauging our rates of mutation and knowing more about our genotype structure and the dynamics of forces at work shaping it, we can postulate the future of humanity and propose programs for our genetic lot that might be better than what would come if we simply "let nature take its course."

MATHEMATICAL MODELS

To discover how populations are constructed and how predictions can be made for the genetic outcome, it is necessary to develop mathematical models determined by known factors. We can measure mutation rates, adaptive values, rates of migration, gene action potential, breeding systems, genetic architecture, population sizes, and so on, amassing experimental data; but unless this information can be generalized into a statement of theory from which deductions can be made concerning the outcome of future experiments, it is not useful.

A model represents a generalization constructed after acquiring some knowledge of the properties of a biological system. From those properties the model helps to deduce the behavior of the entire system or any portion thereof we desire to predict. By its nature, the model is a simplification because it is virtually impossible to incorporate all the complexities and bewildering heterogeneity of the real world. The model focuses attention on relevant and essential elements while omitting the distracting ones; it may thus run the risk of being far from reality. The elements, or properties of the system, that must be estimated—usually by experiment (or postulated for conceptual purposes)—are called *parameters* (Appendix A-1). A sufficient number of them are necessary for a working model. If a parameter of discrete individuals, or objects, takes a decimal value, as it may in an arithmetic mean, or average, of a population, it becomes unrealistic but still has significance to the model as a central tendency of the population. Besides simplifying reality, a model usually aims at a certain level of organization (such as a local population or a species), and while it may incorporate information in terms of parameters obtained at lower levels, it must define those parameters sufficient for the level of focus. For example, the relevent parameters to the genetic equilibrium known as the Hardy-Weinberg law are gene frequencies, the number of alleles at a genic locus, and random mating; we do not need to incorporate the structure of nucleic acid or chromosomes, genic action in protein synthesis, or the physiology of growth and development, although knowledge of those levels of course contributes to final reality.

Natural selection can be demonstrated as a driving force in evolution completely by biological experiments, but mathematical treatment is necessary to demonstrate how far selection can go, how rapidly it goes, what its limitations are, what kinds of genetic systems can be created by selection, and how it may interact with other forces.

Mathematical models may reveal the possibility of forces at work that were entirely unsuspected by the experimental biologist. The classic example is the principle emerging from a restriction of population size or isolation of populations by distance factors, leading to differentiation of gene pools (the "random genetic drift" concept of Sewall Wright). Just how critical these random effects may be in evolution has yet to be shown by the experimentalist, but that they *could* be critical is a mathematical fact, and we owe the principle to theoretic insight gained from models.

Population genetics has been a unique science to some extent because it was developed in the 1920s and 1930s by theoreticians who extended Mendel's and Darwin's principles by rigorous mathematical exploration of their consequences in the dynamics of genotypes within populations—at first via models based on the expectation from breeding systems, and later via models concerned with selection, mutation, and random effects. Only in recent years has empirical knowledge from observing genotypes in real populations, both natural and experimental, emerged and brought into better perspective the vast number of phenomena controlled by genes in populations and the magnitudes of forces acting on those genes.

Experimental population genetics has brought, for example, the demonstration of selective differences between genotypes, their magnitude, the subtlety of selective forces and how they may be aimed at creating new genotypes, magnitudes of integrating forces, stochastic forces, and non-Mendelian phenomena (cytoplasmic heredity and meiotic drive). Further mathematical generalization needs to be constructed taking these new observations into account. Sewall Wright (1960) has stressed the synthesis of observation with theory:

> *The role of the mathematical theory is that of an intermediary between bodies of factual knowledge discovered at two levels, that of the individual and that of the population. It must deduce from the postulates at the level of the individual and from models of population structure what is to be expected in populations, and then modify its postulates and models on the basis of any discrepancies with observation and so on. . . . Evolution is something that happens to* populations [*italics added*], *and without a mathematical theory, connecting the phenomena in populations with those in individuals, there could be no very clear thinking on the subject.*

HISTORICAL EMERGENCE OF POPULATION GENETICS

The student should appreciate that it is by examining development of ideas that we learn how significant problems have been approached, how solutions have been sought in the past, how concepts forming the discipline have been conditioned, and how we may profit by human experience in solving greater problems in the future. From the earliest days of plant and animal breeding, the study of heredity and variation has consistently featured a populational viewpoint; in turn, from studying the consequences of inbreeding and crossbreeding, the major theories governing behavior of genes in populations have developed and now serve our understanding of evolutionary mechanisms and human genetic phenomena. It would be misleading to think of the historical developments as being divided arbitrarily into periods, because all aspects of knowledge were actually developing simultaneously in continual interaction since the rediscovery of Mendel's laws in 1900. The emphasis in terms of significant literature has gone through about three stages: (1) genetic mechanisms in experimental breeding, (2) mathematical deductions derived from those mechanisms together with the principles of evolution synthesizing new population genetic concepts, and (3) genetic observations and empirical phenomena of real populations, reinforcing and extending the synthesis of those concepts (see Mayr, 1959; Wright, 1960; Haldane, 1964). These three stages have intermingled at times, as should be evident in the following historical account. At any one time and even at the present day, all three stages can be discerned—although not to an equal extent—in the population genetics literature.

A few accomplishments in the eighteenth century can be seen as the earliest careful experimentation with significant consequences for genetics and for population analysis.

1. Plant breeding dates from the time of Thomas Fairchild's first artificial plant hybrid (carnation × sweet william) in England in 1717, and Cotton Mather's hybrid (squash × gourd) in America at the same time, through about 50 years of plant breeding in Europe, all culminating in the publication of works by Josef Gottlieb Kölreuter (1761–1766) in Germany describing over 100 experiments in artificial hybridization, especially crosses, backcrosses, and F_2 generations between tobacco species.
2. Livestock breeding was advanced principally by Robert Bakewell in England when he took over the management of the Dishley Estate in 1760. In 35 years he laid the groundwork for development of many modern breeds of livestock, and proved to his satisfaction that inbreeding was the quickest way to fix the phenotype and that it was not always injurious.

Both Charles Darwin and Gregor Mendel were influenced by these early pioneers. Darwin drew attention to the effectiveness of artificial selection on altering the heredity of domesticated animals and plants and translated the process into natural selection as a basic mechanism for organic evolution. Mendel concentrated on the nature of hereditary determination. From those midcentury foundations to the twentieth century, two main lines of thought emerged that later helped form a synthetic basis for modern population genetics: first the quantitative, or biometric school led by Francis Galton in England, and second the inbreeding-crossbreeding group established by Mendel and led by William Bateson in England and William E. Castle in the United States.

Documentation of this progress can be found in Dunn (1965) and Provine (1971). For convenience and brevity, a chronology of the more significant advances in the formative days of population genetics is presented here:

1858. Charles Darwin's and Alfred Russell Wallace's *Essays on Natural Selection.*

1859. Charles Darwin's *Origin of Species* published in England.

1865. Gregor Mendel (Brünn, Austria, now Brno, Czechoslovakia) reported to the Natural History Society on plant hybridization, establishing the unitary nature of heredity determination, gametic segregation, and independent assortment of two or more pairs of hereditary units. His reports in the proceedings of the society (1866), although available to many scientists throughout Europe, were not recognized as significant until their rediscovery in 1900 by Hugo deVries, Carl Correns, and Erich von Tschermak-Seysenegg. Mendel's consideration of the outcome of self-pollination of frequencies of homozygotes and heterozygotes led directly to the theory of inbreeding; that is, he showed in the section of his paper entitled "The Subsequent Generations Bred from Hybrids" that after n generations of selfing (inbreeding), the genotypes' proportions would be $(2^n - 1)AA$: $2Aa$:

$(2^n - 1)aa$. This statement was the first specification of a population's genetic constitution under a system of mating. It should be noted that the statement of genetic equilibrium later known as the Hardy-Weinberg law is an application of the same principle to cross-breeding populations in which mating is at random.

1869. Francis Galton's *Hereditary Genius* was published. Galton, a cousin of Darwin, believed that by studying quantitative characters in populations, a general theory of heredity could emerge. Twenty years later his *Natural Inheritance* was published; its main theme included countless parent-offspring correlations (human stature, for example, was calculated to show about 0.33 correlation, but that figure was later revised to be 0.50); this, of course, indicated incomplete determination of heredity, but Galton pointed out the fundamental fact that "the characteristics of any population that is in harmony with its environments may remain statistically identical during successive generations." To help reconcile these facts, he showed that individuals with extreme differences from the average share with their relatives those differences, although the relatives are usually less extreme. In other words, parents with extreme phenotypes tend to have children closer to the average for the population, or children of extreme parents "regress" toward the population mean. The graphic expression of the linear relation between parents and offspring then came to be known as a "regression."

1894–1899. Karl Pearson, at first an economist but later a student of Galton, published methods for dealing statistically with frequency distributions, especially the standard deviation, variances, chi-square, and the significance of deviations between observed and theoretical outcome in experiments.

1900. The rediscovery of Mendel's contribution by deVries, Correns, and Tschermak was made after each, independently, had confirmed the principles by his own experimental observations. In the same year, Karl Landsteiner found that human blood could be classified according to agglutinating properties of red cells and serum, thus providing one of the best-known human hereditary characters.

1901. William Bateson in England published a translation of Mendel's paper, and also coined the terms *allelomorph*, *homozygote*, and *heterozygote*. The biometricians Galton, Pearson, and W. F. R. Weldon founded the journal *Biometrika*.

1902. Bateson defended Mendelism against attacks by Pearson and the biometricians who continued to assume wrongly that continuity of variation was an expression of "blending" heredity. Pearson attempted to prove that the observed parent-offspring correlations were quantitatively contradictory to the Mendelian expectation. (Four years later G. Udny Yule, a British biometrician, showed that Pearson's conclusions were based on the assumption of complete dominance, so that if gene action was incompletely dominant the Mendelian scheme would give such correlations.) Pearson's obstinancy lasted much longer, however, no doubt owing to personal quarrels with Bateson. Their dispute continued unabated at least until 1910 when the Mendelian nature of continuous heredity was demonstrated clearly.

1903. The *pure line* concept was proposed by Wilhelm Ludwig Johannsen in Denmark (see Chapter 6). Artificial selection for quantitative characters, he found, was effective in changing the mean of a population only as long as the population was genetically heterogeneous. Selection applied to most lines after three generations of selfing was ineffective, and variation observed within such stabilized "pure lines" must have been due to environmental factors. Phenotype and genotype were thereby defined: selection practiced on the phenotype is ineffective in changing the mean of a genetically homozygous line.

Bateson, deVries, Thomas Hunt Morgan, and their colleagues stressed mutation and qualitative gene differences as of prime importance to the evolutionary process. That group known as "mutationists" deemphasized the importance of Darwinian selection, and unfortunately Johannsen's report was immediately misinterpreted by them as showing how ineffective selection might be as a creative process. They pointed out that Johannsen's inbred lines represented genotypes already in existence in the original unselected population that were "merely sorted out by selection." Darwinism was thus dealt a severe blow and did not recover until about 20 years later.

In the same year William Ernest Castle of the United States published experimental data on the inheritance of coat color in mice showing their Mendelian rather than Galtonian (blending) basis and illustrating the principles of "gametic purity." In sections of his paper he explored the outcome of selection against recessive genes and thus made the first correct statement of genetic changes under complete selection against recessives (in fact, he criticized through misunderstanding Yule's attempt to do the same calculation). He went further to consider what would happen if selection ceased and only random mating continued in subsequent generations. His conclusion for populations that "as soon as selection is arrested the race remains stable at the degree of purity then attained" is in essence the same as the familiar Hardy-Weinberg law (Li, 1967*a*; Keeler, 1968).

1904. Pearson rejected Mendelism and at the same time correctly and inadvertently generalized the principle of segregation showing that the F_2 ratio of $\frac{1}{4}AA$: $\frac{1}{2}Aa$: $\frac{1}{4}aa$ will be maintained indefinitely in a randomly breeding large population (see Dunn, 1965; Wright, 1960, 1967). This statement was, like that of Castle's, a predecessor of the genetic equilibrium principle, but this one specifically applied only to a case of equal frequencies at one locus. Yule had actually shown the same rule two years before (see Wright, 1967).

1905. George Harrison Shull (at the Carnegie Institution of Washington Experimental Laboratory, Cold Spring Harbor, Long Island, New York) and Edward Murray East with H. K. Hayes (at the Bussey Institution, Harvard University, Jamaica Plain, Massachusetts) began inbreeding experiments in maize, opening the field of quantitative theory and applied plant breeding, influenced by Johannsen's pure line concept and using biometric ideas of Galton and Pearson.

1906. At the Bussey Institution, Castle and his entomology student, C. W. Woodworth, discovered that drosophila was easy to culture. They brought this little "gift to genetics" to the attention of T. H. Morgan, but continued to study inbreeding, crossbreeding, and selection with drosophila in addition to their selection experiments with hooded rats. In the U.S. Department of Agriculture, George Rommel, chief of the Animal Husbandry

Division, began inbreeding experiments with guinea pigs that were subsequently carried on in 1915 by Sewall Wright, another student of Castle.

1907. East demonstrated the resolution of a heterozygous population of maize into a series of pure lines as a result of artificial selection and inbreeding following Johannsen's concepts.

Ultimately, evolutionary theory was to benefit from another influence. In the same year, Vernon Kellogg published *Darwinism Today*, summarizing theories of evolution before Mendelism caught hold. Kellogg called attention to the studies of J. T. Gulick (1832–1923) on land snails (Achatinellinae) in certain Pacific Islands. Gulick, an American missionary, had found what appeared to be random differentiation of races in similar environments of deep valleys on these islands. Sewall Wright read the Kellogg acount in 1910 and later (1931) incorporated Gulick's ideas into his concept of random genetic drift, sometimes referred to as the Sewall Wright effect, but perhaps more properly (according to Wright) known as the Gulick effect.

1908. This was a crucial year for population genetics, although it was not recognized as such for at least another decade. The major contributions were two statements of genetic equilibrium published independently by Godfrey Harold Hardy, professor of mathematics at Cambridge University in England, and by Wilhelm Weinberg, a physician in Stuttgart, Germany. Hardy's paper, brief and concise, grew from a need for clarification when R. C. Punnett reported to Hardy that Yule had suggested as a criticism of Mendelism that a dominant gene should spread in a population at the expense of its recessive allele because dominants should tend to be distributed in a $\frac{3}{4}:\frac{1}{4}$ ratio. Nothing was farther from the truth. Actually, Punnett did not get his information straight—Yule had not said it that way—but Hardy, though "reluctant to intrude in a discussion concerning matters of which [he had] no expert knowledge," and expecting the simple point he was to make "to have been familiar to biologists," mentioned Pearson's stability of the 1:2:1 ratio and then contributed his own now-familiar statement for gene and zygote frequencies in a random mating population, generalized for a set of alleles at one locus. That paper was Hardy's sole contribution to genetics.

Weinberg, on the other hand, not only worked out the equilibrium principle for its utility in demonstrating a proof of Mendelian heredity in human families, but he also followed up his original contribution in later papers with extensions of the principle to independently assorting nonallelic loci, correlations between close relatives, and methods of partitioning variance between genic and environmental sources. Properly, then, we could say that Weinberg deserves the "father of population genetics" title if anybody does, because he developed for the first time and quite independently a theoretical set of principles applied to populations and derived from Mendelism. In a brief review of Hardy's paper in 1909, Weinberg commented that he himself had already proved the stability of population proportions "and in a simpler manner" than Hardy. Weinberg also laid the foundations for human genetics by realizing that simple Mendelian ratios are not often achieved in human families because of the "unavoidable fact of incomplete selection of such human families" in which normal parents are both heterozygotes (Stern, 1962). Only families with one recessive appearing can be ascertained completely, and he then invented methods of correcting for types of incomplete ascertainment: the sib method, proband method, and the a priori method. According to Stern (1962) "Weinberg's fate bears comparison with that of Mendel. . . . [He] had no colleagues who collaborated with him . . . no personal students . . . knew a few geneticists, among them F. Lenz (who treated the role of consanguinity in the appearance of recessive traits), but he remained outside the fold of most of his scientific contemporaries. . . . His most significant discoveries, those on population genetics, were overlooked and had to be made by others [Fisher and Wright in the early 1920s]."

In 1908 Shull pointed out the decrease in size and vigor when luxurious F_1 hybrids in his field of maize were inbred. Shull's belief that hybrids owed their vigor to their heterozygosity was a concept that led him later (1916) to coin the term *heterosis.* Along with Johannsen and East, he agreed that inbreeding would tend to separate lines into homozygous and weaker genotypes. The following year he suggested that first-generation hybrids between inbred lines could be used as a basis for practical corn breeding.

In Sweden, H. Nilsson-Ehle explained that seed color in wheat, a continuous "blending" character, could be determined by three Mendelian pairs of alleles with additive, nondominant effects ("multiple factors," later called "polygenes" by Kenneth Mather). This discovery helped to reconcile the blending hypothesis of Galton's biometricians with the Mendelian discrete unit of the mutationists, and it led to a basis for quantitative genetic principles and selection theory.

1909. In England, Archibald E. Garrod published *Inborn Errors of Metabolism,* a revision of the Croonian Lectures to the Royal Academy of Medicine delivered the year before. Also, the Galton Laboratory at the University of London was established from a bequest of Sir Francis to be the first laboratory for the study of human heredity.

1910. Thomas Hunt Morgan proposed sex linkage as a mechanism for white eye heredity in drosophila and was led into the concept of linear order of genes on chromosomes.

East summarized the reconciliation between biometric and Mendelian viewpoints, but much bitter antagonism in England prevented the emergence of population genetics there until after 1920.

1911. Sewall Wright heard Shull lecture at Cold Spring Harbor on inbreeding and crossbreeding in maize. "I recall that I was much impressed," said Wright, "though I cannot claim to have appreciated fully the enormous practical importance of his suggestions." This influence on Wright was to guide him in the next decade.

The human A-B-O blood groups of Landsteiner were demonstrated to be entirely hereditary by E. Von Dungern and L. Hirszfeld, although they proposed two independent pairs of alleles as the Mendelian basis for the two antigens A and B. In 1919 the Hirszfelds (Ludwik and Hanka), army physicians in the Balkans during World War I, determined blood groups for many soldiers of diverse race and nationality. Their finding of differences in relative frequencies of blood groups among the races and nationalities represents one of the earliest human population genetics studies. It was not until 1925 that the correct mode of heredity for these blood groups was postulated by Felix Bernstein (a mathematician of Göttingen) when he applied the Hardy-Weinberg principles to the proportions of blood group phenotypes.

1912–1914. Raymond Pearl, a poultry breeder in the United States, working out theoretic consequences of brother-sister mating, devised a coefficient of inbreeding based on the decrease in number of common ancestors with receding generations, but he erroneously concluded that no change would occur in heterozygote frequency. Herbert S. Jennings and Harold D. Fish independently noted the error and published the correct interpretation showing that sib mating essentially leads to a decline in heterozygosity just as selfing does, although at a slower rate. Jennings took note of Hardy's and Pearson's conclusions on the stability of random mating populations, but his statements caught the attention of Sewall Wright who was to make a considerable contribution to the study of inbreeding in the next decade.

1915. Morgan, Bridges, Muller, and Sturtevant published a very important book, *The Mechanism of Mendelian Heredity*.

Wright left an assistantship with Castle to take charge of a long-term experiment on inbreeding and crossbreeding in guinea pigs at the Animal Husbandry Division of the U.S. Department of Agriculture.

In Cambridge, England, Ronald Aylmer Fisher published his first paper on the distribution of the correlation coefficient. He was subsequently occupied with statistical problems suggested by the writings of Pearson and Yule. Punnett published *Mimicry in Butterflies*, which contained a table worked out by H. T. J. Norton giving the amount of selection intensity with generation time required to change gene frequencies in a Mendelian population. In 1926 Tshetverikov made use of that table in analyzing natural populations.

1916. Several contributions later became formative for significant aspects of the field: (1) J. P. Lotsy proposed hybridization ("introgression") as an important evolutionary mechanism, (2) Pearl demonstrated the effectiveness of pedigree selection contrasted with mass selection in poultry, (3) Shull suggested the word *heterosis* to describe the vigor of first-generation hybrids, and (4) Jennings developed a mathematical theory of inbreeding published in the first volume of a new journal, *Genetics*.

1917. E. C. MacDowell reported his selection experiments on drosophila bristle number, interpreting his results as did Johannsen years before and Sturtevant in the following year in terms of accumulation of modifiers by selection.

The Connecticut Agricultural Experiment Station produced the first commercial "crossed corn" by East and Donald F. Jones. The latter explained heterosis as due to linked dominant genes controlling increased vigor. In the following year Jones proposed the "double-cross' system for producing commercial maize, utilizing the benefit of F_1 hybrids ($A \times B$ and $C \times D$) to improve seed, the progeny of the $F_1 \times F_1$.

Wright, at the U.S. Department of Agriculture, used the Hardy-Weinberg principle by comparing observed with expected values based on random mating in a case of color inheritance in cattle, and in the following year by rejecting a single gene hypothesis for the inheritance of human eye color.

1918. Fisher, who had left Cambridge University the previous year to become head of the statistical department of the Rothamsted Experimental Station at Harpenden,

England (see Neyman, 1967), in his new role as statistical consultant, published his first population genetics paper, "The Correlation between Relatives on the Supposition of Mendelian Heredity" (Royal Society of Edinburgh Transactions). From that time to 1943 Fisher published at least 28 contributions to that field (see Moran, 1962). This paper on the genetic effects of inbreeding and assortative mating helped resolve the differences between biometrical and Mendelian schools, largely because he used Pearson's data on human measurements to show that correlations fitted the theory of *particulate* instead of blending heredity. Reasoning of the biometricians was thereby reversed because they had tried to derive all their "blending" arguments from close-relative correlations. Fisher's method was an independent derivation from that of Weinberg, whose work apparently was unknown to Fisher.

In the United States, H. D. King reported results of inbreeding rats for 25 generations and showed that close inbreeding is not necessarily deleterious because fertility and vigor were maintained in some lines. Francis B. Sumner made some instructive observations on the genetic differences within and between natural populations of the wild deermouse, *Peromyscus maniculatus*. J. Schmidt similarly reported racial studies in fishes (*Zoarces* spp.), as did Richard B. Goldschmidt for the gypsy moth (*Lymantria dispar*). These men had great influence in establishing the genetic basis for geographic diversity within species.

The earlier decades of Mendelism thus ended, with the stage set for major advances, the first in terms of mathematical statements derived from breeding observations coupled with Darwinian evolutionary theory and deVries's mutation theory. Confirmation of those generalizations with empirical determination of real populations, natural and experimental, followed in greater degree about a decade later. We should keep in mind that progress in developing the major concepts in this discipline came as a product of interaction between theoreticians and experimentalists. The fullest realization of progress came about after that fact became more obvious to geneticists, ecologists, and evolutionists, at least by the end of the 1950s. The remainder of this historical section is intended to summarize the account of cornerstone publications and other accomplishments whose subject content this book, it is hoped, will elucidate.

1921. The "Systems of Mating" series of papers by Wright generalized theories of inbreeding and crossbreeding. He invented the method of *path coefficients*, which serve to analyze, by subdividing correlations in a causal scheme, the relative contributions of interacting factors on the determination of a measured effect. (See Li, 1975, for a summary of path analysis.) Correlations between relatives under various mating systems were worked out in a general scheme with path coefficients connecting zygotes and gametes. Wright then proceeded to examine the outcome of random sampling of gametes in populations of limited size, a study that led him to generalize the scattering of genetic variability in the concept of "random genetic drift."

Fisher derived expressions for the outcome of selection favoring heterozygotes, or balanced genetic systems. Later those expressions were to be incorporated in the concept of balanced polymorphism defined and described extensively by E. B. Ford in the 1940s.

1924. John Burdon Sanderson Haldane, a reader in biochemistry at Cambridge University, worked out the mathematical theory of selection by considering a single gene subject to natural selection and mutation: frequencies of autosomal dominants, recessives, sex-linked and partially sex-linked genes, including the equilibrium state when new alleles produced by mutation are balanced by elimination through selection in a steady-state population. This work led him to estimate mutation rates for human deleterious genes. Haldane ascribed most of the ideas on this subject to H. T. J. Norton. (The student might recall others listed above who had considered estimating gene frequencies in populations under selection.)

1926. In Moscow, Sergei Sergeevich Tshetverikov published "On Certain Aspects of the Evolutionary Process from the Standpoint of Modern Genetics." That paper formulated clearly, although without the elegance of mathematical refinements, the basic tenets of what later came to be known as the biological, or synthetic, theory of evolution—namely, that mutational variability is the *source* of raw materials for evolution, but does not constitute evolution itself (see Dobzhansky, 1967). Populations in nature, he concluded, absorb mutations "like a sponge" and retain them in heterozygous condition, thereby providing a store of potential variability out of which the population may utilize a portion for its adaptedness (an idea we now include in the concept called by Muller "the genetic load" of a population). Among the geneticists who were stimulated by Tshetverikov was Theodosius Dobzhansky, who received his first drosophila stocks from Tshetverikov. Other people influenced by Tshetverikov were H. J. Muller, who visited the Soviet Union in 1933–1937, N. W. Timofeef-Ressovsky, N. P. Dubinin, and S. M. Gershenson, all of whom were inspired to analyze natural populations for frequencies of mutant alleles and to discuss the mechanisms of natural selection on the genetic potential of populations, or what has been termed "microevolution." Of these geneticists, one of the most outstanding for his major contributions to experimental population genetics has been Dobzhansky, who arrived in the United States in 1927, worked with T. H. Morgan a few years, and was soon appointed to a staff position at California Institute of Technology.

1929. Fisher published *The Genetical Theory of Natural Selection.* His central theme was the determination of the rate of increase (= Malthusian parameter), or "mean selective value" (in Wright's terminology), or Darwinian "fitness" value (in this book's terminology), of genotypes and the bearing of genetic unfixed elements in the population on that parameter. These considerations culminated in his "fundamental theorem of natural selection"—"the rate of increase of fitness of any organism at any time is equal to its genetic variance in fitness at the time" (in which *genetic variance* refers to the "additive component," or linear component of genetic variance, as we shall see in Chapters 6 and 14). In addition, he summarized much evidence on the evolution of dominance, developing his theory of accumulation of dominance modifiers (later criticized by Wright, 1956, 1960). Fisher included important discussions on sexual selection, correcting and improving much of the Darwinian argument, on mimicry, and on eugenics.

Fisher's book, while difficult for biologists in its mathematical and occasionally enigmatic language, did set forth sharply the Mendelian structure of variation in natural populations. His emphasis on the relationships of genetics to natural selection had a tremendously powerful effect on thought in many fields of biology; this book is often taken as a landmark of renewed interest in the study of evolutionary genetics, or the synthesis of modern evolutionary doctrine.

1931. In the United States, Sewall Wright published an equally important contribution: a long paper in *Genetics*, "Evolution in Mendelian Populations." His central problem was an attempt to synthesize a balanced theory of evolution by consideration of the known forces acting on gene frequencies in populations, at least for simple genetic systems. He set out in orderly fashion those parameters expressing the "directed pressures": recurrent mutation, selection, and migration (or hybridization between populations), plus the "non-directed" random fluctuations due to sample size (or effective population size). He concluded that the state of a population's genotype complex represents a balance among these determinate and indeterminate forces. In the course of his exposition he used an analogy of contour diagrams with peaks and valleys to represent a "surface of selective values" ("W") for genotypes within populations. This graphic presentation of the numerous combinational possibilities of forces allowed Wright an efficient device to symbolize his explorations of forces on genotypes among partially isolated and differentiated local populations ("demes"). Two factors stressed and developed by Wright were (1) random drift of gene frequencies and the fluctuations in the "pressures" brought about by small population size and (2) the significance of interpopulational selection (interdeme), especially when a species is subdivided into partially isolated local populations in which local pressures and random fluctuations provide raw material for intergroup selection—the conditions for most rapid evolution. He assigned a considerable role to random drift as a factor to be accounted for in the network of balanced forces, which produced violent criticism from Fisher and his colleagues who accused Wright of substituting drift for natural selection. This misunderstanding lasted through the following two decades and tended to delay in Great Britain any major role for indeterminate (stochastic) factors in the genetics of populations.

1932. Haldane contributed a synthesis of evolutionary doctrine and genetics in *The Causes of Evolution*. He reviewed his 1924 theory of selection and the concepts of selection intensity (and what he later termed "cost" to the population for substituting new mutant adaptive alleles), differentiated between elimination rate by selection and the relative selective value of a genotype, and finally made cogent summaries and remarks about the recent contributions of Wright and Fisher on the subjects of random drift and isolation.

From that time on, all the ingredients for fruition of an exciting enterprise were established. It may seem to many readers that the beginnings of population genetics contrast markedly with those of other branches of genetics, notably microbial and molecular genetics, in that its early years are famous for major theoretical statements by three giant intellects (Fisher, Haldane, and Wright) rather than for experimental observations. However, it should be apparent after this brief summary of cogent events that in fact the observations made by Mendel, Darwin, and the early geneticists constituted the stimulus for those theoreticians and their generalizations. It was not enough for complete theory to rely only on observations of genetics; with greater sophistication in understanding genetic architecture, the demonstration of selective differences between genotypes in real populations, the magnitude of selection, and the subtlety of the dynamic elements controlling the genetic potential of populations in the following decades extended and modified these important early generalizations so that in the present day we have far deeper concepts of genetic dynamics in populations. Many reviews, summaries, symposiums, and compilations of important papers in the field now available are listed in the bibliography, to which the student is encouraged to turn for further details of population genetics history. Particularly noteworthy is the book by Provine (1971).

Through the following pages students should develop insight into population genetics dynamics and thus permit greater appreciation of the current significant contributions being made by experimentalists, agriculturists, and human geneticists to this rapidly expanding field.

PART 2
GENETIC EQUILIBRIUM AND RANDOM MATING

1

GENETIC NETWORK: THE CASTLE-HARDY-WEINBERG CONDITIONS—ONE PAIR OF ALLELES WITHOUT DOMINANCE

For sexually reproducing biparental species of organisms, the sharing of genetic material among individuals is vastly greater than for asexually reproducing forms in which there is only lineage from single-parent individuals. The sexual processes of meiosis (resulting in genetic segregation) and combinations of gametic pairs into zygotes at fertilization make a fundamental network interconnecting related individuals in the community that share in the common "gene pool." This network containing all genotypes potentially or actually present is our center of interest—the *Mendelian population*—and it is upon this network of biparental lineage that natural selection and other evolutionary forces act via individuals and their component genotypes.

In the biological concept of stages in evolutionary divergence, the *species* represents the most inclusive Mendelian population; that is, a genetically closed system, or network, within which breeding is actually or potentially occurring. Interbreeding with another species is rare or prevented by a number of isolating mechanisms. Usually each species is distinguishable by phenotypic criteria from other species. Groupings of larger and more inclusive constellations of populations than a species, such as subgenera, genera, and so on, may show resemblance in some degree, but do not possess common gene pools, do not share the fundamental network of relationship contemporaneously, and consequently do not have the biological essence of species. On the other hand, most species are often not homogeneous from one end of their ranges to the other. For various reasons, such as slow migration, isolation by distance, territoriality, or preferential mating, individuals tend to cluster or become subdivided into subordinate Mendelian populations: subspecies, races, ecotypes, or local populations ("demes"). Each of these subordinate populations shares in the total genetic network of the species over the long run but may be locally characterized in terms of gene frequencies, proportions of genotypes, or chromosomal variants (karyotypes) peculiar to them or at least measurably different from adjacent populations of that species. As the genotype characterizes the individual, so the gene *frequency*, genotype (or zygote) *frequency*, or karyotype *frequency* may characterize the population. Example 1-1 illustrates genotype frequencies of β-chain hemoglobin variants in three populations of humans.

We can characterize the population by its gene (allele) frequencies—its haploid dose of heredity in gametes. Using the data of the Musoma population from Example 1-1 for sickle-cell hemoglobin, we may count each gene by its frequency, totaling 2×100 percent (since each individual is diploid):

	Method 1—Simple Tally of Genes	Gene Frequency
Musoma population	$Hb^A = (2 \times 0.659) + 0.310 = 1.628$	0.814
	$Hb^S = (2 \times 0.031) + 0.310 = 0.372$	0.186
	Total = 2.000	

An equivalent arithmetic method implies the existence of a "gametic pool" in the population from which zygotes will be formed each generation. Mendelian segregation partitions the heterozygote alleles *equally* into the two gametic pools (sperms and eggs), then the two methods become identical:

Method 2—Gametic Pool Contributions: All Homozygote Frequency + $\frac{1}{2}$ Heterozygote Frequency	
Musoma	$Hb^A = 0.659 + 0.155 = 0.814$
	$Hb^S = 0.031 + 0.155 = 0.186$
Gambia	$Hb^A = 0.867 + 0.068 = 0.930$
	$Hb^S = 0.007 + 0.063 = 0.070$
U.S.A.	$Hb^A = 0.918 + 0.040 = 0.958$
	$Hb^S = 0.002 + 0.040 = 0.042$

These populations differ in both gametic (gene) and zygote (genotype) frequencies. Later we shall see that populations may often have identical gene frequencies but different genotype frequencies.

It should be noted that if some non-Mendelian process (such as meiotic drive, segregation distortion, or nondisjunction) disturbs the segregation ratio, the gametic pool may *not* equal the simple tally of genes present. For most cases it can be assumed that the methods are identical, and the gametic pool method has greater utility in algebraic models.

It must be stressed that the genotypes in any generation are *not* transmitted to the next generation as genotypes. Genes alone (haploid doses) are transmitted, not genotypes. The genic continuity between generations depends on the replication ability of the genetic material and the transmission of a *certain fraction* of that material (usually one-half, specified by meiosis, or Mendelian segregation) from each parent to each offspring. Each fertilized egg (zygote) then has a genotype newly determined, and the array of genotypes in a population with its component gene pool (or gametic pool) must be redetermined each generation. Sex cells from a heterozygous (Aa) parent, for example, contain either A or a (except in rare cases of nondisjunction), so that the parent's heterozygosity is not transmitted directly to any child; in fact he or she could easily have no children heterozygous (Aa) for that locus.

If genotypes stabilize in relative frequency—that is, remain constant from generation to generation—we say they are *at equilibrium*. It is an equilibrium determined by biological forces, predominantly breeding behavior, genetic and environmental factors acting on individuals in the population affecting reproductive ability and survival, plus the laws of probability as those laws apply to the sample of genes transmitted in a population of limited size.

From all of the numerous factors that contribute to the determination of genotypes in progeny, it is best to consider one particular component at a time. It is advantageous for learning to concentrate on the many interesting properties of one factor, but we must always keep in mind that separation of factors is a learning device and that real populations are not conveniently so subdivided; instead, they are undoubtedly regulated, modulated, and generally affected by combinations of several forces and complex interactions between those forces.

The one factor to be isolated for study here is *random mating* (*panmixia*), because its consequences follow simple rules (probability for combinations of two gametes at a time). These rules applied to crossfertilizing organisms in populations were actually implicit in Mendel's principles; they were first sketchily mentioned by Castle (1903) and Pearson (1904) as indicating genetic equilibrium. More general statements with greater applicability were made by Hardy and Weinberg in 1908, although they were independently motivated and apparently unknown to each other (see history section of Introduction).

EXAMPLE 1-1

DIFFERENCE BETWEEN POPULATIONS IN GENOTYPE FREQUENCIES

Human populations may differ in the frequency of the abnormal hemoglobin that produces the condition known as sickle-cell anemia. Homozygotes for the abnormal allele (Hb^S) suffer severe anemia when their red blood cells become elongated or filamentous in their veins in response to lowered oxygen tension. Both parents of anemic children are heterozygous (Hb^S/Hb^A), appear healthy, suffer no ill effects, but in laboratory tests their red blood cells sickle when exposed to reducing agents (methylene blue or Janus green) and they are then said to have sickle-cell trait. African populations and American Negro populations differ in frequencies of genotypes for these hemoglobin alleles (data from Allison, 1956; Neel, 1951).

Population	Hb^A/Hb^A	Hb^A/Hb^S	Hb^S/Hb^S	N
Musoma District, Tanzania (infants)	0.659	0.310	0.031	287
Gambia, West Africa (infants)	0.867	0.126	0.007	446
U.S.A., several cities (all ages)	0.918	0.080	0.002	Several thousand

RANDOM MATING

If any individual of one sex is equally likely to mate with any individual of the opposite sex in a population of large enough size so that sampling error is nearly eliminated, and if the genotypes of those individuals are neutral (not influenced by external conditions), then we can say that *panmixia* (random mating) determines the genotype frequencies in the population. Zygotes result from proportional fertilizations just as the probability for two independent events happening in combination equals the product of individual event probabilities (two coins will fall both heads one-half of one-half, or one-fourth, of the time).

Castle, Hardy, and Weinberg demonstrated that this result of constant gene and genotype proportions applies just as well to pairs of alleles of unequal frequency, and that, providing mating is random, the equilibrium of genotype proportions will be established immediately, as in Example 1-2.

In the numerical examples, if we let the probability for any gene $(A) = p$ and the probability for its mutant allele $(a) = q$ in any gamete of either sex so that $p + q = 1.00$, then the zygote frequencies are $(p + q)^2 = 1.00$. Note that *squaring* denotes proportional fertilizations (male gametes × female gametes) when the two sexes have *identical* gene frequencies.

The Castle-Hardy-Weinberg principle will be referred to in this book as "the square law":

$$\text{(sum of gene frequencies)}^2 = \text{genotype (zygote) frequencies}$$
$$(p + q)^2 = p^2 + 2pq + q^2 \tag{1-1}$$

Figure 1-1 illustrates the principle graphically. In each generation all genotypes are essentially restored by random mating (coupling of single doses by chance); that is, no parent passes on a genotype intact to any offspring because the segregation process disjoins that genotype in gamete formation.

A useful simplification of (1-1) can be made as follows: let the ratio $p/q = u$. Then, dividing the component terms of (1-1) by q^2 throughout, it becomes

$$\frac{p^2}{q^2} + \frac{2pq}{q^2} + \frac{q^2}{q^2} = u^2 + 2u + 1 = (u + 1)^2$$

Castle's original equilibrium law was given in terms of this ratio. Thus, the equilibrium populations in which $p = q$, $p = 2q$, $p = 3q$, etc., have the following relative values of genotypes (*AA*: *Aa*: *aa* respectively):

1:2:1
4:4:1
9:6:1
16:8:1, etc.

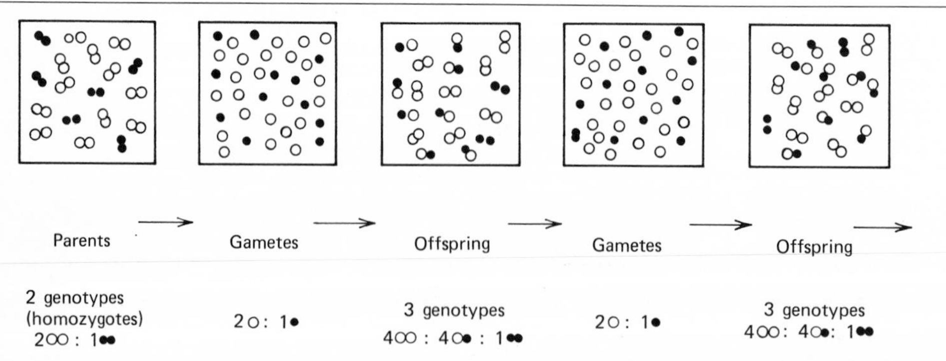

Figure 1-1. Diagrammatic representation of parents' genotypes, their haploid "gametic pool," and offspring after random mating (diploid), then the offspring's "gametic pool," and a second generation. Frequencies are as in Example 1-2. ● = *A*, ○ = *a*.

The early writings of Haldane on natural selection were also formulated in terms of this ratio (u), because selection changes are often easier to describe mathematically in terms of u than in terms of p, q. The student should verify that $p = u/(u + 1)$ and $q = 1/(u + 1)$.

EXAMPLE 1-2

GENETIC EQUILIBRIUM WITH EQUAL ALLELIC FREQUENCIES IN BOTH SEXES

If both sexes have the same genotype frequencies and mating occurs at random (all combinations of ♀♀ × ♂♂ without regard to genotype), equilibrium will be established immediately. Assume one-third of parent females are homozygous AA genotype and two-thirds homozygous aa with males having the same genotype frequencies as below for 1000 of each sex:

$$\text{Parents} \begin{Bmatrix} 333AA\ ♂♂ \\ 667aa\ ♂♂ \end{Bmatrix} \text{and} \begin{Bmatrix} 333AA\ ♀♀ \\ 667aa\ ♀♀ \end{Bmatrix}$$

Parent Mating Proportions and Progeny Genotypes

Parent Males	Parent Females $\frac{1}{3}AA$	Parent Females $\frac{2}{3}aa$
$\frac{1}{3}AA$	$\frac{1}{9}AA$	$\frac{2}{9}Aa$
$\frac{2}{3}aa$	$\frac{2}{9}Aa$	$\frac{4}{9}aa$

Progeny (F_1) totals: $\frac{1}{9}AA : \frac{4}{9}Aa : \frac{4}{9}aa$

Allelic frequencies: $\frac{1}{3}A : \frac{2}{3}a$

Proof of equilibrium if F_1 males mate at random with F_1 females:

F_1 Males	F_1 Females $\frac{1}{9}AA$	F_1 Females $\frac{4}{9}Aa$	F_1 Females $\frac{4}{9}aa$
$\frac{1}{9}AA$	$\frac{1}{81}AA$	$\frac{4}{81}\begin{cases}\frac{2}{81}AA \\ \frac{2}{81}Aa\end{cases}$	$\frac{4}{81}Aa$
$\frac{4}{9}Aa$	$\frac{4}{81}\begin{cases}\frac{2}{81}AA \\ \frac{2}{81}Aa\end{cases}$	$\frac{16}{81}\begin{cases}\frac{4}{81}AA \\ \frac{8}{81}Aa \\ \frac{4}{81}aa\end{cases}$	$\frac{16}{81}\begin{cases}\frac{8}{81}Aa \\ \frac{8}{81}aa\end{cases}$
$\frac{4}{9}aa$	$\frac{4}{81}Aa$	$\frac{16}{81}\begin{cases}\frac{8}{81}Aa \\ \frac{8}{81}aa\end{cases}$	$\frac{16}{81}aa$

Mating frequencies according to (1-2) are fractions at left in this table.

F_2 progeny segregating within each type of mating combination are expressed as fractions to the right of the mating frequencies, except in the corners where all progeny are a single genotype.

$$\text{Total } F_2 \text{ progeny: } \tfrac{9}{81}AA : \tfrac{36}{81}Aa : \tfrac{36}{81}aa = \tfrac{1}{9}AA : \tfrac{4}{9}Aa : \tfrac{4}{9}aa$$

$$F_2 \text{ allelic frequencies: } p_A = \tfrac{1}{9} + \tfrac{2}{9} = \tfrac{1}{3}$$

$$q_a = \tfrac{4}{9} + \tfrac{2}{9} = \tfrac{2}{3}$$

FREQUENCIES OF MATING: GENERAL METHOD RELATING GENOTYPES IN SUCCESSIVE GENERATIONS

In Example 1-2 it is obvious that a 3 × 3 zygote matrix is very cumbersome compared with the 2 × 2 gametic matrix. However complex these fractions for frequencies of mating may appear in the 3 × 3 matrix, it will be essential many times in future discussions to refer to them; consequently we must carry out a detailed determination for them at this time as shown in Table 1-1A. If the sex of the parent is not relevant among the nine possible mating combinations (there being no maternal effects or sex linkage, for example),

TABLE 1-1 *Frequencies of mating at equilibrium as with the data of Examples 1-2 and 1-3*

A

Males	Females		
	p^2_{AA}	$2pq_{Aa}$	q^2_{aa}
p^2_{AA}	p^4	$2p^3q$	p^2q^2
$2pq_{Aa}$	$2p^3q$	$4p^2q^2$	$2pq^3$
q^2_{aa}	p^2q^2	$2pq^3$	q^4

(Where AA, Aa, aa = 3 genotypes in proportions p^2, $2pq$, q^2)

B

Genotype Mating	Frequency of Mating	Offspring AA	Offspring Aa	Offspring aa
$AA \times AA$	p^4	p^4	—	—
$Aa \times AA$	$4p^3q$	$2p^3q$	$2p^3q$	—
$aa \times AA$	$2p^2q^2$	—	$2p^2q^2$	—
$Aa \times Aa$	$4p^2q^2$	p^2q^2	$2p^2q^2$	p^2q^2
$aa \times Aa$	$4pq^3$	—	$2pq^3$	$2pq^3$
$aa \times aa$	q^4	—	—	q^4
	1.00	p^2	$2pq$	q^2

Because $q = 1 - p$, in each column the algebraic sum may be simplified; for example, in the AA offspring column: $p^4 + 2p^3(1 - p) + p^2(1 - 2p + p^2) = p^2$ or, alternatively, factor: $p^2(p^2 + 2pq + q^2) = p^2$.

some of the combinations are duplicates so that they can be reassembled into six as in Table 1-1B.

Frequencies of mating then are represented as

$$(p^2 + 2pq + q^2)^2 = p^4 + 4p^3q + \cdots + q^4 = 1.00 \tag{1-2}$$

Each mating combination contributes to offspring proportionally according to the Mendelian processes with total frequencies again as in (1-1) (see Exercise 1).

APPROACH TO EQUILIBRIUM

If a population is made up of genotypes that do not conform to the square law (that is, zygotes do not occur according to the squares of their allelic frequencies), then *random mating* will produce offspring immediately according to the products of frequencies in the two sexes. It will be apparent that when the sexes have identical allelic frequencies, the "squared" condition in the offspring will hold. In Example 1-2, the parents were homozygotes exclusively but when allowed to mate at random produced progeny conforming to the square law; that is, $(\frac{1}{3}A/A + \frac{2}{3}a/a)^2$ becomes $\frac{1}{9}A/A + \frac{4}{9}A/a + \frac{4}{9}a/a$. In any population in which an arbitrary group of parents is mating at random, gene frequencies can be estimated by the "gametic pool method" (see method 2 mentioned earlier) and the square law applied.

If we observe three genotypes in actual numbers of individuals, let

A = number of AA genotype
B = number of Aa genotype
C = number of aa genotype
N = total number of individuals; then let frequencies be
$D = A/N$, frequency of dominant homozygotes
$H = B/N$, frequency of heterozygotes
$R = C/N$, frequency of recessive homozygotes,

so that their sum totals to 1.00

According to the gametic pool method, gene (allelic) frequencies then are as follows:

$$p = D + \tfrac{1}{2}H, \qquad q = R + \tfrac{1}{2}H \tag{1-3}$$

if frequencies are used or

$$p = \frac{2A + B}{2N}, \qquad q = \frac{2C + B}{2N} \tag{1-3A}$$

if observed numbers are used. (This result is the maximum-likelihood solution; see Appendix A-9(5).)

Provided that the sexes have identical genotype frequencies D, H, and R, the zygote frequencies will be the square of the gametic, or gene, frequencies:

$$\begin{aligned} (D + \tfrac{1}{2}H)^2 &= D \text{ expected} \\ 2(D + \tfrac{1}{2}H)(R + \tfrac{1}{2}H) &= H \text{ expected} \\ (R + \tfrac{1}{2}H)^2 &= R \text{ expected} \end{aligned} \tag{1-4}$$

TABLE 1-2 *Approach to equilibrium in progeny of $D + H + R$ arbitrary parents mating at random, where males and females have identical zygote frequencies (D = "dominant" homozygotes AA, H = heterozygotes Aa, and R = "recessive" homozygotes in relative proportions)*

A

P. Males	P. Females D	H	R
D	D^2	DH	DR
H	DH	H^2	HR
R	DR	HR	R^2

B

Genotype Mating	Frequency of Mating	Offspring AA	Aa	aa
$AA \times AA$	D^2	D^2	—	—
$Aa \times AA$	$2DH$	DH	DH	—
$aa \times AA$	$2DR$	—	$2DR$	—
$Aa \times Aa$	H^2	$\frac{1}{4}H^2$	$\frac{1}{2}H^2$	$\frac{1}{4}H^2$
$Aa \times aa$	$2HR$	—	HR	HR
$aa \times aa$	R^2	—	—	R^2
	1.00	$(D + \frac{1}{2}H)^2$	$2 \cdot (D + \frac{1}{2}H)(R + \frac{1}{2}H)$	$(R + \frac{1}{2}H)^2$

Note the middle column $= 2(DR + \frac{1}{2}DH + \frac{1}{2}HR + \frac{1}{4}H^2)$. Recall that $p = (D + \frac{1}{2}H)$ and $q = (R + \frac{1}{2}H)$.

To "conform to the square law" implies that the following simple conditions must be true: that $H = 2pq$, $D = p^2$, and $R = q^2$, or that $H = 2\sqrt{DR}$, or

$$H^2 = 4DR \qquad \text{(condition for the "square law")} \tag{1-5}$$

Table 1-2 indicates these relationships algebraically, and Example 1-3 illustrates the principle with parental genotypes different from those in Example 1-2. It is a tentative conclusion that if zygote (genotype) frequencies conform to the square law, then mating among parents was *probably* at random and the other conditions of the Hardy-Weinberg equilibrium were probably met.*

* It is nevertheless dangerous to conclude that no selection or other influences are acting on a population just because of conformity with the square law for a single generation. It is safer to score genotype frequencies for at least two successive generations to rule out cases of increments in fitness equal in the two sexes such that fitness of homozygotes is the square of the heterozygote's fitness, for example. See Chapter 14 and the section on increased mating propensity (Chapter 15), where selection is acting but frequencies in a single generation fit a square of gene frequencies. See also the section on conformity with the square law in Chapter 2.

If the parent sexes are not equal in allelic frequencies, the F_1 progeny after random mating are expected to be uniform for autosomal gene frequencies or for any sets of loci independent of the sex-determining system. That is, regardless of any variation in sex ratio, each sex will obtain the *same* average set of heredity from the parents mating at random. In the following generation (F_2) the square law conditions will apply with the allelic frequencies being the average of the frequencies among the original parents:

Let p, q = frequencies of A, a among parent females (where $p + q = 1$); and r, s = frequencies of A, a among parent males (where $r + s = 1$). Then after random mating, all progeny will be $AA = pr$, $Aa = qr + ps$, $aa = qs$. Then the F_1 allelic frequencies will be $p'_A = pr + \frac{1}{2}(qr + ps)$ and $q'_a = qs + \frac{1}{2}(qr + ps)$, where prime (′) refers to F_1. It is easy to see that in terms of the original frequencies the F_1 frequencies will be

$$p'_A = \frac{p + r}{2}, \quad q'_a = \frac{q + s}{2} \tag{1-6}$$

This expression reduces to the equilibrium of allelic frequencies ($p' = p$) when the two sexes have identical allelic frequencies; that is, $p = r$. (See Exercises 3 and 7.)

EXAMPLE 1-3

EQUILIBRIUM ESTABLISHED BY RANDOM MATING

The sexes have equal genotype frequencies, as in Example 1-2. However, we here assume some parents are heterozygotes and employ Table 1-2.

$$\text{Parents} \begin{Bmatrix} 40\ Aa \text{ males} \\ 60\ aa \text{ males} \end{Bmatrix} \text{ and } \begin{Bmatrix} 40\ Aa \text{ females} \\ 60\ aa \text{ females} \end{Bmatrix}$$

Genotype Mating	Frequency of Mating	Offspring AA	Offspring Aa	Offspring aa
$AA \times AA$	0	0	0	0
$Aa \times AA$	0	0	0	0
$aa \times AA$	0	0	0	0
$Aa \times Aa$	0.16	0.04	0.08	0.04
$Aa \times aa$	0.48	0	0.24	0.24
$aa \times aa$	0.36	0	0	0.36
	1.00	0.04	0.32	0.64

$$p = (D + \tfrac{1}{2}H) = 0.20; \qquad q = (R + \tfrac{1}{2}H) = 0.80$$

NOTE: Offspring genotype frequencies can be ascertained by squaring *parents'* gene frequencies:

$$p = 0 + \tfrac{1}{2}(0.40) = 0.20$$
$$q = 0.60 + \tfrac{1}{2}(0.40) = 0.80$$

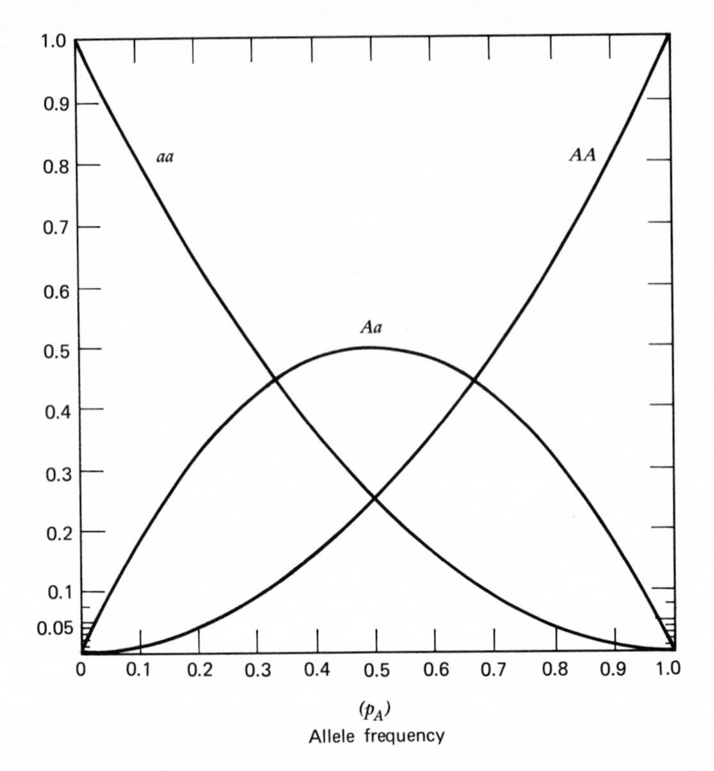

Figure 1-2. Two-coordinate graph of three genotypes for a single pair of alleles as a function of p_A under random mating.

SOME PROPERTIES OF THE GENETIC EQUILIBRIUM IN DIPLOID POPULATIONS

Several important consequences of the binomial distribution follow when allelic frequencies are varied under random mating conditions. In Figure 1-2, changes in p and q values illustrate these relationships for a single pair of alleles at one locus in the conventional two-coordinate graph.* From this graph and the exercises at the end of this chapter the following properties should be evident:

Property 1

With only two doses of each genic locus being considered in a diploid population and a single pair of alleles represented, the maximum heterozygosity attainable is at the

* Representation of genotype frequencies in a triangular coordinate graph is described by C. C. Li (1955, 1976). The base of an equilateral triangle is divided into p,q values and the perpendiculars to the three sides are the relative genotype frequencies. A parabola ($4DR - H^2 = 0$) is described by points representing equilibrium populations, and the projection of points from the parabola to the baseline mark off relative p,q values as segments of the base. See Appendix A-10 for a diagram illustrating use of a triangular coordinate graph.

point where $p = q$, $H = 2pq = 0.50$. If we take the derivative of $H = 2(1 - q)q$ with respect to q we have

$$\frac{dH}{dq} = \frac{d(2q - 2q^2)}{dq} = 2 - 4q$$

If we set the derivative equal to zero to obtain the maximum point (zero rate of change in H as q increases) and solve for q, then $q = 0.50$. (The second derivative is negative and thus indicates a maximum.) Because the two homozygous genotypes (D and R) must be equal under random mating when the gene frequencies are equal, then the heterozygote class can never exceed the sum $D + R$.

Property 2

As p, q values move away from the point of equality, the heterozygote class diminishes and moves from the majority to an intermediate frequency, the point of change being where $H = D$ (or $H = R$, since the curve is symmetric). For example, where $2pq = p^2$, $2p(1 - p) = p^2$, then $p = \frac{2}{3}$, beyond which $p^2 > 2pq > q^2$ (see Exercise 6).

Property 3

As the rarer allele becomes less frequent, heterozygotes carry a much greater proportion of that allele than do homozygotes. If we take the ratio $H/R = 2pq/q^2 = 2p/q$, it is evident that as q approaches zero, the relative size of H compared with the homozygous recessive proportion approaches infinity. Also p approaches 1.0 so that $2pq$ becomes $2q$ effectively, or the heterozygotes become about twice the rare allele's frequency. In other words, when an allele is very rare, it occurs almost exclusively in the heterozygous condition; for example, if a recessive defect has an incidence of one in a million, heterozygous carriers might be expected to occur at a rate of one in five hundred.

Property 4

Among the three parental matings that produce *aa* offspring, namely $Aa \times Aa$, $Aa \times aa$, or $aa \times aa$, the relative proportions of *aa* offspring produced by these parent pairs are in the same proportions as the zygote frequencies expected from the square law (1-1) (referring to Table 1-1B):

Parent Mating	*aa* Genotype Among All Offspring	Relative Frequencies
$AA \times Aa$	p^2q^2	p^2
$Aa \times aa$	$2pq^3$	$2pq$
$aa \times aa$	q^4	q^2
	Total q^2	1.0

Property 5

In Table 1-1B the frequency of mating $Aa \times Aa$ is twice that of $AA \times aa$ (reciprocals included), while the nonequilibrium conditions in Table 1-2 do not indicate such a relationship. This fact of equilibrium was pointed out by Fisher (1918) and by Haldane and Moshinsky (1939), and is true for any genetic equilibrium, whether under inbreeding or random mating; because in $AA \times aa$ matings the homozygotes are replaced by offspring heterozygotes, while in $Aa \times Aa$ only half the offspring are homozygotes (and all other matings produce offspring in the same proportions as their parents), then $Aa \times Aa$ must be twice as frequent as $AA \times aa$ at equilibrium.

As for the nature of the genetic equilibrium determined by random mating, it is worthwhile to point out that this equilibrium is such that if the gene frequencies are changed (perturbated), random mating alone is not a sufficient force to restore the original frequencies. If the gene frequencies are held constant and the genotype (zygote) frequencies altered—for example, by changing the mating system through inbreeding—and then random mating is allowed, the original genotype frequencies will be restored.

The student should recall that there are at least three principal types of equilibrium in populations as in the physical world: (1) dynamic stable, (2) dynamic unstable, and (3) static or neutral. A balance of forces acting on any physical system may hold the system at an equilibrium point. This equilibrium point is *dynamically stable* if, when disturbed (or perturbated), the system tends to return to the original point. If the system tends to diverge from that point when disturbed, it is *dynamically unstable*. If the forces are acting in such a way that they have no relevance to the equilibrium point, the equilibrium is said to be *static* or *neutral*. Physical examples might be clarifying. (1) A freely turning wheel, held by its axle in position with its rim perpendicular to the ground and with a weight attached to its rim, will come to rest with the weight closest to the ground, and if disturbed will return to that same position (stable). (2) If the wheel is positioned so that the weight is perfectly balanced at the exact top center, it will remain balanced unless it is disturbed, when it will begin turning from that position (unstable). (3) The unweighted wheel at rest has the center of gravity in the axle so that when disturbed it may come to rest again but probably not in its original position; here the force of gravity is not relevant to the wheel's final resting position (static or neutral).

The genetic equilibrium determined by random mating can therefore be considered either as a stable-dynamic equilibrium or as a static-neutral one, depending on the point of view either toward genotypes with *constant* p, q values (*stable*: $p^2 + 2pq + q^2$) or toward disturbance of the allelic frequencies themselves (*static*: $p + q$). Random mating restores the "squared" form of genotype frequencies if nonrandom mating had disturbed their distribution. But after disturbing the allelic frequencies by deterministic forces, removal of those forces will not restore the original frequencies.

In summary, the genetic equilibrium described by the Castle-Hardy-Weinberg principle is an expression of genetic conservation: genotypes determined by random mating will be constant from generation to generation as long as allelic frequencies (p, q values) are constant. When the population size is large enough to be insensitive to random sampling and no extrinsic forces (selection, mutation, or migration) are acting on the genotypes present, then allele frequencies and genotypes will be conserved. It should be recalled that before Mendelian heredity was understood, the concept of heredity determiners as "blending pangenes," contributed by the body cells to the sex cells and modifiable by environmental forces, led to the erroneous conclusion that offspring always were intermediate between their parents in hereditary composition; consequently, each generation could only have half the amount of genetic variation of the preceding generation. In a short time, without mutation counteracting the loss of genetic variation, all individuals would be alike on that principle. In fact, the consequence of blending pangenesis was Darwin's main stumbling block in working out his natural selection theory because it eliminated the very hereditary variation needed to make the process of selection effective. The extension of Mendel's discovery of hereditary determiners as self-reproducing units to the random-mating population level, in which each generation forms genotypes anew by proportional fusions of male and female gametes, eliminates once and for all the apparent paradox that bothered Darwin and the early evolutionists. The vast supply of genetic variation available to selection and other evolutionary forces at work on populations is maintained by the Mendelian processes of allelic segregation, and the stable proportions of genotypes are determined by random combination at fertilization in each generation.

EXERCISES

1. A random-breeding population has the following genotype frequencies: 64 percent *AA*: 32 percent *Aa*: 4 percent *aa*. Find p_A and q_a. Construct a table with frequencies of mating for six parent combinations and offspring in the manner of Table 1-1B. Verify that the sums of progeny genotype frequencies equal the parental frequencies.
2. In drosophila, individuals carrying the recessive mutant ebony (*e*) and its wild-type (+) allele are mated in several ways. Give the expected progeny proportions for the following sets of parents (assume equal mating propensity and survival of all genotypes):
   (a) 10 + /*e* ♀♀ × 10 *e*/*e* ♂♂.
   (b) random F_1 ♀♀ × F_1 ♂♂ from (a).
   (c) (5 + /*e* ♀♀ and 5 *e*/*e* ♀♀) × 10 *e*/*e* ♂♂.
   (d) (20 + /*e* ♀♀ and 30 *e*/*e* ♀♀) × (20 + /*e* ♂♂ and 30 *e*/*e* ♂♂).
   In which cases did the progeny genotypes conform to the square law? Why?
3. Suppose matings of (10 + /*e* ♀♀ and 40 *e*/*e* ♀♀) × (40 + /*e* ♂♂ and 10 *e*/*e* ♂♂) are random.
   (a) What would be the genotype and allelic frequencies among the F_1?
   (b) If the F_1 mate at random, what would be the F_2 frequencies, zygotic and allelic?
   (c) If you symbolize the allelic frequencies as *p*,*q* among females but as *r*,*s* among males, how is the F_2 allelic frequency determined algebraically? When do the allelic frequencies become equal between the sexes?
   (d) Can you take the square root of the F_1 homozygote frequencies to estimate allelic frequencies and then square them to obtain parent frequencies? Why or why not?
4. When genotype frequencies (*D*, *H*, *R*) conform to the square law, to what function of *D* and *R* must *H* be equal? Is the relation $H = 1 - (D + R)$ sufficient in expressing the square-law equilibrium? (Hint: Express *D*, *H*, and *R* in terms of allelic frequencies.)
5. Test the following zygotic data for conformity with the square-law equilibrium. (You may use the function obtained in Question 4.) If the data do not conform to equilibrium,

calculate what their frequencies would be after random mating:

	AA	Aa	aa	Totals
a)	50	20	30	100
b)	25	10	1	36
c)	20	20	5	45
d)	9	10	81	100
e)	5625	3750	625	10000

6. Construct a table with the headings as follows:
   p_A, q_a, p^2, $2pq$, q^2, H/R, D/R. Then supply into the second column for q_a (frequencies of a) these values: 0.8, 0.6, 0.5, 0.4, 0.3, 0.2, 0.1, 0.01. Fill in the remainder of the table.
   (a) When is H maximum?
   (b) When does H cease to be in the majority?
   (c) What happens to the H/R ratio as q approaches zero? When $p_A = 0.99$, in which genotype are most of the recessive alleles?
   (d) Graph the H/R and D/R functions of q. Use semilog paper with the log scale on the Y axis and q on the X axis. Which ratio rises faster? Smooth out the lines between your points on the graph so that interpolation is improved.
   (e) Compare Haldane's ratio $u = p/q$ with the H/R column.
   (f) Show that for any row in your constructed table the sum of the three genotypes is $D/R + H/R + 1 = (u + 1)^2$ and $p = u/(u + 1)$, $q = 1/(u + 1)$.
7. Two separate populations of equal size are in equilibrium for the same pair of alleles because of random mating within each. In population I, $p_A = 0.6$, while in population II, $p_A = 0.2$, with $q = 1 - p$ in each population.
   (a) If a random sample of females from one population is crossed to a random sample of males from the other population, what would be the progeny genotype frequencies? If these progeny are then allowed to mate at random, what would the next-generation frequencies be?
   (b) If equal samples of both sexes from each population are mixed so that mating occurs at random in the mixed group, what would be the progeny proportions? What would be the next-generation frequencies if these are then allowed to mate at random?
   (c) In which case would equilibrium occur first?
   (d) How is the principle here similar to the case in Exercise 3 above?
8. Among people of southern Italian or Sicilian ancestry living in Rochester, New York, thalassemia major (Mediterranean anemia, or Cooley's anemia) occurs in about one birth in 2,500; the milder form, thalassemia minor, occurs in about one birth in 25. Can these data help to verify a single genic locus hypothesis as a basis for the heredity of these anemic conditions? (Population size is about 10,000.) Why?
9. A genotype that is presumed to be a simple heterozygote is found to have a frequency of 95 percent in a population. What possible explanations could you offer to account for that frequency?
10. Red cell acid phosphatase alleles p^a and p^b differ in their electrophorectic mobility in

starch gel (Chapter 3). In certain populations of Africa, these alleles have the following frequencies: $p^a = 0.16$, $p^b = 0.84$.

(a) What would be the expected incidence of $p^a p^a$ children in those populations? $p^a p^b$ children? $p^b p^b$ children?

(b) From what types of matings (genotype combination of parents) would $p^a p^a$ children be expected to occur? What would be the relative frequencies of those types of matings?

(c) From those types of matings producing $p^a p^a$ children, estimate the *relative* contributions of each type of family (parental genotype combination) to the incidence of the children ($p^a p^a$). How do these relative contributions compare with your answer to part (a) above?

(d) What relationship do these calculations have with Property 4?

11. A convenient method of representing graphically three genotype frequencies in a population is described by Li (1955*b*), using a tricoordinate system (see Appendix A-10). On a triangular coordinate graph, plot the genotype frequencies you obtained from your table in Exercise 6. Smooth off the parabola to use for reference as the random-mating plot. Verify that the perpendicular to the base cuts off segments that give $p:q$ values. Check the five sets of zygotic data from Exercise 5 to see where those populations would lie (convert actual number to frequencies observed). Which of the five sets lie above, and which below, your random-mating plotted parabola?

12. Harris, et al. (1968, cited in Harris, 1970) described the distribution of 537 English families with parents and offspring typed for phosphoglucomutase allozymes (PGM-1 locus) by finding that certain multiple banding zones on electrophoretic starch gel detected consistent person-to-person differences. Thus, the PGM-1 locus was defined as showing patterns with band 1 only, band 2 only (homozygotes), or both bands 1/2 (heterozygotes). Parent and offspring frequencies for these three presumed genotypes were found as follows:

Parents	Number of Matings	Offspring			Total
		1	1/2	2	
1 × 1	199	392	—	—	392
1 × 1/2	203	207	215	—	422
1 × 2	34	—	71	—	71
1/2 × 1/2	77	35	81	41	157
1/2 × 2	21	—	24	31	55
2 × 2	3	—	—	13	13
Totals	537	634	391	85	1110

(a) Do these PGM-1 patterns indicate a Mendelian explanation? What is the evidence?

(b) Show that the data conform with fundamental Properties 4 and 5. (Hint: for Property 4, first adjust all numbers of children to two per family, for example with 77 matings of 1/2 × 1/2 parents, assume 154 children in a ratio of 34: 80: 40.)

2

UTILITY OF THE GENETIC EQUILIBRIUM SQUARE LAW

It may seem to be the beginning student that conformity to the proportionality square law (Castle-Hardy-Weinberg) constitutes proof of random mating and the establishment of a Mendelian mode of heredity. On further thought, it should become apparent that such agreement of observed and expected data is merely inferential, or indicative of a tentatively plausible explanation for the results, but not proof. Nevertheless, the application of the principle in helping to establish a mode of inheritance has been made many times in the past. In fact, without populational data analysis by this simple rule, we would have far less knowledge of the human genotype than we do. Of course, it is a most useful tool in analysis of population structure both in nature and in the laboratory. But there are countless ways in which genotype frequency data may appear to fit the proportionality rule fortuitously or by particular fit of other hypotheses (see Chapters 5, 14, and 17), so that the investigator must be cautioned against drawing the conclusion of a simple Mendelian explanation and conformity to the conditions of the Castle-Hardy-Weinberg law without further evidence for the Mendelian mode of inheritance or for the randomness of the mating system.

ESTABLISHING A MENDELIAN HYPOTHESIS

Some of the classic cases in which the square law has been applied are worth examining. Weinberg first suggested deriving the expected proportions of parents with particular types of children (dizygotic twins) in order to lend credence to the Mendelian determination. His technique, based on the proportion given in Table 1-1B, was applied to the expected frequencies of twin births among relatives of those mothers who already had multiple-birth offspring. Assumptions for either dominance, incomplete dominance, or recessiveness were made, and the data collected by Weinberg fitted the latter best. While his analysis was not conclusive (if it is due to a single recessive, the genotype must have low penetrance with polygenic modification), the method was important in suggesting a likely mode of inheritance.

Example 2-1 illustrates the application of the square law to an early case of establishing a trait as being probably determined by a single pair of alleles without dominance—namely, the coat colors of shorthorn cattle: roan, red, and white. Wright examined the breeders' data for these coat colors (see Exercise 2). Because each of the six types of mating was recorded, he could estimate proportions of progeny either from a hypothesis of a simple monohybrid roan (red and white alleles without dominance in the heterozygote) or from a two-factor dihybrid hypothesis with dominance in both pairs of loci (*R*-*r* for red-white, *P*-*p* for roan-nonroan). The simpler hypothesis fit the data of progenies better than the dihybrid, but the two hypotheses could not be distinguished when all progenies were pooled (see Chapter 5 on two-factor epistasis).

Classic instances of pointing out the correct genetic theory by use of the square law are those from human genetics in which population genetics principles played a critical role in the first half of the twentieth century. In 1925, Bernstein established the hereditary basis for the *A*, *B*, *O* blood antigens as alleles of a single Mendelian factor by applying the square law to observed blood group frequency data (see Chapter 3 and Example 5-5). More recently, Neel (Neel, 1950, 1951) established the monofactorial basis of thalassemia major (Cooley's anemia) and minor (Cooley's trait) as well as sickle-cell anemia and sickle-cell trait. Thalassemia major (fatal anemia with marked hypochromia of red blood cells had been postulated to be produced by synergistic interaction of two dominant factors, either of which alone would produce the minor form (less severe microcytic anemia). The proponents of this theory had postulated two dominants because of the low consanguinity between parents with thalassemia major children. If this condition had been a recessive, much higher rates of consanguinity would have been expected for such a rare trait (see Chapter 9). But a frequency analysis of Mediterranean-origin peoples both in the United States (Sicilian descendants in Rochester, New York) and in Italy (the Po River Valley and in Sicily) showed thalassemia to be quite common, the minor form reaching nearly 10 percent in the Po Valley.

As for the sickle-cell conditions, it was presumed formerly that anemia could be caused by a dominant allele in heterozygous or in homozygous condition (either *SS* or *Ss*), an implication that an anemic child could have one parent only with the trait. Neel's hypothesis that the minor forms of both these anemias were the heterozygotes with incomplete dominance was established both by careful analysis indicating all parents of thalassemia and of sickle-cell anemia victims to be heterozygotes and by the fit of populational data to the single-locus hypothesis.

TABLE 2-1 *Frequencies (%) of human M-N blood groups (modified from Boyd, 1950, and from Wiener, 1943)*

Population	Place	Number Tested	*M*	*MN*	*N*	p_M	q_N
Eskimo	East Greenland	569	83.5	15.6	0.9	0.913	0.087
American Indians (Navaho)	New Mexico	361	84.5	14.4	1.1	0.917	0.083
American Indians (Pueblo)	New Mexico	140	59.3	32.8	7.9	0.757	0.243
Finns	Karjala	398	45.7	43.2	11.1	0.673	0.327
Russians	Moscow	489	39.9	44.0	16.1	0.619	0.381
Swedes	Sweden	1200	36.1	47.0	16.9	0.596	0.404
Chinese	Hong Kong	1029	33.2	48.6	18.2	0.575	0.425
Japanese	Tokyo	1100	32.4	47.2	20.4	0.560	0.440
Germans	Berlin	8144	29.7	50.7	19.6	0.550	0.450
Belgians	Liege	3100	28.9	50.3	20.8	0.540	0.460
Poles	Poland	600	28.2	49.0	22.8	0.527	0.473
English	London	422	28.7	47.4	23.9	0.524	0.476
Egyptians	Cairo	502	27.8	48.9	23.3	0.522	0.478
Negroes	New York City	730	22.1	49.8	28.1	0.470	0.530
Ainu	Shizunai	504	17.9	50.2	31.9	0.430	0.570
Fijians	Fiji	200	11.0	44.5	44.5	0.332	0.668
Papuans	Papua	200	7.0	24.0	69.0	0.190	0.810
Australian Aborigines	Queensland	372	2.4	30.4	67.2	0.176	0.824

Perhaps one of the most extensive sets of data with a clearly simple genetic basis and a well-established indication of random proportionalities is that of the *M-N* blood group system in humans. There is no doubt about the Mendelian nature of the *M-N* alleles or the random proportionality of genotypes for these alleles. Table 2-1 gives an abbreviated list from Boyd (1950). In the associated *S-s* blood group, antisera testing for *s* is not readily available (Race and Sanger, 1968, 1975), and populational sampling has not been so extensive for *S-s*. The *S-s* factors when combined with *M-N* represent a genetic parallel with the Rh system (*C-D-E*). Many human populations show conformity to the square law for genotypes of the *M-N* blood group (see Exercise 2).

EXAMPLE 2-1

AN EARLY CASE OF THE USE OF THE SQUARE LAW

Wright (1917) established the evidence in favor of a single Mendelian locus as the genetic basis for roan coat color in the shorthorn breed of cattle (roan is the intermingling of white and colored hairs) in contrast with the alternative hypothesis proposed at the time of two loci: a red (*R*-), white (*rr*) pair of alleles with the recessive white epistatic over an independent pair for roan (*P*-), nonroan (*pp*). Supposedly, for example, an *RrPp* (roan) × *RrPp* (roan) mating would produce a ratio of 9 *RrPp* (roan): 3*Rrpp* (red): 4*rrPp* or *rrpp* (white) offspring. Wright used the collected data from herd books to show that a simple hypothesis of a *single* Mendelian locus and random mating fit the data far better than the two-locus hypothesis: Roan was best interpreted as a heterozygote (*Nn*) in which a single dose of the semidominant gene produced a mixture of white and colored hairs with *NN* = homozygous white phenotype and *nn* = red homozygous phenotype.

The crude data from the herd books summarized by Wright gave the following totals:

White	Roan	Red	Total
756	3,780	4,169	8,705

Assuming the single locus (*N-n*) explanation, we let

$$p_N = \text{frequency of white allele} = 0.3040$$

$$q_n = \text{frequency of red allele} = 0.6960$$

Then

	White (*NN*)	Roan (*Nn*)	Red (*nn*)	Total
Observed percentage	8.69	43.42	47.89	100
Expected percentage	9.24	42.31	48.45	100

Thus, the observed frequencies are in good agreement with those expected from random mating and a single pair of alleles without dominance. Consideration of the alternative two-locus explanation can be found in Chapter 5.

CONFORMITY FOR THE WRONG REASON

It is important to realize that many special conditions may imitate the simple square law, which is a genetic application of the binomial (or multinomial) square law (Appendix A-2 and A-4). In populations where matings cannot be controlled by an experimenter, a close fit between observed and expected frequency data does not rule out a more complex genetic basis for phenotypes observed. In Chapter 5, where two or more loci are discussed, cases indistinguishable from the single-locus condition will be described (for example, with dominant epistasis with dominant A = dominant B, and only the genotype *aabb* distinguishable). Also, a fit of the square law does not rule out nonrandom mating or the action of natural selection (see Chapters 14 and 15). Many pitfalls are possible, so that while it is easy to test the square law, when we find agreement between observed frequencies and expected we cannot assume that all questions are solved. Two examples (Example 2-2) illustrate the application of the square law, first to present evidence for a simple recessive genetic interpretation of a most difficult genetic trait, diabetes (Steinberg, 1959), and second to demonstrate that the square law cannot be used to distinguish a genetic from a nongenetic hypothesis (Lilienfeld, 1959). Both of these examples err in two ways, first in the weak assumption (seriously or facetiously made) that the trait is inherited in a simple way, and second in the lack of additional genetic evidence (from pedigree data, for example) confirming what is really a fortuitous case where the incidence of a condition happens to fit the square law.

In spite of these insecurities, a population analysis should always first test the square law, because it is from the *disagreement* between observed and expected values that new opportunities may become evident for discovering determining factors in a population. When the genetic basis for observed variation is known, the square law becomes a valid tool in several instances: either (1) when the equilibrium conditions are not true for the population—for example, when mating is nonrandom or some nonneutral forces are acting on the population, (2) when artifacts seem to imply nonconformity and must be uncovered (such as "stratification" into subpopulations, incomplete ascertainment of inheritance with erroneous incidence information, or wrong genetic theory), or (3) when comparisons are necessary between populations occupying different areas or times (geographic or temporal variation). It must be clear, however, that as long as the genetic basis is not known, the square law must be used with caution in attempting to prove a mode of inheritance.

EXAMPLE 2-2

DOES CONFORMITY WITH THE SQUARE LAW SEPARATE GENETIC FROM NONGENETIC HYPOTHESIS?

Steinberg (1959) and others applied the square law to the published data on diabetes mellitus incidence on the basis of parental occurrence and noted the fit in five out of six sets of data to a simple recessive. See Table A.

The incidence of recessives is expected to occur according to the square law from parents of three types ($Aa \times Aa$, $Aa \times aa$, $aa \times aa$), from property 4 shown earlier. The Steinberg and Wilder data fit expectation from an apparent recessive gene frequency of about $q = 0.10$, while the others range from $q = 0.05$ (Harris) to 0.08 (Pincus and White).

TABLE A *Published incidence of diabetes on the basis of parental occurrence (from Steinberg, 1959).*

	Steinberg and Wilder		Pincus and White		Allen		Harris		Thompson and Watson		von Kries	
Mating	Obs.	Exp.	Obs.	Exp.	Obs.	Exp.	Obs.	Exp.	Obs.	Exp.	Obs.	Exp.
Neither parent diabetic	1589	1588.63	440	440.6	124	122.8	1124	1119.1	1404	1407.9	1137	1134.7
One parent diabetic	370	370.74	80	78.9	17	19.4	109	118.8	223	214.8	160	164.2
Both parents diabetic	22	21.63	3	3.5	2	0.8	8	3.1	4	8.2	8	5.9
X^2 (1 *d.f.*)	0.012		0.10		2.28		8.57*		2.60		1.14	

* $P < 0.01$

The large X^2 for the Harris data is entirely due to the excess of affected offspring where both parents were diabetic.

The inheritance of diabetes mellitus is not so simple, however. According to Clarke (1964), Neel, et al. (1965), and Falconer (1967), there is considerable evidence that (1) the penetrance of the genotype is incomplete (20 to 30 percent), (2) in glucose tolerance tests it is very difficult to separate normal from potentially diabetic persons, and (3) dietary factors (and other environmental influences) are important in cases of late onset. All these factors suggest that if a single gene is concerned, its phenotypic effect must be highly modifiable. Polygenic modification of what may be two essentially different conditions (early and late onset) from alternate genotypes makes more sense than single-gene determination. Thus, it is premature and oversimplified to conclude that this trait is entirely due to a monfactorial recessive, in spite of conformity to the square law from the sources given above.

Lilienfeld (1959) was interested in finding out just how discriminating the square law is between genetic and nongenetic hypotheses. He circulated a questionnaire among the medical students at the University of Buffalo asking whether or not their parents were physicians (or had ever attended a medical school). Results from 261 medical students were as follows:

	No. Medical Students	
Mating	Observed	Expected
Neither parent physician	229	229.98
One parent physician	32	30.04
Both parents physicians	0	0.98

$X^2(1\ d.f.) = 1.11,\ P = 0.30$

Consequently, from the good fit of the data to the square law, one might conclude that the "gene-for-medical-student" is a recessive occurring at a frequency of 0.0613!

Because a zero class was observed (no students had both parents medically trained), Lilienfeld took a more extensive survey from all students at the University of Buffalo, classifying them as to their parental attendance at the same university.

Mating	No. Students	
	Observed	Expected
Neither parent attended U.B.	3640	3630.3
One parent attended U.B.	232	251.3
Both parents attended U.B.	14	4.4

X^2 (1 *d.f.*) $= 22.93, P < 0.01$

The smallest class is too large for the expected (too many "homozygotes") and the middle class too small for a "gene for attending the University of Buffalo" with a q value $= 0.0335$. Quite possibly, this lack of agreement with expected comes from some stratification within the student body analogous to that in genuinely genetic situations.

ARTIFACTS IMPLYING NONCONFORMITY

Stratification, or Subpopulations Differing in Their Gene or Genotype Frequencies

Few populations are completely homogeneous, almost totally randomly mating, and large enough in size to approach the square-law conditions. All sorts of limits to migration, mixing, or intermingling of individuals promote grouping into local populations, or demes; in human populations, it is evident that social and economic barriers exist that prevent or limit marriages between the strata, or classes. If a novice geneticist were to sample a population that is stratified or subdivided into groups with different genotype frequencies, he or she would observe a deficiency of heterozygotes and an excess of homozygotes, even though square-law conditions may hold within the subgroups. This is an illustration of Wahlund's principle, to be discussed under random genetic drift (Chapter 12), and it is demonstrated in Example 2-3. It is introduced here to alert the student to one of the pitfalls likely to arise in sampling real populations when the investigator may make assumptions about uniformity within an observed group of individuals. Thus, an artifact of nonconformity to the square law in pooled totals of divergent subpopulations may be caused by averaging different panmictic units.

Faulty Ascertainment

Completely random samples of a population are seldom attained; many biases on the part of the observer may conspire to prevent a reliable estimate of *incidence* (frequencies at birth, or zygote frequencies) for a genetic condition in a population. Human data are more susceptible than those from experimental organisms because observers who collect data on a particular human genetic condition tend to include a biased large number of cases coming to their attention from institutions or clinics. The various statistical devices invented to compensate for incomplete ascertainment (establishment of ratio and incidence of hereditary conditions) constitute a very large body of knowledge beyond the scope of this book since they do not relate directly to population genetics. (For further discussions of ascertainment, see Neel and Schull, 1954 (Chapter 14); Li, 1961 (Chapter 5); Stern, 1973; Crow, 1965; Smith, 1968.)

EXAMPLE 2-3

THE STRATIFICATION PRINCIPLE

Suppose a large population is actually subdivided into two (or more) subpopulations isolated from each other by a subtle barrier to crossbreeding, such as a behavioral or mental eccentricity in one which the other does not like, and the aversion toward the opposite group is mutually shared. Even though the two subpopulations could interbreed if they could get over their behavioral dissimilarities, they do not. Because they look very much alike, a novice geneticist considers them a single panmictic unit and pools the gene frequency data from both subpopulations as follows:

Observed (400 persons):	$184AA$:	$112Aa$:	$104aa$
Thus $p_A = 0.6$, $q_a = 0.4$ and expected is	$144AA$:	$192Aa$:	$64aa$
The difference, observed-expected, is	$+40$	-80	$+40$

There is a large discrepancy between observed and expected with too many homozygotes and too few heterozygotes. The geneticist draws the erroneous conclusion that the population is not mating randomly at all, but in fact 200 persons have been sampled from each of two isolated subpopulations. The geneticist finally learns by experience that the two groups can be distinguished because one group is thick-skinned and other is thin-skinned. When the data are separated on the basis of skin, the following is found:

Observed (200 thick-skinned):	$164AA$:	$32Aa$:	$4aa$
(200 thin-skinned):	$20AA$:	$80Aa$:	$100aa$

"Obviously," the geneticist says, "now there appears to be random mating within each subpopulation." (The geneticist would find that the variance of gene frequency between subgroups will account for the difference between the pooled genotype frequencies and what would have been the frequencies if all 400 persons had been from a single panmictic unit.)

These data are summarized in Table A. Each set of observed data may be tested for agreement with an expected set by a chi-square as indicated. Thus, there is an artifact of nonconformity with the square law in the pooled totals, caused by averaging divergent but panmictic subgroups.

TABLE A *Averaging of heterogeneous groups: a false impression of nonconformity to the square law, even though each group by itself does conform (see also Table 12-4)*

Subpopulation	Allele Frequency A	q_a		Genotype Incidence AA	AA	aa	Total
I	0.9	0.1	Obs.	164	32	4	200
			Exp. $200(p + q)^2$	162	36	2	200
					$X^2 = 2.47$ (*n.s.*)		
II	0.3	0.7	Obs.	20	80	100	200
			Exp. $200(p + q)^2$	18	84	98	200
					$X^2 = 0.45$ (*n.s.*)		
Average of subpopulations:	$\bar{p} = 0.6$	$\bar{q} = 0.4$	Obs.	184	112	104	400
			Exp. $400(\bar{p} + \bar{q})^2$	144	192	64	400
					$X^2 = 69.44$ ($P < 0.01$)		
Adjustment because of stratification*							
Variance of $q = 0.09$			Exp. 144 + 36	192 − 72	64 + 36		
Expected increment in homozygotes $= (0.09)(400) = 36$			180	120	100		400

* This adjustment by using the variance in q will be discussed under Wahlund's principle in Chapter 12.

CONFIRMATION OF RANDOM MATING

When genotypes have been reliably ascertained, the square law becomes a useful tool in substantiating the nature of the mating system. In both natural and experimental populations, sampling the genotypes' incidence is essential to understand the genetic structure. From a notable series of experiments by Dobzhansky (1947), data from an experimental population of *Drosophila pseudoobscura* maintained over six generations (Example 2-4) illustrate random zygote frequencies each generation, although changes in haploid frequencies took place over the period of time due to natural selection. By checking square-law conditions at each generation, Dobzhansky confirmed the approximate random fertilization of gametes containing the Standard (ST) or Chiricahua (CH) chromosomal arrangements. There was a slight but nonsignificant consistent excess of heterokaryotypes in each sample (which continued in nearly all further samples not given in this example). If consistency is to be tested for significance, however, it must be realized that the change in chromosomal frequencies from the start (20 percent ST: 80 percent CH) to a condition that proved to be equilibrium (70 percent ST: 30 percent CH) makes the pooling of samples a case of "stratification" as illustrated in Example 2-3. First, we may test for conformity to the square law within each sample; then changes in frequencies between samples through time may be tested for statistical significance. The student should realize the various ways in which hypotheses may be applied to these data for estimating "expected" values and some of the tests conventionally used to measure the probability for the "fit" of observed data to expected outcomes. In the next section the data from Example 2-4 are scrutinized for probability of agreement with a few such hypotheses.

EXAMPLE 2-4

RANDOM MATING IN EXPERIMENTAL POPULATIONS BUT CHANGES IN CHROMOSOME FREQUENCIES OVER TIME

In the mid-1940s Dobzhansky began laboratory-population cage experiments with *Drosophila pseudoobscura* taken from natural populations of the species at Piñon Flats on the eastern slopes of Mt. San Jacinto, California. See Example 3-1 for this location. The experimental populations were confined to large boxes, or cages, in which a constant supply of fresh food was provided to allow constant turnover of generations within the cage (a technique originally used by L'Heritier and Teissier in 1934). His objective was to test the validity of selective value (Darwinian "fitness") differences between carriers of arrangements of the third chromosome whose frequencies had displayed seasonal cyclic changes every year in the wild locality. Chromosomal rearrangement due to inversion, it should be noted, can be equated to genic allelism owing to the fact that inversion heterozygotes tend to be crossover suppressors; thus, lowering of recombination between two arrangements tends to make each a genetic unit in an analogous way to a Mendelian segregating allele. Changes of chromosomal arrangement frequencies in the laboratory could be ascertained by sampling eggs at regular intervals from the food surface and growing the larvae under more optimal density conditions in containers outside the cage until the larvae were ready for salivary gland chromosome analysis. From such samples he was able to study the changes in chromosome frequencies in the populations. (Descriptions of these arrangements' frequency changes in experimental populations and the principles of adaptive balanced polymorphism will be found in Chapters 14–17.) Identification of the chromosomal arrangements is done in the third larval instar. In the interval between explanting the eggs and observing the chromosomes there could be differences in growth rates or viabilities of the three karyotypes. Agreement between the observed frequencies in each generation and the expected values on the basis of square-law conditions suggests, however, that the assumption of identity between egg frequencies and late larval frequencies was not at first sight in error to any appreciable extent.

The data in Table A are six samples (for illustration purposes abridged from Dobzhansky, 1947) taken successively at monthly intervals from a population initiated three months previously with 20 percent *ST* (standard) and 80 percent *CH* (Chiricahua) chromosomes among the introduced flies. Whether the eggs were deposited following random fertilizations of gametes was tested by applying the square law to each monthly sample. Note that in each sample, agreement between observed and expected is very good (X^2 is not significant). However, five of the six samples have a slight excess of heterokaryotypes (*ST*/*CH*). One test for the significance of this consistency is shown at the bottom of the table, where the summed observations are considered as if they were from one homogeneous population. Again, there is no serious departure from expectation, although the X^2 of 2.29, with one degree of freedom and a probability of 0.13, hints to the investigator that a real excess of heterokaryotypes might exist. Dobzhansky's original data included 17 samples from populations containing *ST* and *CH* chromosomes. In all but two samples the *ST*/*CH* karyotype was in excess, indicating a slight but consistent

TABLE A *Zygotic and gametic frequencies of Standard (ST) and Chiricahua (CH) arrangements among eggs deposited (i.e., third instar larvae dissected) from an experimental population; samples taken at monthly intervals (modified from Dobzhansky, 1947, population 19)*

Population Sample		Zygote Frequencies				Gametic Frequencies	
		ST/ST	*ST/CH*	*CH/CH*	X^2	$ST_{(p)}$	$CH_{(q)}$
December 1944	Obs.	41	77	32	0.14	0.530	0.470
	Exp.	42.1	74.7	33.1			
	d.	−1.1	+2.3	−1.1			
January 1945	Obs.	56	78	16	2.17	0.633	0.367
	Exp.	60.1	69.8	20.2			
	d.	−4.1	+8.3	−4.2			
February 1945	Obs.	53	75	22	0.33	0.603	0.397
	Exp.	54.6	71.8	23.6			
	d.	−1.6	+3.2	−1.6			
March 1945	Obs.	65	66	19	0.14	0.653	0.347
	Exp.	64.0	68.0	18.0			
	d.	+1.0	−2.0	+1.0			
April 1945	Obs.	61	74	15	1.19	0.653	0.347
	Exp.	64.0	68.0	18.0			
	d.	−3.0	+6.0	−3.0			
June 1945	Obs.	70	71	9	2.70	0.703	0.297
	Exp.	74.2	62.6	13.2			
	d.	−4.2	+8.4	−4.2			
						Average	
						$\bar{p}$	$\bar{q}$
Totals	Obs.	346	441	113		0.6294	0.3706
{Exp.* from average gametic frequencies $(\bar{p}+\bar{q})^2(900)$}		356.5	419.9	123.6			
	d.	−10.5	+21.1	−10.6	$X^2 = 2.29$ ($d.f. = 1$, $P = 0.13$)		

* A more exact method for estimating the expected when samples are pooled and averaged is given in Chapter 12 (Wahlund effect).

parameter not in agreement with the square-law expectation. Whether the excess was due to nonrandom mating among parent flies or to differences in developmental rates or viability from egg stage to larval third instar was not determined. Later evidence has pointed to all these factors to some extent favoring heterokaryotypes (Dobzhansky, 1947, 1961; Spiess, Langer, and Spiess 1966).

TESTS OF SIGNIFICANCE

In making observations on any real population, the investigator must realize that the observed values, even though entirely unbiased (that is, taken completely at random), may deviate from the values expected for at least two reasons: (1) error due to sampling of finite numbers will be the result of "chance" (indeterminate and random factors), while (2) a possible misjudgment in basic premises may invalidate the hypothesis. Tests of significance are designed to distinguish between these two types of deviation, to judge whether the departure from expectation can be assumed "nonsignificant" (due to chance) or "significant," in which case the hypothesis must be reexamined or rejected.

For convenience the pertinent features of these tests are summarized in the Appendix, along with other statistical or mathematical techniques that will be useful throughout simple population analysis. An extensive description of these techniques is beyond the scope of this book. For further understanding in the subjects of biometry, experimental biology design, and the mathematical bases for derivations of statistical tests, the reader is referred to those sources listed in the Appendix.

Two commonly used tests of significance are the chi-square test (χ^2) and variance analysis ("Student's t" is a special application of the latter for a single contrast) (see Appendixes A-7 and A-8). While these tests are mathematically related, the choice of which is more appropriate depends on the nature of the data. In the chi-square test, observed deviations (or squared deviations) are compared to standard deviations (or variances) *fixed by hypothesis.* "Expected" values are calculated by applying some function like the square law to the sample's estimate of a parameter like p, q. In variance analysis (or Student's t), the observed variation itself must be used to furnish parameter limits by its very nature and is not fixed by hypothesis. Usually frequency data lend themselves to the chi-square test or similar tests with a fixed hypothesis, while measurement data are dealt with more frequently by constructing distributions from the experimental observations themselves. Nevertheless, frequency data can be treated as sets of random variables dependent on population size. We shall illustrate first the use of the fixed-hypothesis chi-square method, followed by the use of the normal deviate, or method of t, with p, q as random variables.

USE OF CHI-SQUARE FOR TESTING AGREEMENT WITH THE SQUARE LAW

In Example 2-4, the condition of random mating is tested by applying the binomial square law of the frequencies of ST and CH arrangements for each monthly sample and the average frequencies ($\bar{p}$, $\bar{q}$) for the total of all samples pooled together. Then, because

$$X^2 = \sum \frac{(\text{obs-exp})^2}{\text{exp}} = \sum \frac{(\text{obs})^2}{\text{exp}} - N \qquad \text{(from Appendix A-8c)}$$

if we let A, B, and C = observed numbers of the three karyotypes (or genotypes) ST/ST, ST/CH, and CH/CH, respectively, p, q = estimated single chromosome (or allelic) frequencies, N = number of individuals sampled, as in formula (1-3a), then, using X^2 as an estimation for the theoretical χ^2,

$$X^2 = \frac{(A - p^2N)^2}{p^2N} + \frac{(B - 2pqN)^2}{2pqN} + \frac{(C - q^2N)^2}{q^2N} = \frac{A^2}{p^2N} + \frac{B^2}{2pqN} + \frac{C^2}{q^2N} - N \qquad (2\text{-}1)$$

There is a single degree of freedom because two fixed values, total N and the allelic frequency q, use up two of the three classes counted. Thus, for the June 1945 sample, we would supply the following data (using (2-1)):

$$X^2 = \frac{(70)^2}{(0.494)150} + \frac{(71)^2}{(0.418)150} + \frac{(9)^2}{(0.088)150} - 150 = 2.70$$

For small samples ($N < 100$) or when gene frequency is so low that the result expected for the rarest class (q^2N) is <1.0, it is important to realize that calculation of expected values should be based on a more exact probability test than the square law provides—use of the multinomial probability distribution for three classes. Hogben (1946) and Levene (1949) developed the following formula, whose derivation is given by Li (1955, 1976) for expected values:

Genotypes	AA	Aa	aa	Sum
Expected	$\frac{g_1(g_1 - 1)}{2(2N - 1)}$	$\frac{g_1 g_2}{2N - 1}$	$\frac{g_2(g_2 - 1)}{2(2N - 1)}$	N

(2-2)

where

$g_1 = 2A + B$, $g_2 = 2C + B$ (with A, B, C the observed numbers of genotypes AA, Aa, and aa, respectively)

N = number of individuals sampled, as above

then

$$p = \frac{g_1}{2N},\ q = \frac{g_2}{2N}$$

This formula gives homozygote expected values smaller than those calculated the usual way. For example, using the June 1945 sample from Example 2-4,

	Obs. =	70	71	9	
Usual calculation	Exp. =	74.201	62.597	13.202	$X^2 = 2.70$
Levene's exact	Exp. =	74.097	62.806	13.097	$X^2 = 2.58$

The decrease in chi-square is not very much, although in a borderline case the difference could well be important.

The square-law hypothesis can be tested at each generation (samples of 150 larvae), and if homogeneity between samples is assumed, at the end of the sequence all samples pooled (900 larvae) provide an average overall frequency value ($\bar{p}$, $\bar{q}$). These chi-squares are given in Table A of Example 2-4. (For a relationship of the chi-square to the inbreeding coefficient that measures deficiency of heterozygotes, or excess of homozygotes, compared with the random mating expectation, see Chapters 9 and 12.) All six samples in this population conform to the square law; none of their chi-squares is significant. When all six are pooled there is still no evidence of departure from expectation (random mating). Actually, the samples do tend to have a slight excess of heterozygotes. A more exact method for calculating expected totals will be given in Chapter 12 (Wahlund effect), after we consider the general problem of averaging subpopulations or samples that differ in p, q values; it will be shown that the excess of heterozygotes does become significant when the expected amount of heterozygosity has diminished from application of the stratification principle.

It is important to note that the samples in this population change in relative frequencies of *ST* and *CH*. Now we need to illustrate the testing of two or more samples for the "null hypothesis" that there is no difference between them in the amount of one gene (in this instance, chromosomal arrangement) or genotype (karyotype).

USE OF CHI-SQUARE FOR TESTING DIFFERENCES BETWEEN SAMPLES

In respect to the chromosomal haploid frequencies over the six-month interval (Example 2-4), it is evident that the *ST* arrangement did increase; in fact, the increase was more or less continuous. We can test the significance of the difference in p, q values between any two samples, the first (December 1944) and last (June 1945) being the most relevant contrasts here because they are most divergent. Two simple tests are illustrated in Tables 2-2A and B: (1) a X^2 test for total *ST* versus *CH* ($2N = 300$ chromosomes

TABLE 2-2A *Chi-square tests for haploid chromosome frequency differences between two samples from Table A of Example 2-4*

Sample		*ST*	*CH*	Sum
December 1944	Obs. (O)	82 + 77	77 + 64	300
	Sum (O)	159	141	300
	Exp.* (E)	185	115	300
	$d = (O - E)$	−26	+26	0
	d^2/E	3.65	5.88	9.53
June 1945	Obs. (O)	140 + 71	71 + 18	300
	Sum (O)	211	89	300
	Exp.* (E)	185	115	300
	$d = (O - E)$	+26	−26	0
	d^2/E	3.65	5.88	9.53
Obs. sum.		370	230	600
			Contingency X^2 =	19.06

Degrees of freedom: total items = 4
−1 for grand total
−2 for two marginal totals (one for columns and one for rows)

Therefore, $d.f.$ = 1 remainder

$P < 0.001$ We conclude that a highly significant difference exists between these two samples in frequency of chromosomal arrangements

* Expected calculated from contingency of rows and columns; that is, by proportionality of any row or any column average. See Appendix A-8D for contingency chi-square test.

TABLE 2-2B *Chi-square tests for diploid (zygotic) differences between two samples from Table A of Example 2-4*

Sample		ST/ST	ST/CH	CH/CH	Sum
December 1944	Obs. (O)	41	77	32	150
	Exp.* (E)	55.5	74	20.5	150
	$d = (O - E)$	-14.5	$+3$	$+11.5$	0
	d^2/E	3.79	0.12	6.45	10.36
June 1945	Obs. (O)	70	71	9	150
	Exp.* (E)	55.5	74	20.5	150
	$d = (O - E)$	$+14.5$	-3	-11.5	0
	d^2/E	3.79	0.12	6.45	10.36
Obs. sum.		111	148	41	300
				Contingency X^2 =	20.72

Degrees of freedom: total items = 6
−1 for grand total
−1 for one row total
−2 for two column totals

Therefore, $d.f.$ = 2 remainder

$P < 0.001$ We conclude that these two samples differ significantly in their zygotic frequencies (diploid karyotypes)

* Expected calculated from contingency of rows and columns as in Part A. Also note that by considering zygotic data, the degrees of freedom are increased by 1, but the total X^2 is only slightly increased over that in part A (haploid data).

per sample), and (2) a X^2 test for the zygotic data ($N = 150$ karyotype individuals per sample).

In both tests the expected values are obtained on the assumption that the two samples are homogeneous—that is, they belong to the same population (or in this case no change has taken place in p, q over generation time). Each "box" (haploid or diploid observed number in a sample) is expected to have the same proportion as the frequency for that arrangement or karyotype in the sum of the two samples. For example, in Table 2-2A, ST has a frequency of 370/600 in the total; therefore, ST for December 1944 would be expected to be 370/600 times 300, or 185.0. In the contingency test of Table 2-2A, the haploid frequencies between the two samples are significantly different with $X^2 = 19.1$, with one degree of freedom. In Table 2-2B, diploid frequencies not only are determined by p_{ST}, q_{CH} but also by the square law. Because the chromosome frequencies differ as demonstrated already, the new contingency chi-square is only slightly larger ($X^2 = 20.7$, with two degrees of freedom) due to the small discrepancies between the two samples in their conformity to the totals for three karyotypes. It can readily be seen that separation of the three karyotypes is neither necessary nor as efficient as using the haploid frequencies in demonstrating significance of the sample difference. The homokaryotype differences are most affected by the change in p, q values and make a greater contribution to the X^2 than the heterokaryotype does.

CONFIDENCE LIMITS AND METHODS OF *t* FOR DIFFERENCES IN FREQUENCIES

In contrast with the chi-square test, which requires a hypothesis to fix the expected values, the experimental design often requires that the observed data alone must provide the estimate of true variance and thus the distribution limits for parameter values based on an underlying normal distribution. For example, if any two samples differing in allelic

TABLE 2-3 *Comparison of two samples of size = 150 = N, or 300 chromosomes from Example 2-4, first (A) by estimating confidence limits from binomial standard errors, and second (B) by assuming a null hypothesis calculating the Student's t = d/s* and its probability from the standard error of the difference between samples*

A

Sample	p_{ST}	q_{CH}	$2N$	Variance ($pq/2N$)	Standard Error (s_q)
December 1944	0.5300	0.4700	300	0.0008303	0.0288
June 1945	0.7033	0.2967	300	0.0006960	0.0264
Average ($\bar{p}$, $\bar{q}$)	0.6167	0.3833			

For 95% confidence limits we take $\pm 1.96(s_q)$ as follows:

0.470 ± 0.0564—lower limit $= 0.4136$
0.297 ± 0.0517—upper limit $= 0.3487$
limits then not overlapping

For 99% confidence limits we take $\pm 2.58(s_q)$ as follows:

0.470 ± 0.0743—lower limit $= 0.3957$
0.297 ± 0.0681—upper limit $= 0.3651$
limits then not overlapping

B

Assuming the null hypothesis (no difference between samples), we calculate the Student's $t = d/s$

or

$$t = \frac{(q_1 - q_2)}{\sqrt{\bar{p}\bar{q}\left(\frac{1}{2N_1} + \frac{1}{2N_2}\right)}}$$

$$t = \frac{0.470 - 0.297}{\sqrt{(0.6167)(0.3833)(\frac{1}{150})}} = \frac{+0.173}{0.0397} = +4.36$$

Because these samples could be greater or less than the assumed mean ($\bar{p}$, $\bar{q}$), it is a two-tailed test and d could be positive or negative. Therefore, $P < 0.0001$ that the two samples are drawn from a single population of $\bar{p}$, $\bar{q}$.

* t = ratio of the difference between sample means (d) to the standard error (s) of the difference, or in this case the standard error of the average $\bar{p}$, $\bar{q}$ values, assuming the two samples to be drawn from the same population (null hypothesis). See Appendixes A-5, A-6E, and A-7.

frequencies were to be compared, the investigator should not use one population's frequencies as the "expected" for the other. Consequently, the observer has to estimate parameter limits for each population, relying on an estimate of standard error for confidence limits around the observed frequencies. It is then possible to apply the null hypothesis that there is no difference between them, or to estimate the probability that the two populations are drawn from a single inclusive population. The variance analysis (or its modification, Student's t) is based on the fact that if random factors alone act on allelic frequencies, their distribution will be according to a binomial with a variance of

$$V_q = \frac{pq}{2N} \tag{2-3}$$

where N = number of diploid individuals sampled (see Appendix A-4B, formula (3); A-6E for derivation; and Chapter 12).

In Table 2-3A, confidence limits are indicated for the p, q values of the samples from Example 2-4. From the 95 and 99 percent confidence values, it is evident that the limits are not overlapping, and therefore it is quite unlikely that the two samples were drawn from the same parameter population of average p value. We conclude that they must have changed in chromosome arrangement frequency.

A test that arrives at the same probability as the chi-square in this case is illustrated in Table 2-3B (Student's t): the difference between q values of the two samples is divided by the standard error of the difference. The two samples are assumed to be drawn from the same parameter (null hypothesis), and the average q value ($\bar{q}$) is used for the standard error calculation as if it were the same for both samples (see Appendix A-7). In Table 2-3B, the t value of 4.36 is far too great to be ascribed to chance and the null hypothesis is rejected.

DOMINANCE AND ESTIMATES OF GENE FREQUENCIES

Up to this point we have considered gene frequency analysis only for populations with genotypes distinguishable as AA, Aa, and aa (except for Example 2-2 on diabetes). Estimates of p, q values have been made efficiently by using the principle that $p = D + \frac{1}{2}H$ and $q = R + \frac{1}{2}H$ (formula 1-3).

These are maximum-likelihood estimates (see Appendix A-9). However, with complete dominance, genotype AA is phenotypically identical with Aa; consequently, the efficiency of estimating gene frequency becomes limited if only a single generation is observed. If the population is large and can be considered random mating, and the sampling of phenotypes is done on independent subjects (that is, avoiding counting of close relatives as independent observations), then the estimate of q is easily obtained from $q^2 = C/N = R$, where C = observed number of recessives and R = proportion of recessives. Then

$$q = \sqrt{\frac{C}{N}} = \sqrt{R}* \tag{2-4}$$

* See Haldane's improved estimate in Exercise 9, where

$$q = \sqrt{\frac{4R \cdot N + 1}{4N + 1}} = \sqrt{\frac{4C + 1}{4N + 1}}.$$

with variance

$$V_q = \frac{(1-q^2)^*}{4N} = \frac{pq}{2N} + \frac{p^2}{4N} \tag{2-5}$$

(See Appendix A-9 and Exercise 17 for this derivation.) This variance is larger than that in (2-3) by the second term ($p^2/4N$). When dominance is complete, the standard error of this q estimate (from (2-4)) diverges increasingly from that of the estimate we would make if all genotypes were ascertainable (from (1-3)) as q decreases. If the population is not large and panmictic, and if good reliable genetic evidence for the trait is not available, estimating q in this way is hazardous, because under any other mating system the square root of the recessive frequency will not be q.

With advances in biochemical and immunological techniques in recent years, traits that were completely recessive on the basis of earlier criteria have become "unmasked" in heterozygotes by the demonstration of single allelic products, as discussed in the following chapter. It is reassuring when estimates of genotype frequencies are made before a new testing reagent is available for detecting heterozygotes to find substantial agreement once the new substance is used. Example 2-5 illustrates the frequencies of the human blood group *S-s* in England before and after the discovery and availability of anti-*s*, the serum specifically typing for the "recessive" allele. The less efficient estimate obtained from use of anti-*S* alone based on (2-4) and the later estimate based on ascertainment of *SS* and *Ss* genotypes agree, and thus the population is indicated to be at equilibrium for these alleles, with high probability.

EXAMPLE 2-5

DOMINANCE AND CONFIRMATION OF EQUILIBRIUM AFTER APPLYING A REAGENT TO DETECT HETEROZYGOTES

The human blood antigens *S* and *s*, associated with the *M-N* antigens, were first tested on English populations (1947–1950) with only anti-*S* serum available; so that *SS* and *Ss* genotypes were indistinguishable, and *ss* genotype was *S* negative (Race and Sanger, 1968).

Blood Type	Observed	Expected %	Estimated q
S-	776	$\begin{cases} SS = 10.7 \\ Ss \quad 44.0 \end{cases}$	$q_s = \sqrt{643/1419} = 0.6731 \pm 0.0098$
s (non-*S*)	643	*ss* 45.3	
Total	1419		

By the late 1950s, anti-*s* serum became available, so that in effect the alleles were co-dominant. Another group of English persons was sampled by Cleghorn in 1960 (cited by Race & Sanger).

Blood Type	Observed	Expected %	Estimated q
SS	99	9.5	$q_s = 0.6920 \pm 0.0146$
Ss	418	426	
ss	483	47.9	
Total	1000		

The estimates are not significantly different. This close agreement follows from the fact that the population is probably at equilibrium for these alleles.

PARTIAL PROGENY TESTING WITH DOMINANCE

Without the aid of suitable refinement techniques for characterizing heterozygotes biochemically, immunologically, or morphologically, the investigator must resort either to progeny testing, if the organism is experimental, or to collecting data on relatives, if human genotypes are being analyzed, for accuracy in gene and genotype frequency estimates. We shall first illustrate a method often used for experimental organisms in which progeny testing can be done—but not for all dominant individuals available if the number observed is too large to make sampling practical. An arbitrary sample of the dominants may be tested, leaving a large group classified only according to dominant versus recessive classes. The same method could be applied to cases in human genetics where either new antisera may be discovered or old ones may become exhausted so that four groups are recorded. Appropriately weighted estimates of gene frequency can be made in the manner illustrated in Example 2-6, following the method of Cotterman (1954) for dominants with partial subclassification. If the first classification is to dominant (A-) versus recessive (aa), and a small proportion (k) of A- are progeny tested and subdivided into AA and Aa, we may list the four classes and their expected frequencies as follows:

Phenotypes	Observed	Expected
A-	A	$N(1-k)(1-q^2)$
AA	B	Nkp^2
Aa	C	$Nk \cdot 2pq$
aa	D	Nq^2
Total	N	N

Then:

$$q = \frac{-(2B + C) + \sqrt{(2B + C)^2 + 8N(C + 2D)}}{4N} \qquad (2\text{-}6)$$

$$\text{Variance } q(V_q) = \frac{q(1 - q^2)}{2N(k - kq + 2q)}, \qquad \text{where } k = \frac{B + C}{A + B + C} \qquad (2\text{-}7)$$

Here the information both from the progeny-tested sample of dominants and from the square root of the recessive frequency are weighted by employing the maximum-likelihood method (see Appendix A-9 and Exercise 18) in order to achieve the expected values. In Example 2-6 chi-square tests indicate significant excess of heterozygotes in the drosophila populations; in contrast, the mouse natural populations tended to be deficient in heterozygotes.

EXAMPLE 2-6

PARTIAL SAMPLING OF DOMINANTS IN DROSOPHILA AND MOUSE POPULATIONS

To analyze for gene and genotype frequencies from experimental or natural populations with complete dominance at particular genic loci, it is efficient to progeny test a sample of dominant individuals, separating them into relative frequencies of *AA* and *Aa* genotypes. Final tabulation then consists of counting a large number of dominant and recessive phenotypes with a small subsample of tested dominants.

Hexter (1955) analyzed several laboratory populations of *Drosophila melanogaster* in which six autosomal recessive mutants were run for more than a year in competition with their wild-type alleles with a constant turnover of generations. The following data are from Hexter's low frequency *eyeless*^{2} versus $+ey$ first-generation count as an illustration. One hundred wild-type males were randomly selected each generation and individually mated to three mutant virgin females to determine their genotype. Of these, 93 were ascertained for genotype $+/+$ or $+/ey$:

Genotypes	Observed	Expected
$+/-$	1526	1519.62
$+/+$	31	41.19
$+/ey$	62	51.35
ey/ey	272	278.84
Total	1891	1891.00

Using (2-6) and (2-7)

$$q = \frac{-(124) + \sqrt{(124)^2 + 8(1891)(606)}}{4(1891)}$$

$$q = 0.384 \pm 0.0104$$

Note that this estimate of q is between the square root of R (272/1891), $q = 0.379$, and what one could derive by assuming the 1619 dominants to be subdivided proportionally to the tested $+/+$ and $+/ey$ fractions (1:2); namely, $q = 0.429$. The advantage of the maximum-likelihood estimate here is that the four groups can then be tested with a X^2 for agreement with the square law; $X^2 = 4.9$, with one degree of freedom (since q, k, and the total use up three out of four classes), $P = 0.026$. In the case of all six mutants, Hexter found an excess of heterozygotes during the early generations. In four populations (*eyeless*, *clot*, *hairy*, and *black* mutants versus their wild alleles), random square-law proportions were characteristic of later counts, while in two populations (*thread* and *vestigial* versus $+$) the excess persisted.

Petras (1967*b*) has employed this method in estimating frequencies of coat color alleles from natural populations of the house mouse, *Mus musculus*, inhabiting farm areas in southeastern Michigan. The dominant agouti-white-belly allele (A^w), characterized by absence of the dark apical band on hairs of the ventrum, was found commonly throughout the area. Several white-belly individuals were progeny tested by obtaining at least seven offspring to indicate A^w/A^w or A^w/A genotypes:

G. Lindeman Farm sample (pooled 1959–1962)

Phenotypes	Observed	Expected	
$A^w/-$	49	50.94	
A^w/A^w	4	1.34	
A^w/A	11	14.24	$q = 0.8408 \pm 0.016$
A/A	163	160.48	
Total	227	227.00	

For these data the $X^2 = 6.13$, 1 *d.f.*, $p = 0.014$; however, it should be noted that the expected value for the homozygous dominant class is very small, which makes the greatest contribution to the X^2 and probably excessively so. This result contrasts with the previous illustration from Hexter by displaying a slight excess of homozygotes in this mouse population.

DOMINANCE AND FAMILY DATA

Before the emergence of biochemical and immunological techniques used in ascertaining genotypes, especially for heterozygotes, dominant traits in human populations particularly were studied by family analysis of either two generations (parent-offspring) or a single generation (sib pairs or cousin pairs) in order to estimate genotype frequencies. These methods are rarely met in the literature today except in testing the hereditary components of epidemiological diseases or phenotypes on which only morphological criteria can be relied. It is naturally easier and more efficient for the geneticist to work with genotypes that are simple to ascertain. The modern trend therefore is to find biochemical or immunological criteria for identification of heterozygous carriers of hereditary conditions instead of relying on the more statistical devices worked out for dealing with family data by mathematical geneticists during the 1940s and 1950s. The problem of genetic determination for many conditions (diabetes, for example) remains unsolved, but such statistical devices applied to family data, although rarely conclusive, may be helpful in narrowing down the possible answers significantly. A brief summary is presented here for two relatively uncomplicated methods: Snyder's ratio and the use of cousin incidence in estimation of gene frequencies.

If a hereditary condition is common in the population, so that gene frequencies can be expected to be reasonably high and the sampling adequate (several families unrelated and independently observed), and both parents are known, then the percentage of recessive offspring found in the three types of family ($D \times D$, $D \times R$, and $R \times R$) may be compared with the *population ratios* expected for recessives from these families (Snyder, 1932, 1947). Remember that $D \times D$ matings may be of three kinds, only one of which ($Aa \times Aa$) would produce recessive offspring; $D \times R$ matings may be of two kinds, only one of which ($Aa \times aa$) would produce recessives (recall Table 1-1). The familiar Mendelian ratios would be expected *within* such families, but the population as a whole should produce recessive proportions according to the following ratios:

Snyder's ratios: S_1 = percent recessives from $D \times R$ matings (one parent D), ($AA \times aa$ or $Aa \times aa$ matings). From Table 1-1, dominant progeny $= 2p^2q^2 + 2pq^3 = 2pq^2$; recessive progeny $= 2pq^3$, so that

$$S_1 = \frac{R}{(D + R)} = \frac{2pq^3}{[(2pq^2 + 2pq^3)]} = \frac{q}{(1 + q)} \quad (2\text{-}8)$$

$$\text{with variance } (S_1) = \frac{1 - q^2}{4N(1 + q)^4}$$

S_2 = percent recessives from $D \times D$ matings (both parents D), ($AA \times AA$, $Aa \times AA$, $Aa \times Aa$ matings). From Table 1-1, dominant progeny $= p^2 + 2p^3q + 2p^2q^2 = p^2(1 + 2q)$; recessive progeny $= p^2q^2$, so that

$$S_2 = \frac{R}{(D + R)} = \frac{p^2q^2}{[p^2(1 + 2q) + (p^2q^2)]} = \frac{q^2}{(1 + q)^2} \quad (2\text{-}9)$$

$$\text{with variance } (S_2) = \frac{q^2(1 - q^2)}{N(1 + q)^6}$$

(See Exercise 16 for variance derivations.)

Note that $S_2 = S_1$ squared. If we estimate q from the independent parents or from other persons in a population (but not from the sibs within the families, since sibs will have some genes in common and are then not independent observations), that estimate can be tested for agreement with the observed percent of recessives. Snyder (1932) collected family data on the ability to taste the synthetic substance phenylthiocarbamide (PTC) and found agreement with these ratios. Thus, the hypothesis was confirmed that the ability is dominant (T-) to the lack of ability to taste the substance (tt). Inheritance of blood antigen P is illustrated in Example 2-7 using Snyder's ratios. It must be pointed out, however, that agreement with these population ratios does not discriminate between monofactorial and duplicate multiple-locus recessives (see Li, 1953, 1955*b*, 1961).

The methods for using parent-offspring combinations, sib pairs, or cousin pairs to estimate gene frequencies or to ascertain dominance versus recessiveness for hereditary conditions are at present nearly obsolete. These methods have been devised on the rationale that the simple method of gene frequency estimation by counting all genes in a population is less precise than if it were based on a sample of independent, unrelated individuals. Because close relatives possess some replicates of the same gene in common, their contributions to a frequency count are not at all independent. The reader is urged to consult the following references for illustrations of these methods or for mathematical details: Finney (1948*a*, *b*), Li (1955, 1961), Neel and Schull (1954).

Example 2-8 illustrates use of family data in conjunction with the square law for estimating the gene frequency in cystic fibrosis, a genetic disease that is probably monofactorial and is relatively common in the United States. While the method may be imprecise, at least upper and lower limits for incidence and frequency can be computed to give an order of magnitude. Family grouping in the population then can become a useful tool both to the geneticist and epidemiologist, although in genetically complex situations or where expression may be modified to a great degree by environmental factors, the pitfalls may be insurmountable. Pedigree and relative occurrence analysis is sometimes an indispensable recourse when better genetic tools cannot be found.

EXAMPLE 2-7

SNYDER'S RATIOS ILLUSTRATED WITH HUMAN BLOOD GROUP P

Blood antigen P, discovered by Landsteiner and Levine in 1927, about at the same time as they discovered the M-N groups, was found to be dominant to lack of the antigen. These alleles, previously called $P+$ and $P-$, are now designated P_1 and P_2, respectively (owing to the fact that the P system of blood groups is far more extensively known today and these phenotypes also include positive reaction to an antibody (anti-PP_1) for which a few individuals, are negative). Race and Sanger (1968) list family data on the incidence of P_1 and P_2 phenotypes as recorded by Henningsen (1950) for Danish families.

Parents	No. Families Observed	Offspring P_1	P_2	Total
$P_1 \times P_1$	194	471	53	524
$P_1 \times P_2$	93	169	71	240
$P_2 \times P_2$	17	1*	38	39
Total	304	*extramarital		

If the 608 parents are considered as independent observations, to calculate frequency of P_2,

$$q^2 = \frac{93 + 34}{608} = 0.2089, \qquad \text{or} \quad q = 0.4572$$

Population ratios (Snyder's) then are as follows:

	Observed	Expected	Standard Error
$S_1 =$	0.2958	0.3138	0.01357
$S_2 =$	0.1011	0.0984	0.00469

Substantial agreement indicates that P_2 is recessive (although not necessarily monofactorial).

EXAMPLE 2-8

Cystic fibrosis (mucoviscidosis) appears to segregate in families as an autosomal recessive trait. It is characterized by an abnormality in the secretions of mucous cells of respiratory and gastrointestinal systems, as well as an abnormally high electrolyte (chloride) composition of the secretions from sweat, lacrimal, and salivary glands. It is usually fatal in childhood because abnormally viscous mucus cannot be cleared adequately from lungs and bronchi, so that respiratory pathogens are favored. Incidence of the disease has been reported, for example, as $\frac{1}{3700}$ in Cleveland and at about half that amount in Indianapolis (Hanna, 1965) (see Neel, Shaw, and Schull, 1965, for additional references). Hanna collected data from sibs and cousins of affected children over a two-year period and pointed out that because there was no significant increase in parental consanguinity, the gene must be relatively common in the population. This is a surprising conclusion for a semilethal condition, and possibly it indicates a selective heterozygote advantage or a high mutation rate. Hanna's analysis follows (modified for simplicity).

Both parents of a fibro-cystic proband must be heterozygous (*Cc*) and must have received the recessive allele from a *Cc* grandparent unless a recent mutation to *c* had occurred. Each sibling of the proband's parents (the proband's aunts and uncles) will therefore have a precise probability of being a carrier. The frequency of cystic fibrosis among the first cousins of the proband then gives an estimate of the *Cc* individuals in the population and the frequency of the mutant allele.

Because *cc* can be considered lethal for practical purposes, only homozygous normal (*CC*) and heterozygotes (*Cc*) reach reproductive age.

If we assume constant frequencies in successive generations and concentrate only on known carriers, there are two possible kinds of mating producing parents of probands, namely:

Type 1	Type 2
$CC \times Cc$	$Cc \times Cc$
Known parent Cc ↙ ↘ ?	Known parent Cc ↙ ↘ ?

If we let P = frequency of CC and Q = frequency of Cc,

then P = probability of mating type 1

and Q = probability for mating type 2 (since Cc is common to both types of mating)

Within each family:

Type 1. Probability for any sib of the known parent to be $Cc = \frac{1}{2}$

Type 2. Probability for any sib of the known parent to be $Cc = \frac{2}{3}$

Therefore, total probability for either type of mating to occur and to produce a *Cc* sib of the known parent $= P/2 + 2Q/3 = (3 - 3Q + 4Q)/6 = (1 + Q/3)/2$.

Because the probability for any sib to marry a carrier $= Q$, and the chance of having an affected child $= \frac{1}{4}$, therefore the total probability for a first cousin of the proband to be affected $Q/4 \times (1 + Q/3)/2 = (Q + Q^2/3)/8$.

Because $Q = 2q$ (according to property 3), the chance for a first cousin in terms of gene frequency is $q/4 + q^2/6$. The last term is negligible, and we can estimate that $q = 4R/N$, where R = the number affected among N cousins in a random sample of cousin sibships. Hanna's data for incidence of cystic fibrosis were as follows:

	Children Affected	Total Children	
Probands	95	233	$q = 3(4)/982 = 0.0122$
First cousins	3	982	$s = \sqrt{V_q} = \sqrt{q(4 - q)/N} = 0.00704$

(See Exercise 16 for variance derivation.)

Consequently, the incidence in the population is expected to be from an average of about 15/100,000 up to 68/100,000 maximum (using $q + 1.96s$) for affected children.

SUMMARY

Where controlled matings can be made or where all types of matings can be accounted for, contrasting genetic hypotheses are distinguishable by using the square law. Where genotypes can be distinguished, combinations may be tested for agreement with random combinations (implying a random mating origin). These rules are not secure in reverse, however, because many circumstances may appear in binomial proportions without a genetic basis or with a more complex genetic basis than mere monofactorial determination. Also, artifacts of subgroupings, faulty ascertainment, or wrong genetic theory may imply a nonconformity and lead to erroneous conclusions when in fact random factors could account for the observations.

Testing observations by using a *fixed* hypothesis to calculate expected values are illustrated for conformity to the square law and for differences between samples or populations in allelic frequencies (chi-square). Also, frequency data may be treated as a set of random variables with variance determining confidence limits.

Dominance lowers the efficiency of estimating allelic frequencies. Unless the dominance can be "unmasked" in heterozygotes or progeny tested to ascertain all genotypes, methods of estimation are less reliable and more difficult to acquire substantive data; usually two generations (parent-offspring) or one generation (sib or cousin incidence) must be sampled when populations do not conform to the conditions necessary for Castle-Hardy-Weinberg agreement; that is, where a simple square root of homozygous recessive incidence is likely to be misleading.

EXERCISES

1. A population geneticist must be careful about drawing conclusions from gene frequency or genotype frequency data. Below, statements on the left give information about a population. Conclusions on the right may or may not be justified on the basis of the information given. Consider that successive statements *given* add information to that of the previous statement. For each case, do you agree that the conclusion is justified? Why?

Given	*Is This Conclusion Justified?*
(a) A large population of wild mice exists with 36% of the individuals white and 64% wild type (black).	(a) White is recessive to black.
(b) Mating is known to be random in this population between all coat colors of mice.	(b) White allele's frequency is 0.60.
(c) When crosses are made between mice, their progenies are as follows: wild × wild: all wild-type progeny; white × wild: some progenies all white, some with about half white mice and half wild mice; white × white: some progenies all white, some with $\frac{3}{4}$ white: $\frac{1}{4}$ wild.	(c) You can reasonably be confident in the allele frequencies of this population. What are your estimates of allelic frequencies? Explain.

2. In Example 2-1, Wright reported the total progenies among short-horned cattle for red, white, and roan coat color. He listed (1968, Vol. 1, p. 193 and his Table 9.8) for 6000 short-horn calves from 1000 pairs chosen at random from herd books of 1920 the following incidence of matings (NN = white, Nn = roan, nn = red):

$$NN \times NN = 14 \qquad Nn \times Nn = 1310$$
$$NN \times Nn = 374 \qquad Nn \times nn = 2609$$
$$NN \times nn = 441 \qquad nn \times nn = 1252$$

   Assuming each of these matings is independent, do the frequencies of mating conform to the expected numbers on the basis of random mating among the cattle with genotype frequencies given in Example 2-1? Use Table 1-1, obtain expected frequencies of mating, and multiply by 6000 for expected incidence of matings. Which types of mating are in excess and which types are deficient? Do you think the cattle breeders are favoring particular matings and avoiding others?
3. Tay-Sachs disease is recessive, usually fatal before age four, and characterized by excessive accumulation of lipids in brain ganglia due to lack of the enzyme hexosaminidase A. (Assume random mating and no consanguinity between parents in the following populations.)
   (a) In Sweden, the incidence of Tay-Sachs disease is about 0.000036. What would be the estimated frequency of Swedish persons heterozygous for this gene?
   (b) Since this disease is an effective lethal before maturity, which of the nine types of mating will be missing in the population? What would be their total proportion of all types of matings? What would be the expected frequency of matings producing children with Tay-Sachs disease in Sweden?
   (c) Among Ashkenazic Jews, Tay-Sachs disease occurs at about 0.000225 birth incidence. How much greater is the frequency of the mutant allele among this group of people than among Swedes?
   (d) What would be the expected frequency of matings producing Tay-Sachs offspring among this Jewish group of people (assuming marriages are exclusively intragroup)?
4. (a) Test the observed incidence of the M-N blood groups for conformity with the square law (Castle-Hardy-Weinberg) using a chi-square: Pueblo (American Indians), Navaho, Germans, Belgians, Fijians, and Papuans. Which populations show significant departure from conformity? Which genotypes are in excess or in deficiency?
   (b) Find the 95 percent confidence limits for p_M frequencies in these populations. Use the method of t as given in Table 2-3B and compare Pueblos with Navahos, Germans with Belgians, and Fijians with Papuans for significance of difference in allelic frequencies.
5. In Example 2-4, Table A, if all six samples had constant relative frequencies of zygotes (karyotypes) equal to the six-sample total frequencies of karyotypes, these would be their expected numbers (averages) adding up to 150 (sample size) based on an assumption that their differences are entirely due to chance. (Hint: They will be close to the expected numbers in Table 2-2B.)
   (a) Complete a contingency chi-square test based on these observed numbers for the six samples. Verify that the X^2 value is 26.41. How many degrees of freedom are there? What is the probability for the null hypothesis?

(b) When you compare this X^2 with those in Tables 2-2A and 2-2B, from what source do you ascribe the major contribution to the X^2 value?

6. The drug isoniazid used in chemotherapy of tuberculosis is inactivated in the liver, but human variants differ in the rate of inactivation controlled by allozymes of acetyl transferase. Phenotypes are rapid (r), slow (s), or intermediates (r/s). Dufour, Knight, and Harris (1964) found the following frequencies among three population samples of serum from patients given an oral dose of isoniazid six hours previously:

Population	r	r/s	s	Total
Caucasians	7	37	61	105
Negroes	6	51	59	116
Japanese	108	81	20	209

(a) Test each of these groups for agreement with the random-mating expectation.
(b) Compare the observed allelic frequencies between each pair of populations. Which populations differ significantly in r and s frequencies? Which are apparently alike?
(c) Calculate the average genotype frequencies over these three populations. What are the average allelic frequencies over all three?
(d) If all three populations were not isolated but were actually one random-mating population with the average allelic frequencies you just calculated, what would be the expected genotype frequencies in such a mixed population? How does your expectation compare with the average genotype frequencies you calculated above in (c)?
(e) What can you say about the hazards of averaging over diverse populations?

7. Frequencies of sickle-cell anemia and sickle-cell trait individuals in some human populations are given by Livingstone (1967). In the following sample populations, assume the remainder are normal (Hb^A/Hb^A) homozygotes. All persons were adults of African derivation.

Locality	No. Tested	Hb^A/Hb^S	Hb^S/Hb^S
Dallas	1165	78	7
Philadelphia	1000	74	3
Kampala, Uganda	3362	542	3

(a) Test the observed numbers for agreement with random-mating expectation. If any population frequencies do not fit the square law, what conclusions might seem justified?
(b) Do any of these populations agree in allelic or genotypic frequencies? Pool together the populations that agree and test their pooled frequencies of genotypes for agreement with the square law.

8. The following data are proposed by Falk and Ehrman (*Behavior Genetics*, 1973) to illustrate a spurious case of confirmation for the square law (random mating) by

pooling two groups, neither of which is randomly mating. Given the following two subpopulations:

Subpopulation	Observed			
	AA	Aa	aa	Total
I	40	220	140	400
II	85	150	165	400
Pooled	125	370	305	800

(a) Calculate chi-squares for confirmation for random mating within each subpopulation and then test the pooled data.
(b) Test the difference between these subpopulations for the allelic frequency with a contingency X^2. Then test the difference for genotype frequency. What is the main basis for the difference between these subpopulations?
(c) If the observer geneticist was not aware that the population was subdivided, what might be concluded? What does the observer need to establish before applying the Castle-Hardy-Weinberg principle?

9. Haldane (*Annals of Human Genetics*, 1956) pointed out that formula (2-4) is not quite equal to the true q value of a population, even though the sample of recessives is an unbiased estimate that $R = q^2$. The square root of R is not unbiased for q. A better estimate for small samples is as follows:

$$q = \sqrt{\frac{4C + 1}{4N + 1}}, \text{ with variance } (q) = \frac{1 - q^2}{4N + 1}$$

where C = the number of observed homozygous recessives and N is the sample size. Suppose you make a survey of a trait in which heterozygotes are ascertainable, as in Hb^S/Hb^A sickle-cell trait, but you only score children in families for being "normal" (not anemic) versus anemic (Hb^S/Hb^S). Furthermore, if you score children only from four-child families in which both parents are heterozygous so that you expect an average of 25 percent anemic children, then you might tabulate your data as follows (first entry given in table) (where $N = 4$; C = observed recessive homozygotes):

C	Family Frequency	$\sqrt{\frac{C}{N}}$	$\sqrt{\frac{4C+1}{4N+1}}$
0	81	0	$\sqrt{.0588} = .2425$
1	108	——	——
2	54	——	——
3	12	——	——
4	1	——	——
Total	256	Mean?	Mean?

(a) Fill in the missing calculations (how many significant digits?).
(b) Find the means for the estimate based on (2-4) and the estimate using Haldane's correction. Which is closer to the expected frequency?

10. In a wild small mammal population, a recessive albino form is quite common. Out of 1000 individuals of the species captured in one locality, 162 are albino. Of the wild-type remainder, a sample of 50 is brought back to the laboratory for testing to ascertain heterozygosity by outcrossing to an albino strain. From the outcrossing, 29 are determined to be heterozygous and 21 homozygous wild type.
(a) Estimate the albino allele frequency (q) from these data.
(b) What are the confidence limits of this estimate?
(c) Would you be confident in agreement with the random-mating expectation?

11. Samloff and Townes reported (*Science* 168: 144–145, 1970) on the demonstration of seven electrophoretically distinct pepsinogens in extracts of human gastric mucosa, with the first five being also found excreted in the urine. Pepsinogen 5 (*Pg 5*) was found to be polymorphic in being present or absent in different subjects. Of 931 unrelated Caucasians without known peptic ulcer condition, 132 lacked the *Pg 5* band on the electrophoretic gel and were suspected of having a null allele (designated Pg^b) in homozygous condition, while the remainder were designated Pg^a. Recessive inheritance was tested by studying parents and progenies from 100 matings involving 75 unrelated families (25 families had related spouses), as follows:

Parents	No. Families Observed	Offspring Phenotypes		
		$Pg\ 5^a$	$Pg\ 5^b$	Totals
$Pg^a \times Pg^a$	64	137	11	148
$Pg^a \times Pg^b$	33	65	27	92
$Pg^b \times Pg^b$	3	0	6	6

(a) Calculate the expected values of S_1 and S_2 (Snyder's ratios), using the independent sample of unrelated individuals for estimating q.
(b) What are the observed values of recessive proportions for the two types of family?
(c) Calculate confidence limits for the S_1 and S_2 ratios using square roots of the variances given in formulas (2-8) and (2-9). Are the observed values within the confidence limits?
(d) Also test for agreement between expected and observed data by using a chi-square test for each type of family. What is the probability for the null hypothesis?

12. Using the reasoning of Example 2-8,
(a) If one first cousin out of 1000 first cousins of probands is affected by a recessive trait, what is the recessive allelic frequency and what is its upper confidence limit?
(b) What would be the expected affected frequency among first cousins of probands if the allele frequency is $q = 0.05$?
(c) Why are first cousins advantageous to the estimation of allele frequency?

13. For the following small samples, calculate expected values on the basis of expectation from random mating using the usual method and using Levene's exact probability method:

Population	Genotypes Observed			
	AA	*Aa*	*aa*	Total
1	4	12	24	40
2	2	16	22	40

(a) In which genotypes does the more exact method lower the expected?
(b) What is the difference in computing chi-square between the usual method and the more exact method?
(c) Is X^2 greater when there are too many heterozygous or too few?

14. In Chapter 1, Table 1-1B gives the expected frequencies of single progeny from each of six types of mating. If a family has two or more children, they can be arranged in sib-pairs. For example, if parents are $AA \times Aa$, the probability for producing each of the following sib-pairs is $\frac{1}{4}$: *AA-AA*, *AA-Aa*, *Aa-Aa*. With matings of homozygotes $\times$ homozygotes, all sib-pairs will be alike. Finally, if the mating is $Aa \times Aa$ there will be nine sib-pair combinations in frequencies given by $(\frac{1}{4}AA + \frac{1}{2}Aa + \frac{1}{4}aa)^2$. If the birth order is not important, there will be six types of sib-pairs. The first column (see table below) is given as a model.
(a) Verify the total of the first column.
(b) Fill in spaces in the remaining columns and verify the totals. (Spaces marked with *X* cannot occur.)
(c) If observed sib-pair data on any trait appeared to agree with data expected on the basis of these sib-pair frequencies, what conclusion could you draw about the inheritance of the trait?

Genotypes Mating	Frequency of Mating	Types of Sib-Pairs					
		AA-AA	*Aa-Aa*	*aa-aa*	*AA-Aa*	*AA-aa*	*Aa-aa*
$AA \times AA$	p^4	p^4	*X*	*X*	*X*	*X*	*X*
$Aa \times AA$	$4p^3q$	p^3q	—	*X*	—	*X*	*X*
$aa \times AA$	$2p^2q^2$	*X*	—	*X*	*X*	*X*	*X*
$Aa \times Aa$	$4p^2q^2$	$\frac{1}{4}p^2q^2$	—	—	—	—	—
$aa \times Aa$	$4pq^3$	*X*	—	—	*X*	*X*	—
$aa \times aa$	q^4	*X*	*X*	—	*X*	*X*	*X*
	Totals	$\frac{1}{4}p^2(1+p)^2$	$pq(1+pq)$	$\frac{1}{4}q^2(1+q)^2$	$p^2q(1+p)$	$\frac{1}{2}p^2q^2$	$pq^2(1+q)$
		Sibs Alike			Sibs Unlike		

15. Hopkinson and Harris reported (*Ann. Hum. Gen.* 31: 359–367, 1968) that phosphoglucomutase (PGM) locus-3 alleles were faster migrating on electrophoretic gel and showed independence of the other two PGM loci. In a sample of placentas from single

births to English and to Nigerian mothers, the following incidence of three genotypes for PGM-3 alleles was found:

	PGM-3 Phenotype (Gel Band Pattern)			
	1	1/2	2	Totals
Births to English mothers	315	233	35	583
Births to Nigerian mothers (Yoruba)	32	96	107	235

(a) Calculate frequencies of alleles PGM3-1 and PGM3-2 in these two populations.
(b) These authors also obtained samples of dizygotic twin pairs from both populations. From the observed allelic frequencies within each population above, estimate the expected numbers of sib-pairs for each of the six types given below. Do the data conform to the model of a single locus at the allele frequencies given? Use the totals for columns in the previous exercise to help you estimate the expected.

PGM-3 Phenotypes in Sib-Pairs	Observed from Dizygotic Twins	
	English	Nigerian
Like pairs		
1 1	145	6
1/2 1/2	77	25
2 2	14	23
Unlike pairs		
1 1/2	75	13
1 2	7	4
1/2 2	14	31
Totals	332	102

16. Derivation of the variance of a simple binomial variable that is some function of q is approximated by using the method of small increments ("delta") as employed in elementary differential calculus. Suppose two variables are related such that $Y = f(X)$—for example, Snyder's ratios (2-8 and 2-9) or the proportion of affected cousins (R/N) in Example 2-8, all of which are functions of q. Any small increment in Y is related to a corresponding increment in X as follows:

$$\Delta Y = \left(\frac{dY}{dX}\right)\Delta X \qquad \text{(E-1)}$$

Squaring, $(\Delta Y)^2 = (dY/dX)^2(\Delta X)^2$. Then the expected $(\Delta Y)^2 = V_Y$ (variance of Y), since variance is the average squared deviation. Also, $(\Delta X)^2 = V_X$. Thus, variances of X and Y are related by the square of their derivative function:

$$V_Y = \left(\frac{dY}{dX}\right)^2 V_X \qquad \text{(E-2)}$$

(a) We illustrate this method with finding the variance of $q(V_q)$ from Example 2-8. Let $X = q = 4R/N$, and let $Y = R/N = q/4$ = proportion of affected cousins. From the binomial variance of affected cousins, we have

$$V_{R/N} = \frac{(R/N)(1 - R/N)}{N} = \frac{(q/4)(1 - q/4)}{N} = \frac{q(4 - q)}{16N}$$

Taking the derivative of $R/N = q/4$ with respect to q, $dY/dX = d(R/N)/dq = \frac{1}{4}$. Then using (E-2) above, $q(4 - q)/16N = (\frac{1}{4})^2 V_q$. Thus, $q(4 - q)/N = V_q$, as in Example 2-8.

(b) Use this method to derive the variances of S_1 and S_2 as given after the formulas (2-8) and (2-9) in the text.

17. Derivation of the maximum-likelihood estimate of q in formula (2-4) for two autosomal alleles with dominance can be worked out with the help of reasoning from Appendix A-9. If we observe N individuals in two classes, A dominants and C recessives, the likelihood function will be

$$L = \frac{N!}{A!C!}(1 - q^2)^A(q^2)^C$$

and the log of L will be

$$\log L = \log\left[\frac{N!}{A!C!}\right] + A\log(1 - q^2) + C\log(q^2)$$

taking the derivative of L with respect to q:

$$\frac{d\log L}{dq} = \frac{-2Aq}{(1 - q^2)} + \frac{2C}{q}$$

(a) Set this derivative equal to zero for a maximum, and verify that

$$q = \sqrt{\frac{C}{N}} \tag{2-4}$$

The variance of this estimate of q is obtained by setting the second derivative of the log L function to its negative reciprocal (A-9-6):

$$\frac{d^2(\log L)}{dq^2} = \frac{(1 - q^2)(-2A) + 2Aq(-2q)}{(1 - q^2)^2} + \frac{-2C}{q^2}$$

then we supply expected values for A and C, which are $\exp A = N(1 - q^2)$ and $\exp C = Nq^2$.

(b) Verify that the solution is

$$\frac{d^2(\log L)}{dq^2} = \frac{-4N}{(1 - q^2)}$$

Thus the variance becomes

$$V_q = \frac{1 - q^2}{4N} \tag{2-5}$$

(c) Show that this $V_q = pq/2N + p^2/4N$.

18. Derive the maximum-likelihood estimate for q and variance of q for formulas (2-6) and (2-7) based on the case of sampling a proportion (k) of dominants from a population for progeny testing, as given on page 51. Let the observed numbers of individuals be A, B, C, and D. Then the likelihood function will be as follows:

$$L = \frac{N!}{A!B!C!D!}(1 - k)(1 - q^2)^A \cdot (kp^2)^B \cdot (k2pq)^C \cdot (q^2)^D$$

(a) Set the log of L and take the derivative with respect to q. Verify

$$\frac{d(\log L)}{dq} = \frac{-2Aq}{(1 - q^2)} - \frac{2B}{1 - q} + \frac{C(1 - 2q)}{q(1 - q)} + \frac{2D}{q}$$

(b) Verify formula (2-6).

3

MULTIPLE ALLELES AND GENETIC POLYMORPHISMS

The genetic potential of a population extends beyond the single pair of alleles at one locus into many dimensions. In this chapter the dimension for consideration is the cistron, or functional genic unit including its allelic (mutable site) variations, along with any effectively segregating genetic entity (such as an entire chromosome with crossover suppression). In other words, our attention here will be concentrated on any intrachromosomal genetic unit within which recombination is so low as to be negligible for a short-term population analysis.

Mutation in the broad sense may include any hereditary change from the most refined limit of a change in a single nucleotide base in a codon triplet, through larger variations including regroupings, rearrangements, deletions, or duplications within cistrons or between adjacent ones, up to rearrangements of a chromosome including many loci such as inversions, deletions, duplications, or shifts of loci. After repeated production of mutations at these diverse levels of organization, series of segregating genic units (alleles) or chromosomal arrangements have arisen in nearly all species of organisms. Although the extent of this genetic variation was known to be very large, until the 1950s our appreciation of its magnitude was restricted by methods and techniques largely to morphological and cytological forms that were easily detectable without biochemical analytical devices. More recently, the discoveries in protein and macromolecular structure have very much extended our former estimates of populational genetic potential. In this chapter, after first considering the consequences of randomization for variation within single segregation units (cistrons or regions of restricted crossing over within chromosomes), we shall examine cases of genetic polymorphism and discuss current feelings about the magnitude of genetic potential in this dimension.

EQUILIBRIUM FROM RANDOM MATING AND MULTIPLE ALLELES

As long as the entities of a genetic series can be considered unrecombining (or segregating) units, they may be treated as homozygotes (AA, aa, $a'a'$, $a''a''$, ...) or as heterozygotes (Aa, Aa', Aa'', aa', aa'', $a'a''$, ...) in the diploid state, and we may examine the occurrence of genotypes as an extension of the square law resulting from proportional matings with all its properties derived as before.

As early as 1908, Hardy's statement (also made by Weinberg in 1909) of the proportionality principle included the extension to multiple alleles. The statement is in essence as follows. If A, a, a', ... are alleles of a single genetic locus with proportions of p, q, r, ... in both sexes of a population, then zygotes will be formed proportionally by random mating (random union of gametes) according to the square of their frequencies: $(p + q + r + \cdots)^2 = p^2 + q^2 + r^2 + 2pq + 2qr + 2pr \ldots$ for AA, aa, $a'a'$, Aa, aa', Aa', ..., respectively. In other words, the proportional mating between males and females having identical allelic frequencies results in progeny homozygotes with the squared frequencies summed plus heterozygotes at twice the product of each pair of frequencies, or

$$\left(\sum_i q_i A_i\right)^2 = \sum_i q_i^2 A_i A_i + 2 \sum_{i<j} q_i q_j A_i A_j \tag{3-1}$$

where q_i = frequency of any allele and $\sum q_i = 1$. These relationships may be visualized more clearly by reference to Figure 3-1. A chart graphically depicts relative proportions of genotypes formed by random mating when four alleles are segregating equally in the two sexes.

Female gametes

Male gametes	A (p) 0.1	a (q) 0.4	a' (r) 0.2	a'' (s) 0.3
A (p) 0.1	p^2 0.01 AA	pq 0.04 Aa	pr 0.02 Aa'	ps 0.03 Aa''
a (q) 0.4	pq 0.04 Aa	q^2 0.16 aa	qr 0.08 aa'	qs 0.12 aa''
a' (r) 0.2	pr 0.02 Aa'	qr 0.08 aa'	r^2 0.04 $a'a'$	rs 0.06 $a'a''$
a'' (s) 0.3	ps 0.03 Aa''	qs 0.12 aa''	rs 0.06 $a'a''$	s^2 0.09 $a''a''$

Total genotypes:

$p^2 + q^2 + r^2 + s^2 + 2pq + 2pr + 2ps + 2qr + 2qs + 2rs$

$0.01AA + 0.16aa + 0.04\,a'a' + 0.09a''a'' + 2[0.04Aa + 0.02Aa' + 0.03Aa'' + 0.08aa' + 0.12aa'' + 0.06a'a'']$

Figure 3-1. Genotype frequencies generated by random mating when the sexes have equal frequencies of four alleles with $p = 0.1$, $q = 0.4$, $r = 0.2$, $s = 0.3$.

SOME PROPERTIES OF THE GENETIC EQUILIBRIUM FOR MULTIPLE ALLELES

In parallel with the properties of the binomial distribution for a single pair of alleles, the multiple allelic circumstances and consequences are simply an extension to the square of a multinomial. First the *number of homozygotes* possible is equal to the number of alleles (k), while the *number of heterozygotes* is the number of combinations (C) possible, taking two at a time.

$$\text{No. heterozygotes} = {}_kC_2 = \frac{k^2 - k}{2} = \frac{k(k-1)}{2} \tag{3-2}$$

where k = number of alleles (see Appendix A-2B). For example, in Figure 3-1 with four alleles, there are four homozygotes plus six heterozygotes.*

Second, because the number of heterozygotes (${}_kC_2$) increases geometrically with increasing numbers of alleles, the proportion of heterozygosity in the population can exceed 50 percent, the maximum for a single pair of alleles (see property 1 in Chapter 1). Maximum heterozygosity will be obtained when the allelic frequencies are equal ($p = q = r = s = \cdots$). This fact may be demonstrated by taking the case of three alleles at frequencies p, q, and r: total $H = 2pq + 2pr + 2qr$. Letting $q = 1 - p - r$, $H = 2p - 2pr - 2p^2 + 2r - 2r^2$. Taking partial derivatives, first with respect to p and then with respect to r:

$$\frac{\delta H}{\delta p} = 2 - 2r - 4p$$

$$\frac{\delta H}{\delta r} = 2 - 4r - 2p$$

Setting the total equal to zero:

$$4 - 6r - 6p = 0$$

$$4 = 6(p + r) \text{ or } p + r = \tfrac{2}{3}$$

For the same reason, $p + q = \tfrac{2}{3}$

and $q + r = \tfrac{2}{3}$

Therefore, for maximum H, $p = q = r = \tfrac{1}{3}$ or, in general, with k alleles, maximum heterozygosity will be achieved with each allele at $1/k$ frequency. These relationships can now be arranged as in Table 3-1. Note that the last column "maximum heterozygosity" may be obtained by the following:

* Representation of the six genotypes in a three-allelic system within an equilateral triangular coordinate graph is described by Li (1955, 1961). Perpendiculars from any point in the triangle to the three sides are p, q, r whose sum = 1. Then, by inscribing parallels to the three sides, the areas of three parallelograms and three triangles so formed represent relative proportions of heterozygotes and homozygotes, respectively. Relative lengths along with the baseline represent frequencies of p, q, r. This method can be extended to more than three alleles, as shown by Li (1976) for the blood types A_1, A_2, B, and O. See Appendix A-10.

$$2\,(\text{no. heterozygotes})\left(\frac{1}{k}\right)^2 = 2\left[\frac{k(k-1)}{2}\right]\left(\frac{1}{k}\right)^2 = \frac{k-1}{k} \tag{3-3}$$

It can be seen that with increasing numbers of alleles, heterozygosity may approach a limit of nearly 100 percent, provided alleles are equally frequent.

Third, as $p, q, r \ldots$ values become unequal, total heterozygosity diminishes, but total homozygotes may equal heterozygotes at a number of points. A formula for such points where total H = total $D + R$ is complex, but can be derived from the following expression: $2pq + 2pr + 2qr \cdots = p^2 + q^2 + r^2 + \cdots = \frac{1}{2}$, or $pq + pr + qr = \frac{1}{4}$. For three alleles, letting $r = 1 - p - q$, the expression of p in terms of q for equality between heterozygotes and homozygotes is a quadratic whose solutions are the real roots of

$$p = \frac{(1-q) \pm \sqrt{2q - 3q^2}}{2} \tag{3-4}$$

Note that if $q = \frac{1}{2}$, then $p = \frac{1}{2}$, reducing to the case of a single pair of alleles (property 1 in Chapter 1, where $H = D + R$). Also, q must be $<\frac{2}{3}$, as a limiting condition. For situations with more than three alleles, it may be best to treat k alleles as a smaller number (two or three at the most), distinguishing only major alleles and lumping the remaining $k - 1$ or $k - 2$ alleles to reduce the complexity, as in the following manner:

$$\underset{p^2}{AA} + \underbrace{Aa + Aa'}_{2p(q+r)} + \underbrace{aa + aa' + a'a'}_{(q+r)^2} \tag{3-5}$$

The last term of (3-5) will include all homozygotes and heterozygotes that lack allele $A(p)$. These reductions may be necessary in applying statistical tests or when pooling small frequencies that individually are not critical to the analysis (see Exercise 9).

Fourth, analogous to property 5 in Chapter 1 (see Exercise 7), at equilibrium two different heterozygotes ($Aa \times aa'$) with an allele in common a mate at twice the frequency of matings between the common allele homozygote (aa) and the heterozygote (Aa'). In the latter mating, segregation will restore those two heterozygotes (Aa and aa') in equal amounts, while the $Aa \times aa'$ mating produces aa and Aa' in only half its progeny.

TABLE 3-1 *Multiple alleles and heterozygosity*

No. Alleles = no. homozygotes = k	No. Heterozygotes Possible	Maximum Heterozygosity (when $p = q = r = \cdots$)
2	1	$\frac{1}{2} = 0.50$
3	3	$\frac{2}{3} = 0.67$
4	6	$\frac{3}{4} = 0.75$
5	10	$\frac{4}{5} = 0.80$
$\vdots$	$\vdots$	$\vdots$
k	$\frac{k(k-1)}{2}$	$\frac{(k-1)}{k}$

ESTIMATING $p, q, r, \ldots$

The gametic output of alleles can be estimated in the same manner as with a single pair by the gametic pool method of formula (1-3) provided that all genotypes can be ascertained:

$$p_A = AA + \tfrac{1}{2}(Aa + Aa' + Aa'' \cdots) = p^2 + pq + pq + pr + ps \cdots \qquad (3\text{-}6)$$

For example, using data from Figure 3-1, $p_A = 0.01 + 0.04 + 0.02 + 0.03 = 0.10$. The remaining allelic frequencies may be obtained similarly. Example 3-1 illustrates the principle with data from three natural populations containing four "allelic" chromosomal arrangements from *Drosophila pseudoobscura.* A test of conformation to expectation from the multinomial square law may be applied in the usual way with chi-square (or when sample sizes are very small, utilizing Levene's exact method for expected values*). As for comparing frequencies between populations, either tests on haploid doses or confidence limits with variance of the binomial ($pq/2N$), applied by testing one allele against pooled frequencies of all the others, may be used.

Another important illustration is presented in Example 3-2 where three alleles of hemoglobin (sickle-cell *S*, *C*, and normal *A*) are discussed for certain African populations in which abnormal hemoglobins are in high frequency and for American Negroes with the abnormal types at much lower frequency.

EXAMPLE 3-1

CHROMOSOMAL POLYMORPHISM IN *DROSOPHILA PSEUDOOBSCURA*

From 1932 to the mid-1940s Dobzhansky and his colleagues collected samples from natural populations of *Drosophila pseudoobscura* and its related species, *D. persimilis* and *D. miranda*, from British Columbia to Guatemala and from the Pacific coast to central Texas. Larval salivary glands were analyzed with attention given to the gene arrangement variation on chromosome 3. Flies were attracted to fermenting banana mash in paper cup or bucket traps. Females caught in the wild were most often already inseminated; when brought back to the laboratory they were allowed to lay eggs. The chromosomes of a single F_1 larva from each female were examined. Because each larva is an independent zygote, the arrangement frequencies obtained probably reflect the natural population frequencies. In addition, wild males were usually mated to laboratory standard females with known chromosomal constitution; seven or eight larvae examined were sufficient to establish the most likely zygotic combination for the parent male. Thus, two chromosome-3 arrangements were determined in a single offspring of each wild female or two for each wild male.

Local populations in these species were almost always polymorphic for chromosome-3 arrangements, having up to eight arrangements in many populations of *D. pseudoobscura,*

* Levene's exact method for expected values is in principle the same as for a single pair of alleles (2-2), letting for each pair of alleles $g_1(g_1 - 1)/2(2N - 1)$ = homozygous expected, and $g_1 g_2/(2N - 1)$ = heterozygous expected.

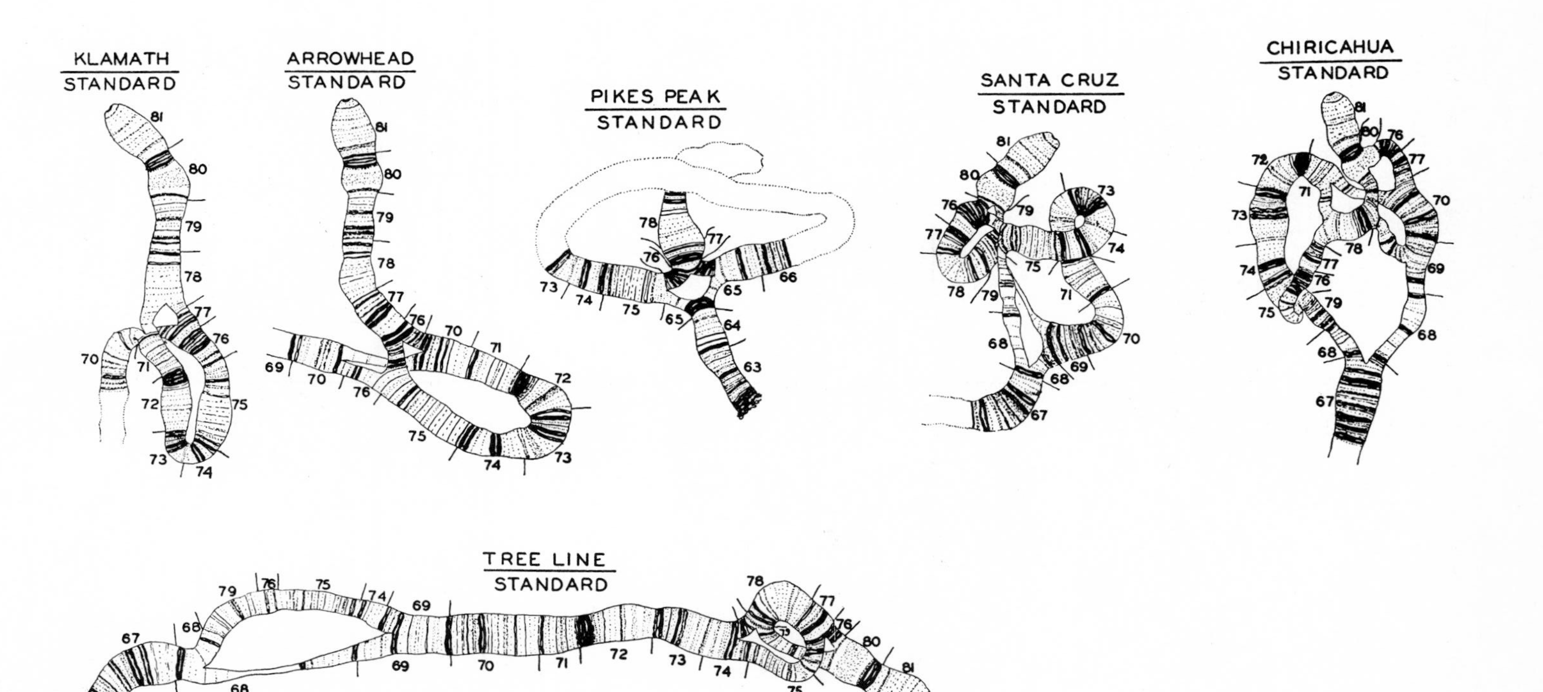

Figure A. Some of the common third-chromosome arrangement heterokaryotypes in *Drosophila pseudoobscura* (*AR*/*ST*, *PP*/*ST*, *SC*/*ST*, *CH*/*ST*, *TL*/*ST*) and in *D. persimilis* (*KL*/*ST*) as they appear in late third instar larval salivary glands and labeled according to the map units specified by Dobzhansky and Sturtevant (from *Genetics*, Vol. 23, 1938, p. 28).

with at least four in high frequency. At present there are 25 known arrangements of chromosome 3 in that species and 12 in its sibling, *D. persimilis.* One arrangement, Standard, is shared by both species (see Dobzhansky, 1970, p. 132; Dobzhansky, et al., 1975). These cytological variants have the advantage of exact detection through the bands of the salivary gland giant chromosomes. A few of the common arrangements are shown in Figure A. However, no morphological or physiological properties are known to be determined by any of them. Phylogenetic relationships determined by single-inversion events are given in Figure B. Beginning in 1939, Dobzhansky concentrated his efforts on seasonal variation within an area limited to the region around Mt. San Jacinto in southern California, especially three sampling stations known as Keen Camp (4300 feet elevation on the western

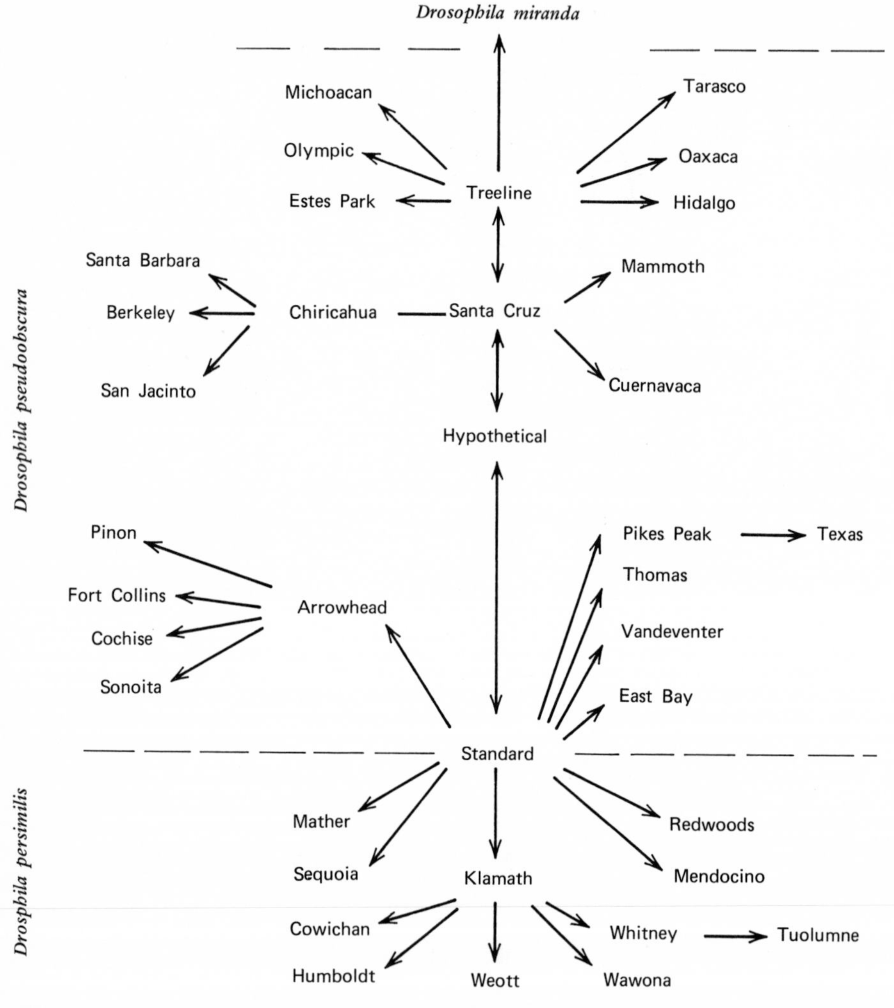

Figure B. Phylogenetic relationships for third-chromosome arrangements in *D. pseudoobscura* and *D. persimilis.* From Anderson, et al. (*Evolution*, Vol. 29, 1975, p. 26) and from Dobzhansky, et al. (*Jour. Heredity*, Vol. 66, 1976, p. 204). Arrows indicate most likely direction of derivation by single inversion events.

slope with severe winters and mild summers in yellow pine forest), Piñon Flat (4000 feet with extremes of temperature and dry conditions in intermediate forest), and Andreas Canyon (800 feet, an oasis in the eastern slope desert near Palm Springs with mild winters and hot dry summers). Distances are given on the map in Figure C (Wright, Dobzhansky, and Hovanitz, 1942; Dobzhansky and Epling, 1944). Data from a typical year (1940) are given in Table A. Note that the *CH* arrangement is highest at Keen and that the *ST* arrangement increases to the east in drier and warmer stations. *AR* is fairly constant for all three stations and also shows little seasonal variation for the arrangements in these localities. *TL* is low but present throughout. Heterokaryotypes (chromosomal arrangement heterozygotes) have high frequency in all localities, although they drop slightly as *ST* and *CH* become more unequal.

The X^2 is calculated in the usual way to test for square-law confirmation. Degrees of freedom will be the number of classes (10) minus 1 for the total and for each estimated independent parameter (any chromosome-3 haploid frequencies), or a total of 6 *d.f.* Note

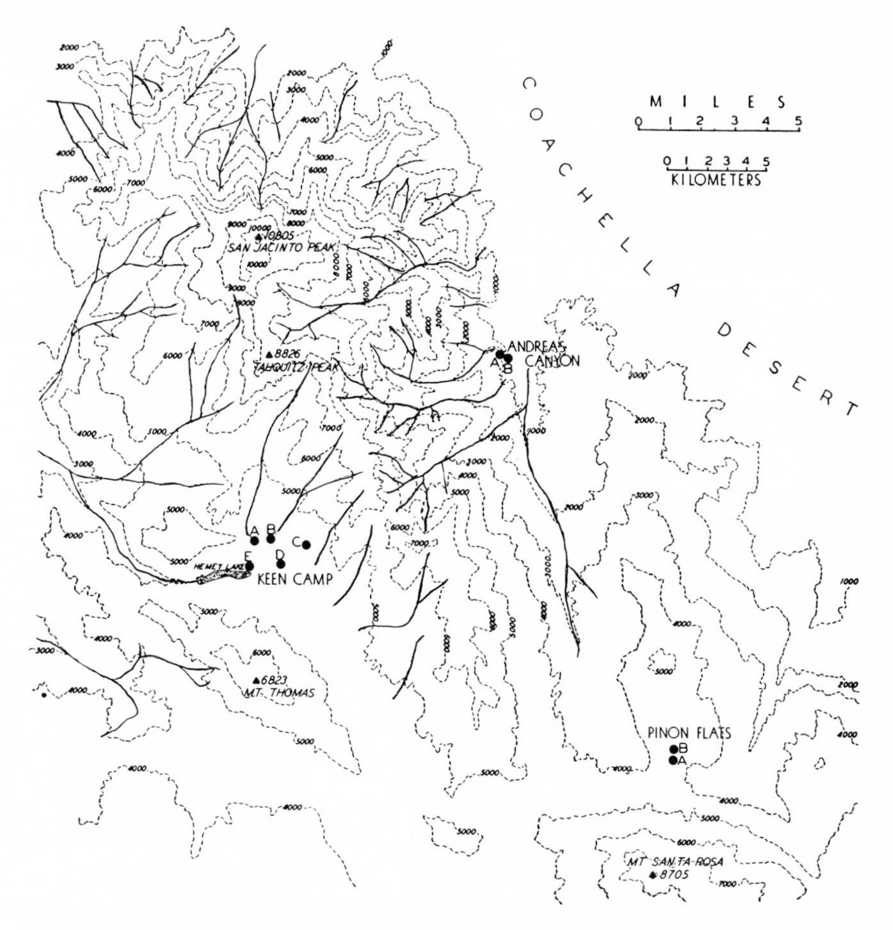

Figure C. Map showing the location of collecting stations on Mt. San Jacinto, California. From Wright, Dobzhansky, and Hovanitz (*Genetics*, Vol. 27, 1942, p. 364).

TABLE A *Numbers of observed chromosome-3 arrangements in* Drosophila pseudoobscura *collected from localities in the vicinity of Mt. San Jacinto, California, in spring of 1940 (from Dobzhansky and Epling, 1944) (ST = Standard, AR = Arrowhead, CH = Chiricahua, TL = Tree Line)*

Locality	ST/ST	AR/AR	CH/CH	TL/TL	ST/AR	ST/CH	ST/TL	AR/CH	AR/TL	CH/TL	Total
Keen Camp	30	11	44	0	53	66	3	48	3	6	264
Piñon Flat	31	11	21	0	40	53	5	37	3	7	208
Andreas Canyon*	89	18	4	1	87	47	12	20	4	2	284

* Expected for Andreas Canyon based on random principle: $(p + q + r + s)^2 \cdot N$ (in order as above).

	92.4	19.0	5.2	0.4	83.9	43.9	11.4	19.9	5.2	2.7	284

$X^2 = 2.18$, $d.f. = 6$, $P = 0.90$.

	Haploid frequencies (%)				
	p_{ST}	q_{AR}	r_{CH}	s_{TL}	% Heterokaryotypes
Keen Camp	34.5	23.8	39.4	2.3	67.8
Piñon Flat	38.5	24.5	33.4	3.6	69.7
Andreas Canyon	57.0	25.9	13.6	3.5	60.6

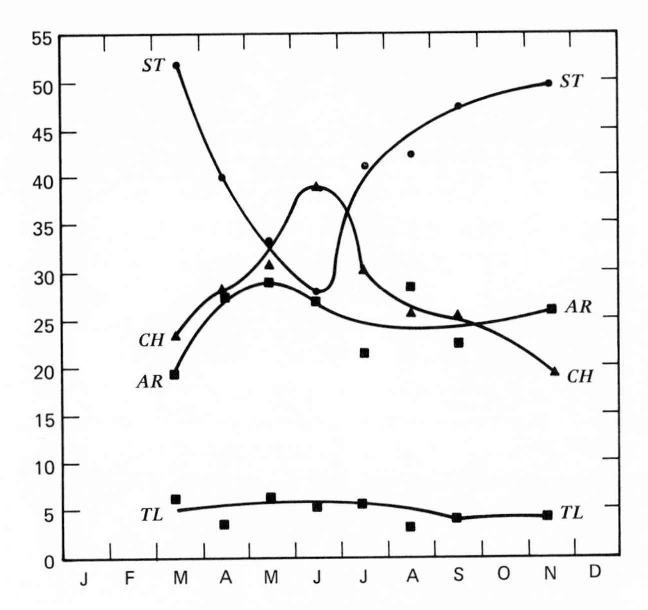

Figure D. Seasonal cycles in chromosomal arrangement frequencies in the population from Piñon Flats. Ordinate in percentage, abscissa in months. Combined data for six years from 1939–1946. From Dobzhansky (*Evolution*, Vol. 1, 1947, p. 6). *ST* = Standard, *AR* = Arrowhead, *CH* = Chiricahua, *TL* = Tree line arrangements.

that there is no significant departure from the square law; apparently panmixia was the case. After several seasons of collecting, Dobzhansky found that a cycle of *ST* versus *CH* frequencies regularly took place at the warmer localities, Piñon and Andreas, with *CH* rising in spring at the expense of *ST*, while the reverse took place in summer and early fall, restoring the higher frequency of *ST* by winter (see Figure D).

These regular changes brought Dobzhansky to a fundamental turning point in his thinking with regard to the forces determining the distribution of these karyotypes in nature. In 1942, Dobzhansky embarked on his laboratory population cage analysis. Some data from it are illustrated in Example 2-4. The constant changes in relative frequencies of gene arrangements within those experimental populations paved the way for understanding the natural distributions to be adaptive and the result of natural selection forces.

EXAMPLE 3-2

HUMAN HEMOGLOBIN POLYMORPHISM

Abnormal hemoglobin first gained attention via the severe anemia of individuals whose erythrocytes "sickled" in their veins or under reduced oxygen tension, as described by Herrick in 1910. This severe hemolytic anemia usually terminates fatally before the age of 20. Pauling, et al., in 1949 were able to separate normal from sickle hemoglobin electrophoretically; in the same year Neel and Beet independently established the hereditary basis of these hemoglobins as due to a single allelic difference. The pioneer analysis was that of Ingram in 1957, who established the single amino acid residue 6 in the β chain to be the site of substitution in sickle cell hemoglobin (valine) with respect to normal glutamic acid. On reexamination of several families in which one parent did not show the sickle trait yet had an anemic child, other variant hemoglobins were discovered: Hb^C and Hb^D. C is a structural allele of S because it substitutes a lysine in the same position, residue 6 of the β chain. Individuals homozygous for hemoglobin C exhibit a mild anemia with hemolysis, not quite as severe as the condition of S homozygotes.

Population data on these and many other hemoglobin variants are now quite extensive (Giblett, 1969). In central Africa most populations either have high frequencies of Hb^S (see Example 1-1) or Hb^C, but in Ghana a large number of tribes have both alleles in substantial amounts (see Table A). Unfortunately, the incidence of newborn children was not ascertained for these populations so that frequencies are undoubtedly lower for the anemias and higher for other genotypes than would be the case without selective action. In contrast, data summarized by Livingstone (1967) for the incidence among American Negro children and adults indicate presence of both types of variant hemoglobin that occur in west Africa but at lower frequencies (Table B).

There is considerable evidence that this hemoglobin polymorphism is maintained by a balance of selective forces, involving susceptibility to falciparum malaria (Hb^A), a subject to be discussed later under selection mechanisms (Chapter 15).

TABLE A *Frequencies of hemoglobin variants in adult Africans from Ghana (6 tribes plus northern territory miscellaneous) (Allison, 1956)* $A(= Hb^A$ *(normal)*, $S = Hb^S$ *(sickle cell)*, $C = Hb^C$ *(C variant)*

	A/A	A/S	A/C	S/S	S/C	C/C	Total
Number	719	199	114	2	5	3	1042
Frequency	0.6900	0.1910	0.1094	0.0019	0.0048	0.0029	

Allelic frequencies

$Hb^A = 0.8402$, $Hb^S = 0.0998$, $Hb^C = 0.0600$

TABLE B *Frequencies of hemoglobin variants in American Negroes from Baltimore (1400), Houston (400), and Philadelphia (1000). The samples are substantially homogeneous (from Livingstone, 1967).*

	A/A	A/S	A/C	S/S	S/C	C/C	Total
Number	2501	213	64	14	4	4	2800
Frequency	0.8932	0.0761	0.0229	0.0059	0.0014	0.0014	

Allelic frequencies

$Hb^A = 0.9427$, $Hb^S = 0.0437$, $Hb^C = 0.0136$

DOMINANCE

As noted in the previous chapter, the efficiency of gene frequency estimation is reduced when heterozygotes and homozygous classes become indistinguishable. With multiple alleles, phenotypes of some heterozygotes may show dominance, complete or incomplete, or there may be codominance between any two alleles. If dominance is complete between three or more alleles in succession, and if gene frequencies must be estimated without recourse to progeny testing, they can be estimated by taking the square root starting with the most recessive class frequency (as in formula 2-4) and proceeding as follows.

If $A > a > a'$ is the order of dominance for three alleles so that all zygotes with A-have the A phenotype, all with a- and without A have the a phenotype, and so on, then under the square law conditions,

let A = proportion $A - /N = p^2 + 2pq + 2pr$

B = proportion $a - /N = q^2 + 2qr$

C = proportion $a'a'/N = r^2$, where N = number of individuals observed

then

$$r = \sqrt{C} \tag{3-7}$$

and, since $B + C = q^2 + 2qr + r^2$,

$$q = q + r - r = \sqrt{B + C} - \sqrt{C} \tag{3-8}$$

and p may be obtained by subtraction from 1.

An example illustrating estimation for a series of three alleles with complete dominance is given in Example 3-3 from data on populations of the snail *Cepaea nemoralis* L. in England. This classic case of genetic polymorphism will be discussed in further detail later under evidence for selective forces maintaining the polymorphism and under random genetic drift. At this stage it is essential to realize that these estimates of gene frequency are first approximations for illustration and that in reality nonrandom mating and selective effects on these snails could make the square-root method of estimation erroneous to a certain extent. Workers on *Cepaea* customarily do not calculate gene frequencies, however, because the phenotype frequencies must reflect the genotypes. In addition, the labor involved in progeny testing would be insurmountable, inasmuch as the mature forms are usually already fertilized in nature.

EXAMPLE 3-3

SHELL COLOR AND BANDING POLYMORPHISM IN *CEPAEA NEMORALIS*

One of the most common and highly variable (polymorphic) land snails in Europe is *Cepaea nemoralis*, which may show changes in proportions of shell colors and banding patterns within a few yards from fields to hedgerows to woodlands (Figure A). How much of this variation may be due to selective pressures (visual predation, for example) and how much to other forces (random genetic drift or gene flow) will be discussed later. The genetics

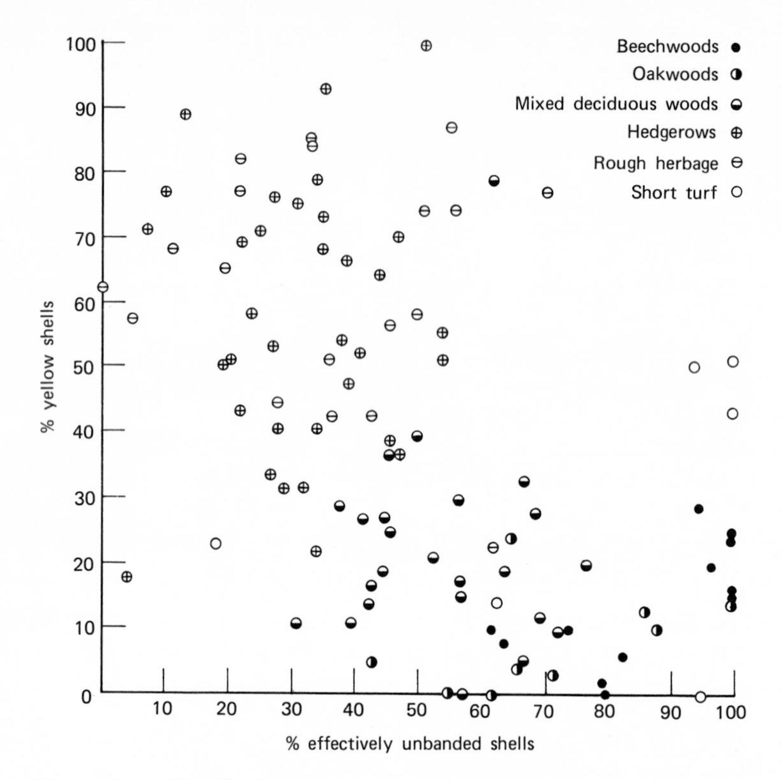

Figure A. Distribution diagram for the proportion of yellow shells as a function of the proportion of unbanded shells in *Cepaea nemoralis*, as well as the type of habitat. From Cain and Sheppard (1954).

of shell color and banding have been demonstrated to be due to two closely linked allelic series, with darker colors dominant to lighter at one locus and unbanded dominant to banded at the other (see references by Cain, B. Clarke, L. Cook, Lamotte, Murray, and Sheppard). The calcareous layers of shell are suffused with violet, pink, or yellow; in combination with the yellowish periostracum, these produce dull brown, a range of red to lighter shades (pinks), or dark to light yellows. The color locus exhibits a hierarchy of dominance in that order, with light yellow being the most recessive. Each color results from a different pigment rather than from a dilution of one pigment. The banding locus also has a series of multiple alleles. A striking case of association between color, banding, and habitat was documented by Cain and Sheppard (1954). Some of the color frequencies are listed below (Table A) for a few pooled or large samples from four different habitats. Brown shells are most common in beechwoods, where there are wide expanses of dark soil; pink ones are mostly in leaf litter on forest floors of mixed hardwoods; yellow ones are in greatest abundance in green turf or hedgerows. While banding is also important, it is omitted here for brevity.

Estimating frequencies of alleles C^B, C^P, C^Y (in descending dominance) from Table A, we may let their frequencies be p, q, r, respectively, and use (3-7) and (3-8). If we allow an oversimplified assumption that the square law applies to the zygotic distribution, then in the beechwoods habitat, for example, phenotypes will be

	Brown	Pink	Yellow
With expected frequencies	$(p^2 + 2pq + 2pr)$	$(q^2 + 2qr)$	(r^2)

then

$$r = \sqrt{\frac{Y}{N}} = \sqrt{\frac{115}{731}} = 0.3966 \text{ for } C^Y$$

$$q + r = \sqrt{\frac{(Y + P)}{N}} = \sqrt{\frac{558}{731}} = 0.8737$$

$$q = 0.4771 \text{ (by subtraction) for } C^P$$

$$p = 0.1263 \text{ (by subtraction) for } C^B$$

Note that agreement with the square law cannot be tested by a chi-square in this case (as it cannot with a single pair of alleles showing dominance), because with three phenotypic classes, two parameter estimates (q and r) plus the total (N) eliminate three degrees of freedom, leaving zero degrees of freedom.

The variance of r (see formula (2-5)) is as follows:

$$V_r = \frac{1 - r^2}{4N} = \frac{(1 - r)r}{2N} + \frac{(1 - r)^2}{4N} \qquad (V_r = 0.000288 \text{ in this example})$$

It is important to note that in those environments where the most recessive homozygote is least common (yellow shells in beechwoods), the gene frequency estimate (incidence in young stages) may be in error more than in the other localities if selective predation tending to eliminate it is most prevalent there. Under those conditions, the estimate made from applying the square-root rule will not be accurate, because heterozygotes may be in greater abundance than calculation from this estimate would imply. While progeny testing for ascertainment of heterozygote frequency is much needed, the technical problem of testing mature individuals caught in nature seems insurmountable since nearly every one is already fertilized. The phenotypic frequency is sufficient in most cases for comparisons between localities and study of selective forces, gene flow, and other evolutionary trends, because relative genotype frequencies ought to reflect allelic (haploid) frequencies, provided the intensity of selection is roughly the same in all localities even though directed against different phenotypes.

TABLE A *Some samples of shell color frequencies in* Cepaea nemoralis (*from Cain and Sheppard, 1954*)

Habitat	Brown (B)	Pink (P)	Yellow (Y)	Total (N)
Beechwoods	0.2367	0.6060	0.1573	731
Deciduous woods	0.0550	0.6751	0.2699	1508
Hedgerows	0.0449	0.3120	0.6431	779
Rough herbage	0.0041	0.2183	0.7776	971

MIXTURE OF DOMINANT AND CODOMINANT ALLELES

Although lack of dominance between certain members of an allelic series seems to improve the efficiency of gene frequency estimation compared with completely dominant alleles, the improvement depends on the number of genotypes ascertainable without progeny testing and the grouping of these genotypes into various sets of dominant or codominant subsets. Unfortunately, along with the improvement in gene frequency estimation, mathematical complexity increases, involving the solutions to maximum-likelihood functions for efficient estimates (Neel and Schull, 1954; Li, 1955).

To point out in an elementary way some of the difficulties, it seems best to illustrate the principles involved with the human A, B, O blood groups because they are so well known. In 1900 Landsteiner found agglutination of red blood cells when suspended in saline and mixed with sera from different donors. These A and B antigens were distributed between four phenotypes (A and B, A not B, B not A, and neither antigen). In those days the blood groups seemed most easily explained like Mendel's two independently assorting pairs of loci (A – a, B – b), according to von Dungern and Hirzfeld in 1911. It was not long, however, before pedigree and population analysis by Bernstein (1925, 1930) clearly indicated that these blood types are determined by multiple alleles of a single locus. When only two antisera are available, anti-A and anti-B, the familiar four serological types with genotypes and expected frequencies are as follows: Let I^{A}, I^{B}, and i represent the isoagglutinogen locus with codominant $I^{A-}I^{B}$ but each dominant over i, in proportions p, q, r, respectively:

Blood group	AB	A	B	O
Genotype	$I^{A}I^{B}$	$I^{A}I^{A}$, $I^{A}i$	$I^{B}I^{B}$, $I^{B}i$	ii
Expected frequency	$2pq$	$p^2 + 2pr$	$q^2 + 2qr$	r^2

On inspection, it is evident that, letting the blood group symbols stand for frequency of each blood type in a random mating population,

$$\mathrm{A} + \mathrm{O} = (p + r)^2$$
$$\mathrm{B} + \mathrm{O} = (q + r)^2$$
$$\mathrm{O} = r^2$$

so that theoretically the following relationships should hold:

$$p = \sqrt{\mathrm{A} + \mathrm{O}} - \sqrt{\mathrm{O}} = 1 - \sqrt{\mathrm{B} + \mathrm{O}}$$
$$= (p + r) - r = 1 - (q + r) \tag{3-9}$$

In similar manner,

$$q = \sqrt{\mathrm{B} + \mathrm{O}} - \sqrt{\mathrm{O}} = 1 - \sqrt{\mathrm{A} + \mathrm{O}}$$
$$= (q + r) - r = 1 - (p + r)$$

However, the frequency of AB has been left out of consideration; therefore, it is not "fully efficient" to utilize these relationships as they stand in estimating allelic frequencies.

When estimating allelic frequencies from observed data, the maximum-likelihood method becomes very complicated (see Neel and Schull, 1954, pp. 189–195), but alternative and simpler solutions have been obtained. For a preliminary estimate, Bernstein (1925) used the expressions on the right in (3-9) while Wiener and others used those on the left in (3-9). Later, adjustments were made on each of these expressions to improve their efficiency, as described by Li (1970*b*). Bernstein's adjustment (1930) is the simpler, and we shall follow it here without proof. It produces a very close approximation to the more complex maximum-likelihood estimation.

We use the symbol (*) for the efficient estimate, including Bernstein's adjustment, and the symbol (′) for the less efficient preliminary estimate.

$$\begin{aligned} p' &= 1 - \sqrt{\mathrm{B} + \mathrm{O}} \\ q' &= 1 - \sqrt{\mathrm{A} + \mathrm{O}} \\ r' &= \sqrt{\mathrm{O}} \end{aligned} \tag{3-10}$$

Let $D = 1 - p' - q' - r'$ (difference between 1 and preliminary estimate total). Then Bernstein's adjustment

$$\begin{aligned} p^* &= p'\left(1 + \frac{D}{2}\right) \\ q^* &= q'\left(1 + \frac{D}{2}\right) \\ r^* &= \left(r' + \frac{D}{2}\right)\left(1 + \frac{D}{2}\right) \end{aligned} \tag{3-10A}$$

The sum $p^* + q^* + r^* = 1 - D^2/4$, not 1 exactly. With the maximum-likelihood estimate the sum of allelic frequencies would equal 1 exactly.

Population data given in Table 3-2 indicate a few of the racial and national frequencies of these blood types throughout the world. The large sample from England (United Kingdom airmen in 1946) may be used for illustration, using (3-10) and (3-10A) and keeping significant digits (plus 1) from the original data cited in Race and Sanger (1968):

Blood Types	Frequency
O	0.4668388
A	0.4171587
B	0.0856045
AB	0.0303980
	1.0000000

From (3-10)

$$\begin{aligned} p' &= 1 - \sqrt{0.5524433} = 0.2567347 \\ q' &= 1 - \sqrt{0.8839975} = 0.0597886 \\ r' &= \sqrt{0.4668388} = 0.6832560 \\ & \qquad\qquad 0.9997793, \quad \text{then } D = 0.0002207 \end{aligned}$$

Then from (3-10A)

$$\begin{aligned} p^* &= 0.25676303 \\ q^* &= 0.05979518 \\ r^* &= 0.68344176 \\ & \quad 0.99999997 \end{aligned}$$

TABLE 3-2 *Frequency of ABO blood groups (in percent) from some representative populations listed in order of increasing amounts of antigen B (data from Boyd, 1950; Li, 1955; Mourant, et al., 1958; Race and Sanger, 1968)*

Population	Locality	No. Tested	AB	B	A	O
American Indians (Navaho)	New Mexico	359	0	0	22.3	77.7
American Indians (Pueblo)	New Mexico	310	0	1.6	20.0	78.4
Eskimo	Cape Farewell, Greenland	484	1.4	3.5	53.8	41.3
Australian Aborignes	Queensland	447	0	3.6	37.8	58.6
English	United Kingdom	190,177	3.04	8.56	41.72	46.68
Melanesians	New Guinea	500	4.8	13.2	44.4	37.6
Germans	Berlin	39,174	6.5	14.5	42.5	36.5
Japanese	Tokyo	29,799	9.7	21.9	38.4	30.1
Abyssinians	Addis Ababa	400	5.0	25.3	26.5	42.8
Chinese	Yunnan	6,000	10.1	27.1	32.0	30.8
Pygmies	Congo	1,032	10.0	29.1	30.3	30.6

The less efficient preliminary estimate may be sufficiently accurate in most cases. The adjusted estimate is improved at the fourth decimal place in the r value but at the fifth place in the other two frequencies. For variances of these ABO frequencies, the reader should consult the two references already cited (Neel and Schull, Li), since the variances are complex. For most purposes, to confirm agreement with the square law it is easier to use the chi-square test.

The ABO locus is more complex than this. It consists of many subgroups. The A group has at least two common subgroups, A_1 and A_2, plus many additional rare subgroups. There are possibly also a few B subgroups. In English populations, relative frequencies for these group alleles are approximately $p_{A_1} = 0.20$ and $p_{A_2} = 0.07$. Both these alleles are dominant to the O allele (i) in I^{A_1}/i and I^{A_2}/i genotypes. Thus, six phenotypes are possible. The same principles apply in estimating allelic frequencies from observed data. The Bernstein adjustment can be made as before with $D/2$. For an example see Li (1961, pp. 50–52).

DETECTING ALLELIC VARIATION

The discoveries in biochemical genetics, especially in primary gene action, have established the colinearity of cistron mutant sites with amino acid sequence of polypeptide products. For example, tryptophane synthetase in bacteria (Yanofsky, et al., 1964; Yanofsky, 1967), hemoglobin in vertebrates (Ingram, 1963), and several other polypeptides (see Jukes, 1966, and Brookhaven Symposia in Biology, 1968, for summaries) have been described in terms of their amino acid residues in primary structure and the mutant sites within structural genic loci that determine them. However, working out these sequences is an exceedingly complex and time-consuming procedure. Powerful techniques of gel electrophoresis and immunological specificity (Smithies, 1955; Beckman, 1966; Harris, 1970; Greenwalt, 1967) allow detailed analysis of molecular variations without dissecting the molecules down to their ultimate residues. We have high-resolving power tools for exploring systematically the genic variation of populations, a far more refined group of tools than was possible to achieve with morphological or cytological criteria alone, and more or less free of the problems of dominance and phenotypic interactions.

Along with the advantage of refinement, however, we must note the limitations of these methods. The specification of a difference between alleles depends on the site or sites of amino acid substitution in the protein product, on the electric charge of the amino acid, and on the molecular size of the protein. Only a fraction of variant alleles will be detectable with these methods, which are restricted to the detection of a net charge difference (electrophoresis) or to binding properties between antigen and antibodies (immunodiffusion). Consequently, many amino acid substitutions may occur in a protein without making a detectable difference, and proteins are by no means equal in that respect; for example, for tryptophane synthetase about seven-ninths of all mutations tested are electrophoretically detectable (Henning and Yanofsky, 1963), for hemoglobin variants about 30 percent of all amino acid substitutions result in a change of charge when electrophoresis is carried out at pH 8.6 (Lehmann and Carrell, 1969), but none of the variations in cytochrome-*c* are separable by that technique even with an extensive search throughout plant and animal kingdoms (Margoliash, personal communication to Lewontin and Hubby, 1966). Consequently, there are no shortcut methods for ascertaining all protein variants without taking them apart in the time-consuming way. We can estimate conservatively that these biochemical tools of genetic refinement uncover on the average no more than 50 percent of all mutant forms in most proteins. Such variants are nonetheless the most efficient phenotypes to be discerned in evaluating the genetic structure of populations.

THE EXTENT OF POLYMORPHISM—ISOALLELIC SYSTEMS

Estimation of any parameter demands achieving a *random sample* of what exists. To find out what is typical of a species's genome, how it is structured, and in how many dimensions, "clearly we need a method that will randomly sample the genome and detect a major proportion of the individual allelic substitutions that are segregating in a population" (Lewontin and Hubby, 1966). Pioneer work in protein variation was mostly inspired by clinical medicine and based on medically oriented research (blood types, hemoglobins, "inborn errors of metabolism," enzyme variants, and serum proteins); these factors were analyzed in the past largely because of their functional significance in disease or in abnormal hereditary conditions. Geneticists and evolutionists have now added to our information on the incidence of allelic variants through surveys of isoallelic systems with no apparent deleterious effects in populations of a wide number of organisms. But most *polymorphisms**

* Defined by Ford (1940) as "the occurrence together in the same locality of two or more discontinuous forms of a species in such proportions that the rarest of them cannot be maintained by recurrent mutation." The "discontinuous forms" in this chapter are the existing alleles within a functional genic unit, or cistron.

have been studied precisely because they are interesting as genetic markers in evolutionary investigation, and we have had no idea how many genetic loci exist that are *not* polymorphic, not genetically variable in most of any species's gene pool. The techniques developed to detect substitutions of amino acids in polypeptides—by electrophoretic methods, for example—are particularly useful because we know enzymes and proteins in general to be determined by one or a few structural genes. Thus, by examining single individuals for electrophoretic mobility of their proteins, it should be possible to detect genic variation from individual to individual at single loci, irrespective of any function allelic variants may have and the forces that may affect their maintenance in populations. We wish to describe the extent of multiple allelic systems including the proportion of the total genome for any species that is monomorphic—that is, genetically determined but not genetically variable—and the proportion that is polymorphic. We may then ask how polymorphic a locus may be for the entire species or for any population within the species. Finally, we may seek to estimate just how heterozygous or homozygous an average individual is likely to be for any population.

These considerations have been vital ones for discussion in many aspects of population genetics theory, but until recently they were only points of pure speculation with widely divergent answers depending on opinions and biases of both experimentalists and theoreticians. Results of random sampling of genomes to make these estimates were published in 1966 independently by Lewontin and Hubby at the University of Chicago (studying 18 loci in *Drosophila pseudoobscura*) and by Harris and associates at University College, London (sampling 10 human blood enzymes). Their work has been extended since that time within both organisms, and much more information from several other organisms has rapidly accumulated to expand our concept for the magnitude of this dimension in genic variation (see Selander, et al., 1969; Selander and Johnson, 1973; Powell, 1975). Examples 3-4 and 3-5 illustrate these two fundamental studies. They agree in demonstrating the vast amount of genic variation in two diverse organisms, both of which are sexually reproducing and outcrossing species. As a conservative estimate, about 50 percent of loci in these genomes have segregating alleles, and there is about 12 percent heterozygosity per individual. Multiple allelic systems are definitely not rare. About 30 percent of the loci have three or more alleles per population! This is truly an astonishing value if it is upheld by extensions of our knowledge encompassing more of the genome.

EXAMPLE 3-4

PROTEIN VARIANTS IN *DROSOPHILA PSEUDOOBSCURA*

With the analytical tool of electrophoretic mobility of proteins through acrylamide gel, Hubby and Lewontin (1966; and Lewontin and Hubby, 1966) surveyed the genomes of six populations of *Drosophila pseudoobscura*. Samples were from central, marginal, and isolated localities in the species distribution. They intended to ascertain the extent of allelic polymorphism "typical" of the genome as a whole, and they chose 18 loci solely because of reliability in detection of bands on the gel (constant and unambiguous stainability). Variants observed were demonstrated to behave as simple Mendelian segregants. The species of drosophila chosen is one of the most common and widespread species of the genus in north

Central America (with one isolate from Colombia, South America) and should be typical of "successful" sexually reproducing organisms. Figure A indicates the distribution of this species and that of its closely related species, *D. persimilis*. Each population was represented by a number of strains, each descended from a single fertilized female caught in nature. Strains had been obtained nine to five years earlier but were often segregating for allelic variants, even though they had been maintained by small mass matings (40–50 pairs on the average) each generation in the laboratory. While separate strains could easily have

Figure A. Known approximate distribution of *Drosophila pseudoobscura* (dashed lines) and *D. persimilis* (dotted lines) in North and South America. For localities, see footnote to Table A.

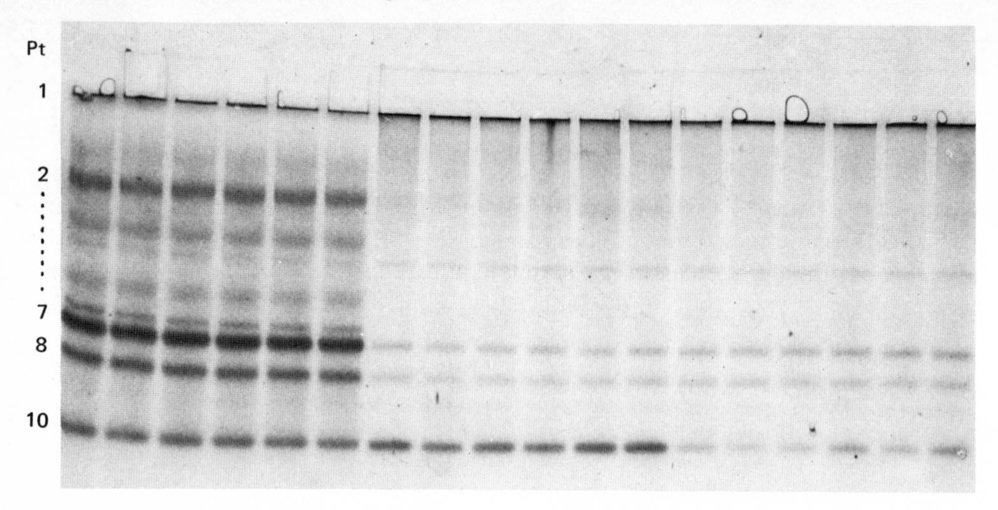

Figure B. Larval proteins (Pt-1 through Pt-10) from *D. pseudoobscura* individuals from acrylamide gel separation and stained with coomasie blue. Proteins 11–13 are not visible in this photograph. Slots from left to right, 1–6 = controls, 7–12 = after treatment with 5 minutes of 57°C, which partially denatures most proteins though Pt-10 is resistant, 13–18 after treatment with 10 minutes of 57°C. Flies from Strawberry Canyon, Cal. (Courtesy R. C. Lewontin.)

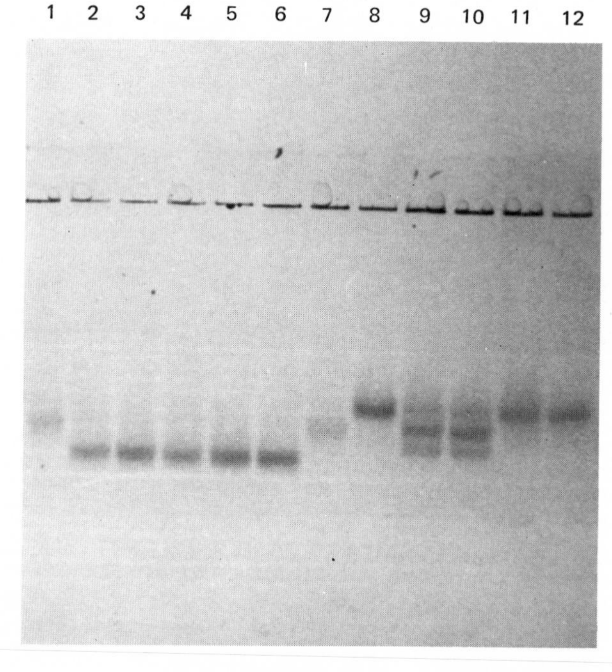

Figure C. Esterase-5 allozymes from *D. pseudoobscura* stained on acrylamide gel. Relative mobility of variants, from left to right slots: 1 = 0.95, 2–6 = 1.00, 7 = 0.95, 8 = 0.85, 9–10 = 0.85/1.00 heterozygotes, 11–12 = 0.85 (0.95 was a known variant). (Courtesy R. C. Lewontin.)

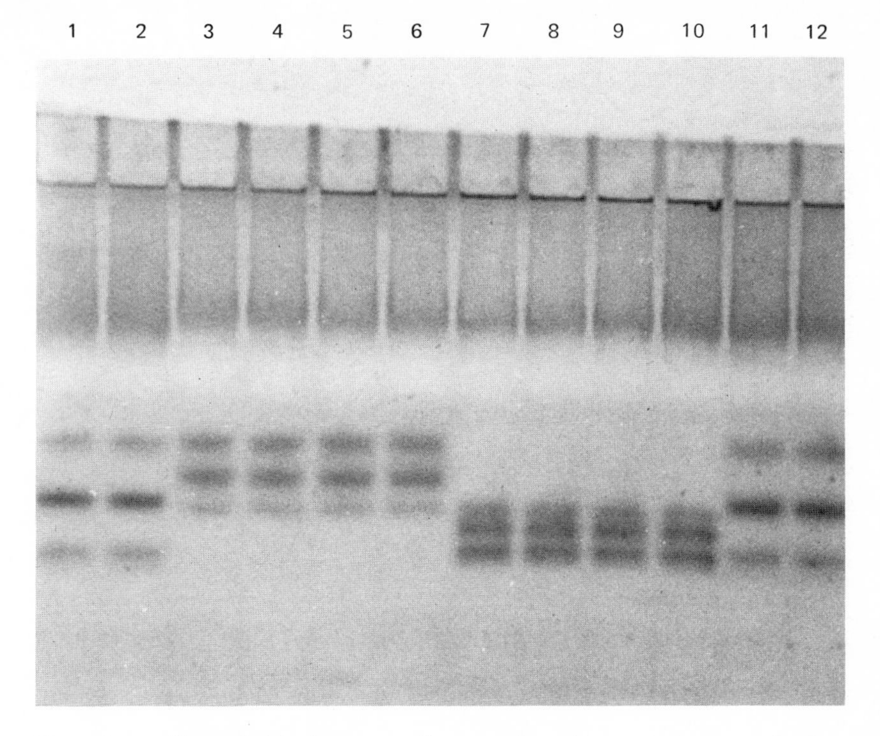

Figure D. Octanol dehydrogenase allozymes from *D. pseudoobscura* stained on acrylamide gel. All individuals heterozygous for variants given in relative mobility, from left to right slots: 1–2 = 1.22/0.86, 3–6 = 1.05/0.86, 7–10 = 1.22/1.05, 11–12 = 1.22/086 repeat. (Courtesy R. C. Lewontin.)

become homozygous by inbreeding, the differences between lines should have preserved some portion of the original population's genetic variety, so that a survey of strains from each population was expected to produce an unbiased estimate for that population.

By 1969–1970 this work was extended to include 10 more genic loci and fresh strains from additional populations (Lewontin, 1974). These 28 loci (17 of them enzymes, with their banding variants now termed "allozymes" by these authors) are listed in Table A with their known alleles and frequencies in five localities. Each allele is designated by its mobility relative to the most common allele, in the case of the allozymes designated as 1.00 or, in the case of the nonenzymatic larval proteins, relative to each other (with the least mobile protein *Pt*-1 at 0.21 up to the fastest protein *Pt*-13 at 1.37) Figure B, C, and D illustrate distributions of larval protein and allozymes *ODH* and *Est*-5 bands in acrylamide gel electrophoresis. From the data of Table A, we may be able to draw many important conclusions about the extent of multiple allelism and polymorphism with regard to each locus and population and for the species as a whole.

NUMBER OF POLYMORPHIC ALLELES AT A LOCUS. More than one allele was found in 17/28, or 61 percent of the loci. Of these, sex-linked esterase-5, with 12 alleles, was most variable, *MDH* with 6, *LAP* and *XDH* with 5 each, four loci had 4 alleles each, three

TABLE A *Allozyme and protein variants at 13 polymorphic loci of* Drosophila pseudoobscura. *From the work of Lewontin, Hubby, Prakash, and Crumpacker, 1966–1970. See Lewontin, 1974, for more complete data. Note localities and supplemental information at end of table. Number of strains in parentheses.*

Locus	Chromosome	Assay	Allele*	Locality** and frequencies of Alleles (%)				
				S.C.(110)	M.V.(120)	P.D.(100)	A.(23)	B.(19)
Esterase-5 (*Est-5*)	*X*	Adult	0.85	0	4	0	0	0
			0.90	0	0	0	2	0
			0.95	12	11	14	3	3
			0.97	0	0	0	3	0
			1.00	42	36	43	29	97
			1.02	1	5	13	11	0
			1.03	8	4	0	0	0
			1.04	1	10	5	15	0
			1.07	19	20	23	26	0
			1.09	1	0	0	0	0
			1.12	13	10	2	5	0
			1.16	2	0	0	6	0
Acid phosphatase-4 (*AP-4*)	*X*	Larva	0.93	0	0	0	3	0
			1.00	100	100	100	86	100
			1.05	0	0	0	11	0
Acetaldehyde oxidase-2	2	Adult	0.90	1	0	0	0	0
			0.93	3	0	0	0	5
			1.00	94	100	100	100	83
			1.02	2	0	0	0	12
Octanol dehydrogenase-1 (*ODH-1*)	2	Adult	0.86	0	1	0	0	0
			1.00	98	96	96	100	100
			1.22	2	3	4	0	0
Protein-7 (*Pt-7*)	2	Larva	0.68	1	0	0	0	0
			0.73	1	1	2	1	5
			0.75	95	95	96	97	92
			0.77	3	4	2	2	3
Protein-8 (*Pt-8*)	2	Larva	0.80	1	1	2	1	87
			0.81	47	41	54	44	10
			0.83	51	58	43	51	3
			0.85	0	0	1	4	0

Xanthine dehydrogenase (*XDH*)	2	Adult	0.90	5	2	1	2	0
			0.92	7	7	5	4	0
			#—					
			0.99	26	30	27	23	0
			1.00	60	58	63	66	100
			1.02	1	3	4	5	0
(# 0.97 allele occurs in southern California, Nevada, and Arizona localities at frequencies of 1–13%)								
α-Amylase-1 (*Amy-1*)	3	Adult and Larva	0.74	3	0	0	0	0
			0.84	29	21	38	12	100
			1.00	68	79	62	88	0
Protein-10 (*Pt-10*)	3	Larva	1.02	1	2	0	1	0
			1.04	61	97	60	94	0
			1.06	38	1	40	5	100
Protein-12 (*Pt-12*)	3†	Larva	1.18	55	94	99	90	100
			1.20	45	6	1	10	0
Malic dehydrogenase (*MDH*)	4	Adult	0.80	0	0	1	0	0
			1.00	97	95	91	96	100
			1.20	3	5	8	4	0
Leucine aminopeptidase (*LAP*)	Autosome (not yet assigned)	Pupa	0.90	1	2	1	4	0
			0.95	5	1	0	2	0
			1.00	89	94	93	87	95
			1.10	5	2	6	5	5
			1.12	0	0	0	1	0
Protein-13 (*Pt-13*)	Autosome (not yet assigned)	Larva	1.23	6	2	3	2	0
			1.30	94	98	96	98	73
			1.37	0	0	1	0	27

* Alleles designated by relative mobility in acrylamide gel.
** Localities: S.C. = Strawberry Canyon near Berkeley, CA; M.V. = Mesa Verde, CO at 7000 feet; P.D. = Palmer Divide (three localities averaged from State Recreation Area, Hardin Ranch, and Nelson Ranch, CO at 6000 feet; A. = Austin, TX; B. = Bogotá, Colombia at 9000 feet.
† Chromosome-3 loci have alleles associated with specific chromosomal arrangements: *Pt-10* allele 1.04 is mostly found in *ST*, *AR*, and *PP* (Standard Phylad), while allele 1.06 characterizes the *SC*, *CH*, *TL*, *EP* (Santa Cruz Phylad). —Amylase similarly has allele 1.00 mostly in *ST* Phylad with allele 0.84 predominantly in the *SC* Phylad (except *CH* has more of 1.00 allele). *PT-12* allele 1.20 is predominantly found in the *ST* arrangement, while the 1.18 allele is mostly in all the other arrangements see Prakash and Lewontin, 1968, 1971).
NOTE: The following proteins and enzymes were found to be monomorphic: glucose-6-phosphate dehydrogenase (G6PD, *X* chrom.) with a single allele except at M.V. where a second rare allele (1.10) occurred, oxidase, larval acid phosphatases-6, -7, -8, larval alkaline phosphatases -4, -6, -7 (rare second alleles were found at single localities not given above), α-glycerophosphate dehydrogenase, and proteins -1, -4, -5, -6, -9, -11.

had 3 alleles, and six had 2 each. Many alleles occurred only once in a single strain; they were suspected as being due to mutation and not falling within the definition of polymorphism (if an allele was found only once in just a single locality, it was not counted toward polymorphism). Examples of such "mutants" might be $Est\text{-}5_{1.09}$, $LAP_{1.12}$, $MDH_{0.80}$, $G6PD_{1.10}$, $Acet\text{-}ox\text{-}2_{0.90}$, $ODH\text{-}1_{0.86}$, $AlkP\text{-}4_{0.93}$, $AlkP\text{-}6_{-}$, and $Pt\text{-}7_{0.68}$. When these "mutants" are eliminated, the remaining 13 loci are polymorphic (46 percent), with 2 to 11 alleles per locus.

AVERAGE NUMBER OF ALLELES PER "DEME" (LOCAL POPULATIONS). Maximum local polymorphism is in *Est*-5 at Strawberry Canyon, with nine alleles. In fact, this locus is polymorphic in every locality sampled. *LAP*, *XDH*, *MDH*, *Acet-ox-2*, *α-Amy-1*, *Pt-7*, *Pt-8*, and *Pt-10* are also quite polymorphic, having from three to five alleles segregating in certain localities. The remaining polymorphic loci have only *local* pairs of alleles.

PROPORTION OF THE SPECIES' GENOME. While nearly 50 percent of the loci exhibit some amount of variation, 10 loci, or 36 percent, show a high degree over more than one locality. This is a much greater fraction of the total genome than was suspected previously to be variable. Since all the biases in the Lewontin-Hubby method conspire on the conservative side, these estimates probably undershoot the parameter values for most populations of this species.

AVERAGE HETEROZYGOSITY PER INDIVIDUAL. If we take the allelic frequencies for each locus in a population, calculate the expected frequencies of heterozygotes on the assumption of random mating, and then average these heterozygote values over all loci *including the monomorphic ones*, which contribute zero heterozygosity, we can estimate the proportion of an individual's genome for which it is heterozygous. (Obviously, the sex-linked esterase-*5* and *AP-4* loci would be included only for females.) When all such calculations are made over all localities, the estimates come out as follows. The proportion of the genome heterozygous per individual is given for the localities with best samples (Prakash, Lewontin, and Hubby, 1969).

Locality	Percent
Strawberry Canyon	14
Mesa Verde	11
Austin	12
Bogotá	4
Mean (excluding Bogotá)	12.3
Grand mean	10.3

Clearly, the isolated population at Bogotá stands out as genetically depauperate, with only about 25 percent of its loci showing any segregation and average heterozygosity less than 5 percent. The remaining three localities, two "marginal" with respect to the species distribution and one (S.C.) central, have about the same high heterozygosity proportion of about 12 percent.

EXAMPLE 3-5

ENZYME VARIANTS IN HUMAN POPULATIONS

Harris and his associates at the Galton Laboratory, University College, London (1966, 1969, 1970) estimated from certain human populations the extent of "genetically determined enzyme diversity among what might be regarded as normal individuals"—that is, a random sample of the human genome without regard for presence or absence of variation at a locus or possible selective advantage or disadvantage for the existing variants in a population. They confined their attention for the most part to enzymes present in the blood (in erythrocytes and in serum) and detected variants by starch gel electrophoresis, chosen solely because sensitive separation and stain technique on the gel was available. By 1969 they had surveyed 19 different enzyme systems representing at least 26 different loci in English, African, and other population samples. Their major results are presented in Table A with frequencies of polymorphic alleles.

In ten cases (red cell acid phosphatase, *PGM-1*, *PGM-3*, E_1, E_2, peptidase *D*, adenosine deaminase, 6-phosphogluconate dehydrogenase, placental alkaline phosphatase, and acetyl transferase) both Caucasians and Africans are polymorphic, although their allelic frequencies vary in some degree. At one locus (adenylate kinase), only Caucasians are polymorphic, while at two loci (peptidase *A* and *PGM-2*), only Africans are. In acetyl transferase, the Japanese population has the opposite frequencies of alleles from both Caucasians and Negroes. In summary, then, 13 out of the 26 loci (50 percent) exhibit polymorphism in at least one population. Of the 26 loci, however, the number of polymorphic alleles per locus is never more than 3 in any one population, in contrast with *Drosophila pseudoobscura*, which has several loci, each with 5 or more alleles in the same population. Finally, we may estimate that the average individual in the English population, for example, is heterozygous at 10.6 percent of his or her loci, remarkably close to the value estimated for *D. pseudoobscura* (12 percent) by Lewontin and Hubby.

As expressed by Harris (1969) for the English population sample,

> *... each of these polymorphisms appears to occur independently of the others ..., by combining frequencies of the most common allele, one finds that the most frequent combination of phenotypes will occur in less than 2 percent of the population .... Furthermore, two randomly selected people in the population would have exactly the same combination of enzyme types about 1 in 200 ..., or approximately 97 percent of people in this population must be heterozygous for at least one of the 10 loci ....*
>
> *If anything, the results must underestimate the true incidence of polymorphism simply because the enzymes were scrutinized for electrophoretic differences ..., and it is quite possible that in some cases polymorphic variation has been missed .... This must surely represent only the tip of the iceberg, and one may plausibly conclude that, in the last analysis, every individual will be found to have a unique enzyme constitution.*

Later Harris and Hopkinson (1972) summarized a more extensive survey: a total of 71 allozyme structural loci, including the original 26, for the English population. Of these, 20 (28 percent) were found to be polymorphic. Average heterozygosity per locus was estimated at 6.7 percent. These estimates are lower than the original, owing to the discovery

TABLE A *Eighteen enzyme systems representing 25 loci in human populations, with incidence of variable alleles* (*autosomal*), *based on the work of Harris and associates* (*1969*). (*See Giblett, 1969, for more extensive populational data.*)

Enzyme	Locus: Alleles	Population Frequencies (%) (sample size in parentheses)						
1. Red cell acid phosphatase				Negroes in				
		British (367)	Tristan da Cunha Islanders (140)	London (23)	Seattle (429)*			
	p^a	36	9	20	25			
	p^b	60	91	80	72			
	p^c	4	0	0	1 (and 1.5% p^r allele)			
2. Phosphoglucomutase						Negroes in		
		British (2115)	Greeks (88)	Iraqi Jews (69)	Turkish Cypriots (243)	London (103)	Nigeria (153)	S. Africa (99)
	PGM-1 1	76	69	67	70	79	76	79
	2	24	31	33	30	21	24	21
	PGM-2 1	100	100	100	100	99	99	97
	2	0	0	0	0	1	1	3
		British (332)	Nigerians (102)					
	PGM-3 1	74	34					
	2	26	66					

3. Pseudocholinesterase

		Canadians (2017)	Negroes (N. America) (666)*
E_1	u (usual)	98	99.5
	a (atypical)	2	0.5

		British (1941)	Tristan da Cunha Islanders (214)	Negroes in Seattle	
				(100)	(317)*
$E_2(C_5)$	+	5	8	1	2
	−	95	92	99	98

4. Peptidase (red cell)

		Europeans (2279)	Negroes in U.K. (293)	Negroes in Nigeria (155)	Negroes in S. Africa (100)
A	1	100	93	90	85
	2	0	7	10	15
B	1	99.9	100	100	100
	2	0.1	0	0	0
C	1	100	100	100	100
D	1	99	95	—	—
	2	1	3		
	3	0	2		

5. Adenosine deaminase

		English (580)	Negroes in London (147)	Indians in London (213)
ADA	1	94	97	89
	2	6	3	11

6. Adenylate kinase

		English (960)	Negroes in Ghana and Nigeria (800)*	Indians in England (132)*
AK	1	95	100	90
	2	5	0	10

TABLE A (*Continued*)

Enzyme	Locus: Alleles	Population Frequencies (%) (sample size in parentheses)		
7. 6-Phosphogluconate dehydrogenase (red cell)		English (4558)	Negroes in Nigeria and Uganda (209)	S. Africa (200)*
	6PGD *a*	98	94	85
	c	2	6	15
8. Placental alkaline phosphatase		English (597) (single births)	Negroes in West Africa and West Indies (94)	
	PL s_1	64	89	
	f_1	27	6	
	i_1	9	4	
9. Acetyl transferase (liver) (isoniazid inactivation)**		Caucasians (105)	Negroes (116)	Japanese (209)
	rapid	24	27	71
	slow	76	73	29
10. Phosphohexoseisomerase		Monomorphic (except for rare variants)		
11. Lactate dehydrogenase		Monomorphic		
12. Malate dehydrogenase		Monomorphic		
13. Isocitrate dehydrogenase		Monomorphic		
14. Red cell hexokinase		Monomorphic		
15. Methaemoglobin reductase		Monomorphic		
17. Pyruvate kinase		Monomorphic		
18. Red cell "oxidase"		Monomorphic		
19. Placental acid phosphatase		Monomorphic (two loci)		

* See Giblett for references under each enzyme.
** Dufour, Knight, and Harris, 1964.

of about three times as many monomorphic as polymorphic loci in the later sample. It must be kept in mind, however, that the criterion used for allozyme variation was simply that of demonstrable electrophoresis difference. Examples of polymorphisms not demonstrable by electrophoresis but by enzymatic activity or enzyme quantity were found in the serum pseudocholinesterase loci E_1 and E_2, as well as in inosine triphosphatase. It is likely, then, that techniques supplementary or additional to electrophoresis (immunological, heat stability, and so forth) will uncover many other polymorphic loci. Harris (1975) now has listed a total of 31 enzyme polymorphisms known for human populations plus 21 non-enzymatic protein polymorphisms.

HUMAN POLYMORPHISMS

There is no intention here to be exhaustive or encyclopedic in listing or discussing the enormous amount and complexities of human allelic variation, but it will serve a useful purpose to illustrate briefly some of the better-known systems of genetic polymorphism in humans. Their concise treatment at this point will indicate further the vastness of genetic variation available to our populations and at the same time provide us with reference material from which to draw illustrations later in this book, particularly when they are discussed in connection with mechanisms responsible for their maintenance.

Serum and Red Cell Enzymes

The autosomal allelic variant systems are discussed in connection with the general problem of heterozygosity incidence and extent of polymorphism illustrated by the work of Harris (1970) and associates in Example 3-5. The sex-linked polymorphisms are singled out for the next chapter.

Hemoglobin

Few proteins have been as thoroughly studied as vertebrate hemoglobin. Knowledge of its structure, especially its sequence of amino acids, has played a key role in our concepts of gene action and evolution of molecules. It is fairly easy to obtain because it is the principal protein of red blood cells, with about 280 million molecules per cell, and it is remarkably similar in all vertebrates, having a molecular weight from about 64,000 to 68,000, with human hemoglobin at 67,000. An exception is the lamprey, which has its hemoglobin existing as a monomer. Hemoglobin then consists of four globin chains (tetramers), each combined with an iron porphyrin heme group; the tetramer molecule can be dissociated into two dimers and thence into monomeric chains under low pH or in solutions containing urea or guanidine. Normal adult human hemoglobin (adult *Hb A*) has two α chains with 141 amino acids each and two β chains with 146 amino acids each, comprising about 97 percent of the total molecules. A minor component (about 3 percent adult $Hb\ A_2$) has two α chains and two δ chains to each molecule. Human embryos and early infants, however, possess fetal hemoglobin (*Hb F*), which has two α chains identical with those in the adult, but the two β chains are replaced by two γ chains. These relationships are diagrammed in Figure 3-2. The amino acid sequence of the four chains can be arranged to indicate maximal homology between chains (Lehmann and Carrell, 1969). While the α chain is five residues

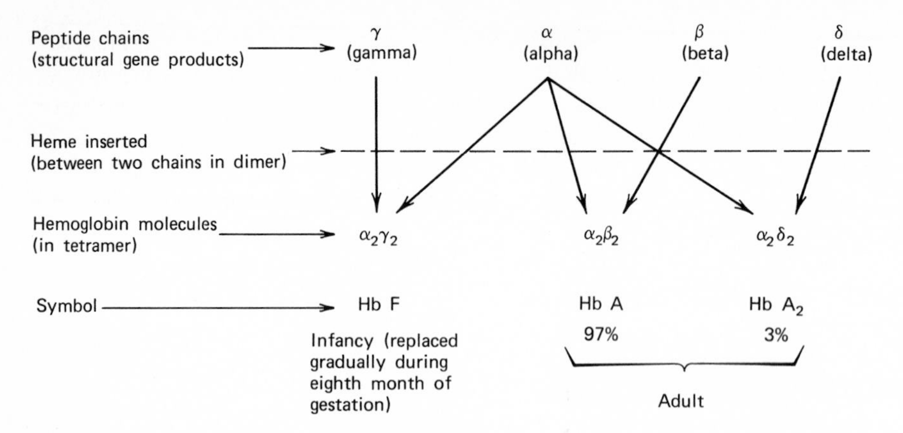

Figure 3-2. Diagram of human hemoglobin components as structural gene products, the four peptide chains—gamma, alpha, beta, and delta.

shorter than the others, it has the same amino acid as β at 64 sites; γ differs from β at 39 sites, while δ differs from β at only 10 sites. Each chain is probably determined by a separate locus, but there is evidence that β and δ are very closely linked; hemoglobin Lepore, which is found sporadically in certain populations (Greece, Boston, and Papua), consists of two normal α chains but varying parts of β and δ chains; that is, one end of a chain is identical with normal β and the other end identical with normal δ, a sequence that could have arisen from unequal crossing over between adjacent cistrons.

Most hemoglobin variants were first discovered by relative mobility in electrophoresis and later characterized by other properties, such as solubility and fingerprinting of tryptic digest peptides. From lists of hemoglobin variants in humans (Livingstone, 1967; Lehmann and Carrell, 1969), there are now known to be at least the following numbers of single amino acid substitutions (several are not yet known by exact substitution) plus a few abnormal chain combinations.

	Number of Variants	
Chain	Substitution Known	Abnormal Combination
α	36	1 ($\alpha 4$)
β	59	2 ($\beta 4 = H$, $\beta + \delta$ = Lepore)
γ	4	1 ($\gamma 4$ = Barts)
δ	5	1 ($\delta 4$; also Lepore)

The majority of these variants are nonpathological, at least to the extent that they do not give their carriers any handicap, and most are also quite rare. However, the following (mentioned in order of their discovery) can be considered polymorphic in that they occur in a few populations at frequencies that are at least one percent.

Hb S (1949). (β 6 glutamic acid → valine) when homozygous, sickle cell anemia. Highest frequencies in central Africa, but found also in Greece, southern Italy, India, and in American Negroes in North, Central, and South America (see Example 3-2).

Hb C (1950). (β 6 glutamic acid → lysine) *C* disease is more mild anemia than sickle cell. West Africa, Mediterranean, and American Negroes (see Example 3-2).

Hb D (1951). A number of hemoglobins with the same electrophoretic mobility (about equal to *Hb S* at pH 8.6 but different at pH 6) but more soluble than *Hb S* and further subdivided after tryptic digestion. With discovery of this heterogeneity, further hemoglobins were given designations from the place of original carrier. Two forms common in parts of India are $Hb\ D_{\text{Punjab}}$ (β 121 glutamic acid → glutamine) and $Hb\ D_{\text{Gujerati}}$ or $Hb\ D_B$ (β trypticase peptide 3 missing), with frequencies of 1 to 4 percent. Homozygotes have mild microcytic anemia.

Hb E (1954). (β 26 glutamic acid → lysine) *E* disease of homozygotes is a mild microcytic anemia. Very common in southeastern Asia, especially in Malaya, Burma, Indonesia, and reaching highest frequency in eastern Thailand at over 50 percent.

Hb K (1956). At least three types (one of which is a rare α chain form). The common form (β 46 glycine → glutamic acid) is found at up to 8 percent among the Berbers of North Africa and is widespread at low frequencies (5 percent or less) in West Africa.

Hb O (1956). Two forms are known. One of these, the $Hb\ O_{\text{Indonesia}}$ (α 116 glutamic acid → lysine), is the only α chain variant common enough to be considered polymorphic, approaching 2 percent of the population on Celebes.

In connection with hemoglobin polymorphisms, it is important to mention also those inherited defects in globin chain synthesis, the *thalassemias* (see Weatherall, 1969; Weatherall and Clegg, 1972; Kabat and Koler, 1975). We are not yet certain whether the genetic basis for these disorders is a regulator or operator locus affecting each hemoglobin chain's structural genic product and its rate of production or whether the disorders represent partial or complete deletions of those structural genes. But the evidence points to close linkage between each structural chain locus and the corresponding thalassemia locus. β thalassemia, the most common form, is a very severe anemia in homozygotes with little or no *Hb A* produced and with *Hb F* and α chain synthesis elevated; in heterozygotes there is variable lowering of β chain synthesis but moderate increase in $Hb\ A_2$ (6 percent) plus *Hb F* (1–5 percent) in adults. Probably the β thalassemias represent a series of alleles that may control varying rates of β-globin synthesis. As a group these β thalassemias are most frequent in regions of high malarial incidence except for tropical Africa, where it is quite rare; for example, in northern Italy and Greece it occurs from 6–14 percent, a bit less in northern Africa and Arab nations, and from 2–5 percent in southeastern Asia. The α-thalassemia, with little or no α chain synthesis, is fatal to homozygotes before birth. It is detectable in heterozygotes by an elevated amount of γ chain (*Hb* Barts) or β chain (*Hb H*) hemoglobin tetramers lacking α chains. This form is therefore difficult to classify as a genotype independent from pure homozygous *Hb* Barts or *Hb H*, but some low amount of α present indicates the defect, which has been found to be fairly common in southeastern Asia. Possibly the *Hb H* disease found among some Chinese and Greeks may be an α thalassemia locus defect.

For review articles on hemoglobins see the following: Allison, 1964; Baglioni, 1967; Giblett, 1969; Ingram, 1963; Jukes, 1966; Lehmann and Carrell, 1969; Livingstone, 1967;

Perutz, 1964; Rucknagel and Neel, 1961; Weatherall and Clegg, 1972; Zuckerkandl, 1964, 1965.

Blood Cell Antigens

Among the first human genotypes to be detected and now perhaps the best known cases of polymorphism in both human and nonhuman vertebrates are those determining red cell antigens. In parallel with the studies of Lewontin and Hubby (Example 3-4) and Harris (Example 3-5) on estimating the average heterozygosity of various organisms, Lewontin (1967) pointed out, on a suggestion by Sheldon Reed at the University of Minnesota, that even though there was no way of detecting monomorphic loci among the blood groups, the bias of being dependent on the presence of genetic variation may be avoided by distinguishing those blood groups with rare mutants alone (where a mutant propositus establishes a family study of a "private" blood group) from the more or less "public" blood group factors (polymorphisms). Blood antigens have been observed since 1900, and new types are now being discovered very infrequently even with constant surveillance for them in blood bank laboratories; therefore, the bias resulting from early discovery of mostly polymorphic systems can be considered very largely diminished. (For reviews, see Race and Sanger, 1968, 1975; Giblett, 1969; Mourant, 1954, 1958.)

Table 3-3 lists all human blood group antigens known to be genetically determined, ordered by year of their discovery as presented by Lewontin (1967) and modified by additional information from Race and Sanger (1968). Frequencies of the most common alleles for each locus are estimated from the white population of Great Britain. For each locus, average heterozygosity is calculated from the gene frequencies including all alleles at that locus under the assumption of random mating within the English population. The next to the last column of the table gives the cumulative average heterozygosity (sum of the previous column entries divided by the number of loci down to that particular blood group) estimated over all loci discovered up to that date. Finally, the last column gives the proportion of all loci known (to that year) that are polymorphic—those loci with alleles of one percent frequency or greater.

Many of the new factors have not been cross-tested for allelism with all other public and private factors, and if any allelism is discovered between them in the future, the estimated polymorphism and heterozygosity would be *increased* because such discoveries would decrease the number of monomorphic loci. It is also important to note that polymorphic loci are still being discovered (four since 1962: *Bg* in 1963, *Cs* and *Do* in 1965, and *Sd* in 1967), and that some factors are monomorphic in English but polymorphic in other populations (*Js* in Africans, *Di* in Mongoloid Asians, *Go* in Africans, and *Co* in Scandinavians).

The asymptotic levels of polymorphism (about 30 percent of loci) and average heterozygosity (about 13.5 percent) were achieved in 1963 and have simply varied around those levels in the succeeding years. It is quite astonishing that these values come so close to estimates of similar nature made for protein variants in drosophila and humans. Perhaps all these types of genic products are some special class (enzymes, antigens, and other proteins) of substances that are maintained at higher polymorphic levels than other types of genic products by natural selection or other evolutionary forces. From these random samples of genic loci, we can take these values as a reasonable, although conservative,

TABLE 3-3 *Human blood groups with their dates of discovery, frequency of most common allele, heterozygosity per locus, cumulative heterozygosity, and proportion of polymorphic loci in the English population (data from Lewontin, 1967, modified and amplified from Race and Sanger, 1968)*

Blood Group	Year	Frequency of Most Common Allele	Heterozygosity at Locus	Cumulative Heterozygosity	Proportion of Loci Polymorphic
1 ABO	1900	0.683	0.478	0.478	1.00
2 MNS	1927	0.389	0.700	0.589	1.00
3 P	1927	0.540	0.497	0.558	1.00
4 Se	1930	0.523	0.499	0.543	1.00
5 Rh	1940	0.407	0.662	0.567	1.00
6 Lu	1945	0.961	0.075	0.485	1.00
7 K	1946	0.936	0.122	0.433	1.00
8 Le	1946	0.815	0.301	0.417	1.00
9 Levay	1946	~1.00	~0	0.373	0.889
10 Jobbins	1947	~1.00	~0	0.337	0.800
11 Fy	1950	0.549	0.520	0.353	0.818
12 Jk	1951	0.514	0.500	0.366	0.833
13 Becker	1951	~1.00	~0	0.337	0.769
14 Ven	1952	~1.00	~0	0.313	0.714
15 Vel	1952	~1.00	~0	0.292	0.667
16 H	1952	~1.00	~0	0.274	0.625
17 Wr	1953	0.999	0.002	0.258	0.588
18 Be	1953	~1.00	~0	0.244	0.556
19 Rm	1954	~1.00	~0	0.231	0.526
20 By	1955	~1.00	~0	0.219	0.500
21 Chr	1955	0.999	0.002	0.209	0.476
22 Di*	1955	~1.00	~0	0.199	0.454
23 Yt	1956	0.995	0.010	0.191	0.434
24 Js*	1958	~1.00	~0	0.183	0.417
25 Sw	1959	0.999	0.002	0.176	0.400
26 Ge	1960	~1.00	~0	0.169	0.384
27 Good	1960	~1.00	~0	0.163	0.370
28 Au	1961	0.576	0.489	0.175	0.393
29 Lan	1961	~1.00	~0	0.168	0.380
30 Bi	1961	~1.00	~0	0.163	0.366
31 Xg	1962	0.644	0.458	0.173	0.387
32 Sm (Bu)	1962	0.999	0.002	0.168	0.375
33 Tr	1962	~1.00	~0	0.163	0.364
34 I	1956 (1960)	~1.00	~0	0.158	0.353
35 Bx	1961	~1.00	~0	0.153	0.343
36 Ht	1962	~1.00	~0	0.149	0.333
37 Go*	1962	~1.00	~0	0.145	0.324
38 LW	1962	~1.00	~0	0.141	0.316
39 Ls	1963	~1.00	~0	0.138	0.308
40 Wb	1963	~1.00	~0	0.134	0.300
41 Bg	1963	0.798	0.403	0.141	0.317
42 Or	1964	~1.00	~0	0.137	0.309
43 Luke	1965	~1.00	~0	0.134	0.302
44 CS	1965	0.840	0.264	0.137	0.318
45 Do	1965	0.580	0.487	0.145	0.333
46 Gf	1966	~1.00	~0	0.142	0.326
47 Gy	1966	~1.00	~0	0.139	0.319
48 Wu	1966	~1.00	~0	0.136	0.312
49 Co*	1967	~1.00	~0	0.133	0.306
50 At	1967	~1.00	~0	0.130	0.300
51 Rd	1967	~1.00	~0	0.128	0.294
52 Sd	1967	0.704	0.417	0.133	0.308
53 To	1967	~1.00	~0	0.131	0.302

* Polymorphic in other populations.

estimate of genetic polymorphism for outcrossing species whose population genetic structure may be similar to that of humans—that is, many species of vertebrates and drosophila.

Blood Serum Proteins

One of the best sources of vertebrate genetic markers is the extensive and complex group of serum proteins, often classified according to the strength of their negative surface potential at a pH of about 8 in moving boundary electrophoresis. The principal constituents can be discerned in the technique designed by Smithies (1959); after an initial separation of major components on filter paper for about 0.015 ml of serum, a second separation

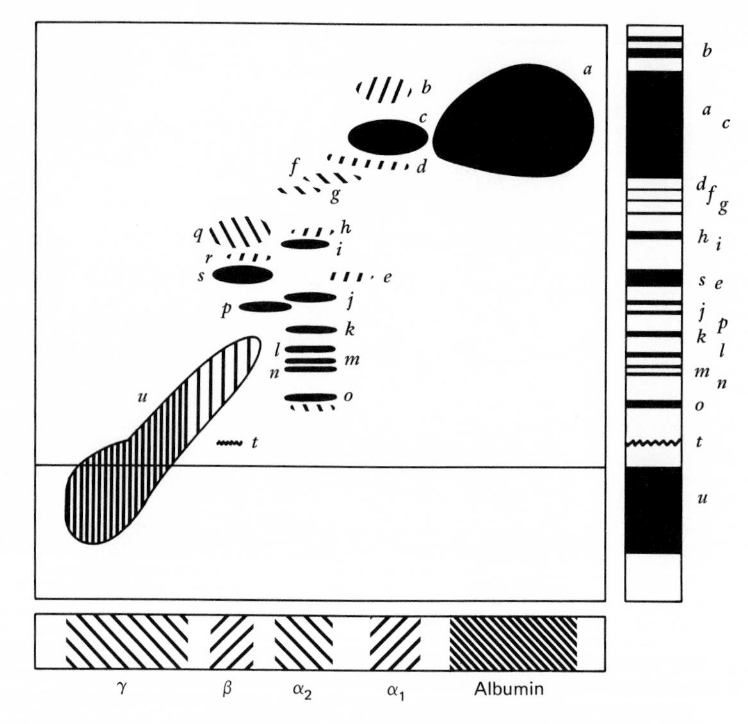

Figure 3-3. Diagrammatic representation of two-dimensional starch gel electrophoresis separation of human serum components with haptoglobin type 2-1 and transferrin type *C*. Below: the filter paper separation into globulins (gamma, beta, alpha-2, alpha-1) and albumin. At right: Starch gel separation in one dimension. Zones are as follows: a = albumin, b = acidic α_1-glycoprotein "orosomucoid," $c = \alpha_1$-antitrypsin; α_2-glycoproteins: d-f-g = *Gc* heterozygote (1-2), h-j-k-l-m-n- = haptoglobins (type 2-1); β-globulins: s = transferrin *C*, t = lipoprotein; γ-globulins = u. (Modified from Smithies, *Advances in Protein Chemistry*. New York: Academic Press, Vol. 14, 1959, p. 83.)

is made by running the material through starch gel at 90 degrees to the original separation, as indicated in Figure 3-3. Serum components thus resolved are, in order of their rate of migration on filter paper (that is, from highest to lowest negative charge), the albumins, α_1, α_2, β, and γ globulins. Many subzones are distinguished in the gel because the rates of migration for these proteins are different in the two media. High resolution between proteins is thereby achieved. Genetic variants of particular subzones may be subsequently studied by electrophoretic, immunological, or combined techniques. Genetic systems that are polymorphic in at least one or a few populations are listed in Table 3-4. (For reviews see

TABLE 3-4 *Principal serum protein genetic systems, polymorphic classified according to fractions from paper electrophoresis* (*condensed largely from Giblett, 1969*)

Serum Fraction	Genetic System	Locus	Number Alleles with ≥ 1% Frequencies	Ethnic Group Sampled
Albumin	Bisalbuminemia	*Alb*	At least 3	No. American Indians
Alpha$_1$ (α_1)	Antitrypsin (protease inhibitor)	*Pi*	4	Scandinavians
	Acid glycoprotein (orosomucoid)		2	Caucasoid Americans Japanese
Alpha$_2$ (α_2)	Haptoglobin	*Hp*	4	Widely sampled
	Group specific component	*Gc*	4	Widely sampled
	Ceruloplasmin	*Cp*	4	American Negroes American Caucasoids American Indians Orientals
	Macroglobulin	X_m	2	American Caucasoids American Negroes Norwegians Easter Islanders
Beta (β)	Transferrin	*Tf*	5	Widely sampled
	Lipoprotein	*Lp*	2	European, American Caucasoids American Negroes Brazilians, Tanzanians, Easter Islanders, Labrador Indians
Gamma (γ)	Immunoglobulin			
	IgG ("heavy chain")	*Gm*	*	Widely sampled
	IgG ("light chain") (also on *IgM* and *IgA* light chains)	*Inv*	3	Brazilian Indians, Negroes, Caucasoids, Japanese, New Guineans

* Inherited as a large series of alleles but possibly as three or more closely linked loci analogous to the Rh system. Possibly each *IgG* subclass is a locus with a set of alleles. At least eight such "alleles" are polymorphic.

Buettner-Janusch, 1970; Giblett, 1969; Greenwalt, 1967; Harris, 1970; Sutton, 1967.) Some details of the more important and widespread variants are the following.

HAPTOGLOBIN. This group of α_2 glycoprotein molecules (discovered in 1938 by Polonovski and Jayle) has the ability to bind oxyhemoglobin into a complex that does not pass into the glomerular filtrate of the kidneys and thus conserves heme iron. Its molecular structure, like the globins of hemoglobin, consists of four polypeptide chains, α_2 and β_2. Variations in the α chain are responsible for the different electrophoretic patterns of intact molecules, the *allotypes*, first described by Smithies (1955) from starch gel separation. The β chain contains a carbohydrate moiety and has the binding site to which the α chains of hemoglobin are bound.

Haptoglobin phenotypes in starch gel electrophoretic bands are generally of three kinds: (1) a single fast-moving band (*Hp* 1-1), (2) a set of multiple slow-moving bands (*Hp* 2-2), and (3) a set with a weak *Hp* 1-1 band plus the slower bands moved into a new position (*Hp* 2-1). These three types were shown by family studies to represent a pair of alleles with *Hp* 2-1 the heterozygote (Figure 3-4). The multiple bands of *Hp* 2 after purification were found not to contradict the genetic doctrine of one gene—one polypeptide—but to represent a series of stable polymers formed by the *Hp* 2 molecule, which is nearly twice the length of the *Hp* 1 molecule (*Hp* 1 has a chain length of 83 amino acids and *Hp* 2 has 142). In the heterozygote, a polymeric series forms, each member of which contains

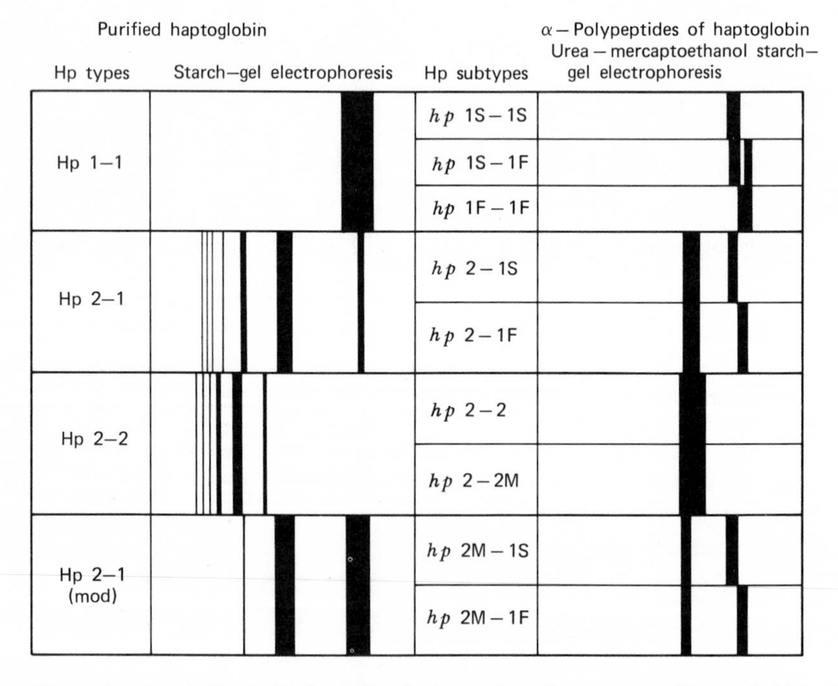

Figure 3-4. A diagram of starch gel electrophoretic patterns of haptoglobin phenotypes and their representative subtypes. From Hsia, *Inborn Errors of Metabolism.* Chicago: Year Book Medical Publishers, 1966.

products of both alleles except for the fast-migrating band, which is a pure *Hp* 1 chain. By breaking down haptoglobin molecules into their component polypeptides by cleavage with urea and mercaptoethanol, two alternative *Hp* 1 chains were revealed: a faster and slower designated $Hp\ 1^F$ and $Hp\ 1^S$, thereby distinguishing six genotypes with *Hp* 2 (Figure 3-4, right column). From amino acid sequence studies, the two polypeptides were found to differ at position 54 with lysine (1^F) or glutamic acid (1^S). The *Hp* 2 gene product contains both of these $Hp\ \alpha^1$ molecules joined together in tandem duplication minus one small fragment of both (from number-70 amino acid of one to about number-12 of the other), so that $Hp\ \alpha^2$ could be designated α^{1F1S}, or the reverse, with a deletion at their junction, possibly having originated by unequal crossing over.

A quantitative variant affecting the concentration of slower-moving polymers, $Hp\ 2^M$ (*Hp* 2-modified), was found to be common in African populations (see Figure 3-4, bottom). Finally, a variant with low haptoglobin or lacking it completely (anhaptoglobinemia), Hp^0, occurred in Nigerians (about 30 percent) and in American Negroes (about 5 percent). Evidence from family studies favored the interpretation that the Hp^0 phenotype is actually $Hp\ 2^M/Hp\ 2^M$ or that $Hp\ 2^M$ may be a controller element affecting the production of the $Hp\ \alpha^2$ chain. (A true amorph Hp^0 allele with no demonstrable gene product probably exists as a rare mutant found in certain European and Japanese families.)

Europeans have remarkably constant frequency of *Hp* 1 (from 35–43 percent), while Africans have higher frequencies but also considerable $Hp\ 2^M$ and anhaptoglobinemia so that the real frequencies of *Hp* 1 may be even higher there. Asian populations have lower *Hp* 1 than Europeans, with lowest values in India (10 percent). In Australian aborigines the frequency is 18 percent. American Indian tribes, in contrast, are highly variable even between tribes in close proximity. $Hp\ 1^F$ and $Hp\ 1^S$ have not been studied as extensively, but from all evidence obtained so far, Mongoloid populations, including American Indians, rarely have the fast allele, while Africans have more of the fast than the slow, and Caucasians have about one-third fast to two-thirds slow.

TRANSFERRIN. This β globulin, a glycoprotein discovered in 1933 by Starkenstein and Harvalik, has the function of selective iron (two ferric ions per protein molecule) transport from plasma to red cell precursors in the hemopoetic tissues. It is a single polypeptide chain with an *N*-terminal amino acid residue of valine and a probable molecular weight of 73,000–76,000. However, the subunit structure and the number of peptides in the molecule are at present unresolved.

Poulik and Smithies noted in 1958 a prominent band as the third in migration rate among the β globulin components. They called it transferrin C. With serum samples from a few Africans and Australian aborigines, they observed a slower band *D*, while a few Caucasians have a faster band they called *B*. After studying several families in which these bands were segregating, they proposed codominant alleles Tf^B, Tf^C, and Tf^D, with Tf^C the most common in all human populations. At present there are known to be at least 18 alleles ordered by rate of band migration relative to the *C* band, although only four variants are truly polymorphic (D_1, D_{Chi}, B_2, B_{0-1}). In each case apparently a single amino acid substitution has occurred, probably from the *C* peptide as a point mutation.

In geographic distribution, TfD_1 appears in two main populations: among aborigines of Australia and New Guinea and among Africans south of the Sahara. American Negroes have a frequency of about 9 percent TfD_1. TfD_{Chi} is common in southeastern Asia and

Mongoloid peoples including American Indians. TfB_2 occurs, at less than 5 percent, among western Europeans and certain isolates of Australian aborigines, while TfB_{0-1} occurs sporadically among certain American Indian tribes (Navajo, and Mexican-Central American people).

GROUP-SPECIFIC COMPONENT (Gc). In addition to haptoglobin, a second α_2 globulin glycoprotein was found to be polymorphic by Hirschfeld in 1958. Its function is yet to be determined, but this *Gc* protein with a molecular weight of about 51,000 is known to be synthesized in the liver and occurs in several body fluids: urine, ascitic, and spinal fluids. *Gc* probably consists of two subunits of 25,000 molecular weight, each held together with disulfide bonds. Amino acid sequences have not been worked out, but at least one substitution has taken place between Gc^1 and Gc^2.

Four alleles out of ten alleles known appear to occur in high enough proportions to be considered polymorphic. The fast-moving Gc^1 and slower Gc^2 seem to occur in nearly all human populations, with the former usually in excess over the latter except for certain Brazilian Indian tribes where Gc^2 may reach up to nearly 70 percent. Gc^2 is fairly constant throughout European populations at about 30 percent, although lower (13 percent) among northern people (Swedes and Finnish Lapps); it is less than 10 percent in frequency among African Negroes and sporadically reaches lowest levels in certain American Indian tribes (Navajo have 2 percent). The faster-moving allele Gc^{Chip} has a 10 percent frequency among the Chippewa tribe, with Gc^2 at 21 percent. The fourth polymorphic allele Gc^{Ab} is present among Australian and New Guinea aborigines in low frequency.

IMMUNOGLOBULINS. Some of the most heterogeneous protein molecules in the serum are the antibody fraction, immunoglobulins, symbolized *Ig*. Certain portions of these *Ig* molecules, known as the *V* portions, are so highly variable in amino acid sequence that each normal individual may possess thousands of different sequences; thereby versatility in antibody synthesis is apparently acquired within the individual. The remainder of the *Ig* molecule—more or less constant in amino acid sequence and known as the *C* portion—is responsible for two genetic polymorphisms: the *Gm* and *Inv* systems of alleles. (For reviews on structure of *Ig* see the following: Edelman, 1970; Edelman and Gally, 1968; Herzenberg, et al., 1968; Hood and Talmage, 1970; Putnam, 1968; Smithies, 1968; Steinberg, 1969.)

Ig (antibody) molecules are synthesized by plasma cells and lymphocytes, usually after an antigenic stimulus. Their heterogeneity would seem at first to make the working out of their structure technically impossible, but *Ig* homogeneity does occur in the tumorous condition discovered by Henry Bence Jones (physician in London in the midnineteenth century) known as multiple myeloma, or malignant proliferation of plasma cells. Each myeloma patient produces a unique set of *Ig* chain subtypes and Bence Jones (*B-J*) proteins (light chains only), so that structure has been worked out basically from them. All immunoglobulins have the same basic structural unit consisting of two heavy (*H*) polypeptide chains (molecular weight 50,000–70,000) and two light (*L*) chains (molecular weight 20,000) held together by disulfide bonds (see Figure 3-5). This basic unit can be fragmented in several ways by using different enzymes; for example, with papain, disulfide bonds are split, separating the *H* chain into two fragments, F_{ab} (antigen binding) containing the entire *L* chain plus the *V* and part of the *C* portion of the *H* chain, and F_c (crystalline) nonantigen-

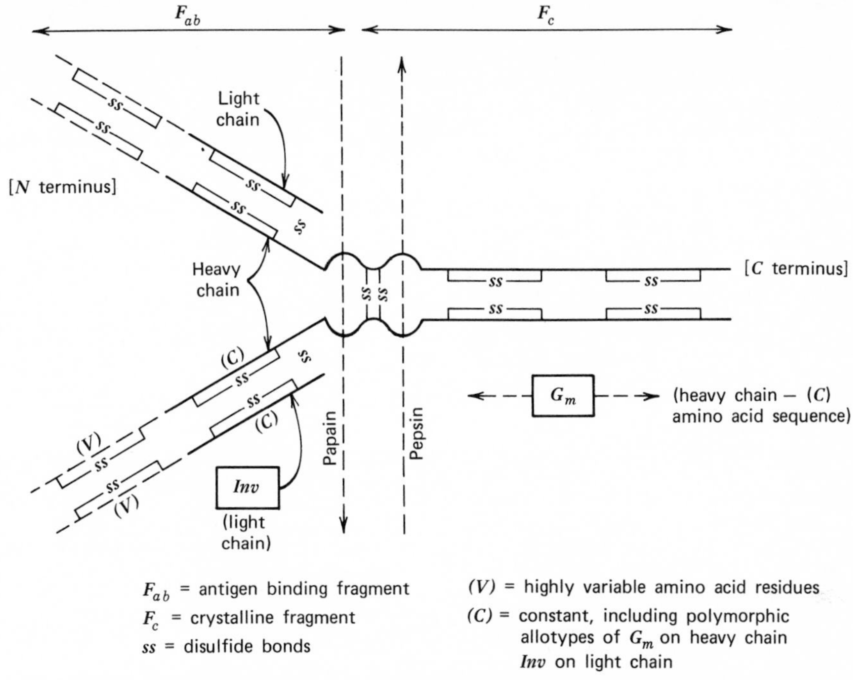

Figure 3-5. A diagram of the *IgG* molecule structure, adapted from several sources including Giblett (1969) and Edelman (1970).

combining remainder of the *H* chain. Other enzymes fragment the molecule, as indicated in Figure 3-5.

The five main classes of *Ig* molecules (*IgG*, *IgA*, *IgM*, *IgD*, and *IgE*) are characterized by particular *H* chain molecular weights, polymeric structures, and biological properties. Those structural specificities common to all members of the species but differentiating classes of molecules are called *isotypes*; those allelic variants that differ within the species as polymorphisms are termed *allotypes*; the protein products of single-cell clones within individuals (such as the particular chains of *B-J* myeloma proteins) are called *idiotypes*. Four subclasses) (isotypes) of *IgG H* chains have been distinguished by means of antisera induced in rabbits or monkeys against the *IgG* of myeloma patients—γ_1, γ_2, γ_3, and γ_4. Two isotypes of *L* chains also are distinguished—κ and λ. Thus, eight combinations of *H* and *L* chains are normally possible with two identical *H* and two identical *L* chains per molecule of *IgG*. *L* chain amino acid residues 1 to 108 are highly variable (*V*), while residues 109 to 214 constitute the constant (*C*) portion. *H* chain amino acid residues have not yet been described completely, although the *V* portion localized on the F_{ab} fragment may resemble its partner in the *L* chain to a great extent, and the sequence of certain fragments associated with *Gm* allotypes has been worked out (Steinberg, 1969).

The *Gm* system of allotypes was discovered in 1956 by Grubb. Rh positive red blood cells coated with "incomplete" anti-*D* antibodies (*IgG*, which reacts with the *rbc* antigen but does not bring about agglutination) were agglutinated by the serum of certain rheumatoid arthritis patients; however, this agglutination was inhibited by the serum of about

60 percent of normal donors. This inhibition was interpreted as follows: the *IgG* in the serum of the donor must contain the specific antigen (*Gm*) that combines with the antibodies (anti-*Gm*) from the rheumatoid arthritis patient, diverting them so that no agglutination occurs. This factor responsible for inhibiting agglutination was found only in the *IgG* (gamma globulin fraction), and it was named *Gm*(*a*). At present, there are 17 allotypes (symbolized as *Gm*(1), *Gm*(2), etc.) characterized according to their antigenicity with "Ragg" (rheumatoid arthritis agglutinant) or with "SNagg" (serum normal agglutinant) from normal donors. In 1965 the World Health Organization (WHO) accepted a notation of all immunoglobulin terms, converting a complex diversity into a useful number system for the *Gm* antigens. They were demonstrated to be inherited as a series of alleles or very closely linked sets of factors in much the same way as the Rh set (*DCE*) is known; that is, a family pedigree may indicate that as many as five, six, or seven *Gm* factors were transmitted together as if determined by a single codominant allele (symbolized with superscripts such as $Gm^{1,5}$, for example) representing either a single-compound locus or such closely linked cistrons that crossing over is exceedingly rare (Steinberg, 1969). Most of these antigens are associated with the F_c fragment, although each is found in one or two particular subclasses (for example, *Gm*(1) is found in γ_1, while *Gm*(23) is in γ_2 and *Gm*(5) is in γ_3); however, *Gm*(3) and *Gm*(17) are associated with the F_{ab} fragment and depend on quarternary structure of *IgG* to be expressed. Consequently, certain authors designate three closely linked but separate cistrons controlling these fragments or subclasses of *H* chains (Kunkel and associates), while others (Steinberg and associates) feel that a single cistron would explain the genetic data. (See Grubb, 1970, for a summary of *IgG* genetic markers.)

Perhaps the most interesting feature of the *Gm* allelic polymorphism is the distinct identification of specific alleles with racial groups (Table 3-5), so that racial mixture and variation within races can be analyzed with greater confidence than with any other known genetic system. Studies on Caucasoid admixture in United States Negroes indicate about 30 percent white gene flow into Negro populations throughout the United States (Reed, 1969) using Causasoid antigens *Gm*(1), *Gm*(1, 2), and *Gm*(5) compared with pure African antigen *Gm*(1, 5). Similar studies on Bushman admixture into South African Negro tribes has been presented by Steinberg. In European populations a cline exists with *Gm*(1) in-

TABLE 3-5 Gm "*alleles*" *commonly found in certain populations* (*from Steinberg, 1969*). *Sample sizes large for first three and Melanesians.*

Population	Alleles (or Multialleles)
Caucasoid	$Gm^{1,17,21}$, $Gm^{1,2,17,21}$, $Gm^{3,5,13,14}$
Negroid	$Gm^{1,5,13,14,17}$, $Gm^{1,5,14,17}$, $Gm^{1,5,6,17}$, $Gm^{1,5,6,14,17}$
Mongoloid	$Gm^{1,17,21}$, $Gm^{1,2,17,21}$, $Gm^{1,13,17}$, $Gm^{1,3,5,13,14}$
Ainu	$Gm^{1,17,21}$, $Gm^{1,13,17}$, $Gm^{2,17,21}$, $Gm^{1,2,17,21}$ (possible Japanese mixture)
Bushmen	$Gm^{1,17,21}$, $Gm^{1,5,17}$, $Gm^{1,13,17}$, $Gm^{1,5,13,14,17}$
Pygmy	$Gm^{1,5,6,17}$, $Gm^{1,5,13,14,17}$ (possible Negroid mixture)
Melanesian	
New Guinea	$Gm^{1,17,21}$, $Gm^{1,2,17,21}$, $Gm^{1,3,5,13,14}$, $Gm^{1,5,13,14,17}$
Bougainville	$Gm^{1,17,21}$, $Gm^{1,2,17,21}$, $Gm^{1,3,5,13,14}$

creasing to the north and east. Other clines and frequencies of alleles are given in Steinberg's review (1969).

The *Inv* system consisting of three allotypes is inherited via three alleles: Inv^1, $Inv^{1,2}$, and Inv^3. The first two were discovered by Ropartz and associates in 1961. They found, in similar fashion to the discovery of *Gm*, that the serum of a donor named Virm. agglutinated Rh+ red cells coated with anti-*D* and that about 19 percent of serum from French donors inhibited the agglutination. Pedigree analysis showed this system to be independent of *Gm* so it was named *InV*, then changed to *Inv*. *Inv* allotypes depend on the κ subtype of the *L* chain and differ by amino acid substitution at residue position 191; *Inv*(1) is associated with leucine and *Inv*(3) with valine at that position. *Inv*(2) has not yet been isolated from *Inv*(1). Their antigenic expression depends on the quarternary structure of the *IgG* molecule because *L* chains alone from normal sera or *B-J* proteins do not manifest *Inv* activity at the usual low concentrations.

In all populations analyzed so far, both *Inv*(1) and *Inv*(3) allotypes exist, with the latter usually predominating. Exceptions with *Inv*(1) reaching high frequencies (greater than 80 percent) occur in certain South American Indian tribes and in Melanesians on the island of Bouganville.

SUMMARY

For the alleles at one locus (cistron) or effectively segregating genetic units, equilibrium of genotypes follows the square law of allelic frequencies as an extension of the Castle-Hardy-Weinberg principle. With increasing numbers of alleles and approach to equal proportions between them, heterozygosity becomes nearly 100 percent for the locus concerned.

Dominance between alleles or mixtures of dominance and codominance introduce problems of estimating allelic frequencies, which are mathematically complex if high accuracy is to be achieved, although convenient approximations are available.

With the discoveries in biochemical genetics in which gene products (polypeptides) can be detected by analytical tools capable of separating single amino acid substitutions or other differences in peptides, the problems of dominance and phenotypic interactions have been reduced. An impressive array of allotypic polymorphisms has been revealed in populations of many organisms. The extent of genic variation in the dimension of single cistrons may now be estimated because both polymorphic and monomorphic loci can be ascertained. Some loci have three or more alleles stablized in populations, and the average individual in many species has about 10–15 percent loci in the heterozygous state.

In human populations, a few polymorphisms (hemoglobins, for example) may appear to have strong selective value when genotypes produce clinical symptoms, although most allotypes of enzymes, serum proteins, and antigens have no known selective value and must at this time be presumed to have neutral fitness. Nevertheless, certain polymorphisms are widespread and at about constant frequencies in most populations sampled (*Hp*-1 and *Hp*-2, *Gc* 1 and *Gc* 2, *Inv* 1 and *Inv* 3), others show predominant alleles (*Tf-C*, Gm^1, *rbc* phosphatase p^b, *PGM*-1, for example), and still others are mostly race- or population-specific both for their occurrence and their frequencies (*Gm* alleles particularly). These specific distributions imply a variety of causes for maintenance of allotypes and suggest that they are not actually neutral.

EXERCISES

1. If mating is at random in a population with three alleles (A, a, and a') whose frequencies are $p_A = \frac{2}{3}$, $q_a = \frac{1}{6}$, $r_{a'} = \frac{1}{6}$,
   (a) What would be the total amount of homozygosity expected?
   (b) What would be the total amount of heterozygosity?
2. (a) How many heterozygotes are possible when a genic locus has four alleles? six alleles? ten alleles?
   (b) What would be the maximum heterozygosity (H) with four alleles, six alleles, or ten alleles in a random-mating population?
   (c) In Example 3-1, frequencies of certain third chromosome arrangements from natural populations of *Drosophila pseudoobscura* are given in Table A. How close to maximum heterozygosity (heterokaryotype percentage) is the observed percentage in the three localities?
3. Suppose five of six alleles in a population are equally frequent ($p = q = r = s = t$) and all are moderately rare at 10 percent frequency, but the sixth allele is common at 50 percent.
   (a) At equilibrium from random mating, what will be the total homozygote frequency? total heterozygote frequency?
   (b) What fraction of total heterozygosity will be due entirely to the five more rare alleles?
4. In a case of complete dominance in a multiple allelic series of three alleles (such as the color, himalayan, albino series in rodents), there are only three phenotypes. In a population of random-mating rodents having these phenotypes, estimation of gene frequencies can be made as follows. Let p = color gene, q = himalayan gene, and r = albino gene, then:

	Color Zygotes	Himalayan Zygotes	Albino Zygotes	Total
	$p^2 + 2pq + 2pr$	$q^2 + 2qr$	r^2	
Observed	2527	1627	1846	6000

   (a) What are the gene frequencies (p, q, and r) in this population?
   (b) Refer to Example 3-3 and calculate the variance and standard error of your estimate of r.
5. From Table 3-2, calculate the expected value and goodness of fit for the A, B, O, and AB blood groups for the following populations assuming the multiple allele basis of inheritance (Bernstein):
   (a) Germans.
   (b) Pygmies.
6. For a population with three alleles, the total amount of homozygosity may equal heterozygosity over a wide array of p,q values (letting $r = 1 - p - q$).
   (a) Solve formula (3-4) for $q = 0$, 0.05, 0.10, 0.20, 0.30, 0.45. Make a graph of these results. What is the higher p value when $q = \frac{1}{6}$?
   (b) Show that for each of these points, total homozygote frequency = total heterozygote frequency.

7. If a population is in equilibrium for three alleles, for every pair of different heterozygotes ($Aa - a'a$, for example) that mates as parents, there must be one pair that produces those heterozygotes ($aa \times Aa'$). However, all progeny from the latter pair will be heterozygotes, while from the former pair only half the progeny will be the same heterozygotes.
   (a) Show algebraically how these frequencies are balanced (see Property 5 in Chapter 1). One set of balanced pairs will be sufficient.
   (b) With the observed frequencies of hemoglobin genotypes in the Ghanian population of Example 3-2, what would be the expected mating frequency of $A/S \times A/C$? Expected mating frequency of $A/A \times S/C$? Do these frequencies represent the equilibrium situation?
8. In Example 3-2, using the incidence of genotypes of hemoglobin alleles given for Ghanians and for Americans of African ancestry, reduce the number of genotypes in both samples by considering three combinations only with regard to the presence-absence of Hb^A allele: A/A, $A/$-, non$A$.
   (a) As in formula (3-5), what will be the expected proportions (from random mating) of these three combinations for the populations given in Tables A and B?
   (b) Do the same by lumping together only with regard to presence or absence of Hb^S. Then with regard to Hb^C. Which groupings make the biggest difference in contrasting the Ghanian with African-American population?
9. As cited by Race and Sanger (1968), Cleghorn, 1960, tested 1000 English persons with antibodies for M, N, S, and s blood types. On the basis of assuming that M-N and S-s are tightly linked and are not known to recombine, we may consider the combination of MN with Ss as defining four allelic states: MS, Ms, NS, and Ns. Cleghorn's sample is as follows:

Blood Type	Observed Number
MS/MS	57
MS/Ms	140
MS/NS	39
MS/Ns } Ms/NS }	224
Ms/Ms	101
Ms/Ns	226
NS/NS	3
NS/Ns	54
Ns/Ns	156
	1000

   (a) On the assumption of random mating and four allelic states, calculate the expected numbers for these blood types.
   (b) Verify a $X^2 = 3.24$. How many degrees of freedom are there? What is the probability for the null hypothesis?
10. A biochemical geneticist takes a survey of 10 blood proteins from a population of 1000 individuals and is able to distinguish these proteins by electrophoretic gel separation. Only a single band on the gel is found for all samples of blood for the proteins 2, 3, 5, 6,

and 9. For the remaining proteins, the geneticist recognizes two or more bands as indicated below with their frequencies among the 1000 samples:

Protein No.	Bands	Frequencies	Protein No.	Bands	Frequencies
1	F	0.10	8	Z	0.001
	S	0.90		W	0.999
4	A	0.30	10	D	0.25
	B	0.10		E	0.25
	C	0.60		F	0.25
				G	0.25
7	X	0.50			
	Y	0.50			

(a) What is meant by polymorphism? Which proteins should be designated as monomorphic or as polymorphic? Why?
(b) For each protein, determine the heterozygote expected frequency assuming random mating.
(c) What would be the frequency expected of persons homozygous for each protein?
(d) How would you calculate the frequency of persons homozygous at all 10 loci? What could be said about the level of heterozygosity in this population?

11. In Example 3-5, estimate average heterozygosity (H) for the English population.
    (a) For the loci with three alleles each, red cell acid phosphatase (p), and placental alkaline phosphatase (Pl).
    (b) For the loci: PGM-1, PGM-2, PGM-3, E-2, ADA, AK, and 6PGD.
    (c) What is the average H for all nine of these loci?
    (d) If we assume that enzymes 10 through 19 are monomorphic in the English population, what would be the expected H for an average genome, assuming these loci to be a random sample of the "English genome."
    (e) How do these estimates of average heterozygosity compare with the estimate based on human blood groups from Table 3-3?
12. From the data in Table A, Example 3-4, on *D. pseudoobscura* allozymes, calculate the expected frequency of heterozygosity (total H) for the following populations: Strawberry Canyon (S.C.), Mesa Verde (M.V.), and Bogotá, Colombia (B.):
    (a) Leucine aminopeptidase.
    (b) Xanthine dehydrogenase.
    (c) Acetaldehyde oxidase-2.
    (d) Protein-10.
    (e) What are the main contrasting features between these populations? Is the same allele always the most frequent one in all populations?
13. Koehn, Milkman, and Mitton (1976) sampled populations of the blue mussel (*Mytilus*), a dioecious bivalve mollusk, along the Atlantic coast, and genotypes for allozymes were determined. For the allozyme leucine aminopeptidase (LAP), they found three alleles to be common at Barnstable, Massachusetts: LAP^{98}, LAP^{96}, and LAP^{94} (in order from fastest to slowest migrating protein variants observed as bands in starch gel electrophoresis). Frequencies of the alleles were found as follows:

LAP	98 (p)	96 (q)	94 (r)
Frequency	0.22	0.28	0.50

However, the heterozygote frequencies were found to be as follows (modified for the sake of this sample):

LAP	98/96	98/94	96/94
Frequency	0.086	0.154	0.196

(a) Do these heterozygote frequencies agree with the expectation from random mating? What is the overall amount of difference from expectation (total H_{obs} − total H_{exp})?

(b) Among other explanations for the observed difference, the authors pointed out that if a "null allele" existed at a low frequency, the data would fit expectation much better. Assume a null allele (LAP^0) recessive to all three protein product alleles exists which produces *no* LAP allozyme. If such an allele existed in this population at a frequency of $s = 0.15$ (and readjusting the three product alleles above so that $p + q + r + s = 1.00$), what would be the expected heterozygote frequencies with the null allele? Expected H for the three alleles that do produce the enzyme?

(c) Could this null allele then explain the observed difference from expectation in part (a), assuming a reasonable sample size?

14. Prakash and Lewontin (1968, 1971) have shown that allozymes linked to the third chromosome in *D. pseudoobscura* include those from three loci: α-amylase, as well as proteins 10 and 12. Frequencies of the chromosomal arrangements found at Strawberry Canyon are given below (from Strickberger and Wills, *Evolution* 20: 592–602, 1966) followed by the frequencies within arrangements of amylase alleles (*amy*) from that population (phylad = the set of arrangements derived from an "ancestral" arrangement by a single or double inversion event, as in Example 3-1 Figure A).

Arrangement	Arrangement Frequency	α-Amylase Alleles	
		0.84	1.00
ST	0.47	0.12	0.88
AR	0.09	0.30	0.70
PP	0.03	—	1.00
ST phylad total	0.59		
SC	0.01	1.00	—
CH	0.19	0.36	0.64
TL	0.14	0.79	0.07*
SC phylad total	0.34**		

* another allele, *0.74*, occurs in TL.

** Plus three rare arrangements, which you may assumed to be exclusively *amy*-0.84 allele.

(a) What are the expected *amy*-allele genotype frequencies within the *ST* phylad? within the *SC* phylad?
(b) What is the average *amy* heterozygosity expected among arrangements within phylads?
(c) How many arrangement heterozygotes (heterokaryotypes) are possible (omitting the rare *SC* phylad arrangements) within phylads? between phylads?
(d) What is the average heterozygosity expected for between phylad heterokaryotypes?

4

SEX LINKAGE

In organisms whose sex determination mechanism depends on a chromosomal difference between the sexes—that is, where one sex is heterogametic and the other homogametic—genes carried on those chromosomes by which the sexes differ determine more genotypes than autosomal genes do. Specifically, the heterogametic sex is usually haploid for an X chromosome and will have zygotes occurring according to allelic frequencies, while the homogametic sex, being diploid, usually fits the same scheme as the autosomal already discussed in previous chapters.

The condition in which females are homogametic (XX) and males heterogametic (XY) is very common, found in most vertebrates including humans, in many insects and other invertebrates including drosophila, and in some dioecious plants (*Lychnis*, for example, in the Pink family). A similar condition in which males simply lack a Y chromosome, whose genic contents may have been translocated into one or more autosomes, occurs to some degree in Orthoptera (short-horned grasshoppers for example), Heteroptera (aphids), and Odonata (dragonflies). Finally, the sexes may be reversed with respect to heterogamy versus homogamy, as in some birds (poultry for example), moths and butterflies, and some lower vertebrates.

In all these cases, the sex chromosome genotypes will be different for the two sexes, and the basic precepts of frequency analysis and equilibria must be considered anew for these systems. For convenience we shall treat the XY sex as the male, with euchromatic

TABLE 4-1 *Frequencies of mating and expected progeny at equilibrium for X-linked alleles*

A

		Females		
		p^2_{AA}	$2pq_{Aa}$	q^2_{aa}
Males	p_A	p^3	$2p^2q$	pq^2
	q_a	p^2q	$2pq^2$	q^3

B

Genotype Mating	Frequency of Mating	Progeny Females AA	Progeny Females Aa	Progeny Females aa	Progeny Males A	Progeny Males a
$AA \times A$	p^3	p^3	—	—	p^3	—
$Aa \times A$	$2p^2q$	p^2q	p^2q	—	p^2q	p^2q
$aa \times A$	pq^2	—	pq^2	—	—	pq^2
$AA \times a$	p^2q	—	p^2q	—	p^2q	—
$Aa \times a$	$2pq^2$	—	pq^2	pq^2	pq^2	pq^2
$aa \times a$	q^3	—	—	q^3	—	q^3
Totals	1.00	p^2 +	$2pq$ +	q^2	p +	q

alleles only on the X chromosome. If mating frequencies are determined randomly and allelic frequencies are the same for the two sexes, it can be shown by an arrangement analogous to Table 1-1 that the population's frequencies are at equilibrium. The six types of matings, their frequencies, and offspring produced are given in Table 4-1. Note that the two sexes in the progeny are treated separately.

In view of the zygotic expectations, it is evident that sex-linked recessives will appear to be far more common among males (q) than among females (q^2)—that is, in a ratio of $1:q$. For example, in European populations red-green color blindness occurs in about 8 percent of males and only about 0.4 percent of females (see Example 4-1). The value for females is less than expected (0.64 percent) because there are actually two closely linked loci (protanopy, or red blindness, and deuteranopy, or green blindness) that are complementary, so that female "trans" heterozygotes (protanopy/deuteranopy) have normal vision. Nevertheless, if sex-linked alleles are at equilibrium in the population, it is expected that frequencies should be equal in the two sexes.

EXAMPLE 4-1

HUMAN RED/GREEN COLORBLINDNESS

One of the earliest sex-linked anomalies to be observed is red/green colorblindness, which is now understood to be probably determined by two separate but closely linked loci, a protanoid (red defect) and a deuteranoid (green defect), and likely to be separated by the locus for G-6-PD (glucose-6-phosphate dehydrogenase) (see Kalmus, 1965; Post, 1962, 1965 for reviews). A person with normal vision who sees white light when certain intensities of blue, red, and green are projected on a screen is said to be trichromatic. Those with color vision defects require either a greater intensity of red (protanomaly) or green (deuteranomaly), the trichromatous anomalies, or they see white light with only two colors projected (dichromatic) and are said to have protanopia (red not distinguishable) or deuteranopia (green not distinguishable). Approximately 60 percent of all colorblind defects are deuteranomalies. Women who are heterozygous for both red and green defects (either in "repulsion" (trans) or in "coupling" (cis) linkage) seem to be normal.

Waaler in 1927 studied a large number of children in Norway for colorblindness, giving the following results cited in Kalmus and Post references:

	Colorblind	Normal	Total	Percent Colorblind
Males	725	8324	9049	8.01
Females	40	9032	9072	0.44

If the colorblindness were due to a single locus and the frequencies of an allele for defective vision were equal in the two sexes, then the number of colorblind females observed would be too small based on the square of the haploid value observed for males; that is, 0.64 percent should be found, or 58 females. In a later analysis Waaler separated the four vision defects as follows:

	Protanopia	Protanomaly	Deuteranopia	Deuteranomaly	Defects Total
Males	80	94	93	458	725
Females	0	3	1	36	40

If we postulate that protanoid and deuteroid defect genes are complementary in heterozygotes, giving normal vision, the excess of normal females could well be explained. The student may check the likelihood of this explanation by solving problem 6 at the end of this chapter.

Post (1962, 1965) has summarized the populational incidence of colorblindness in a wide number of ethnic groups. In general, the "contemporary paleolithic" groups (Eskimo, Australian aborigines, Fiji Islanders, and American Indians) have the lowest incidence, with about 0.5 percent protanoids and 1.5 percent deuteroids. The highest incidence appears to occur among western European males with 2.0 percent protanoids and 6.0 percent deuteranoids. Oriental, African, and racially mixed groups are intermediate in these frequencies. Post has made the interesting suggestion that the explanation for these differences may lie in the fact that hunting and gathering paleolithic cultures were selective against any vision defects, but with the relaxation of such selection in neolithic and modern cultures, mutation rates and possibly some special advantages to color vision anomalies could raise the frequencies of these defects to the observed levels.

APPROACH TO EQUILIBRIUM

If allelic frequencies are not equal in the two sexes, it may be recalled from formula (1-6) that for autosomal genes, the average of the two parent frequencies will be established after random mating among the first-generation progeny. Sex linkage is complicated however by two facts: (1) progeny males' X chromosome allelic frequencies are determined by those of their mothers, and (2) the average frequency of the two parents is not simply half the sum of the parents' frequencies but is the average weighted by the dosage of X chromosomes ($2X♀ + 1X♂$, divided by three). The alternation of X's between the sexes from one generation to the next brings about an oscillating approach to equilibrium, while the weighted average frequency is a constant and will be the ultimate equilibrium value after several generations of random mating.

An illustration of the oscillation and eventual equilibrium is given in Table 4-2 and Figure 4-1. From any generation to the next the following relationships are evident.

Male allelic frequencies depend on those of their mothers. If we adopt the same symbols used in formula (1-6), with p_A, q_a (female allele frequency) and r_A, s_a (male frequency for the same alleles), then progeny males' A frequency becomes

$$r'_A = p_A \tag{4-1}$$

where $'$ refers to progeny.

Each progeny female is determined by contributions of X chromosome doses from both of her parents as with autosomal alleles; therefore, the progeny females will have the mean frequency of their parents, as in formula (1-6).

TABLE 4-2 *An example of the approach to equilibrium for sex-linked genes when allelic frequencies in the two sexes initially are unequal*

	Females					Males		
Generation	p_A	q_a	AA	Aa	aa	r_A	s_a	Difference $(q_a - s_a)$
0	0.40	0.60	0.16	0.48	0.36	0.88	0.12	+0.48
1	0.64	0.36	0.352	0.576	0.072	0.40	0.60	−0.24
2	0.52	0.48	0.256	0.528	0.216	0.64	0.36	+0.12
3	0.58	0.42	0.3328	0.4944	0.1728	0.52	0.48	−0.06
4	0.55	0.45	0.3016	0.4968	0.2016	0.58	0.42	+0.03
5	0.565	0.435	0.3190	0.4920	0.1890	0.55	0.45	−0.015
6	0.5575	0.4425	0.31075	0.4935	0.19575	0.565	0.435	+0.0075
⋮	⋮	⋮	⋮	⋮	⋮	⋮	⋮	⋮
∞	0.56	0.44	0.3136	0.4928	0.1936	0.56	0.44	0

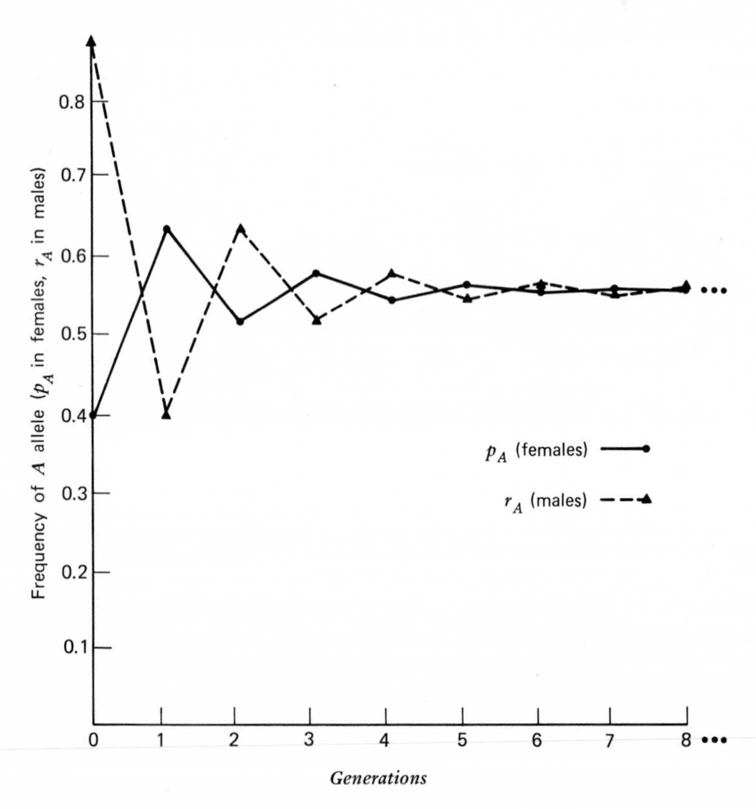

Figure 4-1. An example of approach to equilibrium for sex-linked allele frequencies when the sexes' initial frequencies are unequal, as in Table 4-2.

$$p' = \frac{p + r}{2} \tag{4-2}$$

omitting the A subscript.

The sex difference in frequency in any generation is halved in the following generation, but opposite in sign:

$$p' - r' = \frac{p + r}{2} - p = \frac{-(p - r)}{2} \tag{4-3}$$

The weighted (by X doses) average of any generation is a constant, representing the total dosage of sex-linked alleles initially:

$$\hat{p} = \frac{2p + r}{3} = \frac{2p' + r'}{3} = \cdots \tag{4-4}$$

where $\hat{p}$ refers to equilibrium frequency. In Table 4-2, this value ($\hat{p} = 0.56$) is approached in both sexes and can be calculated from any generation.

From these considerations it is evident that for a single locus under random mating the sexes should be equal in allelic frequencies; if they at any time become unequal due to chance or some arbitrary circumstance, they should return to equality at equilibrium.

ESTIMATING *X*-LINKED ALLELIC FREQUENCIES

When the allelic frequencies are not equal in the two sexes and estimating p, q values must be done on the assumption that they are equal, the maximum-likelihood method should be employed (as illustrated in Neel and Schull, 1954; Crow and Kimura, 1970; Li, 1976). If there is no dominance, so that heterozygotes can be ascertained, the weighted average (by dosage of X chromosomes) is appropriate:

Let N_f = number of females observed

N_m = number of males observed

A, B, C = observed numbers of homozygous dominant, heterozygous, and homozygous recessive females, respectively, so that $A + B + C = N_f = D \cdot N_f + H \cdot N_f + R \cdot N_f$

AY, CY = observed number of dominant and recessive males, respectively so that $AY + CY = N_m = D \cdot N_m + R \cdot N_m$

$$q = \frac{2C + B + CY}{2N_f + N_m} \tag{4-5}$$

with variance of $q(V_q) = q(1 - q)/(2N_f + N_m)$.

However, with dominance, the estimate for females becomes less efficient than that for males, and the maximum-likelihood estimate is as follows:

$$q = \frac{-AY + \sqrt{(AY)^2 + 4(2N_f + N_m)(2C + CY)}}{2(2N_f + N_m)} \quad \text{and} \quad p = 1 - q \tag{4-6}$$

with variance of q $(V_q) = q(1 - q^2)/[4N_f q + N_m(1 + q)]$ (see Neel and Schull or Li for derivation of variance q).

As an illustration, if we assume the frequencies of the $X_m a^-$ allele are equal in the two sexes among Norwegians, as given in Example 4-2, we obtain the estimate of q, using (4-6):

$$q = \frac{-23 + \sqrt{529 + 4(202 + 100)(88 + 77)}}{2(302)} = 0.702$$

$$V_q = 0.0008, \text{ standard error}_q = 0.028$$

An inequality between the sexes in X-linked allelic frequencies may indicate at least one of three important possible causes: (1) a complex genetic situation where more than one genic locus may be concerned with the phenotype (as with red-green colorblindness illustrated in Example 4-1), (2) a sex-limited or sex-influenced difference in effective selective forces (as may be operating in the case of G-6-PD, see Example 4-4), or (3) a nonrandom mating system as would be true if there were selective differences in mating success (illustrated in Example 4-3). It is therefore worthwhile estimating sex-linked allelic frequencies separately for the two sexes, in the event that any of these explanations may be uncovered. Example 4-2 illustrates human population data for the sex-linked serum protein Xm, in which there is a regular excess of the dominant allele in females. Whether this excess reflects a possible complexity of the genic locus (as with red-green colorblindness) or some constant sex-limited selective difference favoring the dominant in females is not yet clear. Data should be tested statistically for sex differences in frequencies whenever possible in calculations of X chromosome allelic frequency values.

EXAMPLE 4-2

ANTIGEN X_m IN HUMAN POPULATIONS

A polymorphism for an antigen in the α_2 macroglobulin portion of human serum was shown in 1966 by Berg and Bearn to differ widely between the sexes and to be sex-linked. Presence of the antigen known as X_m (m refers to macroglobulin) was shown to be dominant ($X_m a+$) over its absence ($X_m a-$). Subsequent linkage studies revealed the X_m locus to be closely linked to the deutan (green defect) locus (Berg and Bearn, 1968). Four populations were sampled by these investigators to determine allelic frequencies, and their data are as follows:

Population	Number Sampled	Number $X_m a+$	$X_m a-$	q
Norwegians	M 100	23	77	0.7700
	F 101	57	44	0.6600
U.S. Caucasians	M 57	15	42	0.7368
	F 67	34	33	0.7018
U.S. Negroes	M 81	25	56	0.6914
	F 151	90	61	0.6356
Easter Islanders	M 66	16	50	0.7576
	F 80	38	42	0.7246

It is apparent that these populations have similar allele frequencies and that the recessive allele (X_m-) in females is less than in males by about 6 percent. Thus, an excess of antigen X_ma in females cannot be explained as due to any developmental bias, because the investigators pointed out that the antigen does not appear in girls often before age 10, and the bias would then be in the opposite direction. Of course, if the locus determining the anitgen is a complex one in which complementation can occur (a situation similar to that between the closely linked protan and deutan colorblindness loci, which in "trans" phase females produce normal vision), and if a least two different alleles produce the negative form, some "trans" phase female heterozygotes could be X_ma+, thus producing an overestimate of the dominant alleles. Far more sampling and knowledge of antigenic products from this locus is needed before final interpretation is possible.

EXAMPLE 4-3

SEX-LINKED RECESSIVES IN DROSOPHILA AND MALE MATING ACTIVITY

Experimental populations of *Drosophila melanogaster* in which one of four different sex-linked recessive mutants were observed for their frequencies relative to their wild type alleles over many generations gave the following results according to Merrell (1953). Starting with equal frequencies of mutant: wild type allele among males and heterozygous females so that $p = q$ in both sexes, he showed that for three of the four mutants (yellow body, cut wing, and raspberry eye color), there was a continuous excess of the mutant allele for males (s = frequency of recessive in males) each generation. The recessive allelic frequency difference ($s - q$) was always positive in the population, if one calculated the females' q value by taking the square root of female homozygous mutant frequency. The fourth mutant (forked bristles) did not show a significant difference between the sexes. A summary of Merrell's data is as follows:

Population	No. Days from Start	Mutant Males Frequency	($s - q$) Average for All Generations
I Yellow vs +	180	0.006	+0.066
II Cut vs +	180	0.003	+0.062
III Raspberry vs +	1351	0.048	+0.026
IV Forked vs +	1209	0.049	+0.004

With sex-linked genes, homozygous recessive female progeny come from only two types of mating (not from three types as with autosomal) in which the male parent is a recessive hemizygote; therefore, if the mutant males fail to mate as often as wild-type males, there will be a deficiency of mutants among females compared with the random-mating situation. For the forked (f) versus its wild-type allele ($+^f$), there were as many cases in which the estimate of q among females was higher than that among males as the reverse, so that the net +0.004 was not significantly greater than zero. This result indicated random mating between those genotypes in that population. Merrell was able to demonstrate in separate mating tests the lowered success of the other three mutant-type males (*y*, *ct*, *rasp*) in rivalry with wild-type males for females. This mating propensity effect for the *yellow* and *cut* mutants was especially noteworthy and was shown to be the chief selective cause for the elimination of those mutant alleles from their respective populations. Forked males were at no such disadvantage, mating equally in rivalry with wild-type males. Females homozygous for all these mutants were equal to wild-type females in receptivity except for the *cut* females, which were less receptive.

SEX-LINKED POLYMORPHISMS IN HUMANS

Genetic polymorphisms are common in which males express a trait predominantly because the trait is recessive and the X chromosome is hemizygous. In cases of anomalous conditions such as colorblindness and G-6-PD deficiency, there may be a considerable effect on relative fitness to survive in particular cultures (Examples 4-1 and 4-4), and frequencies of recessive variants may therefore be retained in populations. Other polymorphisms may exist with variants in high frequencies such as serum protein (Xm in Example 4-2) or blood antigen factors (Xg, for example) that have no apparent effects on reproductive fitness. Xg was discovered in 1962 (see Race and Sanger, 1975) with the positive allele (Xg^a) dominant to its recessive silent allele (Xg) with negative phenotype and found to be probably located on the short arm of the X chromosome because it is not closely linked to any well-known X chromosome trait, although it does appear to be located near the locus for the rare ichthyosis trait. Race and Sanger report the dominant allele is the more common in nearly all human populations sampled, ranging from about 40 percent among Taiwan aborigines to 60 percent on mainland Asia, about 70 percent in Europe, and nearly 80 percent among Navajo Indians. Surprisingly, in a survey of the antigen among various apes and monkeys, Xg^a was found in one species of gibbon, *Hylobates lar lar*, in a sample of 42 at a frequency of 29 percent, so that we may hazard the assumption that this polymorphism is of very long phylogenetic standing.

About 65 known loci remain on the X chromosome of humans, but their mutant forms are so rare that we cannot consider them polymorphisms, such as the familiar hemophilia A and B loci, muscular dystrophy, ocular albinism, and angiokeratoma. No doubt these mutant forms are preserved in the population largely by their mutation rate from normal alleles. It is possible that the more common form of hemophilia (A, at an incidence of 13 per 100,000 males) may have some effect on fertility. Clarke (1959) pointed out that in the English royal family including Queen Victoria, seven heterozygous females carriers produced 37 children compared with six sib females (probably homozygous normal) who produced 28 children. So far such data are too scarce to make a clear distinction at this point for hemophilia A between the higher value of fertility for heterozygous females and that of homozygous normal females. If such a possibility is sustained by future studies, the hypothesis that the hemophilia polymorphism is exclusively mutational would need to be reconsidered.

EXAMPLE 4-4

G-6-PD HUMAN POLYMORPHISM

A most widespread human polymorphism is the series of alleles at the sex-linked locus for glucose-6-phosphate dehydrogenase (G-6-PD). Attention was drawn to deficient variants of this enzyme, especially following World War II, when association was demonstrated between red cell hemolysis and antimalarial drugs (primaquine and pentaquine, for example), antibacterial agents (sulfonamides and sulfanilamides), as well as ingestion of raw *Vicia faba* beans. In all these cases, the "pentose shunt" pathway in the oxidation of glucose-6-phosphate to 6-phosphogluconate and concomitant reduction of the coenzyme NADP (nicotinamide adenine dinucleotide) to NADPH is limiting in red cells with the variant G-6-PD enzyme; their inability to respond to increased demands on this shunt pathway brings about hemolysis and anemia (see Giblett, 1969; Harris, 1970; Kirkman, 1971, for reviews). There are now known to be at least 29 allelic variants distributed as follows with respect to their enzyme activity: (1) six about normal, including the most common allele *B* found in all human races, (2) two with increased activity, (3) thirteen with 50 percent or less enzyme activity but no anemia unless the shunt pathway is stimulated, and (4) eight with low activity and hemolysis in hemizygous males. The alleles that can be considered polymorphic either have normal activity or fall into the third group with no regular anemia.

The G-6-PD molecule is probably a hexamer of subunits with 40,000 molecular weight each, making 240,000 total; in its active state it is associated with its coenzyme NADP. While there is a possibility that two structural loci may control the complete molecule, it is more likely that a single locus does this, although until amino acid sequences are worked out the evidence is not conclusive. The genic locus symbol is *Gd*.

The following alleles can be considered polymorphic; they are separable as electrophoretic components at a pH of 8.6:

B: found in all human populations; can be considered standard.

A: variant common in Negro populations, with about 90 percent activity, slightly faster mobility in electrophoresis than *B*, and differing from *B* in substitution of aspartic acid for asparagine at a single amino acid residue. In Nigeria, about 22 percent of males are Gd^A, while in the United States, about 15–20 percent Negro males are of this genotype.

A−: variant common in Negro populations, with about 15 percent enzyme activity, the same electrophoretic mobility as *A*, and probably a structural variant of *A* but less stable in older red blood cells when demands are put on the shunt pathway. About 22 percent of Nigerian males and 10–15 percent of American males of African descent show this variant.

B−: variant common among Caucasians in the Mediterranean region and in the Orient among Mongoloid peoples, with 0–7 percent enzyme activity, the same electrophoretic mobility as *B*. However, unlike the Negro variants that are expressed in red cells, the *B*− variant enzyme is expressed also in leucocytes, liver, skin, and other tissues; cells of all ages are sensitive to the demands of antimalarial drugs. Also known as $Gd(-)$ Mediterranean allele.

Canton: common in populations of southern China, with about 20 percent enzymatic activity and fast electrophoretic mobility.

Athens: common in populations of Greece, with about 25 percent enzymatic activity and slow electrophoretic mobility.

Along with many of the human hemoglobin variants, the G-6-PD polymorphic variants are highly correlated with the distribution of falciparum malaria. Apparently, parasite proliferation is decreased in G-6-PD-deficient individuals (Livingstone, 1967; Luzzatto et al., 1969). More recently an association between the incidence of Gd^{B-} and red cell acid phosphatase alleles (P^a and P^c) has been found (Bottini, et al., 1971), so that there may be many factors involved in maintaining this sex-linked polymorphism.

SUMMARY

In organisms with one sex heterogametic and the other homogametic for at least one chromosomal pair, there are more genotypes than for pairs of autosomes. With the XY, XX condition with few euchromatic loci on the Y chromosome, the XY sex is effectively haploid.

Each sex's genotypes must be considered separately when population frequencies are determined. Under random mating if allelic frequencies are not equal in the two sexes, an ultimate equilibrium will be achieved via a succession of oscillations in frequency; both sexes are expected to have the same equilibrium frequency ($\hat{p}$) determined as the weighted average dosage of X chromosome frequencies in the population.

Sex-linked polymorphisms are common in human males: colorblindness (two loci), enzyme G-6-PD variants, serum protein Xm variants, and blood group Xg alleles. Hemophila A and B (two loci) are not high enough in frequency to be considered polymorphisms.

EXERCISES

1. Sex linked Bar-eye (B) in drosophila occurred in a mixed culture as follows: males: 30 Bar (B), 70 wild type ($+$); females: 36 Bar (B/B), 48 bean-eye ($B/+$), and 16 wild type ($+/+$). Assume mating between all individuals takes place at random.
   (a) What would be the expected frequencies of eye types in the first progeny generation (G_1) for each sex?
   (b) Determine the expected frequencies of eye types for the next three generations (through G_4) for each sex.
   (c) What will be the expected equilibrium frequency of alleles (B and $+$) and the ultimate zygote frequencies?
2. With sex linkage (XX females and XY males), we let p,q, represent A,a allele frequencies in females and r,s represent the same alleles, respectively, in males in any generation.

(a) Show that the succession of p values (or r values one generation later) will be as follows:

$$p, \quad \frac{p + r}{2}, \quad \frac{3p + r}{4}, \quad \frac{5p + 3r}{8}, \quad \ldots$$

(b) What will be the next term in the sequence?

(c) Check the succession of frequencies of the Bar allele in Exercise 1 to verify that sequence. Also, verify the sequence of Table 4-2.

3. We may use the terminology for successive terms of a recurrent series (see Appendix (A-13) where any given term (u_t) is composed of constant proportions of preceding terms: $u_t = bu_{t-1} + cu_{t-2}$, where b,c, are called *scales of relation*. From (4-2), we have the succession of female frequencies: $u_t = (\frac{1}{2})u_{t-1} + (\frac{1}{2})u_{t-2}$ because the male frequency of one generation back equals the female frequency of two generations back. Thus, $b = \frac{1}{2}$ and $c = \frac{1}{2}$ in this case.

(a) Show that this sequence can also be stated as "any term equals the previous term minus one-half the difference between the frequencies of the sexes."

(b) Verify these scales of relation by taking any two sequences of three generations from Table 4-2 (female q values) and solve simultaneously for b and c. What are the b,c scales for the Bar sequence in Exercise 1?

(c) It is important to obtain a general expression for any term in a sequence without finding all the preceding terms, because we want to know the limiting value of the series, whether it converges on a nonzero point or goes to zero or to infinity. In Appendix A-13, we find that a recurrent series can be expressed as the sum of corresponding terms of two (or more) component geometric series with the same scales of relation and common ratios (λ), so that the general expression for any term is $u_t = x(\lambda_1)^t + y(\lambda_2)^t$, where x and y are the initial terms of two component geometric series. Because

$$\lambda_1 = \frac{b + \sqrt{b^2 + 4c}}{2}$$

and

$$\lambda_2 = \frac{b - \sqrt{b^2 + 4c}}{2}$$

then $\lambda_1 = +1$ and $\lambda_2 = -(\frac{1}{2})$. The general expression for any term in the series then becomes $u_t = x(1)^t + y(-\frac{1}{2})^t$. Find the numerical values for x and y from the first two generations in Table 4-2 of female q values:

$$u_0 = 0.60 = x + y$$
$$u_1 = 0.36 = x + y(-\tfrac{1}{2})$$

Find the limiting value of u_t when $t \to \infty$.

(d) Using this general expression, show that the general limiting value is

$$u_t = \frac{2p + r}{3}$$

4. In a certain population, a sex-linked recessive occurs in 21.16 percent of the females

(a) What percentage of males would you expect to show this trait?

(b) If all the females from this population were mated to recessive males and the population was allowed to breed at random for several generations, what would be the eventual frequency of recessive males?

5. (a) In European populations, about two males in 25 are colorblind (sex-linked). What proportion of females are expected to be colorblind?
(b) If among Norwegian males the frequency of the dominant macroglobulin $X_m a+$ (Example 4-2) is 0.23, what would be the expected frequency of the same phenotype among females? the expected frequency of $X_m a-$ among females?
(c) In the Norwegian sample of Example 4-2, is the difference in q value between the sexes significant at the 5 percent level?

6. In Example 4-1, there was a deficiency of colorblind females (about 0.20 percent) based on estimating their number from the percentage of colorblind males on the assumption of a single locus for colorblindness. After the discovery that there were two main defects (protan and deutan) in colorblindness, it became logical to assume two separate X-linked loci controlling them; thus, "repulsion" (or "trans") heterozygous females could be assumed to exist as well as some doubly defective colorblind males and very rare "coupling" (or *cis*) homozygous colorblind females. Assume the frequencies of protan ($p+$) and deutan ($+d$) mutant chromosomes are as given from the data of Waaler: $\frac{175}{725}$ protan, $\frac{551}{725}$ deutan.
(a) How frequently will $p+/+d$ females be expected to occur in the population?
(b) If these females are normal instead of colorblind (expected under the single locus hypothesis), can the difference between expected and observed for females be accounted for by the two-locus hypothesis?

7. In Nigeria there are three polymorphic variants of G-6-PD—namely, B, A, and $A-$, with the latter two occurring at about 22 percent each in males (Example 4-4).
(a) What would be the expected frequencies of homozygous women for each allele?
(b) What frequencies of women would be heterozygous for the $A/A-$ alleles?
(c) What would be the expected frequency of women who are either homozygous or heterozygous for defective enzymes?

5

TWO LOCI, INDEPENDENT OR LINKED

When two or more loci with allelic variation are considered simultaneously in a random mating population, whether the loci are linked or independently assorting, it is important to describe the haploid combinations of alleles within each gamete. The dimension here for consideration, then, is that of two recombinable genetic entities. We could, for example, describe two nonallelic point loci (two separate cistrons) with at least a pair of alleles in each locus, or we could treat two pairs of entire nonhomologous chromosomes as units if they were distinguishable cytologically or with some distinctive markers segregating on each pair.

EQUILIBRIUM FOR TWO LOCI

As with all genic systems discussed so far, the frequencies of zygotes from allelic frequencies may be specified by the product rule of proportionality. If the proportions p,q represent frequencies of A,a alleles at one locus and u,v the frequencies of B,b alleles at another, then the four gametic combinations of these two loci, at equilibrium if the two loci are unassociated (that is, independently occurring) will be as follows:

$$\begin{aligned} \overline{AB} &= pu \\ \overline{Ab} &= pv \\ \overline{aB} &= qu \\ \overline{ab} &= qv \end{aligned} \tag{5-1}$$

where $\overline{AB}$, etc., represent haploid gametic frequency combinations of alternate alleles between loci.

Therefore, the random union of these four gametic combinations can occur in sixteen ways, as shown in Table 5-1A. It should be noted that there are four different double homozygotes produced by one mating combination each on the diagonal from upper left to lower right, while there are four double heterozygote combinations on the diagonal from upper right to lower left. These four double heterozygotes can of course be produced either by union of "coupling" ($\overline{AB} \times \overline{ab}$) gametes (two) or by "repulsion" ($\overline{Ab} \times \overline{aB}$) gametes (two). All the remainder are single heterozygotes for which there are always two combinations.

Grouping together all identical combinations (without regard for reciprocals) produces the zygote matrix ($\hat{Z}$) of nine genotypes with frequencies shown in Table 5-1B. Note that the rows are constant for genotypes segregating for $A - a$, while the columns represent genotypes with $B - b$. It is well to keep that scheme in mind for uniformity throughout the following discussion.

If gametic combinations are considered as "gamete pools" from the various parental genotypes producing them (in like manner to the "gene pool" concept for single loci), the

TABLE 5-1 *Frequencies of genotypes for two pairs of alleles at equilibrium*

A Matings

Males	Females			
	$pu_{\overline{AB}}$	$pv_{\overline{Ab}}$	$qu_{\overline{aB}}$	$qv_{\overline{ab}}$
$pu_{\overline{AB}}$	p^2u^2 $AABB$	p^2uv $AABb$	pqu^2 $AaBB$	$pquv$ $AaBb$
$pv_{\overline{Ab}}$	p^2uv $AABb$	p^2v^2 $AAbb$	$pquv$ $AaBb$	pqv^2 $Aabb$
$qu_{\overline{aB}}$	pqu^2 $AaBB$	$pquv$ $AaBb$	q^2u^2 $aaBB$	q^2uv $aaBb$
$qv_{\overline{ab}}$	$pquv$ $AaBb$	pqv^2 $Aabb$	q^2uv $aaBb$	q^2v^2 $aabb$

B Zygotes ($\hat{Z}$ = zygotic, or genotype, set at equilibrium)

$$\hat{Z} = \left|\begin{array}{ccc} AABB & 2AABb & AAbb \\ 2AaBB & 4AaBb & 2Aabb \\ aaBB & 2aaBb & aabb \\ \hline BB & Bb & bb \end{array}\right| \begin{array}{c} AA \\ Aa \\ aa \\ \\ \end{array} \text{ with frequencies: } \left|\begin{array}{ccc|c} & & & \sum \\ p^2u^2 & 2p^2uv & p^2v^2 & p^2 \\ 2pqu^2 & 4pquv & 2pqv^2 & 2pq \\ q^2u^2 & 2q^2uv & q^2v^2 & q^2 \\ \hline \sum\ u^2 & 2uv & v^2 & \end{array}\right.$$

C Gametic output of $\hat{Z}$ with independent assortment of nonalleles

$$\begin{aligned}
\overline{AB} &= AABB + \tfrac{1}{2}(2AaBB) + \tfrac{1}{2}(2AABb) + \tfrac{1}{4}(4AaBb) \\
\text{Freq.} &= p^2u^2 + pqu^2 + p^2uv + pquv = pu(pu + qu + pv + qv) = pu \\
\overline{Ab} &= AAbb + \tfrac{1}{2}(2Aabb) + \tfrac{1}{2}(2AABb) + \tfrac{1}{4}(4AaBb) \\
\text{Freq.} &= p^2v^2 + pqv^2 + p^2uv + pquv = pv(pv + qv + pu + qu) = pv \\
\overline{aB} &= aaBB + \tfrac{1}{2}(2aaBb) + \tfrac{1}{2}(2AaBB) + \tfrac{1}{4}(4AaBb) \\
\text{Freq.} &= q^2u^2 + q^2uv + pqu^2 + pquv = qu(qu + qv + pu + pv) = qu \\
\overline{ab} &= aabb + \tfrac{1}{2}(2aaBb) + \tfrac{1}{2}(Aabb) + \tfrac{1}{4}(4AaBb) \\
\text{Freq.} &= q^2v^2 + q^2uv + pqv^2 + pquv = qv(qv + qu + pv + pu) = qv
\end{aligned}$$

$$\hat{g} = \left|\begin{array}{cc} \overline{AB} & \overline{Ab} \\ aB & ab \end{array}\right| = \left|\begin{array}{cc|c} & & \sum \\ pu & pv & p \\ qu & qv & q \\ \hline \sum u & v & \end{array}\right., \quad \text{where } \hat{g} = \text{gametic set at equilibrium}$$

gametic output may be derived by assuming Mendelian independent segregation of alleles for the two loci, as in Table 5-1C. Squaring the gametic output $(pu + pv + qu + qv)^2$ again produces the zygotic array of nine by the matings given in Table 5-1A, indicating an equilibrium condition. Because the two pairs of loci can be considered separately, their independence also can be demonstrated by the product of the two-locus allelic frequencies: $(p + q)^2 \cdot (u + v)^2$, or the product of the row by column totals in Table 5-1B. These relationships are illustrated in Example 5-1 with data similar to the incidence of certain chromosomal variants given in the grasshopper populations of Example 5-2.

EXAMPLE 5-1

A SAMPLE OF TWO-LOCUS COMBINATIONS SIMILAR TO THE NEXT EXAMPLE

The following frequencies of genotypes in which two pairs of alleles are independently assorting are at equilibrium and parallel to the karyotype frequencies of the grasshopper population in Example 5-2. They illustrate the expectations from Table 5-1. Let the observed values be as follows:

	BB	*Bb*	*bb*	Sum
AA	72	284	284	640
Aa	36	142	142	320
aa	4	18	18	40
Sum	112	444	444	1000

So that $p = \frac{4}{5}$, $q = \frac{1}{5}$
$u = \frac{1}{3}$, $v = \frac{2}{3}$

Then the gametic output of this population would be

	Frequency (approx.)
$\overline{AB} = 72 + \frac{1}{2}(284 + 36) + \frac{1}{4}(142) = 267.5$	$\frac{4}{15}$
$\overline{Ab} = 284 + \frac{1}{2}(284 + 142) + \frac{1}{4}(142) = 532.5$	$\frac{8}{15}$
$\overline{aB} = 4 + \frac{1}{2}(36 + 18) + \frac{1}{4}(142) = 66.5$	$\frac{1}{15}$
$\overline{ab} = 18 + \frac{1}{2}(18 + 142) + \frac{1}{4}(142) = 133.5$	$\frac{2}{15}$

Thus, the gametic determinant would be at equilibrium:

(Frequencies)

$$\hat{g} = \begin{vmatrix} \frac{4}{15} & \frac{8}{15} \\ \frac{1}{15} & \frac{2}{15} \end{vmatrix} \quad \text{and} \quad \text{zygotes } (\hat{Z}) = \begin{vmatrix} 0.0711 & 0.2844 & 0.2844 \\ 0.0356 & 0.1422 & 0.1422 \\ 0.0067 & 0.0177 & 0.0177 \end{vmatrix}$$

When multiplied by 1000, these expected zygote frequencies can be seen as in conformity with the observed values above.

EXAMPLE 5-2

NONHOMOLOGOUS CHROMOSOMAL ARRANGEMENTS FROM NATURAL POPULATIONS OF GRASSHOPPERS

When two or more segregating systems are analyzed simultaneously in a population, the possibility arises that there may be interdependency between those systems due to evolutionary forces. While the incidence of the two genetic systems may appear to be independent if only a single population is sampled at a single generation, significant associations sometimes are discerned only after constant sampling from several generations or from more than one population. In the Australian grasshopper *Keyacris* (formerly *Moraba*) *scurra*, White (1957, 1958) described pericentric inversions on two nonhomologous chromosome pairs that at first seemed independently occurring but on further examination from several localities over time showed association (Lewontin and White, 1960). The data presented below illustrate first the principles of gametic output from two-locus systems (which can, of course, apply to whole chromosome arrangements just as for genetic markers). *Keyacris scurra* is found in southeastern Australia as two races that differ in chromosome number: an eastern one (eastern New South Wales and northern Victoria) with eight pairs of chromosomes in females ($2N = 15$ in males, since there is no Y chromosome) and a western (New South Wales) race with nine pairs in females ($2N = 17$ in males). Cytologically, the eastern race has a pair of large metacentric AB chromosomes represented by two pairs of acrocentric elements A and B in the western race, thus accounting for the difference in chromosome number. All the remaining chromosome pairs are designated in order of size CD, EF, and so on. Chromosome CD has a Standard sequence (St), while the inverted sequence Blundell (Bl) has a subterminal centromere. The next smaller chromosome EF also has a metacentric Standard (St) sequence and a sequence called Tidbinbilla (Td), which differs from St by a pericentric inversion.

TABLE A *Observed numbers of male grasshoppers at Royalla B locality collected in 1956–1958 by White*

EF Chromosome	*CD* Chromosome			*EF* Sum	Frequencies
	St/St	*St/Bl*	*Bl/Bl*		*EF* Chromosome
St/St	59	282	231	572	$p_{St} = 0.7878$
St/Td	24	152	150	326	$q_{Td} = 0.2122$
Td/Td	2	14	19	35	
					CD Chromosome
CD Sum	85	448	400	933	
					$u_{St} = 0.3312$
					$v_{Bl} = 0.6688$

In a locality of the Australian Capital Territory called Royalla B, the grasshopper, which had been sparsely distributed throughout the area, was found in large numbers from 1956–1958. Table A lists the pooled numbers of the nine karyotype combinations for standard and inverted sequences for CD and EF chromosomes sampled for those years. There are at least three ways that we can make assumptions about agreement of observed values and expected values from these data. Two of these ways are given in Table B, and the third is presented in Example 17-4.

Assumption 1. From row to row (EF chromosome karyotypes) the columns are in the same proportions (CD chromosome karyotypes); that is, all are independent, based on marginal totals as would be evident from a contingency test. This assumption does *not* necessarily test whether the observed numbers agree with gametic equilibrium and random mating expectations.

TABLE B *Expected numbers based on assumptions 1 and 2 (in parallel)*

Assumption 1			Assumption 2		
52.12	274.65	245.23	63.52	256.53	259.00
29.70	156.54	139.76	34.22	138.19	139.53
3.19	16.81	15.00	4.61	18.61	18.79
		933			933
$X^2 = 5.89$, $d.f. = 4$, $P = 0.20$			$X^2 = 13.72$, $d.f. = 6$, $P = 0.033$		

Assumption 2. From separate chromosome frequencies based on gametic equilibrium and random mating, the observed frequencies agree with the square law. We calculate p_{St}, q_{Td} and u_{St}, v_{Bl} and then perform the product $(p + q)^2(u + v)^2$.

Under assumption 1, a chi-square (contingency) of 5.89, with 4 degrees of freedom (rows $-$ 1) (columns $-$ 1), is not significant ($P = 0.20$).

Under assumption 2, however, the chi-square = 13.72, the degrees of freedom are 6 (9 $-$ 2 constants $-$ 1 total), and it is significant ($P = 0.033$). Therefore, we conclude that although these chromosome combination frequencies did not depart significantly from expectation based on separate chromosome frequencies, there was a significant departure either from random mating or from gametic equilibrium. It is the latter, as indicated by calculating the gametic determinant (D), as given in Table C. As pointed out by Turner (1972), samples from several other grasshopper populations collected by White in New South Wales over a period of years were consistent in having positive gametic determinants of similar magnitude. The likelihood that these karyotypes are adaptively important, with those in excess (under assumption 2) having selective advantages over those that are deficient, thus became evident to Lewontin and White (1960). Example 17-4 will discuss that point further.

TABLE C *Gametic frequencies*

	CD		
EF	*St*	*Bl*	
St	0.26795	0.51983	$D = +0.007045$
Td	0.06324	0.14898	

(using 5-8A): $D_{max} = (0.3312)(0.2122) = +0.070281$

Therefore, observed gametic determinant is 10 percent of maximum.

APPROACH TO GAMETIC EQUILIBRIUM

Each locus will be at equilibrium with respect to its own set of alleles immediately upon random mating, but the gametic combinations of two loci may not be at equilibrium at the outset, although under random mating they will approach the equilibrium condition rapidly if they are independently assorting (not linked or otherwise associated). The principle can be illustrated by the following example:

If we cross two pure lines differing in two independent pairs of alleles using a "no-choice" mating, as in a simple Mendelian dihybrid cross (*AABB* females × *aabb* males), then all progeny (Z_1) are identical *AaBb*. The four gametic combinations will be equal and on inbreeding will produce the familiar zygotic frequency set:

$$Z_2 = \begin{vmatrix} \frac{1}{16} & \frac{2}{16} & \frac{1}{16} \\ \frac{2}{16} & \frac{4}{16} & \frac{2}{16} \\ \frac{1}{16} & \frac{2}{16} & \frac{1}{16} \end{vmatrix}$$

If mating is random from then on, the gametic combinations will be constant at equality (one-fourth each of $\overline{AB}$, $\overline{Ab}$, $\overline{aB}$, $\overline{ab}$).

In contrast, if we cross the same two strains *at random*, not restricting the mating to one type of female × another type of male but giving both sexes in each strain equal opportunity to cross out to the other or mate to its own strain, as follows:

AABB		*aabb*
males and females	×	males and females

TABLE 5-2 *Random matings expected from a parent population containing "coupling" gametes exclusively*

Matings (Generation 0) Gametic Output D

$$Z_o = \begin{vmatrix} \frac{1}{2}AABB \\ \frac{1}{2}aabb \end{vmatrix}_{♀♀} \times \begin{vmatrix} \frac{1}{2}AABB \\ \frac{1}{2}aabb \end{vmatrix}_{♂♂} \qquad g_o = \begin{vmatrix} \frac{1}{2} & 0 \\ 0 & \frac{1}{2} \end{vmatrix} = \frac{1}{4}$$

Progeny (Generation 1) randomly mated

$$Z_1 = \begin{vmatrix} \frac{1}{4} & 0 & 0 \\ 0 & \frac{2}{4} & 0 \\ 0 & 0 & \frac{1}{4} \end{vmatrix} \qquad g_1 = \begin{vmatrix} \frac{3}{8} & \frac{1}{8} \\ \frac{1}{8} & \frac{3}{8} \end{vmatrix} = \frac{1}{8}$$

Subsequent randomly mated progeny

$$Z_2 = \begin{vmatrix} \frac{9}{64} & \frac{6}{64} & \frac{1}{64} \\ \frac{6}{64} & \frac{20}{64} & \frac{6}{64} \\ \frac{1}{64} & \frac{6}{64} & \frac{9}{64} \end{vmatrix} \qquad g_2 = \begin{vmatrix} \frac{5}{16} & \frac{3}{16} \\ \frac{3}{16} & \frac{5}{16} \end{vmatrix} = \frac{1}{16}$$

$$Z_3 = \begin{vmatrix} \vdots \end{vmatrix} \qquad g_3 = \begin{vmatrix} \vdots \end{vmatrix}$$

$$\hat{Z}_\infty = \begin{vmatrix} 1 & 2 & 1 \\ 2 & 4 & 2 \\ 1 & 2 & 1 \end{vmatrix} \qquad g_\infty = \begin{vmatrix} \frac{1}{4} & \frac{1}{4} \\ \frac{1}{4} & \frac{1}{4} \end{vmatrix} = 0$$

$$D_\infty = 0$$

(assuming both sexes equal in numbers), then there are four matings in equal numbers with the following progeny result:

$$Z_1 = \begin{vmatrix} \frac{1}{4} & 0 & 0 \\ 0 & \frac{2}{4} & 0 \\ 0 & 0 & \frac{1}{4} \end{vmatrix} \qquad \tfrac{1}{4}AABB:\tfrac{2}{4}AaBb:\tfrac{1}{4}aabb$$

This process is illustrated in Table 5-2. The parental gametic output for both sexes in the parents was $\frac{1}{2}\overline{AB}:\frac{1}{2}\overline{ab}$ ("coupling" gametes only). Owing to the formation of all four types of gametes in equal numbers from the double heterozygotes, assuming independent assortment, the progeny (Z_1) will have a gametic output of $\frac{3}{8}\overline{AB}:\frac{1}{8}\overline{Ab}:\frac{1}{8}\overline{aB}$; $\frac{3}{8}\overline{ab}$. "Repulsion" gametes ($\overline{Ab}$ and $\overline{aB}$) arise as one-fourth of the double heterozygote frequency. This output can always be determined from the following relationship: for each gametic combination, the sum of frequencies includes all the appropriate double homozygote, plus one-half of each appropriate single heterozygote, plus one-fourth of the double heterozygote frequencies.

If we symbolize the zygotic matrix for any population with order of genotypes as in Table 5-1B:

$$Z_o = \begin{vmatrix} a & b & c \\ d & e & f \\ g & h & i \end{vmatrix}$$

where Z_o = original, or parent, zygotes, then the frequencies of parent gametic output will be (letting $a, b, c, \ldots$ be frequencies):

$$\begin{aligned} \overline{AB}_o &= a + \tfrac{1}{2}(b + d) + \tfrac{1}{4}e \\ \overline{Ab}_o &= c + \tfrac{1}{2}(b + f) + \tfrac{1}{4}e \\ \overline{aB}_o &= g + \tfrac{1}{2}(d + h) + \tfrac{1}{4}e \\ \overline{ab}_o &= i + \tfrac{1}{2}(f + h) + \tfrac{1}{4}e \end{aligned} \qquad \text{or} \qquad g_o = \begin{vmatrix} \overline{AB}_o & \overline{Ab}_o \\ \overline{aB}_o & \overline{ab}_o \end{vmatrix} \tag{5-2}$$

where g_o = original, or parent gametic output (see Example 5-1).

On random mating, these four gametic combination frequencies will be squared according to the proportionality rule so that the progeny zygotes (Z_1) can be expressed in terms of the parental gametic output (omitting o subscripts):

$$Z_1 = \begin{vmatrix} \overline{AB}^2 & 2\overline{AB}\cdot\overline{Ab} & \overline{Ab}^2 \\ 2\overline{AB}\cdot\overline{aB} & 2(\overline{AB}\cdot\overline{ab} + \overline{Ab}\cdot\overline{aB}) & 2\overline{Ab}\cdot\overline{ab} \\ \overline{aB}^2 & 2\overline{aB}\cdot\overline{ab} & \overline{ab}^2 \end{vmatrix} \tag{5-3}$$

Using (5-2) to obtain the gametic output for the progeny (Z_1) in terms of the original gametic output (g_o) gives us the recurrence relation of gamete frequencies from one generation to the next (omitting subscripts o for previous generation):

$$\begin{aligned} \text{For } \overline{AB}_1 &= \overline{AB}^2 + \overline{AB}\cdot\overline{Ab} + \overline{AB}\cdot\overline{aB} + \tfrac{1}{2}(\overline{AB}\cdot\overline{ab} + \overline{Ab}\cdot\overline{aB}) \\ &= \overline{AB}^2 + \overline{AB}\cdot\overline{Ab} + \overline{AB}\cdot\overline{aB} + \overline{AB}\cdot\overline{ab} - \overline{AB}\cdot\overline{ab} + \tfrac{1}{2}(\overline{AB}\cdot\overline{ab} + \overline{Ab}\cdot\overline{aB}) \\ &= \overline{AB}[\overline{AB} + \overline{Ab} + \overline{aB} + \overline{ab}] - \tfrac{1}{2}(\overline{AB}\cdot\overline{ab} - \overline{Ab}\cdot\overline{aB}) \\ \overline{AB}_1 &= \overline{AB} - \tfrac{1}{2}(\overline{AB}\cdot\overline{ab} - \overline{Ab}\cdot\overline{aB}) \end{aligned}$$

where the $\overline{AB}_1$ refers to the frequency of the double dominant gamete in the Z_1 generation.

Similarly,

$$\overline{Ab}_1 = \overline{Ab} + \tfrac{1}{2}(\overline{AB} \cdot \overline{ab} - \overline{Ab} \cdot \overline{aB})$$
$$\overline{aB}_1 = \overline{aB} + \tfrac{1}{2}(\overline{AB} \cdot \overline{ab} - \overline{Ab} \cdot \overline{aB})$$
$$\overline{ab}_1 = \overline{ab} - \tfrac{1}{2}(\overline{AB} \cdot \overline{ab} - \overline{Ab} \cdot \overline{aB})$$

These relationships between successive generations of the example are illustrated in Table 5-2. It should be noted that for each pair of alleles (row or column totals) in the zygote matrices, equilibrium is immediately established; the gametic combinations (A with B) are not at equilibrium, although they approach it fairly rapidly.

The change in each gametic frequency is one-half the difference between the product of the two "coupling" gametic frequencies and the product of the two "repulsion" gamete frequencies.

It may be recognized that this product difference represents a simple determinant of the second order, $wz - xy$ from the form: $\begin{vmatrix} w & x \\ y & z \end{vmatrix} = D$, in which the first diagonal product is positive and the other diagonal negative. The gametic output (5-4) then can be more simply written:

$$\begin{aligned} \overline{AB}_1 &= \overline{AB}_o - \tfrac{1}{2}D_o \\ \overline{Ab}_1 &= \overline{Ab}_o + \tfrac{1}{2}D_o \\ \overline{aB}_1 &= \overline{aB}_o + \tfrac{1}{2}D_o \\ \overline{ab}_1 &= \overline{ab}_o - \tfrac{1}{2}D_o \end{aligned} \qquad (5\text{-}4A)$$

where D_o = gametic output determinant of the previous generation, and AB_o, etc., are the gametic frequencies of the previous generation.

The student should check by determinant algebra that the absolute value of the determinant is reduced by one-half each generation: $D_n = \tfrac{1}{2}D_{n-1}$, until it comes to zero at equilibrium when the "coupling" and "repulsion" gametic products are equal and $\hat{D} = 0$ (see Exercise 2). The rate of change in the AB gamete, for example, is illustrated in Figure 5-1 (lowest curve). D is expected to be zero in large random mating populations on the assumption that there is no association between particular alleles at the two loci. Nevertheless, natural populations and experimental populations are known to display a nonzero D; in Example 5-2, a natural population of grasshoppers has been found to show a slightly significant nonzero D value, and in Example 5-3 certain loci in an experimental drosophila population came to a near-zero D value but other loci remained at nonzero D value.

Another way of expressing the value of the gametic matrix determinant, or lack of independence, is to take the difference between the observed value of any gamete frequency and the appropriate row × column product (expected value) for that frequency, as follows (Kojima and Lewontin, 1970):

$$D = wz - xy = w - (w + y)(w + x)$$

because $z = 1 - (w + y + x)$, or

$$D = \overline{AB} - p_A \cdot u_B \qquad (5\text{-}5)$$

as may be verified with any generation in Table 5-2. This definition of the gametic determinant is the equivalent of the covariance between alleles in the gametic matrix, as the

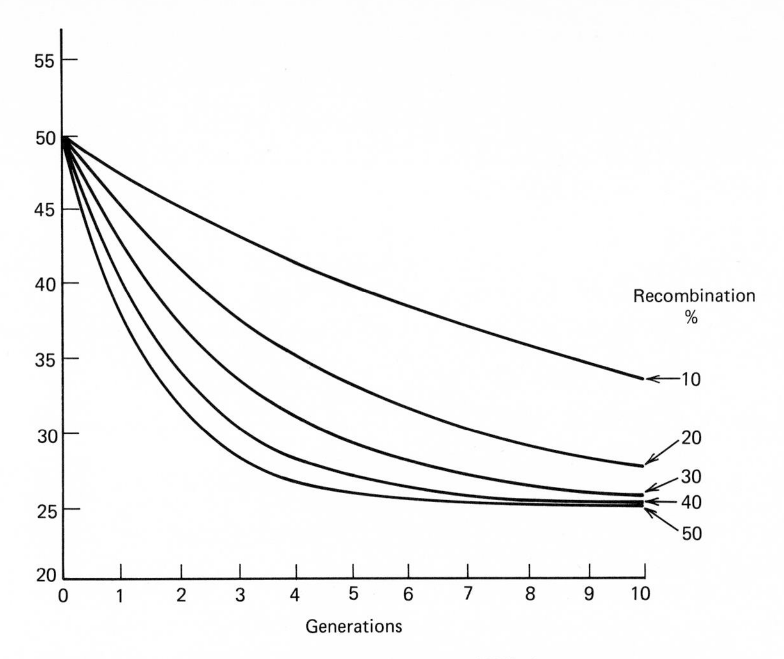

Figure 5-1. Approach to equilibrium for $\overline{AB}$ gametic frequency from initial condition (in Table 5-2) with various amounts of recombination between A and B loci.

student may prove in Exercise 7. It may be extended to a case of three loci, as for A, B, C: $D = \overline{ABC} - p_A \cdot u_B \cdot w_C$, where w_C is the frequency of the third locus allele C and $\overline{ABC}$ is the observed frequency of the triple dominant gamete. Pairs of loci are usually sufficient for our consideration in multilocus systems, and for higher-order departures from equilibria specific to three loci, the reader is referred to Kojima and Lewontin (1970).

LINKAGE EFFECTS

When the approach to equilibrium involves two or more loci occurring on the same chromosome, the rate of change in the gametic set will be slower than with independently assorting loci. The tighter the linkage, the slower the rate of change.

In describing the effect of linkage, it is essential only to realize that linkage makes a difference in the gametic output from the double heterozygotes (*AaBb*); homozygote outputs are obviously unaffected by linkage, and single heterozygotes (*AaBB*, for example) produce single heterozygotes and homozygotes in a 50:50 ratio irrespective of linkage.

If we take note of the mating frequencies in Table 5-1A, it is evident that the double heterozygotes in the diagonal from upper right to lower left are formed either by fusion of two "coupling" ($AB \times ab$) gametes or by two "repulsion" ($Ab \times aB$) gametes. In expression (5-2), with independent assortment, the double heterozygote produces all four gametes in equal numbers so that one-fourth of its frequency is an increment to each gamete combination. We must reconsider separately the fractional contributions from these two types of linked heterozygotes, however, in the case of linkage. The middle term in the Z_1 matrix of

(5-3) then contributes not one-fourth of its total but the following fractions to the gametic output: for $\overline{AB}$, the fourth term of the $\overline{AB}_1$ in computing (5-4):

$$2[\tfrac{1}{2}(1 - c)\overline{AB} \cdot \overline{ab} + \tfrac{1}{2}(c)\overline{Ab} \cdot \overline{aB}]$$

where c = crossover percentage and $1 - c$ = noncrossover percentage, so that

$$\begin{aligned} \overline{AB}_1 &= \overline{AB}^2 + \overline{AB} \cdot \overline{Ab} + \overline{AB} \cdot \overline{aB} + (1 - c)\overline{AB} \cdot \overline{ab} + c\overline{Ab} \cdot \overline{aB} \\ &= \overline{AB}^2 + \overline{AB} \cdot \overline{Ab} + \overline{AB} \cdot \overline{aB} + \overline{AB} \cdot \overline{ab} - c(\overline{AB} \cdot \overline{ab} - \overline{Ab} \cdot \overline{aB}) \\ \overline{AB}_1 &= \overline{AB} - c(\overline{AB} \cdot \overline{ab} - \overline{Ab} \cdot \overline{aB}) = \overline{AB} - cD \end{aligned} \tag{5-6}$$

Similarly

$$\begin{aligned} \overline{Ab}_1 &= \overline{Ab} + cD \\ \overline{aB}_1 &= \overline{aB} + cD \\ \overline{ab}_1 &= \overline{ab} - cD \end{aligned}$$

Because the crossover percentage is always less than 50, the change in gametic output frequencies is always slower than in the case of independent assortment. Figure 5-1 gives the approach to linkage equilibrium for various recombination percentages. The fundamental point to be noted is that irrespective of whether genes are located on nonallelic independently assorting chromosomes or are linked on the same chromosome, the zygotic proportions at equilibrium are entirely determined by the allelic frequencies. Consequently, it is not possible to detect linkage simply by an analysis of gametic frequencies in a population; the "coupling" product must equal the "repulsion" product at equilibrium.*

ASSOCIATION AND CONSTANT NONEQUILIBRIUM ("DISEQUILIBRIUM")

If two traits are associated, or correlated, in a population, it is often erroneously postulated that they are ipso facto linked. But phenotypic association may come about for a number of reasons: a single genic locus may have multiple effects (pleiotropy), two or more genotypes may have common environmental influences, inbreeding may increase certain genotypes' frequencies, a large population may be stratified into diverse subgroups, or particular genic combinations may be favored by selective forces. The latter influence can be an important deterministic one, but it does not depend on linkage between the genic loci concerned because complementary or epistatic effects are certainly not confined to linked loci. Action of selection on two or more genic loci will be discussed in Chapter 17.

Let us here take note of the fact that in populations with conditions conforming to the square law (large-sized N, random mating, and so forth), gametic combinations may experience a constant excess or deficiency of particular combinations whether the loci concerned are linked or not. Because the gametic products should be equal ($\hat{D} = 0$) at equilibrium, a permanent nonzero gametic determinant would indicate that forces are preventing certain gametic combinations from attaining independent levels. This lack of

* For linkage to be detected in human families, segregation ratios among offspring who have at least one doubly heterozygous parent must be ascertained. Methods in human families can be found by consulting Mather (1951), Neel and Schull (1954), or Li (1961).

independence between loci has been called "linkage disequilibrium," constant nonequilibrium, gametic phase unbalance (Jain and Allard, 1966), or "an index of divergence of an arbitrary population from equilibrium" (Li, 1955). A nonzero D may be constant in the absence of linkage, and the first of these possible terms is a misnomer; it is also misleading to speak of a disequilibrium because the allelic (single-locus) frequencies may well be in equilibrium. Nevertheless, the term has now been used so much in this context that it perhaps is preferable and more convenient to retain the term *disequilibrium* than to propose a new one. D is at maximum $= \frac{1}{4}$ when $p = q = u = v = \frac{1}{2}$ and all gametic combinations are either in coupling or in repulsion, as in the parents in Table 5-2. Evidence of a sizable disequilibrium is illustrated in Examples 5-2, 5-3, and 5-4.

The gametic determinant D is affected not only by the amount of association between nonhomologous loci but also simply by allelic frequencies themselves. In Table 5-2 allelic frequencies are held constant and D decreases from its maximum value to zero at linkage equilibrium. Alternatively, we could determine the maximum D for any set of gametic frequencies with maximum "coupling" or "repulsion" for the single-locus gene frequencies given. While D is greatest when all allelic frequencies are equal, it decreases as nonhomologous alleles become rare. This sensitivity of the gametic determinant both to genuine association and to gene frequency changes was pointed out by Lewontin (1964), who expressed the amount of association as a percentage (D') of the maximum value for any given set of gametic frequencies.

A general expression for maximum D (complete association) can be derived as follows. From (5-5) we note $D = \overline{AB} - pu = -\overline{Ab} + pv$, etc. Because $\overline{AB} + \overline{Ab} = p$, then combining $\overline{AB} = D + pu$ and $\overline{Ab} = pv - D$,

$$(pu + D) + (pv - D) = p \qquad (5\text{-}7)$$

For maximum D, either $\overline{AB}$ or $\overline{Ab}$ must be zero—that is, the component in parentheses on right or left in (5-7). If $(pu + D) = 0$, then D must be negative and equal to pu. Conversely, if $(pv - D) = 0$, then D must be positive and equal to pv. The same reasoning is true for the remainder of the 2×2 matrix.

$$\overline{aB} + \overline{ab} = q \qquad \text{then} \qquad (pu - D) + (qv + D) = q \qquad (5\text{-}7A)$$

The expressions for maximum D (gametic association) are then

$$D_{max(1)} = pu, \quad \text{or} \quad -qv \quad \text{(whichever is closer to zero)} \qquad (5\text{-}8)$$

and

$$D_{max(2)} = +pv, \quad \text{or} \quad +qu \quad \text{(whichever is closer to zero)} \qquad (5\text{-}8A)$$

Only in the extreme case (Table 5-2 top) will $D_{max(1)} = D_{max(2)}$. When any inequality exists between p,q or between u,v there will be two D_{max}'s. For example, if $p = 0.2$, $q = 0.8$; $u = 0.4$, $v = 0.6$, maximum D would occur as follows:

(*a*)

0	0.2	0.2
0.4	0.4	0.8
0.4	0.6	

$D_{max(1)} = -0.08$

(*b*)

0.2	0	0.2
0.2	0.6	0.8
0.4	0.6	

$D_{max(2)} = +0.12$

The student should be satisfied that these are indeed maxima and then compare them with the equilibrium state:

$$\begin{vmatrix} 0.08 & 0.12 \\ 0.32 & 0.48 \end{vmatrix}$$

It should be obvious in (5-8, 5-8A) why it is necessary to choose the product closer to zero. If the greater product were chosen, how would the solution to (5-7) be affected?

If we suppose the observed gametic frequencies give a D value intermediate between the equilibrium and D_{max} state, it is evident that the amount of association between loci could be expressed as a percentage of D_{max}. In Example 5-2, observed D for grasshopper chromosomal arrangement combinations was about 10 percent of D_{max}, as given in Table C. In Example 5-3, in Cannon's populations of *Drosophila melanogaster* after 25 generations of random mating, the linked block of three loci ($ss - k - e$) maintained an intense disequilibrium; for *ss* and *e*, for example, the average D of $+0.1034$ is 68 percent of maximum ($D_{max(2)} = +0.152$). As estimated by Lewontin (1964), if lack of recombination in drosophila males is taken into account, the percent of maximum should be 21 after 25 generations for those loci.

EXAMPLE 5-3

LINKED GENOTYPIC ASSOCIATIONS IN EXPERIMENTAL POPULATIONS OF DROSOPHILA

With drosophila, it is technically feasible to observe associations between two or more mutant loci. Experimental populations have been designed by different investigators to focus attention on particular aspects of evolutionary theory. In a test for gametic equilibrium, it sometimes happens that a disequilibrium is found.

Carson (1961) and his students (Cannon, Smathers, and Susman) analyzed several populations of *Drosophila melanogaster* from 1958 to 1963 in which a set of linked markers on chromosome 3 ("sesro") had been introduced along with their wild-type alleles. Their design was to find out whether simple "luxuriance," or the hybrid vigor displayed in crossbred progenies between inbred lines, can contribute to continued biological success of a population. "Sesro" refers to a block of five recessive mutant markers on the third chromosome with the following linkage relationships:

se	centromere	*ss*	*k*	*e*	*ro*
∧	∧	∧	∧	∧	∧
26.0	46.0	58.5	64.0	70.7	91.1

where *se* = sepia eye, *ss* = spineless, *k* = kidney eye, *e* = ebony body, and *ro* = rough eye.

With Carson's base populations, founder flies of one relatively inbred strain (either the "sesro" or wild-type Oregon-R) were established in very crowded "supervial" cultures (modified from the technique of Buzzati-Traverso, 1955). Once the productivity of a base population became approximately stabilized, a single male heterozygous for the five loci (*sesro*/+ + + + +) was introduced into the population to establish one haploid set in "coupling." The populations of interest in the first round of experiments were sampled at regular weekly intervals over a period of between 50 and 100 weeks (about 20 to 40 generations for these crowded cultures). In the two reciprocal types of populations, introduction of the haploid alternate set of coupling alleles produced a significant increase in number of flies and biomas. Whether the base population was homozygous *sesro* or + + + + +, the outcome was similar in that the mutant alleles for the end loci (*se* and *ro*) were retained at higher frequencies (12–20 percent *se* and 50 percent *ro*) than the mutant alleles for the central block (*ss k e*), whose frequency was less than 10 percent.

Cannon (1963) followed Carson's study by founding three replicate populations with these mutants in "repulsion" as follows. Males with three combinations, *se* + + + +,

TABLE A *Gametic combination frequencies from three populations averaged in which the "sesro" markers had been introduced in repulsion (from Cannon, 1963). Total N = 285 for three populations together*

	At Introduction of Markers			After 28 Weeks			After 50 Weeks		
ss e:									
		+	*ss*		+	*ss*		+	*ss*
	+	0.989	0	+	0.903	0.017	+	0.749	0.060
	e	0	0.011	*e*	0.023	0.057	*e*	0.049	0.142
	$D = +0.011$			$D = +0.0510$			$D = +0.1034$		
se ro:									
		+	*se*		+	*se*		+	*se*
	+	0.987	0.006	+	0.890	0.058	+	0.872	0.053
	ro	0.006	0	*ro*	0.052	0	*ro*	0.072	0.003
	$D = -0.00001$			$D = -0.0030$			$D = -0.0012$		
ss ro:									
		+	*ss*		+	*ss*		+	*ss*
	+	0.983	0.011	+	0.877	0.070	+	0.742	0.182
	ro	0.006	0	*ro*	0.048	0.005	*ro*	0.056	0.020
	$D = -0.00007$			$D = +0.0010$			$D = +0.0046$		
se ss:									
		+	*se*		+	*se*		+	*se*
	+	0.983	0.008	+	0.871	0.058	+	0.744	0.054
	ss	0.009	0	*ss*	0.071	0	*ss*	0.200	0.002
	$D = -0.00007$			$D = -0.0041$			$D = -0.0093$		

$\underline{+ + + + ro}$, and $\underline{+ ss\ k\ e\ +}$, were mated to virgin Oregon-R wild-type females. All these mutants became established in the three replicate populations and reached frequencies, over a period of 50 weeks, that were not significantly different between populations—about 5 percent *se*, about 7 percent *ro*, and about 20 percent for the central block of $\underline{ss\ k\ e}$. This outcome was just about the opposite of Carson's result in which the linkage block had been in coupling and in which the end loci became higher in frequency than the central block. At the 28th week and again at the 50th week following introduction of these mutants, Cannon test-crossed males from each population to "sesro" females. The three populations were not significantly different for most of the data, and the gametic frequencies have been averaged for four combinations of interest (out of 10 combinations possible) in Table A. (The mutant *k*, being closely linked with *ss* and *e*, does not differ significantly from them and is omitted.)

There is no significant linkage disequilibrium for the end loci (*se* and *ro*) nor between either of the end loci with any of the central block loci (*ss* is given as representative of the block, which acted very much as a unit). However, the block of $\underline{ss\ k\ e}$ increased in frequency with loss of + + +, as can be seen in the top row of Table A, with a significant linkage disequilibrium. From these data one gets the distinct impression that the trend might continue because the tendency was for the *D* value to increase from close to zero to greater than 0.10.* Unfortunately, there is no apparent reason why the association between certain loci in these "repulsion" populations should be favored while it was not favored in the "coupling" populations. The fact that all populations had been allowed some opportunity to adapt to the laboratory culture conditions before introducing the allelic set (either of mutants or wild-type alleles) suggests that distinctly different genetic complexes must have become established in each case, thus determining a different reaction to the introduced set, favoring or not favoring them.

DOMINANCE

When heterozygous and homozygous classes become indistinguishable because of dominance, the number of phenotypes is reduced, gene and gametic frequencies become more difficult to estimate, and in organisms such as humans in which genetic tests cannot be made easily, some confusion may arise in attempting to fit models of genotype frequencies. With independence of two pairs, or linkage equilibrium, the expected proportions should be according to the products of single-pair frequencies, as in Table 5-3. These proportions occur such that the diagonal products are equal: (*A-B-*)(*aabb*) = (*A-bb*)(*aaB-*). Example 5-4 gives an illustration of dominance in two pairs of factors from snail populations (*Cepaea nemoralis*): for the midbanded–all–banded pair and yellow-pink pair of shell phenotypes, there is independence; but for the unbanded-banded and yellow-pink there is definite association (linkage disequilibrium).

In Example 5-5 the early hypothesis for the inheritance of the A-B-O human blood groups (vonDungern and Hirzfeld, 1911) serves to illustrate how populational data can be utilized in resolving the validity of two alternative genetic hypotheses: that of Bernstein (multiple alleles of a single locus) versus two loci showing dominance. There is no doubt

* Statistical significance of *D* from sampling data has not yet been clearly defined. The matter of utilizing contingency chi-square as a test of gametic association is discussed later in Example 17-6.

TABLE 5-3 *Proportions of phenotypes with dominance and epistasis in two pairs of alleles*

	Basic Form				Simulation of Monofactorial Case When 3 Phenotypes		
A	No epistasis						
	B-	bb	Sum				
A-	$(1-q^2)(1-v^2)$ A-B-	$(1-q^2)v^2$ A-bb	$(1-q^2)$				
aa	$q^2(1-v^2)$ aaB-	q^2v^2 $aabb$	q^2				
Sum	$(1-v^2)$	v^2	1.00				
B	With aa-- epistatic to B-: bb			If $v^2 = 2q/(1+q)$, then			
	B-	bb	Sum		B-	bb	Sum
A-	$(1-q^2)(1-v^2)$	$(1-q^2)v^2$	$(1-q^2)$	A-	$(1-q)^2$	$2q(1-q)$	$(1-q^2)$
aa	— =	—	q^2	aa	— =	—	q^2
Sum	$(1-v^2)$	v^2	1.00	Sum	$\frac{(1-q)}{(1+q)}$	$\frac{2q}{(1+q)}$	1.00
C	With B- epistatic to A-: aa			If $v = 2q/(1+q^2)$, then			
	B-	bb	Sum		B-	bb	Sum
A-	—	$(1-q^2)v^2$	$(1-q^2)$	A-	—	$\frac{4q^2(1-q^2)}{(1+q^2)^2}$	$(1-q^2)$
	‖				‖		
aa	—	q^2v^2	q^2	aa	—	$\frac{4q^4}{(1+q^2)^2}$	q^2
Sum	$(1-v^2)$	v^2	1.00	Sum	$\frac{(1-q^2)^2}{(1+q^2)^2}$	$\frac{4q^2}{(1+q^2)^2}$	1.00

after applying these proportionality principles to the observed data from countless populations that the multiple-allele theory is the valid one (see p. 80), because the lack of fit between observed data and the older hypothesis (two loci) is incontrovertible.

An example of dominance at one locus (leucine aminopeptidase A-O) but codominance at another locus in the same linkage group (esterase-6 F-S) is illustrated in Example 5-6 from the work of Rasmuson, et al., on experimental populations of *Drosophila melanogaster*. The method of estimating gametic frequencies from six phenotypic classes is useful in testing for linkage equilibrium. Apparently in the Rasmuson populations there was no constant linkage disequilibrium, only temporary lack of independence, although there was an excess of heterozygotes at the Est-6 locus.

EXAMPLE 5-4

ASSOCIATION BETWEEN LOCI FOR BANDING AND SHELL COLOR IN *CEPAEA NEMORALIS*

In Example 3-3 some of the shell color polymorphism in the European land snail *Cepaea nemoralis* was described. These color variants represent the series of C alleles for brown, pink, and yellow, and they are often associated with particular habitats. Superimposed upon this ground color is a banding pattern determined by at least five other nonallelic loci. According to the work of Cain, Sheppard, and King (1968) and Carter (1968), the presence (B^B) or absence (B^o) of bands is due to a locus very closely linked to C. B^o is dominant to B^B and epistatic over all other loci that affect banding. The completely banded snail has five black or dark brown longitudinal bands. Another locus not linked to B or C, called U (unifasciata) has a dominant allele (U^3) that suppresses all bands except the middle, or third, band compared with its recessive unmodified (U^-) allele.

The data given in Tables A, B, and C (from Carter, 1968) give samples of phenotypes representing pairs of these loci taken from populations of *C. nemoralis* in the Berkshire Downs (England) during 1960 and 1962. No significant departure from expectation occurs for the C and U loci (Table A), while the B and C loci show linkage disequilibrium (Table C). The excess of banded yellow and unbanded pink shells observed here was also found

TABLE A *Observed number of midbanded* (U^3) *and completely banded* (U^-) *with ground colors yellow* (C^Y) *and pink* (C^P), *omitting the unbanded* (B^o) *snails.*

	1960			1962	
	Yellow	Pink		Yellow	Pink
Midband	503	199	Midband	690	247
All bands	101	38	All bands	138	44
Total		841	Total		1119
	$D = -0.0014$			$D = -0.0030$	

TABLE B *Observed number of bandless* (B^o) *and banded* (B^B) *snails, omitting the ground color effects*

	1960		1962
No bands	344	No bands	235
Midband	702	Midband	937
All bands	139	All bands	182
Total	1185	Total	1354

TABLE C *Observed numbers of bandless (B^o) and banded (B^B) snails, with ground colors, omitting the midband effects*

	1960				1962		
	Yellow	Pink			Yellow	Pink	
B^o	215	129		B^o	150	85	
Banded	604	237		Banded	828	291	
Total			1185	Total			1354
	$D = -0.0192$				$D = -0.0146$		

to be true for populations on the Isles of Scilly by Murray (1966). Table B gives the case of dominant epistatic frequencies between the banding loci (B and U); note that if one were confronted with those three phenotypes without knowing their true genetic basis, one could assume a single pair of alleles with the midband form "heterozygous" but showing an excess over the expected square-law "heterozygote" frequency (a false heterosis).

EXAMPLE 5-5

AN EARLY TWO-LOCUS HYPOTHESIS FOR A-B-O ANTIGENS

During the first decade of this century, just after the rediscovery of Mendel's principles and also the finding by Landsteiner (1900) that there were four phenotypes of people with the A − B antigens (AB, A not B, B not A, and neither A nor B = O), it was very tempting to presume that inheritance of these antigens followed the second law of Mendel. At first it was thought that independent assortment would explain the A-B-O situation with a dominant-recessive pair of alleles for A, not A and a second pair for B, not B, so that O type would be *aabb*, the A type either *AAbb* or *Aabb*, the B type either *aaBB* or *aaBb*, and the AB type any of the remaining four genotypes (*AABB*, *AaBB*, *AABb*, *AaBb*). In 1911 von Dungern and Hirzfeld proposed this method of inheritance for these blood types, but it was not until 1924 that Bernstein correctly determined the mode of heredity to be a single locus with multiple alleles.

If we refer to Table 5-3A, we can see that with two pairs of factors, each pair showing complete dominance, the expected frequency of any one phenotype (corner box in the table) would equal the opposite diagonal product divided by the remaining box. For example,

$$\text{Expected } A\text{-}bb = \frac{(A\text{-}B\text{-})(aabb)}{(aaB\text{-})} \quad \text{or} \quad (1 - q^2)v^2 = \frac{(1 - q^2)(1 - v^2)q^2v^2}{q^2(1 - v^2)}$$

Therefore, for any population in which these blood types have been collected, one can test the observed from expected departure. Let us then take the following frequency data from Japanese populations sampled in 1945 (about 40,000 persons):

Type	Observed Frequency	Expected: Two Loci with Dominance
A	0.384	0.133
B	0.219	0.076
AB	0.097	0.279
O	0.301	0.868

Obviously, there is no "fit" between observed and expected, and the hypothesis of two pairs of loci simply is not correct.

EXAMPLE 5-6

GAMETIC EQUILIBRIUM FOR TWO ALLOZYME LOCI IN EXPERIMENTAL POPULATIONS

Rasmuson, Rasmuson, and Nilson (1967) were able to investigate changes in frequencies of alleles at two loci controlling allozymes in four laboratory populations of *Drosophila melanogaster* by typing each adult fly in a sample by gel electrophoresis for both enzymes: esterase-6 (fast and slow mobility variants with codominance so that *FF*, *FS*, and *SS* are distinguishable) and leucine aminopeptidase (*A* producing a certain band with dominant genotypes *AA* and *Aa*, with *O* absence of any band being recessive *aa* genotype).

Stable equilibria for both loci were formed after 35 generations of observation, although the equilibrium frequencies varied from one population to another and depended on particular founder strains. Approach to linkage equilibrium was rapid between the two loci, and where there was evidence of an excess of heterozygotes (Est-6 locus, *FS*) there was no evidence of interaction between the loci.

The data analysis serves to illustrate the estimation of expected values for a combination of dominance in one pair of alleles and codominance in the other pair. As an example, the final sample from population 1 is given below:

	Est-6			
Lap	*SS*	*FS*	*FF*	Total
A	32	38	5	75
O	14	10	1	25
	46	48	6	100

Our first task is to estimate gametic frequencies $\overline{AS}$, $\overline{AF}$, $\overline{aS}$, $\overline{aF}$. The method illustrated here is efficient in that variances of the estimates have been worked out based on a method by DeGroot and Li (1960). The method of estimation used by Rasmuson, et al., is algebraically longer. From the column totals we estimate the allelic frequencies u_S and v_F for the Est-6 locus:

$$u = \frac{46 + 24}{100} = 0.70 \qquad v = \frac{24 + 6}{100} = 0.30$$

From the row totals we estimate q, the frequency of allele a for the LAP locus:

$$q = \sqrt{0.25} = 0.50$$

Thus, the total allelic frequencies may be represented as marginal totals as follows:

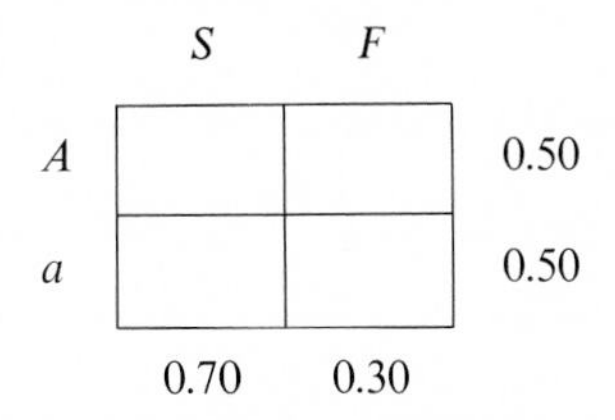

It is evident that only one of the four gametic frequencies is needed to fill up this table.

If we concentrate on the second row of the observed data table (all aa genotype, LAP-O phenotype, we would have no problem of ascertainment. For those genotypes observed as 14SS, O; 10FS, O; 1FF, O, we can estimate the frequencies of gametes $\overline{aS}$ and $\overline{aF}$ that must add up to $q = 0.50$. Thus,

$$\overline{aS} = \frac{14 + 5}{25}(0.50) = 0.38, \qquad \overline{aF} = \frac{5 + 1}{25}(0.50) = 0.12$$

The remaining two gametic frequencies may be obtained by subtraction to fill in the gametic table as follows:

	S	F	Total
A	0.32	0.18	0.50
a	0.38	0.12	0.50
Total	0.70	0.30	1.00

$D = -0.03$

Variances for these gametic estimates (from DeGroot and Li) are as follows:

$$V_{\overline{aS}} = \frac{(\overline{aS})(\overline{aF})}{2RN} + \frac{(\overline{aS})^2 D}{4RN} = 0.001996$$

$$V_{\overline{aF}} = \frac{(\overline{aS})(\overline{aF})}{2RN} + \frac{(\overline{aF})^2 D}{4RN} = 0.001020$$

$$V_{\overline{AS}} = \frac{v(\overline{AS} - \overline{aS})}{2N} + V_{\overline{aS}} = 0.001906$$

$$V_{\overline{AF}} = \frac{u(\overline{AF} - \overline{aF})}{2N} + V_{\overline{aF}} = 0.001230$$

where D = frequency of LAP-A and RN = number of LAP-O individuals observed. Standard errors, respectively, are 0.0447, 0.0319, 0.0436, 0.0351.

Under random mating, expected numbers of the six phenotypes would be as follows, using the estimated gametic frequencies:

	SS	*FS*	*FF*	Total
A	34.56	32.88	7.56	75
O	14.44	9.12	1.44	25
	49.00	49.00	9.00	100

Chi-square = 2.09
d.f. = 2
P = n.s.

Thus, there is no significant departure from the expectation of random mating with the gametic frequencies estimated. There is a D value of -0.03, which is 20 percent of maximum D, but in view of the large standard errors for gametic estimates we cannot be confident in the magnitude of this D estimate. In contrast, if we increase the two coupling gametic frequencies by one standard error (using 0.04 roughly) so that $\overline{AS} = 0.36$ and $\overline{aF} = 0.16$ while we lower the two repulsion gametic frequencies by the same amount making $\overline{AF} = 0.14$ and $\overline{aS} = 0.34$, then $D = +0.01$. Rasmuson, et al., concluded that there was no significant departure from gametic equilibrium for these loci.

EPISTASIS

When there is an inhibitory effect of any allele at one locus over the genotypic expression in the phenotype by a nonallelic locus, the inhibitory locus allele is said to be epistatic, in the narrow sense. For example, the albino condition in mammals usually results from a homozygous recessive that inhibits pigment formation in the entire body, so that when a pair of alleles controlling skin color or coat color is segregating, no segregation is detectable in the albino. Consequently, epistasis restricts the number of phenotypes to less than four if the segregating nonallelic pair of alleles exhibits dominance. By referring to Table 5-3B, if we propose that the *aa* genotype, for example, be an epistatic albino genotype but the *B*-locus be segregating for a color such as black-brown (as in mice or guinea pigs), then the lower row adding up to q^2 would not be subdivided into two phenotypes, and relative frequencies of *B*-:*bb* would have to be estimated from the upper row. With epistasis (recessive) the three phenotypes *A-B-*, *A-bb*, and *aa--* would occur in these proportions: $(1 - q^2)(1 - v^2)\colon (1 - q^2)v^2\colon q^2$.

It may happen that the relative proportions of these phenotypes occur in such a way that their frequencies imitate the proportions of three genotypes with a *single pair of alleles*, as noted by Li (1961). Because $v^2(1 - q^2)/q^2 = v^2(1 + q)(1 - q)/q^2$, we take the last fractional part of the expression as the relative proportion factor by which the H genotype is related to R genotype for a single pair of alleles: we note
(1) the single-locus ratio $H/R = 2q(1 - q)/q^2$, but
(2) the *A-bb*/*aa--* ratio $= v^2(1 + q)(1 - q)/q^2$.
Then equating (1) and (2), $v^2(1 + q) = 2q$ and $v^2 = 2q/(1 + q)$, we obtain the
A-bb frequency $[2q/(1 + q)]\,(1 + q)(1 - q) = 2q(1 - q)$, and the
A-B- frequency $(1 - q^2)\,[1 - 2q/(1 + q)] = (1 - q)^2$.

Thus, the phenotypic frequencies would not be distinguishable from those of a single allelic pair lacking dominance $(1 - q)^2\colon 2q(1 - q)\colon q^2$ (see Table 5-3B). As pointed out by

Li (1961), it cannot always be concluded that just because three phenotypes occur in square-law proportions, the genetic basis is monofactorial. Of course, breeding tests on the *A-B-* type and the *aa--* should distinguish the dihybrid from monohybrid situation. A case in point was illustrated by Wright's analysis of roan coat color in shorthorn cattle (Example 2-1). The herd book data alone could not distinguish between a unit pair of factors determining red:roan:white (with roan heterozygote) from the two-pair factor case with roan:red (*P-*:*pp*) pair independent of a red color:white (*R-*:*rr*) with *rr* (white) epistatic to the *P-p* pair. Wright had to examine particular crosses to distinguish the two contrasting types of heredity and finally to prove the correct mode to be unifactorial.

Dominant alleles may be epistatic, as illustrated in Example 5-4, in which the bandless pattern (B^o locus) of the snail *Cepaea nemoralis* inhibits all banded phenotypes (*U* locus). In similar fashion to the recessive epistatic case, the three phenotypes could simulate the monofactorial case as follows:

From Table 5-3C, with dominant epistasis the double recessive class (*aabb*) has a frequency of q^2v^2, but if the *A-bb* class were in a monofactorial ratio of H/R such that $(1 - q^2)v^2 : q^2v^2$ as $2(1 - qv) : qv$, then $v = 2q/(1 + q^2)$ and the three phenotypic classes would be simulating the monofactorial case (see Exercise 15).

DUPLICATE COMPLEMENTARY FACTORS

When two or more of the phenotypes in Table 5-3 are indistinguishable because nonallelic factors seem to produce equal effects, we say the factors are "duplicate," at least from the point of view of the observer, who cannot analyze their primary genic action. If *A-bb* = *aaB-* in phenotype and if each dominant factor is separately effective so that *A-B-* can be distinguished from them, the frequencies in Table 5-3 in the upper right and lower left would be added. It should be apparent that if the frequencies of the dominant alleles at both loci should be equal—that is, $p_A = u_B$, or $q_a = v_b$—then a population with such frequencies could not be distinguished from the monofactorial case (see Exercise 13). In the simplest case where $p = q$, $u = v$, or $\frac{1}{2}$ for each allele at both loci, the two single dominant phenotypes would be equal and the total ratio would be $\frac{9}{16}$(*A-B-*): $\frac{6}{16}$(*A-bb* + *aaB-*): $\frac{1}{16}$(*aabb*), which could simulate the monofactorial case with $p = \frac{3}{4}$, $q = \frac{1}{4}$.

This form of heredity is often called additive duplicate dominance, because the substitution of either an *A* or *B* allele can bring about equally the increase in phenotype, which is then doubled in the *A-B-* form. For example, these phenotypes might be expected in some quantitative gradation as with depth of pigment in skin or hair.

If duplicate genes also exhibit epistasis, the number of phenotypes is reduced to two. When all allelic frequencies are equal, as in the familiar F_2 generation of a Mendelian cross, the ratios 9:7 (duplicate recessive epistasis), 15:1 (duplicate dominant epistasis), or 13:3 (epistatic dominant duplicate for a recessive) may be exhibited. Any of these hereditary modes could be mistaken for monofactorial cases with simple dominance without suitable progeny testing or pedigree analysis. In discussing the utility of the square law in Chapter 2, it was pointed out for the data on diabetes (Example 2-2) that the square law could fit a monofactorial case and could also fit a two-locus or multiple-locus case, so that population frequencies alone, such as those collected on the basis of one or both parents having the disorder, cannot be used to distinguish these two hereditary modes (one locus versus two loci). It should be obvious that if two diabetic parents had nondiabetic children, the complementary nature of at least two loci would be established, and there appears to be considerable evidence in that direction (see Neel, et al., 1965) (it is not easy to disprove incomplete penetrance in the offspring as an explanation).

All these cases of dominance and phenotypic interactions are included here at some length to impress on the reader that conformity of population frequencies with expectation for a single pair of alleles (square law) is not sufficient proof for monofactorial inheritance. Without progeny testing or study of family pedigrees, the mode of heredity for any phenotypic variant cannot always be ascertained from a single-generation populational analysis.

EXERCISES

1. In a certain population, A-a and B-b alleles at A and B loci, respectively, occur in the following genotype frequencies (grouped as in Table 5-1B) (assume independent assortment between loci).

$$Z = \begin{vmatrix} 0.09 & 0.18 & 0.09 \\ 0.12 & 0.24 & 0.12 \\ 0.04 & 0.08 & 0.04 \end{vmatrix}$$

   (a) What are the allelic and the gametic frequencies?
   (b) What would be the expected genotype frequencies at equilibrium assuming random mating?
   (c) Check the equilibrium by calculating the gametic matrix determinant. What is the "coupling" gametic product? the "repulsion" gametic product?

2. Suppose four completely homozygous genotypes of an experimental organism are introduced into an area where they can mate at random. They are introduced in the following frequencies: $AABB$ = 40 percent, $aaBB$ = 10 percent, $AAbb$ = 20 percent, $aabb$ = 30 percent (assume independent assortment between loci).
   (a) What will be expected for the genotype frequencies of the progeny generation? allelic frequencies?
   (b) What is the parental gametic determinant? What will be the progeny gametic output and gametic determinant?
   (c) Assuming random mating to occur between all genotypes for two more generations, what will be the gametic output over those generations (G_0 = parent generation through G_3)?
   (d) What will be the ultimate genotype frequencies and gametic output after many generations of random mating?

3. One thousand individuals in a population were typed for two traits (A-a and B-b), determined, respectively, by A and B loci. Heterozygotes were distinguishable so that the following nine genotypes were found with these numbers:

$$\begin{vmatrix} 36 & 48 & 16 \\ 72 & 96 & 32 \\ 252 & 336 & 112 \end{vmatrix} \begin{matrix} = 100\ AA \\ = 200\ Aa \\ = 700\ aa \end{matrix}$$

$$\begin{matrix} 360 & 480 & 160 & 1000 \\ BB & Bb & bb & \end{matrix}$$

   (a) Determine if the alleles at *each locus* are at equilibrium. If not, what would the *separate* locus genotype frequencies be?

(b) Determine if gametic sets are at equilibrium. If not, what would each gametic combination frequency be?
(c) Is this array of nine genotypes at equilibrium as it stands? If not, what would be the expected frequencies of these genotypes?
(d) Can you tell whether these loci are linked or not? Why?
(e) How many ways can we estimate *expected* frequencies of genotypes when we have data on two genic loci? List them.

4. If it is known that two genic loci (A-a and B-b) have two alleles each and that the alleles occur at equal frequencies in a certain population (that is, $p_A = q_a = \frac{1}{2}$ and $u_B = v_b = \frac{1}{2}$),
   (a) Can we determine what the *gametic* combination frequencies are based only on these allelic frequencies? What assumptions must you make before you estimate gametic frequencies in a population?
   (b) Suppose you know that the AB gamete in the population has a frequency of $\frac{1}{6}$; knowing the allelic frequencies of each locus, determine the frequency of the remaining three gametic types in the population.
   (c) If mating is at random and the genic loci are independently assorting, what would be the gametic frequencies in the next generation?
   (d) If mating is at random but the loci $A - B$ are linked with 30 percent crossing over, what would be the gametic frequencies in the next generation?
   (e) What will be the limiting gametic frequencies (equilibrium) allowing random mating for a long period of time?
5. Assume loci A and B are linked with 10 percent crossing over in Table 5-2. Starting with the random mating of $AABB$ (♂♂ and ♀♀) with $aabb$ (♂♂ and ♀♀), give the genotype frequencies and gametic output for the Z_1 and Z_2 generations.
6. If original parents included twice as many $AABB$ as $aabb$ individuals, what would be the outcome of random mating for the first two progeny generations? Assume independent assortment first, then 10 percent crossing over.
7. Kojima and Lewontin (1970) pointed out that the parameter D may be defined as the covariance of allelic distribution in the gametes. Look up the covariance in Appendix A-6-9. Consider a 2 × 2 table of allelic frequency association:

	B	b	
A	$\overline{AB} = pu + D$	$\overline{Ab} = pv - D$	p
a	$\overline{aB} = qu - D$	$\overline{ab} = qv + D$	$(1 - p)$
	u	$(1 - u)$	

Assign a quantitative value of 1 to the capital letter alleles A and B, and a value of 0 to the lower-case alleles a and b. The marginal average for A is p and the average for B is u. Letting the A locus be the X variate and the B locus the Y variate, since $CoV = \sum(X)(Y) - (\sum X)(\sum Y)$, then $CoV = \overline{AB}(1 \times 1) + \overline{Ab}(1 \times 0) + \overline{aB}(1 \times 0) + \overline{ab}(0 \times 0) - pu$. Then see formula (5-5).

8. (a) In Exercise 2, marginal allelic frequencies in the gametic matrix were $p_A = 0.6$, $q_a = 0.4$, $u_B = v_b = 0.5$. Fill in the following gametic matrices given that one gametic frequency is zero. Which matrices are not possible, given the marginal

frequencies? What relationship do the gametic determinants (D) have to formula (5-9) or (5-9A)?

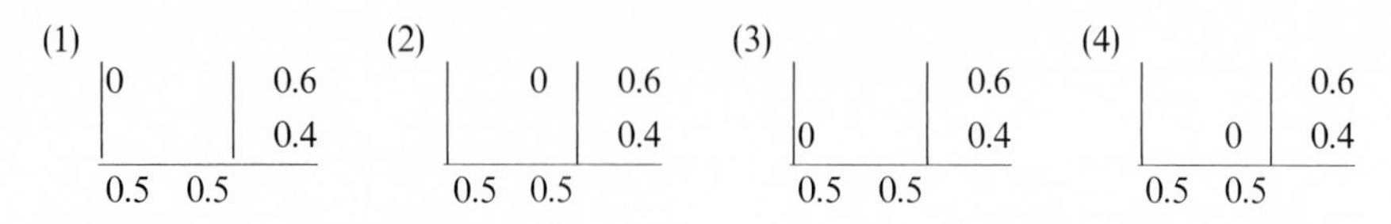

(b) What is D_{max} for the gametic matrix of Exercise 4(b)? D_{max} for the gametic matrix of Exercise 6?

(c) To see the influence of allelic frequency on the D_{max} value, determine D_{max} for the following:
 (1) $p = 0.1, q = 0.9; u = 0.7, v = 0.3$.
 (2) $p = 0.2, q = 0.8; u = 0.6, v = 0.4$.
 (3) Which has the greater D_{max}? Does the change in D_{max} value mean that there is greater disequilibrium in one case than another? How can the disequilibrium be expressed to put cases of allelic changes on an equal basis for comparison?

9. In a population of Pacific Coast hagfish, Ohno, et al. (*Science* 156: 96–98, 1967) reported frequencies of allozymes, fast and slow electrophoretic variants of lactate dehydrogenases (LDH-1 and LDH-5) as follows:

	LDH-1		
LDH-5	*FF*	*FS*	*SS*
FF	14	8	1
FS	5	8	5
SS	5	3	2

(a) If we assume that LDH-1 and LDH-5 are determined by two separate loci, what are the frequencies of alleles at the two loci?

(b) Consider the gametic determinant. What percent of maximum is it?

(c) Determine expected values for the genotypes based on three separate assumptions (as in Example 5-2): (1) row by row and column by column, the proportions are the same (contingency test); (2) based on allelic frequencies, the genotypes occur according to "linkage equilibrium"; (3) gametic frequencies are the result of random mating according to formula (5-3).

(d) Complete chi-squares for each of these three sets of expected values. How many degrees of freedom and what is the probability in each case for goodness of fit?

10. The old theory as to the inheritance of human A-B-O blood groups assumed two pairs of independent factors with dominance in each pair, as illustrated in Example 5-5. Use the data in Table 3-2 for the Germans and the Pygmies to calculate expected values for blood group phenotype frequencies. Which is the more likely mode of inheritance based on these populational data—the Bernstein multiple-allele hypothesis or the two independent pairs of loci?

11. In Example 2-1, Wright estimated the hypothetical frequencies of nine genotypes on the basis of short-horn cattle coat color (red, roan, and white) being due to two independent loci; white (*rr*) was presumed to be recessive to red (*R*-) and epistatic over other colors, while roan (*P*-) was presumed to be dominant to nonroan (*pp*). On the

expectation of equilibrium for two independent loci in this breed of cattle, Wright estimated the separate gene frequencies and then the nine genotype frequencies from the overall breed frequencies: white = 0.08685, roan = 0.43423, red = 0.47892.

(a) Calculate the R-locus and P-locus hypothetical frequencies by using formula (2-4) to obtain recessive allele frequencies (q and v) and genotype frequencies on the expectation from random mating. Then determine expected nine genotype frequencies and set in a 3 × 3 table (as in Table 5-1B).

(b) Refer to Table 5-3B for the relationship that would be expected if the frequencies of phenotypes simulated those of a single locus with an intermediate heterozygote (roan), as postulated by Wright. Assume mating to be at random and estimate v^2 first (nonroan) then q^2 (white).

(c) Does the estimate of v^2 equal $2q/(1 + q)$? Is it possible to separate the monofactorial from the bifactorial interpretation by using these populational data?

(d) Wright resolved the problem of one-locus versus two-loci interpretation by analyzing the progeny data from particular matings, the most important being red × white and roan × roan. Expected and observed percentages for these matings are as follows:

Mating	Progeny Percentage Red	Roan	White
Red × White			
Exp. 2 loci	55.9	21.3	22.8
1 locus	0.0	100.0	0.0
Obs.	2.2	95.9	1.9
Roan × Roan			
Exp. 2 loci	16.7	78.1	5.2
1 locus	25.0	50.0	25.0
Obs.	26.6	52.9	19.5

Check the expected values for the two-loci possibility. There were some errors in the herd books and mistaking extreme roans for white or red to explain the small discrepancies in the single-locus interpretation. Which interpretation is more likely?

12. By determinant algebra, show that $D_t = \frac{1}{2}D_{t-1}$, from formula (5-4A). Proceed as follows:

$$D_1 = \begin{vmatrix} (\overline{AB}_0 - \frac{1}{2}D_0) & (\overline{Ab}_0 + \frac{1}{2}D_0) \\ (\overline{aB}_0 + \frac{1}{2}D_0) & (\overline{ab}_0 - \frac{1}{2}D_0) \end{vmatrix}$$

Perform the inner products and rearrange into four components.

$$= \begin{vmatrix} \overline{AB}_0 & \overline{Ab}_0 \\ \overline{aB}_0 & \overline{ab}_0 \end{vmatrix} + \begin{vmatrix} \overline{AB}_0 & \frac{1}{2}D_0 \\ \overline{aB}_0 & -\frac{1}{2}D_0 \end{vmatrix} + \begin{vmatrix} -\frac{1}{2}D_0 & \overline{Ab}_0 \\ \frac{1}{2}D_0 & \overline{ab}_0 \end{vmatrix} + \begin{vmatrix} -\frac{1}{2}D_0 & +\frac{1}{2}D_0 \\ +\frac{1}{2}D_0 & -\frac{1}{2}D_0 \end{vmatrix}$$

$$= D_0 - \tfrac{1}{2}D_0(\overline{AB}_0 + \overline{aB}_0) - \tfrac{1}{2}D_0(\overline{Ab}_0 + \overline{ab}_0) + 0$$

Complete the simplification.

13. When nonallelic dominants produce duplicate phenotypic effects (for example, if A-bb = aaB-, and if they are distinguishable from A-B-), the frequencies of phenotypes may simulate a monofactorial case when $p_A = u_B$. Suppose hair color depends on two additive dominant independent factors such that blond is *aabb*, light brown can be either A-bb or aaB-, and dark brunette is A-B-. Use the frequencies in Table 5-3A. What would be the frequencies of these hair colors in a random-mating population with $q_a = v_b = 0.10$? Change the ratios of $p{:}q$ and $u{:}v$ in parallel at 0.10 intervals and show that their joint frequencies simulate the monofactorial case expressed in Property 2 in Chapter 1.
14. In house mice, albinism (lack of pigment) is due to homozygosity of a recessive allele at the color (C) locus, usually designated c. The independent B-b locus controls black (B) versus brown (bb) coat color. When an animal is cc we cannot distinguish the B-b genotypes without progeny testing.
    (a) If the albino allele has a frequency of $q_c = 0.20$ and if there are twice as many black mice as brown in a random-mating population of house mice, what would be the expected frequencies of the three coat colors? (Use Table 5-3B.)
    (b) How do these frequencies imitate a monofactorial case?
    (c) If the albino gene has $q_c = \frac{1}{7}$, what frequency of brown mice would bring a simulation of the monofactorial case? (Hint: First find $v^2 = \frac{1}{4}$.) What would the frequencies of the three phenotypes be?
15. In Table 5-3C, if the phenotypes A-bb:$aabb$ are simulating monofactorial heterozygote: recessive homozygote frequencies (H/R), then their ratio will be $2(1 - qv)qv/q^2v^2$. Thus, the simulation would occur if the A-bb frequency equals the simulated heterozygote $2(1 - qv)qv$.
    (a) Verify that $v = 2q/(1 + q^2)$.
    (b) Suppose $q = 0.50$ and $v = 0.80$. What would be the frequencies of phenotypes in Table 5-3C? How do these values simulate the monofactorial case?
    (c) Fill in the frequencies in the right-hand portion of Table 5-3C to verify the indicated relationships.
16. In Example 5-4, Table B, three phenotypes bandless (B^0), midbanded, and all bands present are produced by two loci (B and U) as described, but with B^0 epistatic over the U locus (midband U^3 or U^- all bands). Expected frequencies could be estimated from using Table 5-3C.
    (a) If we take the frequency of B^0 as that in the 1962 collection and equal to $(1 - v^2)$, estimate v and then q^2 and q on the basis of random mating.
    (b) Does the observed set of frequencies for 1962 conform to what would be expected for simulating a single locus with the midband condition as presumed heterozygote?
    (c) If the frequency data do not simulate a single-locus result, can you deduce the fact that these banded phenotypes result from two loci? What possible explanations can you present to account for these frequencies?

6

POLYGENES AND CONTINUOUS VARIATION

All groups of organisms display continuous gradations, or smooth distributions, in a large number of attributes that can be measured either by a scale (for example, dimensions, weights, amounts of pigment, relative disease resistance, mental abilities) or by meristic qualities (for example, numbers of vertebrae, bristles, segments, eggs laid, progeny produced). This kind of variation is the commonest type to be observed. No sharply distinct classes may be discernible between the limits of the distribution's range, and individuals with dimensions or numbers near the center of the distribution are the most frequent sort.

Mendel's success in observing fundamental segregation and assortment ratios depended largely on the discrete, or discontinuous, qualitative features he chose. As long as the distribution of segregating genetic elements in a population produces a continuum, it is simply not possible to analyze it in the usual fashion for single-locus effects. We must utilize largely statistical tools in order to demonstrate numbers of loci controlling the distribution, their potency relative to environmental factors and to each other, their frequencies, and how the population might be affected by changing their frequencies.

In this chapter we review some of the fundamental work forming the basis for our present concept of continuous variate heredity. Polygenic theory as applied to random mating populations will be introduced at an elementary level. We must first explore the expectations of segregation with genetic factors affecting continuous variation, numbers of loci segregating, and the relative magnitudes of their genic action affecting the phenotype in random mating populations. In the following chapter some of the methods used for ascertainment of genetic parameters will be discussed: estimation of genetic variance from controlled crosses, artifical selection response, and parent-offspring correlation.

TURN OF THE CENTURY CONCEPTS

Francis Galton made considerable progress in studying continuous variation in humans throughout the later years of the nineteenth century. By measuring parents and their offspring for a number of traits (as well as traits in many domestic animals), he and his associate, Karl Pearson, clearly showed continuous variate characters such as height, weight, body proportions, and mental traits to be heritable (significant parent-offspring correlation). From these studies, the application of statistical mathematics to biological problems was given a significant beginning, although the biometric school of Galton and Pearson was not able to solve the main problem of hereditary transmission.

A sample of data collected by Galton is given in Table 6-1, showing heights for parents and offspring. It is apparent that while the parent and offspring mean heights are about equal, indicating equilibrium between generations, the main results are that (1) there is a heritable tendency (short parents tend to have short children, and tall have tall), and

TABLE 6-1 *Heights (inches) of parents and offspring in English populations. Female heights were multiplied by 1.08 to make them comparable with male heights. Data collected by Galton (1889), modified from Davenport (1917). Total offspring measured = 928*

Height of Midparent (P)	Median Height of Offspring (O)	Deviation $O - P$
Over 72.5	72.9	[omit (only 4 cases)]
72.5	72.2	−0.3
71.5	69.9	−1.6
70.5	69.5	−1.0
69.5	68.9	−0.6
68.5	68.2	−0.3
67.5	67.6	+0.1
66.5	67.2	+0.7
65.5	66.7	+1.2
64.5	65.8	+1.3

Average parent = 68.4 ± 0.1; average offspring = 68.1 ± 0.1; correlation: $r = +0.45 \pm 0.03$.

(2) extremes among the parents have children who are less extreme—children tend to "regress" toward the populational average. This latter principle, which might better have been termed "retrogression," was stated as follows by Galton: "Individuals differing from the average character of the population produce offspring which, on the average, differ to a lesser degree but in the same direction as their parents." The slope of the moving average between these correlated parents and offspring then came to be called a linear "regression." In the Galton data, the regression slope ($b = 0.63$) and the correlation coefficient ($r = 0.45$) express the main heritable tendency. A better study by Pearson and Lee in 1903 compared the stature of 1078 sons with their fathers and gave a correlation of $r = 0.51$, indicating the same retrogression of children from extreme parents (see Neel and Shull, 1954). But the reasons for the retrograde tendency could be due to at least two main causes: (1) hereditary factors are perhaps not sufficiently determinative to control the phenotype exactly; that is, the nonheritable factors of environment may often have determined the differences of particular individuals; for example, a short person may be short because of low nutrition during childhood in spite of a genotype that could determine greater height if better nutrition had been available during the years of growth. (2) Because only one-half of each parent's heredity is transmitted to each child, the effect of mingling any extreme heredity with the average for the population would tend to bring offspring closer to the population mean. Example 6-1 presents a more extensive illustration of stature in an American population collected by Davenport (1917) and contrasting with Galton in the matter of "retrogression."

Unfortunately, the rivalry between Galton's biometrical followers and the Mendelians led by Bateson and deVries delayed any attempts to discern the true genetic solution to the determination of quantitative-continuous traits. Experiments by two geneticists in Scandinavia and certain Americans in the first two decades of the twentieth century helped to resolve the differences between these two schools. W. Johannsen in Denmark and H. Nilsson-Ehle in Sweden, both reporting their classic work in the same year (1909), performed the essential experiments that served as models for much of modern polygenic theory. Later in the United States, E. M. East, G. H. Shull, and W. E. Castle corroborated their work. In England, R. A. Fisher helped to resolve the differences between the biometricians and Mendelians by establishing the model for multiple-factor heredity, deriving the expected correlations between parents and offspring, producing evidence that these factors could display dominance, and finally attempting to partition the observed phenotypic variance into additive and dominant components.

Johannsen wished to test Darwin's theory of natural selection—that is, how effective selection might be in determining change toward a particular goal by selecting for an increase or decrease in the trait each generation and noting the amount of progress in the direction of the goal. He wondered whether selection applied would act directly on the hereditary makeup or simply on the trait itself to achieve the goal. He chose to experiment with a dwarf bean (Princess variety of *Phaseolus vulgaris*). Starting in 1900, he weighed a heterogeneous lot of beans carefully, noting that this unselected population displayed a continuous distribution of wide range. He divided the seed into classes according to weight, planted them, self-pollinated their flowers to produce the F_1 (524 beans). Lightweight beans tended to produce lighter than average progeny and heavyweight heavier than average. Johannsen then concentrated on 19 lines derived from 19 of the original parent beans, each line having been produced by selfing, and continued with the same method. In 1902 the F_1 beans were planted, selfed, and produced the large F_2 progenies in the 19 lines, as shown by the composite data of Table 6-2.

It was clear that the average weight of the daughter beans was related to that of their mothers. Each line tended to have a characteristic average weight; those derived from the lightest parent beans had lightweight offspring, and the heaviest had heavier offspring. Consequently, the original heterogeneous lot was a genetic mixture, and each pure line (selfed) had sorted out certain genetic determination for weight. But as Johannsen continued to self his lines year after year, he found that, irrespective of the weight of parent beans within each line, progenies of those lines maintained a nearly constant average; we would say that selection for weight was no longer effective in changing the average, or that correlation between parent and offspring tended to be zero. Data from two of the pure lines (line 1 the heaviest and line 19 the lightest) are given in Table 6-3 as representatives for all the lines. Whether plus or minus deviations from each line's mean were chosen as parent phenotypes, no significant differences occurred between daughter averages within lines for the period of six years (1902–1907).

Johannsen ascribed correctly to the breeding method (self-pollination) the mechanism by which the hereditary determiners (he later called them "genes") became uniform ("homozygous," to use Bateson's term) within each *pure line*, a fact Mendel had first proposed. Selection for this continuous variate (weight of bean) was effective at first when applied to the genetically diverse original lot of beans, which must have included a number

TABLE 6-2 *Weights (centigrams) of beans (Princess strain of* Phaseolus vulgaris). *F_1 seeds produced in 1901 were classified by weight before planting. Offspring (F_2) in 1902 were separated by parental line. From Johannsen, 1903*

Weight Groups Parent Beans	Weights of Progeny (1902)																	
	15	20	25	30	35	40	45	50	55	60	65	70	75	80	85	90	N	Average Progeny ± *s.e.*
65–75	—	—	—	2	3	16	37	71	104	105	75	45	19	12	3	2	494	58.47 ± 0.43
55–65	—	—	1	9	14	51	79	103	127	102	66	34	12	6	5	—	609	54.37 ± 0.41
45–55	—	—	4	20	37	101	204	281	234	120	76	34	17	3	1	—	1138	51.45 ± 0.27
35–45	5	6	11	36	139	278	498	584	372	213	69	20	4	3	—	—	2238	48.62 ± 0.18
25–35	—	2	13	37	58	133	189	195	115	71	20	2	—	—	—	—	835	46.83 ± 0.30
15–25	—	—	1	3	12	29	61	38	25	11	—	—	—	—	—	—	180	46.53 ± 0.52
Totals	5	8	30	107	263	608	1068	1278	977	622	306	135	52	24	9	2	5491	50.39 ± 0.13

(524 mother beans in 19 lines)

Correlation: $r = +0.336 \pm 0.008$

TABLE 6-3 *Six generations of selection for the heaviest and lightest beans (Princess variety) within two pure lines (No. 1 heaviest line and No. 19 lightest line). Data summarized by Babcock and Clausen (1927) from Johannsen (1909)*

A Line No. 1

Harvest Years	Total Number of Beans	Mean Weight of Mother Beans of Select Strains		Difference $b - a$	Mean Weight of Progeny Seeds of Select Strains		Difference $b - a$
		a minus	*b* plus		*a* minus	*b* plus	
1902	145	60	70	10	63.15 ± 1.02	64.85 ± 0.76	+1.70 ± 1.27
1903	252	55	80	25	75.19 ± 1.01	70.88 ± 0.89	−4.31 ± 1.35
1904	711	50	87	37	54.59 ± 0.44	56.68 ± 0.36	+2.09 ± 0.57
1905	654	43	73	40	63.55 ± 0.56	63.64 ± 0.41	+0.09 ± 0.69
1906	384	46	84	38	74.38 ± 0.81	73.00 ± 0.72	−1.38 ± 1.08
1907	379	56	81	25	69.07 ± 0.79	67.66 ± 0.75	−1.41 ± 1.09

B Line No. 19

Harvest Years	Total Number of Beans	Mean Weight of Mother Beans of Select Strains		Difference $b - a$	Mean Weight of Progeny Seeds of Select Strains		Difference $b - a$
		a minus	*b* plus		*a* minus	*b* plus	
1902	219	30	40	10	35.83 ± 0.44	34.78 ± 0.38	−1.05 ± 0.58
1903	200	25	42	17	40.21 ± 0.65	41.02 ± 0.43	+0.81 ± 0.78
1904	590	31	43	12	31.39 ± 0.29	32.64 ± 0.21	+1.25 ± 0.36
1905	1657	27	39	12	38.26 ± 0.16	39.15 ± 0.17	+0.89 ± 0.23
1906	1367	30	46	16	37.92 ± 0.22	39.87 ± 0.16	+1.95 ± 0.27
1907	594	24	47	23	37.36 ± 0.30	36.95 ± 0.21	−0.41 ± 0.37

of heterogeneous heredities (he called them "genotypes"). Owing to the fact that the bean plant is naturally self-fertile, it can be presumed that most of the original beans were largely homozygous, although in the first two or three years Johannsen may have decreased even more what little heterozygosity they had. His selection, then, only sorted out, or isolated, originally separate genotypes. This view of selection did not yet appreciate the vast array of genetic variation produced by the Mendelian processes nor did any biologist in those days perceive the subtle specificity of natural selection's action in generating genotypic diversity or in molding it into adapted genotypes that we are beginning to appreciate now. Consequently, Johannsen's results were misinterpreted by anti-Darwinists at the turn of the century as a proof that natural selection cannot be creative! Such erroneous and shortsighted views dealt an unfortunate blow to the theory of evolution by natural selection and our understanding of the adaptive process.

But to return to the nature of hereditary determination, Johannsen pointed out that his results followed the Galtonian regression principle: progeny averages tended to approach the population average away from parental extremes, apparently because environmental factors had been more effective in producing individual differences than genetic determiners (at least for the bean weight character). The phenotype then results from an interaction of hereditary and environmental factors. The genotype may be relatively weak in the case of weight in determination of an exact quantity expressed. Therefore, in order to demonstrate the genotypic control, outcrossing followed by progeny testing, or selection with an observed response from the population in the direction of selection applied, was essential.

While Galton and Johannsen demonstrated that continuous variate heredity was a fact, reconciliation with the Mendelian scheme necessarily awaited the finding of discrete qualitative differences within an otherwise continuous distribution. The Swedish geneticist, Nilsson-Ehle, provided that resolution by finding a trait with distinguishable classes: the depth of red pigment in the glume of wheat (*Triticum vulgare*). A graded series of redness from deep red to white formed a distribution with discontinuous groups that could be easily classified. When a dark red-grained variety was crossed to a white-grained, the F_1 grains were intermediate (no dominance), and by selfing the F_1's an array of distinct color varieties appeared in the F_2 with an approximate ratio of 1:6:15:20:15:6:1 from deep red (1/64) through the lighter classes down to almost white (1/64). When each of the F_2 classes was selfed, the extremes bred true, as did 3/15 of the medium red (15/64) or light red classes. From the 6/64 next to the extremes, segregations of 1:2:1 took place, while segregations into 1:4:6:4:1 arose from 12/15 of the medium or light red classes, which were double heterozygotes. Finally, 8/20 in the intermediate class again produced the entire array as the F_1 had done. Nilsson-Ehle realized that the original parents must have differed by three independent loci with pairs of red:non-red alleles at each locus and that the allele for pigment in each case acted additively on the total phenotype; the more "red" factors present, the greater the depth of color in the grain. We can now realize that *Triticum vulgare*, being a hexaploid, has three genomes in replicate so that presumably there are at least three duplicate loci with homologous alleles affecting the same trait. These triplicate sets of alleles, displaying no dominance in heterozygotes and acting cumulatively on the phenotype, are diagrammed in Figure 6-1. Note that of the 20 combinations with equal dosages of red:non-red, 8 are triply heterozygous and the remainder (12/20) are heterozygous at one locus and "balanced" homozygotes for the other two loci.

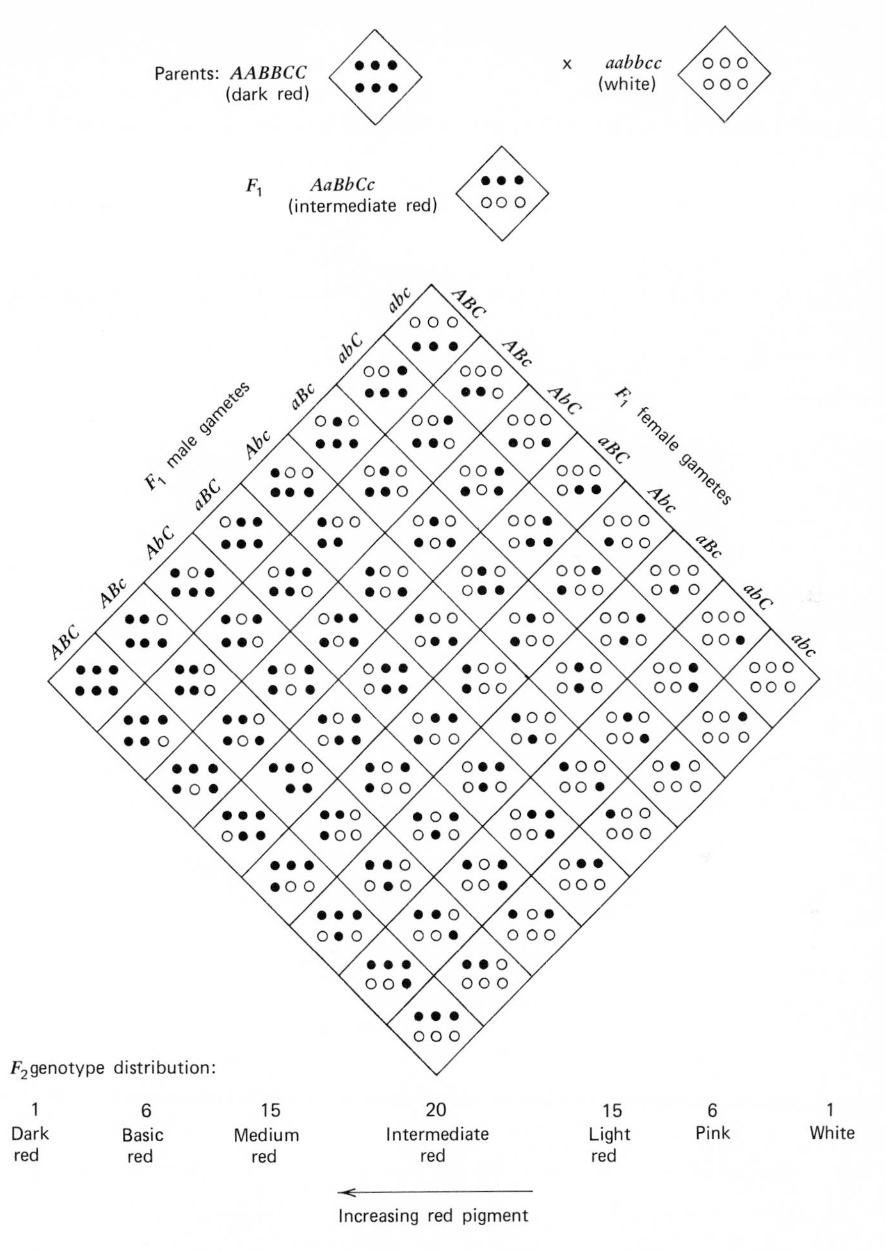

Figure 6-1. Genotypes resulting from crossing two varieties differing in three pairs of independent alleles. Additive gene action according to Nilsson-Ehle's model for color of wheat grains. Note the F_2 has a graded array of seven phenotypes from the 27 genotypes.

EXAMPLE 6-1

A POLYGENIC TRAIT IN HUMAN FAMILIES

In 1917 C. B. Davenport made an extensive survey of human stature, collecting data from 869 American families with 3298 children. To minimize the sex difference problem, he expressed the data in terms of deviations from average height (considered 68 inches for men and 63 inches for women). While Davenport classified matings into 25 categories (tall × tall, tall × medium, very tall × short, etc.), he was unable to distinguish discrete ratios among offspring and was therefore led to consider stature as due to multiple factorial inheritance, with some short phenotypes tending toward slight dominance and tallness tending to be recessive.

His data, summarized in Table A, corroborate Galton on the general heritable nature of the trait but are in contrast with Galton, as well as with Pearson and Lee (1903), in the "retrogression" principle, which seems to have been true only for children of short parents in these American families. On further examination, if the deviation of O-P is considered, it turns out that *all* children tended toward increasing height (except for the small sample of children from very tall × short where O-$P = -0.16$)! Perhaps this is one of the first records of the trend in American populations since the turn of the century for a slow but univeral increase in stature, most evident in the comparison of military servicemen's stature from World Wars I and II.

TABLE A *Matings arranged in order of average deviation (inches) of midparents from medium stature 68 inches with the average deviation of the children's stature within families. Offspring-parent (O-P) indicates the general trend for children. Data from Department (1917). N = numbers of children*

Matings	Average Deviation—Midparents (*P*)—from Population Average	Average Deviation—Children (*O*)—from Population Average	*O-P*	*N*
Very tall × very tall	5.80	6.08	+0.28	105
Very tall × tall	4.45	4.74	+0.29	397
Medium × very tall	3.02	3.45	+0.43	240
Tall × tall	2.93	3.56	+0.63	392
Very tall × short	1.57	1.41	−0.16	58
Medium × tall	1.51	2.01	+0.50	591
Medium × medium	0.03	0.34	+0.31	338
Short × tall	0.03	0.59	+0.56	305
Medium × short	−1.45	−0.38	+1.07	419
Tall × very short	−1.53	−1.28	+0.25	47
Short × short	−2.75	−1.36	+1.39	153
Very short × medium	−3.37	−2.19	+1.18	118
Short × very short	−4.41	−3.01	+1.40	111
Very short × very short	−6.42	−5.33	+1.09	18

Single parent-offspring correlation: $r = 0.52$ (assuming equal variance in both generations).

The increases were greater for children of short parents than of tall, and Davenport was led to believe that some dominant factors may be involved in short stature (as indeed they are in chondrodystrophic dwarfism, for example). Also, variances were greater among children from short parents than among children from exclusively tall parents.

Finally, it should be noted that Davenport showed there had been a highly significant tendency for positive assortative mating: tall × tall, medium × medium, and short × short. Correlation between individual parents and offspring is expected to be 0.50 in a panmictic population under ideal conditions when genotypes are strictly additive and there are no environmental effects. The high value here ($r = 0.52$) is due in part to assortative mating: if parents of similar phenotype share genes for that trait, their offspring will have more genetic agreement than children of random parents.

POLYGENES AND MAJOR GENES

When many duplicate factors (loci), all contributing to a given inherited trait, reinforce each other in a cumulative way, a graded series will be formed. The distribution expected for such a trait is diagrammed in Table 6-4. Provided that each positive, additive, or perhaps multiplicative allele at each locus has the same phenotypic effect, the distribution of

TABLE 6-4 *Genes with cumulative phenotypic effect: distribution of gametes and zygotes in $F_1 \times F_1$ Mendelian progenies.*

Let x = positive phenotypic effect identical for any capital letter allele ($A, B, C, \ldots$), additive above some average value.

y = negative effect for any lower-case letter allele ($a, b, c, \ldots$) with lowering effect below that average value.

One Pair Alleles $Aa \times Aa$ Matings

Gametes $(\frac{1}{2}x + \frac{1}{2}y)$ F_2 zygotes $(\frac{1}{2}x + \frac{1}{2}y)^2$

F_2 phenotypes $2(x):(x)(y):2(y)$

Frequencies $\frac{1}{4}$ $\frac{1}{2}$ $\frac{1}{4}$

Two Pairs Alleles $AaBb \times AaBb$ Matings

If $A = B = x$ and $a = b = y$, then the Ab gamete = aB gamete = xy in additive effect

Gametes $(\frac{1}{4}xx + \frac{2}{4}xy + \frac{1}{4}yy) = (\frac{1}{2}x + \frac{1}{2}y)^2$ F_2 zygotes $[(\frac{1}{2}x + \frac{1}{2}y)^2]^2$ or $(\frac{1}{2}x + \frac{1}{2}y)^4$

F_2 phenotypes $4(x):\frac{3(x)}{1(y)}:\frac{2(x)}{2(y)}:\frac{1(x)}{3(y)}:4(y)$

Frequencies $\frac{1}{16}$ $\frac{4}{16}$ $\frac{6}{16}$ $\frac{4}{16}$ $\frac{1}{16}$

Three Pairs Alleles $AaBbCc \times AaBbCc$ Matings

If $A = B = C = x$ and $a = b = c = y$, then the ABc gamete = AbC = aBC = xxy, and abC = aBc = Abc = xyy in additive effect

Gametes $(\frac{1}{8}xxx + \frac{3}{8}xxy + \frac{3}{8}xyy + \frac{1}{8}yyy) = (\frac{1}{2}x + \frac{1}{2}y)^3$ F_2 zygotes $[(\frac{1}{2}x + \frac{1}{2}y)^3]^2$ or $(\frac{1}{2}x + \frac{1}{2}y)^6$

F_2 phenotypes $6(x):\frac{5(x)}{1(y)}:\frac{4(x)}{2(y)}:\frac{3(x)}{3(y)}:\frac{2(x)}{4(y)}:\frac{1(x)}{5(y)}:6(y)$

Frequencies $\frac{1}{64}$ $\frac{6}{64}$ $\frac{15}{64}$ $\frac{20}{64}$ $\frac{15}{64}$ $\frac{6}{64}$ $\frac{1}{64}$

---

gametic types can be expressed as $(\frac{1}{2}x + \frac{1}{2}y)^n$, where x represents the positive effect allele, y the negative effect (above and below the basic midvalue phenotype halfway between the extremes), and n the number of such loci segregating. Mating of identical genotypes squares the distribution frequencies, and the phenotypic distribution is represented by the familiar binomial of the gametic frequencies squared. It is apparent that with just a few pairs of loci segregating in this way, a nearly normal distribution of phenotypes would occur among the progeny; the completely extreme forms (*all positive* alleles or *all negative* alleles) would then occur at $(1/2)^{2n}$.

This simple cumulative action system probably seldom can be found when continuous variate heredity is analyzed in a population. If we consider the spectrum of genic variation—some traits mostly qualitative, some more quantitative—with allelic differences making large or small phenotypic effects, one locus may be more major in its control than another, so that equality of positive effect is seldom to be found. A degree of dominance will often characterize some alleles more than others, just to make things more complex, while epistatic interactions no doubt distort the simple picture with many traits. Finally, residual variation brought about by numerous small genetic factors plus environmental modification of the phenotype will tend to obliterate any discreteness of categories separating genotypes; consequently, statistical treatment of progeny data to find parameter means and variances becomes the predominant method of analysis. These genetic factors inherited in the Mendelian fashion, usually with similar and cumulative positive effects on the phenotype but individually small in relation to the total variation, were called "polygenes" by Mather (1941), with the assumption that polygenic systems exist in most organisms and make possible the fine adjustment of phenotypic traits to natural selection. Classic illustrations of polygenic traits with additive genic action are those of corolla length inheritance in *Nicotiana* by East (1916) and Hayes (1913), of various quantitative traits in maize by Emerson and East (1913) and Shull (1921), and of traits in rabbits and rats by Castle (1922). Detailed summaries can be found in Mather (1949), Mather and Jinks (1971), Falconer (1960), and Wright (1968, Chapter 15). Among human quantitative traits, fingerprint ridge number (Holt, 1961, 1968) seems to fit the simple additive scheme better than any other trait, while stature and skin color (Harrison and Owen, 1964; Stern, 1970, 1973) display nearly simple additivity with environmental effects and relatively little dominance.

We cannot often say whether a trait is controlled predominantly by numerous polygenes—in a narrow sense, originally conceived to be more or less similar to a simple additive system—or by alleles of major qualitative genes that modify the given trait in a polygenic way. Polygenes may be of prime importance in their own right, as proposed by Mather, but their relationship to traits important to adaptive fitness and the nature of genetic variation controlling adaptive or nonadaptive traits often must be analyzed by the methods based on polygenic assumptions. We shall postpone discussion of the more complex concept of how genotypes are related to fitness until later in this book. It is essential here to derive and discuss expressions from populational models describing polygenic expectations that will have the additional advantage of serving as a basis for quantification of genotypes and phenotypes in general. Many population principles are based on utilization of hypothetical quantity attached to alleles, gametes, genotypes, and phenotypes. Here we describe these principles at an introductory level in regard to randomly mating populations. Further principles will be derived, therefore—especially when nonrandom conditions and selection are applied to the problems of evolution.

In a practical sense, a primary objective for quantification of genotypes is to predict the outcome for breeding improved domestic plant and animal populations. For more general significance, we wish to know the properties of populations and the genetic basis of natural variation, how genetic and nongenetic factors may influence metric characters, whether they are truly polygenic in the narrow sense or are a quantifiable genetic complex affecting an intricate trait such as Darwinian fitness (differential reproduction).

ADDITIVE SYSTEMS: EXPECTED PHENOTYPIC VARIATION

Polygenes may influence quantitative traits via at least three general routes: (1) the frequencies of alleles and genotypes at each locus contributing to the trait, (2) the number of genic loci segregating that contribute to variation in the trait, and (3) the types and strengths of genic action, such as amount of additive or multiplicative effect with each allelic substitution, amount of dominance, and amount of nonhomologous interaction (epistasis in the broad sense) between loci. To describe all these effects simultaneously along with variation from environmental sources would be far too complex. We shall thus derive these principles first from the simplest possible system without dominance. Thence, by introducing dominance, environmental effects, and finally multiple loci, we can express the more general relative influences of genetic and nongenetic factors on the expected distributions of quantitative phenotypes in populations under random mating.

Single Pair of Alleles: Mean and Variance

We proceed with the fundamental principles established by Fisher (1918) concerning quantification of alleles, genotypes, and the components of phenotypic variance. If we limit our attention to a single pair of alleles A_1A_2, the scale of measurement may be represented as in Figure 6-2. The halfway point between the homozygotes represents an exactly intermediate phenotype if dominance is lacking; it is known as the "midparent" position. Each homozygous phenotype then can be considered as deviating from the midparent by an additive value "a" (either $+a$ when substituting A_1 for A_2 from A_1A_2 to A_1A_1 or by $-a$ in the opposite direction). When A_1A_2 exhibits some dominance, its deviation from the midparent is symbolized "d" (either positive or negative depending on whether the "high" or "low" allele displays dominance).

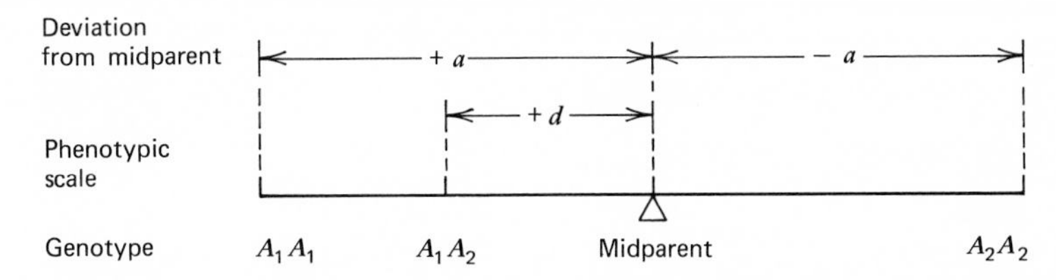

Figure 6-2. Additive scale for quantifying allelic substitutions and phenotypes. The additive effect a is half the distance between homozygotes A_1A_1 and A_2A_2 at the point marked as the midparent position. If the heterozygote A_1A_2 displays dominance, its derivation (d) from the midparent is indicated as a departure from the position expected from exact additivity.

LACK OF DOMINANCE. As an illustration for demonstrating the mean and variance in a randomly mating population, we assume human height to be controlled only by a single pair of alleles with A_1A_1 individuals exactly 72 inches tall, A_1A_2 68 inches, and A_2A_2 64 inches in a hypothetical population. Then the additive effect of the gene A_1 when substituted for A_2, either from shortest genotype to heterozygote or from heterozygote to tallest genotype, would be the same—namely, a 4-inch increment ($a = 4$ in the phenotypic scale, Figure 6-2). Because the heterozygous individuals are at the exact midpoint between homozygotes (midparent), we say there is no dominance, or gene action is simply additive. The arithmetic *mean height* and *variance* in a randomly mating population depend on frequencies of these genotypes, as illustrated in Table 6-5A, cases 1 and 2. Calculation can be simplified either by subtracting a common value (60 inches) from each phenotype, leaving the observed (Y) values as increments above 60 inches or, preferably and more simply, we utilize the scale from Figure 6-2 and give the Y values as deviations from the midparent. Note in case 2 of the table that the mean height will be 70 inches, which is entirely a theoretical parameter because no individual in that population would have that height.

In general, then, the mean ($\bar{Y}$) and variance (σ^2) can be represented as shown in Table 6-5A, cases 3 and 4:

$$\text{Mean without dominance: } \bar{Y} = a(p - q) \tag{6-1}$$

where $\bar{Y}$ is the average deviation from the midparent.

$$\begin{array}{l}\text{Variance in genotypes} \\ \text{(lack of dominance)} \quad : \sigma_G^2 = 2pq\, a^2\end{array} \tag{6-2}$$

where $a =$ half the difference between homozygotes of the phenotypic scale.

PARTIAL DOMINANCE. If we assume the heterozygous phenotype to exhibit some positive dominance, letting A_1A_2 individuals be 70 instead of 68 inches (for example, $d = 2$ in the scale of Figure 6-2) and leaving the homozygous heights as before, then population means and variances would change with gene frequency, as shown in Table 6-5B. At allelic equality ($p = q$), dominance increases both mean and variance compared with the no-dominance condition, since both now have an additional component. However, although the mean is greater when more tall individuals occur (case 2), variance is reduced even though it has more components, because the deviations from the mean are less when the positive allele is more common. A derivation of these fundamental formulas is presented in Table 6-5B.

$$\text{Mean (with dominance): } \bar{Y} = a(p - q) + 2pqd \tag{6-3}$$

where $\bar{Y}$ and a are defined as before and $d =$ deviation of A_1A_2 from midparent.

$$\text{Variance (with dominance): } \sigma_G^2 = 2pq[a + d(q - p)]^2 + 4p^2q^2d^2 \tag{6-4}$$

Note that when $d = 0$, (6-3) and (6-4) reduce to (6-1) and (6-2). Also, if $p = q = \frac{1}{2}$, as expected when crossing two inbred lines, the dominance (d) of the first component is reduced to zero, so that $\sigma_G^2 = \frac{1}{2}a^2 + \frac{1}{4}d^2$. This result will be most useful in estimating these components in the next chapter.

LINEAR VALUES WITH DOMINANCE. It is essential for further understanding of these variance components to derive them in a somewhat different manner. Let us consider what effect an allelic substitution would have if a randomly chosen individual gets

TABLE 6-5 *Mean and genetic variance for quantitative traits.* Y = *phenotypic value,* f = *frequency,* σ_G^2 = *genetic variance*

A Without Dominance (A_1A_2 = midparent)

	Deviation from 60 inches							Deviation from midparent 68 inches							
	Case 1: $p = q = \frac{1}{2}$				*Case 2*: $p = \frac{3}{4}, q = \frac{1}{4}$			*Case 3*: $p = q = \frac{1}{2}$				*Case 4*: General			
Genotype	Y	f	fY	fY^2	f	fY	fY^2	Y	f	fY	fY^2	Y	f	fY	fY^2
A_1A_1	12	$\frac{1}{4}$	3	36	$\frac{9}{16}$	$\frac{27}{4}$	81	+4	$\frac{1}{4}$	1	4	$+a$	p^2	p^2a	p^2a^2
A_1A_2	8	$\frac{1}{2}$	4	32	$\frac{6}{16}$	$\frac{12}{4}$	24	0	$\frac{1}{2}$	0	0	0	$2pq$	0	0
A_2A_2	4	$\frac{1}{4}$	1	4	$\frac{1}{16}$	$\frac{1}{4}$	1	−4	$\frac{1}{4}$	−1	4	$-a$	q^2	$-q^2a$	q^2a^2
$\sum$		1	8	72	1	10	106		1	0	8		1	$a(p-q)$	$p^2a^2 + q^2a^2$

$$\sum f = N, \quad \frac{\sum fY}{N} = \bar{Y}; \quad \sigma^2 = \frac{\sum fY^2 - N\bar{Y}^2}{N}$$

Therefore, (Mean)	(1) $\bar{Y} = 8$ (above 60)	(2) 10 (above 60)	(3) 0 (above 68)	(4) $a(p-q)$ (above midparent)
Variance	$\sigma_G^2 = 8$	$\sigma_G^2 = 6$	$\sigma_G^2 = 8$	$\sigma^2 = 2pqa^2$

General

$$\bar{Y} = p^2a - q^2a = a(p+q)(p-q) = \boxed{a(p-q)}$$

$$\sigma_G^2 = p^2a^2 + q^2a^2 - [a(p-q)]^2 = \boxed{2pqa^2}$$

TABLE 6-5 (*Continued*)

B With Dominance ($a = 4, d = 2$)

	Deviation from 60 inches							Deviation from midparent 68 inches							
	Case 1: $p = q = \frac{1}{2}$				*Case 2*: $p = \frac{3}{4}, q = \frac{1}{4}$			*Case 3*: $p = q = \frac{1}{2}$				*Case 4*: General			
Genotype	Y	f	fY	fY^2	f	fY	fY^2	Y	f	fY	fY^2	Y	f	fY	fY^2
A_1A_1	12	$\frac{1}{4}$	3	36	$\frac{9}{16}$	$\frac{27}{4}$	81	+4	$\frac{1}{4}$	1	4	$+a$	p^2	p^2a	p^2a^2
A_1A_2	10	$\frac{1}{2}$	5	50	$\frac{6}{16}$	$\frac{15}{4}$	$\frac{75}{2}$	+2	$\frac{1}{2}$	1	2	d	$2pq$	$2pqd$	$2pqd^2$
A_2A_2	4	$\frac{1}{4}$	1	4	$\frac{1}{16}$	$\frac{1}{4}$	1	−4	$\frac{1}{4}$	−1	4	$-a$	q^2	$-q^2a$	q^2a^2
$\sum$		1	9	90	1	$\frac{43}{4}$	$\frac{239}{2}$		1	1	10		1	$\bar{Y}$	$\sum fY^2$
Therefore, Mean	(1) $\bar{Y} = 9$ (above 60) or 1 (above midparent)				(2) $10\frac{3}{4}$ (above 60) or $2\frac{3}{4}$ (above midparent)			(3) 1 (above midparent)				(4) $a(p - q) + 2pqd$			
Variance	$\sigma_G^2 = 9$				$\sigma_G^2 = 3\frac{15}{16}$			$\sigma_G^2 = 9$				$\sigma_G^2 = 2pq\{[a + d(q - p]^2 + 2pqd^2\}$			

General

$$\bar{Y} = p^2a + 2pqd - q^2a = \boxed{a(p - q) + 2pqd}$$

$$\begin{aligned}\sigma_G^2 &= p^2a^2 + 2pqd^2 + q^2a^2 - [a(p - q) + 2pqd]^2 \\ &= 2pq[d^2 + a^2 - 2adp + 2adq - 2pqd^2] \\ &= 2pq[a^2 + 2ad(q - p) + d^2(q - p)^2 + 2pqd^2] \\ &= \overbrace{2pq[a + d(q - p)]^2}^{\text{“additive”}} + \overbrace{4p^2q^2d^2}^{\text{“dominant” components}} \\ \sigma_G^2 &= \sigma_A^2 + \sigma_D^2\end{aligned}$$

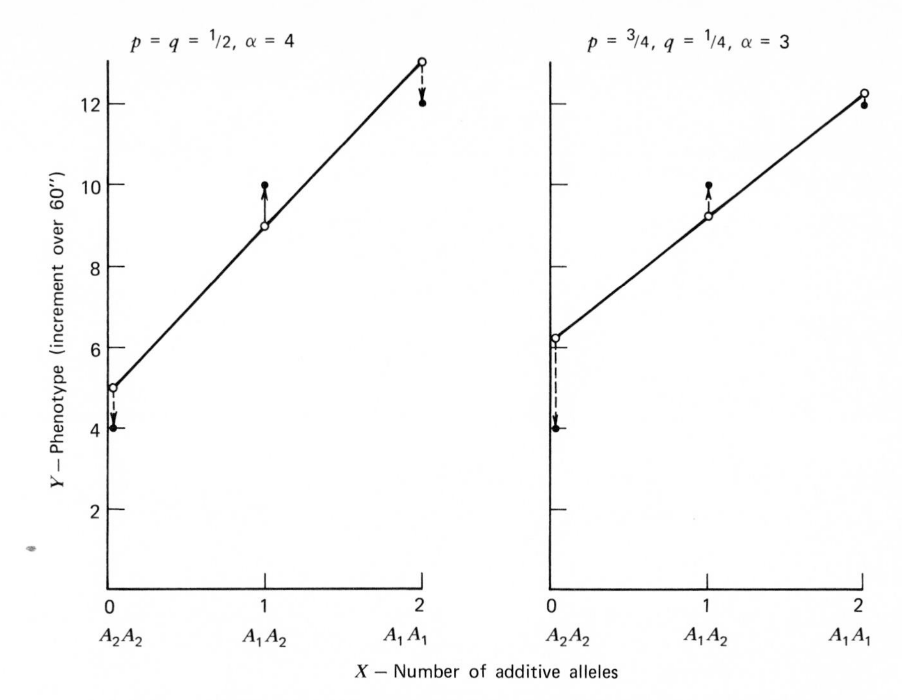

Figure 6-3. Linear values of genotypes with dominance from Table 6-5B, cases 1 and 2. Note the change in slope of the regression with change in allelic frequencies. Linear values = ○, constant phenotypic values = ●.

an A_1 (positive allele) replacing an A_2 allele. What *average* change in phenotype would be expected by doing that? If we are not cognizant of which genotype is receiving the substituted allele, we must consider the situation as a hypothetical one—that is, as if any substitution were equal to any other (like the lack of dominance case), irrespective of whether the change goes from A_2A_2 to A_1A_2 or from A_1A_2 to A_1A_1. The proper "average" additive effect then would amount to the best "fit," or regression slope, of phenotypic values on genotypes for any set of frequencies characterizing the population. Figure 6-3 illustrates regression lines for cases 1 and 2 in Table 6-5B. The slopes of these regression lines can be seen to depend on the *frequencies* of phenotypes in each population; the slope is usually symbolized by b in statistical treatments (see Appendix A-6G), but we shall denote this slope by α, or linear increment.

We can think of α (average phenotypic increment per allelic substitution) as analogous to a in the lack of dominance case; α is dependent on allelic frequencies and on d, while a is a constant value determined by the difference between homozygotes on the phenotypic scale. The analogy is made clearer by completing the algebra for the regression:

Let X = number of positive alleles in each genotype (0, 1, 2)

Y = phenotypic value of each genotype (Y_0, Y_1, Y_2 corresponding to A_2A_2, A_1A_2, A_1A_1)

Then the slope of the regression $\alpha = \dfrac{\sigma_{xy}}{\sigma_x^2} = \dfrac{\sum fXY - \bar{X}\cdot\bar{Y}}{\sum fX^2 - \bar{X}^2}$ (see Appendix A-6G)

With dominance,

Genotype	f	X	Y	fXY
A_1A_1	p^2	2	$+a$	$2p^2a$
A_1A_2	$2pq$	1	d	$2pqd$
A_2A_2	q^2	0	$-a$	0

Then the numerator of the slope (covariance) $\sigma_{XY} = 2p^2a + 2pqd - 2p[a(p-q) + 2pqd]$, where $\bar{X} = 2p$, and $\bar{Y}$ is given from (6-3), which simplifies to $\sigma_{XY} = 2pq[a + d(q-p)]$. Then, since $\sigma_X^2 = 2pq$,

$$\alpha = a + d(q - p) \tag{6-5}$$

which is the bracketed component of (6-4). Therefore, (6-4) can be written as

$$\sigma_G^2 = 2pq\alpha^2 + 4p^2q^2d^2 \tag{6-6}$$

(genetic variance including additive and dominant components). Thus, the genetic variance can be seen as composed of two parts, the additive portion, analogous to (6-2), or "linear component," plus the dominant portion. Using the data from case 2 in Table 6-5B, it is apparent that the total genetic variance can be computed as follows: the slope $\alpha = 4 + 2(-\frac{2}{4}) = 3$ and the genetic variance $\sigma_G^2 = \frac{6}{16}(3)^2 + 4(\frac{9}{16})(\frac{1}{16})(2)^2 = 3\frac{15}{16}$.

The linear (fitted) values (Y_L^0, Y_L^1, Y_L^2) used in Figure 6-3 can be calculated from the linear regression equation [see Appendix A-6G(16)]:

$$Y - \bar{Y} = \alpha(X - \bar{X}), \text{ letting } X = 0 \text{ to obtain the lowest } Y_L^0$$
$$Y_L^0 = \alpha(0 - \bar{X}) + \bar{Y}, \text{ supplying (6-5)}$$
$$= [a + d(q-p)](-2p) + a(p-q) + 2pqd = 2p^2d - pa - qa.$$

Therefore,

$$Y_L^0 = 2p^2d - a \tag{6-7}$$

(lowest linear value as a deviation from the midparent*). Thus, in Figure 6-3 left, the Y_L^0 becomes -3, as a deviation from the midparent, or $+5$ in the scale of that figure. On the right, linear Y values and deviations from them would be as follows:

	f	X	Y	Y_L	D	fY_L	fD
A_1A_1	$\frac{9}{16}$	2	12	$12\frac{1}{4}$	$-\frac{1}{4}$	$\frac{441}{64}$	$-\frac{9}{64}$
A_1A_2	$\frac{6}{16}$	1	10	$9\frac{1}{4}$	$+\frac{3}{4}$	$\frac{222}{64}$	$+\frac{18}{64}$
A_2A_2	$\frac{1}{16}$	0	4	$6\frac{1}{4}$	$-2\frac{1}{4}$	$\frac{25}{64}$	$-\frac{9}{64}$
					$\sum =$	$10\frac{3}{4}$	0

* Linear values may also be expressed as deviations from the population mean (see Exercise 10):

$$Y_L^0 = -2\alpha p,\ Y_L^1 = (q-p)\alpha,\ Y_L^2 = 2q\alpha \tag{6-7A}$$

From (6-7), $Y_L^0 = [2(\frac{3}{4})]^2 - 4 = -1\frac{3}{4}$ (below midparent) $= 6\frac{1}{4}$ (when midparent $= 8$). Then each subsequent Y_L value is increased by 1α and 2α. Note that the mean linear value is equal to the mean phenotypic value ($10\frac{3}{4}$). D, or dominance deviations ($Y_i - Y_L^i$), add up to zero, and they represent the vertical distances between dots in Figure 6-3. It is easy to verify that the variance of these linear values (σ_A^2) $= 3\frac{3}{8}$, while the variance of dominance deviations (σ_D^2) $= \frac{9}{16}$. These data illustrate a fundamental principle of quantitative population genetics—that total genetic variance can be partitioned into additive and dominant components:

$$\sigma_G^2 = \sigma_A^2 + \sigma_D^2 \tag{6-8}$$

where $\sigma_A^2 = 2pq\alpha^2$ and $\sigma_D^2 = 4p^2q^2d^2$ (6-6).

That proportion of total variance due to additive (linear) values is known as the heritability (h^2):

$$h^2 = \frac{\sigma_A^2}{\sigma_P^2} \tag{6-8A}$$

where σ_P^2 = total phenotypic variance.

In this case, where we are not considering environmental variation, total variance is entirely genetic, but the fraction due to additivity (heritability) is

$$h^2 = \frac{3\frac{3}{8}}{3\frac{15}{16}} = \frac{6}{7} = 0.857$$

The next chapter will consider more fully the meaning and utility of the heritability concept.*

A SHORT METHOD FOR VARIANCE COMPONENTS. If we are not concerned with estimating the linear values (Y_L) but only the variance components, a simplification can be used, called by Li (1961) "the method of successive differences," derived from the general regression equation. Let us simply use the phenotypic values (Y_i) instead of converting them into additive effects (a) or dominance deviations (d) to calculate the linear slope (α), where successive values of gene dosage (X) values are 2, 1, and 0 for Y_2, Y_1, and Y_0, respectively:

$$\alpha = \frac{\sum fXY - \bar{X}\bar{Y}}{\sum fX^2 - \bar{X}^2} = \frac{2p^2Y_2 + 2pqY_1 - 2p\bar{Y}}{2pq}$$

$$= \frac{pY_2 + qY_1 - p^2Y_2 - 2pqY_1 - q^2Y_0}{q}$$

which simplifies to

$$\alpha = p(Y_2 - Y_1) + q(Y_1 - Y_0) \tag{6-9}$$

or weighted mean of two differences between successive phenotypes. The student should verify that if $+a$, d, and $-a$ are the Y_2, Y_1, and Y_0 values, respectively, (6-5) will result. In case 2 of Table 6-5B, $\alpha = \frac{3}{4}(2) + \frac{1}{4}(6) = 3$, using (6-9), which is simpler than (6-5) because a and d do not need to be estimated.

* Often in the literature the additive proportion of total phenotypic variance is known as "heritability in the narrow sense," while the proportion of genetic variance (σ_G^2/σ_P^2) is called "heritability in the broad sense." In this text we are concerned only with the former heritability in general; for the latter we prefer to use a term from Falconer (1965): "degree of genetic determination."

Again, the regression equation can be applied; this time the differences from (6-9) are the variables to be considered.

f	X	Y	fXY
p	1	$(Y_2 - Y_1)$	$p(Y_2 - Y_1)$
q	0	$(Y_1 - Y_0)$	0

The slope of the differences (β) then can be determined.

$$\beta = \frac{p(Y_2 - Y_1) - p\alpha}{pq}$$

since $\bar{X} = p$ and $\bar{Y} = \alpha$, given in (6-9) and $\sigma_X^2 = p - p^2 = pq$.

$$\beta = \frac{p(Y_2 - Y_1) - p^2(Y_2 - Y_1) - pq(Y_1 - Y_0)}{pq}$$

which simplifies to

$$\beta = Y_2 - 2Y_1 + Y_0 \tag{6-10}$$

The student should verify that $\beta = a - 2d - a = -2d$ when additive effects and dominance deviations are considered. Note that the slope is negative when the dominant allele gives a positive increment to the phenotype. In case 2 of Table 6-5B, $\beta = 12 - 20 + 4 = -4$.

Li (1961) showed that the total variance can then be represented as follows (compare with 6-6):

$$\sigma_G^2 = 2pq\alpha^2 + p^2q^2\beta^2 \tag{6-11}$$

Using the same example, we have

$$\sigma_G^2 = \tfrac{6}{16}(3)^2 + (\tfrac{3}{16})^2(-4)^2 = 3\tfrac{3}{8} + \tfrac{9}{16} = 3\tfrac{15}{16}, \qquad \text{as given above.}$$

COMPLETE DOMINANCE AND OVERDOMINANCE. With increasing dominance, the $(Y_2 - Y_1)$ difference approaches zero if the positive allele is the dominant one, so that α approaches $p(0) + q(Y_1 - Y_0)$ and β approaches $-(Y_1 - Y_0)$. If we symbolize the $(Y_1 - Y_0) = r$ (recessive effect), then with complete dominance, the linear (additive) component of genetic variance is (using 6-9 and 6-6) $\sigma_A^2 = 2pq^3r^2$; using (6-11), the dominant component is $\sigma_D^2 = p^2q^2r^2$. Relative to each other, the variance components would be $\sigma_A^2 : \sigma_D^2 = 2q : p$; they are dependent on the frequencies of alleles and completely independent of the phenotypic values. As the dominant positive allele increases in a population, the dominant component of variance would become more prominent while high frequency of the recessive would bring the reverse (see Figure 6-4).

Heterosis was described and defined by G. H. Shull in 1908 (see Shull, 1948) as the increased vigor of F_1 hybrids over their inbred parents, and it was visualized genetically as arising from the specific interaction of unlike alleles at "heterotic" loci (others, notably D. F. Jones, attributed the increased vigor to dominance of vigor alleles). It has remained technically difficult to prove for any specific locus the greater expression or magnitude of a single-locus heterozygote (A_1A_2) than the homozygote (A_1A_1), and thus establish

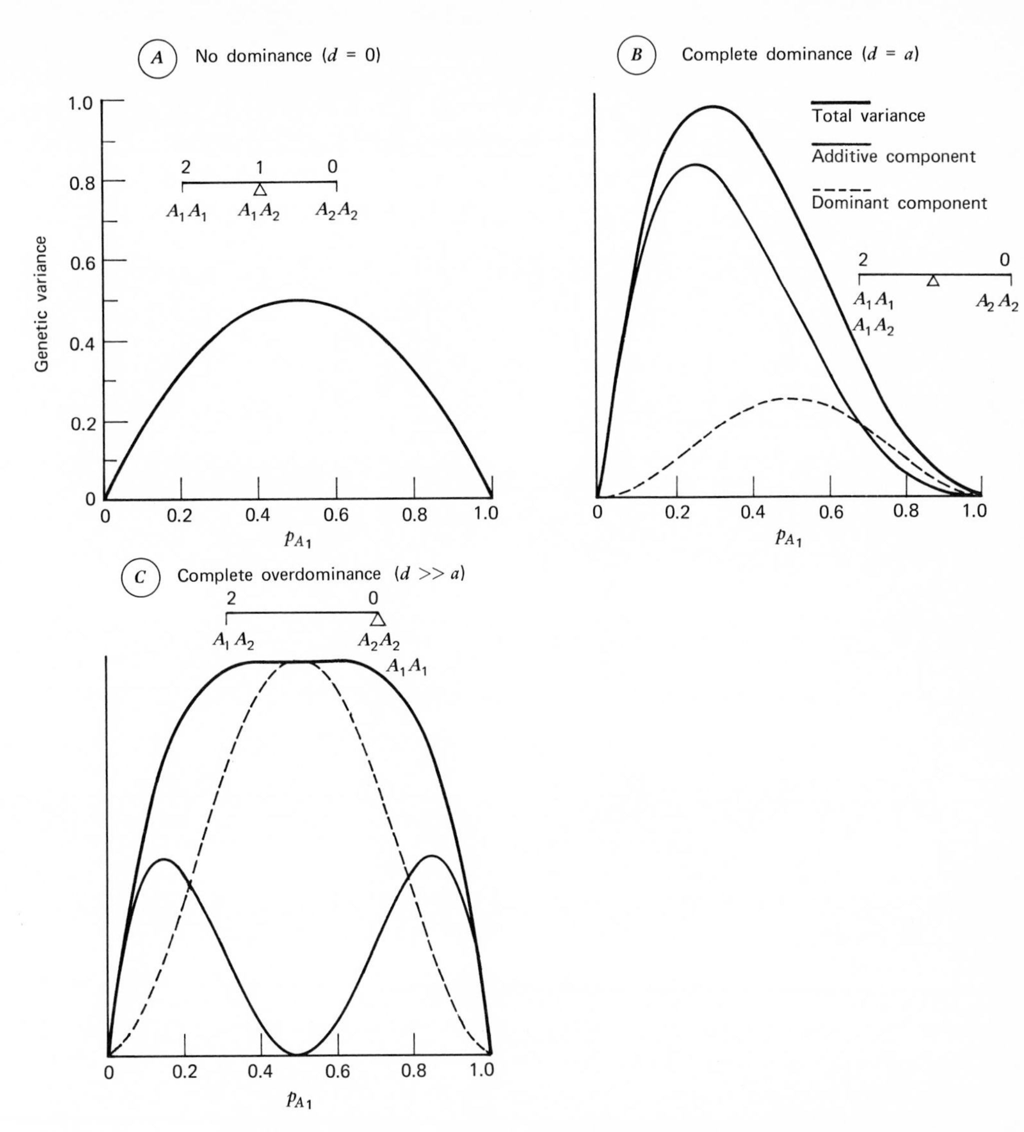

Figure 6-4. Components of genetic variance under different amounts of dominance at a single locus with two alleles. A = no dominance, B = complete dominance, C = complete overdominance and no difference between homozygotes. Total σ_G^2 = heavy solid line, additive component σ_A^2 = thin line, dominance component σ_D^2 = dashed line.

veritable cases of "overdominance," as single-locus heterosis was called. It is highly likely that overdominance in the narrow sense exists, and we recognize that one of the important mechanisms maintaining balanced polymorphisms in populations must be superiority of genically heterozygous combinations (see Chapter 15). Overdominance in the broad sense, as applied to large chromosome blocks such as inversions, or "supergenes" (in drosophila populations, for example), must be considered in any generalized discussion of polygenic theory. Consequently, if we assume the heterozygote surpasses the greater homozygote (or is less than the smaller, in negative overdominance), so that $d > a$ on the phenotypic scale, the linear component of genetic variance becomes smaller and may disappear altogether ($\sigma_A^2 = 0$) under the following conditions: when the differences $(Y_2 - Y_1)$ and $(Y_1 - Y_0)$ are of opposite sign and in proportions $q:p$, then the value of (6-9) becomes zero and the genetic variance is entirely due to dominance. If $\alpha = pq + q(-p)$, or if $p(-q) + pq$, then the additive component is zero. For example, if $p = \frac{3}{4}$, $q = \frac{1}{4}$, and $A_1A_1 = 10$, $A_1A_2 = 13$, $A_2A_2 = 4$, then $\alpha = \frac{3}{4}(-3) + \frac{1}{4}(9) = 0$, and $\sigma_A^2 = 0$, so that $\sigma_G^2 = \sigma_D^2 = 5\frac{1}{16}$. Here the linear heritability would be zero. These relationships are similar to those in Figure 6-4C, in which both homozygotes are illustrated as identical but the heterozygote is superior to both, and the linear component is zero.

MULTIPLE ALLELES. A more general method for representing the additive and dominant effects of each allele at a locus follows from a consideration of the "general" (additive) and "special" (nonadditive) combining abilities of each locus as in a random block design with frequencies in each block representing the relative "pool" for each allele (Falconer, 1960).

If phenotypes can be arranged as follows in a random block design:

	C_1	C_2	C_3	$\cdots$
R_1	Y_{11}	Y_{12}	Y_{13}	$\cdots$
R_2	Y_{21}	Y_{22}	Y_{23}	$\cdots$
R_3	Y_{31}	Y_{32}	Y_{33}	$\cdots$
$\vdots$	$\vdots$	$\vdots$	$\vdots$	

with heterozygotes $Y_{12} = Y_{21}$ in phenotype and in frequency $\frac{1}{2}[(f_{12}) + (f_{21})]$ so that each block of a heterozygote would have its frequency $= q_iq_j$ and each homozygote $= q_{ii}^2$, then each block's value could be represented as $Y_{ij} = \mu + g_i + g_j + s_{ij} + e$, as determined by additive "column and row" effects (g_i and g_j would be deviations of row means or column means from the population mean—that is, $g_i = \bar{R}_i - \mu$ and $g_j = \bar{C}_i - \mu$ are called "general combining abilities") plus special nonadditive interactions (s_{ij}) plus error of sampling (e). Each allele substituted can then be conceived as bringing about a deviation from the population mean (μ) (instead of using the concept of linear regression with increasing dosage in the genotype, which gets to be unworkable with more than a single pair of alleles). The data of case 2 in Table 6-5B can be represented so that the average effect of each allele (average row effect $\bar{R}$) can be estimated from a row × column table, where the number of rows (columns) is the allelic number (k) for the genic locus (in this example, $k = 2$, but the extension to more than two alleles will be apparent).

Phenotype due to Allele	Y_{A_1}	Y_{A_2}	Allelic Mean Effect	p,q	$\bar{R}$
Y_{A_1}	12	10	$\bar{R}_{A_1} = [\frac{27}{4} + \frac{15}{8}]$	$\frac{3}{4}$	$= 11.50$
Frequency	$\frac{9}{16}$	$\frac{3}{16}$			
Y_{A_2}	10	4	$\bar{R}_{A_2} = [\frac{15}{8} + \frac{1}{4}]$	$\frac{1}{4}$	$= 8.50$
Frequency	$\frac{3}{16}$	$\frac{1}{16}$			

The population mean (μ) = 10.75. The deviation due to A_1 is $\bar{R}_{A_1} - \mu = \alpha_1 = 0.75$ (see Falconer, 1960, pp. 119 ff.). The deviation due to A_2 is $\bar{R}_{A_2} - \mu = \alpha_2 = -2.25$, where α_1 and α_2 can be considered "general combining abilities," or g_i for the separate alleles.

The average effect of allelic substitution is the difference between these deviations due to each allele—namely, $\bar{R}_{A_1} - \bar{R}_{A_2} = \alpha_1 - \alpha_2 = 3.00 = \alpha$. Note that the weighted mean of the allelic deviation is zero, as it should be: $p\alpha_1 + q\alpha_2 = \frac{3}{4}(0.75) + \frac{1}{4}(-2.25) = 0$.

The fitted ("linear") values due to additive allelic effects then are as follows from the random block design:

$$
\begin{aligned}
& & & Y_L \\
A_1A_1&: \mu + \alpha_1 + \alpha_1 = 10.75 + (0.75)2 &=& \ 12.25 \\
A_1A_2&: \mu + \alpha_1 + \alpha_2 = 10.75 + 0.75 - 2.25 &=& \ 9.25 \\
A_2A_2&: \mu + \alpha_2 + \alpha_2 = 10.75 + (-2.25)2 &=& \ 6.25
\end{aligned}
$$

Because the deviations for each allele are now given, the additive component of variance can easily be computed as follows for two alleles: $\sigma_A^2 = p^2(4\alpha_1^2) + 2pq(\alpha_1 + \alpha_2)^2 + q^2(4\alpha_2^2)$, which simplifies to

$$\sigma_A^2 = 2(p\alpha_1^2 + q\alpha_2^2) \tag{6-12}$$

because $p\alpha_1 + q\alpha_2 = 0$. In the cited example, $\sigma_A^2 = 2(\frac{3}{4})(0.75)^2 + 2(\frac{1}{4})(-2.25)^2 = 3\frac{3}{8}$, as before.

With more than two alleles, the additive effect of each allele can be conceived in the same way in a $k \times k$ block design (k = number of alleles at the locus in the population). Each allelic mean (row effect $\bar{R}_i$) will have its deviation from the population mean (α_i) so that the total additive component of variance will be

$$\sigma_A^2 = 2\sum q_i\alpha_i^2 \tag{6-13}$$

which is an extension of (6-12) to k alleles. Dominance deviations from the fitted values ($Y_i - Y_L$) when squared and summed would then be the dominant component of variance (see Exercise 17).

ENVIRONMENTAL EFFECT. Nearly always, nongenetic factors influence quantitative phenotypes to such an extent that any gaps between phenotypes that might have existed under completely genetic control in a strictly uniform environment (as we have been assuming above), tend to become obliterated, and a smooth distribution results. Total phenotypic variance will them comprise genetic and environmental components.

By making simplifications for illustration, we may consider a phenotype such as stature controlled by a single pair of alleles without dominance, as in case 1 in Table 6-5A, with $p = q = \frac{1}{2}$. We further assume individuals are distributed among three environments (E_1, E_2, E_3) in relative frequencies of $\frac{1}{4}$, $\frac{2}{4}$, $\frac{1}{4}$. If all genotypes are equally expressed under these environmental influences (if there are no genotype-environmental interactions), so that E_1 (beneficial) tends to add two units to the phenotype, E_2 (average) adds zero units, while E_3 (detrimental) tends to diminish the stature by two units, then we would have an array as in Table 6-6.

When the environmental influences are defined as deviations of each E group from the mean, the sum of those squared deviations is the environmental component (σ_E^2), as in-

TABLE 6-6 *Environmental factor influence on phenotype (single pair alleles without dominance, $p = q = \frac{1}{2}$). No gene-environment interactions. Distribution over environments in ratio of $\frac{1}{4}$ in E_1, $\frac{2}{4}$ in E_2, and $\frac{1}{4}$ in E_3 (mean phenotypes = 8, d = deviation $(Y - \bar{Y})$, σ_P^2 = total phenotypic variance)*

Genotype	Environment	Phenotype	d	f	fd	fd^2
A_1A_1	E_1	14	6	$\frac{1}{16}$	$\frac{3}{8}$	$\frac{9}{4}$
	E_2	12	4	$\frac{2}{16}$	$\frac{4}{8}$	$\frac{8}{4}$
	E_3	10	2	$\frac{1}{16}$	$\frac{1}{8}$	$\frac{1}{4}$
A_1A_2	E_1	10	2	$\frac{2}{16}$	$\frac{2}{8}$	$\frac{2}{4}$
	E_2	8	0	$\frac{4}{16}$	0	0
	E_3	6	-2	$\frac{2}{16}$	$-\frac{2}{8}$	$\frac{2}{4}$
A_2A_2	E_1	6	-2	$\frac{1}{16}$	$-\frac{1}{8}$	$\frac{1}{4}$
	E_2	4	-4	$\frac{2}{16}$	$-\frac{4}{8}$	$\frac{8}{4}$
	E_3	2	-6	$\frac{1}{16}$	$-\frac{3}{8}$	$\frac{9}{4}$
$\sigma_P^2 = 10$, where $\sigma_A^2 = 8$ and $\sigma_E^2 = 2$			Sum	1	0	10

Environment Component

	Effect	f	fe	fe^2
E_1	$+2$	$\frac{1}{4}$	$\frac{1}{2}$	1
E_2	0	$\frac{2}{4}$	0	0
E_3	-2	$\frac{1}{4}$	$-\frac{1}{2}$	1
Sum		1	0	2

e = effect of environment defined as deviation from mean.

dicated at the bottom of the table. Heritability is then 0.80 of case 1 where there were no environmental effects:

$$h^2 = \frac{\sigma_A^2}{\sigma_A^2 + \sigma_E^2} = \frac{\sigma_A^2}{\sigma_P^2}.$$

It is obvious that if only the phenotypic array were given, there would be a smooth distribution symmetrical around the population mean, because an A_1A_1 in the detrimental environment equals an A_1A_2 in the beneficial environment. Unless either the genotypes or environments can be ascertained completely, it would be difficult to separate the two components. In experimental populations or in the field, we usually use special designs to separate the environmental influence; real populations, unfortunately, are not simple, and genotypes are often not independent of environment in their expression so that the problem of separation of the two variance components often becomes intractable.

Two or More Loci

The polygenic concept includes a consideration of many "duplicate" sets of genic variants with cumulative or reinforcing action on the phenotype. We have now examined a single pair of alleles with quantitative effect from a fundamental viewpoint, and we have described population means and variances under varying amounts of dominance and environmental influences under the assumption of random mating. Let us finally allow for the *polygenic* nature of the system—what happens to means and variances when two or more loci act more or less in duplicate fashion as conceived by quantifying alleles and phenotypes.

With the simplest polygenic system consisting of two loci, each locus contributing equally, it can be demonstrated algebraically that the mean and variance of the population consist of the sums of the two separate locus effects. It is then easy to realize that all the more complex cases involving dominance, differing p,q frequencies, etc., must be extensions of this simple case. The total effect will be the sum of all individual locus effects, provided there is no linkage, no special interaction between loci (epistasis), no complex organization, and no environmental interaction. In Table 6-7 the algebraic solution is given for mean and variance of the two-locus case without dominance (as if both loci were acting additively together as in case 1 in Table 6-5A). While the algebra becomes complex, the same type of solution can be made for the more general case with dominance so that we may give

$$\text{Mean: } \bar{Y} = \sum_{i}^{k} [a_i(p_i - q_i) + 2p_i q_i d_i] \quad (6\text{-}14)$$

$$\text{Genetic variance: } \sigma_G^2 = \sum_{i}^{k} [2p_i q_i \alpha_i^2 + 4p_i^2 q_i^2 d_i^2] \quad (6\text{-}15)$$

where the subscript i refers to the parameters for each locus for k loci affecting the phenotype—that is, the sum of (6-3) and (6-6) for all loci.

TABLE 6-7 *Quantitative inheritance of two genic loci contributing equally to the phenotype. Gene action is duplicate additive without dominance:* $A_1 = B_1 = +a$, $A_2 = B_2 = -a$ *where* $+a$, $-a$ *are deviations from midparent. Frequency* $A_1 = p$, $A_2 = q$, $B_1 = u$, $B_2 = v$

Genotype	Phenotype Y	f	fY	fY^2
$A_1A_1B_1B_1$	$+2a$	p^2u^2	$2ap^2u^2$	$4a^2p^2u^2$
$A_1A_2B_1B_1$, $A_1A_1B_1B_2$	$+1a$	$2pqu^2 + 2p^2uv$	$2a(pu)(qu + pv)$	$2a^2(pu)(qu + pv)$
$A_1A_2B_1B_2$, $A_2A_2B_1B_1$, $A_1A_1B_2B_2$	0	$4pquv + q^2u^2 + p^2v^2$	0	0
$A_2A_2B_1B_2$, $A_1A_2B_2B_2$	$-1a$	$2q^2uv, 2pqv^2$	$-2a(qv)(qu + pv)$	$2a^2(qv)(qu + pv)$
$A_2A_2B_2B_2$	$-2a$	q^2v^2	$-2aq^2v^2$	$4a^2q^2v^2$

$$\text{Mean} = \sum fY = 2a[p^2u^2 + (pu - qv)(qu + pv) - q^2v^2]$$
$$= 2a(pu - qv)(pu + qv + pv + qu)$$
$$\bar{Y} = 2a(pu - qv)$$
$\bar{Y} = a(p - q) + a(u - v)$, which is the sum of (1) for each locus.

Variance $= \sum fY^2 - \bar{Y}^2$. Noting that $(\bar{Y})^2 = 4a^2(pu - qv)^2$, when subtracted from the last column above, reduces to the genetic variance as components of two loci.

$$\sigma_G^2 = 2a^2(pq + uv)$$
$$= 2pqa^2 + 2uva^2$$

TABLE 6-8 *Example of two quantitative loci acting additively on the phenotype. A_1A_2 is from Table 6-5B. Y = deviations from midparent. Fractions are frequencies.*

Genotype	Y	Freq.	B_1B_1	B_1B_2	B_2B_2
			3	1	−3
			$\frac{1}{4}$	$\frac{1}{2}$	$\frac{1}{4}$
A_1A_1	4	$\frac{1}{4}$	7 ($\frac{1}{16}$)	5 ($\frac{1}{8}$)	1 ($\frac{1}{16}$)
A_1A_2	2	$\frac{1}{2}$	5 ($\frac{1}{8}$)	3 ($\frac{1}{4}$)	−1 ($\frac{1}{8}$)
A_2A_2	−4	$\frac{1}{4}$	−1 ($\frac{1}{16}$)	−3 ($\frac{1}{8}$)	−7 ($\frac{1}{16}$)

or Combining Phenotypes

Y	f	fY	fY^2
7	$\frac{1}{16}$	$\frac{7}{16}$	$\frac{49}{16}$
5	$\frac{1}{4}$	$\frac{20}{16}$	$\frac{100}{16}$
3	$\frac{1}{4}$	$\frac{12}{16}$	$\frac{36}{16}$
1	$\frac{1}{16}$	$\frac{1}{16}$	$\frac{1}{16}$
−1	$\frac{3}{16}$	$-\frac{3}{16}$	$\frac{3}{16}$
−3	$\frac{1}{8}$	$-\frac{6}{16}$	$\frac{18}{16}$
−7	$\frac{1}{16}$	$-\frac{7}{16}$	$\frac{49}{16}$
Sum	1	$\frac{24}{16}$	$\frac{256}{16}$
		$= 1\frac{1}{2}$	16

Population mean $\bar{Y} = 1\frac{1}{2}$ (above midparent)
Genetic variance $\sigma_G^2 = 16 - (\frac{3}{2})^2 = 16 - \frac{9}{4} = 13\frac{3}{4}$

Considering each locus separately

Locus	a	d	α	β	Mean (above midparent)	Total Variance	Additive Component	Dominance Component
A-a	4	2	4	−4	1	9	8	1
B-b	3	1	3	−2	$\frac{1}{2}$	$4\frac{3}{4}$	$4\frac{1}{2}$	$\frac{1}{4}$
Both loci together					$1\frac{1}{2}$	$13\frac{3}{4}$	$12\frac{1}{2}$	$1\frac{1}{4}$

If the midparent is 68 inches tall, then the population mean is $69\frac{3}{4}$ inches with $\sigma_G^2 = 13\frac{3}{4}$ and heritability = 12.5/13.75 = 0.91.

The numerical example in Table 6-8 uses A_1-A_2 from case 1 in Table 6-5B plus genic locus B_1-B_2. For simplicity of illustration, both loci have alleles at the same relative frequencies ($p = q = \frac{1}{2}$) and only vary in their additive and dominant effects, as shown in the table. It is evident that total phenotypic variance has components due to each locus; these in turn can be subdivided into respective additive and dominant portions. Finally, we can think of the proportion of additive components for both loci constituting the heritability in total of 0.91.

IMPLICATIONS AND LIMITATIONS

It is important to consider some of the implications and limitations of these concepts before proceeding further to estimate polygenic parameters. First, recalling how the additive component of variance has been derived as outlined in this chapter, the student should realize that a significant additive component (heritability) does not imply additivity of genic action in development. On the contrary, considerable dominance, overdominance, or complex nonadditive genic action can easily be the case at the phenotypic level. Additivity as a population parameter depends on average comparisons of homozygotes at opposite ends of the phenotypic scale and heterozygotes relative to them. When average effects of single-allele substitutions are estimated or linear ("breeding") values of genotypes are proposed, they must be considered hypothetical, necessary for the sake of selection and breeding predictions, as will be discussed briefly in the following chapters. Only when non-additive complexities are lacking (zero dominance and epistasis) can we conclude the genic action in a developmental sense (cumulative codominance) to be additive.

Second, with a single locus or two, distributions of genotypes will be discontinuous and often skewed when allelic frequencies are far from equality. The student should note the increase in continuity (diminished qualitative differences) and in normality of distribution as the number of contributing loci with smaller individual effects are found to control a quantitative trait. Even when gene frequencies are far from equal, if each locus's effect is individually highly skewed, cumulative action of many loci tends to normalize the total distribution. For example, if 10 pairs of independent alleles all display complete dominance (*A-a, B-b, . . . , J-j*) so that the phenotypic distribution would be represented as the expansion of $(\frac{3}{4}A + \frac{1}{4}aa)^{10}$, the resulting array could not easily be distinguished from a normal one in practice.

Third, when two or more loci contribute to the phenotypic value, each locus may have the cumulative effect as postulated in the simplest model with more or less independence of genic action, but many polygenic phenomena arise that can be described as "modifier," "suppressor," "inhibitor," or, in general, "epistatic interaction." For two loci, these interactions might be detectable between their linear values, between the dominance deviation effects, or between the linear effect of one locus and dominance of the other. Of course, for more than two loci, interactions between main effects and various combinations of dominant and additive values produce a high-order interaction complexity between linked and unlinked loci. Much of the distinction between simple additive systems and highly complex ones with interacting components can be discerned from the results of selection experiments, as shown in a later chapter. When estimates of genetic variance appear to indicate some heritability but little or no response to selection occurs, nonadditive interactions are suspected. Considerable statistical sophistication and design have been proposed, especially by Kempthorne (1955), Cockerham (1954), and Hayman (1958), using such techniques as diallele crossing to measure components of genetic variance attributable to epistatic interaction. For many traits (those not affecting fitness to any degree), these interactions are often negligible.

The special nonadditive effect of heterosis may be considerable (overdominance) and may be modified via epistatic interactions for traits that do affect fitness and survival. Instead of extending this section to discuss those special cases, we shall discuss them further when we meet them. Suffice it to summarize that the phenotypic variance can be separated into at least the components mentioned and symbolized as $\sigma_P^2 = \sigma_A^2 + \sigma_D^2 + \sigma_I^2 + \sigma_E^2$, where σ_I^2 is the component of genetic variance due to epistatic interactions in the broad sense.

EXERCISES

1. This is an exercise in calculating variance.
   (a) Arrange the following array of Y values in order and group into frequencies: $Y = 6, 9, 2, 3, 5, 6, 10, 9, 7, 6, 1, 6, 12, 7, 8, 4, 3, 5, 7, 6, 4.$

(b) Find the following values: N, $\sum fY$, $\bar{Y}$, $\sum fY^2$, "sum of squares" $[\sum fY^2 - N(\bar{Y})^2]$, variance ($\sigma^2$),and sample variance ($s^2$) (see Appendix A-6 for the distinction between parameter variance and sampling variance).

(c) Confirm the sum of squares by obtaining each deviation and squaring to obtain $\sum f(Y - \bar{Y})^2$.

(d) Calculate the standard deviation and standard error of the mean. What percentage of these numbers lies between ± 1 *s.d.* around the arithmetic mean? What would be the confidence limits for the mean?

2. Suppose that, in mice, body size depends on a pair of alleles, *dwarf* (*dw*) and *wild type* (+), which are strictly additive so that three genotypes at six weeks of age weigh (in grams) approximately as follows:

$$+/+:14, \qquad +/dw:10, \qquad dw/dw:6.$$

(a) What is the additive effect of each "+" substitution?

(b) A population of mice is random mating with frequencies of $p_+ = 0.8$, $q_{dw} = 0.2$, while a second population, also random mating, has frequencies of $p_+ = 0.40$, $q_{dw} = 0.60$. What would be the means and phenotypic variances of each population?

(c) Suppose a large number of females from the first population are mated to an equal number of males from the second population. What would be the expected frequencies of genotypes (or phenotypes), mean weight, and variance for the hybrid population?

(d) If the hybrids from (c) were allowed to mate at random, what would be the expected frequencies, mean weight, and variance of their progeny?

(e) What effect does forced hybridization have on the genetic variance?

(f) What would have been the result if the two populations had been allowed to mate at random instead of forcing the hybridization?

(g) When will equilibrium in frequencies be established—after the forced hybridization (c) or after randomly mating (f)?

3. Suppose that the range in mouse weight from 6–14 grams was controlled by four loci, each of which had an equal effect on the mouse phenotype when allelic substitution took place, so that $A = B = C = D$ as a certain increment over the lower-case letter allele, without dominance as in the previous problem.

(a) What would be the additive effect of each capital-letter allele?

(b) In a random-mating population (such as the first one in Exercise 2), let the frequencies of alleles at each locus be parallel with those at the other loci—that is, $p_A = 0.8 = u_B = w_C = y_D$, with lower-case letter alleles all equal to 0.2—what would be the total phenotypic variance expected for that population?

(c) How does the variance of these several small-effect loci compare with the variance of the corresponding population in Exercise 2 where a major gene has affected body weight? Why has the variance changed in the way it has?

4. Falconer (1960) cited (p. 113) a partially recessive gene in the mouse called *pg* "pygmy." At six weeks of age, it produces the following average weight phenotypes in grams:

$$+/+:14, \qquad +/pg:12, \qquad pg/pg:6.$$

(a) We note that the additive effect of "+" is greater from the 0 to 1 dosage (pygmy to heterozygote) than from 1 to 2 doses (heterozygote to wild type). Thus, the heterozygote is not at the midparent position, but there is some dominance. What are the additive and dominance deviation values here?

(b) What would be the expected mean and variance in a population with $p_+ = 0.8$, $q_{pg} = 0.2$ under random mating?

(c) If $p_+ = q_{pg}$, what would be the mean and variance? If $p_+ = 0.4$, $q_{pg} = 0.6$?

(d) Compare these means and variances with those in Exercise 2(b) where a population has the same allelic frequencies. In what respects does dominance affect means and variances?

5. This is an exercise in calculating covariance, a linear regression slope (b) using the least-squares method, and the correlation coefficient (r) (Appendix A-6-G).

(a) From the following pairs of X,Y numbers, draw a scatter diagram on graph paper:

													Sum
$X =$	1	1	1	2	1	2	4	5	4	5	5	5	36
$Y =$	3	2	4	6	5	8	9	9	13	10	11	16	96

(b) Calculate the regression slope

$$b = \frac{\sum X \cdot Y - N\bar{X} \cdot \bar{Y}}{\sum X^2 - N(\bar{X})^2} \quad \text{or} \quad \frac{\text{covariance}}{\text{variance of } X}.$$

(c) Find the value of the Y intercept (where $X = 0$) and draw the regression line on the graph.

(d) Calculate the correlation coefficient

$$r = \frac{\text{covariance}}{\text{geometric mean variance}} = \frac{\sum XY - N\bar{X}\bar{Y}}{\sqrt{[\sum X^2 - N(\bar{X})^2][\sum Y^2 - N(\bar{Y})^2]}}.$$

6. Use the following phenotypic values for the genotypes listed below:

$$A_1A_1 = 10, \qquad A_1A_2 = 8, \qquad A_2A_2 = 2.$$

(a) Calculate the linear additive effects (α) and dominance deviations for gene frequencies from $p = 0.1$, $q = 0.9$ through to their opposite frequencies by intervals of 0.10. How does the linear (additive effect) change with gene frequency?

(b) How do the linear (additive) and dominant components of the variance change with p,q? What happens to the components when $p = q$?

7. (a) Graph the two components of variance from Exercise 6 over the range of p,q, superimposing them on the same graph to compare their relative sizes.

(b) Let the phenotypic values show complete dominance—let $A_1A_2 = 10$ with the other phenotypes as in Exercise 6. Graph the components of variance, superimposing them on a separate graph for this case. (Use formulas (6-9), (6-10), and (6-11).)

(c) For each of these cases (partial and complete dominance), calculate the total genetic variance and draw in the corresponding curve on each graph.

(d) Where do the maximum points fall (p,q value) for additive and dominant components? for the total genetic variance? Compare your results with Figure 6-4B.

8. The following phenotypes are a case of heterosis:

$$A_1A_1 = 8, \qquad A_1A_2 = 10, \qquad A_2A_2 = 2.$$

   (a) Find the linear additive effect (α), dominance deviation (d), and the additive and dominance components of variance for the following allelic frequencies: $p = 0.9$, 0.8, 0.7, 0.6.
   (b) At what point is the genetic variance entirely due to dominance?
   (c) At what point(s) will the the additive component of variance be maximal?
   (d) Do these maximal points for additive and dominant components fall near the corresponding maximal points in Figure 6-4C? Explain.
   (e) When we find that genetic variance has two components, one additive and one due to dominance, what can we infer about the genetic basis of the phenotypes we observe? In other words, what are the limitations of working backward and drawing conclusions about genotypes from calculations of variance components?
   (f) How does overdominance affect the relative amounts of the additive and dominant variance?
9. For each of the following three phenotypic sets, from formula (6-7) calculate linear (fitted) values (Y_L), and graph similarly to those in Figure 6-3:

	Phenotypic Values				
Set	A_1A_1	A_1A_2	A_2A_2	p	q
I	10	8	2	0.8	0.2
				0.5	0.5
II	10	10	2	0.8	0.2
				0.5	0.5
III	8	10	2	0.8	0.2
				0.5	0.5

   Find the dominance deviations, $\sum f Y_L$, variance of linear values, and sum of squared dominance deviations for each set at each frequency to confirm the components as you did by applying formulas in Exercises 6–8.
10. Linear values may be expressed either as deviations from the midparent, as above in formula (6-7), or as deviations from the population mean, as used by Falconer (1960) and in footnote formula 6-7A. Change a (additive halfway point between homozygotes) into α (linear increment) from (6-5): $a = \alpha - d(q - p)$; then subtract the population mean expressed in terms of α (see Table 6-5B). Thus, show that $Y_L^0 = 2p^2d - [\alpha - d(q - p)] - [\alpha(p - q) + d(p^2 + q^2)] = -2\alpha p$, and for the same reason, $Y_L^1 = (q - p)\alpha$ and $Y_L^2 = 2q\alpha$, the linear values as deviations from the population mean.
11. A cornfield is open-pollinated year after year. Plants range in height 3–7 feet tall in 1-foot intervals only. Assume that height is controlled entirely by two pairs of independent factors (A-a and B-b). A three-foot plant crossed to a 7-foot plant is known to produce 5-foot tall progeny only. Each incremental locus has additivity equal to the other locus.

(a) Arrange the nine genotypes for the cornfield in order and give phenotypic values expected for each genotype on the basis of the information given. Which genotypes might occur among the 5-foot tall plants in the field? What proportion of those could "breed true" if self-pollinated?
(b) If the F_1 plants from a cross of 3 × 7-feet were self-pollinated to produce an F_2, what would be the expected frequency distribution of phenotypes? Graph this distribution.
(c) Let the *A-a* pair of alleles have frequency of $p_A = 0.6$, $q_a = 0.4$ with the *B-b* pair frequencies $u_B = 0.2$, $v_B = 0.8$ in the open-pollinated population. What would be the expected frequencies of the nine genotypes? List their phenotypes separately (from 3 to 7 feet) with their frequencies. Graph the frequency distribution. Which is the most common phenotype?
(d) What is the incremental deviation from the midparent (additive effect) due to a single allelic substitution? Calculate the mean height of the corn in this field with the frequencies of alleles as given in (c). Verify that the mean equals the midparent plus the deviations due to both pairs of factors together.
(e) In this open-pollinated population, what is the genetic variance due to the *A-a* alleles? the genetic variance due to the *B-b*? What is the total variance calculated over all nine genotypes? Verify that the two loci contribute additive components to the total variance.

12. Suppose for the cornfield in Exercise 11 that all nine genotypes are distributed randomly over 10 separate, equal-sized plots of ground. In two plots, soil and moisture are optimal and produce on the average $\frac{1}{2}$-foot greater height than in five "average" plots, while in the three remaining plots, soil is poor so that plants are $\frac{1}{2}$-foot shorter than in the five "average" plots.
(a) What will be the phenotypes and expected frequencies (assuming the allelic frequencies from Exercise 11) of plants in this field? What will be the mean height?
(b) What will be the environmental component of phenotypic variance (σ_E^2) for the entire cornfield population? What will be the total phenotypic variance (σ_P^2)?
(c) What would be the heritability (h^2) for this population? ($h^2 = \sigma_A^2/\sigma_P^2$)
(d) What would be the heritability if all allelic frequencies were equal ($p = q$, $u = v$)?
(e) What does the heritability tell us?

13. In the following population, two pairs of alleles are assorting independently (*A-a* and *B-b*). Each capital letter allele adds to the phenotypic value with an increment of 2 over the base value of the lowest genotype (let *aabb* = 0 in phenotype value). Frequencies of phenotypes are arranged in the following array:

$$\text{Frequencies} \begin{vmatrix} 0.16 & 0.08 & 0.01 \\ 0.32 & 0.16 & 0.02 \\ 0.16 & 0.08 & 0.01 \end{vmatrix} \qquad \text{Phenotypes} \begin{vmatrix} 8 & 6 & 4 \\ 6 & 4 & 2 \\ 4 & 2 & 0 \end{vmatrix}$$

(a) Calculate the phenotypic variance (V_P) using the formula

$$V_P = \frac{\sum f Y^2}{N} - \left(\frac{\sum f Y}{N}\right)^2$$

(verify $V_P = 3.28$).

(b) Determine the frequency of each pair of alleles and the genotypes separately. Calculate the variance of the *A-a* genotypes (right margin) and then the *B-b* (lower margin). What are the contributions of each pair to the total variance? Verify that this total genetic variance conforms to formula (6-15).

14. Assume three loci (*A-a, B-b, C-c*) have a single pair of alleles, each with additive effect on the phenotype. Let the *aabbcc* genotype have 0 increment (= base phenotype value). If the triple heterozygote is selfed (*AaBbCc* × itself), determine the frequency distribution of phenotypes in the array of progeny expected under each of the following conditions of incremental effect. For each array graph the distribution:
    (a) Let each positive allele (capital letter) add a value of 2 to the base value so that $Aa = Bb = Cc = 2$, and let each locus be cumulative so that $AaBb = AaCc = BbCc = 4$, etc.
    (b) Let A be dominant over a so that $AA = Aa = 4$, while *Bb* and *Cc* lack dominance as in (a) above. Thus, $AaBbCc = 8$.
    (c) Let *B-b* also exhibit dominance so that $AaBb = AABB = 8$.
    (d) Let all three loci exhibit dominance so that $AaBbCc = AABBCC = 12$.
    (e) What would the distribution look like if the *A* locus had twice the effect of *B* and *C* so that $Aa = 3$, $AA = 6$, while $Bb = 1.5$, $BB = 3$, and $Cc = 1.5$, $CC = 3$ with dominance lacking (as in (a))?
    (f) What would the distribution look like with dominance at each locus as well as the double effect of *A* locus over *B* and *C* so that $AA = Aa = 6$, while $BB = Bb = CC = Cc = 3$?
    (g) How do the increases in dominance or the relative strengths of separate loci affect the expected distributions of phenotypes?
15. Assume three loci as in the previous problem, but let the *A* locus be strongly incremental while the *B* and *C* loci are weakly diminishing in their phenotypic effects such that:

$$\begin{array}{lll} AA = 18 & BB = -3 & CC = -3 \\ Aa = 10 & Bb = -2 & Cc = -2 \\ aa = 2 & bb = -1 & cc = -1 \end{array}$$

    A cross of *AABBCC* (=12) × *aabbcc* (=0) yields F_1 *AaBbCc* (=6), which is then selfed.
    (a) What will be the distribution of phenotypes in the F_2?
    (b) How would you design an experiment to distinguish the opposite phenotypic effects of the *A-a* locus from the other two loci?
16. Assume three loci as in Exercise 14, but with multiplicative genic action so that each positive allele (capital letter) doubles the base value. To avoid the problem of 0, let $aabbcc = 1$, then $Aabbcc = aaBbcc = aabbCc = 2$, but $AA = BB = CC = 4$. Further, let each locus be multiplicative with each other locus so that $AABBcc = 16$ and $AABBCc = 32$, for example. Describe the frequency distribution of the progeny array from selfing a triple heterozygote (*AaBbCc* × itself). Graph the distribution. How would you "normalize" this distribution by transforming the phenotypic data?
17. Using a random block design when multiple alleles affect a phenotype quantitatively, calculate the allelic substitution effects (α_i), deviations due to each allele, fitted phenotypic linear values, and the additive and dominant components for the following

population with three alleles A, a, and a' under random mating:

Genotype	Phenotype	Frequency
AA	+4	0.25
Aa	+2	0.20
aa	−4	0.04
Aa'	+3	0.30
aa'	−1	0.12
$a'a'$	+1	0.09

18. Li (1970*a*) gives a hypothetical example of a trait, based on average human populational data, which has considerable variation but very little heritability. The trait, family size (number of children per family), is distributed approximately as given below. Assume the number of families with mothers of genotype Aa is twice that of either AA or aa mothers.

Mother's Genotype	No. Children/Family (Y) 0	1	2	3	4	5	6	7	8	No. Families
AA	6	10	9	7	5	5	3	2	3	50
Aa	12	20	18	14	10	10	6	4	6	100
aa	6	10	9	7	5	5	3	2	3	50
Total families	24	40	36	28	20	20	12	8	12	200

(a) Calculate the mean and variance of family size (Y) for each genotype of mother and for the totals. Is there any significant difference between mother's genotype and average family size? What would you say accounts for the variance in family size?

(b) Assume that there really is a slight difference between genotypes in fertility that brings about family size which average as follows: for AA mothers: 3 children; for Aa mothers: 2.5 children; for aa: 2 children. Further, assume $p_A = q_a = 0.5$ and random mating occurring in a population. Find the mean and variance of this trait due to genetic differences. Assuming that the variance obtained in part (a) above is nongenetic, what would be the heritability of this trait for this population?

(c) If mothers' genotypes averaged AA: 2 children, Aa: 3 children, aa: 2 children, and with the same p, q values, what would be the genetic variance? What would be the dominant component of variance? the linear (additive) component? the heritability?

(d) Does a genotypic difference in a trait tell us anything about the extent of the heritability?

7

ESTIMATING POLYGENES

There is a common feeling among biologists that formal genetics is not possible for polygenically determined traits. Certainly, techniques become more indirect and complex when we attempt to ascertain the numbers of controlling loci, individual-locus allelic frequencies in populations, architecture of the genotype, and genic action of development when a trait is predominantly polygenic. A biometric study is supposed to measure genetic variation, but it is comprehensive in outlook, not definitive. Discontinuity of phenotypes must be achieved before genetic dissection can be completely meaningful. Numerous techniques and methods of attack now available in a few genetically well-known organisms have been used with varying degrees of success by achieving discontinuities in what had seemed to be continuously distributed polygenic traits. Some examples of genetic dissection will be described later in this chapter, but it has been achieved only in a few organisms. It is essential first to describe the chief quantitative procedures used in measuring the extent of polygenic variation to obtain the overall view of such traits and what information can be obtained.

From quantitative methods predominantly used in plant and animal breeding programs, we can learn at least the following major information: (1) relative amounts of control by genetic and environmental variables, (2) amount of "additivity" available for selective response (heritability), (3) the basis for selective response (correlation between relatives), (4) the minimum number of segregating independent genetic units affecting the trait, and (5) a basis for drawing inferences about genetic structure with regard to the trait. We can derive this information from considerations of parameters discussed in the last chapter: allelic frequencies, additive, dominant, and epistatic effects of loci, and the numbers of loci controlling the trait. Ascertainment of values can be achieved by employing any or a combination of the following techniques: (1) selection applied toward specific fractions of the population and measurement of responses by genotypes controlling the trait, (2) crosses between pure lines or selected lines differing in the trait and genetic analysis of segregating progenies, and (3) correlations between generations (parent-offspring) or within generations (sibs, half-sibs, or cousins) in a random-mating population. Each technique contributes a few more pieces to the puzzle; none is capable of completing the picture of genetic structure alone, but without applying these analytical tools, the view of polygenic factors in a population will be incomplete.

It is logical perhaps first to consider the main features obtained from artificial selection applied to a population. The remaining techniques will then be considered as adjunct to the selection response (although logically any of these could easily be considered first, followed by the others). In human populations without access to selection or crossing techniques, estimating correlations between relatives is the most feasible technique. We consider here only a simple selection design for illustration of basic principles. Other selection designs will be summarized later, although a complete discussion is beyond the scope of this book.

DIRECTIONAL SELECTION APPLIED AND RESPONSE

From the experiments of Johannsen described in the previous chapter and in all subsequent selection work, it is fundamental that the effectiveness of the population's response in changing its average metric value in the direction of selection applied depends entirely on the amount and nature of the genetic variation available to be selected. A pure line has no genetic variation free to respond. Conversely, an inherited change in average value in the direction of selection applied can only mean that some genetic variation was available for response. This fact is basic to the Darwinian concept of natural selection. (See Chapter 14 for a general discussion on modes of selection, natural selection, and general selection theory.) The amount of response (change each generation in progeny averages) compared with the maximum expected is of interest in order to infer the general nature of the genotypes being selected. To that end, it is important at least to introduce the student to a few highlights in what comprises a large and complex discipline: plant and animal breeding (see the following general references in this field: Lush, 1945; Lerner, 1958; Falconer, 1960; Allard, 1960; Mather, 1949; Mather and Jinks, 1971; Hayes, Immer, and Smith, 1955.

Selection experiments have been designed by geneticists and evolutionists for a variety of aims. Earlier studies were directed toward a demonstration that quantitative traits were in fact controlled by polygenes. Plant and animal breeders have been chiefly interested in just how predictable the outcome of selection response can be, based on estimates of free and potential genic variation available for response and the limits to that response. Population geneticists have used selection techniques to estimate the extent of genic variation controlling quantitative traits (polymorphisms), to measure the precision of selection under different environmental conditions, and to compare different selection schemes for efficiency of response, the nature of genetic structure inferred from the response, plus a methodological approach to analysis of natural selection.

The simplest form of directed selection is accomplished by choosing from a population individuals with particular phenotypes to propagate a lineage, and thus rejecting the remainder. Examples 7-1 and 7-2 demonstrate classic selection experiments illustrating the the main features. If a random-mating population is distributed with regard to a trait (Y), as in Figure 7-1, with mean ($\bar{Y}_P$) and if a certain fraction of phenotypes is selected (cross-hatched) with mean ($\bar{Y}_S$), any genotypic differences that control the distribution, partially or completely, and differentiate the fraction selected will bring about new genotypic frequencies among the progeny (response) of those selected ($\bar{Y}_R$). Ordinarily, we cannot discern actual genotype frequencies, but the rate of progress in the direction applied informs us as to how close to expected progress the genotypic structure of the population will allow.

Genetic Variation Additive

Let us first imagine a hypothetical population in which all variation is determined simply by additive genotypes; allelic dosages for a single locus in gametes (haploid) contribute equally to the phenotype, as in Chapter 6, Table 6-5A. Assume that $p_{A_1} = q_{A_2} = 0.5$ and the phenotypic values $Y = +4, 0, -4$, as in case 3 of that table, with a mean of zero

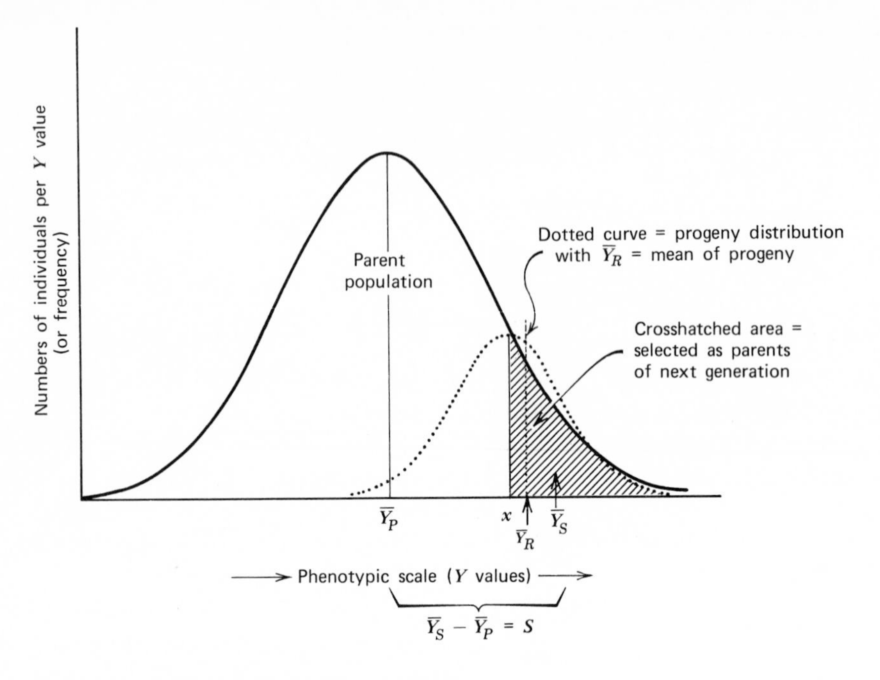

$\bar{Y}_P$ = mean of parent random mating population

$\bar{Y}_S$ = mean of selected parents = height ordinate/area under cross–hatched part of curve

S = deviation ($\bar{Y}_S - \bar{Y}_P$), or *selection differential*

x = point of truncation on phenotypic scale

$\bar{Y}_R$ = mean of progeny from selected parents (depends of heritability)

R = ($\bar{Y}_R - \bar{Y}_P$) deviation of progeny mean from original population mean, or response

Figure 7-1. Selection for a quantitative trait (Y) from a normally distributed population of mean $\bar{Y}_P$. Individuals are selected as parents (with mean of $\bar{Y}_S$) at a point of truncation (x) or greater (crosshatched area under parent distribution). Their progeny distribution (dotted curve) has a mean at $\bar{Y}_R$ at a point on the scale less than the selected parent mean.

($\bar{Y} = 0$). We wish to select for a higher mean value. It would be easy in this case just to select the $+4$ phenotypes and achieve the goal immediately, but for the sake of seeing what happens in the process of selection, let us also select some A_1A_2 genotypes. We select equal amounts of A_1A_1 and A_1A_2 individuals with a mean selected $\bar{Y}_S = +2$. If these selected parents are allowed to mate at random ($p = 0.75, q = 0.25$), it is obvious that their progeny will have the same mean ($\bar{Y}_R = +2$), because random mating simply apportions the same genetic variation into frequencies of genotypes fitting the square law (formula 1-4).

Let S = mean of selected parents as a deviation from the mean of the original population ($\bar{Y}_S - \bar{Y}_P$), known as the *selection differential*. R = mean of progeny as a deviation from the mean of the original population ($\bar{Y}_R - \bar{Y}_P$), known as the *response*.

Then, with simple additivity of genetic variation, we have

$$R = S \quad \text{or} \quad \frac{R}{S} = 1 \tag{7-1}$$

This relationship indicates that progeny phenotypes are completely determined by the parental genotypes—that is, by the *gametic additive values* of the randomly mating group of parents.

When environmental factors influence quantitative traits, the efficiency of selection decreases, because the phenotype is no longer completely determined by the genotype. In Table 7-1 we illustrate simple selection. If we use the example of Table 6-6, in which three environmental effects produce an array superimposed on the additive genetic effects, let us select for increased phenotypic value using all individuals with values greater than the population mean plus half of those with a phenotype equal to that mean. Now selected allelic frequencies are the same as before ($p = 0.75$, $q = 0.25$), and the array of progeny phenotypes will be distributed as shown in the lower portion of Table 7-1. The mean

TABLE 7-1 *Selection for increased phenotypic value with genic variation additive and environmental difference as in Table 6-6. Parents are selected for $Y_i > \bar{Y}_p$ plus $\frac{1}{2}$ of $Y_i = \bar{Y}_p$. Mean phenotype $\bar{Y}_p = 0$, when $p = q = \frac{1}{2}$, as in Table 6-6*

Selected parents ($p = \frac{3}{4}$, $q = \frac{1}{4}$)

Genotype	Environment	Phenotype Deviation ($Y_i - \bar{Y}_P$)	Selected Relative Frequency	fd	
A_1A_1	E_1	+6	1	6	
	E_2	+4	2	8	
	E_3	+2	1	2	$\bar{Y}_S = 20/8 = 2.5$
A_1A_2	E_1	+2	2	4	
	E_2	0	2	0	
			$\sum 8$	$\sum 20$	

Progeny response after random mating of selected parents and growth in three environments

Genotype	Environment	Phenotype Deviation	Relative Frequency	fd	
A_1A_1	E_1	+6	9	54	
	E_2	+4	18	72	
	E_3	+2	9	18	
A_1A_2	E_1	+2	6	12	
	E_2	0	12	0	$\bar{Y}_R = 128/64 = 2.0$
	E_3	−2	6	−12	
A_2A_2	E_1	−2	1	−2	
	E_2	−4	2	−8	
	E_3	−6	1	−6	
			$\sum = 64$	$\sum = 128$	

of the selected parents ($\bar{Y}_S = 2.5$) is decreased in the progeny response ($\bar{Y}_R = 2.0$), owing to the lesser efficiency of selection when environmental influences obscure some of the differences between genotypes. R is less than S, and $R/S = 0.80$, which is the proportion of total variance due to additivity of genotypes (σ_A^2/σ_P^2), or heritability (h^2) in the original population. We speak of the ratio R/S as the realized heritability

$$\frac{R}{S} = h^2 \text{ (realized)} \tag{7-2}$$

or $R = h^2 S$ if we wish to express the response expected for progeny.

As selection is continued generation after generation, gene frequencies change with increasing positive alleles in the direction applied; the genetic variance and total phenotypic variance also decrease, although the environmental component may well remain constant. For example, among these progeny, $\sigma_A^2 = 6$, $\sigma_e^2 = 2$, so that h^2 (progeny) = $\frac{6}{8} = 0.75$. With a constant environmental component, then, the heritability will approach zero as the positive alleles become fixed in the population. This is a fundamental property of a selected population. In spite of this expected decrease in heritability, the actual progress of selection may be a nearly constant linear function for several generations, as illustrated in Examples 7-1 and 7-2. In cases of long response, the environmental component may not be constant, but long-term selection response is more likely due to a slow release of genetic variability by recombination, as will be stressed later.

Selection with Dominance

As with an environmental influence, dominance lowers the efficiency of selection, but environment alone in the above instance did not eliminate the gametic contributions to the phenotype. Within each environment, genotypes were completely ascertainable (although in real populations, environmental effects may often be confounding to a considerable degree). Dominance, being a diploid effect, does reduce differences due to gametic values; consequently, we must rely on the hypothetical linear values ("breeding" values) and increases in α rather than a for estimating *average* gametic effects. As dominance becomes more complete, and at particular allelic frequencies, linear values become slightly less efficient at predicting progeny responses; these estimates are usually close approximations, as we can see from the example given below.

Using case 3 of Table 6-5B, we can select as we did in the simplest case (above) by choosing for parents equal numbers of highest-value (A_1A_1) and intermediate-value (A_1A_2) individuals. In Table 7-2 we proceed as before. The mean of selected parents is now 3.0, but after random mating the progeny mean is 2.75. Then $R/S = (\bar{Y}_R - \bar{Y}_P)/(\bar{Y}_S - \bar{Y}_P) = 1.75/2 = 0.875$ (realized heritability).

Notice that in Table 6-5B, the additive component of variance is 0.889 of total variance, which is slightly larger than the R/S ratio. These values, which are both estimates of h^2, would be equal except that the relation between genotypes of successive generations is not strictly linear when dominance is nearly complete. The progeny response then will be slightly less than that expected from (7-2): $R = h^2S = (0.889)\ 2 = 1.778$ (R expected) compared with $= 1.750$ (R observed). However, the heritability calculated from the population's additive component of genetic variance is sufficiently close for most purposes.

TABLE 7-2 *Selection for increased phenotypic value with partial dominance as in Table 6-5B, case 3 ($\overline{Y}_p = 1$, mean of population; parents selected as equal amounts of A_1A_1 and A_1A_2)*

Selected parents ($p = \frac{3}{4}$, $q = \frac{1}{4}$)

Genotype	Phenotype Deviation ($Y_i - \overline{Y}_P$)	Selected Relative Frequency	*fd*	
A_1A_1	+4	1	4	$\overline{Y}_S = 3$
A_1A_2	+2	1	2	
		2	6	

Progeny response

Genotype	Phenotype Deviation ($Y_i - \overline{Y}_P$)	Relative Frequency	*fd*	
A_1A_1	+4	9	36	
A_1A_2	+2	6	12	$\overline{Y}_R = 2.75$
A_2A_2	−4	1	−4	
		16	44	

$S = \overline{Y}_S - \overline{Y}_P = 2$
$R = \overline{Y}_R - \overline{Y}_P = 1.75$
$R/S = 0.875$

From Table 6-5B, when $p = q = \frac{1}{2}$, in parent population: $\sigma_A^2/\sigma_G^2 = \frac{8}{9} = 0.889$ (amount of variance due to linear values, or additivity); in offspring: $\sigma_A^2/\sigma_G^2 = 3\frac{3}{8}/3\frac{15}{16} = \frac{6}{7} = 0.8571$

Also, it should be noted that the heritability changes as before with change in frequency of alleles. In the illustrated case, h^2 will diminish as p approaches 0.75, q 0.25; the student should note that it will not continue to diminish as in the case of the constant component from environmental variance.

Generalized Response

In the illustrations given, we have supposed a single pair of alleles to be controlling the variation. Obviously, such phenotypic values then would not be normally distributed but would instead be discontinuous. With increasing numbers of loci controlling the trait (polygenes) and summation of genic effects being the case (formulas 6-14, 6-15), phenotypic distributions would be expected to become more normal (or they can be made so if scaling transformations are necessary; see Appendix A-11). The population's phenotypic distribution can be considered in general similar to that in Figure 7-1, provided many genic loci contribute to the phenotype. It is also evident in studying the design that the magnitude of the selection differential depends not only on the proportion of the population selected (thus, the mean of selected parents as a deviation from the population) but also on the phenotypic variation of the trait (standard deviation). In Figure 7-2, we suppose all individuals in the population beyond a point of truncation to be selected. In parts A and B of Figure 7-2, the right half of each curve is based on 20 percent selection, as illustrated in Example 7-3, in which realized heritability was compared with that expected from determinations of additive variance. In Figure 7-2A, we assume in the left half of the curve that one unit of phenotype marks a point of truncation for S, the selection differential, such that, given a standard deviation of just that amount, 15.9 percent of the population would be selected. Thus, within a population of constant variance, the smaller the percent

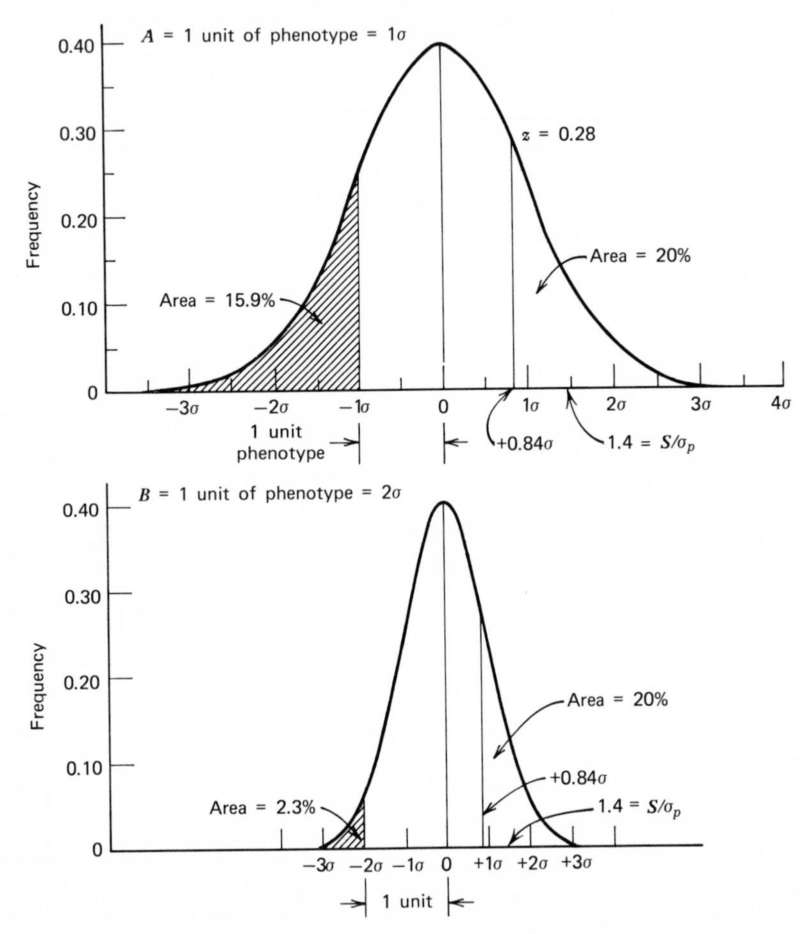

Figure 7-2. Diagram to show how S (selection differential) depends on the proportion of the population selected and on the phenotypic variation (σ_P) of the trait. A. 1 unit of phenotype $= 1\sigma$; B. 1 unit of phenotype $= 2\sigma$. Note that the proportion of the population selected is greater (less intense selection) when the population is more variable. The left side of each curve shows the comparison while the right side is from Example 7-3.

selected, the greater S will be (at 20 percent of selection, S would be 0.84 of a unit). In Figure 7-2B, the phenotypic variance is less so that the same phenotypic unit comprises two standard deviations. Thus, choosing the same point of truncation phenotypically will select only 2.3 percent of the population. However, by choosing S measured in standard deviations instead of phenotypic units as on the right side of both curves (at 1.4σ), the proportion of 20 percent is kept constant. Therefore, a more generalized statement of selection differential and response can be expressed by including the phenotypic standard deviation σ_P. Both S and R may be expressed as units of σ_P; then, in "standardized" form, (7-2) may be expressed more generally as:

$$\frac{R}{\sigma_P} = \frac{h^2 S}{\sigma_P} \qquad (7\text{-}3)$$

The standardized selection differential (S/σ_P) is often called the *intensity of selection* (i), and it measures by how many standard deviations the mean of the selected individuals exceeds the mean of the population before selection. Formula (7-3) then may be written

$$R = i\sigma_P h^2 \qquad (7\text{-}4)$$

Thus, $h^2 = R/i\sigma_P$.

As pointed out by Falconer (1960, 1965), i depends only on the proportion (p) selected and is then related to the normal distribution as follows: $i = z/p$, where $z =$ height of the ordinate at the point of truncation in the normal curve, shown in Figure 7-2. For example, $i = 1.40 = 0.280/0.20$ for the case of 20 percent selected in Example 7-3. That intensity corresponds to a normal deviate (τ) $= 0.8416$. These points are included in Figure 7-2. (Falconer, 1965, Appendix, includes a table of these relationships.)

In Example 7-3, Clayton, Morris, and Robertson were able to achieve expected response in selection for increased abdominal bristle number in *Drosophila melanogaster* when using a proportion selected of 20 percent, or $i = 1.40$, as well as when using a larger proportion of 26.7 percent, or $i = 1.24$. With an average response of 2.52 bristles per generation for the first five generations of selection, and with a mean phenotypic standard deviation (both sexes) $\sigma_P = 3.35$, those authors could estimate (using 7-4) realized heritability $h^2 = 2.52/(1.4)(3.35) = 0.54$, which was very close to their prediction based on parent-offspring covariance (see the next section).

EXAMPLE 7-1

AN EARLY SELECTION EXPERIMENT FOR POLYGENES

Kenneth Mather (1941) was concerned with the demonstration of polygenic variation as a primary mode of heredity and a ubiquitous feature of sexually reproducing populations. He chose to analyze abdominal bristle number in *Drosophila melanogaster* (combined number of chaetae on the ventral side of the fourth and fifth abdominal segments in both sexes), because it was a trait with apparently simple polygenic features: a nearly normal distribution in most strains and approximate additivity among hybrids and segregants of strains. He set about testing strains by crossing those that differed in bristle number: first, yellow (y) × forked (f)-infra-bar (B^iB^i), and second, bar (BB) × Oregon wild-type (+) which was later subdivided into two sublines. In all cases, F_1 progeny were counted and inbred in two or three sets of cultures to provide F_2 segregants as progenitors of selected lines. Among F_2, the two highest and lowest females and males out of 20 were selected to establish a high and a low line. (Selected flies each generation were always intercrossed between sets of sibs to avoid extreme inbreeding.) F_3 flies then comprised the first selected progeny (generation 1).

The outcome of selection is summarized in Figure A. As a frequent feature of quantitative traits, there is considerable fluctuation in each line largely due to contemporaneous environmental causes; consequently, some of that "environmental noise" was eliminated by plotting *differences* between high and low lines instead of the actual line means. Also, the sexes were averaged to help reduce unimportant variation.

In the cross $y \times fB^iB^i$, a quick response occurred in the first two selected generations followed by a nonsignificant gradual rise until termination due to sterility at the eighth generation. In the first $BB \times +$ cross, there was an initial response followed by a period of stability until the fourth generation. In turn, there was a second and larger advance with selection up to the eighth generation followed by stability. A repeat of the same cross ($BB \times$

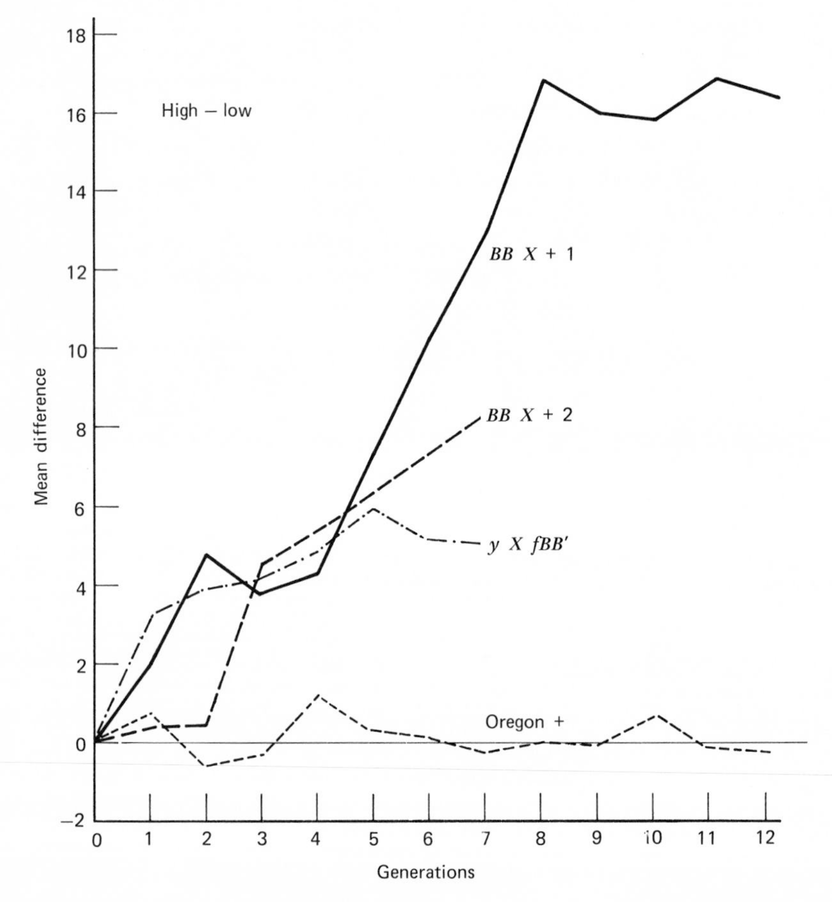

Figure A. Mean differences between high and low selection lines plotted by generations of selection. G_O = the F_2 progeny from the three crosses. From Mather (1941).

TABLE A *Means and variances of bristle numbers in the BB × + (Oregon) 1 cross and high and low lines after 12 generations of selection (from Mather, 1941)*

Strains		Original Strains and Cross		Selected Strains and Cross	
		Mean	Variance	Mean	Variance
Bar (*BB*)	♂♂	36.06	6.0065	43.10	4.1667
	♀♀	43.48	8.2708	49.13	7.1137
Wild type (+)	♂♂	39.88	6.5922	29.91	6.2909
	♀♀	44.59	9.0700	35.79	12.7310
F_1 progeny:	♂♂	37.83	5.7710	35.86	6.6869
	♀♀	43.40	4.9080	40.34	8.3389
F_2 progeny:	♂♂	38.35	6.8201	37.92	16.8555
	♀♀	44.20	7.0436	42.80	18.0826

+2) gave a delayed response two generations after the start of selection, followed by gradual increase in difference until termination. As a check on the possibility that rapid mutation could be effective in providing variation for sudden advances, the inbred Oregon + line was selected, but no response took place. These results were interpreted as follows: (1) immediate responses were due to whole chromosome segregation effects, (2) a plateau followed by increased response followed release of variation by intrachromosomal recombination (as in *BB* × +1, generations 4 to 8), and (3) a corollary to the latter, that polygenes must largely exist in balanced combinations (*AbCd*/*aBcD*, and so forth).

Some of the data demonstrating how much variation can be uncovered and concentrated by the selection process are given in Table A for the *BB* × +1 lines. Several points may be noted:

1. Looking at the means, the bar parent strain was lower in bristle number than the parent wild type, but after selection the high strain became homozygous bar while the low eliminated the bar; the bristle number had to be considered resulting from a high-determining bar-X chromosome, with low-determining autosomes in the original bar strain but high-determining autosomes and low-X in the wild-type Oregon strain. The recombinational events then must have been autosomal.
2. Both F_1 and F_2 means are intermediate not only between the original strains but also between the selected strains.
3. Females always have more bristles and higher variance (except original strain F_1 females) than males.
4. Most important, there is little evidence of additive genetic variance from the *original* cross (F_2 increment of variance over the F_1), and the number of "effective segregating factors" is nearly zero for the original females.

After selection, however, with the great difference between strains, the large additive variance component is greatly increased, and the number of effective factors ($k = 2.3$) is about what should be expected for *D. melanogaster* with three large chromosomes if each chromosome acts as an effective factor. It can then be inferred that factors determining bristle number were balanced similarly (repulsion linkages), with about the same amount of total plus and minus factors but distributed differently in both original strains. Since no response came from selecting the inbred Oregon + line, mutation could be ruled out as a likely source of free variation, and crossing over must be assumed as the source of released, or redistributed, plus and minus factors.

EXAMPLE 7-2

FURTHER DEMONSTRATION OF POLYGENES BY LONG-TERM SELECTION

An extensive set of selection lines was reported by Mather and Harrison in 1949, laying the groundwork for a careful check on polygenic theory. The same trait as in Example 7-1 (abdominal bristle number) was analyzed from a cross between two wild-type strains (Oregon inbred and Samarkand) of *Drosophila melanogaster*. Selection was practiced for both high and low numbers in several lines, from the F_2 generation through 138 generations. (Two high or two low flies of each sex were selected from 20 per culture each generation; two cultures were maintained in each line to provide for crossing within lines and avoidance of extreme inbreeding.) From Figure A, the main features for the first 61 generations can be seen. Bristle number fell erratically for 35 generations in the low line, at which time sterility caused the line to die out. A mass culture (1) begun at the 20th generation was stable for low bristle number. Further selection for either high or low number from the mass culture was usually effective, but these lines died out as a result of sterility (inability of low bristle number males to inseminate females).

The high line showed a progressive, nearly constant rise in the selected trait but a decrease in fertility until the 20th generation, when the number of flies was so small that selection was abandoned and resort was made to mass culture in the hope that natural selection would restore fertility. The bristle number of that mass line (3) fell back 80 percent of the way to the original parental mean level within five generations of mass mating. (This tendency to return to the unselected level on relaxation of selection is known as populational homeostasis; see Lerner (1954) and Chapter 19.) Reselection was begun from line 3, and in four generations the bristle number level of generation 20 was recovered but now without loss of fertility. Both that reselected line (8) and its mass-mated unselected line (7) were stable at about the same level from then on. This correlation between fertility and bristle number is interpreted as the result of crossovers between high number factors and low fertility, while the quick initial responses and reselection responses must mean that the three major chromosomes are acting substantially as units. (The decrease in fertility could be the result of a loss of relational balance following crossing over in the selected lines, but separately segregating fertility loci are perhaps more likely.)

Table A lists average bristle number and the results of a chromosomal assay (see *Locating Polygenes* section) for certain representative lines among the 10 selected. Note that the low selected strain (1) is not very different from the lower parent (*Sk*), with minus

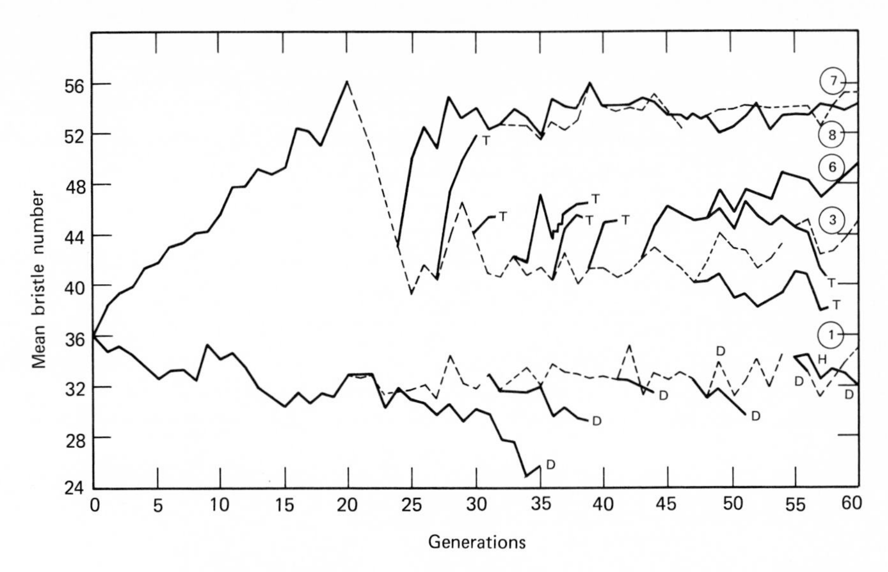

Figure A. Response to selection of mean abdominal bristle number for the first 61 generations of five lines (continued subsequently onward to G_{110}, not shown). Selected lines = solid; mass-mated unselected lines = dashes; T = line terminated; D = died out from sterility. Line 1 was selected in the down direction except at H when it was selected in a high direction. Lines 3 and 6 were selected in a low direction at G_{48} and G_{49}, respectively, and then terminated in that direction at G_{58}. From Mather and Harrison (1949).

TABLE A

Line	Average Bristle Number (females)	Chromosome Assay (net deviations from tester in backcross)						
		Main Effects			Interactions			
		X	2	3	X-2	X-3	2-3	X-2-3
Parents								
Samarkand	37.63	−1.06	−1.30	−3.24	−0.67	−0.83	0.53	0.00
Oregon	43.53	0.79	0.54	−2.44	−0.15	−0.14	0.58	0.04
Selected								
1	36.00	0.25	−1.05	−3.42	0.54	0.60	0.42	−0.69
3	44.35	0.94	0.33	−2.20	−0.64	−0.29	0.27	−0.06
7	59.53	2.75	3.01	0.66	−0.62	0.47	0.72	−0.42
8	63.35	2.62	3.23	1.72	−0.86	1.36	0.60	−0.58
9-2	70.25	2.96	2.76	5.65	−0.37	1.13	0.12	0.23

modifiers on chromosome 3 predominantly. Line 3 (original mass-mated) is virtually at the level of the other parent (*Or*). The high selected lines have responses on each of the major chromosomes, with the third chromosome making the largest response in the highest line (9-2), which had been derived from a cross between lines 3 and 7. None of the interactions is significant except those between the X chromosome and the third in the two highest lines, where presumably some epistatic effect must have been selected.

To estimate relative dominance and the minimum number of "effective factors" to account for the differences in bristle number levels, crosses were made between some of the mass-mated lines, which had become stabilized at particular levels. For example, the upper level line 7 (mean bristle number of 53) was crossed to the lower level line 3 (mean bristle number of 43). Their F_1 progeny was almost exactly intermediate (47.5 bristles), with a slight reciprocal effect indicating some X chromosome determination. With estimates of variance, the number of effective factors (see section in text) segregating in the F_2 were calculated as follows:

Average variance of parents (lines 7, 3) and their F_1 = 8.00 (estimate of environmental variance component)

Average variance of F_2 = 13.14; therefore, the F_2's estimated genetic component of variance = 5.14.

Then, using text formula (7-19), we have $k = D^2/8V_A = (10)^2/8(5.14) = 2.43$ effective factors, a value very close to the number found in Mather's earlier experiments (Example 7-1). These results are what would be expected for whole chromosomes acting as units of segregation in *D. melanogaster*.

EXAMPLE 7-3

AN EXPERIMENTAL CHECK ON POLYGENIC THEORY

The relationship between prediction of selected response (based on heritability measurements from covariances between relatives) and the realized heritability was determined by Clayton, Morris, and Robertson in 1956. The trait, abdominal bristle number in *Drosophila melanogaster*, was used to examine existing theory and how long response would continue according to initial predictions. Their base population was a laboratory population derived from wild flies captured in Kaduna, West Africa (1949), in which the mean bristle number (sum of fourth and fifth abdominal sternites) was stabilized for about 60 generations before the experiments were begun: males = 31.4 ($\sigma = 3.03$), females = 39.2 ($\sigma = 3.54$), which are slightly lower values than those in the strains studied by Mather and Harrison.

Heritability was estimated from the random mating base population.

(1) Offspring-Parent Regression

Samples were obtained from the base population and about 15–20 females were mated to each male as indicated in Table A-1 until 27 males had been used. Females scored for bristle number were allowed separately to produce progeny, two males and two females, which were then scored. Within each sire, mothers and their offspring were regressed: daughter − dam, $b = 0.269$ and son − dam, $b = 0.241$ (after correction for smaller bristle numbers in males). Since offspring-parent regression is equal to one-half the heritability, the two estimates were pooled to give $h^2 = 0.51$.

For estimates of heritability from half-sib and full-sib correlations, the first generation of family selection lines was utilized.

TABLE A

(1) Offspring-Parent Regression

	Scored Mothers	Scored Offspring	
Replicated 1♂ × several times	#1♀	2♀♀ + 2♂♂	
	#2♀	⋮ ⋮	$b_{\text{daughter: dam}} = 0.269$
	⋮		
	#20♀		$b_{\text{son: dam}} = 0.241$ (corrected for difference between sexes)
Total of 27♂♂ from base population	X_i	Y_i or Y_j	

(2) Half-Sib Correlation

First generation 6 selected lines (3 high + 3 low) — 1st line (high) · · · 6th line (low) ↓ Offspring

		Offspring	
Within each line 1♂ ×	#1♀	5♀♀ + 5♂♂	(same for each selected line)
	#2♀	⋮ ⋮	
	⋮		
	#10♀		

(Total 60 families for 6 lines)

Daughters from ♀#1 = half sibs to daughters from ♀#2, etc. (within sires)

Covariance ($\sum dx\,dy$) for *paired* half-sibs = variance between mean daughter (or mean son) of each dam ($\sum d_x^2$) = variance between 60 families

(3) Full-Sib Correlation

First generation 6 selected lines (3 high + 3 low) — 1st line (high) · · · 6th line (low) Offspring

		Offspring	
Within each line	1♂ × 1♀	3♀♀ + 3♂♂	(same for each selected line)
	2♂ × 2♀	⋮ ⋮	
	⋮		
	20♂ × 20♀		

Covariance ($\sum dx\,dy$) for paired full sibs = variance between 120 families since within lines the parents are full sibs ($\sum d_x^2$)

(2) Half-Sib Correlation

The half-sib family selection experiment consisted of three high and three low lines, in each of which were ten families. Each family was sired by a single male mated to ten females as indicated in Table A-2. (In order to score only half-sibs, the progeny of each female has to be separated from that of other females; these authors did not separate each female's progeny, so that when scoring progeny, five females and five males per sire, there is one chance in ten that two individuals could be full-sibs provided all females were equally fertile. In the authors' calculations they corrected for the increased relationship, but their heritability estimate did *not* include the correction. We shall proceed as if all half-sibs had been separated.) Each parent set was transferred to two bottles to obtain an estimate of minor environmental variance.

Heritability was then estimated from analysis of variance, since the variance between families of half-sibs within lines equals the covariance of half-sibs (see Exercise 8 to this chapter).

Source of Variation	*d.f.*	Mean Square	Variance Component
Between families within lines*	54	39.51**	1.256
Between bottles within families	60	14.39†	0.527
Residual	1020	9.21	

* With 6 lines, 60 families of half-sibs, losing one *d.f.* per line.
† $0.005 > p > 0.001$.
** $p < 0.001$.

Note that the differences between bottles made up only about 5 percent of total variance (10.903).

$$h^2 = 4\left[\frac{1.256}{(1.256 + 9.12)}\right] = 0.484 \qquad \text{(within bottles, heritability from half-sibs)}$$

(3) Full-Sib Correlation

The full-sib selection experiment came from three high and three low lines as with the half-sib lines, and in each line twenty families were scored as indicated in Table A-3. Each parental pair was transferred to two vials for scoring of progeny (three of each sex per vial) for minor environmental variance. Heritability was again estimated from analysis of variance components.

Source of Variation	*d.f.*	Mean Square	Variance Component
Between families within lines*	114	46.40**	3.08
Between vials within families	120	9.71	0.20
Residual	1080	8.54	

* With 6 lines, 120 families of full-sibs, losing one *d.f.* per line.
** $p < 0.001$.

Again, the minor variance is small and this time not significant.

$$h^2 = 2\left[\frac{3.08}{(3.08 + 8.54)}\right] = 0.53$$

All three estimates of heritability therefore agree in being not very different from an overall estimate of 0.52.

Let us simply follow the response to selection in which individuals (not closely related as in sib selection or family selection experiments) were selected at different intensities for five generations, and thereafter selection was relaxed for 19 generations. The results are shown in Figure A where *H*1, *H*2 . . . 5, *L*1, *L*2 . . . 5 represent five high and five low lines,

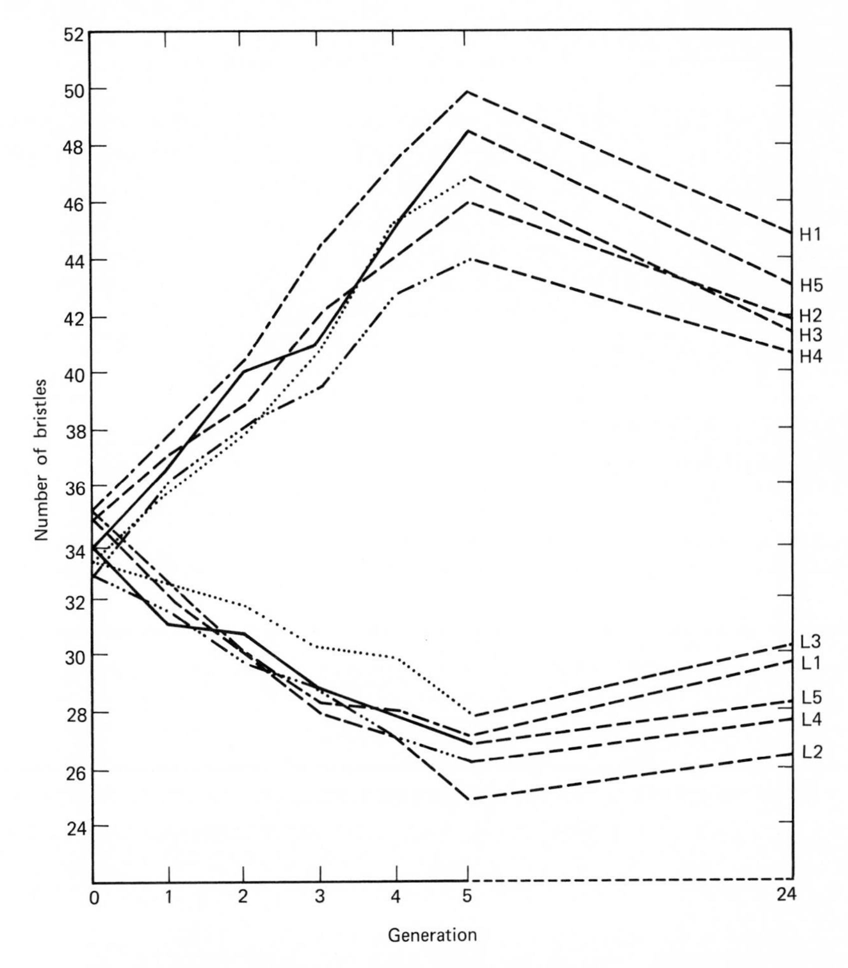

Figure A. Replicated responses to two-way individual selection for abdominal bristle number in *Drosophila melanogaster*. *H*1, *H*2, . . . , *L*4, *L*5 represent five high-bristle-number and five low-bristle-number lines, respectively. Selection was relaxed at generation 5 for 19 more generations. Data from Clayton, Morris, and Robertson (1957), modified from Kojima and Kelleher (1963).

respectively. From 100 flies of each sex scored, 20 of each sex were selected in each line per generation and mass mated. The average response was very marked and very close to that expected on the basis of heritability estimates derived from close relative covariance, as evidenced from the data below:

		Changes per Generation in 5 Generations		
		Observed		
Proportion Selected	i	Up	Down	Expected (from h^2 estimates)
20/100	1.40	2.52	1.48	2.42
20/75	1.24	2.20	1.26	2.14

From the text formula (7-2) put into standardized form (7-4), we can see the realized $h^2 = 0.54$ for 20 percent selected in the upward direction is very close to prediction, although the downward expectation is not achieved (h^2 realized downward $= 0.32$). It might be remembered that Mather and Harrison also found their downward selection was not as efficient as upward. (It was not possible to say whether natural selection was opposing the artificial or whether genetic variation was used up more quickly in the downward direction; asymmetric response to selection may be due to a number of different causes.)

COVARIANCE BETWEEN PARENTS AND OFFSPRING

It has no doubt occurred to the reader that an essential quantitative relationship between parents and offspring (expressed in the concept of heritability) must determine the selection response. This relationship can be derived and described as an additional population property known as the *covariance* (see Appendix A-6D and G) with its derived forms, the *regression* of offspring on parents and the *correlation* between relatives. In practice, we wish to know the extent of the additive genetic component not only to make predictions for the outcome of selection but also to estimate the magnitude of polygenic factors and the genetic structure of the population. Perhaps the first objective belongs more to the practical breeder and the second more to the evolutionist or to the human geneticist. At any rate, by means of computing the covariance between relatives (particularly parent-offspring), we can estimate the additive component of variance, because it is seldom possible in practice to ascertain genotypes for polygenic traits. As another point of view, the regression of offspring on parents will help to explain why the additive component of genetic variance can be used to predict the progress expected from selection.

The simplest to derive and most commonly used *covariance* is that between a single parent and offspring. We shall then proceed to describe the relation between the mean of two parents and their offspring and, finally, the relation between the average phenotypic value of the parent population and their linear ("breeding") values that predicts their offspring's response to selection. By this procedure the student should see that there are several ways of looking at covariances between generations or between groups within

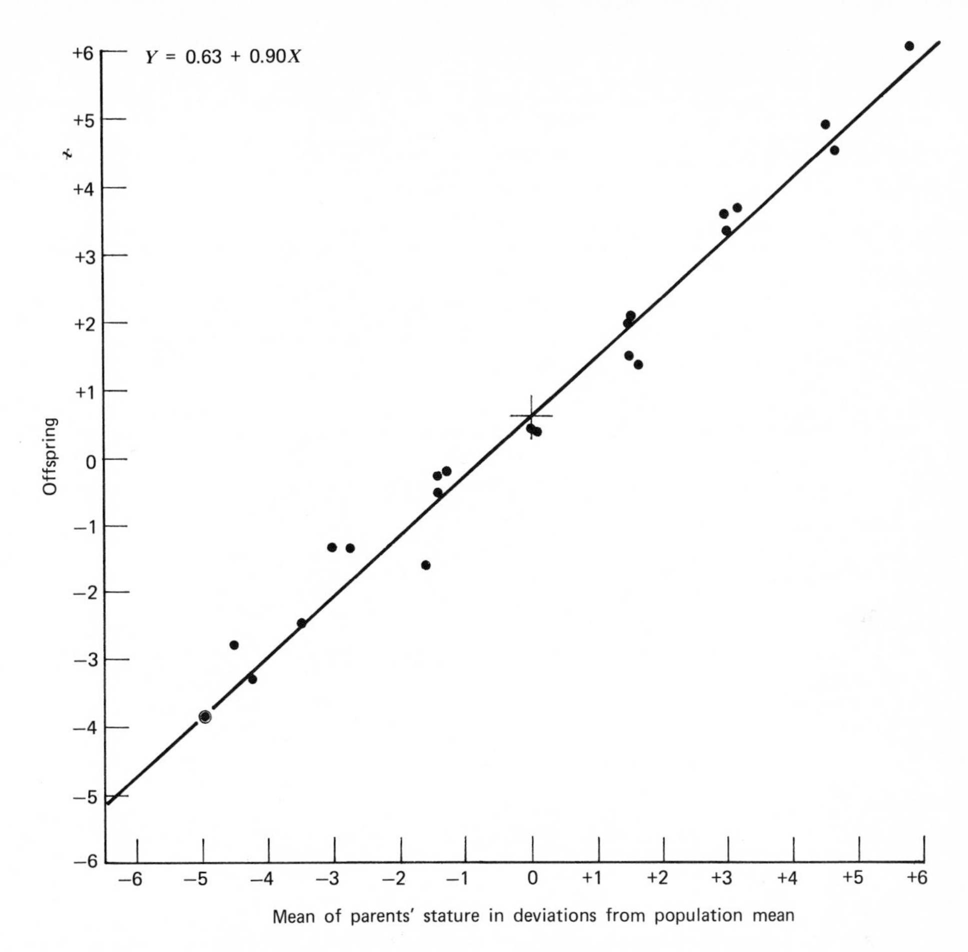

Figure 7-3. Regression of offspring on parents for human stature from Davenport's data given in Example 6-1 and plotted from his Table 5. Stature is indicated in deviations from the population mean of 68″ (males) and 63″ (females).

populations. At the same time, the student may learn by increasing complexity another aspect of the heritability concept. In Figure 7-3, the regression of offspring on two-parent means illustrates the principle from data collected by Davenport for stature in American families (Example 6-1).

Covariance of Single Parent-Offspring

For just one parent, only half the genotype determines each offspring's genotype (autosomal traits in general, although including sex linkage for mother-daughter). The correlation between genotypes of any parent and offspring ("coefficient of relationship") can be obtained by considering genic contributions of each type of parent to offspring in a table of joint distribution for parent-offspring pairs (Table 7-3). Offspring frequencies from each type of parent may be reasoned out as follows: in upper left, to have an A_1A_1 child,

TABLE 7-3 *Joint distribution of single parent and offspring and their covariance in a random mating population (Y = phenotypic value of parent, Y' = phenotypic value of offspring)*

A Genic covariance (additive)

Parent			Offspring					
Genotype	Y	Frequency	Y′	A_1A_1 Frequency	Y′	A_1A_2 Frequency	Y′	A_2A_2 Frequency
A_1A_1	+1	p^2	+1	p^3	0	p^2q		
A_1A_2	0	$2pq$	+1	p^2q	0	pq	−1	pq^2
A_2A_2	−1	q^2			0	pq^2	−1	q^3

$$\begin{aligned}\sigma_{PO} = \sigma_{YY'} &= 1^2 \cdot p^3 + 1^2 \cdot q^3 - (p-q)^2,\\ &= p^3 - p^2 + q^3 - q^2 + 2pq\\ &= 2pq - p^2q - q^2p = pq(2 - p - q)\\ &= pq = \tfrac{1}{2}\sigma_A^2\end{aligned}$$

since at equilibrium $\sum Y = \sum Y'$, $\bar{Y} = \bar{Y}' = (p - q)$, given by formula (6-1) and Table 6-5A.

B Diploid covariance (including dominance)

Parent			Offspring					
Genotype	Y	Frequency	Y′	A_1A_1 Frequency	Y′	A_1A_2 Frequency	Y′	A_2A_2 Frequency
A_1A_1	$+a$	p^2	$+a$	p^3	d	p^2q		
A_1A_2	d	$2pq$	$+a$	p^2q	d	pq	$-a$	pq^2
A_2A_2	$-a$	q^2			d	pq^2	$-a$	q^3

$$\begin{aligned}\sigma_{PO} = \sigma_{YY'} &= \textstyle\sum Y \cdot Y' - \bar{Y} \cdot \bar{Y}' \quad \text{combining products gives}\\ &= a^2p^3 + 2adp^2q + d^2pq - 2adpq^2 + a^2q^3 - [a(p-q) + 2dpq]^2\\ &= a^2[p^3 + q^3 - (p-q)^2] - 2adpq(p-q) + d^2(pq - 4p^2q^2)\\ &= a^2pq + 2adpq(q-p) + d^2[pq(q-p)^2]\\ &= pq[a + d(q-p)]^2 = pq\alpha^2\end{aligned}$$

Therefore, $\sigma_{PO} = \frac{1}{2}\sigma_A^2$

an A_1A_1 mother must receive an A_1 gamete (frequency p) from the child's father; to have an A_1A_2 child, an A_1A_1 mother must receive an A_2 gamete (frequency q); to have an A_1A_1 child, an A_1A_2 mother must receive an A_1 gamete, but only half of the children produced would be A_1A_1 and so forth. The covariance (sum of products) is derived first in the upper portion of the table (A) for complete additivity (genic covariance) and in the lower portion (B) for dominance (diploid covariance), which is the more general case, using the scale symbols of Chapter 6.

The covariances ($\sigma_{YY'}$) between parent-offspring phenotypic values in both cases have the property that they are just one-half of the additive variance component:

$$\sigma_{PO} = \tfrac{1}{2}\sigma_A^2 \qquad \text{(covariance parent-offspring)} \qquad (7\text{-}5)$$

Dominance has no effect on this relationship. (Why?) This relationship basically expresses why selection is effective when genetic variance is significantly greater than zero: the regression (b) of offspring (dependent variate) on parent (independent variate) then becomes simply the ratio of covariance/parent variance:

$$b_{O:P} = \left(\frac{1}{2}\right)\frac{\sigma_A^2}{\sigma_P^2} \quad \text{(regression offspring: parent)} \tag{7-6}$$

Finally, parent-offspring correlation, or the ratio of covariance to the geometric mean of their variances ($\sqrt{\sigma_O^2 \cdot \sigma_P^2}$), is the same, because at equilibrium parents and offspring are expected to have equal variances:

$$r_{PO} = \left(\frac{1}{2}\right)\frac{\sigma_A^2}{\sigma_P^2} \quad \text{(correlation between single parent-offspring)} \tag{7-7}$$

We may now take note of the fact that because heritability is defined as the proportion of additivity (σ_A^2/σ_P^2), both the regression and the correlation between a single parent and the offspring estimate one-half the heritability:

$$b_{O:P} = r_{PO} = \tfrac{1}{2}h^2 \tag{7-8}$$

Examples 7-3 and 7-4 illustrate the use of realized heritability and offspring-parent regression estimated heritability for populations of drosophila. The first example measures a morphological trait, while the latter uses behavioral traits—geotaxis and phototaxis.

The correlation formula may be viewed in a fundamental way that may be helpful for more understanding. First conceived by Fisher (1918) and later independently developed by Wright in path coefficient analysis, the concept that a correlation may consist of various steps, or components, is very useful because it follows from the product rule of independent probabilities. Figure 7-4 (modified from Li, 1961) illustrates the three steps between the parent's phenotype (Y) and the offspring's (Y'). The main genetic step is that between a haploid gametic set in parent and offspring (X $\rightarrow$ X$'$ being the genic effect quantified going from parent to offspring). Because the variance of genic dosages should be equal in both generations at equilibrium, $\sigma_X^2 = \sigma_{X'}^2 = 2pq$ and the covariance $\sigma_{XX'} = pq$ (Table 7-3A), correlation between X's (genic values of parent and offspring) is given by:

$$r_{XX'} = \frac{pq}{2pq} = \frac{1}{2} \tag{7-9}$$

which is independent of gene frequencies.

Finally, within either generation the correlation (r_{XY}) between the individual's haploid genic dosage (X) and phenotype (Y) is a function of the linear value (Y_L) and phenotypic value (Y). Recalling that $\alpha = \sigma_{XY}/\sigma_X^2$ and $\sigma_A^2 = \alpha^2 \cdot 2pq = \alpha^2 \cdot \sigma_X^2$ (formula 6-8), then

$$r_{XY} = \frac{\sigma_{XY}}{\sigma_X \cdot \sigma_Y} = \frac{\alpha\sigma_X^2}{\sigma_X \cdot \sigma_Y} = \frac{\alpha\sigma_X}{\sigma_Y} = \frac{\sigma_A}{\sigma_Y} = \sqrt{\text{heritability}} \tag{7-10}$$

Therefore, the correlation between parent and offspring phenotypes $r_{YY'} = r_{PO}$ is the triple product:

$$r_{XY} \cdot r_{XX'} \cdot r_{X'Y'} = \frac{\sigma_A}{\sigma_Y} \cdot \frac{1}{2} \cdot \frac{\sigma_A}{\sigma_Y} = \left(\frac{1}{2}\right)\frac{\sigma_A^2}{\sigma_Y^2} = \left(\frac{1}{2}\right)\frac{\sigma_A^2}{\sigma_P^2}$$

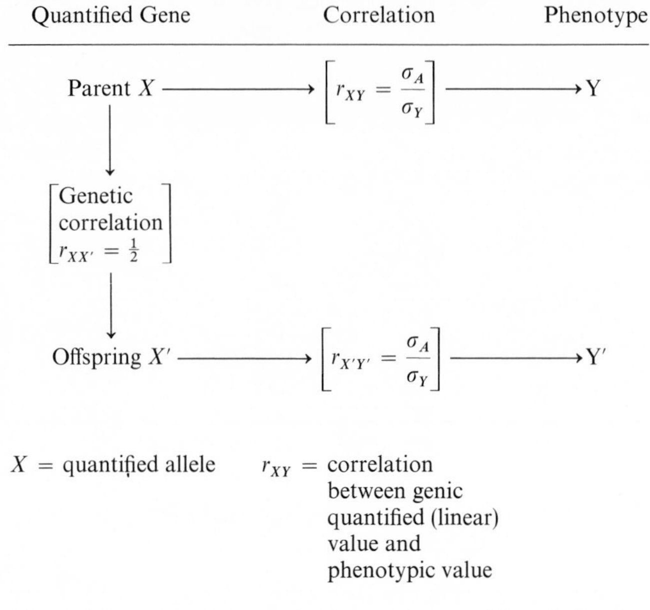

Figure 7-4. Total correlation between parent (Y) and offspring (Y′) phenotypes via gene determination ($X \rightarrow X'$); r_{XY} = correlation between linear (additive) value and phenotypic value. Modified from Li (1961).

It is important to realize that (7-9) is a constant fundamental Mendelian property relating parents to offspring and is frequently called the "coefficient of relationship." This coefficient measures the degree of genetic relationship. For full sibs it is also equal to $\frac{1}{2}$, for double first cousins it is $\frac{1}{4}$, and for ordinary first cousins it is $\frac{1}{8}$. That relationship is based on the probability that these relatives share "identical genes" (see Chapter 8 and following). The relationship of the linear (additive) component to phenotypic value (7-10) brings about the variation in the parent-offspring correlation.

Covariance of Two-Parent-Mean and Offspring

If the two sexes do not differ on the average (or if they are different but can be equalized by adjustment of scale), the relation of average between mother-father and their average offspring may be used to estimate heritability. Intuitively, the student should realize that using information from both parents is more efficient than using only one's. Algebraically, the estimate of covariance for parent average and offspring is a little more complex than for a single parent, but the covariance comes out the same as before, as given in Table 7-4:

$$\sigma_{\bar{P}O} = \tfrac{1}{2}\sigma_A^2 \qquad \text{(covariance two-parent mean-offspring)} \qquad (7\text{-}11)$$

which is identical with (7-5). However, regression and correlation are *not* the same. For the regression, the variance of the two-parent mean values is half the variance of parent

TABLE 7-4 *Joint distribution and covariance of parental mean values and mean offspring within families in random mating populations (using frequencies from Table 1-1B)*

Distribution

Parents			Offspring			
Genotypes	Mean Y	Frequency Mating	A_1A_1 $(+a)$	A_1A_2 (d)	A_2A_2 $(-a)$	Mean Y′
$A_1A_1 \times A_1A_1$	$+a$	p^4	all	—	—	$+a$
$A_1A_2 \times A_1A_1$	$\frac{1}{2}(a+d)$	$4p^3q$	$\frac{1}{2}$	$\frac{1}{2}$	—	$\frac{1}{2}(a+d)$
$A_2A_2 \times A_1A_1$	0	$2p^2q^2$	—	all	—	d
$A_1A_2 \times A_1A_2$	d	$4p^2q^2$	$\frac{1}{4}$	$\frac{1}{2}$	$\frac{1}{4}$	$\frac{1}{2}d$
$A_2A_2 \times A_1A_2$	$\frac{1}{2}(-a+d)$	$4pq^3$	—	$\frac{1}{2}$	$\frac{1}{2}$	$\frac{1}{2}(-a+d)$
$A_2A_2 \times A_2A_2$	$-a$	q^4	—	—	all	$-a$

Covariance

$$\sigma_{\bar{P}O} = \sigma_{YY'} = \sum Y \cdot Y' - \bar{Y} \cdot \bar{Y}'$$ (where Y value is the two-parent mean and Y′ is offspring mean within a family)

First, taking products only

$$a^2p^4 + (a+d)^2p^3q + 2d^2p^2q^2 + (-a+d)^2pq^3 + a^2q^4$$

Collecting like terms

$$a^2(p^3 + q^3) + 2adpq(p-q) + d^2pq$$

Subtracting the mean squared $(\bar{Y} \cdot \bar{Y}')$

$$-[a^2(p-q)^2 + 4a(p-q)dpq + 4d^2p^2q^2] \quad \text{gives covariance}$$

$$\sigma_{\bar{P}O} = a^2pq - 2adpq(p-q) + d^2pq(p-q)^2 = pq[a + d(q-p)]^2 = pq\alpha^2$$

values, so that the denominator of the regression coefficient is as follows: each parent contributes a half value toward the mean if Y_M and Y_F stand for each parent's phenotypic value; the parent mean is $\frac{1}{2}(Y_M + Y_F)$, and the variance of this sum, assuming mother and father values are uncorrelated, is

$$\sigma^2_{1/2Y_M} + \sigma^2_{1/2Y_F} = 2 \cdot \sigma^2_{1/2Y} = 2(\tfrac{1}{4})\sigma^2_Y = \tfrac{1}{2}\sigma^2_Y$$

where $\sigma^2_{1/2Y_{M,F}}$ refers to the variance of $\frac{1}{2}$ the mother's or father's phenotypic value. (See Appendix A.6 on multiplying values by a constant and variance of a sum.) Thus, the regression of offspring on two-parent mean is equal to the heritability.

$$b_{O\bar{P}} = \frac{\frac{1}{2}\sigma^2_A}{\frac{1}{2}\sigma^2_Y} = \frac{\sigma^2_A}{\sigma^2_P} = h^2 \tag{7-12}$$

However, the correlation between offspring and the two-parent mean is not equal to the regression here but is greater (as might be expected with more information from two parents than from one parent).

$$r_{\bar{P}O} = \left(\frac{1}{2}\right)\frac{\sigma_A^2}{\sigma_{\bar{P}} \cdot \sigma_O} \tag{7-13}$$

since $\sigma_{\bar{P}} = 0.707\ \sigma_P$, where $\sigma_{\bar{P}}$ = standard deviation of two-parent mean values and σ_O = standard deviation of offspring within families.

Covariance of Selected Parents and Offspring

When a fraction of a population is selected for carrying on a line, the relation between the average parental phenotype values and consequent linear ("breeding") values can be used to predict the offspring response. Covariance between phenotypic and linear values is for the student to verify (see Exercise 7) for the relation

$$\sigma_{\mathrm{YY}_L} = 2pq\alpha^2 = \sigma_A^2 \tag{7-14}$$

(Covariance of phenotypic and linear values equals the additive component of genetic variance.) Therefore, because σ_{Y}^2 is phenotypic variance, the regression of linear values on phenotypes is

$$b_{\mathrm{Y}_L:\mathrm{Y}} = \frac{2pq\alpha^2}{\sigma_{\mathrm{Y}}^2} = h^2 \quad \text{(heritability)} \tag{7-15}$$

Consequently, predictions of progeny response are based on linear values from the parents selected. This relationship provides the basis for *realized heritability* in a selection program.

It is important for the student to realize a slight technicality in most selection experiments that is not exactly equivalent to this theoretical prediction. Linear values are based on the supposition of random mating in the population *before* any selection is done. In actual selection experiments, virgin females and males are usually chosen assortatively first and then mated so that it is the selected group of parents whose variance is being estimated. However, the selected parents' variance is reduced by the same amount as the offspring-parent covariance, so that the regression of offspring on selected parents is not affected to any degree; therefore, after obtaining a reliable estimate of heritability from the parent generation, progeny response can be predicted from (7-1): $R = h^2S$. Theoretically, the prediction of response is valid for only a single generation—gene frequencies, genetic variance components, and heritability will change each generation. Usually, however, response is nearly constant for at least a few generations, as illustrated in the examples of this chapter. It is then desirable to estimate heritability from parent-offspring correlations or from other relatives (see the exercises and Example 7-3 for estimates from half-sibs and full sibs) and predict the magnitude of response expected.

EXAMPLE 7-4

SELECTION FOR POLYGENES DETERMINING BEHAVIORAL TRAITS

Many behavioral traits in drosophila have been found to respond to linear selection, especially geotaxis and phototaxis (movement as a reaction to gravity and to light, respectively). In 1958, while at Columbia University, Jerry Hirsch constructed an apparatus comprised of a series of plexiglass tubes bifurcating in T units at regular intervals to form a maze in which flies starting at one end in a single tube must choose successively 15 times (either up or down in geotaxis or to light or dark in phototaxis) in order to end up in one of the 16 final collecting tubes (see Hirsch, 1967, for summary). Collecting tubes were numbered increasing toward the positive side in each test, and the flies assorted themselves into positions after a few hours according to their reactions to the environmental gradient. The data of assigned scores were treated as any quantitative character.

Response to geotaxis was excellent starting with a genetically heterogenous population of *Drosophila melanogaster* (Erlenmeyer-Kimling, Hirsch, and Weiss, 1962). From a sample of 200 flies of each sex tested each generation, progeny from the most extreme scoring 10 pairs of parent flies were usually selected. After about 20 generations, positive and negative lines had become well separated with a mean score difference of about six points. Both lines responded similarly with the positive line first coming to a plateau at about the eighth generation; the negative lines continued to respond until about the 25th generation. Later the positive line again increased so that by the 48th generation, maximum separation was attained with 96 percent of individuals in the two lines getting the score for which they had been selected. Similar results were obtained in selection for phototaxis by Hirsch and Boudreau (1958) and Hadler (1964).

A chromosome assay carried out (as in Table 7-5) on the geotactic lines after 28 generations of selection (Table A) by Hirsch and Erlenmeyer-Kimling (1962) indicated that the unselected population chromosomes had the *X* and second chromosomes tending toward positive scores compared with the tester strain, while chromosome 3 was more negative. Selection for positive geotaxis brought about little change for the second chromosome, possibly increased modifiers on the *X*, and had a greater effect on the third, while selection for negative geotaxis caused changes on all three chromosomes more than did positive selection. None of the interactions was important. Also, from crossing oppositely

TABLE A *Chromosome assay main effects and interactions (in deviations from tester strain) for X, second, and third chromosomes from lines selected for geotactic scores (from Hirsch and Erlenmeyer-Kimling, 1962)*

	Average Score Coefficient						
	Main Effects			Interactions			
Line	*X*	2	3	*X*-2	*X*-3	2-3	*X*-2-3
Positive	1.39*	1.81*	0.12	−0.16	−0.10	−0.12	−0.14
± Standard error	0.13	0.14	0.12	0.13	0.18	0.11	0.08
Unselected	1.03*	1.74*	−0.29	0.05	−0.07	0.03	0.00
± Standard error	0.21	0.12	0.17	0.10	0.13	0.17	0.13
Negative	0.47**	0.33	−1.08**	−0.12	0.14	0.06	0.06
± Standard error	0.17	0.20	0.16	0.11	0.13	0.11	0.13

* Significantly different from the tester stock $p < 0.05$, $d.f. = 17$.
** Significantly different from the tester stock $p < 0.05$, $d.f. = 18$.

selected lines and calculating F_1 and F_2 variances at different generations, an increase in F_2 variance with successive generations indicated continual gain of genetic differences between lines.

After 52 generations, selection for geotaxis was reversed for 12 generations, breeding the most negative pairs in the positive line and most positive in the negative line. By the 65th generation these reverse lines had rapidly returned to a neutral average score. Consequently, it was concluded that a considerable genetic reservoir of additive factors must be available in the base population with all major chromosomes contributing to the determination of the geotactic sign, but even after 52 generations of selection plus the unconscious inbreeding that certainly occurred, a large additive genetic reservoir remained: the selected strains had not become homozygous (fixed).

With the techniques of Hirsch and his colleagues in mind, a populational analysis of *D. pseudoobscura* was begun by Dobzhansky and Spassky in 1965 to determine both geotactic and phototactic scores. A Piñon Flats population containing *AR* and *CH* arrangements of the third chromosome at equal frequencies constituted the base population. Out of every 300 pairs of flies tested each generation, 25 were selected with the most extreme scores. Lines diverged quickly, reaching a maximal separation at about the 18th generation in the phototactic experiment and at the 26th generation in the geotactic experiment. After a few generations at a plateau, lines were relaxed from selection by using randomly chosen flies mated in population cages. From Figures A and B it is evident that response to selection was nearly symmetrical in both sets of selected lines; after relaxation of selection, there was a

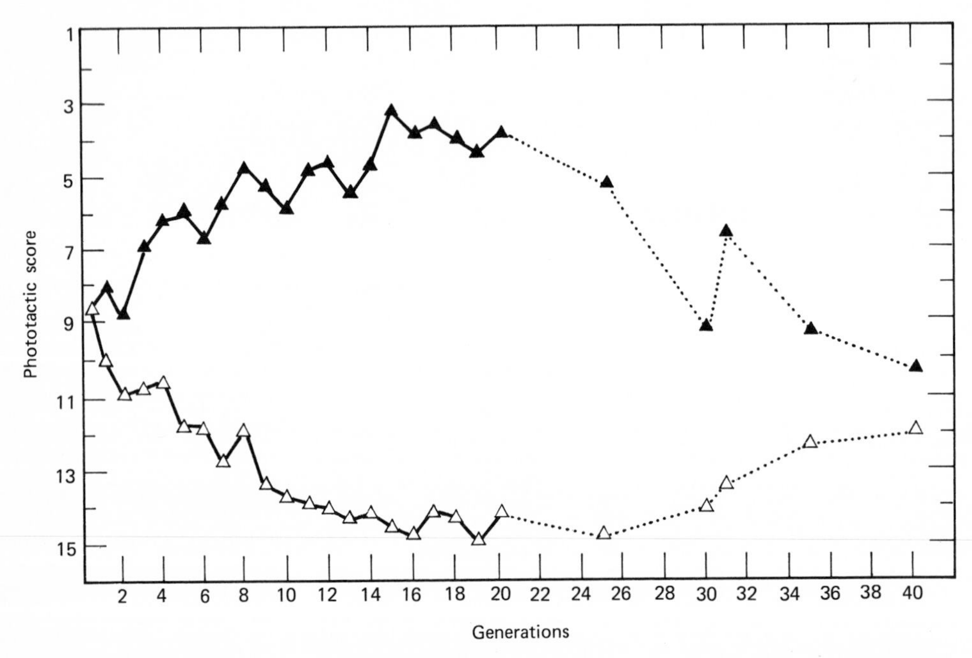

Figure A. Selection for phototaxis. Open triangles = in positive direction; solid triangles = in negative direction; dotted line = selection relaxed. From Dobzhansky and Spassky (1969).

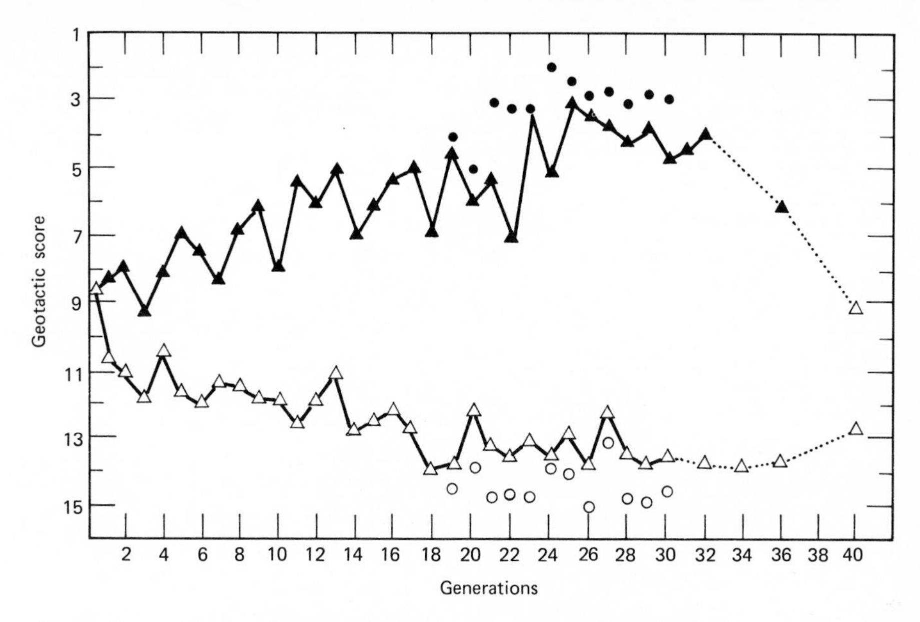

Figure B. Selection for geotaxis. Symbols same as in Figure A. Circle = mean scores on retests of 100 "best" flies. From Dobzhansky and Spassky (1969).

strong tendency for restoration of original scores (except for the positive geotactic line, which did not change significantly). The tendency was noteworthy for the lines with negative scores. These responses parallel very closely those observed by Hirsch. Of course, relaxation of selection differs from reverse selection in that natural selection alone would be responsible for changes in the former, while the latter is directed consciously by the observer.

Heritability was estimated (Richmond, 1969) in two ways. First, the regression of response on cumulative selection differential over the first 15 generations of selection gave the following results:

	Realised Heritability	
Selection Line	Phototaxis	Geotaxis
Positive ♀♀	0.100 ± 0.009	0.028 ± 0.006
Positive ♂♂	0.100 ± 0.008	0.021 ± 0.007
Negative ♀♀	0.090 ± 0.011	0.024 ± 0.011
Negative ♂♂	0.063 ± 0.010	0.034 ± 0.009

It is clear that the reaction to light is more heritable than that to gravity. Both are rather low heritabilities with little difference between the sexes.

The second method was by regression of male and female progeny on the mean parent scores done on flies from the base population before selection. The following regression coefficients equal the heritability estimate:

	Regression (b) Offspring on Parent Mean $= h^2$	
Progeny Sex	Phototaxis	Geotaxis
♀♀	0.167 ± 0.044	0.175 ± 0.044
♂♂	0.154 ± 0.067	0.055 ± 0.038

These estimates are clearly higher in every case, except possibly for geotaxis males, than the realized heritabilities from the selection of 15 generations. It is remarkable that such good responses took place under selection when the heritabilities were clearly so low. The high intensity of selection was in part responsible for the response observed; but the lower than expected realized heritability could result if, for example, the selection response was higher at first but leveled out after early generations so that by averaging over 15 generations the mean response would be biased by the low-response later generations.

In both positive geotactic and phototactic lines, the *AR* arrangement increased to about 85 percent at the expense of *CH* with selection, remaining at about that frequency after relaxation of selection for positive geotaxis. In the negative lines, *AR* did not increase as much (74 percent) in the phototactic line, and it decreased to 32 percent in the geotactic line. The latter exceptional chromosomal constitution was reversed following the relaxation of selection. In unselected populations kept in the laboratory, *AR–CH* from Piñon Flats reach equilibrium at a level of 70–80 percent *AR*, so it is quite likely that the negative geotactic line was indeed showing a correlated response to selection that favored *CH* chromosomes. As with the Hirsch experiments, a populational homeostasis, or tendency to return to the original score, was clearly in evidence. The average neutrality of the natural populations must then be an adaptive trait characterized by a relationally balanced genotype ready to respond to environmental forces quickly and easily.

ESTIMATING GENETIC VARIANCE FROM CROSSES

Once a particular genetic system has been selected, usually in "up" and "down" lines, it is often convenient to cross these opposite lines, measure their progenies in F_1, F_2, backcrosses (B_1 and B_2), and other generations derived from the cross, in order to estimate additivity, dominance, interaction, and environmental components of variance. These estimates can often be assumed to have some bearing on the nature of the genetic system of the original base population before selection, although we should always feel cautioned in drawing such conclusions from particular responses in selected lines, because the number of possible genetic systems that can be fixed in lines could be large indeed; the diversity of genetic pathways for particular phenotypes undoubtedly can be great in most sizable populations where random mating has been the natural system. It would be erroneous to conclude that the selection scheme had uncovered the only genetic system available to produce the response.

The student should also be warned about drawing conclusions of additivity, dominance, and other components of variance from progenies derived by crossing selected lines. It can be assumed that most gene frequencies determining the differences between the "up" and "down" lines would be equalized ($p = q = 0.50$) by the crossing. Such allelic frequencies may or may not be close to those in the original base population from which the lines were selected, and because additive, dominant, and interaction components of variance depend on gene frequencies, they will be ascertained with those qualifications in mind. Nevertheless, it is useful from at least two standpoints to proceed with crosses; first, we can estimate a minimum number of "effective factors," or "segregation index," of independently assorting loci (Wright, 1952) by which the selected lines differ, and, second, relative amounts of dominance and additivity and thus heritability can be calculated (although realizing that gene frequencies may be quite different in the base population). Therefore, by making a few simplifying assumptions, much useful information can be obtained from crossing lines and analyzing the variance of resulting progenies. The following abridged and elementary description is intended only to be introductory to acquaint the student with a few fundamental procedures. (For more complete discussion of problems and methods, consult Falconer, 1960; Mather and Jinks, 1971; or Wright, 1968.) The assumptions made can often be tested by experimental designs, which are beyond the scope of this book.

Let us assume lines have been selected for a sufficient number of generations to achieve a plateaued pair of opposite lines. We may further assume that homozygosity has been achieved at some loci for opposite alleles (how could an assumption of homozygosity be tested?), so that on crossing, allelic frequencies for those loci will approximate 50 percent. The variation inherent in the selected lines and in their F_1 progeny we shall assume to be exclusively nonheritable (except for mutation). Consequently, we may use their variances (properly weighted) as estimations of the environmental component (σ_E^2) when analyzing later generations (F_2 and backcrosses, for example), provided we assume that no genotype-environmental interactions in those generations would change the total variance substantially. Now we can proceed to estimate the additive and dominant components of genetic variance with F_2 and backcross progenies.

In the F_2 progeny, the contributions to the genetic variance made by each segregating locus can be expressed by (6-6). An F_2 simulates a random mating array with $p = q$, in which the additive component $2pq[a + d(q - p)]^2$ simplifies to $2pqa^2$, because the second term is zero. Provided that the loci concerned are more or less independent genetically and phenotypically (not closely linked or interacting to any extent), the total variance given by k loci will be the sum (as in 6-15):

$$\sigma_{F_2}^2 = \tfrac{1}{2}\sum a^2 + \tfrac{1}{4}\sum d^2 + \sigma_E^2 \qquad (7\text{-}16)$$

In backcrosses of the F_1 to either of the line parents, there will be a 50-50 split for homozygotes and heterozygotes with respect to each locus in which the parent lines differ. On the midparent scale (Figure 6-2), the means of backcross progenies will be as follows:

B_1 (backcross to "up" parent): $\frac{1}{2}(a + d)$

B_2 (backcross to "down" parent): $\frac{1}{2}(d - a)$

Contributions to variance in the two backcrosses then are as follows (see Exercise 10):

$$\sigma^2_{B_1} = \tfrac{1}{4}\sum(a - d)^2 + \sigma^2_E \tag{7-17}$$

and

$$\sigma^2_{B_2} = \tfrac{1}{4}\sum(a + d)^2 + \sigma^2_E$$

If these two backcross variances are summed, we obtain

$$\sigma^2_{B_1} + \sigma^2_{B_2} = \tfrac{1}{2}\sum a^2 + \tfrac{1}{2}\sum d^2 + 2\sigma^2_E \tag{7-18}$$

If we compare (7-18) with (7-16), it is evident that an estimate of additive and dominant components could be made as the simultaneous solution for the two equations: The difference (7-18) minus (7-16) estimates $\frac{1}{4}d^2 + \sigma^2_E$ from which the dominant component of genetic variance can be easily obtained.

In Example 7-5, with some simplifying conditions added to the original estimates made by the authors, this calculation is illustrated. (Barnes and Kearsey, 1970, have included weights to accommodate homozygote × heterozygote interactions, which we have omitted.) Of course, assumptions of linkage and interactions will generally lower the total additive variance estimates for the backcrosses (and for F_3 means, which are also omitted here). The use of more generations from the crosses (F_3 from selfing or from random F_2 pairs) has been explicitly described by Mather and Jinks (1971) together with their variance and covariance component estimates. All these further crosses can improve the refinement of additive, dominant, and interaction components with additional equations and weights for their simultaneous solution. These are beyond the scope of this book, and the reader is referred to the paper of Barnes and Kearsey (1970) for an illustration of these refinements.

Finally, overall dominance between alleles selected may be considered from the standpoint of deviation between the F_1 mean and the midparent (halfway between selected line parents). Some allelic differences between the lines will, of course, produce more dominant effects than others, and the deviation of F_1 from midparent represents a resultant, or preponderance in one direction, of all individual locus dominant effects. This resultant in fact measures the ratio $\sum d/\sum a$ for the k loci and has been termed the "potence ratio" by Mather (1949); it represents the net dominance of one parental genic set over the other. In Example 7-5 it is negative, indicating preponderance of dominance in the low direction. It is important that we should have no preconceptions that all genes of like effect are concentrated exclusively in one parent or the other or that for each locus all the d effects have the same sign (unidirectionality). Suppose in reality that the F_1 would equal the midparent; then there would be an apparent complete lack of dominance, but there could be full dominance at each locus if there were an exact balance between plus and minus dominant effects (that is, with $A(+)$ dominant over $a(-)$, but $B(-)$ dominant over $b(+)$). Nevertheless, the potence ratio is a useful indication, especially when comparing selected lines, because the genetic response to selection will depend on overall dominance relationships for the loci with free variation. For example, in Example 7-5, the lower number of sternopleural bristles, being dominant overall, may be an indication that there is a selective advantage to flies with lower numbers of bristles in natural populations, a fact that was reinforced when the same authors tested flies of varying bristle number for their study of Darwinian fitness in random mating populations (Barnes and Kearsey, 1970*b*).

EXAMPLE 7-5

GENETIC ANALYSIS OF A METRICAL TRAIT FROM A DROSOPHILA POPULATION

At the University of Birmingham, Barnes and Kearsey (1970) postulated that stabilizing selection is able to maintain variation for characters that show no dominance if the effects of different loci are unequal. To ascertain the extent of genetic variation for an additive quantitative trait (sternopleural bristle number), they developed a program to analyze wild and laboratory populations of *Drosophila melanogaster*. Their base population was derived from flies caught in nature at Austin, Texas in 1965 and was maintained in a laboratory population cage (size about 2500 flies) as the source of study.

The mean and variance of bristle number (sum of right and left sides of the fly) among females in the base population is given in Table A, top row. From the low variance of the trait, it is clear that either the genotype controlling the trait was highly homozygous or, if heterozygous, a remarkable balance with genetic stability had been maintained in the population. That the latter was the case was evidenced from the selection response quickly ensuing. In the low direction, bristle count dropped linearly for about 12 generations and plateaued at about 12 bristles. In the high direction, the count responded much faster for 18 generations to a plateau at about 44 bristles. Realized heritability from the generations in which selection response was about linear was 0.814 ± 0.018. Heritability calculated from regression of offspring on parental mean in the original population was 0.593 ± 0.169. These estimates are not significantly different. Selection was relaxed in certain later generations, at which time bristle numbers tended to return to the original number.

At the 34th generation a genetic analysis was done by crossing high and low selected lines, raising F_1, F_2, and both backcrosses. Means and variances of these tests are given in Table A. There is some dominance for the low bristle number when we compare the F_1 mean (20.41) with the midparent position (25.08); this fact can account for the better response in the upward direction. Barnes and Kearsey fitted their crossing data to a genetic model including additive, dominant, and digenic interaction deviations. For the sake of

TABLE A *Sternopleural bristle numbers (females) for original Texas population, selected lines at G_{34} and crosses between lines (from Barnes and Kearsey, 1970)*

Lines and Progenies	N	Average Bristles	Average Variance Within Families
Original population		18.40	3.13
High line (G_{34})	120	37.58	10.91
Low line (G_{34})	90	13.53	3.39
F_1(H and L)	188	20.41	4.52
F_2	240	22.06	42.05
Backcross × H	470	27.86	44.91
Backcross × L	371	17.54	7.91

simplicity we shall omit the latter, and illustrate the method of estimating additive and dominant components of genetic variance from these data, by allowing a few minor approximations.

Let us estimate V_E (environmental component) $= 4$ (approximately) from the low line parent and F_1 variances within families. (The higher parent's variance value may have been due in addition to some continuing segregation plus some lack of developmental homeostasis in a relatively divergent-from-normal morphotype.)

Then, we may let $\sigma^2_{F_2} = \frac{1}{2}A + \frac{1}{4}D + V_E = 42.05$, where $A = \sum a_i^2$ and $D = \sum d_i^2$ and $V_E =$ environmental component.

$$\sigma^2_{B_1} + \sigma^2_{B_2} = \tfrac{1}{2}A + \tfrac{1}{2}D + 2V_E = 52.82$$

$$\text{or} \begin{cases} \tfrac{1}{2}A + \tfrac{1}{4}D = 38.05 \\ \tfrac{1}{2}A + \tfrac{1}{2}D = 44.82 \end{cases}$$

simultaneously solving for D

$$\tfrac{1}{4}D = 6.77$$

$$D = 27.08$$

$$\text{and } A = 62.56$$

Because $\sigma^2_A = \frac{1}{2}\sum a_i^2$, then heritability $= 31.28/42.05 = 74.39$ percent in the F_2. In other words, about 75 percent of the observed phenotypic variance in this F_2 is from the additive component.

Finally, we may calculate the minimum estimate of effective factors ($k =$ segregating units) using (7-19).

$$k = \frac{(P_1 - P_2)^2}{8(\sigma^2_A)} = \frac{(24.05)^2}{8(31.28)} = 2.31$$

This estimate again is close to what would be expected if the major chromosomes in this species were acting as entire units, segregating in the F_2. An indication of recombinants can be implied in the increased variance of the backcross progeny from the higher parent (44.91).

Obviously, however, the excellent selection response must mean that several loci were involved. The authors made some interesting simplifying assumptions and proceeded to calculate minimum effective factors based on the randomly mating base population from which the selected lines were drawn, using the realized heritability. We shall proceed in like manner using some approximations to simplify the calculations.

In the base population, the mean ($\bar{Y}$) and additive component of variance (σ^2_A) are as follows:

from (6-14) $$\bar{Y} = k[\bar{a}_i(\bar{p}_i - \bar{q}_i) + 2\bar{p}_i\bar{q}_i\bar{d}_i] + (m)$$

from (6-15) $$\sigma^2_A = 2\bar{p}_i\bar{q}_i \cdot k(\bar{\alpha}_i)^2$$

where $(m) =$ midparent value and $\bar{p}_i$, $\bar{q}_i$ are average allelic frequencies over all loci.

Barnes and Kearsey made the assumptions that (1) all additive and dominant effects were equal for all loci ($a_i = a_j \cdots$) and ($d_i = d_j \cdots$), and (2) allelic frequencies (dominant low bristle alleles at p frequency) were equal for all loci.

Using the original population data, we can estimate the constant p frequency with the following quadratic expression from (6-14) for the mean:

$$18.40 = [12 - 24q - 9.32q + 9.32q^2] + 25.08$$
$$q = 0.696$$
$$p = 0.303$$

using $k\bar{a} = \frac{1}{2}$ difference between selected parents $= 12$ and $k\bar{d} = \frac{1}{2}$ difference between F_1 and midparent $= -4.66$ and changing p to q, compared with the authors' more refined estimate of $p = 0.296$.

With this estimate of bristle average allelic frequency, we may solve for k (effective factor number). The additive component of variance (σ_A^2) may be computed using the observed variance in the original population ($\sigma_P^2 = 3.13$) times the heritability realized by selection ($h^2 = 0.814$) to give $\sigma_A^2 = 2.547$. Average α can be estimated $[\bar{a} + \bar{d}(q - p)] = 12 - 4.66\ (0.370) = 10.2758$. Thus, we have $2.547 = 2(0.315)(9.685)k\ [10.2758]^2$, or $k = 17.9$ (compared with the authors' more refined estimate of $k = 16$, approximately).

Barnes and Kearsey make note of the fact that inequality of additive or dominant effects would result in k being an underestimate. For the condition of inequality $p_i \neq p_j \neq \ldots$, the general results have not been examined but "some special cases considered numerically suggest that k may be overestimated but to a negligible extent." Consequently, there must be in the order of 16–18 factors available for the selection response from the base population. On the other hand, it is possible that owing to the fact that relaxed selection caused quick tendency to return to the normal number, perhaps many of these factors were linked in a disequilibrium originally so that effectively there were fewer factors until recombination released some of them against natural selection tendencies. Barnes and Kearsey went on to examine the adaptive values of different bristle numbers, but that story will be postponed until Chapter 15 (Kearsey and Barnes, 1970).

SCALE TRANSFORMATIONS AND ADDITIVITY

While qualitative discontinuous traits have easily distinguishable classes and thus have no ambiguity due to scale of measurement, quantitative traits can only be described in terms of means, variances and other statistical relationships that do depend on the scale. A choice of scale appropriate to the trait being measured is one of the most essential steps in experimental work on quantitative characters. In the example illustrated in Figure 7-5, with original measurements, their logs, and their square roots compared, it is evident that conclusions about additivity, dominance, and interactions may be quite different depending on the scale chosen. Consequently, even a brief consideration of scale effects and the more desirable ways to deal with them is necessary. A few of the more useful scale transformations are listed in Appendix A-11.

For the purposes of the genetic analysis when crossing strains differing in a quantitative trait, our objective can be summarized by the following points according to Wright (1952):

1. Genetic strains tested and their F_1 should be as uniform in variance as possible. We must assume that genic action in producing the phenotype is taking place on

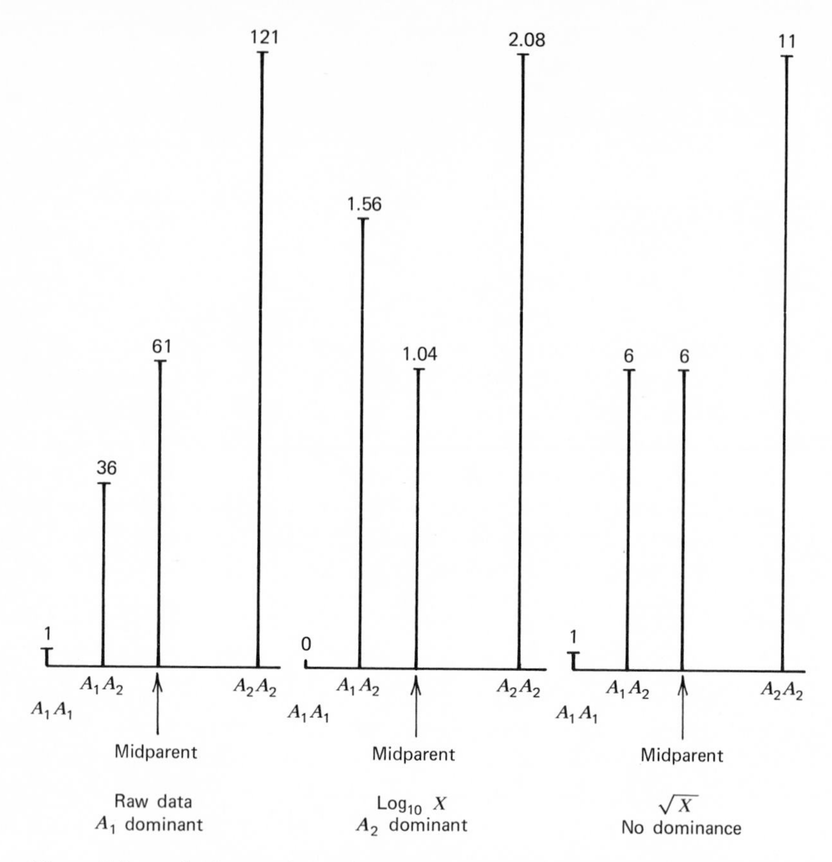

Figure 7-5. Scale transformation from raw data showing dominance of one allele: $\log_{10} X$ makes opposite allele "dominant" and $\sqrt{X}$ eliminates dominance. Modified from Lerner *The Genetic Basis of Selection* (New York: John Wiley & Sons, 1958).

the same scale irrespective of genotype. Consequently, the coefficient of variation (σ/m) should be equal within sampling error for parents and F_1.

2. Distributions within classes, within generations, should be normal.
3. Genetic and environmental effects should be determined from an additive scale (for example, the $\sqrt{X}$ scale in Figure 7-5) to minimize nonadditive interactions and other "difficult" complexities as much as possible.
4. Relationships between generations and crosses (P_1, P_2, F_1, F_2, and backcrosses) should conform to additive criteria.

Of course, it is clear that attempts at scale transformation for additivity may be in complex genetic cases very much a pious hope; each genic locus may be acting on its own scale independently of other loci. It is important to realize that in spite of averaging out possibly diverse effects for additivity, the limitations of these devices include the possibility of arriving at conclusions about genic action being additive. Actually, the statistical design here outlined must be admittedly empirical until individual genic loci can be ascertained or specifically measured. Nevertheless, as empirical genetic systems for present purposes without the techniques for genetic "dissection" of each locus available, we can utilize these points to considerable advantage as long as we keep those limitations in mind.

Additivity of parents and crosses (the last two of Wright's points above) can be tested, or a confirmation of generation means to the additive-dominant model made, according to Mather's scaling tests (Mather, 1949; Mather and Jinks, 1971), provided that parents, F_1, F_2, and backcrosses to each parent have been recorded.

First, if we consider only the backcross to one parent and the F_1 from the original cross, it is evident that the backcross mean (B_1), which should have a 50:50 distribution of AA:Aa genotypes at each locus, will fall at the halfway point between F_1 mean and P_1 mean, or $B_1 = \frac{1}{2}(F_1 + P_1)$, whence $2B_1 = F_1 + P_1$ and $A = 2B_1 - F_1 - P_1$, where A is expected to be zero. Similarly, for the backcross mean to the other parent (B_2), $B = 2B_2 - F_1 - P_2$, where B is expected to be zero.

If we consider using the F_2, F_1, and parent means, we realize from (6-3), when $p = q = 0.5$, that $F_2 = \frac{1}{2}d$, that is, one-half the deviation of the F_1 from the midparent. Then, letting the midparent $= \frac{1}{2}(P_1 + P_2)$, the F_2 deviation from midparent in terms of F_1 is $F_2 - \frac{1}{2}(P_1 + P_2) = \frac{1}{2}[F_1 - \frac{1}{2}(P_1 + P_2)]$, whence $4F_2 = 2F_1 + P_1 + P_2$ and $C = 4F_2 - 2F_1 - P_1 - P_2$, where C is expected to be zero.

Variances given by Mather for these scaling test coefficients (A, B, and C) can then be estimated, and square roots taken for standard errors and tests of significance from zero can be obtained.

$V_A = 4V_{B_1} + V_{P_1} + V_{F_1}$ (variance of A in terms of variance of means from B_1, P_1, and F_1)

$V_B = 4V_{B_1} + V_{P_2} + V_{F_1}$ (variance of B)

$V_C = 16V_{F_2} + 4V_{F_1} + V_{P_1} + V_{P_2}$ (variance of C)

For the data in Example 7-5, as an illustration, we obtain

$$\text{A} = -2.27 \pm 0.70$$
$$\text{B} = +1.14 \pm 0.39$$
$$\text{C} = -3.70 \pm 1.74$$

All of these are significantly different from zero in absolute value, so we must surmise that the bristle numbers alone are not sufficiently additive to fit the simplest model. Actually, Barnes and Kearsey did not change their scale, but a square root transformation might have improved the additivity in view of the dominance of the lower values. See Exercise 11 for another example. The main point is to obtain a scale that gives the best fit of these coefficients to zero and then proceed with all future analyses using that scale. It is important to realize, however, that the particular transformation used (logs, antilogs, roots, probits, reciprocals, etc.) does not imply anything about individual genic action in producing the phenotype. It is merely a device to permit approximation of results to an additive scale for mathematical facility, estimations of heritability, numbers of effective factors, and predictions of selection outcome. We can know a gene's action only from isolating a locus in a design aimed at its developmental activity.

ESTIMATING NUMBER OF "EFFECTIVE FACTORS" (SEGREGATION UNITS)

When crosses of selected lines have been analyzed and the additive component of genetic variance obtained, we may estimate the number of independently segregating genetic units (k) that could account for the difference between selected parents and the variance in F_2 and other generations. It must be apparent at the outset, however, that most of the simplifying assumptions needed to proceed here conspire to make the estimate a minimum, as we shall see, and the combined sets of assumptions may prove so far from reality that much of the calculation (even with greater refinements allowing for narrowing the assumptions) may seem a waste of time. In the light of recent successes in locating polygenes (to be mentioned in the last section of this chapter), these estimations of effective factor numbers may seem useless, yet they give us minimum order of magnitude and an expectation for early responses to selection. Locating polygenes with special linkage markers, as mentioned below, can only be attained in a very few organisms where such markers are available for testing and locating. For most polygenic traits in most organisms, these advantages are far from attained or attainable in the near future.

With an estimate of the additive component of genetic variance from an F_2 between selected lines, we may make the following assumptions for estimating the minimum number of independent segregating factors (probably not single genic loci) that may account for the difference between selected parents:

1. All the "plus" alleles of the k factors are in one line and all the "minus" alleles in the other—at least those alleles relevant to the difference between selected lines.
2. Additive (a) values are equal for each k factor: $a_A = a_B = a_C \cdots = a_k$, where $A \cdots k$ = each factor or segregating unit.
3. The linkage between factors is loose enough not to distort segregation. Nonallelic interactions are not significant.
4. Dominance is moderate or lacking at all factor loci, so that the dominant component of genetic variance will be small or zero.

With these assumptions in mind, the total additive component for the F_2 generation is

$$\sigma_A^2 = 2pq\, a_A^2 + 2uv\, a_B^2 \cdots = \tfrac{1}{2}a_A^2 + \tfrac{1}{2}a_B^2 \cdots \tfrac{1}{2}a_k^2 = \tfrac{1}{2}ka^2$$

The difference between selected parent values ($P_1 - P_2$, or D) measures twice the additive effect (a) for each factor so that $D = \sum 2a_i = 2ak$, and $a = D/2k$.

Therefore, in terms of difference between selected parents, the additive variance component can be expressed as

$$\sigma_A^2 = \frac{k}{2}\left[\frac{D}{2k}\right]^2 = \frac{D^2}{8k}$$

Solving for k,

$$k = \frac{D^2}{8\sigma_A^2} \tag{7-19}$$

(minimum estimate of segregation units, or effective factors).

This relationship was worked out by Wright in 1921 for W. E. Castle, who had wished to estimate the number of factors segregating for mean weight in his strains of rabbits. Castle estimated segregating units of 3, 14, and 22 in three different crosses between strains of rabbits. Wright (1934) applied the formula to progeny from a cross between four-toed and three-toed strains of guinea pigs to obtain an estimate of four units. A number of subsequent estimates made on many other organisms have been summarized by Mather (1949) and Mather and Jinks (1971). We can apply this principle to data illustrating the estimation from Examples 7-1, 7-2, and 7-5 concerned with factors determining abdominal bristle or sternopleural bristle numbers in drosophila. Without knowledge of linkage and other restrictions to our simplifying assumptions (above), we should note that the number of segregating units (effective factors) in these examples, estimated from F_2 variances, appear to be the same order of magnitude as the major chromosome number in *D. melanogaster*. In other words, with no crossing over in males plus the probability of linkage between many loci affecting the traits measured, the differences between selected lines amount to whole chromosomes acting as the units that determine the observed variation.

We might look briefly without mathematical rigor at the consequences of removing the simplifying assumptions we have made. What happens to the estimate if each assumption cannot be substantiated?

1. If the plus and minus factors are not equally distributed between selected lines, then the difference $(P_1 - P_2)$ would be less than $2\sum a_i$, and k would be underestimated. This would be the case if there were two loci concerned, for example, with one line $+-/++$ and the other $-+/--$, with a preponderance of alleles in the right direction but not homozygosity at both loci.

2. If the the additive value for each k factor is not equal to that of the other factors, inequality of increments will lead to a variance in increments that will be an additional component of the denominator and thus lower the estimate of k.

3. It can be seen intuitively that if two factors are tightly linked, recombination will not be sufficient from moderate samples of F_2, backcrosses, or F_3 to separate them, and they will be spuriously reckoned as one locus.

Mather (1949) and Mather and Jinks (1971) have described at considerable length that when backcrosses and F_3 are utilized, the effects of dominance, linkage (in the order of 0.10 to 0.30), and nonallelic interactions may be separated to refine the estimate of additive component in the denominator. Of course, inability to separate those elements lowers the estimate as before. As stated by Wright (1952), "Unfortunately very close linkage, which is the most important from the standpoint of ultimate release of variability in balanced systems, is also not likely to be revealed statistically by changes that occur in a single generation." In short, the number of segregating units must be thought of not in terms of single loci since they probably are not "final units," but they are useful in predicting rates of advance under selection for a few generations while initial additivity remains without recourse to recombination for newly produced linkages with nonallelic interactions.

In Example 7-1, Mather demonstrated low genetic variance when original strains were crossed and analyzed for bristle number, but there was considerable genetic variance from the selected strains. That fact plus the initial temporary plateau (at generations 2-4) indicated to Mather that in the original strains polygenes were probably "balanced" in repulsion linkages

$$\left(\frac{+-+-\cdots}{-+-+\cdots}\right).$$

It appeared that very little variation was made available by recombination at first. By the fourth generation of selection, however, there was presumably a release of variability by crossing over, so that the temporary plateau was broken and much more response was achieved. After 12 generations of selection, the amount of genetic variance had nearly tripled over that characterizing the original strains. In Example 7-2, a more impressive demonstration was made for increased selection response following release of variability and uncoupling from a depression of fertility in the high bristle number line when stringent selection was relaxed and genetic recovery had taken place. These results led to an explicit concept of genetic structure and the expectation of response from selection: (1) polygenes are often "relationally balanced" in wild populations, and (2) additive genetic variance for a trait is converted into linear response in the direction of applied selection either until the variance component reaches zero or until the genetic system determining the population's fitness becomes so unbalanced or disturbed that fitness is lowered excessively. This latter expectation would occur if genic loci affecting reproductive potential were multitudinous and tightly linked to loci determining the selected trait. See Chapters 17 and 19 (populational homeostasis).

LOCATING POLYGENES

At present we have a limited knowledge of the architectural organization and properties of polygenic systems, but in a few organisms such as drosophila, with a small number of chromosomes, useful marker loci available to delimit chromosomes, and linked blocks within chromosomes, it has been possible to indicate in some degree the number and kinds of loci that control a few quantitative traits; in fact, we can trace down with some accuracy location and specification of the "effective factors." It has been useful to narrow down the controlling regions, first beginning with whole chromosome distribution. We shall briefly outline the main methods customarily employed in such analysis.

Chromosome Assay

The conventional technique for locating any mutant to linkage group by dominant markers with crossover suppressors can be used to assay selected lines (or any wild population chromosomes that differ from the dominant marker in the quantitative trait) for relative contributions to the phenotype for each chromosome marked. Mather and Harrison (1949) (see Example 7-2) employed the technique assaying for X, second, and third major chromosomes with about 90 percent efficiency (that is, recombination from chromosome 4 and a low amount of double crossing over with possible environmental effects lowering the efficiency by about 10 percent). Essentially, the technique would be as follows for any three linkage groups. Let A, B, C = dominant markers (usually lethal as homozygotes and linked with crossover suppressors) on chromosomes 1, 2, and 3, respectively. Let $+^1$, $+^2$, $+^3$ = wild-type alleles on homologues from the selected line being assayed. Then

parents: *selected line* $(+^1/+^1;\ +^2/+^2;\ +^3/+^3) \times (A/Y;\ B/+;\ C/+)$,
(females) (males with dominant markers)

and F_1 females $(A/+^1;\ B/+^2;\ C/+^3)$ backcrossed $\times$ selected line males $(+^1/Y;\ +^2/+^2;\ +^3/+^3)$ produce eight phenotypic classes of progeny, as given in Table 7-5. The phenotype selected has been measured in the line, in the dominant marker strain, and in the F_1. The F_1 differs from the heterozygous dominant marker in that the wild-type homologues of the lab stock $(+;\ +;\ +)$ are replaced by selected line chromosomes. Therefore, if a comparison between F_1 and the laboratory marker shows a net significant difference, dominance of selected polygenes over the marker homologues would be indicated. One limitation of the chromosome assay method is that the greater the amount of dominance of selected alleles over those on the marker, the less will be the difference between classes 1 and 8 among backcross progeny. If completely recessive, the selected alleles will produce the equivalent of $\sum a_i$ additive effect for each chromosome, while partial dominance would produce an intermediate value. Presumably, if selection had concentrated plus alleles on the "up" line on any particular chromosome, they would be measureable as an increment in double $(+'/+' \cdots)$ dose over the marker heterozygote $(A/+' \cdots)$ for that chromosome, provided they are recessive or only partially dominant. Consequently, the relative effects of the major chromosomes can be ascertained by the usual process of calculating the net sums, as indicated in Table 7-5 for main effects.

The particular comparisons to be made between phenotypes to obtain main effects and interactions is due to R. A. Fisher and is a standard three-factor analysis (described by Mather, 1965). Letting A be the phenotype (bristle number, for example) of the individuals

TABLE 7-5 *Coefficients for chromosome assay from backcross progenies with three pairs of selected versus marked homologues (A, B, C = markers on chromosomes 1, 2, and 3, respectively. $+_i$ = selected wild-type homologue)*

	Phenotype			Combinations of Coefficients Measuring						
				Main Effects			Interactions			
Class of Progeny	Chromosome 1	2	3	1	2	3	1-2	1-3	2-3	1-2-3
1	A	B	C	−	−	−	+	+	+	−
2	A	B	$+_3$	−	−	+	+	−	−	+
3	A	$+_2$	C	−	+	−	−	+	−	+
4	A	$+_2$	$+_3$	−	+	+	−	−	+	−
5	$+_1$	B	C	+	−	−	−	−	+	+
6	$+_1$	B	$+_3$	+	−	+	−	+	−	−
7	$+_1$	$+_2$	C	+	+	−	+	−	−	−
8	$+_1$	$+_2$	$+_3$	+	+	+	+	+	+	+

marked by the A tester and 1 the phenotype of those lacking A (selected chromosome homologue of A), then we have the expected coefficients for the eight classes of backcross progeny giving the main effect of the first chromosome by expanding the expression $(1 - A)(1 + B)(1 + C)$; that is, the contrast for A versus its homologue only, independent of the second and third chromosomes. The net sum of the selected homologue minus A is given in the fith column of Table 7-5, $- ABC - AB - AC - A + BC + B + C + 1$. In like manner, the main effects of second and third chromosomes are found from the expansions of $(1 + A)(1 - B)(1 + C)$ and $(1 + A)(1 + B)(1 - C)$, respectively. First-order interactions depend on the joint action of two main effects, so for the first and second chromosome interaction, with the third being not considered, the joint contrast will be $(1 - A)(1 - B)(1 + C) = + ABC + AB - AC - A - BC - B + C + 1$. Likewise, the remaining first-order interactions will be expansions of $(1 - A)(1 + B)(1 - C)$ for first and third chromosomes and $(1 + A)(1 - B)(1 - C)$ for second and third jointly. Finally, the seventh contrast, or second-order interaction, measures the joint departure from simple additivity of all three chromosomes considered together—that is, the expansion of triple contrast $(1 - A)(1 - B)(1 - C)$ given in the last column of the table. These functions are orthogonal to one another, and the seven sets of values are independent so that they may be interpreted as separate entities.

In Example 7-2, Mather and Harrison, using a refinement of this method, found all major chromosomes in their 10 selected lines to have accumulated some alleles affecting bristle number (polygenes). The lowest lines tended to have minus modifiers very much as they were distributed in the lower parent, while the high lines responded mostly by concentrating plus modifiers, with especially the third chromosome markedly altered in contrast with the unselected low parent. These modifiers displayed little dominance so that the efficiency of measurement over the tester marker was quite good. Also, selection had produced a positive interaction (epistasis) between the X and third chromosomes in the highest lines. A second notable example of this chromosomal assay method is cited in Example 7-4 on geotactic scores of drosophila selected by Hirsch and his colleagues (1967); as with bristle numbers, these behavioral control effects were polygenic on all major chromosomes, with no significant interactions, but with predominantly X and second chromosomes responding in selection for positive geotaxis (downward behavior) and with the third chromosome responding for negative (upward behavior).

Intrachromosomal Location

It is enormously more time consuming than the chromosomal assay to pinpoint polygenic activity or a specific polygene locus; however, locating a few polygenes of moderately strong effect has been successful. An indication of the approximate number and kinds of loci that control a morphological feature (bristle number) and a component of fitness (viability) was first made by Breese and Mather (1957, 1960) by separating linked segments of particular chromosomes from selected lines of drosophila. The more exact location and specification of "effective factors" was accomplished by Thoday and his colleagues in 1962 when analyzing sternopleural bristle number from populations of disruptively selected design (see Chapter 16). The methods utilized by these groups of workers differ sufficiently to give a brief description of each. The first (Breese and Mather) employs segments of a selected "wild-type" chromosome without the influence of marker mutants. The second technique (Thoday) requires some dominance of the metric trait polygenes from the selected chromosome over the tester homologue in order to be ascertained. The first method, although more time consuming than the second, perhaps can be considered more exhaustive because the dominance condition of utility in Thoday's method may prevent ascertainment in cases of complete recessives.

Breese and Mather found from their chromosomal assays that two contrasting selected lines (a high line from a cross between numbers 8 × 9 of Mather and Harrison, Example 7-2, and a low line of Harrison's) differed mostly on chromsome 3. They proceeded to construct strains with known segments of chromosome 3, first by crossing both high and low lines to a recessive marker stock containing eight mutants ($\underline{ru\ h\ st\ p\ cu\ sr\ e\ ca}$), backcrossing heterozygous F_1 females of each to compound mutant males so that recombinant chromosomes consisting of portions of the selected wild-type chromosome could be recognized ($\underline{+^H\ +^H\ st\ p\ cu\ sr\ e\ ca}$ or $\underline{ru\ h\ +^L\ +^L\ +^L\ +^L\ +^L\ +^L}$, for example), crossing these complementary recombinants to produce wild-type daughters, again mated to compound mutant males so that completely wild-type recombinant progeny would have a compound high-low chromosome with portions from each selected line ($\underline{+^H\ +^H\ +^L\ +^L\ +^L\ +^L\ +^L\ +^L}$). These compound chromosomes were then "stored" in balanced condition over a dominant marker with crossover suppressor to be used subsequently for a diallelic analysis between all such compound and original parental chromosomes. No marker genes were left to confuse ascertainment due to pleiotropic or dominant effects in the wild-type progenies of these balanced strains. Each of the six regions delimited approximately by crossovers between marker loci was then tested for contributions to additive, dominant, and interaction effects with the genetic background controlled. All six regions showed some activity toward production of phenotype, and most showed some dominance for the lower bristle number, although activity was greatest in the segment nearest the centromere. Breese and Mather interpreted their findings as evidence for polygenic activity throughout the chromosome with at least six loci involved. (Viability was a more complex matter than bristle number, and we shall postpone discussion of its genetic architecture until later in this book.)

The second method (utilized by Thoday) depended on the following principles (paraphrasing Thoday, 1961). Suppose that a chromosome gives a higher metric value (from partial dominance) than that given by a homologous chromosome marked with recessive markers a and b. Let us arbitrarily designate the selected high strain as H while the ab/ab strain is relatively low (L). We may then testcross an H/ab heterozygote by the homozygous ab/ab stock and assay the four types of progeny for the metric trait: $++/ab$, ab/ab, $a+/ab$, and $+b/ab$. First, if there is only one HL locus, then the F_1 tested individuals are either $H++/Lab$ (if to the left of a), $++H/abL$ (if to the right of b), or $+H+/aLb$ (if between the markers). If the locus is closely linked to the markers so as to be little affected by recombination to left or to right in the first two instances, *recombinants* (neglecting double crossovers) between $a - b$ will indicate the location of the locus. For example, if HL is to the left of a, then $a+$ will be low and $+b$ will be high and parental types will be variable depending on crossover distance, $++$ being H but occasionally L by crossing over and ab being L but occasionally H. However, if the HL locus is between the markers, recombinants of $a - b$ (again ignoring double crossovers) will be heterogeneous, there being two types depending on linkage distance from respective markers to the HL site. The first step in locating particular polygenic activity, then, is to search for a pair of markers that will give this latter result—that is, the expectation from markers on opposite sides of the HL locus.

With the possibility of more than one HL locus between markers $a - b$, again the recombinant classes $a+$ and $+b$ would produce varying proportions of H and L, as follows (assuming two loci between markers): $+HH+/aLLb$ heterozygous F_1 back-crossed × ab/ab will produce uniformly HH or LL for parental classes but varying proportions of HH, HL, and LL for the recombinant classes. Thus, the difference between the single-locus and multilocus result is indicated: with one locus, recombinant classes produce only two types, H and L, while two loci produce three classes, HH, HL (or LH), and LL—that is, a high, an intermediate, and a low class. The technical problem is then resolved into progeny testing a sufficient number of recombinants to assess not only the number of quantitatively different classes but also their frequency to obtain map distances.

Just such tests were employed by Gibson and Thoday (1962) to discover and locate relevant loci on the second chromosome and by Wolstenholme and Thoday (1963) for loci on the third chromosome from a polymorphic population that had been "selected disruptively" for sternopleural bristle number with positive assortative mating (see Chapter 16). With detailed knowledge of these quantitative loci, these authors were in a position to investigate the genetic architecture of the population involved as well as developmental effects and specific properties of each locus. Three loci were then postulated as a minimum for the second and four loci for the third chromosome.

General Remarks

An important review of the situation should be considered at this point. Perhaps finding such genes of marked effect, which have been termed "oligogenes" by Thoday and others to imply phenotypic control by relatively few loci, could explain much of the observed variation instead of invoking the large number of very small-effect "polygenes" originally proposed by Mather. However, the number of loci influencing response to selection for bristle number, for example, may be much larger than the number *directly* affecting that trait. Most of the selection may have been determined by a few loci of relatively pre-ponderant effect. Probably "locatable polygenes" are a nonrandom sample comprising the most easily found and effective portion of the genotype from a continuous spectrum that ranges down to "genes of vanishingly small effect" (Thoday). Our very techniques of selection may tend to screen out small-effect genes and draw attention to loci with more major effects. Certainly, in cases of some techniques ("genetic assimilation," see Chapter 19), major genes often emerge as most influential, but modifiers are often found to alter slightly or markedly in subsequent selection the trait which the major genes seem to control. The simple assumptions we had to make to estimate numbers of loci biometrically seldom apply to most quantitative traits; however, some traits such as body size or "fitness" must be influenced by an enormous number of genes, and the loci concerned probably have about equal effects.

The hope of studying specific loci controlling continuous variation lies first in making the character discontinuous and second in reducing the number of loci segregating in any progeny. New techniques available for getting nearer the biochemical gene action level and redefining the characters in terms closer to the gene's primary action offer con-siderable promise. For example, Spickett, Shire, and Stewart (1967) have taken a continuous character (body weight in mice) and redefined the character in terms of quantity of a particular hormone per unit weight of hormone-producing tissue, thus breaking down the continuous character into a single-locus effect.

By aiming at particular relevant genes, by "genetically dissecting" a complex character with chromosomal markers, or by selecting for certain components of the character and excluding other components, an accounting for polygenic characters in terms of important loci can be made. Drosophila, with its small number of chromosomes and numerous other advantages, provides a means of studying the chromosomal distribution of effective factors and the extent of interactions between these factors. Fraser, Nassar, Scowcroft, and associates (1965–1966), working with a strongly "canalized" character (scutellar bristles, which are normally four in number), have had considerable success in allotting particular portions of selection response to the major chromosomes and interactions between factors. Other notable analyses for chromosomal allotment of selection responses are to be found in the work of King (1965) on DDT resistance, Mohler (1965–1967) on cross-veinless polygenes, Kearsey and Kojima (1967) on body weight and egg hatchability in drosophila, and Wehrhahn and Allard (1965) on heading time between varieties of wheat. Scowcroft (1966) has summarized the following important points to be ascertained from subdivision of these genetic changes, not fully realized from most selection studies or from analysis of genetic variance components in this field:

1. Interactions between segments on different chromosomes affect the character both in unselected populations and in unselected individuals.
2. Selection can alter interaction patterns and can also induce and utilize favorable interactions between nonhomologous chromosome segments.
3. The numerous lines similarly selected for the same trait, even though each line is derived from a single wild individual within a Mendelian population, are quite genetically heterogeneous; that is, genes responsible for the metric traits selected appear to be scattered throughout the genome, and selection is opportunistic in utilizing particular portions of the genome in eliciting a response.

Further discussion of selection and its consequences on genetic systems, precision in producing response, correlated responses, different selection modes, genetic assimilation, and inferences on the action of natural selection will be found in Chapters 14–19. A helpful summary review of artificial selection, its predictability, and its responses can be found in Lee and Parsons (1968).

SOME REMINDERS, COMPLEXITIES, AND MISCONCEPTIONS IN HUMAN POLYGENIC TRAITS

At the end of Chapter 6, the student was reminded of a few of the limitations involved when estimating polygenic parameters. Even at this introductory level, we have an intuitive sense that precision is much reduced when so many simplifying assumptions need to be made in order to estimate genetic variance components, numbers of segregating loci, predictions for selection outcome, and locations of polygenes. But it is much more difficult to attain precision for those parameters in human populations. Of course, what is needed from a genetics point of view is a human metric trait with "good additivity" to be a reliable start for basic models from which more complex traits and models of their genetics may in turn be derived. With the diversity of interests among many sciences, each with its goal of satisfactory information about human metric traits, it is vital to make a good start, but few traits can meet the ideal. Consequently, nongeneticist investigators in anthropology, medicine, psychology, sociology, and education may rely on models and concepts applied to animal and plant breeding techniques, not realizing perhaps fully enough some of the major limitations of metric trait genetics to prevent some misjudgment and consequent controversy. That has been the case particularly with certain articles by nongeneticists on the genetics of intelligence quotient (IQ) and mental ability. To gain some perspective, in this section we shall consider first the sort of data that are achievable with the most ideally additive human polygenic trait found so far in humans—fingerprint ridge count. We shall then proceed to consider some of the data documented on mental ability (IQ) and the controversy over its heritability and genetic differences between groups.

Precision is best achieved in human polygenic studies with estimates of genetic variance components made from parent-offspring, sib-sib, and other close relative correlations. Reliability for genetic determination, populational differences, and frequency analysis is increased for traits free of environmental influences. One human trait that seems to be nearly ideal, with little environmental influence and no nonrandom components as yet detected, is fingerprint ridge count (Holt, 1961, 1968; Penrose, 1969). The reported correlations are as follows:

Finger Ridge Count Correlations (from Holt, 1961)

Between	r observed	r theoretical (additivity)
Monozygotic twins	0.95	1.00
Dizygotic twins	0.49	0.50
Sibs	0.50	0.50
Parents	0.05	0
Parent mean-child	0.66	0.70 (see formula 7-13)
Mother-child	0.48	0.50
Father-child	0.49	0.50
Left-right hand (same individual)	0.94	1.00

Apparently, all correlations are very close to those expected with complete additivity of gene action. Now that these facts are established, comparisons between groups or populations for this trait can have substantial significance for gene frequency and genetic difference analysis.

Another human trait with good additivity, mentioned in the previous chapter, is stature. The significant amount of assortative mating for stature tends to inflate the additive variance component because more homozygotes will be produced under that mating system than under random mating. In Example 6-1, we made note of a correlation between parent-offspring for stature amounting to $r = 0.52$ when there is about 20 percent assortative mating in the population sampled. The expected value under random mating would be

(1 − 0.20), or 80 percent of that, $r = 0.41$ ($h^2 = 0.82$). Stature, then, can be considered a reasonably good trait for genetic difference comparisons between groups, but caution must be maintained in view of obvious environmental influences.

For these traits, heritability must be close to 1 because dominance, nonadditive interactions, and environmental influences are relatively slight in determining the phenotype. Within families, the phenotypic distributions from most of these traits seem to fulfill expected values based on additive genic action. What environmental influences can alter fingerprint ridges or how they may do so is still a matter of conjecture, though we know that sex, body size, and contributing features affect them; therefore, like all heritable traits, fingerprint ridges have not developed in an environmental vacuum. Stature is unquestionably easily influenced by nutrition, exercise, and other environmental factors. Consequently, we cannot say that phenotypic differences in these traits, even though having high heritability in most populations, are exclusively genically determined. Of course, with increasing nonadditive genic action at loci controlling any trait, heritability will depend on gene frequencies as well as genotype-environmental interactions. It cannot be overemphasized that while heritability expresses how to predict selection outcomes and the amount of similarity to be expected between relatives, it is a populational attribute, *not* a constant feature of any trait. These concepts must be kept clearly in perspective when discussing traits that are less easy to describe and define than these morphological ones, especially when we analyze those contributing to behavior such as mental ability, IQ, or "intelligence."

A second method of estimating heritability is that employing monozygotic and dizygotic twin data—data particularly gathered for studying behavioral and mental traits. Because the variance observed between members of monozygotic twin pairs should be entirely environmental, a broad heritability estimate (degree of genetic determination) can be made from the ratio $(V_{DZ} - V_{MZ})/V_{DZ}$, where V = variance between members of twin pairs, MZ = monozygotic, and DZ = dizygotic. It can be shown that this ratio is equivalent to a comparison of intraclass correlation coefficients as follows:

$$H = \frac{r_{MZ} - r_{DZ}}{1 - r_{DZ}} = \frac{V_{DZ} - V_{MZ}}{V_{DZ}}$$

where H = heritability in the broad sense, or degree of genetic determination

r = intraclass correlation coefficient between members of twin pairs

This relationship was used originally by Newman, Freeman, and Holzinger in 1937 when testing IQ among twin pairs. (For their method and derivation of the formula, consult Neel and Schull, 1954.) They found that for the standard Binet IQ test, the correlations were 0.881 for MZ twins, 0.631 for DZ twins, from which $H = 0.678$, meaning that mental ability as exemplified by the Binet test showed about twice as much hereditary determination as environmental. (A modern treatment of twin data analysis is given by Jinks and Fulker, 1971.)

A more extensive attempt to estimate heritability of mental traits was accomplished by Cyril Burt and his colleagues (1958) for over 1000 pairs of siblings, their parents, and other close relatives in a borough of London, using Fisher's variance analysis. While Burt made a more complete assessment of mental ability than could be indicated by test scores alone, the variance components given below are for comparison with the Holzinger, et al., estimate. A summary of the variance component values based only on crude test scores is given below. (Unfortunately, a reexamination of Burt's data-collecting and recording techniques (Jensen, 1974) has revealed some inconsistencies and ambiguities that weaken their usefulness.)

Source	Variance Values Crude Test Scores (from Burt, 1958)	
Genetic component		
Fixable by selection, or additive	40.5	
Nonfixable (non additive)	16.7	Total genetic 77.1
Assortative mating	19.9	
Environmental		
Systematic	10.6	
Random	5.9	Total nongenetic 22.9
Unreliability	6.4	

The degree of genetic determination for mental test scores for these two studies is in substantial agreement. Note, however, that the significant amount of assortative mating (parents tend to marry within social and intellectual strata) and the dominant component in Burt's study reduce the additive component to about 40 percent (h^2). The broad-sense heritability (77 percent) includes all genetic factors from nonadditive sources plus any increase in genetic relationships due to mating of similar genotypes—in fact, a pool of all genetic factors inclusive.

These data from Burt have been utilized to draw conclusions that may be more full of emotion than objectivity. An explosive controversy has developed with wide sociological connotations over the heritability of between-group differences in IQ, starting mainly from the publication of an article by A. R. Jensen in 1969 in the *Harvard Educational Review*, in which it was suggested that large-scale educational attempts to raise IQs of underprivileged children, white and black, were failing because of high heritability of IQ and that social class and racial differences in mean IQ were due largely to differences in the genic distributions of these groups. The question at issue was summarized well in a review of the argument by Scarr-Salapatek (1971*a*, 1974). See also Loehlin, Lindzey, and Spuhler (1975).

Such ideas as Jensen's could easily develop into reinforcement of pernicious racial prejudice in our society, and yet the main point he raised (evidence for genetic differences in mental ability between social and ethnic strata) had never been carefully measured. While IQ differences have been documented for some time between economic and social groups, it was a major jump to the conclusion that the group differences were predominantly due to a genetic determination. A study reported two years later by Scarr-Salapatek (1971*b*), done in Philadelphia public schools using nearly 1000 pairs of twins among black and white children, indicated that there was a much lower proportion of genetic variance for aptitude among lower-class than among middle-class children in both black and white groups. These estimations are of heritability in the broad sense; we know that heritability is a populational statistic, a ratio whose denominator includes nongenetic components, and the lowered value in the disadvantaged group comes about no doubt from greater environmental influences there. Another study done on children from interracial (black × white) matings throughout the United States (Willerman, Nayler, and Myrianthopoulos, 1970) indicated that the race and marital status of the mother made a difference in IQ, with male children from Negro unmarried mothers showing the lowest scores. Social disadvantage in prenatal and postnatal development can substantially lower IQ and reduce genotype-phenotype correspondence. Consequently, average IQ differences between social strata may be considerably larger than any genotypic differences they may have.

The appeal of attempting to explain our present social evolution by assuming genetic differences between social strata surfaced in 1971 with a popular article in the *Atlantic* by a psychologist at Harvard University, Richard Herrnstein. He enthusiastically espoused IQ testing and went on to assert and to reason that "social standing will be based to some extent on inherited differences among people," because when arbitrary barriers on social mobility are removed, as in the class system of European and North American countries, most of the people with high IQs gradually rise to the top and most of those with low ones fall to the bottom of the social ladder so that IQ differences between top and bottom become more and more genetically differentiated. In contrast, under a caste system where barriers prevent mobility between castes, genetic variability within each caste is more likely to be preserved. In countries where racial groups are barred from moving up the ladder, genetic differentiation of social groups should not be expected.

While details of the controversy (both the racially "hot" point of Jensen and the more "responsible" one of Herrnstein) are beyond the scope of this book, the student should be aware of the likely flaws in their arguments, especially when derived from attempting to apply heritability estimates of mental ability. Jensen relied heavily on data supplied by Burt taken from the relatively uniform English population. Of course, we cannot legitimately extrapolate from that population to other populations, especially to our highly heterogeneous one in the United States. The heritability value emphasized by Jensen and also utilized by Herrnstein was high (about 0.80)—a value that certainly includes many components contributed by nonadditive genetic effects and nonrandom mating, leaving only about half the total (40 percent) available to respond to Herrnstein's proposed selection for social level. We know that the "additive" variance component may well have a dominant and epistatic component when allelic frequencies are not equal at all loci, and we do not have the faintest idea of the nature of the genotype that determines IQ. Therefore, the heritability could easily move to zero with slight change in allelic frequencies if much of the determination is from overdominant (heterotic) loci (See Exercise 8 in Chapter 6). To make Herrnstein's provision for change in genotype with mobility between social classes, it is necessary to propose that the additivity measured will produce a good response to selection forces. We do not know whether high IQ results from homozygote concentration of positive alleles, from high levels of heterozygosity, or from special genetic combinations, both epistatic and overdominant. If the high-IQ genotype is a highly heterozygous "balanced" one, all the selection one could muster in an all-out eugenics program would never create a permanent, uniformly stable (homozygous) high-IQ genotype. Segregation for lower IQ (or different mental abilities if we speak more generally) would always continue. At the other end of the scale, segregation of higher IQ from low-mentality parents would probably also occur. When genetic variation is *unfixable*, selection may bring about a quick plateau at a not very high level. No doubt that process has already taken place over the past five or ten generations in our society. If a major fraction of human genotypes acting on mental traits can be attributed to heterozygosity, it would be very hazardous to predict the outcome of selection purely from parent-offspring heritability estimates. Realized selection progress is often much less than that predicted (see especially Examples 7-4 and 7-5).

Throughout this discussion the student may ask how useful the heritability concept may be for human mental trait predictability. Heritability estimates for plant and animal breeding predict selection gains to be expected, but what use are they to human predictions? Scarr-Salapatek (1971) points out that they may not be entirely without merit; they may be useful

> *. . . as indicators of the effects to be expected from various types of intervention programs. If, for example, IQ tests . . . show low heritabilities in a population, then it is probable that simply providing better environments which now exist will improve average performance in that population. If h^2 is high but environments sampled in that population are largely unfavorable, then (again) simple environmental improvement will probably change the mean phenotypic level. If h^2 is high and the environments sampled are largely favorable, then novel environmental manipulations are probably required to change phenotypes, and eugenic programs may be advocated.*

In short, "neither intelligence nor h^2 estimates are fixed." Those who fear a high heritability of intelligence because they think that it means intelligence thereby becomes genetically fixed (more homozygous) and phenotypically unchanging misunderstand the concept completely. And they certainly have no comprehension of probably the most complex, highly heterozygous, highly balanced (in Mather's sense of linked complexes) genetic system of all organisms. The notion that we shall run out of genetic variability in the human population or that any group or race might do so is as far from reality as that the sun will run out of energy tomorrow. Certainly, euthenic policies (improving environmental conditions) have not been explored or put into operation to any extent. A genuine concerted effort to improve conditions for underprivileged students with efficacious tutoring and intervention programs to improve the lot of children from disadvantaged parents has not yet been tried. For further detailed discussions and evidence on the inheritance of human mental abilities, see Robinson (1970), Lerner (1972), and Loehlin, Lindzey, and Spuhler (1975).

EXERCISES

1. Suppose an open-pollinated cornfield, as in Exercise 11, Chapter 6, has plants ranging in height from 3 to 7 feet at 1-foot intervals only, with height controlled by two pairs of equally additive independent factors (1 foot = additive effect of each allele). Let allelic frequencies all be equal ($p = q$, $u = v$) with the population mean at 5 feet.
   (a) If only 5-foot plants are selected for propagation and allowed to cross-pollinate at random, what would be the expected mean height and variance of the progeny corn?
   (b) If a random sample of plants taller than the mean (6 and 7-foot plants) is selected for propagation, first determine the expected mean of the selected plants ($\bar{Y}_S$), their gametic output, then the mean height and variance of their progeny plants ($\bar{Y}_R$).
   (c) When phenotypes are completely determined by additive genotypes, what is the expected progeny mean (response) from a given selected parent mean value? What are the values of R, S, and h^2 in this case?

2. Nine genotypes of corn plants are distributed over ten separate plots of ground as in Exercise 12, Chapter 6, and allelic frequencies all equal ($p = q$, $u = v$)—that is, two plots that add $\frac{1}{2}$ foot in height, five plots that are "average," and three plots that diminish the height by $\frac{1}{2}$ foot.
   (a) What will be the frequency distribution of those plants greater than the mean height? What would be their mean? Let the value be $\bar{Y}_S$ and assume that all plants greater than the mean are selected for propagation.
   (b) Find the gametic output of the selected plants from (a). What would be the frequency distribution of progeny plants and their heights? What would be the expected mean of the progeny ($\bar{Y}_R$)?
   (c) What will be the values of R, S and h^2? How does the R/S ratio compare with the proportion of additive variance out of the total variance, as you computed in Exercise 12(d), Chapter 6?
3. Suppose a randomly breeding population has three genotypes with the following phenotype values and frequencies:

Genotype	Phenotype Value	Frequency
AA	8	0.16
Aa	7	0.48
aa	1	0.36

   (a) What is the mean and genetic variance of this population? What are the additive and dominant components of the genetic variance? What would be the heritability (additive proportion of the variance)?
   (b) Suppose this population is an organism that produces few progeny per family unit, but it is desirable to raise the mean value of the population by selecting for high-value phenotypes. The experimenter selects all the *value 8* individuals plus one-half of the available 7's (or 40 percent of the population counted). What would be the mean of that selected group?
   (c) After random mating is allowed among the selected group, what would be the mean of their progeny? What would be the expected progeny mean based on the heritability of the original population and the selected parents' mean?
4. In Table 6-8, both *A-a* and *B-b* pairs of alleles display partial dominance.
   (a) If a random sample of individuals whose phenotype is greater than the population mean is selected for propagation, what will be the mean ($\bar{Y}_S$)?
   (b) After random mating of these selected parents, what would be the expected progeny frequency distribution for the nine genotypes produced? What will be the expected progeny mean ($\bar{Y}_R$)?
   (c) What are the values of the ratio R/S? How does this realized heritability compare with the expected h^2 given at the bottom of Table 6-8? Should these heritability values be identical? Why or why not?
5. Mather (Example 7-1) did not make heritability estimates of abdominal bristle number for the inbred strains (BB and Oregon +) or for the F_2 progeny following the crossing of those strains, yet his selection response was very significant.

(a) How could Mather have predicted this selection response? How much genetic variance was there in the original strain F_2 progeny?
(b) What sort of genotype probably existed for his parental strains in order to account for the low genetic variance in the original strain F_2 progeny?
(c) How much genetic variance was there in the F_2 from the selected line cross? What might account for the difference between this result and that from the unselected strain cross?
(d) Estimate the number of effective factors segregating in the F_2 of the cross between selected lines (use formula 7-19).
(e) Contrast Mather's base population with Clayton, Morris, and Robertson's Kaduna base population (Example 7-3) as to general genetic structure controlling the trait.

6. Example 6-1 and Figure 7-1 give the relationship between the two-parents' mean stature (measured as a deviation from the population mean) and their offspring's average stature.
(a) From the data given below, obtain the regression coefficient (b) and the product moment correlation coefficient (r) from the 24 groups of parents-offspring given by Davenport:

Parents' Mean (X)	Offspring Average (Y)
$\sum X = -0.088000$	$\sum Y = 15.052000$
$\bar{X} = -0.003666$	$\bar{Y} = 0.627166$
$\sum X^2 = 228.324478$	$\sum Y^2 = 197.014362$
Var. $X = 9.513504$	Var. $Y = 7.815592$

$$\sum XY = 204.7495$$

Verify that $b = 0.8970$, Y intercept $= 0.6305$, $r = 0.9896$, $\sigma_r = 0.0043$.
(b) Do you believe the regression coefficient is the best estimate of the heritability of stature from these data? Do you think it might be an overestimate or an underestimate? Why?

7. The relation between the average parent phenotype value (Y) and the parent's linear value (Y_L) determines the amount of progeny response to selection. In Chapter 6, Y and Y_L were usually expressed in terms of deviation from the midparent. Algebraically, it is easier to work out the covariance (σ_{YY_L}) and thus the heritability expression if Y and Y_L are put into deviations from the population mean ($\bar{Y}$) instead, since thereby $\sigma_{YY_L} = \sum d_y \cdot d_{Y_L}$ directly.
(a) First verify the deviations from the population mean $\bar{Y} = a(p - q) + 2dpq$ for each Y_i phenotypic value as given below. (Hint: Use formula (6-5) converting the additive effect a into the linear increment α and see Exercise 10, Chapter 6.)

Genotype	Frequency	Phenotypic Value Y	Linear Value Y_L	Sum of Products fYY_L
AA	p^2	$2q(\alpha - qd)$	$2q\alpha$	$p^2[4q^2\alpha(\alpha - qd)]$
Aa	$2pq$	$(q - p)\alpha + 2pqd$	$(q - p)\alpha$	$2pq[(q - p)^2\alpha^2 + 2pqd(q - p)\alpha]$
aa	q_2	$-2p(\alpha + pd)$	$-2p\alpha$	$q^2[4p^2\alpha(\alpha + pd)]$

(b) Obtain the total sum of products (covariance) as follows. First, taking only products with d, find that they sum to zero. Of the remaining products, factor out $2pq\alpha^2$ to obtain $\sigma_{YY_L} = 2pq\alpha^2$, which is the covariance between phenotypic and linear values.
(c) What is the significance of the expression: $2pq\alpha^2/\sigma^2$?
(d) Verify these relationships with the data from the linear values in Figure 6-3B or Table 6-5B. Show that $h^2 = \sigma_{YY_L}/\sigma_Y^2$.
(e) Because each offspring receives one-half of its parent's genotype, one-half of each linear value above equals the mean value for offspring of that genotype. Show the relationship of this result to formula (6-5).

8. In Example 7-3, an estimate of covariance between close relatives (half-sibs) can be obtained by the hypothetical assumption that "nonadditive" and "additive" effects on the phenotype are independent of each other, as in the following model. A genetic trait with phenotype y_{ij} in the jth member of the ith group (a sibship, for example) is made up of three components:

$$y_{ij} = \mu + g_i + s_{ij}$$

where μ = grand mean over all groups, g_i = group additive effect, and s_{ij} = special nonadditive effect. Both effects are expressed as deviations from μ, so that

$$\sum_i g_i = 0, \qquad \sum_i \sum_j s_{ij} = 0$$

Then variance $V_y = V_{y_{ij}} = V_g + V_s$.

The expected (E) covariance of any two members of one group $CoV(y_{ij}y_{ik})$ is the mean product of the array of such products; that is,

$$CoV(y_{ij}y_{ik}) = E[(g_i + s_{ij})(g_i + s_{ik})]$$

(where E[. . .] is the expected value overall)

$$= E(g_i^2) + E(g_i s_{ij}) + E(g_i s_{ik}) + E(s_{ij}s_{ik})$$

If all g_i and s_{ij} and s_{ik} are independent of each other, the last three terms in this expression will be zero. Thus, $E(g_i^2) = V_g = CoV(y_{ij}y_{ik})$; in words, the variance in group effects (group means expressed as deviations) equals the average covariance between members of a group. In Example 7-3, the covariance between half-sibs or between full sibs was estimated from the variance component "between families."

(a) Half-sibs are offspring with a single parent in common. Thus, the mean phenotypic value of a group of half-sibs from a single parent (all other parents being different individuals) is one-half the linear value of the common parent. Using linear values as deviations from the population mean as in the previous exercise and realizing that one-half the linear value of the parent equals the value of the offspring, show that the variance of the means of half-sib families is

$$\begin{aligned} V(\text{half-sib families}) &= p^2q^2\alpha^2 + 2pq \cdot \tfrac{1}{4}(q - p)^2\alpha^2 + q^2p^2d^2 \\ &= \tfrac{1}{2}pq\alpha^2 \\ &= \tfrac{1}{4}\sigma_A^2 \end{aligned}$$

This is also the expected covariance between half-sibs; however, the long method for working out the covariance between half-sib pairs may be tried by setting up a table of half-sibs pairs and their frequencies based on a general plan such as that shown by Li (1955*b*, pp. 19–22) for full sibs.

(b) With full sibs, the covariance is the variance of full-sib families. Using Table 7-4, the last column on the right gives offspring mean values. Those may be converted into deviations from the population mean (as in Exercise 7), squared and summed to produce the mean square, or variance between full-sib families. Show that

$$V(\text{full-sib families}) = pq\alpha^2 + d^2p^2q^2$$
$$= \tfrac{1}{2}\sigma_A^2 + \tfrac{1}{4}\sigma_D^2$$

This is the expected covariance between full sibs.

9. In 1911, East (cited in Mather, 1949) made crosses between varieties of tobacco (*Nicotiana longiflora*) that differed in length of the corolla tube. Lengths of corolla (mm) are given below for parent strain means, F_1 and F_2 progeny derived by selfing of F_1:

		Strain 383	Strain 330
Parents			
	Mean	40.37 mm	93.11 mm
	Variance	3.03	5.64
Sample N		211	169

	F_1	F_2
Mean	63.53 mm	68.76 mm
Variance	8.62	42.37
N	173	444

(a) What is the midparent value? Is there evidence of any dominance?

(b) If the average variances of parents + F_1 are taken as an approximation for environmental component of variance, estimate the effective number of factors segregating using formula (7-19).

(c) From F_2 plants, East raised F_3 progenies. Average F_3 progeny regressed on F_2 plant corollas gave a coefficient of $b = 0.77$. If there were no dominance and the estimate of environmental variance could be assumed approximately as above, what would be your estimate of h^2 for this F_2? If you use the environmental component as equal to the F_1 variance, what would be your estimate for h^2? Does there appear to be any significant amount of nonadditive (dominant) component or is the F_2 variance made of chiefly of additive genetic + environmental components?

10. Verify the contributions to variance by backcrossing to each parent as given in formula (7-17) for any single locus by which selected parent lines may differ. Hint: Start with the following:

	Genotype	Frequency	Y Value
For B_1	AA	$\frac{1}{2}$	$+a$
	Aa	$\frac{1}{2}$	d
For B_2	Aa	$\frac{1}{2}$	d
	aa	$\frac{1}{2}$	$-a$

11. Kessler (1969) tested mating speed in *Drosophila pseudoobscura* not only by recording the number mating during this time of observation but also by weighing the number mating in 5-minute intervals by the reciprocal of time so that faster maters were given greatest weight (Spiess and Spiess's "mating index," 1966). After isolating two selected lines that differed in mating, Kessler made Mendelian crosses to test for additivity of both mating frequency and mating speed (index). Using the additive scales of Mather (*A*, *B*, *C*), test the following columns of data from Kessler's crosses to determine which scale transformation provides the best "additivity":

No. Observed Mating Chambers	Tests	Average % Mated (1)	Arcsin Transformation of % Mated (2)	Mean Mating Index (3)	Log of Mean Mating Index (4)	Square Root of Mean Mating Index (5)
P_1	16	77.4 ± 1.7	2.16 ± 0.05	11.87 ± 0.52	1.069 ± 0.020	3.43 ± 0.07
P_2	16	19.7 ± 2.0	0.91 ± 0.05	2.23 ± 0.53	0.331 ± 0.030	1.48 ± 0.05
F_1	22	76.0 ± 1.7	2.13 ± 0.05	11.11 ± 0.52	1.035 ± 0.020	3.31 ± 0.08
F_2	24	51.8 ± 3.6	1.61 ± 0.08	7.04 ± 0.56	0.810 ± 0.040	2.60 ± 0.11
BC_1	27	59.4 ± 3.2	1.78 ± 0.07	8.72 ± 0.61	0.916 ± 0.028	2.91 ± 0.10
BC_2	27	43.9 ± 3.5	1.44 ± 0.08	5.60 ± 0.54	0.693 ± 0.045	2.30 ± 0.11

12. Sternopleural bristle number is an "additive" trait in drosophila with most strains of *D. melanogaster* averaging from 8–10 bristles on each side. A high bristle number strain (Canton-S) was crossed to a low strain (Lausanne-S). Intermediate F_1's were backcrossed to each parent line as well as inbred to produce F_2 flies. The following means and variances were obtained by counting bristles on one side only for 100 of each sex per generation:

	♀♀		♂♂	
	Mean	Variance	Mean	Variance
P_1	8.82	0.885	8.34	0.841
P_2	12.46	1.192	12.12	1.102
F_1	10.12	1.056	9.48	1.000
F_2	9.09	1.165	8.89	1.283
BC_1	9.22	1.002	8.76	1.134
BC_2	11.06	2.400	10.59	2.022

(a) Using Mather's scales (*A*, *B*, *C*), would you say these data show complete additivity throughout?
(b) Is there any evidence of dominance? Which parental type tends to be dominant?
(c) Is there any evidence that either or both parent strains is genetically non-homogeneous?
(d) Estimate the additive component of variance from the F_2 generation. Do each sex separately. What would you use as an estimate of the environmental component? Why?

(e) What possible genetic or nongenetic factors could be assumed to account for the large variance in the backcross to the larger parent? How might such factors help to explain the Mather scale results?

(f) Wright (1952, 1968) has shown that when there is a degree of dominance (assuming it to be constant over all loci affecting the trait), the number of segregating factors (k) may be estimated by modifying formula (7-19) as follows. Let h be the proportion of overall heterozygosity effect: $h = (d + a)/2a$ so that h ranges from 0 to 1, and if there is no dominance $h = 0.5$. Then $k = D^2/8s_G^2\,[1.5 - 2h(1 - h)]$, where s_G^2 is the sample estimate of the genetic component of variance. Determine h and verify that $k = 5.4$ segregating loci.

13. Selection for sternopleural bristle number was performed on the *D. melanogaster* F_2 generation obtained as a separate sample from the crosses described in Exercise 12. Each generation, $\frac{10}{100}$ flies of each sex were selected as the highest (H line) or lowest (L line). A nonselected control was also maintained by mating 10 pairs at random each generation. Data were collected as follows over four generations in males (females were similar) (s^2 = sample variance of total progeny):

	H Line				*L* Line				Control Line	
Gen.	$\bar{Y}_P$	s^2	$\bar{Y}_s$	$\bar{Y}_R$	$\bar{Y}_P$	s^2	$\bar{Y}_s$	$\bar{Y}_R$	$\bar{Y}_P$	s^2
F_2	9.05	1.133	11.20	10.71	9.05	1.133	7.50	9.04	9.05	1.133
F_3	10.71	2.006	13.60	11.60	9.04	0.907	7.60	8.96	9.42	0.913
F_4	11.60	2.202	14.10	12.03	8.96	0.907	7.70	9.03	9.77	1.532
F_5	12.03	2.330	—	—	9.03	0.898	—	—	10.23	2.017

(a) Determine the selection differential (S_d) and the response (R) for each generation in each selected line.

(b) How do these data compare with the control values? What do you expect the control values to do? Should the selection differentials and responses be corrected in any way utilizing the control values?

(c) Find the cumulative selection differentials and responses, and find the slope of the regression that is an estimate of heritability. Which is more appropriate for this estimate, the crude values or the values corrected by comparing with the controls?

14. (a) What is the value of H (degree of genetic determination) for the fingerprint data from MZ and DZ twins?

(b) Assuming heritability for a trait is high, discuss whether it is legitimate to equate within-group h^2 with between-group heritability. You might begin by trying to define the latter first.

15. Herrnstein (1971) stated the following on the heredity of IQ: "the more advantageous we make the circumstances of life, the more certainly will intellectual differences be inherited." Do you agree? What is he referring to? How would you alter his statement to express what he was perhaps trying to point out? Do you agree with his concept?

PART 3

NONRANDOM MATING: CONSEQUENCES FOR THE GENOTYPE

8

INBREEDING, EXTREME AND SIMPLE: SELFING

The main consequence of random mating (panmixia) on genic element frequencies is genetic equilibrium determined by the familiar random proportionality rule. That is an idealized situation. The conditions set forth for panmixia in the first chapter of the previous section include the specification that individuals must mate without bias based on phenotypic resemblance, relatedness, or physical distance between them. Obviously, in real populations of sexually reproducing organisms, individuals form mating pairs with violation of the randomness condition to a considerable degree and in a variety of ways. Even in large and well-dispersed creatures such as houseflies, planktonic crustaceans, marine fishes, grasses, and human beings, where mates could be expected to be unrelated and unassociated, there may well be correlation between parents, morphological resemblance, common origin, ecological preference, or biases of many sorts. Proximity between mates may increase the likelihood that male and female are related by descent, cause some preference over more distant individuals, or result in preference for either morphological or behavioral tendencies. Families, tribes, cohorts, and groupings, social or economic, bring about stratification into subunits almost universally in natural populations including humans. Under these nonrandom conditions, genic elements (alleles, gametic combinations, chromosomal variants, and the like) will not form zygotes independently by simple proportionality rules, and we must consider how the genic and genotypic pools of the population can change according to the mating system.

In this section, we isolate for consideration the genetic consequences of nonrandom mating for a fundamental understanding of the main effects. It is important first to look at the general rules of the mating system both in order to see the overall evolutionary results and to make predictions at an elementary level in experimental or natural populations. We proceed by considering a population of genotypes as an entity that practices nonrandom mating and how the genic elements in the population as a whole become distributed with time and continued mating. For illustration of this process, we begin with the simplest and most extreme form of inbreeding—self-fertilization or selfing; then, via less extreme and less simple inbreeding systems (sib mating and more distant relatives), we consider principles that may apply more generally to populations, with the assumption that no other forces are acting on the genotypes. An alternative mating system of interest—assortative or preferential mating—will be examined briefly for comparison. Finally, because the main effects of nonrandom mating are directed toward genotype distributions, we look into the important generalization of heterozygosity loss with limitation of population size (numbers in the mating unit), the dispersion of genotypes, and the random fixation of genic factors in restricted populations.

These general considerations were established by Wright in his early definitive series of papers on mating systems (1921), from which he derived much of the reasoning concerning correlations between parents, offspring, and close relatives. Much of his theoretical derivation was based on a method known as path coefficient analysis (for a concise treatment of inbreeding using path coefficient terminology, consult Li, 1955*b*, 1975). It is based on correlation between an effect and multiple causes, analyzed in terms of relative causation or the components of multiple correlation (analogous to components of variance in analysis of variance). An alternative method of analysis based on probability was developed first by Haldane and Moshinsky (1939) and later extended by Malécot (1948). We have found this method simpler to describe than path analysis, although we utilize much of the fundamental terminology and reasoning of Wright in the following discussion. Path analysis is of much wider applicability than the probability method, and through his path coefficient analytical method, Wright subsequently pointed out the widespread nature of nonrandom mating systems in natural, laboratory, and breeding populations; these systems can have very significant evolutionary consequences on population structure (1964, 1965, 1969).

INBREEDING: GENERAL CONSIDERATIONS

In real populations, individuals are related; that is, they are likely to share some genes by common descent from one or a few ancestors. In a conventional illustration of a two-sex species such as humans, every individual has two parents, four grandparents, eight great grandparents, and so on, so that at t generations back the individual has 2^t ancestors. Obviously, at about 25 generations or more before ours there would not have been enough people on earth to add up to the vast number of ancestors if everybody were independently descended separately; we must therefore be much more interrelated than it would appear.

The fundamental genetic consequence of two individuals having a common ancestor is that both may carry replicates of a single allelic form of that ancestor's genetic makeup. If the two individuals then mate and produce offspring, it is possible for them to bring about the double-dose (homozygote) condition of that ancestral allele. In most basic terms, if an organism capable of self-fertilization is a heterozygote (Aa) and produces gametes in equal proportions ($A{:}a$), that organism in its growth has replicated its genic substance many times following its origin as a zygote. The "identical" DNA representing either allele will be replicated countless times into the nuclei of all cells of that organism including its gametic tissue. Any of that individual's gametes containing A will be *identical* with every other A gamete because it will be an exact replicate (except for mutation, which we shall discount for the time being). Also, all a gametes will be identical with each other. On selfing, the $\frac{1}{4}AA$ progeny will be *identical* homozygotes, as will be the $\frac{1}{4}aa$, or $\frac{1}{2}$ of the total progeny will be identical in this sense because they represent the double-dose condition of what existed in single (haploid) dose in the parent. We shall thus find inbreeding to be a mechanism for making "identical" diploids via sexual reproduction and the Mendelian segregation process.

Therefore, there appear to be two kinds of homozygotes: (1) *identical* by descent as just outlined, and (2) *random combinations* from unrelated parents. In a sense, these may be only theoretically distinguishable, since the latter type of homozygote actually could contain two alleles that may well be identical. However, the unique feature of inbreeding is that it is a mating system that creates the former type of homozygote. When we know that a progeny or population has resulted from mating between related parents, we can estimate how many identical homozygotes have arisen. That general feature of inbreeding—to create identical homozygotes out of nonidentical (either from heterozygotes or some random type of genic combination)—is the fundamental consideration to be kept in mind throughout the following discussion.

SELFING: EXTREME AND SIMPLE

Although natural populations that are exclusively selfing are no doubt quite rare, many (monoecious) plant species are capable of it or are predominantly selfing (Baker, 1959, refers to these as "habitual inbreeders") but occasionally outcrossing—for example, some of the grasses such as oats (*Avena*) and *Agropyron*, shepherd's purse (*Capsella*), chickweed (*Stellaria*), and many crop plants such as wheat, rice, barley, peanuts, soybeans, tobacco, tomatoes, peas, beans, cotton, and citrus fruits, although wild ancestral forms may be cross-fertilizing predominantly. Stebbins (1974) pointed out that this predominantly selfing system of mating is probably the most common type among annual flowering species. A few hermaphroditic animals, autogamous ciliates, parasitic flatworms, certain freshwater snails, and some crustaceans may be capable of selfing, but for animals in general it is certainly a rarity.

Whether the population is capable of selfing or not, we may consider the consequences of selfing as if mating within the same genotype were taking place. Mendel in 1865 was the first to consider the relative frequencies of genotypes resulting from continued selfing, which he expressed as if each plant furnished just four seeds each generation, starting to self from the F_1 strain hybrid (Aa):

Generation After Selfing	AA	Aa	aa	Ratios AA:Aa:aa
1	1	2	1	1:2:1
2	6	4	6	3:2:3
3	28	8	28	7:2:7
4	120	16	120	15:2:15
5	496	32	496	31:2:31
t				$2^t - 1:2:2^t - 1$

It is easy to see that proportions of the three genotypes in each generation are obtained from Mendel's last three columns. These results follow from the segregation of the heterozygote into one-half perpetuating itself plus one-half forming "identical" homozygotes (AA and aa). By the 10th generation of selfing, the heterozygote proportion in the population will be $(\frac{1}{2})^{10}$, or $\frac{1}{1024}$, and with continued selfing it will soon be reduced to zero in any finite population. We now should proceed to cases with different genetic base populations to illustrate that with selfing (or, as we shall see, any regular system of mating that we may consider) in simple genetic situations, the general effect of inbreeding is similar in terms of proportional change from heterozygotes to homozygotes.

Selfing, Unequal p, q Frequencies

As an illustration, Table 8-1A starts with $p = 0.4$, $q = 0.6$ in which the parent (base population) is panmictic. If all genotypes are selfed exclusively, then all proportions of

TABLE 8-1 *The main effects of selfing, or mating of each genotype to itself*

A *Single genic locus, unequal p,q values. Let $p = 0.4$, $q = 0.6$ with random mating original parent generation (G_0), which is selfed exclusively to produce progeny ($G_1 \ldots$) in successive generations (t = generation)*

Generation	AA	Aa	aa	$(\frac{1}{2})^n$	$(\frac{1}{2})^n H_0$ H Proportion
0	0.16	0.48	0.36	1	$0.48 = (\frac{1}{2})^0(0.48)$
1	0.28	0.24	0.48	$\frac{1}{2}$	$0.24 = (\frac{1}{2})^1(0.48)$
2	0.34	0.12	0.54	$\frac{1}{4}$	$0.12 = (\frac{1}{2})^2(0.48)$
⋮	↓	↓	↓		
$t = \infty$	0.40	0	0.60	$(\frac{1}{2})^\infty$	$0 \leftarrow (\frac{1}{2})^\infty(0.48)$

Recall that $p = D + \frac{1}{2}H$

B *Single genic locus with multiple alleles. Let $p = q = r$ and original parent generation (G_0) to be exclusively heterozygotes in equal numbers ($H_0 = 1.00$), which are then selfed exclusively*

Generation	AA	Aa	aa	Aa'	aa'	$a'a'$	H Proportion
0	0	$\frac{1}{3}$	0	$\frac{1}{3}$	$\frac{1}{3}$	0	$1 = (\frac{1}{2})^0$
1	$\frac{1}{6}$	$\frac{1}{6}$	$\frac{1}{6}$	$\frac{1}{6}$	$\frac{1}{6}$	$\frac{1}{6}$	$\frac{3}{6} = \frac{1}{2} = (\frac{1}{2})^1$
2	$\frac{3}{12}$	$\frac{1}{12}$	$\frac{3}{12}$	$\frac{1}{12}$	$\frac{1}{12}$	$\frac{3}{12}$	$\frac{3}{12} = \frac{1}{4} = (\frac{1}{2})^2$
⋮	↓	↓	↓	↓	↓	↓	
$t = \infty$	$\frac{1}{3}$	0	$\frac{1}{3}$	0	0	$\frac{1}{3}$	$0 \leftarrow (\frac{1}{2})^\infty$

C *Two pairs of alleles (independent loci). Let $p = q$, $u = v$ and original parent be only a dihybrid ($AaBb$) selfed and progeny selfed from then on*

Generation		BB	Bb	bb	H Proportion
	AA	0	0	0	
0	Aa	0	1	0	1
	aa	0	0	0	
		$\frac{1}{16}$	$\frac{2}{16}$	$\frac{1}{16}$	
1		$\frac{2}{16}$	$\frac{4}{16}$	$\frac{2}{16}$	(Counting both loci): $\frac{16}{32} = \frac{1}{2} = (\frac{1}{2})^1$
		$\frac{1}{16}$	$\frac{2}{16}$	$\frac{1}{16}$	
		$\frac{9}{64}$	$\frac{6}{64}$	$\frac{9}{64}$	
2		$\frac{6}{64}$	$\frac{4}{64}$	$\frac{6}{64}$	$\frac{32}{128} = \frac{1}{4} = (\frac{1}{2})^2$
		$\frac{9}{64}$	$\frac{6}{64}$	$\frac{9}{64}$	
⋮		↓	↓	↓	
		$\frac{1}{4}$	0	$\frac{1}{4}$	
$t = \infty$		0	0	0	$0 \leftarrow (\frac{1}{2})^\infty$
		$\frac{1}{4}$	0	$\frac{1}{4}$	

homozygotes are perpetuated, and new identical homozygotes are added to their respective totals. In one generation of selfing, heterozygotes have given up one-half of their proportion to become identical homozygotes equally distributed to *AA* and *aa*: $0.48Aa \rightarrow [0.24Aa + 0.12AA + 0.12aa]$. If we assume that the new identical homozygotes are indistinguishable from the random original homozygotes in the parent generation, then the total proportions are as given in the second row of the table. Each generation heterozygotes are reduced by one-half until after many generations of selfing only homozygotes remain at relative frequencies equal to the original p, q values. The student should recall that according to (1-3), $p = D + \frac{1}{2}H$ and $q = R + \frac{1}{2}H$. Essentially, selfing has converted a diploid $(p^2 + 2pq + q^2)$ population into a $(p + q)$ diploid population. No change has occurred at the haploid level. Allelic frequencies have not been altered by the mating system; only genotype frequencies have been redistributed.

Selfing, Multiple Alleles

In Table 8-1B, we let three alleles be equal to give maximum heterozygosity, and in fact instead of equilibrium conditions we start with heterozygotes exclusively for a change. Again, with selfing one generation, one-half the heterozygote proportion remains, and after another generation one-quarter. Ultimately, the genotypes will be all identical homozygotes in this case, and their frequencies will equal the p, q, and r of the base population.

The relationship of recurring $H_t = \frac{1}{2}H_{t-1}$—the heterozygosity of any generation equals one-half of that in the previous generation—is an intrinsic property of the mating system, in this case selfing. The amount of H depends on the original amount in the base population and on the number of generations of inbreeding. The population can be thought of as splitting into isolates because no individual crosses with any other in selfing, until ultimately it is composed exclusively of isolated homozygotes in relative frequencies of $p, q, r, \ldots$.

Selfing, Two Independent Loci

While the details look more complex when two independent pairs of alleles are considered, the final outcome is substantially the same. In Table 8-1C, we begin selfing a double heterozygote (*AaBb*). Remembering basic Mendelian expectations, we find that genotype to be reduced to one-fourth while single heterozygotes (*AABb*, *AaBB*, *aaBb*, *Aabb*) *increase* from zero to one-half of the total. The double heterozygote is reduced twice as fast as single heterozygotes in the single-locus case $[(AaBb)_t = \frac{1}{4} \cdot (AaBb)_{t-1}]$, but the single heterozygotes do not decrease as fast as before because they are partially supplied by the segregating double heterozygote. The student should verify the zygote matrix in Table 8-1C for each generation.

We may now consider this case of two independent loci from three points of view: (1) *one locus at a time*: the proportion of H decreases just as before (either *Aa* or *Bb* goes down by $\frac{1}{2}$ each generation), (2) *for both loci*: the proportion of *heterozygous* loci decreases in the same way—for example, in generation 2—and we may suppose 64 individuals, or 128 loci (two loci per individual), out of which 32 loci are heterozygous (either *Aa* or *Bb* or both); that is $\frac{1}{4}H$, as the following tabulation indicates:

Number of Individuals	Proportion of Individuals	Loci	
		$D + R$	H
36 homozygotes at 2 loci	0.5625	72	0
24 homozygotes at 1 locus, heterozygotes at the other	0.3750	24	24
4 double heterozygotes at both loci	0.0625	0	8
		96	32

Finally, (3) the *average individual* (*genotype*) is $\frac{1}{4}$ heterozygous: 0.5625 are zero H, 0.375 are $\frac{1}{2}H$, and 0.0625 are completely heterozygous, or $1H$. Thus, their weighted average is $0 \times 0.5625 + 0.5 \times 0.375 + 1 \times 0.0625 = 0.2500$.

In summary, then, the following recurrence relationship of heterozygosity for selfing follows for all simple genetic systems considering either a single locus, proportion of loci, or average genotype in the population:

$$H_t = \tfrac{1}{2}H_{t-1} = (\tfrac{1}{2})^t \cdot H_0 \tag{8-1}$$

where subscripts refer to generation (0 = original, t = tth, etc.). Then, as t generations $\to \infty$, $H \to 0$, and homozygotes for a single locus approach allelic frequencies ($p, q, r \ldots$) or for multiple-loci at gametic frequencies (pu, pv, qu, qv).

INBREEDING COEFFICIENT *F*

As the selfing mating system, if continued, reduces nonidentical genotypes toward zero, it increases identical homozygotes by converting a regular fraction of nonidentical genotypes each generation. The proportion of nonidentical genotypes (heterozygotes) converted into identicals each generation can be thought of as the probability that any two gametes forming a zygote will be identical, or as Wright termed the correlation between uniting gametes, the "fixation" coefficient F. In Table 8-1A we consider the proportions of H and identical $D + R$ in the following way:

Generation	H	Identical $D + R$	F (cumulative)
0	$2pq = H_0$	0	0
1	$pq = \frac{1}{2}H_0$	$\frac{1}{2}H_0$	$\frac{1}{2}$
2	$pq/2 = \frac{1}{4}H_0$	$\frac{3}{4}H_0$	$\frac{3}{4}$
3	$pq/4 = \frac{1}{8}H_0$	$\frac{7}{8}H_0$	$\frac{7}{8}$
t	$(\frac{1}{2})^t \cdot H_0$	$[1 - (\frac{1}{2})^t]H_0$	$[1 - (\frac{1}{2})^t]$

The proportion of original heterozygotes fixed by the inbreeding system, or the coefficient $[1 - (\frac{1}{2})^t]$, is the unique feature of that mating system.

We may look at how the F coefficient is derived more simply by taking two successive generations of the H recurrence series and expressing the generations all in terms of the first generation as follows:

$$H_1 = \tfrac{1}{2}H_0 = H_0 - \tfrac{1}{2}H_0$$

and for the same reason

$$H_2 = H_1 - \tfrac{1}{2}H_1$$

so that putting all in terms of H_0, we have

$$H_2 = H_0 - \tfrac{1}{2}H_0 - \tfrac{1}{2}(H_0 - \tfrac{1}{2}H_0)$$

or

$$H_2 = H_0 - \tfrac{3}{4}H_0$$

the fraction of H_0 remaining in H_2

$$H_2 = H_0(1 - \tfrac{3}{4})$$

In other words, for any generation, heterozygotes equal the fraction of original heterozygotes not yet made identical.

$$H_t = H_0(1 - F_t) \tag{8-2}$$

This relationship between identical homozygotes and heterozygotes is always true in any generation; whatever the F values may be, we can always determine how many heterozygotes there are compared with the base population from which the inbreeding was started. In the cases given in Table 8-1B and C where the original heterozygosity was complete $H_0 = 1$, and the heterozygote frequency in any generation is simply $(1 - F)$ for that generation. We shall see when various inbreeding systems are considered that "F," the probability of identical homozygotes, is determined directly by the mating system. Consequently, our primary aim is to determine that coefficient and from it derive the heterozygosity frequency (H_t) subsequently by using (8-2). In the next chapter we shall examine the increment in F each generation.

GENERALIZED ZYGOTIC DISTRIBUTION UNDER INBREEDING

While it is easy to see the change in heterozygotes and total identical homozygotes, the distribution among the latter (AA and aa) is not so evident. From the numerical example of Table 8-1A we see that from the parent generation to the F_1 an equal increment of 0.12 has been given to each homozygote, while twice that amount (0.24) has been lost from Aa. We may consider in steps, then, the apportionment into identical homozygotes (shown in Table 8-2):

1. One-half of the new identical homozygotes are AA, one-half aa; apportioned equally to each box (upper left or lower right);
2. These sum up to F for the entire set, the fraction of *identical* homozygotes—
3. Out of the original fraction of heterozygotes ($2pq$).

Putting all probabilities together, the increment in each homozygote will be $\tfrac{1}{2}(F)(2pq) = Fpq$. In the case illustrated, $F = \tfrac{1}{2}$, $pq = 0.24$, so that the fractional increment is 0.12. The following generation F increases to $\tfrac{3}{4}$, so that the net increment over two generations is

TABLE 8-2 *Distribution of zygotes under inbreeding (from Table 8-1)*

A *Apportionment of genotypes from G_0 to G_1*

	A	a
A	0.16 + 0.12 = 0.28	0.24 − 0.12 = 0.12
a	0.24 − 0.12 = 0.12	0.36 + 0.12 = 0.48
$\sum =$	0.40	0.60

B *Generalized zygotes in generation following inbreeding*

	A	a
A	$p^2 + Fpq$	$pq - Fpq$
a	$pq - Fpq$	$q^2 + Fpq$
$\sum =$	p	q

where F = probability of identical homozygotes in the generation

0.18, and so forth. At the same time, of course, the heterozygotes have lost twice the Fpq amount. The generalized distribution of genotypes then can be expressed as follows:

$$\begin{array}{ccccc} AA & : & Aa & : & aa \\ [p^2 + Fpq] & : & [2pq(1 - F)] & : & [q^2 + Fpq] \end{array} \tag{8-3}$$

Note that this new expression is a greater generalization than the Castle-Hardy-Weinberg square law, which is the special case for distribution of genotypes under random mating. Because each mating system will determine F, we can always estimate genotype proportions for any system of mating provided we know the base population distribution and the number of generations the inbreeding system has been going on.

A more formalistic way of looking at the genotypic distribution (8-3) is to consider the four combinations in Table 8-2B to be derived from two components, a random, nonidentical (1-F) and a nonrandom identical (F) component, under selfing, as follows:

	Random Component	Identical Homozygote Component		
Generation	$(1 - F)$	New F +	F from G_{t-1} =	Cumulative F
0	$1(p + q)^2$	0	0	0
1	$(1 - \frac{1}{2})(p + q)^2$	$\frac{1}{2}(p + q)^2$ +	0	$= \frac{1}{2}(p + q)^2$
2	$(1 - \frac{3}{4})(p + q)^2$	$\frac{1}{4}(p + q)^2$ +	$\frac{1}{2}(p + q)^2$	$= \frac{3}{4}(p + q)^2$
3	$(1 - \frac{7}{8})(p + q)^2$	$\frac{1}{8}(p + q)^2$ +	$\frac{3}{4}(p + q)^2$	$= \frac{7}{8}(p + q)^2$

The student should realize that the formation of *identical* genotypes is not only confined to the heterozygotes. By representing the populational components above as $1 - F$

and F, we wish to emphasize that *all* genotypes must go through the same process under inbreeding—the duplication and doubling of DNA strands via two separate gametes from an ancestor to a descendant individual. If the genotype is already a homozygote, the doubling process changes nothing for the case of selfing, so that the total proportion of identical homozygotes is simply cumulative, as shown above. (Under sib mating, discussed in the next chapter, or any other inbreeding, however, the accumulation will not be so simple. An identical homozygote may mate to a heterozygote or to a different homozygote.) Thus, in any generation, genotype AA includes a random component $p^2(1 - F)$ plus an identical component $F(p^2 + pq)$ thus:

$$p^2(1 - F) + Fp = p^2 - p^2F + pF = p^2 + Fpq.$$

Similarly, aa includes $\quad q^2(1 - F) + qF = q^2 + Fpq.$

Finally, Aa is entirely random $\quad = 2pq(1 - F).$

GENETIC VARIANCE UNDER INBREEDING

When we know the distribution of genotypes under inbreeding, we may look at the quantitative phenotypic array in the population as a whole to note the increment in genetic variance that takes place, owing to the redistribution of phenotypes. If we consider the simplest quantitative condition (lack of dominance) on the phenotypic scale (Chapter 6) and a single pair of alleles, so that D (A_1A_1 homozygote) has a value of $+a$, H (heterozygote) zero, and R (A_2A_2) $-a$, then the genetic variance under any nonrandom state will be:

$$\sigma_G^2 = a^2D + a^2R - [a(D - R)]^2 = a^2[D + R - (D - R)^2]$$

Then if we supply the zygote distribution from (8-3), we have

$$\sigma_G^2 = a^2[p^2 + Fpq + q^2 + Fpq - (p^2 + Fpq - q^2 - Fpq)^2]$$
$$= a^2[p^2 + 2Fpq + q^2 - (p - q)^2]$$

which simplifies to

$$\sigma_G^2 = a^2 \cdot 2pq(1 + F) \tag{8-4}$$

The genetic variance has been increased by the fraction F of itself over that under random mating (formula 6-2). The same will be true for all loci without dominance according to our conclusions of the previous sections, and we may take (8-4) as a generalization for an equilibrium system. (However, recent considerations of multiple-locus systems will make this calculation less simple, although still approximately true, because multiple heterozygotes and homozygotes may occur in excess over the expectation from the product of their separate locus probabilities.)

The entire population as a whole under inbreeding has increased variance compared with the random-mating population, but the student must realize that within any isolate of that population (the selfed individual, a sib pair, cousin pair, and so on), the variance is being reduced by the same proportion (F), and once homozygosity is attained in a pure line or isolate, the genetic variance becomes zero in that pure line or isolate. The consequences of inbreeding within isolates or lines is examined in Chapter 12.

EQUILIBRIUM WITH SELFING AND RANDOM MATING COMBINED

It is evident that under continued selfing (or any regular inbreeding system), identical homozygotes will continue to increase each generation, though at a gradually slower rate—that is, in an infinite series (discussed in the next chapter in more detail). Any time mating occurs exclusively at random just once in an inbred population, genotype proportions would be restored at square-law frequencies, as pointed out in Chapter 1. The student can easily see with a little reflection that if both inbreeding and some amount of outcrossing take place in "just the right amounts," genotypes could be retained at equilibrium frequencies somewhere between the completely random condition and the completely inbred proportions. If the conditions for equilibrium are met (especially property 5 in Chapter 1), we may obtain an expression for these "just right amounts" of inbreeding and random mating.

First, using formula (1-5), under random mating, $H = 2pq$, $H^2 = 4p^2q^2 = 4DR$, or $4DR - H^2 = 0$. Under inbreeding, this expression will be >0, as can be seen from the following when we set substitute genotypes from (8-3):

$$4DR - H^2 = 4(p^2 + Fpq)(q^2 + Fpq) - [2pq(1 - F)]^2$$
$$= 4Fp^3q + 4Fpq^3 + 8Fp^2q^2$$

on simplification, this becomes

$$4DR - H^2 = 4Fpq \qquad (8\text{-}5)$$

Property 5 (Chapter 1) included the fact that at equilibrium in any population, new *Aa* (heterozygotes) would arise from matings of $AA \times aa$, while matings between heterozygotes ($Aa \times Aa$) reproduce just half their number in the same genotype (*Aa*). The other half become homozygotes; therefore, equilibrium between these genotypes exists only when this relationship between mating frequencies occurs:

$$2[AA \times aa] = [Aa \times Aa]$$

If we look at the frequencies of mating in the population, it is convenient to consider that there are two types of mating going on: a fraction (s) of the population is selfed (or inbred in some other way) plus a fraction of randomly mated ($1 - s$). Nine matings can be represented as in Table 8-3, a modification of Table 1-2A. By applying the equilibrium

TABLE 8-3 *Mating frequencies of genotypes with a fraction of the population selfed (s) plus a fraction randomly mating (1 − s).*

	D	H	R	
D	$(1 - s)D^2 + sD$	$(1 - s)DH$	$(1 - s)DR$	D
H	$(1 - s)DH$	$(1 - s)H^2 + sH$	$(1 - s)HR$	H
R	$(1 - s)DR$	$(1 - s)HR$	$(1 - s)R^2 + sR$	R
Total	D	H	R	1.00

property 5 for the equality between matings, we have

$$4(1 - s)DR = (1 - s)H^2 + sH$$

which simplifies to

$$4DR - H^2 = \frac{sH}{(1 - s)} \tag{8-6}$$

Because the left side of (8-6) is the same as that of (8-5), we may substitute the right sides to solve for F in terms of s. That is, the amount of inbreeding that increases the expression (8-5) above zero is counterbalanced by a ratio between selfed and outcrossed portions of the population.

$$4Fpq = \frac{sH}{(1 - s)}$$

because $H = 2pq(1 - F)$, we have

$$4Fpq = \frac{s}{(1 - s)}[2pq(1 - F)]$$

Then, solving for F,

$$F = \frac{s}{2 - s} \tag{8-7}$$

equivalent inbreeding coefficient when a portion of the population is selfed (s) and a portion outcrossed $(1 - s)$.

Example 8-1 illustrates the situation in a population that is predominantly selfed and yet retains a high degree of genic polymorphism at apparent equilibrium. In a wide number of self-pollinated species of grasses, R. W. Allard at the University of California, Davis, and his colleagues have described the genic architecture of populations, many of which parallel the case given in this example. (For further reference to this field, see Allard, 1965, 1975; Allard, Jain, and Workman, 1968; Hamrick and Allard, 1975.) In the example, a discrepancy exists between the calculated F value based on the known fraction of selfed pollen in the population and the F value estimated from observed genotype frequencies. From the known fraction of selfing (0.986), $F = 0.972$, using (8-7), which is far greater than the average inbreeding coefficient estimated from the observed frequency of heterozygotes—namely, $F = 0.75$, using (8-2) rearranged. These facts strongly indicate an excess of heterozygotes for four of the six genic loci sampled; for example, $H(\text{expected}) = 0.013$, $H(\text{observed}) = 0.11$, for E_4 locus. This result is interpreted by the investigators as evidence for balanced polymorphism (see Chapters 14–18).

The approach to the equilibrium state from a completely random-mating situation can be derived by consideration of two components, a random and an inbreeding portion of each genotype: $(1 - s)(p^2 + 2pq + q^2) + s(p + q)$. We may consider in the following succession of generations just the heterozygote frequencies following the first generation array. Let us assume selfing and random mating in a population originally randomly breeding and then proceed as in Table 8-4 with consideration of just the heterozygote frequency at equilibrium. From one generation to the next, the heterozygote frequency is reduced in the selfing (s) fraction always by $s/2$, as shown by the transition from G_1 to G_2. In G_2 after algebraic simplification, the H reduction by the $s/2$ fraction produces a geometric

TABLE 8-4 *Approach to equilibrium for a population with selfing and random mating combined*

Generation	Genotypes		
	AA	Aa	aa
G_0	p^2	$2pq$	q^2
G_1	$(1-s)p^2 + s\left(p^2 + \frac{pq}{2}\right) = p^2 + \frac{spq}{2}$	$(1-s)2pq + spq = 2pq\left(1 - \frac{s}{2}\right)$	$(1-s)q^2 + s\left(q^2 + \frac{pq}{2}\right) = q^2 + \frac{spq}{2}$
G_2		Heterozygotes only $(1-s)2pq + spq\left(1 - \frac{s}{2}\right)$	
		$= 2pq\left[1 - \frac{s}{2} - \left(\frac{s}{2}\right)^2\right]$	
⋮			
G_t		$H_t = 2pq\left[1 - \frac{s}{2} - \left(\frac{s}{2}\right)^2 - \left(\frac{s}{2}\right)^3 \cdots \left(\frac{s}{2}\right)^t\right]$	
At equilibrium, $t \to \infty$,		$H_t = 2pq\left[\frac{1-s}{1-s/2}\right] = 2pq\left[1 - \frac{s}{2-s}\right]$	

$F = \frac{s}{2-s}$ under selfing and random mating combined; thus, $H_t = 2pq\,(1 - F_t)$ as in (8-2) and (8-3) series so that at the tth generation H_t can be expressed as the net sum of terms in that geometric series at the limit ($t \to \infty$). This relationship makes the limiting case the equivalent of (8-2) when the net inbreeding coefficient is the ratio $s/(2 - s)$.

The smaller the amount of selfing (s), the closer to the simple random square law at equilibrium but the slower the approach to the equilibrium state (8-3), if selfing is imposed on a random-mating population. Of course, in any real population of limited size, the equilibrium state would not be detectably different from one very nearly approaching that state.

EXAMPLE 8-1

GENOTYPES IN A PREDOMINANTLY SELFED POPULATION

Allard (1975) and his colleagues surveyed natural populations of certain grasses (particularly wild oats, *Avena fatua* L. and *A. barbata* Brot.) for enzyme polymorphism over a wide area of California. These species of oats are principally selfing and would be expected to exist as homogeneous and homozygous populations, but many contain substantial amounts of allozyme variability. Populations from the more moist habitats in central California are characterized not only by considerable heterozygosity but also by a great diversity of genotypes if about a half-dozen genic loci are assayed together among individual plants. Furthermore, allozyme allelic frequencies may differ considerably when populations only a few miles apart are compared.

A particularly variable population of *A. barbata* was described by Marshall and Allard (1970) from a locality they designated *CSA* near the city of Calistoga in Napa County. It was scored for six loci by starch gel electrophoresis of leaf tissue after germination of seedlings established from seeds of single wild plants: four esterase loci (E_1, E_4, E_7, and E_{10}), a phosphatase locus (P_5), and an anodal peroxidase locus (APX_5). Two loci (E_1 and E_7) showed only homozygotes of one allele, while the remaining loci were polymorphic with the genotype frequencies, as given in Table A.

To determine whether the observed level of heterozygosity would be consistent with levels of outcrossing in the population, the proportion of selfing was estimated for each locus from the proportion of heterozygous progeny among homozygotes pollinated in nature. The proportion of outcrossed pollen amounted to 0.008–0.027 in this population and averaged 0.014. From formula (8-7), the equivalent amount of inbreeding in the population should be

$$F = \frac{s}{2 - s} = \frac{0.986}{1.014}$$

$$F = 0.972$$

However, the observed F level is much lower than that expected for all four loci. If we take each locus in turn, an estimate of F can be calculated from (8-2), as follows: By rearranging (8-2), we have $F = 1 - (H_t/H_0)$, where H_0 = heterozygote frequency under random mating and H_t = observed heterozygote frequency with inbreeding.

TABLE A *Genotype frequencies for six genic loci in a population of* Avena barbata *(Brot.) from CSA locality (Napa County, California) (from Marshall and Allard, 1970)*

Locus	Genotype	Frequencies	Sample Size
E_1	11	0	
	12	0	54
	22	1	
E_4	11	0.30	
	12	0.11	54
	22	0.59	
E_7	11	0	
	12	0	54
	22	1	
E_{10}	11	0.46	
	12	0.13	85
	22	0.41	
P_5	11	0.40	
	12	0.15	86
	22	0.45	
APX_5	11	0.48	
	12	0.11	86
	22	0.41	

In the case of E_4 genotypes, for example, estimated $F = 1 - \left[\frac{0.11}{0.458}\right] = 0.76$

also from E_{10}, $F = 0.74$

from P_5, $F = 0.70$

and from APX_5, $F = 0.78$

All these estimates are reasonably close and can be considered uniform, so that their average $F = 0.75$ indicates a much lower level than that expected on the basis of self-pollination observed. Consequently, the authors were confident that a highly significant excess of heterozygotes occurred in the population, probably owing to a selective advantage.

COMPLEXITIES WITH TWO LOCI UNDER SELFING

When more than one locus is considered under selfing, the situation can be more complex than that illustrated in Table 8-1C either if the original base population was not at gametic equilibrium (Chapter 5) when selfed or if selfing and random mating are combined. If selfing is carried out on a population in which the double heterozygotes are not equally divided between coupling and repulsion genotypes ($AB/ab \neq Ab/aB$), the limiting state for the four homozygotes will not be as stated above (under selfing, two independent loci, page 242)—$AB/AB \neq pu$, and so forth—but there will be a positive association to increase (or to decrease) coupling at the expense of repulsion gametes and thus the total homozygotes at the limiting state. Bennett and Binet (1956) examined the outcome of selfing and selfing combined with random mating for the two-factor case, which we may summarize briefly as follows without derivation (also see Workman and Allard, 1962).

After complete selfing to the limiting state, the frequency of the coupling gamete ($\overline{AB}$) has a positive increment dependent on the recombination frequency (c) and on the double heterozygote difference (repulsion minus coupling genotypes) in the base population, so that the coupling homozygote (AB/AB) would have this frequency:

$$\frac{\overline{AB}}{\overline{AB}_{\infty}} = \overline{AB}_{\infty} = \overline{AB}_0 + \frac{c}{1 + 2c}\left(\frac{\overline{Ab}_0}{\overline{aB}_0} - \frac{\overline{AB}_0}{\overline{ab}_0}\right)$$

The other gametes (genotypes) can be calculated by remembering that the two coupling forms will have the same increment. This fact is true even if A and B loci are not linked.

Finally, when both selfing and random mating are combined with relative frequencies of s and $(1 - s)$, respectively, linked factors become associated to an extent that depends not only on the proportion of selfing but also on the magnitude of recombination. Even when the loci are independently assorting ($c = 0.5$), double homozygotes and double heterozygotes will show positive association and single heterozygotes negative association. That is, these associations are the excesses of the final equilibrium frequencies over the expected frequencies of genotypes based on the product of the single-locus probabilities. In general, the excess of double heterozygotes (ω) for any recombination frequency (c) would be as follows: $\omega = \overline{AaBb}_{\infty} - \overline{Aa}_{\infty} \cdot \overline{Bb}_{\infty}$

$$\omega = \frac{16s(1 - s)}{(4 - s)(2 - s)^2}\left[\frac{2 - s - 4c(1 - c)(1 - s)}{4 - 2s + 4sc(1 - c)}\right] pquv$$

Under random mating, this excess diminishes to zero similarly to the gametic determinant (D), although more rapidly; under varying proportions of selfing and random mating, this excess will be maximum when the selfing proportion is approximately 70 percent. (See Allard, Jain, and Workman (1968) for graphic representation of these nonrandom effects.) There is then a tendency for double heterozygotes to become concentrated in a few members of a population that is selfing and randomly mating. Double homozygotes will also be in excess at about one-fourth of ω, but there will be a deficiency of single heterozygotes at about -2ω. Of course, the net results for a single locus are not altered by these considerations.

EXERCISES

1. Given the following genotype frequencies from the parent generation G_0, calculate the expected genotype frequencies through the G_4 and the limiting frequencies if only self-fertilization occurs:

	AA	Aa'	Aa	$a'a'$	$a'a$	aa
(a)	0.05	0.20	0.10	0.06	0.28	0.31
(b)	0.04	0.12	0.20	0.09	0.30	0.25
(c)	0	0	0.40	0	0.60	0

2. Given the following sets of nine genotype frequencies for two loci from parent generation G_0, arranged as in Table 8-1C, calculate the expected genotype frequencies through the G_2 and limiting frequencies if selfing alone is the method of reproduction.

(a) $\begin{vmatrix} 1 & 4 & 4 \\ 4 & 10 & 4 \\ 4 & 4 & 1 \end{vmatrix} = 36$

(b) $\begin{vmatrix} 25 & 0 & 0 \\ 0 & 50 & 0 \\ 0 & 0 & 25 \end{vmatrix} = 100$

(c) $\begin{vmatrix} 0 & 0 & 0 \\ 25 & 50 & 25 \\ 0 & 0 & 0 \end{vmatrix} = 100$

3. If a single heterozygous plant is selfed and its descendants are selfed continuously thereafter, how many generations would it take to achieve approximately 98.4 percent homozygotes?
4. If a small isolated population of a self-fertilizing species consisted only of six AA and two Aa individuals, what percentage of the population would be heterozygous after five generations of selfing? What are expected to be the genotype frequencies in that generation? allelic frequencies and genotypes in the population?
5. In a selfing population, the following genotype frequencies were observed: 8AA: 24Aa: 68aa. From formula (8-3), estimate F and the number of generations of selfing that might have produced these frequencies.
6. Assume a cornfield such as the one in Exercise 1, Chapter 7, with plants ranging in height from 3–7 feet at 1-foot intervals and height controlled by two pairs of factors that are equally additive and equal in allelic frequencies.
   (a) Give the expected genetic variance for this cornfield if these corn plants are selfed over four generations, using formula (8-4).
   (b) If the A-a pair of alleles had $p = 0.2$, $q = 0.8$ and the B-b pair had $u = 0.4$, $v = 0.6$, what would be the expected genetic variance after three generations of selfing?
   (c) Suppose the additive effect of the A-a pair is 1 foot as before, but the additive effect of B is to add 2 feet over b. What would be the expected genetic variance after three generations of selfing? What factors affect the amount of genetic variance?
7. What would be the expected genetic variance among the progeny of plants selfed from a single $AaBb$ plant (assuming the additive effect of $A = B = 1$ unit). On continued selfing for a second generation, what will be the expected average genetic variance within progenies? What happens to genetic variance within inbred lines?
8. Verify the following identity: if each homozygous frequency under inbreeding is

divided by its allelic frequency, the sum of all such ratios will be as follows:

$$\frac{p^2 + Fpq}{p} + \frac{q^2 + Fpq}{q} = 1 + F$$

9. Start with a single genotype *AaBbCc* in a population undergoing selfing. In G_1, $\frac{1}{8}$ of the progeny will be triple heterozygotes like the parent, $\frac{3}{8}$ of the progeny will be double heterozygotes (*AaBBCc*, *Aa Bb CC*, etc.), $\frac{3}{8}$ of the progeny will be single heterozygotes (*AaBBCC*, etc.), and $\frac{1}{8}$ of the progeny will be zero heterozygotes (*AABBCC*, *aaBBCC*, etc.). Complete filling in the spaces in the table below to show the distribution of genotypes under selfing with different amounts of heterozygosity, total heterozygous loci, and percent heterozygosity over all loci:

$G(t)$ of Selfing	Heterozygotes None	Single	Double	Triple	Total $= 2^{3t}$ Individual Types	Total Individual Heterozygous Loci out of →	All Loci	$\%H$
0	0	0	0	1	$1 = 2^0$	3	3	100
1	1	3	3	1	$8 = 2^3$	12	24	50
2	27	___	___	___	$64 = 2^6$	48	192	___
3	___	___	___	___	___	___	___	___
⋮								
Limit	___	___	___	___	___	___	___	___

10. If a corn plant is heterozygous for four closely linked genes with the genotype *AbCd*/*aBcD*, what is the probability that on self-fertilization a plant will be produced homozygous at all four loci,
    (a) On the assumption of zero crossing over between any loci?
    (b) On the assumption of 1 percent crossing over between each locus and its neighbor locus with no multiple crossovers occurring?
    (c) With the assumptions of (b), how long (generations) would it take to achieve a possible *ABCD*/*ABCD* homozygote?
11. In Example 8-1, from the known amount of selfing, average F was calculated to be $F = 0.986$. Average homozygote frequency for the four polymorphic loci was 0.875.
    (a) Why is this latter frequency not considered to be the estimate of F?
    (b) What genotype frequencies would you have expected to find in a population of these wild oats knowing the amount of selfing taking place? Why?
    (c) How could you account for the difference between the E_1–E_7 loci and the other loci?
12. A certain plant population capable of selfing has the following frequencies of genotypes:

AA	*Aa*	*aa*
12	16	72

    (a) What level of inbreeding can the population be said to have?
    (b) If random mating as well as selfing takes place in this population, what amounts of the two types of mating could be said to be taking place?

(c) What assumptions must be made about other forces at work on this population before you can put confidence into the calculations you have made about relative amounts of inbreeding and random mating?

13. Construct a pair of curves on two-coordinate graph paper as in Figure 1-2. Let the X axis $= p_A$ and the Y axis $=$ genotype frequency. First draw curves for genotypes under random mating, then use formula (8-3) and draw curves for genotypes *aa* and *Aa* under selfing at G_1, G_2, and G_5. Describe the changes in the homozygous and heterozygous frequency curves.

9

INBREEDING, LESS EXTREME: SIBS AND MORE DISTANT RELATIVES; CONSEQUENCES FOR PHENOTYPES

For sexual organisms that are not monoecious or hermaphroditic, the mating systems that bring about formation of identical homozygotes most rapidly are sib mating and parent-offspring mating. We shall being by considering only the former since the latter is in principle very similar. Once we see the main points for sib mating, other regular systems of inbreeding will appear to be less complicated. Sib mating is most widely used by animal breeders and geneticists in general for production of homozygosity, but it is doubtful whether any natural population exhibits sib mating on a regular or exclusive basis; at least there are none on record. Owing to the dispersion of young in most wild populations, probably little sib mating takes place in nature, although there are few if any genetic mechanisms to prevent it specifically. Genetic mechanisms of pollen self-sterility, for example, prevent selfing in many flowering plants.

Nevertheless, in our consideration of general principles with inbreeding, we must explore in some detail what the consequences of sib mating can be for genotype distribution, because we shall see a number of useful generalities emerge thereby. First, we proceed through a realistic but lengthy method for visualizing the consequences of sib mating for a single generation. Because proceeding further with that method would be quite long, complex, and inefficient, we then summarize the probability method of analysis of inbreeding (based on Malécot, 1948) in order to generalize the changes in genotype distribution by repeated application of the probabilities through successive generations. The consequences for genotype distribution can then be examined for more distant relative matings, inbreeding effects of common ancestors, and mixed mating systems.

SIB MATING, ONE GENERATION

If an entire population practices sib mating, each sibship (pair of brother-sister) would be an isolate in the same sense that the individual that selfed is an isolate, or closed system. Each family resulting from a brother-sister mating can be considered a unit, and the population then must be considered as an assemblage of numerous brother-sister pairs for the following generation. But all matings within a family provide as many new isolates as there are sibships possible; if each brother-sister pair among the progeny from a sibship were not isolated from the progenies of all other sibships of a family, the following generation would include cousins as well as sibs. (A first cousin is the child of a full sib of one's parent.) In addition, for simplicity, we must also assume that all sibships have equal numbers of male and female progeny, that all sib pairs have equal progeny, and that all progeny survive. In other words, we need to generate probabilities for genotypes each generation by the mating within sib pairs only in order to see the general outcome, which we may then compare with selfing.

TABLE 9-1 *Basic effect of sib mating for one generation (assume all four alleles unlike at single locus in parent generation)*

Generation	Genotypes	
G_0 (Parent single pair)	$A_1A_2 \times A_3A_4$	$F = 0$
G_1 (Brothers and sisters)	A_1A_3, A_1A_4, A_2A_3, A_2A_4	$F = 0$

Sibs mate at random in 4 × 4 sibships

	A_1A_3	A_1A_4	A_2A_3	A_2A_4
A_1A_3	$\frac{1}{4}A_1A_1$ $\frac{2}{4}A_1A_3$ $\frac{1}{4}A_3A_3$	$\frac{1}{4}$	$\frac{1}{4}$	0
A_1A_4	$\frac{1}{4}$	$\frac{1}{2}$	0	$\frac{1}{4}$
A_2A_3	$\frac{1}{4}$	0	$\frac{1}{2}$	$\frac{1}{4}$
A_2A_4	0	$\frac{1}{4}$	$\frac{1}{4}$	$\frac{1}{2}$

sibship

$\boxed{\frac{1}{2}}$ $= \frac{1}{2}$ of progeny in sibship are identical as illustrated in upper left box

(Fractions indicate proportion of progeny identical within sibship)

G_2 Progeny from all 16 sibships include: $\frac{4}{16}(\frac{1}{2}) + \frac{8}{16}(\frac{1}{4}) + \frac{4}{16}(0)$, identical homozygotes $= \frac{1}{4}(A_1A_1$, A_2A_2, A_3A_3, $A_4A_4)$; thus, $F = \frac{1}{4}$.

In its simplest form the emergence of identical homozygotes by sib mating can be envisaged as in Table 9-1. We assume that a parental pair has a locus with all alleles unlike ($A_1A_2 \times A_3A_4$). All progeny are sibs, which can in turn become paired (brother × sister) in all 16 possible combinations. Progeny of these brothers × sisters will be identical homozygotes one-fourth of the time, having replicates of each allele from the grandparents in double dose. The diagonal from upper left to lower right supplies half of each sibship with identical progeny, the opposite diagonal zero identical progeny, and all other sibships produce one-quarter of their progeny as identical homozygotes. Because each sibship is one-sixteenth of the total, a final tally amounts to one-quarter identical homozygotes resulting from sib mating. This is a general result from sib mating; one-fourth of non-identical genotypes will always be made identical by this mating system ($F = 0.25$ in progeny of sib matings). This fact can be applied to single loci, average loci in the genome, or the average individual's genotype, as we concluded from considerations in the last chapter.

It is also important to consider the outcome of sib mating from a random-mating base population. Suppose we choose parents as random pairs from a large population in which zygotes fit the square law; nine mating combinations occur, or six genetically different parental pairs (assuming reciprocals to be equal) in frequencies as shown in Table 9-2 (the same as those in Chapter 1, Table 1-1B). We may conclude, of course, that for both parents

TABLE 9-2 *Sib mating among progeny within families from a panmictic population of parents*

A

Pairs of (G_0) parents Parents ↓ →	$p^2(AA)$	$2pq(Aa)$	$q^2(aa)$
$p^2(AA)$	① $p^4[AA]$	② $2p^3q\begin{bmatrix}AA\\Aa\end{bmatrix}$	③ $p^2q^2[Aa]$
$2pq(Aa)$	④ $2p^3q\begin{bmatrix}AA\\Aa\end{bmatrix}$	⑤ $4p^2q^2\begin{bmatrix}AA\ \frac{1}{4}\\Aa\ \frac{2}{4}\\aa\ \frac{1}{4}\end{bmatrix}$	⑥ $2pq^3\begin{bmatrix}Aa\\aa\end{bmatrix}$
$q^2(aa)$	⑦ $p^2q^2[Aa]$	⑧ $2pq^3\begin{bmatrix}Aa\\aa\end{bmatrix}$	⑨ $q^4[aa]$

G_1 progeny = sibs in bracketed sibships for 9 different types of families

B If G_1 *only* brother × sister pairs reproduce, there will be 6 types of sibship with genotype mating frequencies derived as follows (omitting 6, 8, and 9)

Sibship	Frequency	Matings $G_1 \times G_1$ within Sibships
①		
$[AA]$	p^4	$AA \times AA$
② and 4		
$\begin{bmatrix}AA\\Aa\end{bmatrix}$	$4p^3q$	$AA \times AA\ \frac{1}{4}$ $AA \times Aa\ \frac{2}{4}$ $Aa \times Aa\ \frac{1}{4}$
③ and ⑦		
$[Aa]$	$2p^2q^2$	$Aa \times Aa$
⑤		
$\begin{bmatrix}AA\ \frac{1}{4}\\Aa\ \frac{2}{4}\\aa\ \frac{1}{4}\end{bmatrix}$	$4p^2q^2$	$AA \times AA\ \frac{1}{16}$ $AA \times Aa\ \frac{4}{16}$ $AA \times aa\ \frac{2}{16}$ $Aa \times Aa\ \frac{4}{16}$ $Aa \times aa\ \frac{4}{16}$ $aa \times aa\ \frac{1}{16}$

C Tally of $G_1 \times G_1$

Matings	Frequency	G_2 Offspring AA	Aa	aa
$AA \times AA =$	$p^4 + p^3q + (\frac{1}{4})p^2q^2$	1	0	0
$AA \times Aa =$	$2p^3q + p^2q^2$	$\frac{1}{2}$	$\frac{1}{2}$	0
$AA \times aa =$	$(\frac{1}{2})p^2q^2$	0	1	0
$Aa \times Aa =$	$p^3q + pq^3 + 3p^2q^2$	$\frac{1}{4}$	$\frac{1}{2}$	$\frac{1}{4}$
$Aa \times aa =$	$2pq^3 + p^2q^2$	0	$\frac{1}{2}$	$\frac{1}{2}$
$aa \times aa =$	$q^4 + pq^3 + (\frac{1}{4})p^2q^2$	0	0	1

TABLE 9-2 *(Continued)*

D G_2 *offspring totals* (from column × rows above)

$$AA = p^4 + p^3q + (\tfrac{1}{4})p^2q^2 + p^3q + (\tfrac{1}{2})p^2q^2 + (\tfrac{1}{4})p^3q + (\tfrac{1}{4})pq^3 + (\tfrac{3}{4})p^2q^2{*}$$

$$= p^2 + \frac{pq}{4}$$

$$Aa = p^3q + \frac{p^2q^2}{2} + \frac{p^2q^2}{2} + \frac{p^3q}{2} + \frac{pq^3}{2} + \frac{3p^2q^2}{2} + pq^3 + \frac{p^2q^2{*}}{2}$$

$$= 2pq - \frac{pq}{2}$$

$$aa = q^2 + \frac{pq}{4}$$

* In simplifying, refer to Table 1-1B, Chapter 1, where

$$p^2 = p^4 + 2p^3q + p^2q^2$$
$$2pq = 2p^3q + 4p^2q^2 + 2pq^3$$
$$q^2 = p^2q^2 + 2pq^3 + q^4$$

Pool together like terms and subtract appropriately.

and their random progeny (brothers and sisters to be paired in sibships), no change has taken place in genotype frequencies.

The six different parental pairs produce six different types of sibship, four of which are outlined in Table 9-2B for illustration. Visualizing that within each sibship enough brother × sister pairs could exist to include all possible combinations for pair matings at random, we list those matings and within-sibship frequencies separately and finally their resulting progeny in Table 9-2C and D. In the G_2 generation, progeny following the mating of sibs then occurs as follows:

$$AA = p^2 + \frac{pq}{4}, \quad Aa = 2pq - \frac{pq}{2}, \quad aa = q^2 + \frac{pq}{4}$$

The newly created identical homozygotes are partioned equally as one-fourth the gene frequency product, so that $F = \frac{1}{4}$, as in the simpler case of Table 9-1. The student should recall the incremental fraction Fpq added to each homozygote as in Table 8-2 and formula (8-3). F is then the coefficient of pq. (See Exercise 2 for the outcome for sib mating from the special case of a Mendelian F_2 to F_3 when original parents may be defined as identical homozygotes.)

At least three points should be obvious here. First, by comparison with selfing, it takes a generation to produce sibs from random pairs of parents, and the increment in identical homozygotes takes an extra generation; in other words, identical genic dosages must come from two separate individuals instead of from one. Second, the inbreeding coefficient is half that in selfing. Third, the complexity of working out the inbreeding coefficient for just the progeny of sibs is so great as to give us apprehension about going beyond the second generation, let alone working out a recurrence relation for successive generations or learning general results for more remote relative mating systems. What is needed is a simpler and more direct method, which we shall now describe.

RECURRENCE OF *F* IN SELFING

The briefest and simplest method of working out the successive increments in identical genes per generation is the probability method (Malécot, 1948). We shall go through the main steps here for what has already been shown by longer methods with selfing and sib mating so the student may gain confidence in the outcome before proceeding.

First, we note that the symbol F refers to the probability that two identical alleles *have been transmitted* from parent(s) to offspring. It will be useful to introduce another symbol f in lower case to represent the probability that a parent *transmits* two identical alleles or that each of the two parents transmits one allele identical to that from the other parent. It is easier to consider two gametes transmitting identical alleles than to consider the zygote as having received them. (Malécot has called this probability the "coefficient de parenté," and Falconer called it "coancestry," a term that does not convey to the student as much meaning as the term "transmission," which we shall use.) Thus, the probability (F_t) for an individual in the tth generation to be an identical homozygote is the same as the probability that the two gametes transmitted from its parent or parents were identical (f_{PP} if both genes came from a single parent P or f_{PQ} if one gene came from one parent P and the other one from parent Q). In Figure 9-1A, assume the population consists of selfed individuals, each parent having both male (square) and female (circle) elements and with genotype A_1A_2; then the probability that P (parent) transmits identical genes to X (offspring), one via a male gamete and one via a female gamete, is as follows:

1. If A_1 is not identical with A_2, then half of the time both gametes will be either A_1 and A_1 or A_2 and A_2—that is, $\frac{1}{2}(1 - F_{t-1})$, as indicated in the previous chapter.

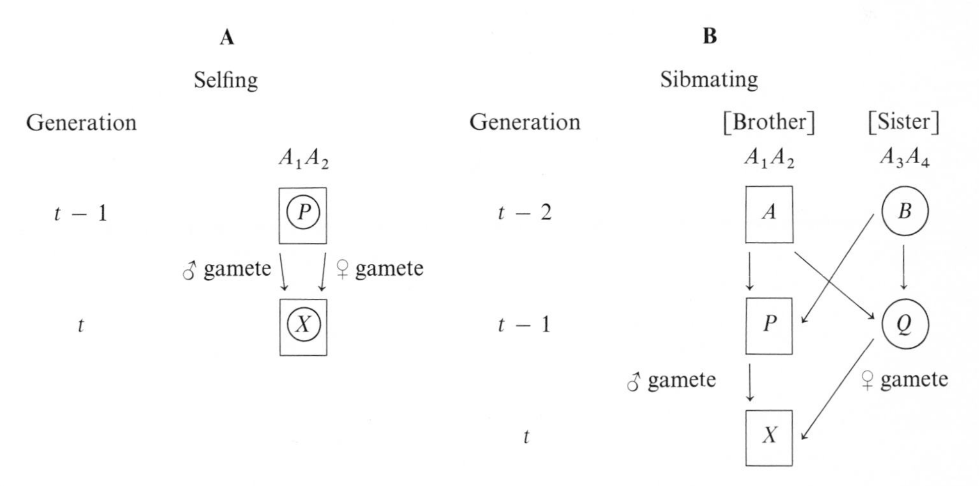

Figure 9-1. Diagram for selfing (A) and sibmating (B) indicating the transmission paths of gametes from parents to offspring X in the t generation. A_i represents any genic allele.

2. If A_1 is already identical with A_2 because of previous inbreeding, then both gametes will be identical already—that is, F_{t-1}.

Then the total probability for any individual following selfing to be identical equals the probability that the parent transmitted identical gametes.

$$F_t = f_{PP} = \tfrac{1}{2}(1 - F_{t-1}) + F_{t-1} = \tfrac{1}{2} + \tfrac{1}{2}F_{t-1} = \frac{1 + F_{t-1}}{2} \qquad (9\text{-}1)$$

(recurrence of F in selfing).

The student should verify that the successive increase in F and decrease in H follows (from a panmictic base population, for example):

Generation	0	1	2	3	4	⋯	t
F coefficient	0	$\frac{1}{2}$	$\frac{3}{4}$	$\frac{7}{8}$	$\frac{15}{16}$	⋯	$\frac{1 + F_{t-1}}{2}$
Heterozygotes	$2pq$	$2pq/2$	$2pq/4$	$2pq/8$	$2pq/16$	⋯	$H_0(1 - F_t)$

Also, it is important to convert the F recurrence statement into the H recurrence algebraically by recalling from formula (8-2):

$$H_t = H_0(1 - F_t) = H_0\left[1 - \left(\frac{1 + F_{t-1}}{2}\right)\right] = \frac{H_0(1 - F_{t-1})}{2}$$

Therefore,

$$H_t = \frac{H_{t-1}}{2}$$

because the numerator above $= H$ in the parent generation (recurrence of H in selfing identical to (8-1)).

RECURRENCE OF *F* IN SIB MATING

The diagram for sib mating in Figure 9-1B parallels that for selfing with the grandparent generation added; of course, it differs from selfing in that the individual offspring X has two parents P and Q. We may presume that the grandparents A and B possess four alleles, any of which can become identical homozygotes in X. The two gametes from P and Q could have originated in four different ways: (1) a single allele (say, A_1) could have been transmitted from ancestor A to both his son P and daughter Q via the two arrows from A; the probability for this is symbolized f_{AA} (both gametes transmitted from same grandparent); (2) a single allele could have been transmitted from grandmother via the two arrows from B to both her son and daughter (f_{BB}); (3) one allele from A to P and one allele from B to Q via outside arrows (f_{AB}); or (4) one allele from A to Q and one from B to P via inside arrows (f_{BA}). In general, for sib mating, the inbreeding coefficient of X offspring is

$$F_t = f_{PQ} = \tfrac{1}{4}[f_{AA} + f_{BB} + f_{AB} + f_{BA}]$$

That is, the path of transmission (gamete probabilities f_{PQ}) from parent sibs to their offspring is divisible into four equally probable paths of transmission from the grandparents to parents.

The student should note that the first two component paths from generation $t - 2$ to $t - 1$ are the probabilities for *one* ancestor to transmit identical genes via *two* offspring, called condition 1. The second two paths are the probabilities for *two* ancestors separately to transmit identical genes via *one* offspring each, called condition 2. Let us then consider the recurrence of F by assuming that a population has been entirely sib mating for a few generations so that some of the alleles at generation $t - 2$ will be identical and some also at $t - 1$ generation because of previous inbreeding. Then we must consider that A_1 may be identical with A_2 or A_3 with A_4. Taking condition 1 first, we see:

1. If A_1 is not identical with A_2, then half of the time both gametes will be either A_1 and A_1 or A_2 and A_2—that is, their probability to be identical in transmission $= \frac{1}{2}(1 - F_{t-2})$. If A_1 is already identical with A_2, then the transmission probability is F_{t-2}. The situation is then the same as that for selfing, just as if individual A had achieved self-fertilization! The probability for A to transmit identical genes to P and Q is the selfing equivalent for A's generation:

$$f_{AA} = \tfrac{1}{2}(1 - F_{t-2}) + F_{t-2} = \frac{1 + F_{t-2}}{2}$$

The same is true for B's transmission to P and $Q = f_{BB}$. The total probability for condition $1 = f_{AA} + f_{BB} = 2[(1 + F_{t-2})/2]$. Then for condition 2, with the grandparents already sibs, we see:

2. If either A_1 or A_2 is identical with either A_3 or A_4, the probability of transmission f_{AB} or $f_{BA} = 2f_{AB}$, because under a regular system of mating all paths from one generation to the next should be equal. Therefore, the path $A \rightarrow P = A \rightarrow Q$ because all individuals in any generation should have the same probability of being identical. So also $B \rightarrow Q = B \rightarrow P$. Consequently, while the diagram implies identical genes going to *different* individuals P and Q, we can assume that each has received identical genes from grandparents A and B with the same probability. Therefore, $2f_{AB} = 2F_{t-1}$; that is, for example, the probability that the outer arrows are identical is the equivalent of arrows $A \rightarrow P$ and $B \rightarrow P$ being identical, so that P has a probability (F_{t-1}) of being homozygous if the grandparents share identical genes. The same is true by noting the identity of the arrows to Q, so all transmission from the grandparents via separate parents leads to identical homozygote increment in the parents' generation just as that from the parents does for the offspring X.

All four transmission probabilities are equal, so the average inbreeding coefficient of the offspring under a regular system of sib mating is

$$F_t = f_{PQ} = \tfrac{1}{4}\left[2\left(\frac{1 + F_{t-2}}{2}\right) + 2F_{t-1}\right] = \tfrac{1}{4} + \tfrac{1}{2}F_{t-1} + \tfrac{1}{4}F_{t-2} \qquad (9\text{-}2)$$

(recurrence of F in sib mating). The recurrent series for increasing F and decreasing H from a panmictic base population then is as follows:

Generation	0	1	2	3	4	...
F coefficient	0	0	$\frac{1}{4}$	$\frac{3}{8}$	$\frac{8}{16}$	...
Heterozygotes (x $2pq$)	1	$\frac{2}{2}$	$\frac{3}{4}$	$\frac{5}{8}$	$\frac{8}{16}$	...

The H series is known as a *Fibonacci sequence* in which each term has an element that is the sum of two (or more) previous terms. (See Appendix A-13 on recurrent series.) In this case, while the denominator increases by doubling, the numerator is a Fibonacci sequence (if we make the second term $\frac{2}{2}$). The equivalent of selfing one generation ($H = \frac{1}{2}$) is accomplished after three generations of sib mating.

Again, the student should convert the F recurrence statement into the H series algebraically by use of formula (8-2):

$$\begin{aligned} H_t &= H_0(1 - F_t) = H_0[1 - (\tfrac{1}{4} + \tfrac{1}{2}F_{t-1} + \tfrac{1}{4}F_{t-2})] \\ &= H_0(\tfrac{1}{2} - \tfrac{1}{2}F_{t-1} + \tfrac{1}{4} - \tfrac{1}{4}F_{t-2}) \\ &= H_0[\tfrac{1}{2}(1 - F_{t-1}) + \tfrac{1}{4}(1 - F_{t-2})] \end{aligned}$$

and substituting in (8-2),

$$H = \tfrac{1}{2}H_{t-1} + \tfrac{1}{4}H_{t-2} \tag{9-3}$$

(recurrence of H with sib mating).

INBREEDING FROM ONE OR TWO COMMON ANCESTORS

Before proceeding to look at regular systems of mating for more remote relatives than sibs, we can easily see an important generalization at this point: the descendants of any single ancestor may each have received identical genes (copies of one DNA strand in the ancestor).

If ordinary cousins mate, what is the inbreeding coefficient of their child? In Figure 9-2A an ordinary first-cousin mating is diagrammed. Using the same reasoning as before, we can obtain the transmission probability of identical genes from grandfather E to A and B: $f_{EE} = (1 + F_{t-3})/2$, or the equivalent of selfing for E. Owing to Mendelian segregation from A to P, B to Q, and from the two parents (cousins) to X, the probability for transmission will be for identical genes from E:

$$f_{PQ} = \left(\frac{1}{2}\right)^4 \left[\frac{1 + F_{t-3}}{2}\right]$$

because there are four paths of transmission where the probability is one half from any parent to his child.

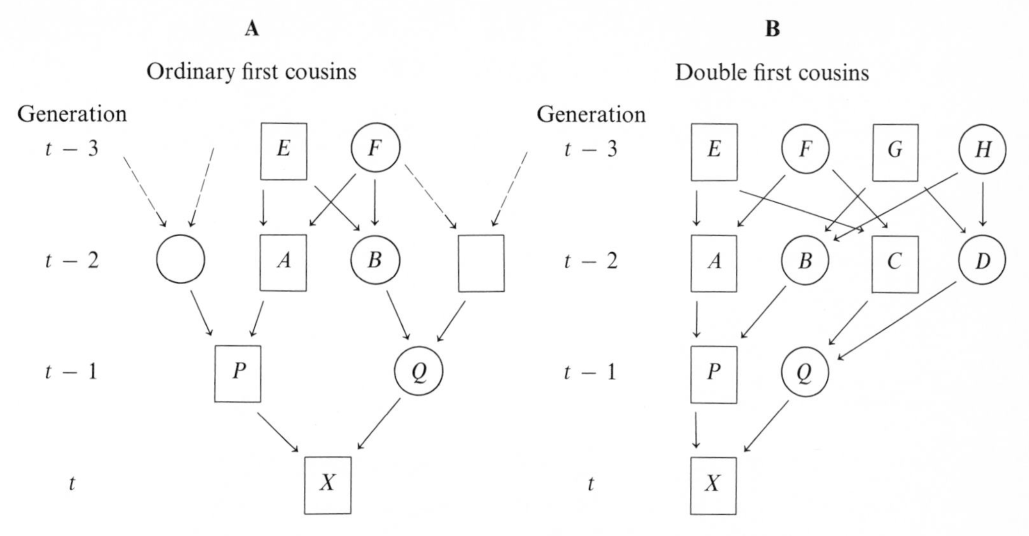

Figure 9-2. Diagram for ordinary first cousin mating (A) and double first cousin mating (B). P and Q are first cousins. Other symbols parallel with Figure 9-1.

The same is true for transmission from grandmother F so that total probability for the child X of ordinary first cousins is

$$F_X = f_{PQ} = 2\left(\frac{1}{2}\right)^4\left[\frac{1 + F_{t-3}}{2}\right] = \frac{1}{16}[1 + F_{t-3}]$$

Note that the exponent of $(1/2)$ is the total number of parent to offspring paths *from the child of one common ancestor downward* to X in succession including all paths from all children of that common ancestor to X. In the case illustrated, the paths to be added are $A \to P$, $P \to X$, $B \to Q$, $Q \to X$ from either ancestor E or F. These four paths together constitute a *compound path.* It is important to realize that there are *not* eight paths but four that are available twice, so that the total probability is simply the sum of compound paths for each common ancestor; the identical homozygote can arise from *either* grandparent. In counting the total paths, we may start at X and go back in generations until we stop short at the *child* of the common ancestor because the transmission of two identical gametes from the ancestor to two of his children is given by the selfing statement $[(1 + F)/2]$ for the ancestor's generation. Under most circumstances, ordinary first cousins will have grandparents who can be considered unrelated—that is, with $F = 0$. Consequently, the coefficient for children of ordinary first cousins will usually be $F = \frac{1}{16} = 0.0625$.

In summary, the F coefficient of any child from parents who are related via one or more common ancestors is as follows:

$$F_X = \sum\left(\frac{1}{2}\right)^t\left[\frac{1 + F_A}{2}\right] \tag{9-4}$$

where F_A = inbreeding coefficient of any common ancestor (A); t = number of parent to child paths from each *child of the common ancestor* to the descendant child (X); $\sum$ = summation of compound path probabilities from all common ancestors to descendant (X).

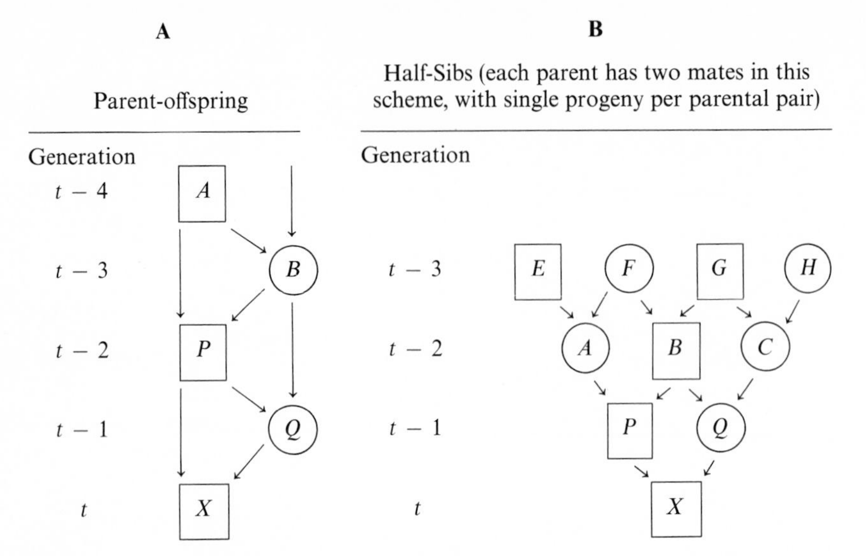

Figure 9-3. Diagram for parent × offspring (A) and a possible half-sib (B) mating scheme. Symbols as in Figure 9-1.

RECURRENCE OF *F* IN OTHER TYPES OF CLOSE RELATIVES

Plant or animal breeders often use a mating system such as parent × offspring, half-sibs, or other close relative matings. To gain some confidence in the probability method of reasoning, the student may refer to diagrams in Figure 9-3 of regular systems of parent × offspring mating and one of the many possible half-sib mating schemes. Exercises 5 and 6 are intended for practice in verifying the F recurrence series given here.

Parent-offspring

$$F_t = \tfrac{1}{4} + \tfrac{1}{2}F_{t-1} + \tfrac{1}{4}F_{t-2} \tag{9-5}$$

(Note: same as for sib mating.)

Half-sib mating

$$F_t = \tfrac{1}{8} + \tfrac{3}{4}F_{t-1} + \tfrac{1}{8}F_{t-2} \tag{9-6}$$

RECURRENCE OF *F* FOR DOUBLE FIRST COUSINS

Double first cousins are those whose four parents belong to two separate sibships; that is, they are cousins via both of the parental lines. We shall consider a bit more in detail a population mating exclusively by double first cousins; this will be helpful when the rate of heterozygosity loss is considered later in Chapter 11. In Figure 9-2B, P and Q are double first cousins. Note that members of the same sex in the $t - 2$ generation are siblings (A and C, B and D). With such a mating structure, all matings are confined to double first cousins, and with perpetuation of that structure an isolate of size 4 may be considered through time; just as with sibs, the isolate is of size 2 and with selfing of size 1. As with those regular systems of mating, the population must be considered as breaking up each generation into sufficient cousinships to generate probabilities for all genotypes each generation.

Basically, the same rules apply for the statement of inbreeding coefficient for any individual X at generation t, as we had for sibs; that is, the paths of transmission (gamete probabilities) from parent cousins to their offspring are divisible into four equally probable paths from grandparents to parents:

$$F_t = f_{PQ} = \tfrac{1}{4}[f_{AC} + f_{BD} + f_{AD} + f_{BC}]$$

What this statement means essentially is that the individual could receive identical genes via those grandparents who are sibs ($A - C$, $B - D$) or via those who are cousins ($A - D$, $B - C$) if we presume that this system of mating has been going on for some generations before the $t - 3$ generation.

Taking the two sib pairs first, we have worked out the sib mating recurrence of F in (9-2), and we can simply recall the statement for F_{PQ} in the sib scheme and use that for f_{AC} and f_{BD} but put the generations back by one. Thus, for the two transmission probabilities from the sib grandparents,

$$f_{AC} + f_{BD} = 2(\tfrac{1}{4} + \tfrac{1}{2}F_{t-2} + \tfrac{1}{4}F_{t-3})$$

Taking the two cousin pairs next, we find the situation much simpler. Paths from $A \rightarrow P$ and $D \rightarrow Q$ have the same probability as $A \rightarrow P$ and $B \rightarrow P$ so that they measure the probability for P to be an identical homozygote (F_{t-1}); therefore,

$$f_{AD} + f_{BC} = f_{AB} + f_{CD} = 2F_{t-1}$$

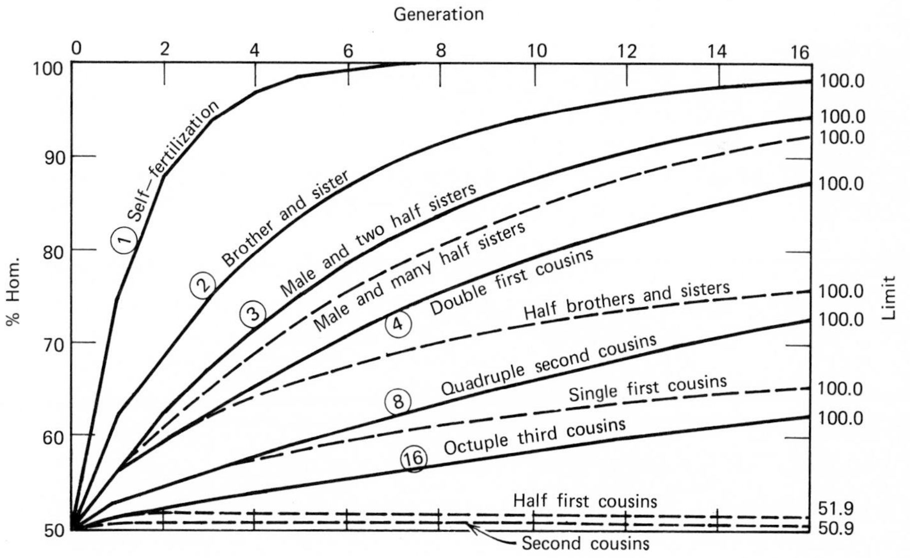

Figure 9-4. The percentage of homozygosis in successive generations of regular close relative matings. Solid lines refer to those systems of mating that contain constant numbers of individuals in isolates characterizing the population. Dashed lines are systems obtaining genes from outside the close relative line. The limit of homozygosis after a large number of generations is given at the right margin (from Wright, *Genetics*, Vol. 6, No. 172, 1921, Fig. 5).

Now we can go back to the main statement of the four equally probable paths into P and Q, converting their total as follows:

$$F_t = f_{PQ} = \tfrac{1}{4}[2(\tfrac{1}{4} + \tfrac{1}{2}F_{t-2} + \tfrac{1}{4}F_{t-3}) + 2F_{t-1}]$$

which simplifies to

$$F_t = \tfrac{1}{8} + \tfrac{1}{2}F_{t-1} + \tfrac{1}{4}F_{t-2} + \tfrac{1}{8}F_{t-3} \qquad (9\text{-}7)$$

(recurrence of F in double first cousin mating). If starting from a panmictic population, it would take three generations (first to get sibs, then to get cousins) before progeny of cousin matings could be produced. Recurrence of F would then be as follows:

Generation	0	1	2	3	4	⋯
F coefficient	0	0	0	$\frac{1}{8}$	$\frac{3}{16}$	⋯

These increases in homozygosity are graphed for many regular systems of mating in Figure 9-4.

INCREMENTS IN *F*

In the regular systems of mating in which matings occur within isolates (selfing, sibs, double first cousins, and so forth), F will always increase until its limit of 1. The student should verify this fact by taking, for any two successive generations, the F statement in formula (9-2) and then subtracting the earlier generation (F_{t-1}) from the later (F_t) to see what the increment is. For example, with sib mating,

$$F_t - F_{t-1} = \Delta F = \tfrac{1}{2}F_{t-1} - \tfrac{1}{4}F_{t-2} - \tfrac{1}{4}F_{t-3}$$

By setting the increment equal to zero, we see that the only solution would be to set $F_{t-1} = 1$ as well as the previous generation F's because they will always be <1 until then. Consequently, F will continue to increase until the limit of 1.0 is reached. Exercise 8 will give the student some practice in this concept. Increment in F will be generalized in Chapter 11.

It is important to realize, however, that when matings take place outside the isolate as they do with ordinary first cousins, second cousins, half first cousins, and more distant relatives, a regular system of inbreeding can occur that includes some input of genotypes from beyond the isolate. With ordinary first cousins, F increases very slowly, but it does have an eventual limiting value of 1:

Ordinary first cousins, F	0	0.0625	0.0935	⋯	0.312	⋯	1.0
Generations of cousin mating	0	1	2	⋯	15	⋯	∞

However, with second cousins (individuals with two out of eight grandparents who are sibs), genotypes enter the pedigree by outcrossing sufficiently for an equilibrium to occur, and F does not increase to unity. If second cousins only are mating in a population, the F recurrence is

$$F_t = \tfrac{1}{64} + \tfrac{1}{8}F_{t-2} + \tfrac{1}{32}F_{t-3} + \tfrac{1}{64}F_{t-4}$$

If successive generations of this recurrence relation are taken, an increment in F statement (ΔF) would quickly approach zero with every generation F equal to the limiting value $F = \frac{1}{64} + F(\frac{11}{64}) = \frac{1}{53}$ (see Exercise 9).

The same is true with any more remote relative matings than second cousins when *sources of genotypes come from outside the isolate.* Equilibrium occurs at values below complete homozygosis and with a constant F value. Biologically, this result is essentially like a combination of close relative mating and random mating in that an equilibrium of genotypes can be established with values given by formula (8-3).

SIB MATING AND RANDOM MATING COMBINED

Intuitively, the student should realize that close inbreeding occurring regularly in combination with random mating in a population can lead to an equilibrium of genotypes. We may proceed to analyze the combination of sib + random mating in parallel with that discussed previously under selfing combined with random mating. The student may then gain a bit of confidence in using these concepts extended beyond the simplest selfing with random-mating situation, even though the basic models are oversimplified compared with natural populations.

Let us assume there is a constant amount of sib mating in each generation combined with random mating. Let the fraction s_b be the part of the population practicing sib mating and $1 - s_b$ the part randomly mating. From Chapter 8, we recall that only the two types of mating $AA \times aa$ and $Aa \times Aa$ need to be reckoned with, the latter having to be twice the former for the equilibrium condition. It is convenient to consider a mating scheme like that in Table 8-3. The random-mating fraction is easily considered as all nine combinations of products $(D + H + R)^2(1 - s_b)$. However, the sib-mating fraction must be considered from the standpoint of the sibships in which the relevant matings can occur. In Table 9-2B, the only sibship that can generate matings of $AA \times aa$ is in box 5 (middle), and only in $\frac{2}{16}$ of that sibship will such matings take place. In the combined sib and random mating population, then, $AA \times aa$ will take place in frequency:

$$2DR(1 - s_b) + \tfrac{1}{8}H^2 s_b$$

For matings of $Aa \times Aa$, the sib mating fraction can be generated from $\frac{1}{4}$ of sibships 2, 4, 6, and 8—that is, $\frac{1}{4}(2DH) + \frac{1}{4}(2RH)$ in terms of their parent matings; plus all of sibships 3 and 7—that is, $2DR$ in terms of parent matings; and $\frac{1}{4}$ of sibship 5, or $\frac{1}{4}H^2$. The total $Aa \times Aa$ for combined sib and random mating population then will be in frequency:

$$H^2(1 - s_b) + [\tfrac{1}{4}(2DH) + \tfrac{1}{4}(2RH) + \tfrac{1}{4}H^2 + 2DR]s_b$$

For the equilibrium condition, we set double the $[AA \times aa]$ matings equal to the $Aa \times Aa$:

$$4DR(1 - s_b) + \tfrac{2}{8}H^2 s_b = H^2(1 - s_b) + \left[\frac{DH}{2} + \frac{RH}{2} + \frac{H^2}{4} + 2DR\right]s_b$$

which simplifies to

$$4DR(1 - s_b) = H^2(1 - s_b) + \left(\frac{DH}{2} + \frac{RH}{2} + 2DR\right)s_b$$

The portion in parentheses on the right side simplifies to $[H(1 - H)/2 + 2DR]s_b$ because $D + R = 1 - H$.

Also, at equilibrium, $2DR = \frac{1}{2}H^2$ (because of property 5, Chapter 1) so that

$$4DR - H^2 = \frac{Hs_b}{2(1 - s_b)}$$

The left side of equilibrium is greater than zero by the factor $4Fpq$ (8-5), and replacing H with its form from (8-3), we have

$$4Fpq = \frac{2pq(1 - F)s_b}{2(1 - s_b)}$$

so that $4F/(1 - F) = s_b/(1 - s_b)$; therefore,

$$F = \frac{s_b}{4 - 3s_b} \tag{9-8}$$

Equilibrium F coefficient when sib mating and random mating are combined.

It may be noted that the same value of F could result from several combinations of inbreeding and random mating. We may be interested in the equivalent amounts of sib mating and selfing in such an equilibrium population. For example, if F is observed to be 0.20, it could be that one-third of the population is selfing and two-thirds random mating or, as an equivalent, half the population is sib mating and half random mating; the same value of F could result from either type of inbreeding combined with random mating or some combination of all three.

ESTIMATING *F* FROM GENOTYPE FREQUENCIES

We cannot specify what sorts of matings are actually occurring if we estimate the inbreeding coefficient from genotype data. There is no shortcut to ascertainment of the mating structure in the population; it must be reckoned by observation. Of course, we can say what equivalent types of matings might have produced a given deficiency of heterozygotes, but it is very important to be cautious about the determination of the genotypic array. (Recall in Chapter 1 that simply because of agreement between the "square law" and the observed frequencies of genotypes, we cannot conclude necessarily that mating was at random.)

Example 9-1 illustrates some genotype frequency data from biochemically polymorphic populations of various rodents in which there is an excess of homozygotes and deficiency of heterozygotes as *could* be produced by a constant amount of inbreeding with a fraction of random mating. Without knowing the breeding structure of the populations concerned, we cannot conclude exactly the causation of the discrepancy. The same result can be produced by (1) subdivision of the population into small breeding units (Wahlund effect, defined in Chapter 2 as the stratification effect and delineated more fully in Chapter 12), (2) presence of one or more "silent" recessive alleles, (3) assortative mating for homozygotes, or (4) selection favoring homozygotes in particular ways (Chapter 14). At this point we can only realize that various forms of inbreeding combined with random mating would give the net estimate of F illustrated in the example. Ascertainment of real breeding structure must include knowledge of the rodents' migration, habits, and family unit organization; also, knowledge of what genotypes exist and their selective value differences must precede a decision about the causation of the genotype frequency arrays observed. It is important to have recorded diploid frequencies, even though we may be short of determining the exact cause of the heterozygote deficiency.

We should be cautious about interpreting the deviations from the square law (any significant F value), and we should also realize that many evolutionary forces may conspire to reduce F to zero, thereby appearing to bring genotypes into confirmation of square-law frequencies. For example, selection may favor heterozygotes and thus counteract their reduction by inbreeding. Objective criteria for ascertainment of the breeding structure in the population are essential. Forces that may bring genotype frequencies into agreement with the square law have been discussed by Workman (1969), and they will be reviewed in Chapters 14 and 15.

It is useful to point out that the coefficient F, when applied simply to describe the deviation from panmixia (as in formula 8-3) without regard for cause of the deviation, has a relationship to the chi-square used in the goodness-of-fit test for conformity with the square law (2-1). If the genotype frequencies fit the model expressed by (8-3), then the chi-square is exactly equal to F^2N, a result derived by Li and Horvitz (1953) and reviewed by Workman. See Table 9-3 and Exercise 10 for this derivation.

Estimated F (symbolized F_{est}) may be calculated either from solving (8-2) for F from observed/expected ratio for heterozygotes, as in Example 8-1,

$$F_{est} = 1 - \frac{H_n}{H_0} \tag{9-9}$$

(see Example 9-1, Table A, for standard error of F_{est}), or from the chi-square test for agreement with the square law.

$$F_{est} = \sqrt{\frac{X^2}{N}} \tag{9-10}$$

For (9-10) all three genotypes are used. With multiple alleles, it has been shown by Li and Horvitz that the chi-square may be used with k alleles as well.

$$F_{est} = \sqrt{\frac{X^2}{N(k-1)}} \tag{9-11}$$

TABLE 9-3 *Calculation of chi-square based on observed genotypic proportions with inbreeding and those expected from the square law (modified from Workman, 1969)*

Genotype	Observed (O)	Expected (E)	$(O - E)^2/E$
AA	$(p^2 + Fpq)N$	p^2N	q^2F^2N
Aa	$(2pq - 2pqF)N$	$2pqN$	$2pqF^2N$
aa	$(q^2 + Fpq)N$	q^2N	p^2F^2N
Totals	N	N	$F^2N = X^2$

EXAMPLE 9-1

GENOTYPES IN SOME NATURAL POPULATIONS OF RODENTS

Genetic biochemical variation is now widely known for a number of rodent populations (see Berry and Southern, 1970), most notable of which are those of house mice *Mus musculus*, studied extensively by Selander and colleagues, Petras, and others; deermice *Peromyscus maniculatus*, studied by Rasmussen; and voles *Microtus* spp. studied by Tamarin and Krebs. Zygotic (genotype) data have been recorded often for these populations with some evidence that excess of homozygotes and deficiency of heterozygotes occurs frequently. Interpretation of these genotype frequencies in terms of population structure remains a matter of conjecture.

Rasmussen (1964, 1970) sampled natural populations of the woodland deermouse, *Peromyscus maniculatus gracilis* (Le Conte), from northern Michigan, typing individuals for two antigenic characteristics (designated *Pm* − A and *Pm* − B) defining three phenotypes A, AB, and B as codominant. Three locality samplings gave the observed incidence of phenotypes with sexes separated as given in Table A.

Rasmussen pointed out that all heterogeneity chi-squares for these data were nonsignificant. (Do you agree? What about the constantly greater deficiency of heterozygotes among females than among males?) He then pooled the totals making the assumption of homogeneity among the samples. Do the pooled data fit a random-mating assumption of a single pair of alleles? Rasmussen estimated an $F = 0.291$ (we estimate 0.240) for these

TABLE A *Incidence of antigenic phenotypes (Pm locus) in Peromyscus (from Rasmussen, 1964)*

Locality	Sex	N	Antigenic Type (Pm) A	AB	B	Est. $F \pm s.e.$*
Munising (Alger County)	F	35	16	12	7	0.266 ± 0.167*
	M	26	13	10	3	0.097 ± 0.205
Huron Mts. #1 (Marquette County)	F	52	23	19	10	0.221 ± 0.138
	M	34	16	13	5	0.146 ± 0.174
Huron Mts. #2 (Marquette County)	F	45	27	11	7	0.391 ± 0.145
	M	35	22	10	3	0.190 ± 0.197
Pooled		227	117	75	35	

* Standard errors of F calculated from $\sqrt{\text{variance } F}$, where

$$\sigma_F^2 = \frac{(1 - F)[2pq(1 + F) + F(2 - F)(q - p)^2]}{2pqn},$$

the inverse of the information matrix of Fisher's maximum-likelihood function (see appendix to Rasmussen, 1964).

pooled data using formula (8-3). (What is the estimated F for the two sexes separately?) That estimate fitted the results much better, so Rasmussen postulated the most likely structure of the populations as being made up of partially inbred local populations with some spatial structure (territoriality, localized breeding places, or home ranges).

This finding has been criticized by Selander (1970) as follows: "the possibility of occurrence of a 'silent' allele (nonreactive) cannot be excluded." As with the human A, B, O blood antigens, a third allele lacking antigenic activity, Pm^0, if at low enough frequency to make silent allele homozygotes rare, would increase the frequency of the homozygous Pm – A and Pm – B types so that the frequency model could be as follows (letting r = frequency of silent allele):

Phenotypes	A	AB	B	O
Frequency	$(p^2 + 2pr)$	$2pq$	$(q^2 + 2qr)$	r^2

If we estimate the p, q, r values from the pooled data using formula (3-9) and Bernstein's adjustment (3-10A), we obtain approximately $p = 0.642$, $q = 0.298$, $r = 0.060$, which may be applied to the total pooled data for a chi-square test. We obtain $X^2 = 4.27$; $p = 0.04$, Selander suggests an r value of 0.08 (with $p = 0.64$, $q = 0.28$), which indeed comes closer to expectation, and thus there is no need to invoke inbreeding ($F = 0$). Without definite knowledge of the existence of a silent allele, of course, the problem cannot be resolved.

Biochemical data on other genotypes from other rodent populations cannot constitute proof for the original Rasmussen observations, but we could be on more certain ground about the level of inbreeding if more loci examined for genotype frequencies had similar excesses of homozygotes and deficiencies of heterozygotes. Results have proved to be equivocal.

Petras (1967*a*,*b*) reported heterozygote deficiencies in populations of house mice (*Mus musculus*) collected over a three-year period on five adjacent farms near Ann Arbor, Michigan, as indicated below:

Protein	Locus	Estimated F
Prealbumin serum esterase	Es – 2	0.18 ± 0.063
Hemoglobin β chain	Hb	0.13 ± 0.062

Also, a coat color (agouti/white belly-agouti) gave an F estimate of 0.147 (see Example 2-6). The hemoglobin locus gives a borderline significant F, while the coat color data come from a sample with too few of the dominant phenotypes to draw clear-cut decisions. Selander (1970) criticized the Petras conclusions of heterozygote deficiency by pointing out that "the period over which collections were made is so long that seasonal and secular changes in allele frequencies may have contributed to the observed heterozygote deficiency . . ." because of stratification (Wahlund effect). To counter that criticism, we may go back to Petras's data for Es – 2 and consider only the observations from those samples in which the allelic frequencies lie within one standard error of the mean p, q value within farms over the years, that is $\bar{p} = 0.30 \pm 0.05$. Ten out of fifteen samples lie within those limits, and their pooled incidence values are

Genotype	$Es-2^a/Es-2^a$	$Es-2^a/Es-2^b$	$Es-2^b/Es-2^b$	Total
Incidence	27	76	106	209

Clearly, it may be verified that heterozygotes are deficient and that the data fit the relationship (8-3) $F_{est} = 0.154$, which is not significantly different from the estimate given above. It must be added, however, that when the hemoglobin genotypes are similarly viewed (only $\frac{8}{14}$ samples within $\pm 1\sigma_p$), the frequencies fit the random-mating square law (to be verified by the student).

Genotype	Hb^d/Hb^d	Hb^d/Hb^s	Hb^s/Hb^s	Total
Incidence	12	46	64	122

There is no compelling reason for attributing this set of data to the effects of inbreeding. We need to know more about the selective values, however. If the heterozygotes for hemoglobin have in fact some *selective* advantage, they may actually be as deficient as the Es $-$ 2 heterozygotes, which could be more selectively neutral. We need to know more about the selective value of these genotypes. Petras, like Rasmussen, attributed his results to small effective breeding unit size.

Selander (1970) has made extensive samplings of house mice throughout many parts of Texas. For the Es $-$ 3 (hemolysate) and Hb (hemoglobin) loci, he reported widely differing amounts of heterozygote deficiency and excess values among samples of varying size. The distribution of estimated F values around the mean $\bar{F} = 0$ was nearly normal for small samples ($n < 70$ from small barns and sheds where populations are capable of housing only a few mice), while in larger barns (n from 70 to 130), $\bar{F}$ tended to become positive at about 0.10 to 0.20. Selander attributed these values for the larger populations to subdivision and semiisolation of breeding units within large barns rather than to close inbreeding (see also Selander et al., 1971, and Wheeler and Selander, 1972, and see Chapter 12).

RARE RECESSIVE PROPORTIONAL INCREASE UNDER INBREEDING

When the breeding system includes random mating sufficient to hold F constant under the equilibrium conditions we have examined, the increment in homozygotes over their panmictic frequency (Fpq) will be proportionally greater as the allele's frequency becomes less. The quantity $q^2 + Fpq$ relative to q^2 depends not only on the level of inbreeding but also on gene frequency. Let us examine the consequences of reducing q. The ratio of recessives under inbreeding (R_i) to those under outbreeding or random mating (R_o) will always be greater than one.

$$\frac{R_i}{R_o} = \frac{q^2 + Fpq}{q^2} = 1 + \frac{Fp}{q} \qquad (9\text{-}12)$$

(relative proportion of recessives between inbred and outbred population for given q value). Therefore, as gene frequencies approach equality, inbreeding makes the least difference in terms of generating identical recessives; on the other hand, as a recessive becomes rare, the

TABLE 9-4 *Relative proportions of recessive from inbred (R_i) as compared with randomly bred parents (R_o) for two inbreeding intensities (single locus with two alleles) (from Li, 1955)*

Allelic Frequency (q)	Panmictic (q^2) Recessives R_o	Relative Proportion R_i/R_o with	
		$F = \frac{1}{16}$	$F = \frac{1}{32}$
0.500	0.2500	1.06	1.03
0.400	0.1600	1.09	1.05
0.200	0.0400	1.25	1.125
0.100	0.0100	1.56	1.28
0.010	0.0001	7.19	4.09
0.001	0.000001	63.44	32.22

difference from that expected under random mating becomes larger. This relationship is summarized by the data of Table 9-4.

In human populations, this consideration is important. Most close inbreeding in humans comes about through first-cousin marriages. If a recessive is rare, its incidence may be principally from the progeny of first-cousin marriages. We know $F = \frac{1}{16}$ for children of first cousins, so that using (9-12), we have $R_i/R_o = 1 + (\frac{1}{16})p/q$. Certainly, if the rare recessive is a deleterious condition, the "harmful" effect of cousin marriages would be in evidence.

For the population as a whole with mixtures of cousin and random matings, we may use reasoning similar to that employed in working out equilibriums for mixed selfing-random or sib-random matings in order to determine the relative effect of first-cousin matings on the population. Letting c be the frequency of first-cousin marriages in the population and $(1 - c)$ be the frequency of random mating, R_i and R_o being the conditional probabilities for having a recessive child under inbreeding or random mating, respectively, we can represent the offspring (dominants + recessives) as follows (modified from Li, 1961):

	Parents: Cousins	Parents: Unrelated	Total
Recessives	cR_i	$(1 - c)R_o$	$R_o + c(R_i - R_o)$
Dominants	$c(1 - R_i)$	$(1 - c)(1 - R_o)$	$(1 - R_o) - c(R_i - R_o)$
Totals	c	$1 - c$	1.0

If we were to ascertain a recessive child in the population from a survey of a genetic disorder, we would more than likely find the child had come from parents who are related. The proportion of children from first cousins can be worked out from the relationships given above; letting k = proportion of first-cousin parents among all parents with recessive (aa) offspring, we have

$$k = \frac{cR_i}{R_o + c(R_i - R_o)} = \frac{c(1 + 15q)}{cp + 16q} \tag{9-12A}$$

(proportion of recessive offspring from first-cousin marriages out of all marriages). For example, where $c = q = 1$ percent, $k = 6.77$ percent, nearly seven times greater than c (amount of cousin mating). As q diminishes, k will increase much more. (This formula was worked out by Weinberg in 1920, but publicized by Dahlberg in 1929.)

It should be noted that finding consanguinity among parents of homozygous children is an important criterion for recessive inheritance when the trait is rare. Cystic fibrosis of the pancreas (see Example 2-8) appears to be recessive; but contrary to the above, there is no significant increase in parental consanguinity for that disorder when q is greater than 1 percent. Perhaps the condition arises from interaction of more than one locus (see Exercises 11 and 12), but when q increases above 2 or 3 percent, it becomes difficult to indicate a significantly higher proportion of children from consanguineous than from marriages between unrelated persons.

INBREEDING DEPRESSION

Among the consequences of inbreeding for the genotypes of a population, two main effects often take place, in fact so frequently as to be considered erroneously as primary effects. These effects are (1) the average lowering of metric values (depression), especially of traits pertaining to fitness (reproduction and survival) and the emergence of deleterious recessives among inbred progenies, and (2) the dispersion of allelic frequencies among inbred isolates. (Discussion of this second effect of distribution among isolates is postponed until Chapter 12, when we turn our attention to changes within isolates and stochastic elements in populations.) Both of these well-known effects are not *primary*; the breeding system has only one direct effect—the change in distribution of genotypes through formation of identical homozygotes. Depression of metric values and dispersion of allelic frequencies follow as secondary accompaniments of the changes in homozygotes and heterozygote frequencies.

The conclusion that depression is not a general concomitant of inbreeding was first established by King (1918–1921), who inbred lines of rats by sib matings for 70 generations without finding any degeneration of reproductive capacity. In 1924, Hyde studied fertility in 17 lines of *Drosophila melanogaster* and found that high fertility was established in certain lines; hence, depression could not be considered as invariably associating with inbreeding. Sang (1964) found that inbred lines did not differ substantially from hybrids in nutritional requirements. The most extensive evidence for the incidence of "normal" viability (percent survival) homozygotes comes from several surveys of concealed viability modifiers in natural populations of *Drosophila* (see Example 9-3 and Dobzhansky, 1970, for detailed discussion). By using dominant markers with crossover suppressors for specific chromosomes, it is technically easy to produce complete homozygotes that can be compared with heterozygotes for relative viability. Depending on the species and the chromosome, normal-viability homozygotes were found to occur from a frequency of about 1 percent up to about 25 percent, and a few chromosomes (2 percent or less) even boosted the viability significantly above normal. Some inbred lines, then, can be normal and stabilized.

Nevertheless, the emergence of rare recessive homozygotes that are deleterious to fitness and have a general depressing effect as an accompaniment to inbreeding informs us that in Mendelian populations the "normal" wild-type composite set of genotypes may contain a hidden recessive deleterious "genetic load." The rate of depression with inbreeding can also inform us about overall dominance of the genotypes contributing to the trait we are measuring.

In the last chapter we considered the increment in genetic variance with inbreeding but did not mention what happens to the population mean. Actually, it was not a serious omission for the case considered where dominance was lacking, since no change will take place in the population mean with the inbreeding unless dominance affects the trait being measured. The contribution of each genic locus affecting the trait can be expressed as follows by recalling formulas (6-3) and (8-3): letting $\bar{Y}_F$ = phenotype mean under inbreeding and $\bar{Y}_o$ = phenotype mean under random mating, we have

$$\bar{Y}_F = a(p - q) + 2pqd(1 - F) = \bar{Y}_o - 2pqdF \qquad (9\text{-}13)$$

which the student should verify (Exercise 13). If several loci affect the trait in such a way that the genotypic values act independently and additively and the allelic frequencies are more or less equal from one locus to another, the population mean is given by summation of the separate locus contributions:

$$\bar{Y}_F = \bar{Y}_o - 2F\overline{pq}\sum d \qquad (9\text{-}13\text{A})$$

where $\overline{pq}$ is the average frequency genic product over the loci affecting the trait. The minus term in this expression indicates the change of metric mean on inbreeding. Obviously, with zero dominance (intermediate heterozygotes at midparent), no depression will take place with inbreeding. With dominance, however, change in the population mean will occur in the direction of the more recessive alleles. Most recessives in wild populations and human populations are metrically reduced compared to "normal," or wild-type, and especially so for traits associated with fitness. Therefore, there will be a depression, maximally if inbreeding is begun when allelic frequencies are about equal. As F increases, the depression in the mean should be linear, provided that the genetic situation is relatively simple with independence of loci and little epistatic interaction.

Example 9-2 cites inbreeding depression data for some traits from human populations. The data indicate how difficult it is to analyze the effects of inbreeding when complex environmental factors may mimic the depressing effect of inbreeding, especially when the amount of inbreeding in a population is correlated with socioeconomic status. Example 9-3 illustrates inbreeding depression in viability in drosophila. Falconer (1960, p. 249) and Lerner (1958, pp. 92–94) list several traits in domestic animals and in drosophila showing approximate linear depression with inbreeding. Latter and Robertson (1962) measured the decline in a competitive reproduction index by putting inbred lines in competition with a tester genotype as a standard; they found the index to decline about 2.7 percent for every 1 percent increase in F. Bowman and Falconer (1960) recorded litter size in 20 inbred strains of mice and found litter size to decline 0.56 progeny for each 10 percent increase in F, but 17 of the 20 lines were lost when F reached 0.70. The remaining three lines were shown to have fixed favorable dominants with high fertility, a result similar to that of Hyde (1924), already mentioned (see also Carson, 1967, for a brief review). Experimental results such as these demonstrate that if an experimental organism is being inbred by carrying several sublines simultaneously and if F increases to the point that high probability for deleterious homozygous recessive is reached, reproduction may diminish markedly in certain lines with the danger of loss to those lines. Surviving lines are likely to harbor genotypes that have resisted the inbreeding depression and will thus be a selected group in which the linearity will no longer apply. Consequently, inbreeding depression as an indication of dominance has a logical basis only during the early stages of inbreeding.

If the separate low-value inbred lines are crossed and random mating is restored, there will be immediate recovery from inbreeding depression with eventual restoration of the base population mean—that is, a hybrid vigor or *heterosis*. If nonallelic recessives had become established at high frequencies in the inbred lines and if the lines were different in allelic frequencies, intercrosses between the lines will at first produce greater frequencies of heterozygotes than under random mating if the lines were very different in allelic frequencies (see Exercise 2 in Chapter 6). If many loci contribute to overdominance, the heterotic effect may be maximal in the first hybrid generation but will fall off with random mating among the hybrids to an equilibrium value in the subsequent generation. Detailed discussion of heterosis and its importance to the evolutionary dynamics of populations is postponed to Chapters 14–19.

EXAMPLE 9-2

INBREEDING DEPRESSION IN HUMAN FAMILIES

Consanguinity (close relationship between mates) is rare in most human populations. A few close relationships have been recorded like brother-sister, father-daughter, uncle-niece, aunt-nephew, or double first cousins (see Levitan and Montague, 1971). More often, consanguineous marriages are those between ordinary first cousins, first cousins once removed, or second cousins. Some exceptionally high frequencies of such marriages can be found in India (Andra-Pradesh) with about 8 percent uncle-niece, aunt-nephew marriages and with 20–30 percent first cousin marriages; Guinea (West Africa) with 16 percent first cousin marriages; Japan (*Kakure* in Hirado) with 12 percent first cousin marriages and an overall 15 percent of marriages consanguineous; there is an incidence of about 2 percent first cousin marriages in Brazil, French Canada, Italy, Spain, and in some isolates such as those of religious sects in the United States (Hutterites and Dunkers), certain Amerind tribes (Rama Navajo in New Mexico), in Israel and Jordan (Samaritans), and certain islanders (Tristan da Cunha, Aeolians). Cavalli-Sforza and Bodmer (1971), Table 7.3, gives an extensive list of consanguinity incidence in humans.

Inbreeding depression has been examined by a few authors; it seems to be linear with increasing F for some traits in certain populations, but when allowance is made for socio-economic status the effect often becomes nonsignificant. Lower-income and less-educated "deprived" levels of society where inbreeding tends to be more common often show lower metric values for many traits so that the inbreeding depression of values may be confounded with that arising from environmental factors. Barrai, et al. (1964) analyzed stature and chest girth of males aged 20 years born in Parma, Italy, between 1892–1911. Chest circumference decreased linearly with F (after adjustment for socioeconomic status and other factors), suggesting some dominance in genes for larger chest size. Stature did not show a linear relationship with F, and there was a peak in stature for progeny of third cousins. On the other hand, Mange (1964) studied effects of parental inbreeding (that is, one or both parents were children of close relatives) and consanguinity on stature in a Hutterite isolate and found negative regression of stature on F (adult male offspring of first cousins averaging 2.2 cm shorter than controls and female offspring 4.8 cm shorter) but no significant (though negative) effect from parental inbreeding. Apparently, the Italian population sampled by Barrai is stratified for stature in certain partially inbred groups, while the Hutterites are more uniform and therefore a more reliable group for genetic conclusions.

A very extensive series of studies on inbreeding in human populations has been done on Japanese populations by Schull, Neel, and their colleagues (1965, 1970) from the Department of Human Genetics, University of Michigan Medical School. Their earlier study on children born to residents of Hiroshima and Nagasaki following World War II (1948–1953) included a variety of data, of which we shall first mention the inbreeding depression and in Example 9-4 the mortality and extreme detrimental genetic effects. For anthropometric data (weight, height, girth of head, chest, and so forth) from inbred and noninbred individuals, in general, offspring from first cousin marriages showed inbreeding depression at about 0.5 percent of the outbred value. There was considerable variation from one population to another in their earlier survey, so these authors undertook a detailed study on a more closed population (the island of Hirado off the west coast of Kyushu, Japan) where there was more social uniformity and consanguinity was high. Japan has unusually favorable facilities for studying consanguinity largely because official family records (*koseki*) provide accurate information on genetic relationships independent of that obtained through personal interviews. Effects of parental consanguinity upon children and the inbreeding level of parents were independently ascertained. Inbreeding depression was estimated (Neel, et al., 1970) on seven anthropometric variables (span, height, head measurements, etc.), on tapping rate (speed of tapping telegraph key), blood pressure, IQ, and school performance (the last two variables were not measured in parents).

Socioeconomic status may confound consanguinity effects. Three main religious groups on Hirado are Buddhists, Kakure ((“hidden Christians,” descendants of sixteenth-century converts to Catholicism who practiced their religion in secret during years of religious persecution), and Catholics. Farming and fishing are the livelihoods of more than half of the island’s residents. Formal education beyond elementary school is rare, although illiteracy is virtually unknown. For these groups the amounts of consanguinity and inbreeding differ as indicated in the following data on 10,530 marriages recorded between 1880 and 1960 (from Schull, Nagano, Yamamoto, and Komatsu, 1970):

Percents of consanguineous marriages on Hirado, Japan, distributed by relationship between husband and wife (C = cousins)

Unrelated	Closer than First C	1st C	$1\frac{1}{2}$ C	2nd C	Greater than Second C	Between 1st-2nd C	Unknown	Total
84.9	1.5	5.5	2.0	3.1	1.5	0.9	0.5	100.0

Proportions of consanguineous marriages within religious groupings on Hirado

	Buddhists	Catholics	Kakure	Others	Total Consanguineous
Prop.	0.145	0.082	0.272	0.141	0.147
N	8,209	1,226	761	261	10,457

Catholics had the lowest and Kakure the highest frequencies of inbreeding. Socioeconomic status was scored by taking account of parental occupation, parental education level, and annual income with highest income, educational level, and more professional occupation with top scores. Out of 18 correlation coefficients between inbreeding effects and socioeconomic status among nonfarmers, 17 were found to be negative although all small on the order of -0.06. Examination of such variables singly would be likely to miss their cumulative impact and would fail to expose the complexity of environmental variables and their possible association with inbreeding. By making allowances for these extraneous variables, the following general consanguinity effects were determined:

1. Anthropometric (seven measures). With increased F of children, no single measure showed significant regression, although all of the seven traits were negative. For their parents, six of the seven regressions were negative. Thus, these data were in conformity with those from the earlier Hiroshima-Nagasaki data in showing slight depression, but the regression was less in magnitude, presumably because of a smaller sample in Hirado. There was no significant regression from inbreeding in the parents.

2. Blood pressure. None of the regressions for inbreeding were significant.

3. Tapping rate. Regressions for inbreeding and tapping rate were negative although not significant for the children and significant for a larger sample of adults.

4. IQ. None of the regressions for inbreeding were significant, although negative and in agreement with the authors' earlier data on Hiroshima and Nagasaki.

5. School performance. Consanguinity effects were consistently negative again, but not significant and similar to the earlier data.

No effects of inbreeding among the parents upon their childrens' performance variables were consistent. That is, if a parent was an offspring of a first-cousin marriage, that was of no consequence to the metric values of his children; likewise, there were no maternal or paternal influences due to parental level in inbreeding.

In total, then, the evidence for inbreeding depression parallels that found by these authors in their larger study, but the level is low though consistent in all cases and therefore probably real.

EXAMPLE 9-3

INBREEDING DEPRESSION ON VIABILITY IN DROSOPHILA

When we try to assess the possible depression in any character as a function of inbreeding, we obtain an overall effect of increased homozygosity throughout the genome instead of from separate chromosomes. Inbreeding depression in drosophila is easily

TABLE A *Percentage of preadult survival with increasing F in representative drosophila species (modified from Mettler and Gregg, 1969)*

Species	F (inbreeding coefficient) 0	$\frac{1}{16}$	$\frac{1}{8}$	$\frac{1}{4}$
D. ananassae*	90.8	75.0	—	65.5
D. willistoni**	84.3	—	76.0	63.5
D. pseudoobscura*	90.2	83.2	—	80.5
D. pseudoobscura†	86.9	—	77.5	73.8
D. arizonensis‡	77.0	73.3	69.1	66.2
D. mojavensis‡	62.4	46.1	48.4	44.1

* From Stone, et al. (1963); population of *pseudoobscura* from Arizona and population of *ananassae* from Ponape (Caroline Islands).
** From Malogolowkin-Cohen, et al. (1964); populations from Venezuela, British Guiana, and Trinidad.
† From Dobzhansky, et al. (1963).
‡ From Mettler, Moyer, and Kojima (1966); laboratory populations established from Sonora, Mexico, progenitor population.

measured by obtaining a sample of wild females from a natural population. Nearly all wild females, being already inseminated when coming to food sources in nature, can be isolated singly in food vials to obtain first-generation progeny. Sibs or other related individuals can be mated in such a way as to produce cousins in later generations. Table A gives representative data from such methods for egg to adult survival percentages as functions of the degree of inbreeding in six species of drosophila. In general, the technique is to collect eggs with a certain expected F value, plant them on a food surface, and count the number of adults emerging. If the survival amount is plotted as a function of F, we record at the $F = 0$ point the fraction surviving owing to random genetic and environmental causes and at the $F = 1$ point, by extrapolation from the sib and cousin values, the expected lowest survival owing to homozygosity for segregating low viability factors throughout the genome. The difference between random genetic deaths at the $F = 0$ point and those at various levels of inbreeding measures the increment in genetic deaths (1 − survival fraction) due to increased homozygosity for these detrimental factors.

In Figure A some of the regressions are indicated (two populations of *D. pseudoobscura* and one of *D. willistoni*). It is clear that even low amounts of inbreeding depress the viability. If extrapolated linearly to the maximum level ($F = 1$), the expected survival would vary from a very low level in the tropical species to less drastic levels in the temperate species. However, such extrapolation might be in considerable error for a number of reasons: first, linear extrapolation implies an additive increment in detrimental homozygotes with each increased level of inbreeding, and it is likely that some chromosomes may contain more than a one detrimental allele. Lethals at two loci cannot have more effect than one

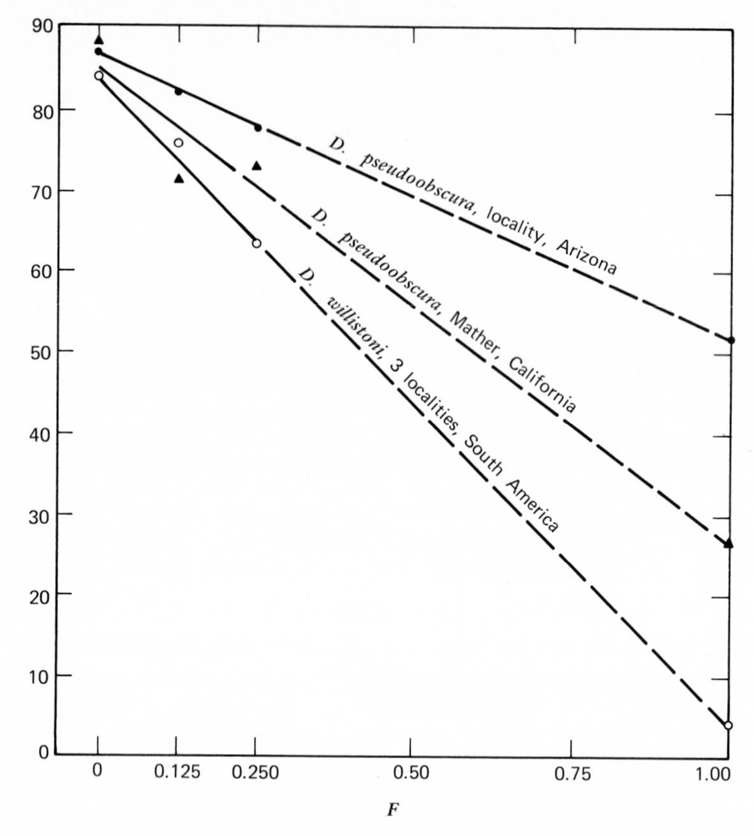

Figure A. Viability (preadult survival percentage) as a function of the degree of inbreeding for two populations of *Drosophila pseudoobscura* and one of *D. willistoni* (from Dobzhansky, et al., 1963).

lethal, for example. There may well be two or more loci with a detrimental allele each, considered later in the text under lethal equivalents. From a more refined method of regression analysis (*A* and *B* statistics outlined in Chapter 18), the average number of lethal equivalents that would result from doubling the haploid set of chromosomes for these two drosophila species is as follows:

Population	Lethal Equivalents at $F = 1$	Reference
D. pseudoobscura		
Mather, Cal.	0.742	Dobzhansky, et al., 1963
Arizona, Locality *C*	0.472	
D. willistoni		
Venezuela	1.052	Malagolowkin-Cohen, et al., 1964
British Guiana	1.100	
Trinidad	1.148	

GENETIC LOAD CONCEPT

It is evident that inbreeding exposes deleterious alleles faster than random mating because it reveals all variation hidden by dominance of "normal" alleles. What sorts of deleterious alleles there are, their distribution, and how and why they come into existence and then either become lost, maintained, or increased in populations are major problems in population genetics.

During the 1940s and 1950s when artificial radiation exposure (X-ray, gamma, and other ionizing radiation in particular) became a major concern, Hermann J. Muller emphasized strongly the increased mutation rates in all organisms exposed and the consequent increased reservoir of damaged genetic material in populations. In a classic paper entitled "Our Load of Mutations" (1950), he attempted to estimate the total mutational load in humans. While all genetic variation stored in heterozygosity is certainly not deleterious under all circumstances, in large measure much of the variation concealed as recessive does constitute a burden, or "load," on the population in the sense that many genotypes include alleles that would lower the reproductive ability of their carriers if made homozygous. In human populations, doubtless much suffering in terms of individual loss is determined by the occurrence of homozygotes for detrimental alleles within families, but the sense of the words *load* and *burden* give an anthropomorphic connotation that is not experienced emotionally by nonhuman organisms to any extent. This genetic load becomes expressed predominantly by inbreeding, and some examples of its magnitude are given in Examples 9-3, 9-4, and 9-5; further discussion of its implications under natural selection are deferred to Chapter 18. Of course, the examples of inbreeding depression considered above illustrate a portion of the expressed genetic load; however, the more drastic genetic effects that bring about severe loss are customarily thought of as genetic load because their phenotypic effects are so definite: lethals, semilethals, or sterility factors.

In 1956, Morton, Crow, and Muller published an article that attracted much attention (further discussed under selection and genetic loads in Chapter 18, since it is pivotal to the concept): "An Estimate of the Mutational Damage in Man from Data on Consanguineous Marriages." Some of their data are presented in Example 9-4 on the incidence of drastic genetic loads in human populations. It is worthwhile for our present purposes to give the basic incidence of drastic factors and withhold discussion of their implications to the theory of their maintenance in populations until we have completed chapters on selection. In addition to the studies mentioned in Example 9-4 on mortality incidence, the student should be cognizant of significant parallel data on such other detrimental conditions as mental defectiveness, deaf-mutism, blindness, and other hereditary diseases, summaries of which can be found in Penrose (1963); Dewey, Barrai, Morton, and Mi (1965); Cavalli-Sforza and Bodmer (1971); Sheba International Symposium (1974); Crawford and Workman (1973).

Since the pioneering work of Tshetverikov in 1926, Timofeef-Ressovsky in 1927, and Dubinin in 1934, populations of drosophila have been screened by inbreeding techniques of various sorts to uncover recessive mutants in wild populations (see Dobzhansky, 1970, pp. 109–125, for a summary). In the early days the technique of brother-sister matings of first-generation progenies from wild-caught flies (Spencer, 1947) was used, and later the tool of dominant markers with linked crossover suppressors was introduced to create completely homozygous wild chromosomes in just two or three generations. One of these, the marker chromosome method patterned after the *Muller ClB* (crossover suppressor-lethal-Bar) technique for sex-linked recessives, is outlined in Example 9-5, Figure A, in terms of autosomal recessives. It is precise especially in screening for lethals and for viability and fertility modifiers, since individuals with the nondominant-marker ("wild") phenotype in G_3 would be homozygous for one of the wild chromosomes from the progenitor fly being screened. Example 9-5 gives some representative results of screening populations for incidence of genetic load factors affecting viability in four drosophila species. A typical frequency distribution of viabilities is illustrated in Figure B of the example; it is bimodal with a high frequency of lethals, a few semilethals, a considerable amount of viability depressors ("subvitals"), some "normal" chromosomes, and a few "supervitals" as defined in the example.

To avoid making the oversimplified assumption that any chromosome so designated by these assays would have only a single viability factor responsible for the homozygous phenotype, we must consider the distribution of such factors over a sample of wild chromosomes. There may be two or more lethals on the same chromosome. There may be semilethals, quarter-lethals, or subvitals in abundance. Loci with "normal" alleles can interact epistatically to lower or to increase viability compared with their individual locus action on the phenotype (for example, "synthetic lethals"; see Chapter 17 on multilocus interactions). To help reduce the error involved in the simplified assumptions of extrapolating inbreeding depression linearly and in drawing conclusions about single-locus effects on viability from chromosomal assays, we can make a correction by the method of "lethal equivalents." This correction will help estimate the number of detrimental factors in the total genome or on a particular chromosome, depending on the technique used. However, caution is necessary: the complexities of epistatic interactions become technically difficult, and the method alone is not effective in accounting for interactions.

According to Morton, Crow, and Muller (1956), "A lethal equivalent is a group of mutant genes of such number that, if dispersed in different individuals [and made homozygous], they would cause on the average one death, e.g., one lethal mutant, or two mutants each with 50 percent probability of causing death, etc." For example, suppose we observe 20 percent wild chromosomes with lethal effect and 80 percent normal in homozygous condition by the assay method in Example 9-5. It is likely that the lethal factors are distributed among the wild chromosomes as a Poisson: 80 percent have no lethal, some fraction will have one lethal, a smaller fraction will have two lethals, and so forth. Since the first term (zero term in the Poisson distribution) is e^{-m}, we can find the average frequency (m) of *lethal equivalents* as if they were dispersed each to a single chromosome by solving for m: $e^{-m} = 0.80$, then $m = 0.223$ lethal equivalents. The difference between 0.223 and 0.200 lethals observed is then due to the fact that some chromosomes carry two or more lethals.

Application of this method to human mortality data will be postponed to Chapter 18. The method as applied toward drosophila chromosomes, however, was worked out by Greenberg and Crow (1960) and Temin (1966), and we may illustrate it as follows. In a chromosome assay of *D. pseudoobscura*, assume that heterozygous viability is "normal" = 1.00, while the lethal frequency for a major autosome is 0.186 and the average viability of all homozygotes relative to heterozygous viability is 0.642. If the chromosomes with lethal effect were the only ones to be reckoned with in producing the lethality, then we would simply have $e^{-m} = (1 - 0.186) = 0.814$, so that $m = 0.206$ lethal equivalents.

However, with a distribution of other deleterious factors over most of the chromosomes, as illustrated in Example 9-5, it is less restrictive to include the semilethals and subvitals as well. Those "detrimentals" could in various cumulative ways act as lethal factors when combined. Certainly, they ought not to be excluded, yet there may be considerable variation in just how they may interact to produce lethality. Greenberg and Crow then use the overall homozygote viability, letting $e^{-m} = 0.642$, so that $m = 0.443$ lethal equivalents.

In Example 9-6, we illustrate the method of lethal equivalents from a chromosomal assay of *D. pseudoobscura* and of *D. willistoni*. If the heterozygote's viability is estimated separately with respect to a constant (dominant marker lacking the wild chromosome), then the expression for total genetic load (T) in lethal equivalents would be, since $m = -\log_e$:

$$T = -ln[V_{ho}] - (-ln[V_{ht}]) = ln[V_{ht}] - ln[V_{ho}] \tag{9-14}$$

where $ln = \log_e$, V_{ho} = average viability of all homozygotes relative to a constant genotype

V_{ht} = average viability of heterozygotes relative to the same constant genotype

It is apparent that for the three major autosomes of *D. pseudoobscura* from this sample, the estimate of total lethal equivalents of 1.429 is three times greater than the estimate from extrapolation of sib and cousin inbreeding depression seen in Example 9-3 (Arizona locality *C*). In *D. willistoni*, the total load from a chromosomal assay (1.46 ± 0.13) is significantly greater than the estimate based on regression of viability on F (1.09 ± 0.06). While some of the difference could be attributed to considerable sampling error in the latter data (regression of viability on F) at low F values, undoubtedly the greater cause of discrepancy is due to the fact that the regression of survival (or mortality) on F is based on the assumption that the effects of harmful genes are additive and that a linear relationship should be in evidence. Disproportionally higher mortality in complete homozygotes than would be expected from additive lowering in compounds of subvitals, for example, must signify that many low viability factors do not act additively but probably synergistically (epistatically). Suppose that two subvitals each lower viability by 10 percent when made homozygous compared with outcrossed controls; if both occur together in coupling homozygotes, they might be expected to lower viability by 20 percent (additive), which is still above semilethality, but they might produce a complete lethal instead. As pointed out by Dobzhansky, Spassky, and Tidwell (1963), with their experiments described in Example 9-6,

> *. . . the discrepancies between estimates of the genetic load obtained with different methods do not necessarily invalidate these methods . . . . The determination of even the order of magnitude of certain genetic parameters, such as the genetic load, is at present desirable. It should be kept in mind that the usefulness of the oversimplified assumptions which we are forced to make in our work in order to do any work at all may be only temporary.*

But we must learn to give up simple assumptions when evidence appears to contradict them. Cases of "synthetic lethals" and other epistatic interactions (see Chapter 17) between factors controlling fitness traits make the expectation from linear extrapolation of inbreeding depression on F to be quite seriously in error.

In any case, there is universal evidence of a recessive genetic load in all these populations (flies, mice, and humans), as revealed by inbreeding. It follows that genotypes of these organisms must be highly heterozygous on the average for alleles or genic complexes that are selectively advantageous to the individual's survival and reproductive capacity. Further incidence data for deleterious genetic factors in populations may be found in Crumpacker (1967); Spassky, Dobzhansky, and Anderson (1965); Wills (1966, 1968); Marinkovic (1967); Magalhães, et al. (1965); Temin (1966); Wallace (1966, 1970); Stone, et al. (1968); Levene, et al. (1965); Watanabe and Oshima (1970).

EXAMPLE 9-4

MORTALITY FROM INBREEDING IN HUMAN POPULATIONS

Inbreeding data from human populations have been extensively used to measure the magnitude of the genetic load. Morton, Crow, and Muller (1956) compiled mortality data from several sources, all of which showed similar trends. From their most reliable source (Sutter and Tabah, 1953), we may summarize the following: consanguineous marriages were recorded from Catholic marriage dispensations issued during 1919–1925 in France; the investigators visited most of the families, taking histories of births, deaths, and noting conspicuous abnormalities. For controls, they obtained similar information from town clerks for unrelated parents married during the same period and selected without regard to their medical history or fertility. The child mortality data of Sutter and Tabah are summarized below.

	From Marriages of		
	First Cousins	Second Cousins	Not Related
Stillbirths and neonatal deaths	$\frac{69}{743} = 0.0929$	$\frac{34}{549} = 0.0619$	$\frac{108}{2745} = 0.0393$
Infantile and juvenile deaths	$\frac{96}{674} = 0.1424$	$\frac{49}{515} = 0.0951$	$\frac{198}{2637} = 0.0751$

It can be noted that the increasing proportions of lethality parallel changes in the expected inbreeding coefficient (F) for progeny homozygotes from these marriages. Morton, et al., extrapolated from these increments of mortality to what the lethal equivalents would be for a completely homozygous ($F = 1$) population. Without using their more sophisticated regression of mortality on F (which involves weighting by the natural log of survivors), we may see in a simpler rough way that if, for example, the children of first cousins have an average inbred coefficient of $\frac{1}{16}$ and the average increment in mortality from the controls for these children is approximately 0.06, then 16 times that amount would be 0.96, or nearly one lethal equivalent. Actually, by the more exact method of calculation, these authors estimated 1.5 to 2.5 as the number of lethal equivalents per gamete in this population.

Schull and Neel (1965) were able to collect considerable mortality data (stillbirths plus death before age 21, nonaccidental) from consanguineous marriages among the inhabitants of Hiroshima and Nagasaki, Japan, as well as from Hirado, Japan (see Example

9-2). In both cases, these authors were able to separate data from different socioeconomic groups, because correlation between inbreeding and deprivation may account for a portion of greater mortality with inbreeding. While separation was important in the large cities of Japan, the Hirado data were not much affected by including or excluding the socio-economic status as far as mortality was concerned.

Percent mortality of children from various types of marriages are given below for Schull and Neel's earlier study (numbers in parentheses = families observed).

City	From Marriages of			
	First Cousins	$1\frac{1}{2}$ Cousins	Second Cousins	Unrelated
Hiroshima	6.12(532)	7.18(192)	4.43(230)	3.55(1384)
Nagasaki	5.25(826)	4.94(263)	3.18(330)	3.42(2078)

In the later study (Schull, Nagano, Yamamoto, and Komatsu, 1970) on Hirado, the data were presented as regressions of mortality on F, both for parents and children of consanguineous marriages. Only the mortality data of inbred children were significant. Stillbirths were short of significance perhaps because of poor recall by mothers of their preparturition losses. At any rate, the overall effect of increasing F on mortality among conceptions surviving at least 21 weeks of gestation but dying before age 21 years gave a regression of 0.7703 ± 0.1620, which is highly significant. Thus, if a regression of mortality on levels of F were to be extrapolated to the level of complete inbreeding ($F = 1$), there would be 77 percent mortality minus about 3.5 percent at the control (unrelated parents), or roughly 73.5 percent expected mortality due to homozygosity for lethal equivalents in the total genome. This is a lower estimate for the Japanese population than that calculated by Morton, Crow and Muller for the Caucasian population data from France, a fact noted by Neel and Schull (1968).

EXAMPLE 9-5

RECESSIVE DETRIMENTAL FACTORS REVEALED BY INBREEDING IN DROSOPHILA

Techniques have been perfected for chromosomal assay in drosophila by using dominant markers (usually homozygous lethal) linked with crossover suppressors (inversions) in order to bring about complete homozygosity of any chromosome so marked—in a far more efficient way than could be achieved by inbreeding alone without markers. In Figure A, it is essential to realize that the marker must have each homologue marked (often a nonallelic dominant is used instead of the recessive on the homologue as shown) in order to avoid ambiguity in generation 2. (Why should the first-generation hybrid single male be backcrossed to the marker strain and not simply crossed immediately to its sibs, thus shortening the process by one generation?) If the wild chromosome being tested possesses any recessive lethal, visible mutant, or sterility allele, all nondominant flies in the third generation will have the identical genotype in homozygous condition

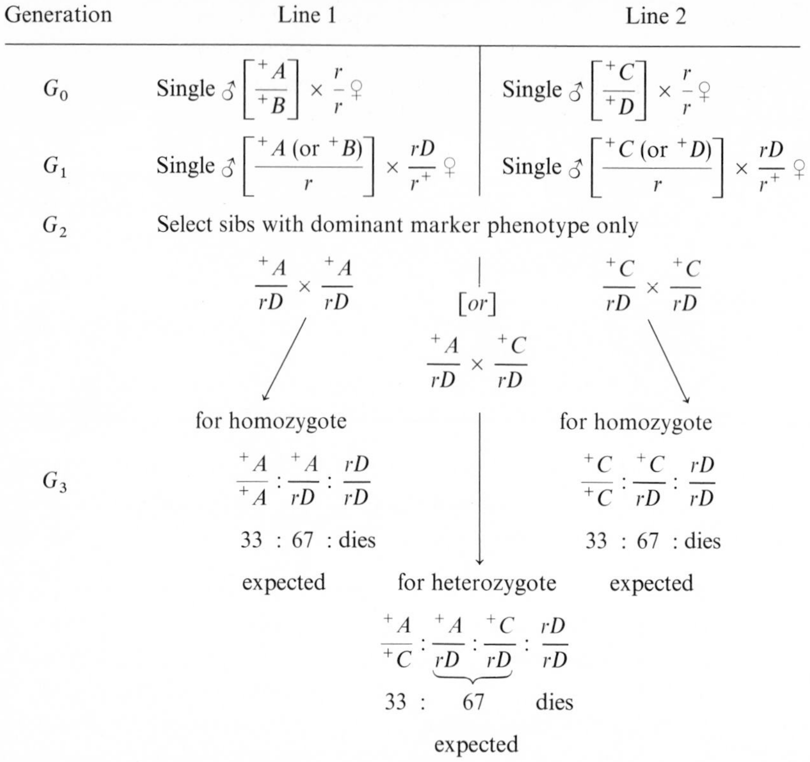

Generation	Line 1	Line 2
G_0	Single ♂ $\left[\frac{^+A}{^+B}\right] \times \frac{r}{r}$ ♀	Single ♂ $\left[\frac{^+C}{^+D}\right] \times \frac{r}{r}$ ♀
G_1	Single ♂ $\left[\frac{^+A \text{ (or } ^+B)}{r}\right] \times \frac{rD}{r^+}$ ♀	Single ♂ $\left[\frac{^+C \text{ (or } ^+D)}{r}\right] \times \frac{rD}{r^+}$ ♀
G_2	Select sibs with dominant marker phenotype only	

$$\frac{^+A}{rD} \times \frac{^+A}{rD} \qquad [or] \quad \frac{^+A}{rD} \times \frac{^+C}{rD} \qquad \frac{^+C}{rD} \times \frac{^+C}{rD}$$

for homozygote (Line 1):

G_3: $\frac{^+A}{^+A} : \frac{^+A}{rD} : \frac{rD}{rD}$

33 : 67 : dies

expected

for homozygote (Line 2):

$\frac{^+C}{^+C} : \frac{^+C}{rD} : \frac{rD}{rD}$

33 : 67 : dies

expected

for heterozygote

$$\frac{^+A}{^+C} : \underbrace{\frac{^+A}{rD} : \frac{^+C}{rD}} : \frac{rD}{rD}$$

33 : 67 dies

expected

Figure A. Technique for revealing recessives in drosophila with a dominant marker (D) mutant (lethal as a homozygote), a crossover suppressor (heavy band on D chromosome), and a recessive marker (r). Wild type chromosomes to be tested for homozygote or heterozygote phenotype are $^+$A, $^+$B, $^+$C, $^+$D.

(just as at the limit of inbreeding but circumventing the long process of sibmating, which was the only technique available before the marker technique was developed). Heterozygotes for wild chromosomes can simply be produced by crossing line 1 × line 2 at G_2.

In the G_3, if 100 flies are counted, normal viability is expected to produce 33 wild types: 67 dominant marker phenotypes. From such assays for viability of wild chromosomes, viability levels have been defined as follows:

1. Lethal = zero wild-type flies.
2. Semilethal = from 1 to 50 percent of expectation; that is, from 1 to 16 wild-type flies per 100 counted in G_3.
3. Quasi-normal = greater than 50 percent of expectation; that is, >16 wild-type flies per 100 counted.

A typical distribution of viabilities measured relative to average outcrossed (heterozygote) viability is shown for *D. prosaltans* in Figure B.

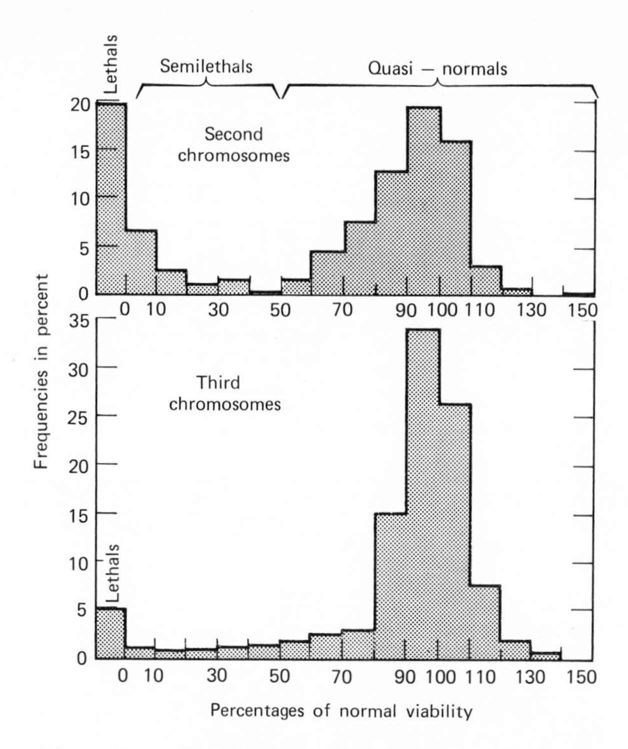

Figure B. Distribution of homozygous chromosomes' viabilities as a percentage of normal (expected 33% wild in G_3) from natural populations of *D. prosaltans*. Ordinates = frequencies in percent of chromosomes sampled (from Dobzhansky and Spassky, 1954).

Further subdivision of the quasi-normal group was proposed by Wallace and Madden (1953), using the following definitions and illustrated in Figure C:

1. Normal = average number of wild-type characterized by random heterozygotes and including the $\pm 2\sigma$ confidence limits around the average.
2. Subvital = homozygotes and/or heterozygotes falling below the -2σ confidence limit from the average.
3. Supervitals = those falling above the $+2\sigma$ confidence limit.

Some representative results of assays for viability and fertility modifiers expressed in homozygotes by Dobzhansky and his colleagues from natural populations of four drosophila species are given in Table A. Subvitals, it should be noted, constitute more than half chromosomes from natural populations of all the widespread species (*D. pseudoobscura*, *D. persimilis*, and *D. willistoni*), although their frequencies are lower in the species *D. prosaltans*, which forms less dense and widely scattered populations in New World tropics.

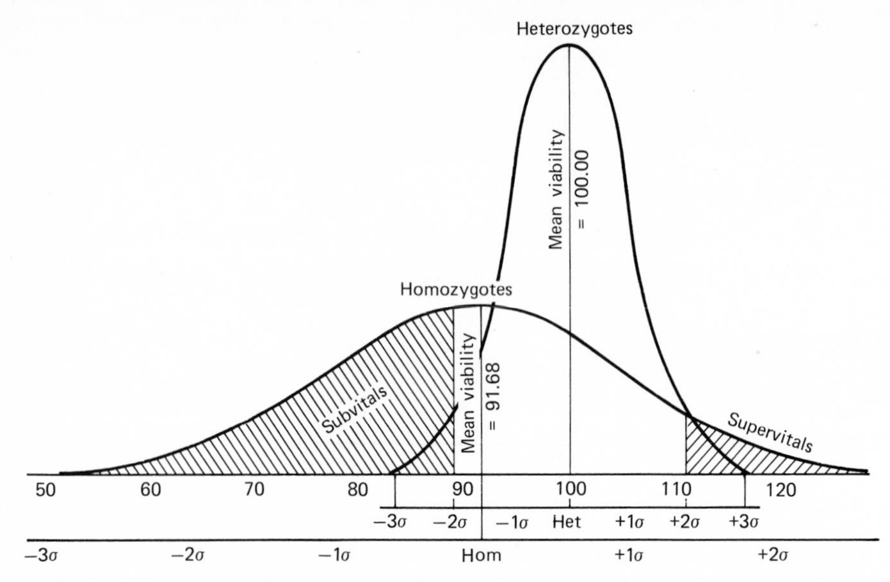

Figure C. Estimated distribution of subvital, normal (or quasi-normal), and supervital viabilities in homozygotes and heterozygotes of *D. prosaltans'* second chromosomes, all measured in relation to a constant dominant marker. Under the abscissa (percentages of normal viability) are shown multiples of the standard deviations for homozygotes (bottom) and for heterozygotes (middle) (from Dobzhansky and Spassky, 1954, based on Wallace and Madden, 1953).

TABLE A *Percentage of homozygous chromosomes tested for recessive modifiers of viability and fertility*

Drosophila Species	Chromosome	Lethals and Sublethals	Subvitals	Supervitals	Female Sterility	Male Sterility
*D. pseudoobscura** (Mather, Cal.)	2	33.0	62.6	<0.1	10.6	8.3
	3	25.0	58.7	<0.1	13.6	10.5
	4	22.7	51.8	<0.1	4.3	11.8
*D. persimilis** (Mather, Cal.)	2	25.5	49.8	0.2	18.3	13.2
	3	22.7	61.7	2.1	14.3	15.7
	4	28.1	70.7	0.3	18.3	8.4
*D. prosaltans*** (Brazil)	2	32.6	33.4	<0.1	9.2	11.0
	3	9.5	14.5	3.0	6.6	4.2
D. willistoni† (Venezuela, British Guiana, and Trinidad)	2	38.8	57.5	<0.1	40.5	64.8
	3	34.7	47.1	1.0	40.5	66.7

* From Dobzhansky and Spassky, 1953; Sankaranarayanan, 1965.
** From Dobzhansky and Spassky, 1954.
† From Malogolowkin-Cohen, et al., 1964.

Other drastic detrimental factors, lethals, semilethals, and sterility factors tend to have highest frequencies among the three densely populated species.

EXAMPLE 9-6

LETHAL EQUIVALENT ESTIMATES IN DROSOPHILA

Samples of *D. pseudoobscura* taken from Mather, California, and from certain localities in the Chiricahua Mountains of Arizona by Dobzhansky, Spassky, and Tidwell (1963) were assayed with dominant marker techniques on all autosomes for wild homozygote and heterozygote viabilities, as defined in Example 9-5 but measured relative to half the dominant marker class (D/wild class in G_3, Figure A, Example 9-5). Mean viability on the scale of wild-type counted out of 100 flies and the viability relative to half the D/wild class are given in Table A(upper). For each chromosome, the total load in lethal equivalents was estimated by using formula (9-14) as the difference between the natural log of heterozygote viability and that of the homozygote chromosome. For example, the second chromosome homozygotes have a total load of $T = ln[0.955] - ln[0.642] = 0.394$. For the three autosomes combined, the total load is 1.43 lethal equivalents per gamete.

For the tropical species *D. willistoni*, the two autosomes were similarly surveyed by Malogolowkin-Cohen, et al. (1964). Viabilities, their standard errors, and estimated total

TABLE A *Autosomal assays for mean viability, viability relative to a constant marker, and lethal equivalents estimate in two species of drosophila*

Chromosome	Mean Viability (average wild-type observed/100 counted)	Viability Relative to $[D/+]/2$	Lethal Equivalents (T)
D. pseudoobscura (Dobzhansky, Spassky, and Tidwell, 1963)			
2 homozygote	22.36 ± 1.17	0.642 ± 0.035	0.394
2 heterozygote	32.24 ± 0.36	0.955 ± 0.011	
3 homozygote	21.65 ± 1.22	0.678 ± 0.038	0.461
3 heterozygote	32.98 ± 0.32	0.987 ± 0.017	
4 homozygote	18.81 ± 1.18	0.520 ± 0.034	0.574
4 heterozygote	30.96 ± 0.29	0.923 ± 0.016	
		Total lethal equivalents	1.429
D. willistoni (Malogolowkin-Cohen, et al., 1964)*			
2 homozygote	21.05 ± 1.58	0.53	0.77 ± 0.10
2 heterozygote	36.43 ± 0.53	1.09	
3 homozygote	20.20 ± 1.24	0.51	0.69 ± 0.08
3 heterozygote	$33.53 - 0.34$	1.01	
		Total lethal equivalents	1.46 ± 0.13

* Viability (V) of homozygotes (or wild-type heterozygotes) was twice the wild type proportion (G_3 of Figure A in Example 9-5) relative to the proportion of dominant markers; that is, $V = 2P/(1 - P)$, where P = proportion of wild-type in G_3.

loads are presented in Table A(lower). As these authors pointed out, the total lethal equivalents from this species of 1.46 is contributed by lethals, semilethals, and subvitals (present in 48, 33, and 19 percent, respectively, for the second chromosome and 18, 47, and 34 percent, respectively, for the third chromosome).

The student should compare these lethal equivalents with the estimates made from regression of viability depression on F in Example 9-3. Note how much greater the estimate is from this direct method of homozygote analysis than by inference based on extrapolation from low F to high.

EXERCISES

1. In a panmictic population with $p = 0.8$, $q = 0.2$, sibs are mated within families only.
   (a) What will be the expected frequencies of genotypes (AA, Aa, aa) among the progeny of these sibs? Use the relationships in Table 9-2D.
   (b) Using Table 9-2B, what would be the frequencies of the six different types of sibship in G_1?
   (c) If the G_1 sibships 2 and 4 $\begin{bmatrix} AA \\ Aa \end{bmatrix}$ only are perpetuated by mating G_2 sibs, what would be the expected distribution of genotypes among their G_3 progeny? (Hint: Set up each type of G_2 sibship first with relative frequency of each.)
2. A Mendelian cross is made between $AA \times aa$ ("identical homozygotes"). Thus, $F = 1$ in the parent (G_0) generation, but $F = 0$ in the G_1.
   (a) If sibs only are mated, what will be the frequency of identical homozygotes in the G_2 and G_3? You may wish to use Table 9-1 as a guide, but with three genotypes instead of four.
   (b) This is a special case at the start, but once having arrived at G_2, what values will be expected for F if sibs only are mated in succeeding generations?
   (c) Since the Mendelian G_2 is the exact equivalent of a panmictic population with $p = q$, predict the genotype frequencies for the G_4 after sibmating the G_3.
3. Complete the blanks in the following table of recurrence for F and H under complete selfing or sib mating from a panmictic base population. Use formulas (9-1) and (9-1A) for selfing and (9-2) and (9-3) for sib mating.

Generation G_t	Selfing Only F	Selfing Only H (coeff. $2pq$)	Sibmating Only F	Sibmating Only H (coeff. $2pq$)
0	0	1 ($2pq$)	0	1 ($2pq$)
1	$\frac{1}{2}$	$\frac{1}{2}$	0	1
2	$\frac{3}{4}$	$\frac{1}{4}$	$\frac{1}{4}$	$\frac{3}{4}$
3	___	___	___	___
4	___	___	___	___
5	___	___	___	___
⋮				
∞				

4. Derive the F coefficient for the children of an ordinary sib mating by the method of summing all paths to a common ancestor. How does the result differ from the recurrence relationship for a continuous sib-mating F?
5. Using Figure 9-3A and the reasoning from probability of identical gene transmission in a regular system of mating, derive the F coefficient recurrence for parent-offspring matings (formula (9-5)). Remember that you are going to express the transmission probability into F—the probability that the individual has received identical genes. Hint: First work out the probability for transmission of P to X and Q (f_{PP}). What is the probability for Q's identical gene to go to X? Total probability for identical genes to go P to X and P to Q to X? Second, what is the transmission probability for P and B to transmit identical genes to Q (f_{PB})? Fill in the four ways in which gametes can be transmitting identical genes to the parents P and Q.
6. (a) Using Figure 9-3B and the reasoning from probability of identical gene transmission in a regular system of mating, derive the F coefficient for the type of half-sib matings illustrated in formula (9-6).
   (b) What fraction in formula (9-6) is the probability for identical genes in the progeny of ordinary half-sibs (no ancestors related)?
   (c) With the type of half-sib mating system illustrated, what would be the F and H recurrence sequences for three generations of half-sib matings?
   (d) Diagram a pedigree of half-sib matings, enlarging on Figure 9-3B in such a way that X would mate to Y, a half-sib. Diagram in such a way that parents of half-sibs form a *circular* system of mates.
7. What would be the F coefficient for X in the following pedigrees (first assume all ancestors' $F = 0$)?

(a)

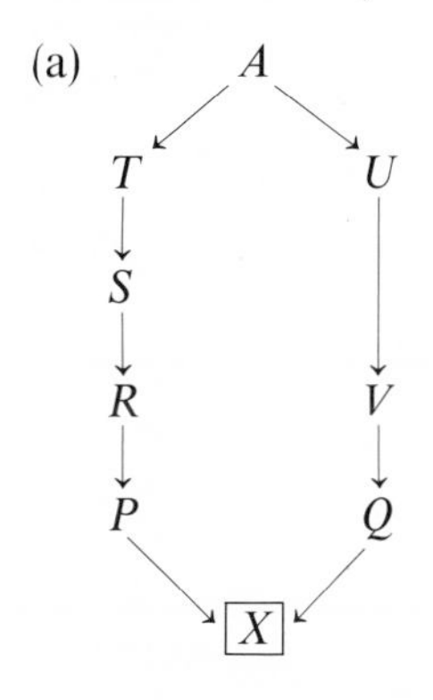

(b)

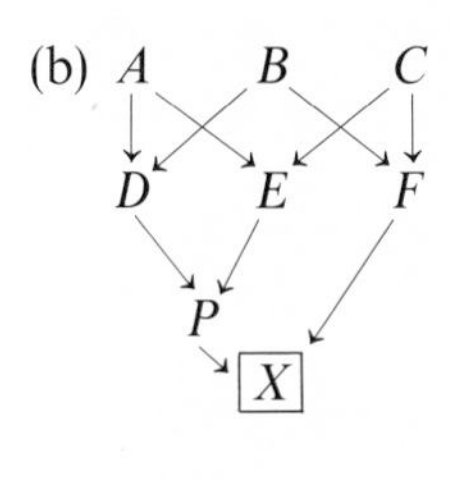

   (c) Assuming original common ancestors (A) in both pedigrees to have $F = 0.4$, what will be F_X in (a) and (b).
8. Find the general increment in F (ΔF) over generations in the following regular systems of mating:
   (a) Selfing.
   (b) Sib mating.
   (c) Double first cousin mating.

   First, work out the recurrence relation of ΔF by application of two generations of the F recurrence formula and subtract (for example), for selfing:

$$\begin{cases} F_t = \frac{1}{2} + \frac{1}{2}F_{t-1} \\ F_{t-1} = \frac{1}{2} + \frac{1}{2}F_{t-2} \end{cases}$$
$$\Delta F = \tfrac{1}{2}F_{t-1} - \tfrac{1}{2}F_{t-2}$$

Then find the amount of ΔF over the first six generations for each system of mating.

9. Using the recurrence relation for second cousin mating (p. 266), work out the limit of ΔF (when it becomes zero, or when $F_t = F_{t-1} = F_{t-2} \cdots$). First apply two generations and subtract as in the previous exercise. Then, show the limiting value of F by letting all F values be equal.
10. When all three genotypes can be estimated, F_{est} measures the departure from square-law expectation, specifically using formulas (9-9), (9-10), or (9-11), whichever may apply more easily. Formula (9-10) is also derived as follows. From formula (8-5), we have $4DR - H^2 = 4Fpq$, which gives $4DR - H^2/(2D + H)(H + 2R) = F$. Note the resemblance of this relation to the contingency chi-square when data can be arranged in a 2×2 table:

a	$\frac{1}{2}b$
$\frac{1}{2}b$	c

$$X^2 = \frac{(4ac - b^2)^2 \cdot N}{(2a + b)^2(b + 2c)^2}.$$

Thus, because genotypes for a pair of alleles can be arranged in this way, we have $X^2 = (F_{est})^2(N)$.

(a) Arrange the data from Table A of Example 1, top line (females from Munising), $16A{:}12AB{:}7B$. Calculate the chi-square for contingency and show $F = \sqrt{X^2/N}$, according to formula (9-10).
(b) Show that the relationships of Table 9-3 are illustrated by these data as well. Use the F value calculated and supply the values for p, q, and N into the table.

11. Use formula (9-12) to estimate the gene frequency of cystic fibrosis from first cousin data as given in Chapter 2, Example 2-8. In the Cleveland area, the incidence of cystic fibrosis was given as $\frac{1}{3700}$. Assume this frequency to be entirely from nonrelated parents (R_0), while $\frac{3}{982}$ were found from first cousin marriages (R_i).
(a) Verify that q (estimated) $= 0.006$.
(b) If the incidence recorded comes entirely from families where parents were first cousins, what would be the estimate of q?

12. If the proportion of matings from first cousins is 2 percent in a population and a recessive has $q = 0.01$, then the rest of the population is randomly mating.
(a) What will be the total frequency of aa genotypes? Compare with a population that is entirely random mating.
(b) What will be the relative risk of having a recessive child from a first cousin mating compared with the risk for unrelated parents?
(c) If q_a is 0.10, what happens to the proportion of first cousin parents among all parents with recessive offspring (k), assuming the amount of first cousin matings remains at 2 percent?

13. Inbreeding depression depends on the dominance of incremental quantitative alleles. We may derive formula (9-13) by combining formula (8-3) with the quantitative phenotypic values (Y_i) with dominance used in Chapter 6:

Genotype	Frequency	Y	Product
AA	$p^2 + pqF$	$+a$	$ap^2 + apqF$
Aa	$2pq - 2pqF$	d	$d2pq - d2pqF$
aa	$q^2 + pqF$	$-a$	$-aq^2 - apqF$
		Sum	

(a) Complete the algebraic sum and show that formula (9-13) is its derivation after simplifying with formula (6-3).

(b) Show that inbreeding depression depends on gene frequencies by taking the data of Table 6-5B (Examples 1 and 2) and assume sibs have mated to produce offspring.

10

PHENOTYPIC ASSORTATIVE MATING

In the broad sense, any mating set not determined by random combination of its elements but by some grouping or class to which mates belong can be said to result from nonrandom, or assortative, mating. Inbreeding, then, could be considered "assortative" because common ancestral relationship (consanguinity) determines the grouping. Ordinarily, in population genetic terminology, we separate inbreeding as a special case where grouping is by genotype derived via common ancestor relationships. We speak of assortative mating in the sense of grouping according to phenotypic criteria: that is, mating is considered assortative when individuals of opposite sex have a tendency to mate by "choice" (preference) or to mate in particular combinations (as when flowers of a single color tend to be pollinated by an insect vector). We shall consider this phenotypic assortative mating and its consequences for genotypes in populations in this chapter.

Terminology can often be confusing, and it will be helpful to clarify some of the ways in which mating associations can occur. An elementary chart is provided in Table 10-1 to illustrate and help define a few general types of mating. Let us assume that a constant equal number of two phenotypes (A and B) is available for mating ($n_A = n_B = 32$ in both sexes) and that any individual only mates once (monogamy). In Table 10-1A, mating is at random, because from the row and column totals we would estimate the internal distribution of the set (equal fourths) based on the equal proportions of A:B, provided both sexes mate totally. If some individuals in one or both sexes do not mate but if the matings that do occur are at random, we might have the outcomes in Table 10-1C. In the three cases illustrated, we know that from marginal totals with A:B proportions we can predict the mating frequency in each combination (internal distribution). A simple $2 \times 2\chi^2$ test would indicate a perfect fit with expectation (Tables 10-1A and C).

Table 10-1B gives three examples of assortative mating. In case 1, positive mating, or homogamy, is where more mating occurs between individuals with phenotypic resemblance than between dissimilar ones. We say the case illustrated is "incomplete assortative mating" because some heterogamy does take place. Case 2 illustrates incomplete assortative mating, but the tendencies for A and B phenotypes to mate may not be equal, so that the "preference" of B for B is not equally strong with the "preference" of A for A. A net loss of total mating then may ensue, eight individuals of each sex not mating in the case illustrated. Finally, case 3 represents negative assortative (incomplete) mating, or heterogamy, with matings more frequently between unlike individuals.

Complete positive assortative mating is illustrated in Table 10-1D. Complete negative assortative mating would be the reverse, with all matings between A $\times$ B. The distribution of matings within these sets (Tables 10-1B and D) cannot be predicted from expectations based on marginal totals; thus, we say these matings are nonrandom.

In Table 10-1C, A and B types are not equal in mating ability. This inequality may be considered a result of Darwinian selection. The internal distribution is random in that

TABLE 10-1 *General types of mating—assume two phenotypes of parent (A and B) occurring in equal numbers, 32 of each, with sexes equal*

A Random Mating

Males	Females *A*	*B*	$\sum$
A	16	16	32
B	16	16	32
$\sum$	32	32	= 64 total matings

B Incomplete Assortative Mating

(1) Positive-symmetrical

Males	Females *A*	*B*	$\sum$
A	24	8	32
B	8	24	32
$\sum$	32	32	= 64

(2) Positive-asymmetrical

Males	Females *A*	*B*	$\sum$
A	24	8	32
B	8	16	24
$\sum$	32	24	= 56

(3) Negative-symmetrical

Males	Females *A*	*B*	$\sum$
A	8	24	32
B	24	8	32
$\sum$	32	32	= 64

C Differential Mating Ability

(1) One sex (♂)

Males	Females *A*	*B*	$\sum$
A	16	16	32
B	8	8	16
$\sum$	24	24	= 48

(2) Both sexes

Males	Females *A*	*B*	$\sum$
A	24	8	32
B	8	3	11
$\sum$	32	11	= 43

D Complete Positive Assortative Mating (=Selfing)

Males	Females *A*	*B*	$\sum$
A	32	0	32
B	0	32	32
$\sum$	32	32	= 64

we can predict it from marginal frequencies. In case 1, B males mate less than A, and the two kinds of female mate equally. In case 2, there is a mating ability difference in both sexes by the same amount. In this case, the student might ask how it is different from the positive asymmetrical assortative case in the row above (B-2). Obviously, the difference lies in the B × B value and its relationship to the other three combinations. The simple contingency X^2 will fit the data almost perfectly in (C-2), indicating random mating, but certain "isolation indexes" are in common usage to measure relative amounts of homogamic versus random versus heterogamic mating; they can indicate some positive homogamy. For example, an isolation index (I) mentioned by Parsons (1967) is the following (from Malogolowkin-Cohen, et al., 1965):

$$I = \frac{(A, A) + (B, B) - (A, B) - (B, A)}{N}$$

where the letters in parentheses represent the row, column observed values in the 2 × 2 table and N = total matings observed.

$$V_I = \frac{4pq}{N}$$

where p = frequency of homogamic matings, q = frequency of heterogamic matings, and V_I = variance of index. Applying this index to random mating cases (Tables 10-1A and 10-1C-1) generally gives $I = 0$, while homogamy gives positive fractions up to $I = 1.0$ for complete positive assortative mating and heterogamy to $I = -1.0$ for complete negative assortative mating. Now the student may verify that the case in Table 10-1B-2 gives $I = \frac{3}{7} = 0.429$, $V_I = 0.058$, *s.e.* = 0.241, but the random case where both sexes have lower mating activity in (C-2) gives $I = \frac{11}{43} = 0.256$, $V_I = 0.022$, *s.e.* = 0.147, even though the X^2 is nearly zero. It is no wonder, then, that once in a while some confusion may arise about whether assortative mating may be taking place.

Schaffer (1968) illustrated the use of multiple-choice mating of A and B types in both sexes in the problem of distinguishing between random (nondiscriminatory) and assortative (discriminatory) mating. He estimated the expected random outcome by assuming the sexes have the same relative mating activities for types A and B. For example, with the case of Table 10-1B-2, the relative mating ability of A compared to B is 4:3, so that if the sexes are equal in their ability, the expected values would be A × A = $(\frac{4}{7})^2 = 0.327$, or 18.3 matings, instead of 24 observed. Schaffer devised an index of discrimination that measures the amount of deviation between observed and random mating frequencies (see Exercise 1). He also pointed out that the isolation index I is not a good measure of assortative tendency if only one sex shows discrimination because it only "goes halfway towards excluding nondiscriminatory mating differences." Thus, we would not agree that the case in (C-2) indicates clearly assortative mating, even though the isolation index indicates some homogamic tendency; it is one-sided. Some might argue that (C-2) represents asymmetrical assortative mating, with A × A being "preferred" over the other combinations; but the randomness can be shown by internal distribution from marginal totals, so the case is not one of assortative mating as it stands.

Extension of these definitions to more than two phenotypes (or genotypes) in mating sets can be made with an $n \times n$ factor analysis.

In Example 10-1, where Wallace observed mating differences between sex ratio (*SR*) and standard (*ST*) *X*-chromosome *D. pseudoobscura*, it first appears that *ST* males × *ST*/*ST* females seem to show a "preference"; however, *SR* males simply mated less, as did *SR*/*SR* females, with both karyotypes of the opposite sex, so that a simple mating propensity difference between the two karyotypes is sufficient to explain the data. Caution is the better part of wisdom in view of possible anthropomorphic biases when measuring mating or behavioral traits in nonhuman organisms; therefore, it should be clear to the student that much empirical observation and thought in experimental design is necessary before we can be certain of evidence for assortative mating (see Exercise 1).

In this chapter, we emphasize the type of mating that does not depend on *selection* for or against any types, but only on the actual set of matings taking place. As pointed out by Lewontin, Kirk, and Crow (1968), three concepts should be distinguished because they have three different effects on genetic structure of populations: (1) selective mating, (2) phenotypic assortative mating, and (3) inbreeding. The first (selective) involves differential reproduction (differences in Darwinian fitness between genotypes or phenotypes); that is, differential contributions made to the next generation's gene pool, a net gain to one type and loss to another, as in Table 10-1C. The second (assortative), while dependent on phenotypic grouping in mating, does not primarily involve loss or gain to any group, and no change in gene frequencies need result (as will be shown in this chapter). In Table 10-1B, the symmetrical cases have equal matings by both phenotypes in both sexes so that there is no selective change; however, the asymmetrical case could be considered a mixture of assortative and selective processes. Finally, the third concept (inbreeding) depends on genotypic grouping as a result of relationship by descent rather than on phenotype and does not result primarily in gene frequency change for the population as a whole (as described in Chapters 8 and 9). Mating can, of course, consist of a mixture of these processes, but analysis can only be made through discernment of what the primary processes are and weighting their relative contributions to determine the outcome to the population.

EXAMPLE 10-1

ASSORTATIVE MATING OR DIFFERENTIAL MATING ABILITY?

One of the earliest reports of mating control by chromosomal arrangements (presumably because the genic contents of different arrangements contain alleles that affect mating activity differentially) is that of Wallace (1948) of the "sex ratio" set of triple inversions (*SR*) in the right arm of the *X* chromosome in *Drosophila pseudoobscura*. He measured several fitness traits that might have been influencing the outcome of changes in relative frequency of the *SR* and *ST* (standard) arrangements. One was the insemination proportion of females by *SR* and *ST* males in "male choice" experiments; that is, 10 males of one karyotype in a single vial with 20 females, 10 of each karyotype. At the end of two hours, females were dissected to ascertain which had mated. When using homokaryotype females only cultured at 25°C, Wallace found the following relative frequencies of matings (sample sizes were 85 to 91 per mating combination, with total N mated $= 193$ out of

347 tested) (see Exercise 1 for the observed data):

Males	Females		
	ST/*ST*	*SR*/*SR*	Average Mating Frequency (Males)
ST	0.404	0.254	0.658
SR	0.197	0.145	0.342
Average mating frequency (females)	0.601	0.399	1.000

Wallace (1968, pp. 124–125) assigned this outcome to a case of positive assortative mating (see Spiess, 1970, pp. 344–346). *SR* males mated less with both types of female, and *SR*/*SR* females also mated less with both types of male, in parallel with Table 10-1C-2. As a matter of fact, the tests were not "multiple choice," in which females would have had the opportunity to accept or reject the two kinds of male, so that the results can more simply be explained as due to relative sexual activity, not to preferential behavior.

COMPLETE POSITIVE ASSORTATIVE MATING, SINGLE PAIR OF ALLELES

With no dominance (*Aa* intermediate between *AA* and *aa*), it is clear that the consequence of mating exclusively like to like produces a result identical with selfing. (If all genotypes are distinguishable so that $AaBB \neq AABb$, for example, obviously the same statement would be true for multiple loci.) Therefore, at the extreme, the consequences for the genotype under positive assortative mating are identical with inbreeding. That is not the case, however, when dominance makes heterozygotes indistinguishable from one of the homozygotes, when mating is incompletely positive, or when polygenes act in an additive way on the phenotype. Let us first examine the complication of dominance.

If dominants ($AA + Aa$) mate only with dominants and recessives only with recessives, the matings may be represented as below:

	AA	*Aa*	*aa*
AA	Dominant ×		—
Aa		dominant	—
aa	—	—	Recessive × recessive

The recessives only mate to their own genotype; thus, the set of matings between *AA* and *Aa* can be considered a separate subset complementary to the recessive subset. Let frequencies of *AA*, *Aa*, and *aa* genotypes be symbolized as *D*, *H*, and *R*, respectively (as in Table 1-2A). Matings of $D \times R$ and $H \times R$ will not occur. The $R \times R$ matings will occur at a frequency of *R*, while the subset of dominants by dominants will occur at $(1 - R)$ frequency, or $(D + H)^2$. As an example, we begin with a population of $D = \frac{1}{4}$, $H = \frac{2}{4}$,

TABLE 10-2 *Complete positive assortative mating with dominance, single pair of alleles*

Mating Frequency	Progeny D	Progeny H	Progeny R
D^2 } $= (1 - R)$	$D^2/(1 - R)$	—	—
$2DH$ } $= (1 - R)$	$DH/(1 - R)$	$DH/(1 - R)$	—
H^2 } $= (1 - R)$	$(H^2/4)/(1 - R)$	$(H^2/2)/(1 - R)$	$(H^2/4)/(1 - R)$
R	—	—	R
Total 1.00 =	$\dfrac{(D + H/2)^2}{(1 - R)}$ +	$\dfrac{H(D + H/2)}{(1 - R)}$ +	$\dfrac{(H/2)^2 + R(1 - R)}{(1 - R)}$
=	$\dfrac{p^2}{1 - R}$ +	$\dfrac{pH}{1 - R}$ +	$\dfrac{q^2 + R(p - q)}{1 - R}$

$R = \frac{1}{4}$. Then $R \times R$ matings will occur at one-fourth frequency while the remaining $\frac{3}{4}$ of the matings will be $(\frac{1}{4} + \frac{2}{4})^2$. Thus, for the latter set of matings: $(\frac{1}{4} + \frac{2}{4})^2 \div \frac{3}{4} = \frac{1}{12} D \times D$, $\frac{1}{3} D \times H$, $\frac{1}{3} H \times H$ matings. Progeny genotypes from all matings then will be as follows:

Mating Parents	Progeny D	Progeny H	Progeny R
$D \times D$	$\frac{1}{12}$	—	—
$D \times H$	$\frac{1}{6}$	$\frac{1}{6}$	—
$H \times H$	$\frac{1}{12}$	$\frac{1}{6}$	$\frac{1}{12}$
$R \times R$	—	—	$\frac{1}{4}$
Totals	$\frac{1}{3}$	$\frac{1}{3}$	$\frac{1}{3}$

Thus, there will be a gain in both homozygotes and loss in heterozygotes from G_0 to G_1. These relationships are symbolized in Table 10-2.

As with inbreeding, there is no change in allelic frequencies under positive assortative mating (with or without dominance). We may calculate the expected p value by taking the sum of dominant homozygote plus half the heterozygote frequencies among the progeny.

$$p_1 = \frac{(D + H/2)^2}{1 - R} + \frac{H(D + H/2)}{2(1 - R)} = \frac{D + H/2}{1 - R}[D + H]$$

Since $1 - R = D + H$, then cancelling, $D + H/2 = p_0$.

The change in genotypes involves changes similar to those of inbreeding. Heterozygote loss may be estimated as follows:

$$H_t = \frac{H_{t-1}(D_{t-1} + H_{t-1}/2)}{1 - R_{t-1}} = \frac{pH_{t-1}}{1 - R_{t-1}}$$

or

$$H_t = \frac{2pH_{t-1}}{2p + H_{t-1}} \tag{10-1}$$

(recurrence of H under complete positive assortative mating). By using this recurrence series for the example given above where $H_0 = \frac{1}{2}$, the student may easily note the simple series of H:

$$\tfrac{1}{2}, \tfrac{1}{3}, \tfrac{1}{4}, \tfrac{1}{5} \cdots 0 \qquad \text{as } n \to \infty$$

This series, whose terms are the reciprocals of an arithmetic series, is called a harmonic series (see also selection against recessive lethals, Chapter 14). Thus, after t generations of continued complete assortative mating, heterozygosity will be in terms of the original generation H_0,

$$H_t = \frac{2pH_0}{2p + tH_0} \tag{10-2}$$

Clearly, because the denominator increases continually, the outcome will be zero as a limiting value. When the original population is panmictic, then (10-2) becomes simply

$$H_t = \frac{2pq}{1 + tq} \tag{10-3}$$

The student should note that here the rate of change depends on allelic frequency, which is greater when the recessive allele is common than when it is rare. This progression is in marked contrast with the effect of inbreeding, where the amount of relationship between parents alone determines the rate of change in zygosity. In fact, this result for assortative mating resembles the changes of selection more than the results of inbreeding in the dependence on allelic frequency. For the entire population practicing that type of mating, no change will take place in p,q if the size of the population is large enough to disregard sampling error. Approach to the limiting state is slow, with the faster changes occurring in the earliest generations from the start of the assortative mating system (Exercise 3).

INCOMPLETE POSITIVE ASSORTATIVE MATING, SINGLE PAIR OF ALLELES

With some heterogamy, the set of matings must be weighted accordingly, and the rate of change per generation will be slower than for the complete positive case as given above. Wright (1921, *Systems of Mating III*) worked out by path coefficient methods a general formula for predicting changes in genotype frequencies depending on the correlation between mates (m). An equivalent method to be considered is based on the same principle we used when discussing mixtures of inbreeding and random mating (p. 246 and p. 267)—imagining the population as divided into randomly mating and assortatively mating fractions, as described by Crow and Felsenstein (1968). For example, in Table 10-1B, we can visualize the positive symmetrical case to be made up of a random-mating fraction with all four combinations equal (eight per combination) plus A × A and B × B to be worth 16 each. In this case, half the matings would be at random and half completely assortative.

Let us represent the fraction of the population that is assortatively mating as m and the random-mating fraction as $(1 - m)$. Then using the contributions to each genotype from Table 10-2, we see that the genotype frequencies in next generation will be as follows:

$$D_t = (1 - m)p^2 + \frac{mp^2}{1 - R_{t-1}}$$

$$H_t = 2(1 - m)pq + \frac{2mpH_{t-1}}{2p + H_{t-1}} \quad \text{(using (10-1) here for the } H_t \text{ second term)} \tag{10-4}$$

$$R_t = (1 - m)q^2 + m\left[\frac{q^2 + R_{t-1}(p - q)}{1 - R_{t-1}}\right]$$

With any value of m except 1, the population will come to nonzero equilibrium for heterozygosity. Setting $H_t = H_{t-1} = H$ gives $H(2p + H) = 2(1 - m)pq + 2mpH$, which simplifies to

$$\left(\frac{H}{2}\right)^2 + p^2(1 - m)\frac{H}{2} - p^2q(1 - m) = 0$$

with quadratic solution, the positive root is

$$H = p^2(m - 1) + p\sqrt{p^2(1 - m)^2 + 4q(1 - m)} \tag{10-5}$$

Sequences of H from initial $H = \frac{1}{2}$ for various genetic situations and three fractions of assortative mating ($m = 1$, or 0.8, or 0.5) are given in Table 10-3. As an example, the student may verify that in the third column from the right the sequence ending with $H = 0.390$ is at its limit if half the population were mating at random and half assortative—for example, with initial $D = \frac{1}{4}$, $H = \frac{1}{2}$, and $R = \frac{1}{4}$. It is noteworthy that the equilibrium is attained fairly rapidly, usually in less than 10 generations. The slight amount of heterogamy offsets the trend toward complete homozygosity; in fact, it might come as a surprise to some students that very little fixation takes place in spite of rather high fractions of assortative mating

TWO PAIRS OF ALLELES, POSITIVE ASSORTATIVE MATING

When correlation between mates is based on resemblance of traits that depend on two or more genetic loci, genotypic changes become very complicated (unless all genotypes are distinguishable, in which case assortment would be equivalent to selfing). Wright (1921) considered the simple case of polygenic additivity (no dominance but additivity with $A = B \ldots$, and $AaBB \cdots = AABb \ldots$, as in Chapter 6). Of nine genotypes with two loci, there are five phenotypes on which preferences could be made (Table 6-7). While Wright carried out the details of four generations of complete assortative mating, he admitted that "the work is rather tedious even in this simple case." He then proceeded to analyze the composition of the population by path coefficient analysis, and his successive estimates of heterozygosity are given in Table 10-3. Wright's derivation has been redescribed by Crow and Felsenstein (1968) somewhat more lucidly than the original. However, instead of heterozygosity, we shall give the recurrence of F (fixation, or correlation between

TABLE 10-3 *Percentage of heterozygosis under positive assortative mating (k = number of loci segregating) (modified from Wright, 1921, Systems of Mating III)*

Generation	Complete Assortative Mating					Incomplete Assortative Mating					
	$m = 1.00$					$m = 0.80$			$m = 0.50$		
	$k = 1$ Dominance	$k = 1$ No Dominance	$k = 2$* No Dominance	$k = 4$* No Dominance	$k = 10$* No Dominance	$k = 1$ Dominance	$k = 1$ No Dominance	$k = 4$* No Dominance	$k = 1$ Dominance	$k = 1$ No Dominance	$k = 4$* No Dominance
0	0.500	0.500	0.500	0.500	0.500	0.500	0.500	0.500	0.500	0.500	0.500
1	0.333	0.250	0.375	0.438	0.475	0.367	0.300	0.450	0.417	0.375	0.469
2	0.250	0.125	0.312	0.406	0.462	0.315	0.220	0.430	0.397	0.344	0.461
3	0.200	0.063	0.266	0.379	0.451	0.291	0.188	0.414	0.392	0.336	0.456
4	0.167	0.031	0.227	0.354	0.439	0.281	0.175	0.402	0.391	0.334	0.453
5	0.143	0.016	0.193	0.330	0.428	0.275	0.170	0.391	0.391	0.333	0.450
10	0.083	0.001	0.088	0.233	0.376	0.270	0.167	0.357	0.390	0.333	0.445
15	0.059	0.000	0.040	0.165	0.330	0.270	0.167	0.343	0.390	0.333	0.444
⋮											
∞	0	0	0	0	0	0.270	0.167	0.333	0.390	0.333	0.444

* With additive effects of nonhomologous loci ($A = B = C \cdots$) so that $AaBBCC \cdots = AABbCC \cdots = AABBCc \cdots$ in the polygenic model (Chapter 6).

gametes) for the case of the cumulative polygenic model without derivation (see Wright, 1969, vol. 2., pp. 273–284).

Given k loci and genetically determined correlation between mates (m) as we have used it above (fraction of population undergoing assortative mating),

$$F_t = \frac{m}{2k}[1 + kF_{t-1} + (k - 1)s_{t-1}] \tag{10-6}$$

where $s_t = (F_{t-1} + s_{t-1})/2$ and s is the correlation between nonhomologous genes in the same gamete of parents for any generation. For example, the third column in Table 10-3 (where k loci $= 2$, no dominance) the sequence can be obtained as follows: with positive assortative mating complete ($m = 1$), we have by generations (with initial $H = \frac{1}{2}$):

Generations	0	1	2	3	...	
F	0	$\frac{1}{4}$	$\frac{3}{8}$	$\frac{15}{32}$	...	
s	0	0	$\frac{1}{8}$	$\frac{1}{4}$	...	
H	$\frac{1}{2}$	$\frac{3}{8}$	$\frac{5}{16}$	$\frac{17}{64}$	...	using (8-2)

One important contrast with the composition of an inbreeding population is that under complete assortative mating with a polygenic trait, the ultimate composition will consist entirely of the two extreme phenotypes: *AABB* and *aabb* instead of all four homozygotes (*AABB*, *AAbb*, *aaBB*, *aabb*). The student may intuitively see this result after doing Exercise 6, because it can be seen that each phenotypic set is slowly converted into the extremes only.

More complex genetic systems lead to similar results, ending in fixation ultimately, if assortative mating is completely positive, but coming to an equilibrium (nontrivial) if it is not completely positive. Furthermore, in real populations, the mating pattern may depend on a number of other factors: monogamy versus polygamy, sex ratio, unisexual preference (asymmetrical cases), and many other variables. For further discussion, see references by O'Donald (1960), Parsons (1962, 1967), Scudo and Karlin (1969), Karlin and Scudo (1969), and Watterson (1959).

GENERAL EFFECTS AND EXAMPLES OF POSITIVE ASSORTATIVE MATING

Example 10-2 presents a case of positive assortative mating from the experimental literature. In addition, successful evocation of isolation between mutants in certain populations of *D. melanogaster* that constituted single gene pools has been achieved independently by a number of investigators. Wallace (1954) was able to effect sexual preference and isolation between straw and sepia mutants by simply destroying wild-type hybrids between them for more than 70 generations; he found a marked preference for intrasepia matings, although tests for sexual preferences did not always bring positive results. Knight, Robertson, and Waddington (1956) described bringing about partial sexual isolation by using ebony and vestigial mutants in a similar technique to that of Wallace. Hybrid wild-type flies declined, and after 20–30 generations an increase in vestigial homogamy (one-sided) occurred followed by both mutants to show homogamy. Light intensity is known to affect mating behavior of ebony flies, and when a container distributed the light intensity more uniformly, ebony mutants displayed as much homogamy as vestigial.

Later Crossley (1974) repeated those experiments, modifying and improving the Knight, Robertson, and Waddington technique and in addition testing for preferential mating throughout the experiments (see Manning, 1965). Consequently, it must be taken as established that isolation (homogamy) within a gene pool can be achieved. Achievement of similar results by "disruptive selection" by Thoday and Gibson will be described in Chapter 16. Also, de Souza, da Cunha, and dos Santos, 1972, reported homogamic mating between two morphs of *D. willistoni* that differ in their sites of pupation in population cages.

In wild populations other than those of drosophila, homogamy and character displacement may be achieved by behavioral mechanisms that involve a conditioning or "imprinting" process. In flowering plants pollinated by animal vectors, Levin and Kerster (1970; Levin, 1972) pointed out that pollen vectors (bees, butterflies, hummingbirds) tend to forage within small areas and move between neighboring plants; if plants are polymorphic for flower color or other corolla expressions (*Phlox*, *Linanthus*, *Leavenworthia*), the pollinators tend to lessen their search time and energy expenditure by learning the features of a common corolla type and consistently visiting that type with reduced tendency to outcross (visit differing types of flower). These habits bring about a disadvantage to the lower-frequency flower type. Color divergence then may well be increased by pollinators' discrimination.

Imprinting, or preference for a phenotype learned during its rearing by a young animal (birds and mammals particularly), will lead to a form of positive assortative mating. Parsons (1967) and Seiger (1967) list some well-known examples of imprinting in geese, skuas, ducks, pigeons, and mice. Apparently, if both sexes are imprinted on the phenotypes of their parents, and if both parents are identical, their offspring will mate only with that phenotype; if parents are different, offspring could mate with either phenotype. In the case of simple monofactorial dominance, it has been shown (Seiger, 1967) that the outcome is very similar to that of complete positive assortative mating, increasing homogamy, and isolation of homozygous genotypes.

Example 10-3 summarizes instances from human populations. The level of correlation between mates—and thus grouping of mating units based on phenotypic traits in both wild and in human populations—is often significant, with implications that the trends could result in increasing diversity, genetically uniform groups in a patchwork distribution, and perhaps semi- or complete isolation. In fact, assortative mating combined with inbreeding can be a very powerful force leading in that direction. Genetic variance, of course, will increase for the population as a whole essentially as described by formula (8-4). Stratification of the population may well be the major contributor to maintaining or increasing variance in many human populations, especially in modern civilized ones where the forces of natural selection have been greatly reduced (Lewontin, Kirk, and Crow, 1968). Any assortment by phenotypes can be a source of bias in estimates of heritability because it tends to inflate the genetic variance, as mentioned in Chapter 9.

Evidence for assortative mating is still far from overwhelming in the literature. When investigators do present evidence, at least two pitfalls must be avoided or ruled out before genetic conclusions can be drawn. First, phenotypic traits, selected by mates, may have little or no genetic basis, or low heritability, so that no changes in distribution of genotypes will result from the mating assortment. This is no doubt the case with the caste system of India (as pointed out by Dobzhansky, 1962; Garrison, Anderson, and Reed, 1968) as well as homogamy for social characteristics in general. Dobzhansky said of the vast "experiment" going on for hundreds of years in India: "It attempted to breed varieties of men genetically specialized in the performance of certain functions. To all appearances, such a specialization has not been achieved." (Of course the factors of inbreeding and subdivision of the population's gene pool must bring about changes leading to increasing genetic diversity in India but independently of the traits by which the castes are delimited.)

A second pitfall has to do with the gathering of data. If two or more groups within which mating is random but between which mating is reduced are combined for measurement of some trait without regard to the grouping, a spurious correlation will be computed (see Exercise 10). For example, if two ethnic groups are measured for some trait such as stature, the groupings will result in overall correlation, when in fact there may be none for the trait if it were not for the grouping on a basis independent of the trait. Alternatively, pooling measurements from two or more time periods by assuming no difference (for example, stature in 1900 and 1970) would also lead to spurious correlation.

EXAMPLE 10-2

PHENOTYPIC POSITIVE ASSORTATIVE MATING IN DROSOPHILA

A general tendency for homogamic mating within populations has been looked for in a few species of drosophila, with the view that incipient sexual isolation between gene pools might arise from such ethological reinforcement tendencies. More often techniques have been aimed at testing different races or geographically distinct groups within species to ascertain levels of isolation; however, objectives from this information are aimed at finding out what isolation has already been achieved following separation of gene pools, as a product rather than a cause of genetic divergence. Intrapopulational homogamy is occasionally found (see Manning, 1965; Parsons, 1967; Spiess, 1968, pp. 199–203; Spieth, 1968, pp. 181–191; Petit and Ehrman, 1969, pp. 178–180).

To measure the level of any homogamy in a population, either of two techniques seems useful: (1) using some metric trait to estimate the correlation between mates for the trait, or (2) selecting for assortment by discarding progeny from random matings, thereby increasing homogamy and demonstrating genetic variance for ethological preferences in the base population. Using the first technique, Parsons (1965) chose a single strain of *Drosophila melanogaster* (Canton-S), raised the flies under three larval density levels to provide more phenotypic variability than otherwise might be the case, and then allowed them to mate in lots of 40 virgins of each sex within each density level. As pairs mated, they were aspirated out of the mating chamber and their sternopleural bristle numbers recorded. After about half the flies had mated, the remaining unmated flies were stored separately for scoring. Then correlation coefficients were calculated as given in Table A for the mated pairs and the unmated remainder, which were arbitrarily matched.

For low larval density and mixed density, significant correlation coefficients close to $r = 0.20$ were obtained for bristle numbers of mated pairs. At high density level, the correlation was less but still significant. Unmated pairs had correlations not significantly different from zero. In drosophila, it may be that assortative mating may be a direct effect of fly size: smaller flies have fewer bristles. Wing areas may be effective in male mating

TABLE A *Sternopleural chaeta numbers for mated and unmated flies and correlation coefficients between members of pairs for chaeta number (modified from Parsons, 1965)*

Larval Competition Level	n	Chaeta Numbers Females Mean ± *S.E.*	Chaeta Numbers Males Mean ± *S.E.*	Correlation Coefficient Between Members of Pairs	Probability That Correlation Coefficient Differs from 0
1. Low					
Mated	212	23.86 ± 2.34	22.87 ± 2.16	0.206	<0.01
Unmated	247	23.86 ± 2.09	22.94 ± 2.03	−0.038	0.5–0.6
2. High					
Mated	568	20.17 ± 1.79	19.81 ± 1.85	0.109	<0.01
Unmated	610	20.12 ± 1.68	19.67 ± 1.84	−0.049	0.2–0.3
3. Mixed					
Mated	172	22.01 ± 2.40	21.17 ± 2.20	0.200	<0.01
Unmated	239	21.62 ± 2.45	21.00 ± 2.28	−0.005	>0.9

success (Ewing, 1964). There may also be behavioral differences between flies of different sizes leading to minor courtship and receptivity modifications. At least these data do present evidence that positive assortative mating exists in unselected populations, and, as mentioned by Parsons, "It may well be that the evolution of positive assortative mating during divergence is a very general phenomenon."

EXAMPLE 10-3

CORRELATION BETWEEN MATES IN HUMAN POPULATIONS

Humans usually take a great deal of time choosing the person with whom they intend to spend the rest of their lives (monogamous cultures). Widely quoted clichés such as "people are made for each other" or "opposites attract" point out how commonly the conscious choice of marriage partners occurs in general.

Pearson and Lee in 1903 measured stature, span of arms, forearm length, and similar characteristics and took note of the correlation coefficients between marriage partners. The coefficients were about 0.20, a value that has been confirmed in many but not in all populations. Davenport's data (Example 6-1) indicate the same tendency.

At the Fourth Princeton Conference on Population Genetics and Demography held in 1967, Spuhler (1968) summarized the degree of correlation (product-moment for continuous variables or contingency for discrete variables) in regard to 105 physical traits in several population samples, compiled from diverse authors beginning with Pearson and Lee's observations to the data of the conference. In addition, Spuhler reported the results of an Ann Arbor, Michigan, adult population study made in 1951–1954 on assortative marriages and fertility. A random sample of 1054 persons participated in the study. Correlations between members of each married pair were calculated for 43 physical measurements, of which 14 (32.6 percent) were not significantly different from zero, while 29 were significantly positive (5 at the 5 percent level of significance and 24 at the 1 percent level). This finding includes strong evidence for assortative mating for body size, although the intensity is not of high degree (0.20–0.30). About half of the traits having to do with size of head and face failed to show significant correlations, while seven traits including ear length and head breadth and head height were significantly correlated. Of course, age had the highest correlation (0.94). Body breadth and circumferences were significantly correlated for the most part (0.20–0.30).

Spuhler (1968) summarized his compilation from diverse authors by stating that one-third of the correlations fell in the range from 0.10–0.20. Homogamy coefficients higher than 0.3 were rare, at least in European peoples. There was some tendency for higher marital associations for physical traits among populations from the Mediterranean region than among those residents of northern or western Europe. Only two negative correlations were significant (number of illnesses and pulse after exercise). For the two populations non-European sampled (Ramaha Navaho Indians and the Japanese), no significant correlations were found for stature; in fact, for the Indians, with the exception of correlation for age ($r = 0.99$) none of the remaining 39 correlations were significant. Consequently, assortative mating for physical measurements can only be considered as specifically true for particular populations.

In measurement for assortative tendencies for behavioral traits and intelligence, however, the situation is much more positive. At the Princeton conference, Kiser (1968) and Garrison, Anderson, and Reed (1968) reported data on correlations for educational attainment and resultant fertility of marriages. The latter authors reviewed past studies on intelligence associations between spouses as well as sociological "traits" (religious affiliation, cultural background, social participation, and the like). Correlations for standard IQ tests ranged from $r = 0.33$–0.55 (cited in Garrison, et al., 1968). In educational attainment (highest grade in school) Garrison, Anderson, and Reed introduced a statistic G useful for discrete alternatives (polytomies) that are not normally distributed continua. This G statistic proved more useful than the usual contingency X^2; the G statistic lies between $+1$ and -1, resembling the product-moment correlation coefficient; it is the difference between the probability of selecting a pair of married couples in which the order of scores for husbands is the *same* as the scores for wives and the probability of selecting a pair in which the order for husbands is *different* from the order for wives. Over 100,000 marriages in Minnesota (1965–1966) gave $G = 0.69$ approximately for educational attainment, and in a larger sample (5 percent of the entire United States population) G ranged from a low average 0.608 ± 0.003 for southern nonwhite to 0.664 ± 0.0008 for southern whites. The value of $r = 0.624 \pm 0.0004$ for the latter group by comparison with G, although r is not interpretable because the underlying assumption of normality of a continuum is not met (see Table A). These authors showed that positively assortative marriages for educational attainment resulted in more children than negatively assortative marriages because a larger proportion of the latter are childless. There is a striking drop in all four G values for childless couples (row 2), and especially an increase up to the third row for couples producing children. Finally, in the fourth row, G was calculated using the actual numbers of children produced by the couples, who were arranged by educational attainment. When G is higher in this row than in the rows above it (as 0.7002 for southern whites), these authors concluded that assortative marriages resulted in production of more

TABLE A *G statistic (tendency for association) with 95 percent confidence limits for educational attainment of husbands and wives, age cohort of 35–44 years based on a sample (5 percent) of the United States population, taken from United States Bureau of the Census, 1964 (reported by Garrison, Anderson, and Reed, 1968)*

	G Statistic for Population			
Category Studied	White	Nonwhite	Southern White	Southern Nonwhite
All couples	0.6172 ± 0.0005	0.6104 ± 0.0017	0.6640 ± 0.0008	0.6083 ± 0.0026
Childless couples	0.5600 ± 0.0016	0.5507 ± 0.0039	0.6006 ± 0.0030	0.5607 ± 0.0060
Couples with children	0.6244 ± 0.0005	0.6258 ± 0.0018	0.6714 ± 0.0008	0.6184 ± 0.0020
Couples by number of children	0.6488 ± 0.0003	0.6336 ± 0.0010	0.7002 ± 0.0005	0.6052 ± 0.0015

children per marriage than when the marriages were less positively assortative. It is evident that in the case of southern nonwhites, the tendency toward assortative marriage for educational attainment and the association of assortative marriage with greater reproduction has weakened, compared with the other groups. Educational attainment may tend to keep some groups stratified if the educational level has heritability of a significant amount.

Nonrandom marriages between ethnic groups brought together by migration to Australia were tabulated by Hatt and Parsons (1965) on the basis of surname origins. Marriages were more often between individuals carrying the same surname derived from the same ethnic group than from different groups (English, Scottish, or Irish origin). (See Exercise 13 for data and calculating the departure from expectation based on random mating.) Assortative mating based on cultural origin is not likely to have much heritability, although any heritable differences that do exist between these groups might be perpetuated in this way.

Finally, it may be pointed out that patterns of marriage may be assortative for handicapped people, especially those with similar defects. A striking social influence occurs among deaf persons, who tend to marry others with the same defect (see Ehrman, 1972*a*, pp. 128–131 for a summary). However, there is little or no evidence for people with mental disorders to marry assortatively. Garrison, et al. (1968), in surveying relatives of psychotic patients (scoring husbands and wives for psychosis, psychoneurosis, or personality disorders), found that their *G* statistic was 0.04—not significantly different from zero. Also, they cite Kallmann (1953), who did not find evidence for assortative marriage of schizophrenics. However, Reed and Reed (1965) pointed out that for about 600 mental retardates who married, 9.4 percent of their spouses were also retarded, or six times the expectation from random marriages.

NEGATIVE ASSORTATIVE MATING

When mating is disassortative, unlike phenotypes mate. Let us consider a single pair of alleles without dominance where three genotypes are distinguishable; if exclusively disassortative, only the matings of $D \times H$, $H \times R$, and $D \times R$ will occur. It is apparent,

TABLE 10-4 *Negative assortative mating with single pair of alleles, three genotypes phenotypically distinguishable*

A If parent generation has $\frac{1}{4}D$, $\frac{2}{4}H$, $\frac{1}{4}R$, then matings occur as follows

	$\frac{1}{4}D$	$\frac{1}{2}H$	$\frac{1}{4}R$
$\frac{1}{4}D$	—	$\frac{1}{8}DH$	$\frac{1}{16}DR$
$\frac{1}{2}H$	$\frac{1}{8}HD$	—	$\frac{1}{8}HR$
$\frac{1}{4}R$	$\frac{1}{16}RD$	$\frac{1}{8}RH$	—

Total (T) $= 1 - D^2 - H^2 - R^2 = \frac{5}{8}$

B

Mating (Ignoring Reciprocals)	Frequency	Progeny D	Progeny H	Progeny R
$D \times H$	$\frac{2}{8}(\frac{8}{5}) = \frac{2}{5} = 2DH/T$	$\frac{1}{5} = DH/T$	$\frac{1}{5} = DH/T$	—
$D \times R$	$\frac{2}{16}(\frac{8}{5}) = \frac{1}{5} = 2DR/T$	—	$\frac{1}{5} = 2DR/T$	—
$H \times R$	$\frac{2}{8}(\frac{8}{5}) = \frac{2}{5} = 2HR/T$	—	$\frac{1}{5} = HR/T$	$\frac{1}{5} = HR/T$
	Totals	$\frac{1}{5}$	$\frac{3}{5}$	$\frac{1}{5}$

C Successive Generations

Generation	D	H	R	T	H/T
0	0.250	0.500	0.250	0.625	0.800
1	0.200	0.600	0.200	0.560	1.0714
2	0.2143	0.5714	0.2143	0.5816	0.9825
3	0.2105	0.5789	0.2105	0.5762	1.0048
4	0.2115	0.5769	0.2115	0.5777	0.9987
⋮	⋮	⋮	⋮	⋮	⋮
∞	0.2113	0.5774	0.2113	0.5774	1.0000

since the first two types of mating will not change the distribution in their progeny ($AA \times Aa$ and $Aa \times aa$ lead to equal frequencies of the same genotypes), that the $D \times R$ matings will increase the amount of heterozygosity in the following generation. In Table 10-4A, assuming a parent generation of $\frac{1}{4}AA$: $\frac{2}{4}Aa$: $\frac{1}{4}aa$, we calculate proportional matings as indicated. Three types of mating do not occur, and relative frequencies of the other three types of mating must be "normalized" (shown in Table 10-4B) by dividing by the total (T). The successive generations can then be calculated by noting that the homozygotes in successive generations become proportional to the ratio of H/T, as noted by these relationships:

$$D_t = D_{t-1}\left(\frac{H_{t-1}}{T_{t-1}}\right)$$

$$R_t = R_{t-1}\left(\frac{H_{t-1}}{T_{t-1}}\right)$$

while the heterozygotes can always be obtained by subtraction:

$$H_t = 1 - D_t - R_t$$

Consequently, if the same mating system continues, the ratio D/R does not change $D_t/R_t = D_{t-1}/R_{t-1}$), and the H/T ratio can be used to calculate the next generation's homozygote values:

$$\frac{H_{t-1}}{T_{t-1}} = \frac{D_t}{D_{t-1}} = \frac{R_t}{R_{t-1}} \qquad (10\text{-}7)$$

Then D or R is easily calculated, as given in Table 10-4C. For all practical purposes, it is evident that the genotypes come rapidly to an equilibrium with higher heterozygosity than initially present. This result is generally true for negative assortative mating, whether complete or incomplete.

If dominance makes D indistinguishable from H, then the complete negative assortative outcome is very simple: only dominant × recessive phenotypes would mate. By using the same initial genotype frequencies as before for comparison, we can see that the mating system immediately produces an equilibrium. If $D + H = \frac{3}{4}$ and $R = \frac{1}{4}$, mating is entirely $\{D + H\} \times R$. Then $D \times R = 2(\frac{1}{4})(\frac{1}{4}) = \frac{1}{8}H$ progeny and $H \times R = 2(\frac{1}{2})(\frac{1}{4}) = \frac{1}{8}H + \frac{1}{8}R$ progeny, or total $\frac{2}{3}H{:}\frac{1}{3}R$ progeny. From then on, no dominant homozygotes occur; all matings will be $H \times R$, yielding $\frac{1}{2}H : \frac{1}{2}R$ as long as that system of mating continues.

In nature, at least a few authenticated cases of this type of mating system can be identified. Heterostyly in many flowering plants (primrose *Primula vulgaris*, loosestrife *Lythrum salicaria*, and shepherd's purse *Capsella grandiflora*) is the condition in which the length of style is controlled by a single genetic unit. In "thrum" flowers, the style is so short that the stigma is halfway down the corolla tube and the anthers are confined to the top of the tube. In "pin," the style bears the stigma at the mouth of the corolla tube with the anthers half-way down (see Ford, 1964, for a summary of the genetics and incidence of this condition in populations of *Primula*, in particular). The genetic unit, which consists of a "supergene" of at least seven tightly linked loci controlling anther height, style length, and pollen tube growth, can be represented as *S*- for thrum (dominant) and *ss* for pin. The population then reproduces by *Ss* × *ss*; *SS* plants occur very seldom if at all because pollen with the *S* allele from thrum plants does not grow a pollen tube well in thrum styles. Except for occasional crossovers within the *S* supergene, at less than one percent of segregants, populations of these plants contain a nearly equal frequency of *Ss*:*ss* plants. The latter are often in excess, however, because *ss* pin plants can self-fertilize.

Mating can be disassortative for genotype in the cases of genetic self-sterility in many flowering plants (discovered in *Nicotiana* by East and Mangelsdorf, 1925; described for *Oenothera organensis* by East in 1939, and for red clover by Williams in 1947; summarized by Lewis, 1949, and by Crowe, 1964). Self-sterility alleles prevent selfing because of a physiological growth inhibition of any pollen grain growing on the stigma of its parent plant; pollen of any allelic type (S_1) can form a tube only on stigmas of plants with non-S_1 constitution (S_2S_3, for example). Thus, a large allelic series must exist in the population to keep pollen loss to a minimum. Such a population, of course, must maintain at least three alleles, and their zygote frequencies would be likely to stabilize at one-third each, consisting entirely of heterozygotes.

Frequency-dependent mating systems in which rare phenotypes may gain an advantage in mating over more common phenotypes can be considered effectively as negative assortative systems (see Example 10-4). They can lead to increased heterozygosity over random mating, and balanced polymorphism may result. It is more proper to consider these nonrandom mating systems under selection, however, because the criterion of differential reproduction usually fits, and a change in the gene pool for the next generation ensues (see Chapter 15).

It must also be pointed out that the sex-determining cytogenetic mechanism of $X:Y$ chromosomes and heterogametes in bisexual populations ensures that all matings are between XX and XY individuals. The entire genetic sex determination in such populations can be considered an equilibrium condition for negative assortative mating.

Beyond these instances, there is little evidence for negative assortative mating in natural populations (see Example 10-4 for a case in moths and see Thorneycroft, 1968, for a case in white-throated sparrows). In fact, it is quite likely that it is more extensive than has appeared until now. In contrast with the tendency under positive assortative mating, the major effect will be to maintain genotypes at equilibrium without fixation and to lower phenotypic variance (lower extremes' frequencies) or to stabilize heterozygosity at a frequency higher than that under random mating. Genetic potential will then be increased or stabilized by storage of genetic variability in heterozygotes.

EXAMPLE 10-4

DISASSORTATIVE MATING IN MOTHS

One of the few examples of negative assortative mating recorded in the experimental genetics literature is that described by Sheppard (1951, 1952, 1953) in the diurnal scarlet tiger moth *Panaxia dominula* (L.), taken from a population at Cothill, Berkshire, England. Fisher and Ford carried out much of the original investigation on this population, in particular from 1939 to 1947, describing population size and the frequencies of a genetic unit (probably a supergene) known as the *medionigra* unit since the common variant in the population is the heterozygote known as variety *medionigra* Cockayne. Homozygotes for the unit are known as var. *bimacula* Cockayne. Frequencies of this genetic unit were high (about 10 percent) in the earlier years of collections (1940); then its frequency diminished (down to 6 percent in 1946 and less than 4 percent subsequently until 1952).

On a suggestion from Kettlewell that *medionigra* individuals tended to mate nonrandomly, Sheppard made tests of matings between all three genotypes by setting up cages with three individuals in each, two of one sex and different genotypes with a single individual of opposite sex, so that the single individual had a "choice" of mates including one the same genotype as itself. If there was no assortative pairing, there would be an even chance of either mating taking place. Data for a total of 150 mated individuals are given in Table A, arranged by the sex with "choice."

With female "choice" the heterogamy is striking and highly significant (rows 1 and 2): heterogamous matings total 61–29 homogamous, a ratio of greater than 2:1. The consistency of the trend is most apparent when we note that the female partner seems to prefer a partner of opposite type.

With male "choice," in row 3 (*dominula* male × *dominula* and *medionigra* females) the same trend occurs. In row 4, *medionigra* females have a slight edge over *dominula* females, and it appears that *medionigra* females simply mate more frequently than *dominula*. The tests with *bimacula* females are too small to be utilized.

TABLE A *Results of first matings in single sex "choice" experiments with three varities of* Panaxia dominula *(modified from Sheppard, 1952) (dom. = var. dominula, medio. = var. medionigra, bimac. = var. bimacula)*

Female "Choice"

		♀dom.	♀medio.
(1)	♂ dom.	11	14
	♂ medio.	22	13
		♀medio.	♀bimac.
(2)	♂ medio.	2	15
	♂ bimac.	10	3

Male "Choice"

	♀dom.	♀medio.	♀dom.	♀bimac.
(3) ♂ dom.	8	20	2	1
	♀dom.	♀medio.	♀medio.	♀bimac.
(4) ♂ medio.	12	14	2	0
			♀medio.	♀bimac.
(5) ♂ bimac.	—	—	0	1

In later tests for fertility, Sheppard (1953, Table 4) indicated that *medionigra* females were equal to *dominula* in percent fertile matings (about 72 percent for each of 83 matings with the former variety and for 45 with the latter), but *medionigra* females laid a slightly greater average number of fertile eggs than did *dominula* (about 307 eggs per fertile clutch for the former variety to 303 for the latter). *Medionigra* males, however, produced 20 percent fewer fertile matings than *dominula* males.

If the tendency for heterogamy is actually controlled by the female of this species in wild populations, it would certainly be an important factor in maintaining the polymorphism observed. Sheppard stated (1952) that "preliminary observations suggest that the disassortative pairing is controlled by the female. On several occasions she has been observed to reject actively a male of her own genotype, and then has at once accepted a male of a different form. The observations suggest that the courtship between like genotypes is longer than between unlike forms. . . . Moreover, because the males, but not the females, mate more than once, the *medionigra* males will have a better chance of mating than the wild type males if the gene frequency is low." See also Ford (1964) and Sheppard and Cook (1962) for summaries of work on *Panaxia dominula*. The latter reference lists a total of 199 choice matings in which 126 resulted in unlike matings and only 73 in matings of like genotypes, a highly significant preponderance of disassortative matings.

EXERCISES

1. The observed data for Wallace's experiment in Example 10-1 on mating proportions between *ST* and *SR* males and females gave the following:

♀ ×	♂	No. Mated/No. Put In
ST	*ST*	78/91
SR	*ST*	49/85
ST	*SR*	38/85
SR	*SR*	28/86

   (a) Construct a 2 × 2 table of these data and calculate a contingency chi-square. Are the four observed types of mating randomly distributed? Why? (Note: to simplify the calculation, readjust for unequal numbers put in.)
   (b) What is the isolation index value I?
   (c) Which is more likely, preferential mating or random mating assortment with differential mating activity? Explain your decision.

2. Given an initial population with genotype frequencies of $AA = 0.1$, $Aa = 0.4$, $aa = 0.5$, assume complete dominance of A- allele and completely positive assortative mating.
   (a) What will be the progeny genotype frequencies for two generations?
   (b) From Table 10-2, show that $(H/2)^2 + R(1 - R) = q^2 + R(p - q)$ from your data and then algebraically by converting H and R symbols into p or q equivalents and simplifying.
   (c) Also verify that the total progeny proportions (bottom of Table 10-2) add up to 1.00 by using your data and also algebraically by using the relationship $1 - R = D + H$.

3. (a) Calculate the values of H over six generations for the case in the previous exercise (0.10, 0.50, 0.40). Then do the same for an initial genotype set of (0.40, 0.50, 0.10), assuming complete dominance and positive assortative mating, as before.
   (b) Make a graph of change in H for initial q values you obtained in part (a). If the initial genotype frequencies conformed to the square law, how would the change in H appear on the graph?
   (c) Compare the proportional loss in H for the two initial frequencies in (a) with that for the initial random-mating frequencies. Is the drop in H greater from square-law frequencies or from the nonrandom frequencies of part (a)?

4. (a) Given initial square law conditions with $p = q = 0.5$, complete dominance for the A-allele, and symmetrical positive assortative mating at 50 percent. Verify the sequence of H given in the appropriate column of Table 10-3 over four generations (use formula (10-4)). Verify the equilibrium H value using (10-5).
   (b) In Table 10-1B where the positive symmetrical case (1) is incompletely assortative, assume A is a dominant phenotype but completely heterozygous (Aa) while B is

the recessive (aa). What would be the sequence of H over four generations assuming the same amount of assortative mating takes place each generation? What will be the final value of H?

5. (a) Go back to Chapter 8. If there is no dominance, and matings are positively assortative, would the form of Tables 8-3 and 8-4 apply? By using those tables, see whether the sequence of H given in Table 10-3 for the cases of "no dominance" with one locus ($k = 1$) for $m = 1, 0.80$, and 0.50 would be similar to the use of s in Chapter 8, at the same values. (Assume square-law conditions in G_0.)
   (b) Do you suppose the case of two loci in Table 10-3, under $m = 1.00$ complete assortative mating, would be similar in sequence to that with selfing? Why?
6. (a) Derive the two-locus case as follows for complete positive assortative mating where phenotypes are due to simple additive action ($AABb = AaBB$, etc.) so that there are five phenotypes. Assume $p = q = u = v = 0.5$ and square-law conditions at G_0. Verify that G_1 genotypes will occur as follows:

	BB	Bb	bb	Totals
AA	13	10	7	30
Aa	10	16	10	36
aa	7	10	13	30
Totals	30	36	30	96

   (Hint: First arrange the five phenotypes in order with their initial frequencies, then allow matings within each phenotype (subset), using gametic matrix (5-2) as an aid.)
   (b) What is the value of H? Is it the same as that given in Table 10-3 for the appropriate column?
   (c) Find the G_2 genotypes' frequencies similarly. What is happening to the corner genotype frequencies in the zygotic matrix? How different is this result from the outcome of selfing?
7. In formula (10-6) s is the correlation between nonhomologous genes in the same gamete of the parents of any given generation. To verify that $s = \frac{1}{8} = 0.125$, we may proceed as follows. Given the phenotypic effect of each dominant allele (A, B) to be 1.00, with recessive allele to be zero in additive value, than their correlation (r_{AB}) is given by covariance between A and B divided by the geometric mean of their variances in G_1. Let A be the X variate and B be the Y, then

$$r_{AB} = \frac{\sum f(XY) - (\sum fX)(\sum fY)}{\sqrt{\sigma_X^2 \cdot \sigma_Y^2}}$$

In G_1 the following will be the case:

Gamete	X_A	Y_B	f
AB	1	1	0.28125
Ab	1	0	0.21875
aB	0	1	0.21875
ab	0	0	0.28125

(a) First verify that the covariance = 0.03125.
(b) Verify that the variance of X_A = variance of Y_B = 0.25.
(c) Find $s = 0.125$ as given on page 303.
(d) Note that F as defined by Wright is the correlation *between uniting gametes.* Verify $F = 0.25$ in G_1 zygotes using your data from Exercise 6.

8. Fisher (1918) found that under assortative mating, the correlation between mates (m) inflates the correlation between a single parent and offspring (quantitative traits r_{PO}) (formula (7-7)) as follows:

$$r_{PO} = \left(\frac{1+m}{2}\right)(h^2), \qquad \text{where } h^2 = \frac{\sigma_A^2}{\sigma_P^2}$$

For human stature, Pearson and Lee (1903) found for English university students $r_{PO} = 0.5066$, while the correlation between their parents for stature was $m = 0.2803$. Since under a completely random-mating system $r_{PO} = \frac{1}{2}h^2$ (7-8), it is clear that the observed parent-offspring correlation might well be too large due to assortative mating for stature. Using Fisher's formula, estimate the heritability of stature.

9. It was shown in Exercise 7, Chapter 7, that a parent has an expected phenotypic linear value Y_L given by the product of the heritability × his phenotypic observed value $Y_L = h^2(Y)$. When there is phenotypic correlation between mates (m), the other parent has an expected phenotype of $Y'_L = m(h^2)(Y)$. Then the mean of the two parents is

$$\bar{Y} = \frac{Y_L + Y'_L}{2} = \frac{h^2 Y(1+m)}{2}$$

How would positive assortative mating change the regression of offspring on means of two parents as given in formula (7-12)?

10. A pitfall that may lead to a spurious recording of assortative mating results from combining data from two or more heterogeneous groups, a procedure that could produce a correlation between mates. Suppose there are populations I and II between which there is complete isolation, but they occupy the same area and ordinarily cannot be distinguished (like the thin-and thick-skinned populations of Example 2-3). Assume population I has $p_A = 0.6$, $q_a = 0.4$, while population II has $p_A = 0.2$, $q_a = 0.8$ with random mating within each population. Furthermore, assume genotypes are additive so that $AA = +1$, $Aa = 0$, and $aa = -1$ in phenotype.
(a) Set up a table with nine kinds of matings in the following manner:

Matings		Population I			Population II			
♀♀	♂♂	X	Y	f	X	Y	f	Average Frequency
AA	AA	1	1	0.1296	1	1	0.0016	0.0656
AA	Aa	1	0	0.1296	1	0	0.0028	0.0928
⋮	⋮	⋮	⋮	⋮	⋮	⋮	⋮	⋮

(b) Determine the correlation between mates within populations I and II. Verify that $m = 0$ in each population.

(c) If we assume the two populations to be equally frequent, use the average frequencies of mating over both populations, not realizing that they are reproductively isolated. What is the correlation between mates you find? Verify that m (spurious) $= 0.20$.

(d) In Example 10-2, Parsons found about 20 percent correlation between flies mating on the basis of chaeta number. Do you think this was a case of lumping heterogeneous groups together? Why?

11. In the Table 10-4 sequence of negative assortative mating genotypes,

(a) Can formula (10-7) be used to calculate the relation between G_0 and G_1? Do you use T before or after mating takes place? Why or why not? What exactly is it that T (total) refers to in Table 10-4A?

(b) Suppose the initial population has $D = 0.36$, $H = 0.48$, and $R = 0.16$ and all matings are negatively assortative. Find the genotype frequencies over the next four generations and then estimate their ultimate frequencies on the assumption of continued negative assortative mating.

(c) How does the limiting value of H compare with the initial H_0?

12. In Table 10-4C, the total (T) each generation oscillates similarly to the successive terms in the recurrent series of sex-linked allele frequencies in Exercise 3, Chapter 4. From the general expression for any term in a sequence, determine the limiting value for the T series by first finding two component geometric series with the same scales of relation.

13. Hatt and Parsons (1965) tabulated the frequencies of marriages between ethnic groups that had immigrated to Australia; individuals were assigned to ethnic group on the basis of surname origins. Six possible types of marriage between English (E), Scottish (S), and Irish (I) surnames are given with their frequencies below.

Marriage	Observed Frequency	Prob. Random	Expected Frequency
$E \times E$	141	p^2	138.785
$E \times S$	148	$2pq$	151.354
$E \times I$	100	$2pr$	——
$S \times S$	50	q^2	——
$S \times I$	41	$2qr$	——
$I \times I$	26	r^2	——
Totals	506	1.00	506.000

(a) Fill in the expected frequency blanks for the remaining 4 marriage types.

(b) What is the rationale for letting a nationality (ethnic) group be tested as if due to single alleles at one genic locus in estimating expected probabilities?

(c) Which marriage types are in excess of expectation based on random mating? Which deficient? Is there any evidence for assortative mating?

11

GENERAL TRENDS UNDER INBREEDING

In the past three chapters, we have examined populations in their entirety by considering them as broken up into an infinity of isolates. Isolate sizes were $N = 1$ under complete selfing, $N = 2$ under sib mating, $N = 4$ under double first cousin mating, and so forth for more distant relatives. For regular systems in which all matings are confined to isolates, with no outcrossing between isolates, allelic frequencies for the population as a whole are not expected to change. These considerations hold under positive phenotypic assortative mating also, as long as it is complete and symmetrical.

It is important to reflect briefly on the general trends for genotype distribution (heterozygote versus homozygote frequencies)—especially the rate of change per generation—first in the limiting case where only the most distant relatives may be mating and then in the average case for a population in which all types of mating can occur including selfing.

REMOTEST RELATIVES ONLY

Heterozygosity recurrence statements for close relative mating systems were derived in Chapters 8 and 9 and are summarized below to illustrate the general trend.

Relationship	H_t Recurrence
Selfing	$\frac{1}{2}H_{t-1}$
Sib mating	$\frac{1}{2}H_{t-1} + \frac{1}{4}H_{t-2}$
Double first cousins	$\frac{1}{2}H_{t-1} + \frac{1}{4}H_{t-2} + \frac{1}{8}H_{t-3}$
Quadruple second cousins	$\frac{1}{2}H_{t-1} + \frac{1}{4}H_{t-2} + \frac{1}{8}H_{t-3} + \frac{1}{16}H_{t-4}$

However, each of these recurrence statements is expressed in terms of additive fractions of previous generations' heterozygosity. We need to express the change in heterozygosity ΔH. We begin by expressing the H recurrence as a *difference*, which can easily be done merely by subtracting successive terms as follows.

In the case of selfing, two successive generations are expressed as

$$\begin{cases} H_t = \frac{1}{2}H_{t-1} \\ H_{t-1} = \frac{1}{2}H_{t-2} \end{cases}$$

If we first divide the lower by 2 and then subtract we obtain

$$H_t = H_{t-1} - \frac{1}{4}H_{t-2}$$

The same simultaneous solution of two successive generations for any of these regular systems of mating can be rewritten as follows (for H_t):

Selfing $\qquad H_{t-1} - \frac{1}{4}H_{t-2}$

Sib mating	$H_{t-1} - \frac{1}{8}H_{t-3}$
Double first cousins	$H_{t-1} - \frac{1}{16}H_{t-4}$
Quadruple second cousins	$H_{t-1} - \frac{1}{32}H_{t-3}$

From these considerations, Table 11-1 can be built to see the relationship between isolate size and the rate of heterozygosity loss. Because the isolate size is a multiple of 2, we can visualize a population with only the most remote relatives being parents as a limiting case (k very large). The loss in H will have the coefficient $(\frac{1}{2})^{2+k}$, which is $(\frac{1}{4})$ times the reciprocal of N (since $N = 2^k$). The final expression for change in H, namely $\Delta H = -(1/4N)$. $H_{t-(2+k)}$, tells us the very slowest rate by which heterozygosity decreases each generation. When N becomes very large, the loss each generation will be almost the same as that of the previous generation; so that even when $(2 + k)$ generations may be very far back, the heterozygosity may be nearly the same as it was just one generation back. The final expression becomes approximately

$$H_t = H_{t-1} - \frac{1}{4N} \cdot H_{t-1} = H_{t-1}\left(1 - \frac{1}{4N}\right) \tag{11-1}$$

(recurrence in H when only remotest relatives are mating).

Another way of expressing the change in heterozygosity is by percentage of H retained per generation. It is easy to see that for selfing the loss is 50 percent of the heterozygosity in any generation. It is not so easy, however, to see the limiting percentage lost or retained with sibs and more distant relatives. If we examine the recurrence of H for sibs (9-3), the coefficients are as follows:

H (sibs)	$\frac{1}{1}$	$\frac{2}{2}$	$\frac{3}{4}$	$\frac{5}{8}$	$\frac{8}{16}$	$\frac{13}{32}$	$\frac{21}{64}$
Generations (t)	0	1	2	3	4	5	6

TABLE 11-1 *Relationship between isolate size and rate of change in heterozygosity. Matings exclusively by the types of relatives indicated (from Wright, 1921, 1951)*

Relatives Mating	Isolate Size (N)	Heterozygosity in tth Generation	Limiting Percentage Change $(H_t/H_{t-1}) = \lambda$ After Several Generations
Self-fertilization	$1 = 2^0$	$H_{t-1} - (\frac{1}{2})^2 H_{t-2}$	0.500
Sibs (full)	$2 = 2^1$	$H_{t-1} - (\frac{1}{2})^3 H_{t-3}$	0.809
Double first cousins	$4 = 2^2$	$H_{t-1} - (\frac{1}{2})^4 H_{t-4}$	0.920
Quadruple second cousins	$8 = 2^3$	$H_{t-1} - (\frac{1}{2})^5 H_{t-5}$	0.965
Octuple third cousins	$16 = 2^4$	$H_{t-1} - (\frac{1}{2})^6 H_{t-6}$	0.983
More distant cousins	$N = 2^k$	$H_{t-1} - (\frac{1}{2})^{2+k} \cdot H_{t-(2+k)}$ $= H_{t-1} - \frac{1}{4N} \cdot H_{t-(2+k)}$	$\frac{4N-1}{4N}$ (approx.)

When we consider the proportion of H retained, the rate of change in percentage of H_t will oscillate about some limiting value, as follows:

Letting λ = percent of each previous generation H (that is, $\lambda = H_t/H_{t-1}$) we would have, for example, at $G_3: \lambda = (\frac{5}{8})(\frac{4}{3}) = \frac{5}{6} = 0.833$. We may then list the λ values for successive terms of H under sib mating as follows:

λ	1	1	0.75	0.833	0.800	0.8125	0.8077	...
G	0	1	2	3	4	5	6	...

The limiting value of λ can be obtained from assuming that it becomes a constant ratio ultimately in the recurrent sequence of H. In that case, $\lambda_{t-1} = \lambda_t = \cdots$. Because $H_t = \frac{1}{2}H_{t-1} + \frac{1}{4}H_{t-2}$, dividing through by H_{t-1}, we would obtain $\lambda = \frac{1}{2} + 1/4\lambda$. Therefore, this expression becomes equivalent to the quadratic equation $\lambda^2 - \lambda/2 - \frac{1}{4} = 0$, which has a positive root $\lambda = 0.8090$, which is the limiting percentage of H retained each generation under sib mating. The complement $1 - \lambda$, then, is the percentage loss in H each generation.

In turn, the limiting equation for double first cousin mating is a cubic: $\lambda^3 - \lambda^2/2 - \lambda/4 - \frac{1}{8} = 0$, whose solution for largest positive root is $\lambda = 0.91964$. Other limiting values are given in the right column of Table 11-1 (see Appendix A-13 on recurrent series).

Thus, an inevitable though very slow loss of heterozygosity will take place even when extreme inbreeding is avoided.

AVERAGE POPULATION ("IDEALIZED") WITH ALL RELATIVE MATINGS POSSIBLE

Progeny can be produced with all possible relationships between uniting gametes, including selfing, with organisms such as monoecious plants that are wind pollinated or with hermaphroditic animals such as certain marine forms that may shed sex cells into the sea where fertilization takes place. A population of such an organism where opposite sex gametes have equal chances of fusing with each other irrespective of relationship among donors of the gametes illustrates the average rate of change in heterozygosity in the clearest and simplest way; we call such a population structure "idealized." We may then reason out the probabilities for uniting gametes to have identical genes and deduce the average change in heterozygosity expected by the following analysis.

By referring to Table 11-2, we may visualize an infinite pool of gametes to which each parent (N = total parents) contributes equally. If each parent is heterozygous for a different

TABLE 11-2 *Diploid monoecious (hermaphroditic) species population of size = N with gametic pool = 2N*

Individuals	First	Second	Third	...	Nth
Male gametes	A_1A_2	A_3A_4	A_5A_6	...	$A_{2N-1}A_{2N}$
Female gametes	A_1A_2	A_3A_4	A_5A_6	...	$A_{2N-1}A_{2N}$

A_i = any allele of a locus.

pair of alleles, as implied by the numbering in the table, then it ought to be evident that any random pair of opposite-sex gametes will have $1/2N$ chance for carrying identical genes. Obviously, two gametes have $(1 - 1/2N)$ chances for carrying nonidentical genes. However, if there has been any previous inbreeding, some of the latter pairs $(1 - 1/2N)$ in the population may be identical with one another; the frequency for this is symbolized for the parent generation as F_{t-1}. Consequently, a random pair not made identical by the present fusion of gametes but by a previous inbreeding will be identical $(1 - 1/2N)F_{t-1}$ proportion of the time.

$$\text{In summary, } P\,\{\text{random pairs to be identical}\} = \frac{1}{2N}$$

$$P\,\left\{\begin{array}{l}\text{remainder pairs to be identical}\\ \text{because of previous inbreeding}\end{array}\right\} = \left(1 - \frac{1}{2N}\right)F_{t-1}$$

The sum of these two probabilities is the inbreeding coefficient for the tth generation of the population:

$$F_t = \frac{1}{2N} + \left(1 - \frac{1}{2N}\right)F_{t-1}$$

On rearrangement, this gives the result in terms of an increment in homozygosity:

$$F_t = F_{t-1} + \frac{1 - F_{t-1}}{2N} \tag{11-2}$$

Students may think of F_{t-1} as symbolizing the proportion of actual identical homozygotes among the parents, but that aspect does not complete the meaning of the symbol. From Chapter 9 it should be recalled that under any regular system of mating, the probability for transmission of identical genes to single-parent individuals is expected to be the same as that for transmission to two separate individuals ($A \to P = A \to Q$ in Figure 9-1B, for example). The concept to keep in mind is the "infinite pool of gametes" in the model; it makes no difference whether two genes identical through previous inbreeding are contributed to the pool, each via separate or both via a single individual; that is, $F_{t-1} = f_{AB} = f_{AA} = f_{BB}$, to use the symbols of Chapter 9, where A and B are separate individuals in the $t - 2$ generation.

The right-hand increment in (11-2) represents the fraction of nonidentical genes that become newly "inbred" each generation, symbolized as ΔF.

$$\Delta F = \frac{1}{2N} = \frac{F_t - F_{t-1}}{1 - F_{t-1}} \tag{11-3}$$

(rate of inbreeding increment as a proportion of panmictic fraction when all types of mating exist). The middle expression $(1/2N)$ is exact for the idealized model population described. Theoretically, this rate of homozygosity increment will become a constant after some generations if the populational model remains the same. Of course, under pure selfing, (11-3) reduces to the familiar rate of $\Delta F = \frac{1}{2}$ when $N = 1$.

In order to see how much change may take place over many generations, it will be simpler to calculate time with the complementary expressions for heterozygosity. Recalling from (8-2) that $1 - F_t = H_t/H_0$ and $F_t = 1 - H_t/H_0$ in any generation, we may substitute and work out the decrease expected in heterozygosity (nonidentical genotypes) from (11-2).

$$1 - \frac{H_t}{H_0} = 1 - \frac{H_{t-1}}{H_0} + \frac{H_{t-1}}{2NH_0}$$

Then,

$$H_t = H_{t-1} - \frac{1}{2N} \cdot H_{t-1} = H_{t-1}\left(1 - \frac{1}{2N}\right) \tag{11-4}$$

(recurrence in H in average idealized population). Compared with (11-1), we see that random union of gametes in a population of size N brings a rate of decrease in heterozygosity twice that when inbreeding was minimized: $-(1/2N)H$ compared with $-(1/4N)H$.

The relative amount of heterozygosity retained as a percent of each generation H value is then, since $\lambda = H_t/H_{t-1}$,

$$\lambda = 1 - \frac{1}{2N} = 1 - \Delta F \tag{11-5}$$

(percentage retained of H per generation in an idealized population).

The ratio H_t/H_0, called the *panmictic index* by Wright (1951), expresses the amount of heterozygosis relative to that in the foundation stock and is usually symbolized as $P = 1 - F$. Then, (11-2) can be rewritten in terms of the panmictic index.

$$1 - P_t = 1 - P_{t-1} + \frac{P_{t-1}}{2N}$$

On rearrangement, this gives

$$P_t = P_{t-1}\left(1 - \frac{1}{2N}\right) \tag{11-6}$$

or, more generally,

$$P_t = P_{t-1}(1 - \Delta F)$$

This statement is the equivalent of (11-4) but is simply expressed as heterozygosity relative to the panmictic original state. For each generation with respect to the previous one, the same relationship holds, so that putting the equivalent of two generations back we have

$$P_t = P_{t-2}\left(1 - \frac{1}{2N}\right)^2$$

and going back to the base population,

$$P_t = P_0\left(1 - \frac{1}{2N}\right)^t \tag{11-7}$$

The panmictic original population, by definition, has a panmictic index of $P_0 = 1$, so that for the idealized population,

$$\frac{H_t}{H_0} = \left(1 - \frac{1}{2N}\right)^t$$

or

$$H_t = H_0\left(1 - \frac{1}{2N}\right)^t \tag{11-8}$$

In order to keep these concepts in mind, the student should complete Exercise 2, which compares P, F, ΔH, λ, and ΔF. It is especially important to see the relationships $P = 1 - F$ and $\Delta F = 1 - \lambda$.

AVERAGE POPULATION WITHOUT SELFING

In most animal populations and in many plants, selfing cannot take place. We then must assume that gametes may combine at random but no two come from any single parent to a single offspring. In order words, the closest relatives possible could be either sibs or half-sibs in the population. Two gametes which can fuse in zygote formation then will have the following two possible probabilities for doing so (refer to Figures 9-1B and 9-3B):

$$P\{\text{two gametes come from one individual in } (t-2) \text{ generation}\} = \frac{1}{N}$$

$$P\{\text{two gametes come from different individuals in } (t-2) \text{ generation}\}$$

$$= 1 - \frac{1}{N} = \frac{N-1}{N}$$

In the first case, the probability for transmission of identical genes is $f_{AA} = (1 + F_{t-2})/2$, where A is any ancestor in the $(t - 2)$ generation. In the second case, the transmission probability is $f_{AB} = F_{t-1}$, where A and B are different individuals in the $(t - 2)$ generation. The total probability for identical genes to occur in the t generation will be as follows:

$$F_t = \frac{1}{N}\left(\frac{1 + F_{t-2}}{2}\right) + \frac{N-1}{N} F_{t-1}$$

This reduces to

$$F_t = \frac{1 - 2F_{t-1} + F_{t-2}}{2N} + F_{t-1} \tag{11-9}$$

Because $F_t = 1 - H_t/H_0$, then

$$H_t = H_{t-1} - \frac{2H_{t-1} - H_{t-2}}{2N}$$

$$H_t = \frac{N-1}{N} H_{t-1} + \frac{1}{2N} H_{t-2} \tag{11-10}$$

(recurrence of H when selfing is avoided). When $N = 2$, as it does for sibmating, (11-10) reduces to the familiar formula (9-3).

For the ultimate constant rate of retention in H per generation, we divide (11-10) through by H_{t-1} to obtain $\lambda = (N - 1)/N + 1/(2N\lambda)$, which becomes $\lambda^2 - ((N - 1)/N)\lambda$

$-1/(2N) = 0$. The applicable positive root then is

$$\lambda = \frac{N - 1 + \sqrt{N^2 + 1}}{2N} \tag{11-11}$$

(ultimate rate of retention in H per generation). A good approximation for ΔF is

$$\Delta F = 1 - \lambda = \frac{1}{2N + 1} \tag{11-12}$$

(approximate rate of increase in F). For populations with a moderate to large size ($N > 25$), we may use $\Delta F = 1/(2N)$.

UNEQUAL SEXES AND INBREEDING EFFECTIVE NUMBER

Results in the previous section are based on the assumption that the N parents are equally divided between males and females ($N♂ = N♀ = N/2$). This situation is still idealized because even in monogamous species it often happens that more than one mate may occur for each parent, and in polygamous species several females may be mated to a single male, or in a few cases a single female to more than one male. The "effective" number of parents (N_e)—that is, those individuals contributing gametes to the next generation—will not be a simple arithmetic average or total of the parental number. We must consider the following facts in order to estimate N_e.

If the population size is constant from generation to generation ($N_0 = N_1 = N_2 \cdots$), then each parent must contribute two haploid doses on the average to the next generation. If the sexes are unequal, of course, the lesser sex will contribute more than two on the average while the greater sex will contribute less, but the probability for any two genes to be contributed to the next generation will be $(\frac{1}{2})(\frac{1}{2}) = \frac{1}{4}$. The probability that both genes have been contributed by the same male is $(1/N♂)(\frac{1}{4})$. Similarly, for any individual female to contribute two genes is $(1/N♀)(\frac{1}{4})$. Therefore, the probability that any two genes have come from the same individual regardless of sex is the sum: $1/(4N♂) + 1/(4N♀) = 1/N_e$ (harmonic mean of the sexes). Solving for the effective number, we have

$$N_e = \frac{4N♂ \cdot N♀}{N♂ + N♀} \tag{11-13}$$

where $N♂$ = number of actual parent males and $N♀$ = number of actual parent females.

If the sexes are equal ($N♂ = N♀ = N/2$), then (11-13) reduces to N. In Exercise 6, it is apparent that the effective population size depends on how close to equality the sexes are. When near equality (monogamy), N_e is close to the total of the two, but with smaller and smaller minority of one sex (males usually), the N_e approaches the arithmetic average of each sex's number and then falls below the average approaching the lower number.

Under conditions of inequality between the sexes, the effective number N_e must be substituted into (11-10) in place of N to estimate the change in H per generation, or into (11-11) for rate of loss in H after several generations. For moderate numbers of $N♂$ and $N♀$, the rate of change in H can be approximated at very nearly the same rate as (11-10) and (11-11); in fact, (11-5) can be considered as an approximate rate for most populations.

VARIATION IN N AND INBREEDING EFFECTIVE NUMBER

Over many generations, population size (effective parents) may not be constant in most real populations. The panmictic index $(1 - 1/2N)$ (11-8) will not be constant so that each generation must be considered.

$$\frac{H_t}{H_0} = \left(1 - \frac{1}{2N_0}\right)\left(1 - \frac{1}{2N_1}\right)\left(1 - \frac{1}{2N_2}\right)\cdots\left(1 - \frac{1}{2N_{t-1}}\right)$$

$$= \prod_{i=0}^{t-1}\left(1 - \frac{1}{2N_i}\right)$$

(where N_i refers to population size of each generation).

However, a population of constant effective size (N_e) could achieve an equivalent magnitude of heterozygosity loss; consequently, the generation-by-generation expression above may be equated to $(1 - (1/2N_e))$ and solved for N_e. While an exact solution may be algebraically quite a chore, a good approximation can be arrived at by realizing that if the N is fairly moderate in size (> 10), the product terms of the long expression will be very small, approaching zero, and can be ignored. Approximately, then,

$$1 - \frac{t}{2N_e} = 1 - \sum \frac{1}{2N_i}$$

and N_e becomes the harmonic mean of generation by generation population sizes.

$$\frac{1}{N_e} = \frac{1}{t}\sum \frac{1}{N_i} \quad \text{or} \quad N_e = \frac{t}{\sum\left(\frac{1}{N_i}\right)} \tag{11-14}$$

Thus, generations with smallest numbers have most effect on decreasing heterozygosity (H) (see Exercise 8). It should also be kept in mind that an increment in inbreeding in one generation due to depletion of the breeding numbers in the population cannot be counteracted by an expansion in numbers later on; the rate of change in H will be slowed by an expansion but not reversed. Only outcrossing, mutation, selection favoring increased heterozygosity, or other forces of similar nature can restore the level of heterozygosity lost through random processes including the inbreeding aspect we are considering.

VARIATION IN PARENT GAMETIC CONTRIBUTION AND INBREEDING EFFECTIVE NUMBER

Each parent may not contribute equally to the next generation. We have assumed up to now that the random sample of gametes from parents implied that each parent was equally effective in producing an offspring, but in most real populations the number of surviving offspring must vary considerably from one parent to another; that is, some parents' gametes become established as offspring more than others. This variation in gametic contribution leads to reduction in effective population size, as first demonstrated by Wright in 1938. (The following derivation is based on that of Crow and Kimura, 1970, from Wright.)

In a population of monoecious diploids, "idealized" with ability to self-fertilize, let k_i be the number of successful gametes from the ith parent in the $t - 1$ generation; then,

$$\bar{k} = \frac{\sum_i^{N_0} k_1}{N_0} = \text{average number of successful gametes from any parent, where } N_0 = \text{number of monoecious parent individuals.}$$

$N_0\bar{k} = 2N_1$ = number of successful gametes—those producing N_1 progeny.

With selfing possible when gametes combine at random, the number of pairs that can form from any one parent is $k(k - 1)/2$, and the total number of pairs from single parents throughout the population is $\sum k(k - 1)/2$. In like manner, the total number of pairs of gametes from all parents in the population is $N_0\bar{k}(N_0\bar{k} - 1)/2$. Therefore, the probability P for two gametes uniting at random to come from the same parent is

$$P\{2 \text{ gametes from same parent}\} = \frac{\sum k(k-1)}{N_0\bar{k}(N_0\bar{k} - 1)} = \frac{\sum k^2 - \sum k}{N_0\bar{k}(N_0\bar{k} - 1)}$$

The distribution of k gametes among the parents must be scrutinized; consequently, we rewrite the numerator of the above probability in terms of variance in k as follows:

$$\sigma_k^2 = \frac{\sum(k - \bar{k})^2}{N_0} = \frac{\sum k^2 - N_0\bar{k}^2}{N_0}$$

Rearranging gives $\sum k^2 = \sigma_k^2(N_0) + N_0\bar{k}^2$, which represents the first term of the numerator in the probability expression above. The second term is $\sum k = N_0\bar{k} = 2N_1$, and we can rewrite the probability as follows:

$$P\{2 \text{ gametes from same parent}\} = \frac{\sigma_k^2(N_0) + N_0\bar{k}^2 - N_0\bar{k}}{N_0\bar{k}(N_0\bar{k} - 1)}$$

This simplifies to

$$\left[\frac{\sigma_k^2}{\bar{k}} + (\bar{k} - 1)\right] \cdot \frac{1}{N_0\bar{k} - 1}$$

Finally, we recall that the proportion of gamete pairs drawn from the same parent is $1/N_e$ (see the derivation of (11-13)). Therefore, the effective population size can be expressed as the reciprocal of the above probability:

$$N_e = \frac{1}{P} = \frac{N_0\bar{k} - 1}{(\sigma_k^2/\bar{k}) + \bar{k} - 1} = \frac{2N_1 - 1}{(\sigma_k^2/\bar{k}) + \bar{k} - 1} \tag{11-15}$$

(general effective number as a function of gametic distribution).

If the population size is constant over generations, then the average gametic contribution of parents must be $\bar{k} = 2$, and $N_0 = N_1$. On substitution into (11-15), the effective number becomes

$$N_e = \frac{4N - 2}{\sigma_k^2 + 2} \tag{11-16}$$

If N is large and constant with no selection acting, the distribution of k will approach a Poisson (Appendix A-4C) so that mean and variance will be equal ($\sigma_k^2 = 2$), and (11-16) becomes $N_e = N$ approximately.

If we consider a population with separate sexes (dioecious), the derivation of (11-15) is complicated by the inclusion of three generations and two separate numbers for males and females. For brevity we shall omit the derivation (given by Crow and Kimura, 1970, pp. 349–351) and merely point out that the numerator of (11-15) is altered to $2N - 2$ instead of $2N - 1$, a difference that would be important only if N is small—about 10 or less.

It is a bit surprising to note that, under special circumstances such as those used by breeders where the exact number of parents can be controlled and each parent contributes equal numbers of progeny, the effective number can be greater than the actual number of parents. When all parents contribute equally, the variance of k is zero and (11-16) becomes $N_e = 2N - 1$—nearly twice as large as the number of parents! In such a case, the rate of inbreeding would be half that in the idealized population of the same size. Furthermore, by *avoiding* all close relative matings and keeping the parents chosen equally in this way, the breeder can reduce the rate of inbreeding maximally without outcrossing.

EMPIRICAL ESTIMATION OF N_e

The rate of fixation (ΔF) can be related to population effective size via (11-3) or (11-12); thus, we may wish to estimate N_e in real populations using these relationships. It would not be difficult to program a computer to simulate the rate of fixation for any population size, or to determine a sample size for any given rate of fixation, but it is important *biologically* to determine just how different the effective number of parents may be from the apparent size of the population. Usually when all sorts of related or unrelated individuals occur together, their contributions of genic material to the next generation may range all the way from zero to a very large fraction of the progeny. Some individuals may not mate at all or if they mate may have few offspring. For particular populations with behavioral properties peculiar to each—mating characteristics, viabilities, and fertilities—it could be difficult to generalize the N_e for various species of organisms. Nevertheless, we need to know methods of approach and orders of magnitude for N_e under given conditions in order to study the changes expected under the influence of evolutionary forces (to be discussed in the next chapters).

We shall first look at an example illustrating procedures based on the inbreeding coefficient (rate of fixation). Another major aspect of limited population size, however, is that of dispersing gene frequencies; this aspect will be examined in the next chapter under *variance effective number* since the amount of dispersion (σ_k^2) is an equally useful parameter, as are methods based on lethal allelism (Chapter 18) to help us measure N_e. Example 11-1 presents some data collected from laboratory populations of drosophila where N_e varied from less than half the actual population size (parent number) up to about 80 percent of the census number. These estimates were based on the relationship in (11-12). Crow and Morton (1955) give estimates of N_e based on the relationship in (11-15); see Example 11-2 for their methods and data.

EXAMPLE 11-1

INBREEDING EFFECTIVE NUMBER IN A DROSOPHILA EXPERIMENT

Laboratory measurement of random sampling effects has been successfully accomplished many times, and estimation of N_e from the constant rate of fixation has been made most efficiently by using genetic mutants that are least likely to have selective effects in organisms such as drosophila. The best models for illustrating the overall inbreeding rate of fixation can be found in the work of Kerr and Wright (1954), Merrell (1953), Crow and Morton (1955), and Buri (1956). (In most of these studies, chance fluctuations in gene frequencies were a parameter, but the aspect of dispersion between small populations is deferred to Chapter 12.) In this example, we illustrate the method of Buri, which was especially clear for present purposes and which was modeled after the experiments of Kerr and Wright.

Buri was fortunate to utilize a pair of alleles of brown eye color (bw^{75} and *bw*) in *D. melanogaster*, which not only are distinguishable phenotypically in the heterozygote from both homozygotes segregating (an unusual feature in drosophila where most mutants are either recessive to wild type or if dominant are often lethal in homozygote), but also both alleles showed little selective differences in most tests so that they could be considered as neutral in fitness. With these *brown* alleles segregating in a scarlet eye color (*st*) genetic background, eye colors of the *bw* alleles were as follows: *bw*/*bw*; *st*/*st* = white eye . . . bw^{75}/bw; *st*/*st* = light orange eye . . . and bw^{75}/bw^{75}; *st*/*st* = reddish orange eye.

After some initial tests for viability and competitive productivity of these three genotypes, Buri concluded that they were reasonably equal in fitness, and he set up a large number of cultures with small controlled breeding size ($N = 16$). Each initial culture founded a line in which each successive generation was initiated from a random sample of uniform size (eight males and eight females) taken from among the flies of the preceding generation. Thus, the form was that of a fairly large population subdivided into many completely isolated units. The entire experiment was replicated into two series, but for this illustration we shall merely concentrate on Series I. (In Series I, cultures were made in 35-ml vials, while in Series II cultures had more food available in 60-ml specimen bottles.) On each successive generation, the samples of 16 flies were taken exactly in the same manner—on the 12th day after mating (or 2 days after emergence of first flies in the culture). Flies were aged 2 days with sexes separated and virgin before mating to perpetuate the series.

Series I was begun with 107 vial cultures, all with heterozygotes bw^{75}/bw; *st*/*st* and 8 pairs per culture. Successive generations and the distribution of bw^{75} numbers from zero (all *bw*/*bw*) up to 32 (all 16 flies bw^{75}/bw^{75}) are presented in Table A. Note that there is a nearly symmetrical spread increasingly wide among these cultures until at the fourth generation one is fixed at the right end (bw^{75} homozygous in all 16 flies), while at the opposite end (*bw* in all) fixation begins at the sixth generation. By the 19th generation there is near equality of fixation at opposite ends of the distribution, which is generally flat in between. Newly fixed cultures are accumulated in the columns to the left and right of the main data. (Compare these distributions with those illustrated in Example 12-2.) On the

TABLE A *Distributions of numbers of the bw^{75} allele among the cultures from the initial to the 19th generation (modified from Series I of Buri, 1956)*

Genera-tion	Total Fixed *bw*	Number of bw^{75} Genes																																	Total Fixed bw^{75}	Total Newly Fixed	Total Unfixed in Preceding Generation	Fixation Rate
		0	1	2	3	4	5	6	7	8	9	10	11	12	13	14	15	16	17	18	19	20	21	22	23	24	25	26	27	28	29	30	31	32				
0																			107																			
1									1	0	0	0	2	5	8	10	13	22	9	14	14	7	1	1														
2				1	0	0	0	2	3	1	1	4	8	7	7	2	15	9	8	13	7	5	3	1	2	4	1	0	2	0	1							
3					2	0	1	1	2	3	5	3	4	9	8	8	9	8	6	7	4	7	2	5	2	2	2	2	3	0	0	1	1					
4					2	0	2	1	5	4	6	8	6	6	2	5	8	9	3	3	7	4	3	3	6	4	2	2	2	0	1	2	0	1	1	1	107	0.0093
5				1	1	2	3	5	3	6	5	1	6	4	5	4	4	5	10	7	5	3	4	3	2	1	4	5	3	1	0	2	0	1	2	1	106	0.0094
6	1	1	1	2	1	2	2	3	6	2	10	7	2	4	4	7	4	9	4	2	1	4	6	7	0	4	0	3	3	2	0	1	0	1	3	2	105	0.0190
7	3	2	2	1	0	0	1	5	5	7	5	2	4	3	5	12	5	5	5	4	5	4	1	3	1	2	5	1	2	7	1	1	0	0	3	2	103	0.0194
8	5	2	0	1	1	2	3	4	4	2	10	3	4	3	4	5	8	6	6	3	2	2	5	3	1	3	1	0	0	3	2	3	3	2	5	4	101	0.0396
9	5	0	1	2	5	3	2	1	3	5	3	7	7	5	3	4	2	5	4	2	1	4	4	3	3	1	3	3	2	3	1	1	3	1	6	1	97	0.0103
10	7	2	2	4	3	5	0	3	1	4	2	3	2	4	5	3	4	3	6	7	4	5	2	1	5	1	2	2	1	2	2	2	2	2	8	4	96	0.0417
11	11	4	1	1	3	6	3	5	2	1	2	1	3	2	2	3	5	4	2	3	5	3	4	5	5	2	2	0	2	2	2	2	3	2	10	6	92	0.0652
12	12	1	1	0	4	6	2	2	4	2	2	1	3	0	6	3	5	1	4	4	2	4	5	3	2	4	3	0	1	2	1	0	1	7	17	8	86	0.0930
13	12	0	4	3	3	4	0	3	3	2	4	1	3	3	0	2	1	7	5	4	3	4	2	1	2	1	2	2	2	2	1	2	1	1	18	1	78	0.0128
14	14	3	6	1	5	2	0	3	1	2	2	4	1	3	3	3	1	6	3	5	2	2	0	0	1	2	2	3	2	1	2	1	0	3	21	6	77	0.0779
15	18	3	4	4	2	3	3	0	0	1	1	5	2	5	3	1	4	4	1	1	0	2	2	0	1	2	1	6	0	2	2	1	3	2	23	5	71	0.0704
16	23	5	0	2	4	1	3	2	1	1	2	1	2	2	7	1	2	2	3	3	2	2	0	1	0	2	0	2	0	4	3	1	3	2	25	7	66	0.1061
17	26	3	0	1	4	1	2	4	1	4	1	0	0	3	3	3	2	3	1	2	2	2	0	2	2	1	2	1	2	1	2	2	1	1	26	4	59	0.0678
18	27	1	3	1	3	1	4	2	0	0	3	1	1	1	4	2	3	1	3	2	0	2	2	3	0	3	1	3	0	0	1	1	2	2	28	3	55	0.0545
19	30	3	2	0	3	1	2	0	5	2	3	1	1	1	5	2	0	0	2	0	2	2	3	2	1	2	2	1	0	0	0	0	4	0	28	3	52	0.0577

far right are the "total newly fixed" isolates. When the latter numbers are divided by the total unfixed in the preceding generation, we have the last column, or fixation rate ΔF.

From about the 10th generation on, the average rate of fixation approaches a constant, although one could argue that beginning by the eighth generation ($\Delta F = 0.0396$), a fairly constant rate has been achieved. It may become a bit arbitrary where to draw the line as to the achievement of constancy, so we may calculate the effective size (N_e) based on upper and lower limits. If we use (11-12) based on the average rate of fixation for the last 12 generations (from G_8–G_{19}), the estimate is: average $\Delta F = 0.05808$ (G_8–G_{19}) therefore, from (11-12), $N_e = 8.11$.

Buri, basing his estimate principally on the dispersion of allelic frequencies (examined in the next chapter), concluded that N_e in Series I was 9.0. However, if we take the average ΔF farther along, say from the 11th generation on, it comes to 0.0673, which estimates $N_e = 7.0$. In any case, the student may realize that the rate of fixation is faster than would be expected from a population of 16 parents with 8 males and 8 females contributing equally to the next generation; in fact, about half or less of the parents are effective in producing progeny from this analysis.

EXAMPLE 11-2

EFFECTIVE BREEDING NUMBER FROM PROGENY DISTRIBUTIONS

The ratio $\sigma_k^2/\bar{k}$ appears in formula (11-15) for estimating the effective number of parents in a population. It is a measure of the degree of departure from idealized conditions, which would be the case if both mean number of progeny and their variance were equal to 2 (Poisson distribution of progeny per parent in a stable population size), thereby making the ratio equal to 1. Since the ratio is capable of direct measurement, counting offspring per parent can provide information to predict effective population size. Data have been collected by Crow and Morton (1955) and discussed further by Crow (1954) indicating that this ratio is commonly greater than unity so that the effective number is less than the census number of parents.

Two of the methods employed to determine this ratio by using drosophila were (1) adult progeny and (2) egg production. In method 1, virgin females marked with single mutants (for example, *cn*, *bw*, *e*, etc.) were mated to multiple recessive males (*cn* – *bw* – *e*) in food vials so that after progeny emergence each female's surviving progeny could be distinguished. After emerging progeny were counted for 10 days, the mean and variance of progeny number per maternal strain parent were determined (total of 90 sets of progenies with from 8–32 vials per set). Similarly, progenies from male parents were counted using single mutant males mated to multiple recessive females (19 sets of progenies with from 11–35 vials per set). Summarized results are given at the top of Table A with the calculation of N_e/N, the proportion of effective size for each sex in the experiment. In method 2, females of each tested strain were mated to multiple-marker males as before but placed in vials containing slanted food containing lampblack so that egg counting would be facilitated. Eggs were counted and parents transferred every 24 hours to fresh vials. Means and variance over mean ratio for individual female egg production are given in the third row

TABLE A *Mean progeny number ($\bar{k}$), ratio ($\sigma_k^2/\bar{k}$), and proportion of effective number to actual number for Drosophila, Lymnaea (snail), and human females. Data from Crow and Morton (1955) and Crow (1954), who give references for data on Lymnaea and humans as well.*

	Mean Progeny $\bar{k}$	Ratio $\sigma_k^2/\bar{k}$	Proportion N_e/N
Drosophila			
Females—adult progeny	13.9	4.73	0.77
Males—adult progeny	17.9	11.35	0.63
Females—egg production (daily)	37.2	11.45	0.67
Lymnaea	390.9	133.6	0.75
Human females			
New South Wales	6.2	2.61	0.79
England	3.5	2.54	0.69
United States (born 1839)	5.5	1.28	0.94
United States (born 1866)	3.0	1.93	0.76

of Table A. Note that mean and variance/mean ratio are greater for males than for females; as a result, N_e/N is smaller for males than for females.

Data also are presented by these authors from the literature on productivity for the snail *Lymnaea* and for human females. In early all cases, the variance in productivity between individuals lowers the effective breeding size to about 70–80 percent of the census size, the exception being women in the early nineteenth century, whose variance/progeny ratio is closest to that of the idealized population (Poisson distribution with variance equal to mean progeny).

GENERAL CONSIDERATIONS

Most of the ideas presented in this chapter were first obtained by Wright (1931) using path coefficient methods and later summarized in his Galton Lecture (1951). They represent some of the most fundamental conclusions in population genetics. Heterozygosis inevitably must fall provided *gametic union is random.* Obviously, it is thus expected that populations should tend toward genetic uniformity over long periods of time provided that no forces are available to counteract that trend, so that the mere fact that most populations have not "stagnated" genetically indicates that a number of forces—some of considerable magnitude—must be at work to maintain genetic variation in populations.

The loss of heterozygotes in populations of finite size is not simply due to occasional inbreeding, as the student might conjecture. When inbreeding is specifically avoided, H decreases, although at half the rate (11-1) achieved by random union (11-4). We shall see in the next chapter that without consideration of inbreeding specifically, mere random distribution of parents in small populations will bring about the same rate of heterozygosity decrease. The dispersion of genotypes and distribution of gene and genotype frequencies from one small population to another will concern us as another aspect of the same phenomenon, but it has important consequences for evolution of new gene pools, races, and species.

EXERCISES

1. Give the successive amounts of heterozygosity (H) for a population that starts with $p = 0.6$, $q = 0.4$ at random-mating proportions in G_0 and reproduces only by double first cousins once they are produced in the G_2 up to G_7.
   (a) First find the H recurrence series by using the additive components from previous generations. Then use expression (11-1) expressed as a loss in H per generation.
   (b) What is the percentage change (λ) in H per generation? What happens to this λ value each generation? Do successive values diverge, converge, or oscillate constantly?
2. Construct the table below to show the consequences of sib mating for five generations, from G_0 to G_6, and the limiting value, using the headings as given below and initial $H_0 = \frac{1}{2}$ (a few key entries are given).

G	H	$P = H_t/H_0$	$F = (H_0 - H_t)/H_0$	$\Delta H = H_t - H_{t-1}$	$\lambda = H_t/H_{t-1}$	$\Delta F = (F_t - F_{t-1})/(1 - F_{t-1})$
0	$\frac{1}{2}$	1	0	0	1	0
1	$\frac{2}{4}$	$\frac{2}{2}$	0	0	1	0
2	$\frac{3}{8}$	$\frac{3}{4}$	$\frac{1}{4}$	$-\frac{1}{8}$	$\frac{3}{4} = 0.75$	$\frac{1}{4} = 0.25$
3						
4						
5						
6						
⋮						
Limit					0.809	0.191

3. Show that ΔH does not equal ΔF though of opposite sign, (decrease in $H \neq$ increase in F), but they are proportional because $H_t/H_0 = 1 - F$, so that

$$\frac{-\Delta H_t}{H_{t-1}} = \frac{\Delta F}{1 - F_{t-1}}$$

   which is different from $\Delta F/F_{t-1}$ (percent change in F). Use numerical examples from your table constructed in the previous exercise and also show this algebraically.
4. Find the amount of heterozygosity expected (H) over three generations if a population at G_0 has $H = 0.5$ and it consists of just 10 hermaphroditic individuals capable of selfing. What would be the panmictic index and amount of H at G_{100} if the population size remains constant each generation?
5. If the population in Exercise 4 consisted only of animals that could not self but could reproduce by all other types of relative matings and if the population consisted of exactly five males and five females all equally fertile, what would be the expected H over three generations? If the population is perpetuated each generation with five pairs only, what will be the ultimate rate of loss in H per generation?

6. Find N_e for the following numbers of males and females:

(a)

♀♀ =	5,	6,	7,	8,	9
♂♂ =	5,	4,	3,	2,	1
Total =	10	10	10	10	10.

(b) Show that formula (11-13) reduces to N as the sexes become equal.
(c) Graph this series of N_e as the sex ratio (♀♀/♂♂) increases.
(d) Double the total to 20 and let the sex ratio change (10:10, 11:9, . . . , 19:1). How does N_e change as the sexes become more unequal?
(e) When the number of males is just 1, what will be the maximum N_e?

7. Suppose a population of ten individuals is made up of eight females and two males and that each male mates with four females to perpetuate the population each generation. How will the change in H compare with that you calculated in Exercise 5 where the sexes were equal? What will be the ultimate rate of loss in H per generation if the same mating system is constant for many generations?

8. A population of dioecious organisms fluctuates in size from one generation to the next because of a seasonal cycle as follows:

	Winter	Spring	Summer	Fall	Winter	Spring
♀♀	12	20	70	40	15	15
♂♂	8	20	30	10	5	15
Totals	20	40	100	50	20	30

Assume that each generation any individual is equally capable of breeding relative to other individuals of the same sex, and that each season has a new generation.
(a) Find the N_e for each generation (season).
(b) What will be the expected panmictic index over these six generations?
(c) What would be the expected panmictic index if all generations were the same size at the highest N_e? What would it be if all were at the lowest N_e? How do these expectations compare with the value you found with the cyclic population sizes in part (b)?

9. For sex-linked genes, the recurrence of H among the XX (homogametic sex) analogous to formulas (11-4) and (11-10) for autosomally linked genes was given by Wright (*Proc. Nat. Acad. Sci.* 19: 411–419, 1933) as follows:

$$H_t = H_{t-1} - \frac{N_♀ + 1}{8N_♀}(2H_{t-1} - H_{t-2}) + \frac{(N_♀ - 1)(N_♂ - 1)}{8N_♀ N_♂}(2H_{t-2} - H_{t-3})$$

Assume the rate of change in H becomes constant ultimately so that $H_t = H_{t-1} \cdots$.
(a) Solve the λ equation to obtain the decrease in percentage of H.
(b) Verify that

$$\Delta H = \frac{N_♀ + 2N_♂}{9N_♀ N_♂} \text{ approximately.}$$

(c) What will be the rate of change in H for sex-linked genes if the sexes are equal and monogamous?

10. Assume a population of six individuals to be monoecious (hermaphroditic) and individuals to be capable of selfing.
    (a) The distribution of progeny produced by random breeding including selfing of these six parent individuals is as follows:

Individual	1	2	3	4	5	6
Progeny	20	40	100	50	20	10

    Use formula (11-15) to estimate the effective population size (N_e). Note: $N_0 = 6$, $N_1 = 240$, so that $\sum k = 480$—total successful gametes.
    (b) Assume that all except six progeny fail to survive in the struggle for existence so that each individual's relative contribution to the total of $N = 6$ is the contribution to surviving (successful) gametes, assuming that there are no differences among individuals for their progeny's survival. For example, individual 1 makes a 0.5 relative contribution to the total of 6. What will be the effective population size (N_e)? How does that compare with the estimate made previously with all progeny surviving?
    (c) Assume all six parents produced exactly 40 progeny, all of which survive. What will be N_e?
    (d) Assume individuals 1, 2, and 5 are sterile in having no surviving progeny while individual 3 has 160 progeny and individuals 4 and 6 each have 40 progeny. What will be N_e?
    (e) Assume as in (b) that just six progeny survive and that fertile parent individuals make relative contributions to the total of six. What would be the N_e?
    (f) What can you say about how the distribution of gametes affects change in effective population size?
11. (a) From Example 11-1, Table A, find the mean fixation rate (ΔF) for generations from G_4–G_{10}.
    (b) For these 16 generations, what is the mean rate of loss in heterozygosity as a percentage of H—that is, λ?
    (c) In the last generation (G_{19}), is it possible to estimate H_{19} and F_{19} for the entire array of 107 lines (the population) from the data given? Does the ΔF you calculated above tell you how much the total homozygosity is for the population as a whole? What further assumptions would you have to make about the *unfixed* cultures before you could estimate F for the population as a whole from the data given?
12. From Example 11-2, use Table A data for $\bar{k}$ and the *variance/mean* k ratio to verify the N_e/N proportions for drosophila adult progenies for females and males, for Lymnaea, and for women from New South Wales. In the ratio column, which human population was most likely to have progenies distributed according to the Poisson? What kind of a progeny distribution do you suspect for *Lymnaea*?

PART 4
FORCES CHANGING GENE FREQUENCIES

INTRODUCTION TO PART 4

The root of biological evolution consists of changes in gene frequencies with one allele (or set of alleles) displacing another over time or through space. For any group of organisms, adaptation cannot be achieved nor can the differentiation of any population take place, whether for functional significance or neutral (nonselective) genetic displacement, without fundamental change in proportions of alleles. Sometimes the limits of complete loss or fixation are reached, although it is probably more often the case that stable equilibriums are attained with many genotypes at nonfixation points. New chromosomal combinations, genotypes, linkages, and higher orders of genetic organization then become potentially available for natural selection and other forces to utilize in formation of new ecotypes, races, or species. In short, the evolutionary process consists primarily of the transformation of genic variation that arises from mutation and recombination into spatially or temporally differentiated groups of organisms which can be described at least in principle according to their allelic proportions.

In the previous sections of this book the conservative nature of the Mendelian process has been emphasized. Chiefly via the continuity of the genetic material and the mechanisms of reproduction, the status quo is preserved, keeping a stable organization of the gene pool. In sexually reproducing biparental populations, the determinacy of random combinations of zygotes from gametic frequencies (Castle-Hardy-Weinberg equilibrium, or square law) has been evidenced along with the basic properties of genetic systems. The interrelatedness of genic combinations manufactured by the sexual process (meiosis and joining of gametes to form zygotes) makes a fundamental network interconnecting and integrating the units (individuals and breeding groups) of the population. Change in components of that network takes place first at the gene level via mutation; then, by random sampling and—most critically—by selection and gene flow, the population's genetic organization is controlled. While the mating system, if it becomes nonrandom, can change what we might call the texture of the hereditary network in a population by altering the way in which genes combine into genotypes, the gene pool *in its entirety* is expected to be unaffected in total allelic dosages by redistribution of the enclosed genotypes. Actually, however, such redistribution over space (between breeding units, or isolates) and time (generations within units) can easily lead to gene frequency change from unit to unit. Both nonrandom mating and the accidents of sampling bring about a dispersion of gene frequencies among subpopulations, providing an "array of probabilities" upon which the deterministic forces can subsequently act. Allelic frequencies as well as all sorts of genotype changes may spread as an array among small populations or breeding units, providing the flexibility of genetic possibilities vital for efficient evolutionary change.

In this section we turn our attention to gene frequency change (alteration of q values) predominantly. Gene frequencies may change with direction and velocity under the influence of *vectorial*, or *deterministic*, processes. Superimposed on such predictable (in principle) changes are fluctuations, which are determinate in amplitude but without direction or velocity—that is *random*, or *stochastic*, processes. We begin by looking at the latter type of process first in an elementary and fundamental way, since our considera-tion of nonrandom mating in the past section of this book has led naturally to a description of random elements in the dispersion of gene frequencies. We should realize when we turn to the vectorial processes that the stochastic elements should be taken into account if we wish to approximate reality. Finally, we consider briefly the interaction of these processes to provide the more realistic view of genes in populations than we would achieve by consideration of determinate processes alone.

Rarely in nature can we observe these factors in isolation; most often each may be counteracted by another, perhaps in ways tending toward a balance between them. We are considering the fundamental processes separately only as an educational and analytical device. As stressed by Wright (1955), "conditions in nature are often such as to bring about a state of poise among opposing tendencies on which an indefinitely continuing evolu-tionary process depends . . . ." Evolution results when one force or another gains an edge over others, when the balance is perturbated, and when directional change in gene frequencies at a fundamental level takes place.

12

RANDOM GENIC CHANGES IN POPULATIONS OF LIMITED SIZE

For a large population considered as a gene pool in its entirety, no change in allelic frequencies is expected to occur, irrespective of the mating system, as was discussed in the last section. Nevertheless, *within* any isolate (subgroup) of the population, or line of descent, under any mating system, the sampling of gametes will not be constant and may vary considerably in space or time. If our point of view then is focused on the isolate, the sample, or the line of descent, it will be evident that gene frequencies fluctuate randomly with considerable amplitude in small breeding units, but with less amplitude the larger the unit.

Whether alleles identical by descent or simply random pairs of alleles are considered, the same dispersion will take place by increasing the variance of gene frequencies along with the random fixation and loss of alleles. Inbreeding, nonrandom mating, and sampling of gametes are actually all different ways of looking at the same phenomenon, although it has been convenient to look at them separately. In this chapter we single out the aspect of reproduction in a population that can be described as random sampling of gametes either in time, from generation to generation, or in space, from one isolate to another. We assume the absence of all other forces affecting gene frequency changes.

In reality, few populations are large enough to be considered infinite in size as required for a perfect fit of the square law equilibrium. Most are not only limited in numbers but their gene pools are usually broken up by barriers (geographic, ecological, behavioral, etc.) into small breeding units, fragmented in various ways into "patchwork," "stepping stones," "islands," or local concentrations. Rarely does a species occupy an area in continuum of uniform density. (For references on distributions and abundance of organisms, see Connell, Mertz, and Murdock, 1970; Andrewartha, 1961; Wilson and Bossert, 1971.) From one local breeding unit to the next, gene frequencies may vary considerably, especially if gene flow between units is limited (for example, the small mammal breeding units in Example 9-1).

RANDOM FLUCTUATION IN ALLELIC FREQUENCIES: SAMPLING OF GAMETES

If we follow any line of descent from generation to generation, random fluctuation in q values takes place. For example, if a population is maintained by sib mating, allelic frequencies within sibships will become diverse from each generation to the next until a sibship becomes completely "fixed." Starting from a random-mating base population, when offspring from sibs arise and sibships are formed, q values will spread among sibships reaching limits of fixation or loss with intermediate frequencies in between (Table 9-2).

One can trace changes in q values by taking any line of descent at random to see the zigzag nature of the process.

The "random walk" of gene frequencies with generations need not be confined to inbred lines, of course. As a matter of fact, the possibilities for change are more restricted with inbreeding than with random mating in small populations: for example, if a sibship has $\frac{1}{2}AA$:$\frac{1}{2}Aa$ individuals, three possible matings are expected to occur in a ratio of $\frac{1}{4}$:$\frac{2}{4}$:$\frac{1}{4}$ (Table 9-2B, 2 and 4), while if a small population containing $q_a = 0.25$ (equivalent of the above sibship) were to be mating at random, nine types of matings could occur including $aa \times aa$ at a frequency of $q^4 = 0.0039$. This random variation in q in small populations has been called "random genetic drift" by Wright (1955), in contrast to what Wright points out is a "steady drift" in gene frequencies due to the directed or deterministic processes. As long as q lies between $1/(2N)$ and $(2N - 1)/2N$, q can drift in either direction, becoming any value after a few generations of random mating within a small population. Once a gene frequency attains 0 or 1, however, it has reached a "dead end" that is irreversible without mutation or hybridization between breeding units. Thus, homozygous breeding units will gradually accumulate just as in the expectation from inbreeding.

In Example 12-1, an illustration of "random walk" in q values is given from the simulation of random drift by a computer. The essential ingredients for the simulation are: (1) Starting with equal proportions of two alleles in the simplest cases, we assume that a very large population initially (G_0) mating at random will be reduced to a limited size (N) at a "bottleneck season" (winter in temperate zones or dry season in the tropics, for example). This small population of survivors we call G_1 *initial*. (2) In the favorable season, we assume rapid expansion in numbers of progeny from the N survivors and panmixis among them, all of which we designate as constituting G_1. (3) On the return of the bottleneck season and reduction of population size to N (we assume to be constant), a sample of $2N$ gametes is randomly obtained from the large gene pool; zygotes obtained by chance combinations of these gametes we designate G_2 *initial*; all progenies produced by these survivors in turn in the favorable season again constitute G_2. Chance alone determines which allele (A or a) of the panmictic gene pool becomes included in the survivors' sample (G_1 *initial* or G_2 *initial*.), and most of the time the small surviving group will have new allelic frequencies ($p' + q'$). Successive application of the stochastic element (random drawing of $2N$ gametes each generation) produces a random walk of q values.

EXAMPLE 12-1

SAMPLING OF GAMETES AND RANDOM WALK OF q

A "random walk" of q values ensues when the size of a gametic sample is limited by chance each generation. Some physical models devised in the past that produce a random walk and thus an array of q among isolates have the common attribute of sampling from a binomial distribution simulating diploid genotypes determined at random (for example, Dubinin and Romaschoff's bowl of marbles (1932) described by Dobzhansky in his 1941 edition of *Genetics and the Origin of Species*, or the pairs-of-beads model described by Moody in 1947). A good classroom demonstration devised by House (1953) omits zygotic frequencies and focuses on random q value changes; with the assumption of panmixia

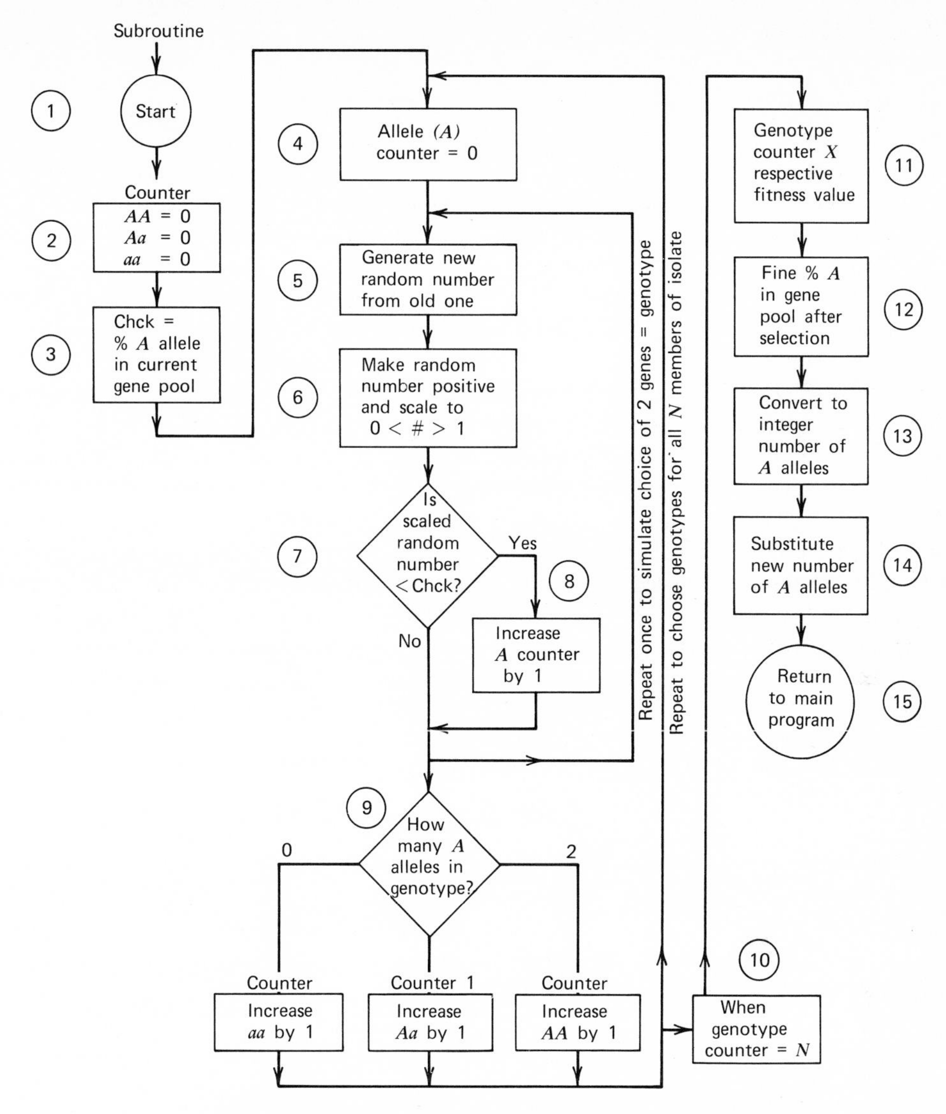

Figure A. Flow chart for computer program subroutine that produces a "random walk" of gene frequency over *one* generation within a single isolate line under the model described in Examples 12-1 and 12-2. At start (step 1), the G_0 array of A gametes + a gametes [or the number of such gametes ($N_A + N_a = 2N$) in the gene pool of any generation after G_0] is introduced and set as a proportion of A (step 3). The "counter" for the current generation is set at zero (step 2). At step 5, a set of random digits is generated. If $A = 20$ percent at G_0, 20 percent of the random digits expected to fall at or below 2/10 will be considered as A gametes (step 7) and will be counted (step 8). Random digits above 2/10 will be a gametes. A second time around the first loop will be counted as a diploid genotype (step 9). When $2N$ gametes have been counted (step 10), exit from the loop proceeds via a selection subroutine (steps 11–14 used in Chapter 14 examples) back to the main program. Under random drift, fitness values of genotypes are set at $1 = 1 = 1$. Successive application of this subroutine is done sequentially over 100 generations or until fixation at $p = 0, 1$. In the main program, the N_A is recorded in a two-dimensional frequency distribution generation by generation (Example 12-2).

within isolates between generations of "bottleneck limitations," it is superfluous to include zygotic frequencies in the model. Random gametic sampling is the only essential ingredient, and constant isolate size at the bottleneck season is the only necessary assumption.

To simulate a sample of gametes in which alleles A and a occur at p and q frequencies, House used containers (cans or boxes) partially filled with many small beads of two colors. For demonstration of random walk, the student must have available as many containers as will be needed for all possible unfixed $p - q$ values ranging from $1/(2N)$ up to $(2N - 1)/2N$. For example, if the "Noah's Ark" case is presumed, $N = 2$, then three containers must be available, one with 100 beads at $25A{:}75a$, one with 50:50, and one with the reverse of the first, $75A{:}25a$. If $N = 4$, there would need to be seven containers with 200 beads, each in ratios ranging from 25:175 to the reverse. The student starts a run with an assumption of the G_0 population, usually at $p = q = 0.5$ and then draws a sample of $2N$ gametes by shaking the 50:50 container and observing a $2N$ sample. For example, if $N = 2$, four beads may be drawn at random. A container is chosen according to the ratio observed, with the same proportion of $A{:}a$. The student shakes that container, again samples $2N$ beads, and continues until all $2N$ beads in any sample are identical (fixation or loss). Tabulation of this random walk by generations can then be scrutinized for an entire class.

An alternative method is to employ a table of random numbers to imitate the procedure of sampling. If we assume the "Noah's Ark" case ($N = 2$) with four genes per isolate, we

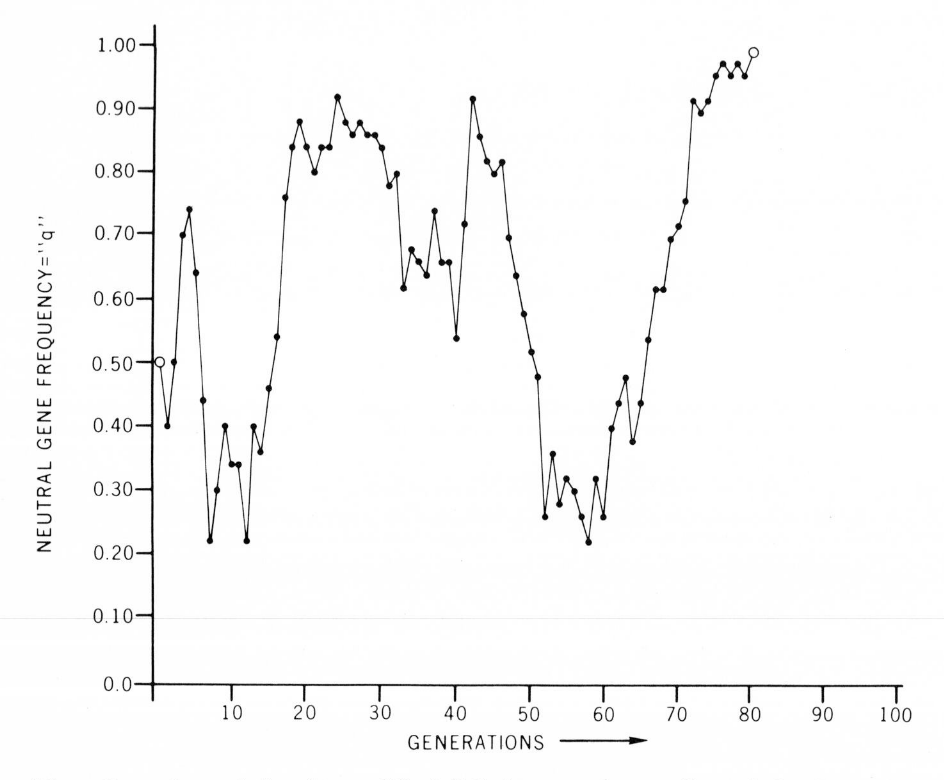

Figure B. A sample "random walk" of allelic frequency in a small population ($2N = 20$) generated by IBM 370 computer simulation at U.I.C.C.

may start from an initial generation at $p = q = 0.5$; simulating the frequencies is accomplished by taking any random number from 0 to 4 inclusive for A allele and 5 to 9 for a. If we choose four digits *at random*, we shall have our gametic sample (similarly, we might choose 10 digits to give $2N = 10$, and so forth). Now, if we obtain $3A:1a$ in the sample, our simulation of 75 percent A:25 percent a must be made by choosing digits in a manner such that 0 to 6 could count as A, with 8 to 9 as a. The digit 7 would be an A if followed by 0–4 as a second digit but it would be an a if followed by 5–9.

With the extensive availability of high-speed digital computers, the mechanical simulation of House's model and drawing random digits are superseded, but the older methods are still useful in the classroom where computers are not available. The essential element of the computer program includes the generation of a set of random numbers each generation from which a new set of gametes and zygotes is drawn. (Computer programs were worked out for the IBM 370 by Paul Garst in 1973 for random genetic drift simulation in our laboratory and are available on request.) The computer program flow chart for a single generation is shown in Figure A. A random walk given in Figure B with $2N = 20$ illustrates the main results: fixation or elimination may take place as late as nearly 100 generations after starting the process or as early as one generation. Trends may occur that apparently proceed in one direction for a long time or may be reversed. Fifty such populations simulated on the computer all came to fixation of A (27 populations) or a (23 populations) by no later than the 94th generation and no earlier than the 6th.

DISPERSION: THE NOAH'S ARK CASE.

When a large population is fragmented into an array of small-size isolated subpopulations either by imposition of geographic barriers or by seasonal reduction, the distribution of allelic frequencies over the array of isolates is most instructive. If all isolates so produced are of equal size for convenience (the simplest model), the probability distribution of isolates in the next generation (G_1) will be given by the binominal expansion $(p + q)^{2N}$ for isolates with q values ranging from 0, $1/(2N)$, $2/(2N)$, $3/(2N)$, ... $(2N - 2)/2N$, $(2N - 1)/2N$, to 1. Now the offspring (G_1) in turn we may assume breed at random, generating zygotes in an expanded progeny according to the square law within each isolate according to its specific $(p_i + q_i)^2$ frequencies. Again, when the reduction to limited size takes place producing an array of G_2 isolates, sampling of gametes according to the binomial expansion within each isolate must be considered.

The simplest description of this dispersion process as a result of limiting population size can be more clearly understood by considering the smallest size for the bisexually reproducing species—that in which a single pair perpetuates the species in each isolate. If we generate the array of possible two-individual isolates ($N_i = 2$) at a bottleneck season with an expansion of numbers by random mating within each isolate, followed by another bottleneck season with an expansion of numbers by random mating within each isolate, followed again by another bottleneck reduction down to the single-pair size, we have the model of random genetic drift in its simple form, as represented in Figure 12-1.

In Table 12-1 the spreading of the variability for the "Noah's Ark" model is given in terms of probabilities for gametic frequencies in isolates of size $N = 2$, starting with an initial pair, which we assume to be $Aa \times Aa$ in order to establish $p = q = \frac{1}{2}$, in

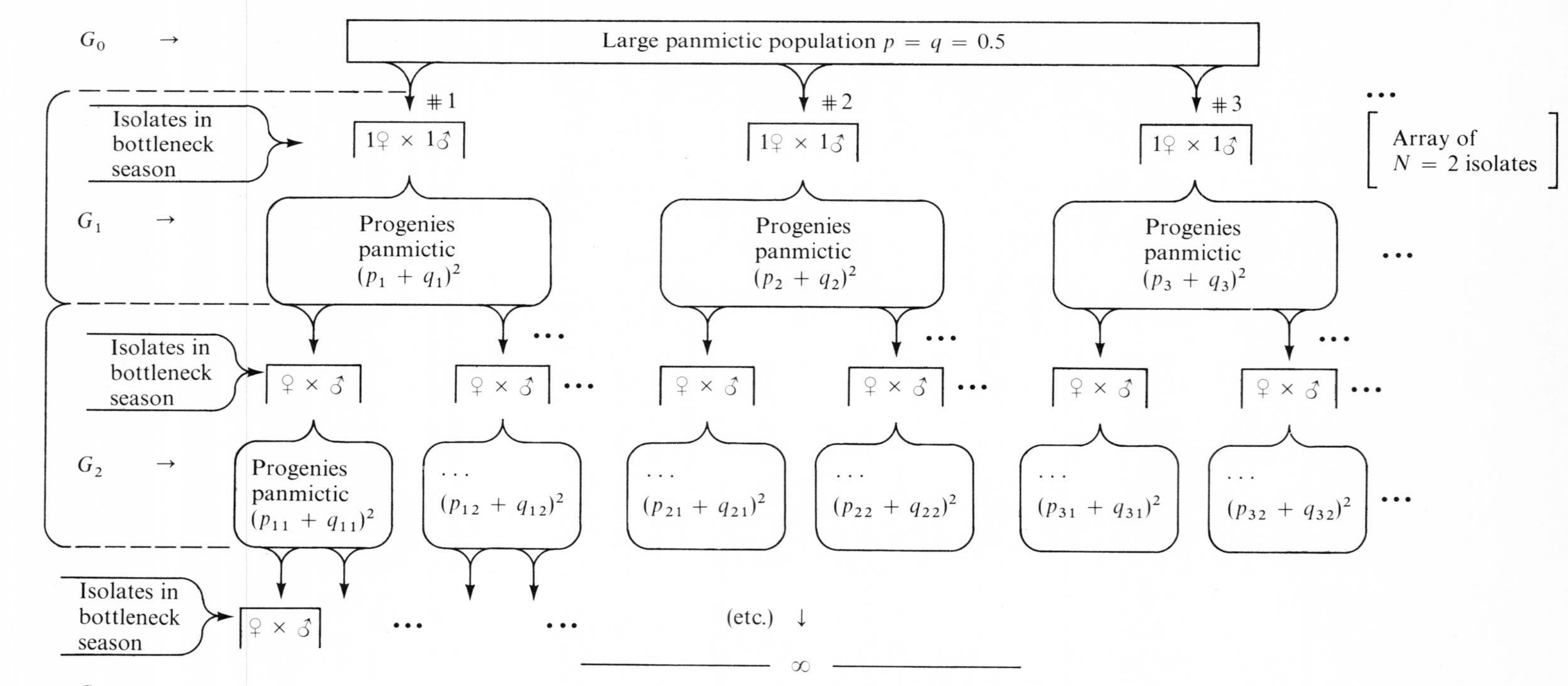

Figure 12-1. Diagram of the "Noah's Ark Case", $N = 2$, in a single model of random genetic drift. Each generation at the bottleneck season, isolates which survive are single pairs (1 ♀ × 1♂) taken at random from a large panmictic population. Within each isolate, the parental pair produces a large progeny that is panmictic in the favorable season. Thus the original population breaks up into a large number of isolates, each of which in turn continues to propagate again and become subdivided in the bottleneck season until an infinite number of isolates and progenies would exist. Triple dots imply repeated isolates of single pairs and their panmictic progeny at each generation.

TABLE 12-1 *Spreading of the genetic variability among isolates with two individuals each: the "Noah's Ark" case. Model: 1♂ × 1♀ produce a large number of progeny, which then mate at random. In the difficult season, all individuals are eliminated except 1♂ and 1♀ to reproduce the next generation. Consider each pair surviving to be an isolate. G_o = all Aa (variance of q between isolates in G_1 and G_2 uses q itself as the Y variate)*

G_1 Isolates	Frequency	q_a Within	G_2 Isolates (q_2 Within)					G_2 Totals
			0.00	0.25	0.50	0.75	1.00	$(p + q)^{2N}$
AA and *AA*	$\frac{1}{16}$	0.00	$\frac{1}{16}$	—	—	—	—	$\frac{1}{16}(1 + 0)^4$
AA and *Aa*	$\frac{4}{16}$	0.25	$\frac{1}{4}(\frac{81}{256})$	$\frac{1}{4}(\frac{108}{256})$	$\frac{1}{4}(\frac{54}{256})$	$\frac{1}{4}(\frac{12}{256})$	$\frac{1}{4}(\frac{1}{256})$	$\frac{1}{4}(\frac{3}{4} + \frac{1}{4})^4$
Aa and *Aa* / *AA* and *aa*	$\left.\begin{matrix}\frac{4}{16}\\ \frac{2}{16}\end{matrix}\right\}\frac{6}{16}$	0.50	$\frac{3}{8}(\frac{1}{16})$	$\frac{3}{8}(\frac{4}{16})$	$\frac{3}{8}(\frac{6}{16})$	$\frac{3}{8}(\frac{4}{16})$	$\frac{3}{8}(\frac{1}{16})$	$\frac{3}{8}(\frac{1}{2} + \frac{1}{2})^4$
Aa and *aa*	$\frac{4}{16}$	0.75	$\frac{1}{4}(\frac{1}{256})$	$\frac{1}{4}(\frac{12}{256})$	$\frac{1}{4}(\frac{54}{256})$	$\frac{1}{4}(\frac{108}{256})$	$\frac{1}{4}(\frac{81}{256})$	$\frac{1}{4}(\frac{1}{4} + \frac{3}{4})^4$
aa and *aa*	$\frac{1}{16}$	1.00	—	—	—	—	$\frac{1}{16}$	$\frac{1}{16}(0 + 1)^4$
G_1	$(H = \frac{1}{2})$ $(\sigma_q^2 = \frac{1}{16})$	G_2 Frequencies	$\frac{170}{1024}$	$\frac{216}{1024}$	$\frac{252}{1024}$	$\frac{216}{1024}$	$\frac{170}{1024}$	$(H = \frac{3}{8})$ $(\sigma_q^2 = \frac{7}{64})$

generating G_0. Progeny from this pair can be visualized as produced in large numbers. In turn they mate at random, producing a panmictic population. Now we suppose the bottleneck season should eliminate all but two individuals to perpetuate the species in any isolate. The possible isolates of that size will occur as in the left column with probabilities as given in the second column. We may call these isolates G_1 with q values according to the third column, even though at least a generation of random mating may have intervened between G_0 and G_1. Of course, this large random-mating intervention is simply a device to engender the gamete sample without bias and to avoid the separation of the two types of 50:50 isolate (*Aa* × *Aa*) and (*AA* × *aa*).

When the process is repeated, single pairs produce large random-mating progenies as before; these are cut down by the bottleneck season to single-pair isolates for G_2. The $\frac{2}{16}$ of the population fixed in G_1 remains, but all of the unfixed isolates again are dispersed according to the binomial distribution of their respective gametic frequencies. On the extreme right of each row in Table 12-1 is given the binomial expressing the distribution within each G_1 type of isolate into G_2 isolates. Note that each term in parentheses represents $(p_i + q_i)^{2N}$, or distribution within G_1 progenies, while the coefficient fraction before each parenthesis is the proportion of G_1 isolates being dispersed. The total G_2 isolated frequencies then are the sums of fractions downward in the table. Isolates where all four gametes are identical have reached fixation. Unfixed classes will continue to spread the same way as long as that mating system and "seasonal" reduction to $N = 2$ continue. Table 12-2 itemizes successive frequencies of these isolates for a few more generations. It is apparent that within a short time the distribution of isolates becomes flattened across the three unfixed classes,

TABLE 12-2 *Successive frequencies of isolates in the "Noah's Ark" case continuing from Table 12-1*

	Isolates (q_a Within)				
Generation	0.00	0.25	0.50	0.75	1.00
0	—	—	1.00	—	—
1	0.0625	0.2500	0.3750	0.2500	0.0625
2	0.1660	0.2109	0.2461	0.2109	0.1660
3	0.2490	0.1604	0.1813	0.1604	0.2490
5	0.3587	0.0904	0.1017	0.0904	0.3587
8	0.4404	0.0381	0.0429	0.0381	0.4404
∞	0.50	0	0	0	0.50

which are uniformly low in frequency throughout; in other words, the probability for any unfixed isolate is about equal to any other. Isolates with a fixed or lost allele continue to increase, making the distribution U-shaped. Ultimately, all isolates will be either "fixed" or "lost" in frequencies of p and q, respectively.

The "Noah's Ark" case is the most extreme form of dispersion, or spreading of genetic variability; the rate of fixation and loss is greatest. Of course, the same process occurs in

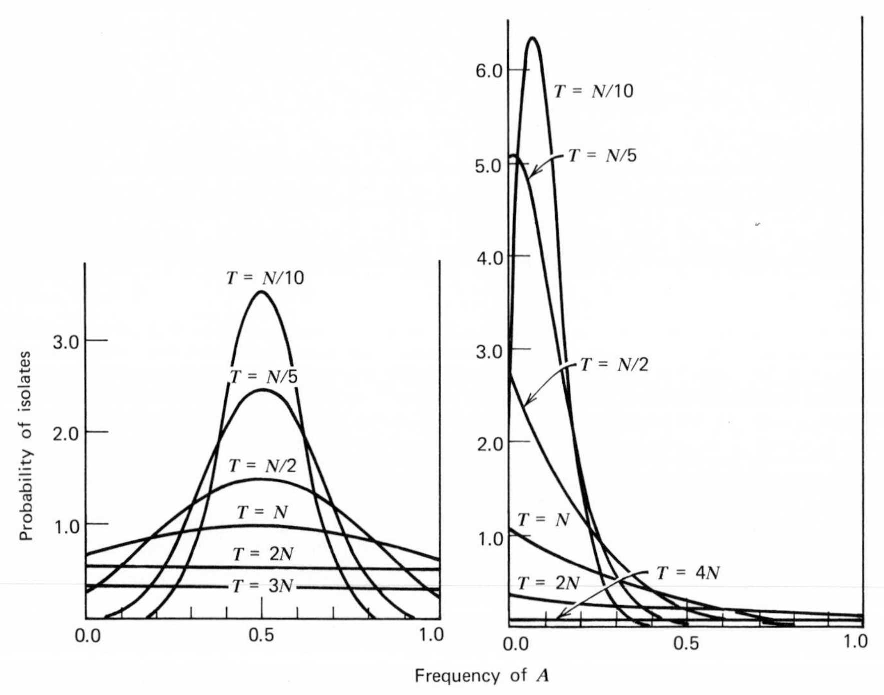

Figure 12-2. The probability density distribution of unfixed isolates of limited size (N) due to random sampling of gametes each generation, over t generations, assuming deterministic forces to be zero. At left, initial $p_0 = q_0$ 0.5, while at right, $p_0 = 0.1$, $q_0 = 0.9$ (from Kimura, 1955).

any population of limited size if the size limitation continues for more than a single generation or two. From each generation to the next, the dispersion is simply the probability array of $(p_i + q_i)^{2N}$, where N is the effective number of parent individuals and the assumptions of random mating within each isolate are reasonably approximated. Example 12-2 gives data produced by random genetic drift simulated by a high-speed digital computer for a succession of generations with $2N = 20$ and the other conditions as set forth for the "Noah's Ark" case. Note that by the $2N$th generation, about half the isolates are unfixed and have equal expectation (average $\frac{1}{40}$ per class of unfixed isolate) and that the remaining half of the isolates are fixed. Rate of fixation (ΔF per generation) for the isolate size in the whole population is given in the example, which confirms the expectation from $N = 10$. Distributions of isolates under random genetic drift with generations related to effective population size ($N_e \geq 10$), as initial p value is either 0.5 or 0.1, are illustrated in Figure 12-2.

EXAMPLE 12-2

COMPUTER SIMULATION OF ISOLATE DISTRIBUTIONS

A distribution of isolates, or dispersion array, generation by generation from the binomial sampling of gametes within isolates (each being of constant size = N individuals) can be simulated by a computer program with three main ingredients: (1) the random number generator as in Example 12-1, (2) a counter for accumulating the individual genotypes in each isolate, and (3) the printout of the isolates array each generation. Figure A shows a sample set of distribution changes for 30 generations (G_1, G_3, G_5, G_{10}, G_{20}, and

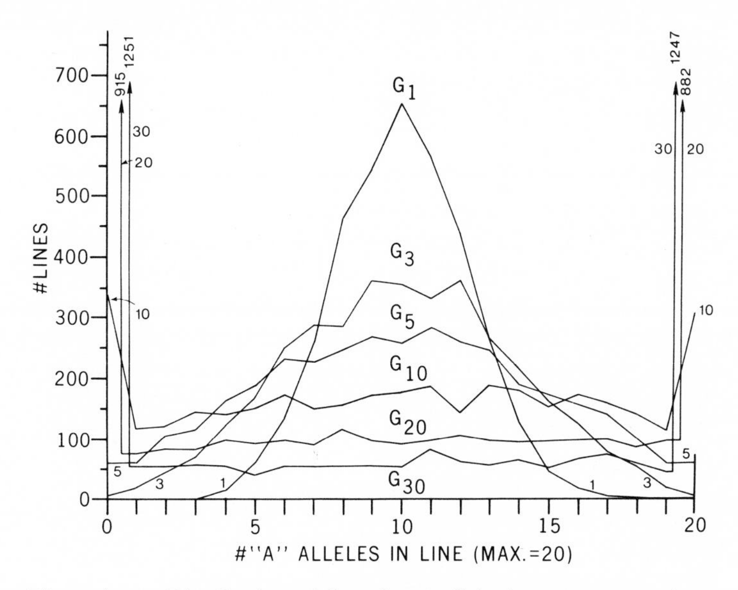

Figure A. Distribution of lines by A allele frequency at various generations starting from G_0 with $p = q = 0.50$ in 10 Aa parents, based on 3600 IBM computer-generated lines of size $N = 10$ over 30 generations.

G_{30}) from the computer simulation of the case of $2N = 20$. The model population is subdivided into 3600 isolates of size $N = 10$ each. Generations 1–5 illustrate the initial spreading of p and q values with successive flattening of the isolate distribution array with fixed ($A = 1.0$) or lost ($A = 0.0$) cases beginning to occur by generation 3. Fixed isolates accumulate in the next few generations so that by the 20th ($2N$th) generation, a U-shaped distribution occurs with all isolates containing unfixed alleles about equally frequent. That flat condition persists from then on when the total number of unfixed isolates becomes very small, but it is evident that even in this limited illustration the probability for any isolate to have a particular p,q value (other than 1 or 0) is equal to any other. Rates of fixation (ΔF) were variable for the first 10 generations; however, for the next 10 generations (G_{10}–G_{20}), average increment in F was $\overline{\Delta F} = 0.047001$. By applying formula (11-3), we obtain $2N = 21.28$, or $N = 10.6$, which is very close to the isolate size used by the computer to generate the array, as it should be. The student is asked to make a similar calculation from the increment in σ_q^2 in Exercise 3.

VARIANCE OF GENE FREQUENCY

At the start of the process, the expected dispersion could be described by the binomial sampling variance. If $2N$ gametes that combine to form G_1 are drawn at random from the parental generation, the sampling variance between isolates will be (Appendix A-4B and A-6E) as follows:

$$\sigma_q^2 = \frac{pq}{2N} \tag{12-1}$$

where $q = (1 - p)$ in the parent generation and $p =$ all remaining alleles collectively. We assume all isolates to be identical in size and the sampling to be from a supply of parent gametes, which can be regarded as infinite.

In the "Noah's Ark" case, the gametic number is too small to approach a continuous distribution of samples; there are only five classes of isolates. We cannot predict accurately the confidence limits of q values based on the standard error (σ_q) because the discontinuities make the distribution nonnormal. For example, at the start of Table 12-1 at G_o, if we use $\sigma_q = \sqrt{pq/2N} = \sqrt{(0.5)^2/4} = 0.25$, then the confidence limits of q would be 0.50 ± 0.25 for 68.3 percent of the time if the binomial were distributed as a normally continuous variable (see Appendix A-5). Instead, we find the G_1 isolates within those limits 87.5 percent of the time. With larger isolate sizes ($2N > 100$), however, the distribution approaches normality, and the standard error of q can be used to estimate confidence limits from generation to generation or from isolate to isolate. For small or intermediate-sized samples, the binomial probabilities must be calculated as illustrated in Appendix 4-1, Table A. Tables giving confidence limits for binomials are available in standard statistical handbooks. (See, for example Table W in Rohlf and Sokal's *Statistical Tables.*) It is fundamentally important to realize that the magnitude of random fluctuations may become large in small-sized populations. If $2N = 100$, then $\sigma_q = 0.05$, while if $2N = 1{,}000{,}000$, then $\sigma_q = 0.0005$.

"Accidents of sampling" may be viewed as the amount of random fluctuation for a single pair of alleles from one generation to the next or from one isolate to another among equal-sized isolates in space within a time level. Alternatively, it may be viewed as the distri-

TABLE 12-3 *Variance of q components in the "Noah's Ark" case in G_2 (Table 12-1)*

G_1 Isolate (q_a Within)	G_1 Frequency (f)	Variance (σ_q^2) Within G_2	$f \cdot \sigma_q^2$
0	$\frac{1}{16}$	0	0
0.25	$\frac{1}{4}$	$\frac{3}{64}$	$\frac{3}{256}$
0.50	$\frac{3}{8}$	$\frac{1}{16}$	$\frac{3}{128}$
0.75	$\frac{1}{4}$	$\frac{3}{64}$	$\frac{3}{256}$
1.00	$\frac{1}{16}$	0	0
	$G_1\sigma_q^2 = \frac{1}{16}$		$\sum f\sigma_q^2 = \frac{3}{64}$ due to dispersion within G_1 isolates into G_2 progenies

Therefore, total variance in $q(\sigma_q^2)$ for G_2 isolates $= \frac{1}{16}$ (component from G_1) plus $\frac{3}{64}$ (component from G_2 dispersion) $= \frac{7}{64}$

bution of a large number of loci whose frequencies were 0.50 in a parent population and then became diversified among progeny genotypes, some loci having higher and some having lower allelic frequencies as a result of random sampling from the parents' genomes. As the effective size of the genetic sample increases toward infinity, fluctuations expected in q tend toward zero, and gene frequencies are stabilized as expressed by the square-law equilibrium.

In successive generations the variance of q between isolates is compounded. In G_1 $\sigma_q^2 = \frac{1}{16}$ (Table 12-1). However, variances will differ in G_2 depending on the q value of each G_1 progenitor isolate; for example, within G_2 isolates derived from G_1 isolates with ratio $25A{:}75a$ or the reverse, the variance in q will be $\frac{3}{64}$. We must then take account of the contribution to total G_2 variance in q due to the dispersion from G_1 unfixed isolates into G_2 arrays. This component of variance is the increment ($\Delta\sigma_q^2$), while the remaining component is that produced by the original dispersion from G_0 to G_1. These computations of the two components are outlined in Table 12-3; $\frac{3}{64}$ is due to dispersion from G_1 unfixed isolates to G_2 arrays, while $\frac{4}{64}$ is due to the original G_1 array. The student should verify the within-isolate variances and the total variance in q among isolates for the G_2.

This numerical example should serve to illustrate the following relationship for the variance in q for the general case of any generation (t) in terms of previous generations. Derivation of the compounding of variance between generations was worked out by Wright in 1942 and described by Crow (1954) and by Crow and Kimura (1970).

$$(\sigma_q^2)_t = \frac{pq}{2N} + \left(1 - \frac{1}{2N}\right)(\sigma_q^2)_{t-1} \tag{12-2}$$

(variance of q in two successive generations). We can let the $(1 - 1/(2N)) = x$ and then work out successively previous generations for the recurrence relationship of the variance.

$$(\sigma_q^2)_t = \frac{pq}{2N} + x(\sigma_q^2)_{t-1}$$

substituting in terms of G_{t-2},

$$= \frac{pq}{2N} + x\left[\frac{pq}{2N} + x(\sigma_q^2)_{t-2}\right]$$

$$= \frac{pq}{2N}[1 + x] + x^2(\sigma_q^2)_{t-2}$$

Thus, if all generations are included with size of $2N$ after panmixia, we would have

$$(\sigma_q^2)_t = \frac{pq}{2N}[1 + x + x^2 \cdots x^{t-1}] + x^t(\sigma_q^2)_0$$

Since we may regard the initial generation (G_0) with zero variance in q as in Table 12-3, the last term can be considered zero.

Furthermore, the quantity in brackets $[1 + x + x^2 \cdots x^{t-1}] = (1 - x^t)/(1 - x)$. Replacing the x with $(1 - 1/(2N))$ and performing the algebra produces the variance in q for any generation:

$$(\sigma_q^2)_t = pq\left[1 - \left(1 - \frac{1}{2N}\right)^t\right] \tag{12-3}$$

or approximately $pq(1 - e^{-t/2N})$. After an infinite number of generations, the variance converges to $(\sigma_q^2)_\infty = pq$ (see Exercises 4 and 6).

ESTIMATING EFFECTIVE POPULATION SIZE

The fact that the rate of change in variance of q depends on sample size, or effective N, gives us the opportunity to estimate N_e. The magnitude of dispersion within isolates from one generation to the next is inversely proportional to $2N_e$ and directly proportional to the gametic product (pq). As an alternate method for estimating the effective size in real populations to that illustrated under methods of inbreeding (Example 11-1), we may use (12-1), rearranged: $2N_e = pq/\sigma_q^2$, provided that we can observe the arrays of gene frequencies from any parent generation to its offspring isolates. In Example 12-3, the method is illustrated for making that estimation. It is clear that the two methods—one based on rate of fixation under inbreeding ("inbreeding effective number," as in Chapter 11), the other based on dispersion of gene frequencies between isolates each generation from donor isolate to recipient ("variance effective number," as designated by Crow and Kimura, 1970)—give equivalent results. In fact, they should produce the same result in simple situations, as was shown by Wright (1931).

It is also possible to estimate N_e from the increment of variance in q between isolates each generation if we consider the total population array using (12-2) rearranged to solve for N:

$$2N = \frac{pq - (\sigma_q^2)_{t-1}}{(\sigma_q^2)_t - (\sigma_q^2)_{t-1}}$$

where p,q = gene frequencies of the $t - 1$ (parent) generation. In Example 12-2, Figure A, the variances for generation 1 through 5 from a computer simulation can be used for the drift of isolates with known size of $N = 10$ just to confirm the effective size. We obtain the calculations as given by the following:

Generations	pq in $(t-1)$	V_t	V_{t-1}	Estimate of $2N$
1–2	0.24999	0.025067	0.013901	21.14
2–3	0.24995	0.037489	0.025067	18.10
3–4	0.24999	0.048708	0.037489	18.94
4–5	0.24998	0.058026	0.048708	21.60

These four pairs of generations give an average value for $2N = 19.93$, or approximately $N = 10$. Thus, when we control the model by computer simulation, we may gain confidence in this method of estimating effective population size.

EXAMPLE 12-3

ESTIMATING EFFECTIVE POPULATION SIZE

From Example 11-1 it was demonstrated by Buri that the rate of fixation could be used to estimate the effective number of gametes in small populations. Estimates of $2N_e$ can also be made if the dispersion of each of the unfixed isolates (from 1–31 bw^{75} genes constituting an isolate) is measured in each generation. (In Buri's series I, he found that $2N_e$ was not completely independent of q with $2N_e$ being greater for the lower frequencies, but we shall disregard that obstacle and use only the portion of his data that illustrates the point of utility; while his series II was free of that effect, the rate of fixation was not so consistent for that series.)

The five isolates closest to the midpoint in the array of bw^{75}: bw distribution are illustrated below. In the absence of selective differences and assuming random mating plus about equal productivity of parents, the mean variance produced in recipient isolate classes by dispersion from each parent isolate over the generations of the experiment could be combined; these are given from Buri's series I data as a portion of his Table 10 with gene frequencies from midrange.

Isolate No. of bw^{75} Alleles	p	pq	Observed Recipient Variance	Estimated $2N_e$
14	0.4375	0.24609	0.0112	21.97
15	0.46875	0.24902	0.0146	17.06
16	0.5000	0.25000	0.0131	19.08
17	0.53125	0.24902	0.0151	16.60
18	0.5625	0.24609	0.0164	15.01
				Average $2N_e = 17.94$

With Buri including all 31 unfixed isolate classes, his average estimate of $2N_e$ was 17.86 for the series I experiment. The isolates with gene frequencies nearest the middle of the array are very close to that average. On comparison with the estimate made from the rate of fixation (Example 11-1), we see that the two estimates (N_e) are very close; the simplified estimate (N_e) based on

$\frac{1 - \Delta F}{\Delta F}$	$\frac{pq}{\sigma_q^2}$
8.11	8.97

Further, Buri compared the rate of decrease in heterozygosis expected on the basis of (11-10) using $2N_e = 18$ and found fair agreement with observed drop in heterozygosity. The data then analyzed either method to estimate an effective gametic number at slightly greater than 50 percent of the census number of parents in each isolate.

In Buri's series II, where developing fly larvae were less crowded than in series I, effective population size was increased. His estimate of $2N_e$ based on observed recipient variance in q was 23 for series II.

RELATIONSHIP BETWEEN ACCUMULATED VARIANCE IN q AND INBREEDING COEFFICIENT

From Chapter 11 (11-7 and 11-8), we noted that in considering the decrease in heterozygosity per generation, the panmictic index (P_t) decreases at the rate of $(1 - 1/(2N))$. With the consideration of dispersion in the "Noah's Ark" case (Table 12-1), a tally of heterozygotes in G_1 and G_2 gives $\frac{1}{2}$ and $\frac{3}{8}$, respectively, so that heterozygosity descends at the same rate due simply to chance sampling of gametes (irrespective of the "identity" of genes). We may equate the two ways of describing the same phenomenon by considering (11-7) with (12-3) as follows:

$$P_t = \left(1 - \frac{1}{2N}\right)^t$$

Because $P_t = 1 - F_t$, we have from (12-3)

$$(\sigma_q^2)_t = pq[1 - (1 - F_t)]$$

so that if we consider average gene frequencies

$$(\sigma_q^2)_t = \overline{pq}F_t \qquad (12\text{-}4)$$

where $\overline{p}, \overline{q}$ are the average allelic frequencies over all isolates of the population.

In Table 12-1, we note that G_2 is identical with the progeny of sibs in that $P_2 = H_2/H_1 = \frac{3}{4}$, $F_2 = \frac{1}{4}$. Exercise 5 indicates that the variance of q between sibships in G_2 will not be the same as that in the "Noah's Ark" case, however, since the latter random mating within isolates produces slightly greater dispersion than does sib mating. (Note also in Table 12-1 that initial $H_0 = 100$ percent, which is a special case producing H_1 at 50 percent immediately; if we consider H_1 to be the equivalent of the panmictic maximum H_0, then the panmictic index (P_2) will be identical with that of P_1 from an initial random-mating population.) Nevertheless, restriction of the breeding size has an effect that can be described either by distribution of gene frequencies or as an effect of inbreeding. Both effects produce random fixation and loss of alleles and can therefore be considered simply as two aspects of the same process.

WAHLUND'S FORMULA AND BREEDING STRUCTURE

Whenever we examine real populations, we usually find them *not* to be large units approaching the model envisaged by the square law of Castle-Hardy-Weinberg. It is far more common to find populations subdivided by various factors (barriers to interbreeding, ecological or ethological preferences within groups, socioeconomic factors in human populations, and so forth) into subpopulations that are often partially isolated from each other. If gene frequencies differ between subpopulations, which we shall consider isolates, the distribution of p,q values among those groups will bring about a simulation of inbreeding when the entire group of subpopulations is considered together as a unit.

The simplest model is to visualize a large species population subdivided into k isolates, each equal in size to the others, and each random mating within itself. In Chapter 2, we mentioned that stratification into subgroups with different allelic frequencies produces a deficiency of heterozygotes and excess of homozygotes compared with a single unit in which all subgroups are considered together as a random-mating entirety. Example 2-3 demonstrated the outcome. Table 12-4 expresses the frequencies of the subgroups (isolates) from Example 2-3 for convenience and reexamination.

Variance in q between k isolates is described as follows:

$$\sigma_q^2 = \frac{\sum(q - \bar{q})^2}{k} = \frac{\sum q^2}{k} - \left(\frac{\sum q}{k}\right)^2 \tag{12-5}$$

Using a line over each symbol to represent the average, we may rewrite the expression for variance in q as follows:

$$\sigma_q^2 = \overline{q^2} - (\bar{q})^2 \tag{12-5A}$$

that is, the variance in q is the difference between the average homozygote (aa) frequency observed among subgroups and the homozygote frequency expected by squaring the

TABLE 12-4 *Allelic and genotype frequencies for two subpopulations ($k = 2$ isolates), assuming equal size of each, as subdivisions of large population*

Subpopulation "Isolate"	Allelic Frequency		Genotype Frequencies		
	p	q	p^2	$2pq$	q^2
I	0.9	0.1	0.81	0.18	0.01
II	0.3	0.7	0.09	0.42	0.49
Average	0.6	0.4	0.45	0.30	0.25
$(\bar{p} + \bar{q})^2$			0.36	0.48	0.16
Difference			+0.09	−0.18	+0.09
σ_q^2 between isolates = 0.09					

average q value of all subgroups. If subgroups are unequal, weights (w) are most easily assigned to each relative to their proportional sizes so that $\sum w = 1$, and the mean and variance of q will be $\overline{q} = \sum w_i q_i$ and $\sigma_q^2 = \sum w_i q_i^2 - (\overline{q})^2$. Subgroup size, then, causes no problem.

With (12-5A) rearranged in terms of the average homozygote frequency observed among subgroups, we obtain

$$\begin{aligned} \overline{q^2} &= (\overline{q})^2 + \sigma_q^2 \\ 2\overline{pq} &= 2\overline{p}\overline{q} - 2\sigma_q^2 \\ \overline{p^2} &= (\overline{p})^2 + \sigma_q^2 \end{aligned} \tag{12-6}$$

These relationships are illustrated in Table 12-4 numerically. They were first noted by Wahlund (1928). (See Exercise 7 for derivation of the heterozygote statement.)

The student may have already noted the resemblance of Wahlund's expression to the distribution of genotypes under inbreeding at equilibrium (8-3). The increment for homozygotes and deficiency in heterozygotes can be recognized as parallel. The Wahlund proportions are the equivalent of inbred changes in the following way:

$$\sigma_q^2 \text{ (between groups)} = F\overline{p}\overline{q} \tag{12-7}$$

This expression is equivalent to (12-4).

We may conclude that by subdividing the population, the genotypes occur in frequencies as a simulation of a certain degree of inbreeding even though each subgroup (isolate) is random mating within itself. As far as the total population is concerned, there are too many homozygotes and too few heterozygotes compared with the panmictic entire undivided population. We might rearrange the expression (12-7) in terms of F, the "equivalent of inbreeding" as if the entire population were propagated by mating close relatives:

$$F = \frac{\sigma_q^2}{\overline{p}\overline{q}} \tag{12-7A}$$

In Table 12-4, the data would indicate $F = 0.09/0.24 = 0.375$, which is the equivalent of two generations of sib mating. We should note that the symbol F is used here to include a new concept. Many authors use a different symbol (f) for this Wahlund "equivalent of inbreeding," but we shall retain this F and specify its meaning as it develops new application to populations.

In Example 9-1, samples of mouse populations were either heterogeneous from one locale to another or from one generation to another, so that there was often a deficiency of heterozygotes. Without exact knowledge of breeding structure in these populations, it was not possible to distinguish close inbreeding from subdivision and isolation between breeding units, but both could conceivably contribute to the final deficiency. Without knowledge of exact relationship between mates in a population, then, it is hazardous to draw conclusions about breeding structure from an empirical estimation of F.

Multiple alleles in a subdivided population make the Wahlund effect more complicated. Nei (1965) and Li (1969) examined some of the problems ensuing from consideration of multiple alleles with subdivision. Briefly we can summarize (from Li) that with respect to multiple alleles, "each allele's frequency has its own variance among the isolates and there is no unique index or coefficient to measure the degree of association between uniting gametes as caused by the subdivision of the population." While it should be clear that the variance of p equals that of q with a single pair of alleles, with three or more alleles, there will be different covariances for each pair, and the variances of each allele among the isolates will differ from one to another. Consequently, heterozygote frequencies are not all changed by the same amount by subdivision; some may decrease and some increase or remain the same as that of a random-breeding total population, since the covariance of frequencies of alleles A, a, a' . . . may be negative, positive, or zero. For example, if a large population is subdivided into isolates with three alleles at a genic locus at frequencies p_i, q_i, r_i in the ith isolate, with random mating within each isolate to give $(p_i + q_i + r_i)^2$ zygotes, the genotype frequencies over all isolates will be as follows:

	A	a	a'
A	$p^2 + \sigma_A^2$	$pq + \sigma_{Aa}$	$pr + \sigma_{Aa'}$
a	$pq + \sigma_{Aa}$	$q^2 + \sigma_a^2$	$qr + \sigma_{aa'}$
a'	$pr + \sigma_{Aa'}$	$qr + \sigma_{aa'}$	$r^2 + \sigma_{a'}^2$

where $\sigma_A^2, \sigma_a^2, \sigma_{a'}^2$ = variance between isolates for each allele, and $\sigma_{Aa}, \sigma_{Aa'}, \sigma_{aa'}$ = covariance for each pair of alleles.

Thus, all homozygotes will be in excess frequency by an amount equal to their variances between isolates, but heterozygotes will have increments amounting to their covariances, which can be negative, positive, or zero. How much deviation from panmixis there will be owing to the Wahlund effect does not seem to have a simple index (Li, 1969) to apply to the entire population as with a single pair of alleles (see Exercise 10).

HIERARCHICAL POPULATION STRUCTURE

We may gain a better understanding of populational structure and an interpretation for the Wahlund effect when we consider a subdivided population and the probability for homozygosity compared with that of a panmictic population, as noted by Wright (1940, 1943, 1951, 1965). Assume a total population (T) to be divided into isolated subpopulations (S), or "demes," within which gametes unite to produce individuals (I). There are at least three ways for us to consider relative homozygosity or identity between gametes:

1. F_{IS} = homozygosity of an individual (I) due to identical alleles derived from its own subdivision, or the correlation between uniting gametes within the subpopulation of which the progeny individual (I) is a member. For Table 12-4, F_{IS} would be zero because each isolate is randomly mated within itself. In like manner, this probability will depend only on the particular mating system within the subdivision. See Exercise 11 and formula (9-9).

2. F_{ST} = correlation between random gametes within subdivisions relative to gametes of the total population. When subpopulations differ in p, q values, this probability is the equivalent of (12-7A), due to the Wahlund effect. For Table 12-4, F_{ST} = 0.375, as given before. It could also be due to reduction of population size among isolates. See Exercise 12.

3. F_{IT} = correlation between uniting gametes relative to gametes in the total population. Since F_{IS} = zero for Table 12-4, $F_{IT} = F_{ST}$.

When we combine the increments of homozygosity due to either inbreeding or random sampling within subdivisions with effects of subdivision, the principle can be illustrated as follows. Let us consider the two subpopulations of Table 12-4 as an example of combining the effects of subdivision and inbreeding. Suppose that in both subpopulations I and II inbreeding has been going on at the same level and has achieved an $F_{IS} = 0.40$ in both. The genotype frequencies will then be, using (8-3),

	AA	Aa	aa	F
I	0.846	0.108	0.046	0.4
II	0.174	0.252	0.574	0.4
Entire population	0.510	0.180	0.310	0.625

It is evident that inbreeding within demes combined with the Wahlund effect increases homozygosity overall; however, the total F is not the simple sum of $F_{IS} + F_{ST}$ (that is, $0.4 + 0.375$). The key to understanding the relationship is to recall that these amounts of homozygosity and heterozygosity have been defined relative to their expected frequencies in random-mating populations. The panmictic index, Wright's P, or $H_t/H_0 = 1 - F$, as given in Chapter 11, is useful here; it is defined in terms of heterozygosity relative to that in a random-mating population. It is easier to state the principle as follows: the probability for an individual to be heterozygous relative to random mating $(1 - F_{IT})$ is the product of two probabilities—that the individual has not come from the union of two genes identical by descent from a common ancestor within the subgroup $(1 - F_{IS})$ *and* that the individual has not been produced by random combination of gametes that are identical because of some more remote relationship $(1 - F_{ST})$. Thus, $(1 - F_{IT}) = (1 - F_{ST})(1 - F_{IS}) = P_{IT} = P_{ST} \cdot P_{IS}$, where P is the panmictic index. Solving this expression for F_{IT}, we obtain

$$F_{IT} = F_{ST} + F_{IS}(1 - F_{ST}) \tag{12-8}$$

In other words, total probability for being homozygous in the overall population is accounted for by two components: random sampling of gametes within subpopulations (Wahlund effect) plus the fraction of the nonhomozygous remainder resulting from consanguinity or nonrandom mating. In the above example, using (12-8), we obtain $0.625 = 0.375 + (0.4)(0.625)$.

Utility of this principle can be extended to other mechanisms that decrease heterozygosity via random sampling as in the case of population size reduction. As an illustration, suppose 10 wild mice (5♂♂ and 5♀♀) are mated and isolated on an island where their progeny can breed at random. Assume the island is inundated by a hurricane once a year so that all progeny mice are drowned except five pairs that propagate in the following season. For five years in a row, hurricanes come along and reduce the population in the same way to five pairs. In the 6th year when the population is again increasing, a geneticist captures a sibling pair of mice from the island to breed in the laboratory. What would be the inbreeding coefficient of the progeny (F_{IT}) from the sib pair? From (12-8) we may estimate the inbreeding coefficient for these mice as follows: first we must find $(1 - F_{ST})$ after five generations of random sampling of $2N = 20$ gametes. Since in any succession of one generation to the next,

$$F_{ST} = \frac{\sigma_q^2}{\overline{pq}} = \frac{\overline{pq}}{2N} \cdot \frac{1}{\overline{pq}} = \frac{1}{2N}$$

and $P_{ST} = (1 - 1/(2N))$, after five generations in a row, the 10 mice surviving hurricanes on the island would have $(1 - 1/(20))^5 = 0.7738$ probability for not being homozygous. (Note that this result is equivalent to the panmictic index as given by (11-7).) Thus, $F_{ST} = 0.2262$ at the sixth generation.

For the progeny of a sib pair, $F_{IS} = 0.25$, so that putting all these probabilities together as an illustration of (12-8), we obtain

$$F_{IT} = 0.2262 + 0.25(0.7738) = 0.4197$$

Thus, the progeny from the sib pair would be expected to be homozygous at about 42 percent of their genic loci instead of 25 percent expected under constant large population size. (See Exercise 13 for the case of no expansion of the population between generations.)

These two illustrations should serve to emphasize two aspects of random sampling of gametes that lead to increased correlation* between them and thus to increased homozygosity: (1) if sampling is done over two or more subpopulations that differ in allelic frequencies, the Wahlund effect contributes to F_{ST}, and (2) if population size has been reduced for successive generations, the spreading of variability by the stochastic factor each generation contributes to F_{ST}. Consequently, from a practical standpoint, the population geneticist must be aware of these additional contributions to changes in genotype frequencies that in turn produce gene frequency random walk as well. With some gene flow to counteract the isolation of gene pools or with less drastic reduction of numbers per generation, these random elements will be less effective, but will nevertheless be present in real populations to be considered when estimating expected genic changes.

GENOTYPE DISTRIBUTIONS IN REAL POPULATIONS

To what extent accidents of sampling are responsible for observed changes in natural or experimental populations or how the present distribution of genetic elements in those populations is affected by random gametic combination is exceedingly unclear and problematical. The less the strength of deterministic forces (selection, gene flow, and recurrent mutation) on genotypes, the more likely that chance alone *can* be responsible for observed changes and distributions. At this point we must reiterate, purely and simply, that if we look only at the sampling of gametes, three major immediate effects are possible: (1) dispersion of allelic frequencies among isolates, (2) fixation of particular alleles due to continued population size reduction, and (3) chance establishment of genotypes (the "founder principle") in isolates. (After consideration of the main vectorial forces, we shall discuss how these and other random processes such as fluctuations in the deterministic forces may have evolutionary consequences.) The point emphasized here is that while these chance

* Wright (1965) points out that correlation can be negative under disassortative mating within subpopulations so that both F_{IS} and F_{IT} could be negative, although F_{ST} must always be positive (see Exercise 12).

factors can and do operate on real populations, it becomes an immensely difficult task to ascertain in any particular case whether observed changes and distributions of genetic elements have resulted from the action of random factors, from the vectorial forces or, more likely, from some combination of the two.

Of course, whenever there is a puzzle to solve, controversy ensues and bias tends to appear. Unfortunately, it is often easier to make judgments without complete objectivity than to withhold judgment until all facts are known, so that the importance of random genetic drift to evolutionary situations has been variously interpreted or misinterpreted by biologists, anthropologists, paleontologists, and mathematicians when "needed" to help explain or argue about possible mechanisms at work in populations. Those biologists trained in Darwinian selection traditions often have minimized stochastic elements (for example, Ford, 1964), while the more mathematically inclined have emphasized the importance of chance (Crow and Kimura, 1970). Controversy between well-known evolutionists over the interpretation of genetic distributions as due to either random or selective factors is exemplified by three arguments: (1) Fisher and Ford (1947) versus Wright (1948) on the interpretation of changes in frequencies of color varieties in the scarlet tiger moth *Panaxia dominula* (see also Example 10-4); (2) Ford (1964, 1975) and Dowdeswell, et al. (1955, 1960) versus Dobzhansky and Pavlovsky (1957) on the interpretation of distribution and changes in wing-spotting patterns of the meadow brown butterfly *Maniola jurtina* on the Scilly Islands; (3) Cain and Sheppard (1950–1968) versus Lamotte (1959) on shell pattern distributions in the snails *Cepaea nemoralis* and *C. hortensis*.

A clear demonstration of the patchy distribution expected from random genetic drift in small semiisolated subgroups (mating units or demes) of a large population was found by Selander (1970) when he trapped house mice (*Mus musculus*) in a grid pattern throughout the interior of a large barn (Hildreth farm of dimensions 192 × 48 ft) in Texas. From the mice living under the barn floor, nearly 380 were typed for genotypes of Esterase-3 (*MM*, *MS*, and *SS*). The distribution of these genotypes is shown in Figure 12-3. It hardly seems possible that such a distribution could be accounted for by selective differences between one part of the barn and another. The mosaic can only be due to small breeding unit size, inbreeding within units as mentioned in Example 9-1, or both.

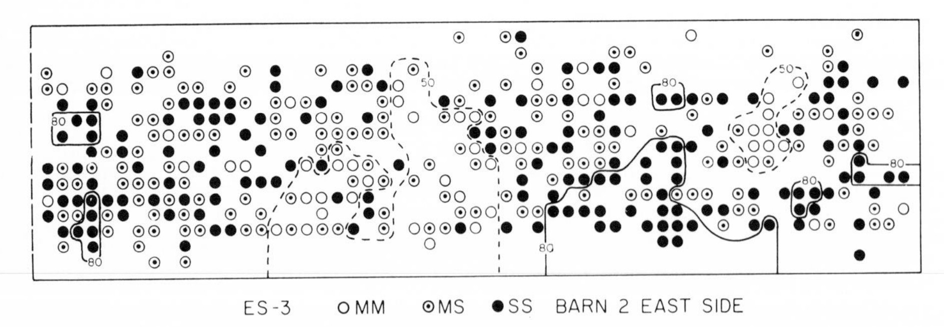

Figure 12-3. Esterase-3 genotypes of mice collected in a grid pattern on the east side of barn 2 at Hildreth Farm, Texas. Symbols as indicated M = medium and S = slow alleles in electrophoretic gel. Contours mark isofrequencies of 80 percent S (solid line) or 50 percent (dashed line). Area 192 ft × 48 ft (from Selander, 1970).

Wright (1955, 1960, 1967) has repeatedly insisted that continued drift in small isolated populations will be nonadaptive and will usually lead to extinction following genic fixation. More significantly, random drift has two aspects: (1) preventing too strict adherence to the influence of the directed vectorial pressures that can limit genetic plasticity, and (2) providing the condition for most rapid evolution—namely, that the population totally consist of partially discontinuous subpopulations among which allelic frequencies must be dispersed. He considers that, by subdivision, random processes and local selection *jointly* bring about differentiation of gene pools; in turn, following differential growth and dispersion of subpopulations (demes), interdemic selection may well achieve far greater success in an adaptive strategy that meets the challenge of common aspects of the environment than could be achieved by intrademic selection alone.

FOUNDER EFFECTS, FLUSHES AND CRASHES

Perhaps the greatest utility of the dispersive process described in this chapter is the proliferation of diverse gene and genotype frequencies—the propagation and multiplication of genic combinations that would not be likely to occur in single large populations under limitations of strictly deterministic forces. The changes then engendered may well lead occasionally to "genetic revolution" of the population's gene pool (Mayr's "founder principle") and to speciation (Carson's "flush and crash principle") (see Chapter 19).

The founder principle enunciated, analyzed, and documented by Mayr (1942, 1954, 1963) "designates the establishment of a new population by a few original founders (in an extreme case by a single fertilized female) that carry only a small fraction of the total genetic variation of the parental population." He stressed the serious loss of genetic variability through inbreeding depression after the founding takes place—the "primary poverty of genetic variability"—followed by a "genetic revolution," or reconstitution of balanced genetic systems with newly acquired allelic frequencies. Most illustrations given by Mayr, however, show *not* how depauperate the genetic variability is in cases of small founding populations, but rather the disparate or unique gene frequencies that characterize them. The latter aspect of the founder principle is undoubtedly true, but the former, the "poverty of genetic variability," may well be argued. For example, in a general discussion on the nature of colonizing species, Lewontin (1965) took exception to the genetic "poverty" aspect of the founder principle and stressed the change in allelic frequencies as a more proper point of emphasis for that principle:

> *If there is a colonization by a single fertilized female, there will be loss of genes and a radical change in gene frequencies at loci where alleles are at intermediate frequencies. But the one thing that will not happen is a profound change in the total amount of genetic variation available [to selection] .... If you cannot select some quantitative trait in a population which is founded by a single fertilized female, then you almost certainly cannot select for it in the original parental population either. So you must distinguish very carefully between the drift of gene frequencies with the loss of rare genes and the essential preservation of the total picture of variability .... If there is a small colonization, the gene frequencies of all the genes may change radically.... [That fact] doesn't make the population much less selectable.*

(See also Example 16-4, in which several selection experiments were carried out on the progeny of a single pair of wild-caught drosophila.)

Nei, et al (1975) explored mathematically and with computer simulation the problem of genic variability reduction when a population goes through a bottleneck, followed by population size expansion. New mutations occurring during the expansion phase restore much variability (see Chapter 13). The amount of reduction in average heterozygosity per locus during a bottleneck season depends not only on the "size of the bottleneck" but also on the rate of population growth afterward. With rapid population growth after the bottleneck, the reduction in average heterozygosity was found to be rather small. On the other hand, the loss in the average number of alleles per locus is profoundly affected by bottleneck size but not so much by subsequent population growth. This difference in the effect of the bottleneck on multiple alleles versus one pair of alleles is derived from the fact that random genetic drift eliminates low-frequency alleles most easily. However, when population size is restored, the average number of alleles increases faster than average heterozygosity.

Examples of isolated or semiisolated populations with radically altered genotype or allele frequencies compared with those of their presumed progenitor populations and therefore likely to be founded by a few individuals abound in the literature. Summaries are presented particularly in Mayr (1963), Ford (1964), Grant (1963), Dobzhansky (1970), and Baker and Stebbins (1965). The earliest classic cases of genetic heterogeneity among populations between which there was little or no migration and which seemed to have all the diagnostic characteristics of random establishment of genotypes and the founder principle were those recorded by Gulick in the nineteenth century and described by Kellogg in 1907 for land snails (Achatinellinae) on various Pacific islands. Later, in 1932, Crampton extended Gulick's observations for species of *Partula* on Tahiti, Moorea, Society Islands, Marianas, Guam, and Saipan. Certainly, Wright was much influenced by these early observations when he developed his concept of random genetic drift. Many of the snail species were restricted to single islands, and often single species were endemic to one or two valleys of particular islands, so that the establishment of genotypes by small numbers of founding individuals seemed to be a very likely explanation for their heterogeneous distribution. As to the argument that the local differentiation of these snail populations has been due to selective neutrality of their features and random fixation of genotypes brought about by continued small population sizes, it can be pointed out that more recent observations have put in doubt the hypotheses of both selective neutrality and small population size. In 1962 and 1967, Clarke and Murray collected *Partula* spp. on the island of Moorea extensively and on a fine scale. They pointed out (1971) that patterns of gene frequencies of *P. suturalis* resemble those found in *Cepaea nemoralis* by Cain and Currey (1963) and are designated as "area effects" (see Example 3-3). Random genetic drift is argued against as the sole cause of the distribution by the fact that the areas and population sizes are too large for much random differentiation and that there are significant associations among factors such as altitude, density of species, and mean shell lengths. They point out, however, that small founder groups arriving in any region could have established differences that have persisted between populations (see also Ford, 1964, for discussion, pp. 168–171).

An experimental verification of the greater genetic diversity among populations with small numbers of founders than with large numbers is given in Example 12-4. Any particular sample of a small number of gametes taken from a large source population can be expected to have a wide variety of p_i, q_i values so that from one sample to another a dispersion of p

and q will be established. The populations' final genetic compositions were a result of interaction between acquiring a particular gametic sample due to chance and the subsequent action of selection on the set of genotypes derived from that sample.

In human populations, anthropologists and human geneticists have documented a large number of cases that are best explained by the founder principle. From the observations on blood group frequency differences, as summarized by Boyd (1950), cases with unusual gene frequencies, such as those of Greenland Eskimos (high M, low N), Australian aborigines (low M, and high N), Amerinds of the Blackfeet and Blood tribes with high A compared with other Amerinds (low A), and Basques with high rh (*cde*) are likely to be accounted for by this principle (see also Mourant, 1954; Race and Sanger, 1968, 1975; Giblett, 1969; and Spuhler, 1972). Small population isolates founded by a few individuals with unique genetic frequencies differing from their progenitor populations have been described as follows:

1. Glass (1954) studied the religious isolate known as the Dunkers of Pennsylvania (German Baptist Brethren)—a sect founded by 27 families in the early eighteenth century who had immigrated from the Rhineland. Because they are genetically isolated by strict marriage customs from mixing with those outside their sect, their gene frequencies are unique, especially in having high blood types A (I^A frequency = 0.60 compared with about 0.40 for the United States or Germany), M (p_M = 0.66 compared with 0.55 for outsiders), and lower rh (p_{cde} = 0.11 compared with 0.15).

2. McKusick, et al. (1964) described a noteworthy case among the Old Order Amish of Lancaster County, Pennsylvania, a group of about 8000 descended from just three couples who immigrated to America in 1770. They possess an Ellis-vanCreveld syndrome at a frequency of about 0.07, but that recessive is extremely rare in other humans.

3. Steinberg, et al. (1967) summarized observations on the Hutterites of South Dakota (descendants of Tyrolean Anabaptist Protestants who immigrated to the United States in the 1870s)—a group of about 9000 descended from not more than 92 immigrants. Frequencies of M-N blood type alleles among the subgroups (colonies) of these people varied from 0.36 M to 0.90 M, as the distribution of one locus was illustrated. Equally great differences were also recorded for ABO, Rh, and Kell loci. This divergence in the M-N locus frequencies has been exceeded only by Australian aborigines and Eskimos.

4. Livingstone (1969) listed several deleterious recessives and a few dominants that have attained uniquely high frequencies in particular populations:

Hereditary Condition	Population
von Willebrand's disease	Åland, Finland
Porphyria	Sweden and South Africa
Tyrosinemia	Chicoutimi, Quebec
Tay-Sachs disease	Eastern European Jews
Hb^S	Brandywine, Southeast Maryland

Because deleterious alleles are eliminated from large populations by selection, variations in their frequencies in major human populations today could be due to the founder effect. Thus, chance establishment of an increased frequency considerably above that which could be accounted for by a balance between mutation and selection (Chapter 18) would result in a state of transient polymorphism for these deleterious alleles since they would be slowly eliminated if not counteracted by factors favoring the polymorphism (see Chapter 14). In whatever way we may account for the origin of the allele's frequency in those particular subgroups today, the uniqueness in a human isolate is most likely to have been brought about by particular founders.

5. Marked differentiation among human isolates and islanders has been recorded by Morton and Yamamoto (1973), Buettner-Janusch, et al. (1973), Steinberg and Morton (1973), and Yamamoto and Fu (1973) for a number of blood markers. They all came to similar conclusions: reduction of gene flow among tribal or island groups has increased genetic diversity among them. These data can be compared with those collected by Cavalli-Sforza, et al. (1964) on genetic differentiation among villages in Italy to the west of the city of Parma; among the parish-villages in the mountains, the F value (σ_q^2/pq) for various blood group alleles was about 0.03, while it diminished gradually with lower elevation to about 0.002 near the city of Parma. Heterogeneity between gene frequencies in the upper Parma valley villages is evidenced from the data of Table 12-5, collected by Cavalli-Sforza, et al. (1964). From the chi-squares, there is significant heterogeneity for all alleles except one in the ABO system, and the Wahlund effect is considerable. Evidently, genetic differentiation by sampling within isolated or semiisolated human populations, whether tribes, clans, or "civilized" villages, has been and is commonly true.

Carson (1967, 1968, 1975) emphasized the importance of the dispersive process to evolutionary change, especially as a result of sudden populational increases, or "flushes," often followed by reduction in numbers to low levels, or "crashes." While many populations are structured into subunits, the oscillation in size—or, more important, the occasional stimulation to greater size that is not cyclic but sporadic (flush)—can be of significant consequence to the establishment of new genetic combinations, leading to new ecotypes, races, or species formation. Flushes might follow from the invasions or introductions of founding individuals into new territory in which selective forces, which had been maintaining balanced numbers in the progenitor population's "home ground," could be reduced or eliminated. Colonizations by nonnative weeds, pests, and parasites across wide areas are well known. Alternatively, a flush may be stimulated by introduction of alien genotypes to a well-adapted local population via some interpopulational hybridity and resulting "luxuriance" of fertility and productivity. Experimental demonstration of hybrid vigor as seen in crossbred F_1's and the contribution of this vigor to the continued biological success of a group has been achieved by Carson and his students using drosophila (see Example 5-3): "Thus a population of mutant individuals held under strictly uniform environmental conditions was maintained by natural selection at an equilibrated size. Following introduction of one haploid set of wild type autosomes, the size of the population trebled in three generations and has remained essentially so since, simultaneously retaining all five

TABLE 12-5 *Heterogeneity chi-squares between allelic frequencies for three blood group loci among 37 parishes in the upper Parma Valley (66 villages with 15,000 persons) (from Cavalli-Sforza, Barrai, and Edwards, 1964)*

Blood Group Alleles	X^2	*df*	*p*	$\frac{\sigma_q^2}{\bar{p}\bar{q}}$
M-N	102.76	36	<0.01	0.048
A_1	98.74	33	<0.01	0.052
A_2	40.23	33	≈ 0.20	—
B	76.08	33	<0.01	0.034
O	110.25	33	<0.01	0.061
r	81.77	33	<0.01	0.038
R_1	75.46	33	<0.01	0.033
R_2	54.54	33	≈ 0.02	0.017
rh remainder	80.01	33	<0.01	0.037
Mean				0.0356 ± 0.006

mutants at substantial frequencies in the experimental populations." The biomass of the populations after the hybridization exceeded the level of the wild-type control population, which had been used as a donor. Hybrid vigor may well be exploited in a population.

As envisioned by Carson (1968), "the crest of the population flush is characterized by a multitude of recombinant genotypes. Under the conditions that prevailed during the log phase of growth, relatively few recombinants would have been subjected to selective elimination. . . . Populations at the crest would accordingly be characterized by extraordinarily great polygenic variability including many recombinants with relatively low fitness." Relationally balanced genetic systems (as discussed by Mather) would recombine to form new combinations for further selection. Recombination might lead to a breakdown of genetic complexes controlling high fitness, but in any case the population at the crest of the flush may become more vulnerable to density-dependent factors so that adjustment is lacking among genome recombinants and a crash in numbers usually ensues, especially if the population does not disperse or migrate outward. On the other hand, with dispersal, colonizations into marginal localities could result in founding effects with nonadaptive or neutral genetic changes that would bring about extensive populational differentiation. Genetic data now being collected on the extraordinary drosophila fauna of the Hawaiian archipelago indicate, more clearly than the instances of the Pacific snails mentioned above, that most of the several hundred endemic species were derived ultimately from just a few ancestral individuals—founders that may well have gone through the flush, recombination explosion of the gene pool, and crash stages in order to bring about such diversity. Finally, key dispersive events may easily follow depletion of a population, leaving remnants, isolated demes, or a mosaic of groups with gene flow reduced between them. These events undoubtedly have greatest significance for further evolution of new gene pools. (See Chapter 19 for a discussion of founding and speciation.)

EXAMPLE 12-4

FOUNDER EFFECT PRODUCES DIVERGENCE OF GENE FREQUENCIES

Dobzhansky and Pavlovsky (1957) described a set of experiments in which the determinacy of outcome of selection between two chromosomal arrangements of *Drosophila pseudoobscura* (*AR* and *PP*) was more uniform and predictable for populations with large numbers of founders than for those with just a few founders each generation.

Twenty replicate experimental populations were kept in a uniform environment for about 18 months. The foundation strains of all the populations consisted of second-generation hybrids between flies descended from a Texas population with *PP* arrangement of chromosome 3 and flies of California (Mather) origin with the *AR* arrangement in the same chromosome. Ten populations were founded by 4000 flies initially, while another 10 populations were started with only 20 founders. Thereafter flies were allowed to breed at random and grow to the maximum population size for the cage conditions, which were identical as far as possible. While $AR = PP = 50$ percent each in all 20 populations at the start, the *PP* arrangement varied, as shown in Figure A, with mean frequency and variance, as given in Table A, at five- and eighteen-month intervals, with a variance about four times greater for the populations initiated with the small number of founders than for the variance of frequencies among the large populations. Also, it should be noted that the "large" populations uniformly showed a steady decrease in *PP* for the time of sampling, while the "small" populations in three cases appeared to be returning to the original *PP*

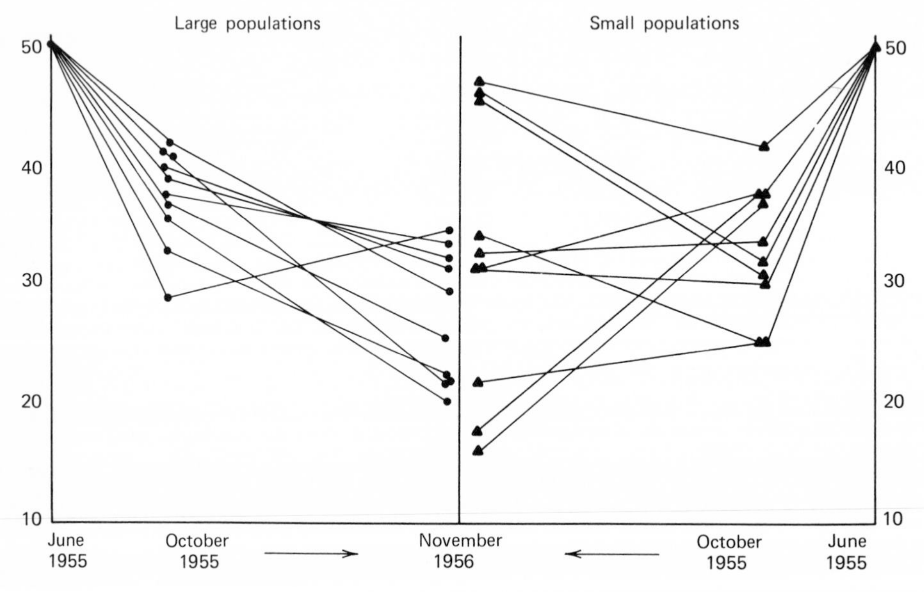

Figure A. The frequencies in percentage of *PP* chromosomes in 20 replicate experimental populations of mixed geographic origin (Texas by California); populations founded by large initial size samples ($N = 4000$) on the left and by small initial size samples ($N = 20$) on the right (from Dobzhansky and Pavlovsky, 1957).

TABLE A *Frequencies in percentage of PP chromosomes in the 20 experimental populations founded by large numbers (4000) and by small numbers (20) at 5 months and at 18 months after initiation. Variance is between the 10 replicate population PP frequencies (from Dobzhansky and Pavlovsky, 1957)*

Large Populations			Small Populations		
No.	Oct. 1955	Nov. 1956	No.	Oct. 1955	Nov. 1956
145	39.3	31.7	155	37.7	18.0
146	42.3	29.0	156	30.7	32.0
147	29.3	34.7	157	31.0	46.0
148	38.0	34.0	158	32.3	46.7
149	33.3	22.7	159	34.3	32.7
150	36.0	20.3	160	41.7	47.3
151	40.3	32.0	161	37.3	16.3
152	41.0	22.3	162	25.3	34.3
153	37.0	25.7	163	37.7	32.0
154	42.0	22.0	164	25.3	22.0
Mean	37.85	27.44	Mean	33.33	32.73
Variance	15.30	26.96	Variance	26.73	118.91

frequency. Heterogeneity is much greater in those populations started with the smaller sample of genetic variability, presumably because the founder effect may have produced a heterogeneity of sample genomes with which selection could act to change the *AR-PP* frequencies. Segments of genome hybridity between Texas and California races must be smaller in the 20-fly founder populations than in the larger founder populations. It may reasonably be inferred that evolutionary changes involving sampling of the genome in founding effects, coupled with subsequent selective changes controlled by the conditions within which the genotypes of the population interact, are common in nature.

This experimental result has been criticized by Clarke and Williamson (1958) as "not satisfactory evidence of genetic drift" in experimental populations, because in the "small" populations there seemed to be three possible points of equilibrium (at 20, 33, and 47 percent *PP*), which "would be indeterminate in the sense that we do not know the genotypes of the originators of each population." In fact, that is precisely what Dobzhansky and Pavlovsky have shown: the founding effect creates divergent genotype frequencies but *not* regular and consistent reduction of genic variation. We must consider that some of the small populations were provided with more, and some with less, polygenic balanced genomes from either original population. The more Texas-like the founding sample, the more Texas-like the outcome (comparing favorably with *AR-PP* pure Texas origin experimental populations) and similarly for those of California origin. With more nearly equal genome segments of hybridity, the *AR*:*PP* outcome would be less predictable because the interpopulation hybrid genome was not coadapted by selection history (see Chapter 19).

Powell and Richmond (1974) studied six laboratory populations of *D. paulistorum*. Two were initiated with over 100 independently derived lines from a natural population, and four were initiated with about six independently derived lines. These lines were segregating for allozymes (F and S) of the tetrazolium oxidase locus (To) on the X chromosome. Over a period of 700 or 900 days, the frequencies of alleles in the 100-line populations remained very stable, while the frequencies of the same alleles in the 6-line populations were very erratic. Thus, the founder effect was similar to that observed by Dobzhansky and Pavlovsky, but the difference in behavior in the case of observing a single locus (To) must be caused by the incorporation of the amount of genetic background variation (linked complexes or epistatic interactions) likely to affect the stable equilibrium of this locus.

EXERCISES

1. Use Table 12-1 for the "Noah's Ark" case and the G_2 frequencies of five types of isolates. Expand the binomial of within-isolate $p_i - q_i$ values, total up the G_3 isolate frequencies, and confirm these data from G_3 of Table 12-2.
2. (a) If a large panmictic diploid population has $p_o = q_o = 0.5$, what would be the frequency distribution of q values for samples of 10 individuals per sample?
   (b) If $p_o = 0.8$, $q = 0.2$, what would be the frequency distribution of samples with 10 individuals each?
   (c) If samples contained 50 individuals each, after $p_o = q_o = 0.5$ as in (a), what percentage of all such samples would occur between the q values of 0.45 to 0.55? What would be the 95 percent limits for q among samples of $N = 50$?
3. A computer simulation of random genetic drift is done on a population similar to that in Example 12-2. Assume 500 isolates of size $N = 10$ and initially $p = q = 0.5$ with G_0 all heterozygotes. For the first 10 generations, the following are isolates that either lose A or fix A:

	No. Isolates	
G	Zero A	All A
1	0	0
2	0	0
3	2	1
4	5	3
5	8	2
6	6	6
7	11	8
8	6	6
9	9	12
10	6	8

   Calculate the rate of fixation ΔF and estimate N_e using formula (11-3).
4. A computer simulation of random drift is done using 8000 isolates all of constant size (N) and an initial $p = 0.1250$, using the same program as that in Example 12-2.
   (a) From the variances in p_A between isolates each generation, use formula (12-2) rearranged to find N from the following data:

G	p_A	Variance of p_A
1	0.1263	0.0134
2	0.1264	0.0254
3	0.1256	0.0355
4	0.1242	0.0444
5	0.1248	0.0527

(b) Show that the expected limit of variance of p_A between isolates $= pq$ by employing the general expression for variance from the following (letting $A = Y$ variable with a value of 1 and $a = 0$):

	Y	Y^2	f
A	1	1	p
a	0	0	q

5. (a) Compute σ_q^2 between sibships in G_2 for Table 9-2. Verify that $\sigma_q^2 = \frac{3}{32} = 0.09375$.
(b) How does that value compare with G_2 in the "Noah's Ark" case where $N = 2$?
(c) Use formula (12-7) to show that for sibmating $\sigma_q^2 = F\overline{pq}$. Why is there a difference between the result under sibmating compared with the "Noah's Ark" model?
6. With a computer simulation of random drift such that constant isolate size has $2N = 8$, the distributions of isolates by the number of A alleles present is given below at five-generation intervals following the start from an initially large population.

	Number of A Alleles per Isolate								
G	0	1	2	3	4	5	6	7	8
5	5516	537	564	446	335	238	175	105	84
10	6403	179	221	196	174	158	149	119	401
15	6723	81	105	94	78	107	95	78	639
20	6887	39	47	51	33	55	42	42	804
⋮									
40	7029	3	2	4	4	3	4	4	947
⋮									
60	7038	0	0	0	0	0	0	0	962

(a) What would you say was the original allelic frequency likely to be for the initial large population from which isolates were formed?
(b) How would you describe the distribution of isolates by allelic number over these generations given?
(c) At what generation does it seem that the probability for all isolates with unfixed frequencies will be about equal?
(d) What is the variance in q between isolates at the G_5? at G_{20}? at G_{60}?
7. A large population is subdivided into small isolated groups ("demes") within which mating is random. The following table represents five demes differing in p,q values (assume equal-sized demes):

Deme	p	q	p^2	$2pq$	q^2
A	0.9	0.1	0.81	—	—
B	0.7	0.3	0.49	—	—
C	0.6	0.4	0.36	—	—
D	0.2	0.8	—	—	—
E	0.1	0.9	—	—	—
Average	—	—	—	—	—

(a) Fill in the blanks of the table and verify that the *deviation* of $\overline{q^2} - (\bar{q})^2$ (average homozygote frequency − expected panmictic homozygote frequency) is equal to the variance in q between demes.
(b) What would be the F-coefficient equivalent for this population (F_{ST})?
(c) In Wahlund's expression, the average homozygote observed overall is always greater than the expected panmictic homozygote frequency—that is, $\overline{p^2} > (\bar{p})^2$. Show this to be true for a case with k subgroups by expanding as follows: let $k = 2$, then $= 3$, then the general case (fundamental inequality for variance):

$$\frac{\sum p_i^2}{k} > \left(\frac{\sum p_i}{k}\right)^2$$

8. When several samples are averaged together, Wahlund's principle may be used for estimating expected genotype frequencies. In Chapter 2, Example 2-4 listed six samples of *Drosophila pseudoobscura* chromosómal arrangements over a period of time in which the arrangement frequencies changed. Thus, estimation of frequencies expected from the square law was based on the average frequencies $\overline{p_{ST}}$ and $\overline{q_{CH}}$.
(a) Verify the variance of p between samples as 0.00281.
(b) For the totals of the six samples in Example 2-4 observed—$ST/ST = 346$, $ST/CH = 441$, $CH/CH = 113$—first obtain expected frequencies based on random mating as if there had been no change from $\overline{p_{ST}} = 0.6292$, $\overline{q_{CH}} = 0.3708$. Then correct the expected values using the *var.* p and the Wahlund expression.
(c) Obtain expected karyotype frequencies and expected numbers (total 900). Verify a chi-square of 3.52. What would be the probability that rejection of the null hypothesis (no departure from random mating) would be in error?
(d) What is your opinion about whether heterokaryotypes (ST/CH) are significantly in excess or not in these samples?
9. Given the following two subgroups (demes) I and II (assume equal sizes) within large populations (a) and (b) below with the genotype frequencies indicated, find F within demes, F_{IS}, F_{ST}, and F_{IT}:

		AA	Aa	aa
(a)	I	0.855	0.090	0.055
	II	0.153	0.294	0.553
(b)	I	0.810	0.180	0.010
	II	0.258	0.084	0.658

(c) In what respects do the (a) and (b) populations differ from the population in Table 12-4 and that on p. 354.

10. A population is subdivided into two equal isolates, I and II, which differ in frequencies at a multiple allelic locus (A, a, a'). Gametic combinations occur at random within these two isolates according to the following 3 × 3 diagram:

		A	a	a'
A	I	0.25	0.15	0.10
	II	0.01	0.03	0.06
a	I	0.15	0.09	0.06
	II	0.03	0.09	0.18
a'	I	0.10	0.06	0.04
	II	0.06	0.18	0.36

(a) Find the allele frequencies of these two isolates and the variance of each frequency (between isolates). (Verify, for example, that $\sigma_A^2 = 0.04$.) Find the covariances of each pair of alleles' frequencies. (Verify, for example, that $\sigma_{Aa'} = -0.04$.)

(b) Show that the average genotype frequencies in the population differ from the expected panmictic frequencies (if there were no subdivision) by the amounts given on p. 353 (from Li, 1969).

(c) Nei (1965) showed that when a population is subdivided and the genotypes result from multiple alleles, F_{ST} (Wahlund effect deviation from panmixis) $= -CoV_{ij}/\bar{p}_i\bar{p}_j$, where $\bar{p}_i$ and $\bar{p}_j$ are mean frequencies of alleles A_i and A_j, and CoV_{ij} is the covariance of their frequencies. Find the value of this F_{ST} for each pair of alleles. What can you say about a single index of Wahlund effect in a case of multiple alleles?

11. Wright defines F as the correlation between uniting gametes in a population. If we give each allele a quantitative value (let $A = 1$, $a = 0$), it is easy to illustrate the calculation of correlation between uniting gametes for subpopulation I on p. 354 as follows:

Progeny Genotype	Parent Gametes X(♀)	Y(♂)	Frequency (f)		fXY	fX	fY
AA	1	1		0.846	0.846	0.846	0.846
Aa	1	0	0.108	0.054	0	0.054	0
	0	1		0.054	0	0	0.054
aa	0	0		0.046	0	0	0
Sum				1.000	0.846	0.900	0.900

Covariance, $\sigma_{A-a} = 0.846 - (0.9)^2 = 0.036$—Variance, $\sigma_A^2 = \sigma_a^2 = 0.90 - 0.81 = 0.09$

Therefore, correlation between uniting gametes is $F = \sigma_{A-a}/(\sigma_A)(\sigma_a) = 0.036/0.090 = 0.4$ (as given on p. 354).

(a) Verify the correlation (F) for subpopulation II on the same page.
(b) Find the F value as a correlation between uniting gametes for the following genotype frequencies: $AA = 0.132$, $Aa = 0.336$, $aa = 0.532$.

12. Wright (1965) defines F-statistics as follows for a subdivided population:

F_{IS} = "average correlation between uniting gametes over all subdivisions relative to those of their own subdivision"

F_{ST} = "correlation between random gametes within subdivisions relative to gametes of the total population"

F_{IT} = "correlation between random gametes over all the population"

(a) Find F_{IS} and F_{IT} by the correlation method for the following genotypes and their frequencies in a pair of equal-sized subdivisions of a population:

Subdivision	AA	Aa	aa	
I	0.846	0.108	0.046	(Note: same as in
II	0.132	0.336	0.532	Exercise 11)
Whole population	0.489	0.222	0.289	

(b) Wright's definition of F_{ST} at first seems difficult to understand because the correlation between random gametes within any subpopulation must be zero by definition (try testing the F for genotype under the square law.) Then the correlation Wright refers to must employ the average genotype frequencies over the population on the assumption of panmixia within subdivisions. Calculate the F_{ST} by the correlation method for the data in Table 12-4.
(c) What is F_{ST} for the data above in part (a) of this exercise? Why are these two F_{ST}'s (parts a and b) alike (or different)?
(d) Verify that

$$F_{IS} = \frac{\Sigma\left(1 - \dfrac{H_i}{2p_iq_i}\right)}{k}$$

where i refers to values within each subdivision in k subdivisions.

$$F_{ST} = \frac{\sigma_Q^2}{\bar{p}\bar{q}}$$

where σ_q^2 = variance in q between subdivisions, and line over frequencies = average for subdivisions.

$$F_{IT} = \left(1 - \frac{\bar{H}}{2\bar{p}\bar{q}}\right)$$

(e) Show how F_{IS} and/or F_{IT} could be negative.

13. Referring to the case of a population of ten mice (five pairs) isolated on an island and swept by a hurricane each generation, we might have assumed a longer generation time for these mice and possibly some predation against them so that the mice would not be able to produce a large random-mating progeny between cataclysms. Let us assume the mice are only able to maintain exactly the same population size (replacement) of ten each generation. What correction would we need to make in the calculation of F_{IT} after five generations on the island? Explain.
14. Allozymes controlled by 27 genic loci were studied in a perennial self-incompatible herb of prairies in Illinois, namely *Liatris cylindracea* (Schaal, 1974, cited by Levin and Kerster, 1974). Fifteen loci proved to be polymorphic. A population was divided into 66 square meter sections, and the bulbous corm of 60 plants was collected from each section. *F*-statistics were calculated on plants within and among sections with the following representative estimates on the polymorphic loci:

	Locus	F_{IS}	F_{ST}	F_{IT}
Highest F_{IS}	Est-2	0.506	p.219	0.614
Lowest F_{IS}	Per^-	0.100	0.014	0.112
Highest F_{ST}	Ap-1	0.467	0.224	0.586
Lowest F_{ST}	Est-3	0.429	0.009	0.434
Average F		0.407	0.069	0.426

(a) What can you say about the relative amounts of diversity within versus between sections?
(b) If the mating system alone were responsible for the amount of homozygosity in this population, what would be expected from genic locus to locus for F values?
(c) What factors do you suppose might account for the genetic structure of this population?

15. In Example 12-4, Table A, variances between replicate populations are given for those descended from an initially large sample and from small sample founders. Do these variances have any relationship whatsoever to the expected increment variance in q between isolates described by formulas (12-2) and (12-3)? Why or why not?
16. To what factors would you attribute the heterogeneity between allelic frequencies among the villages of the Parma Valley in Italy (Table 12-5)? How do you interpret the right-hand column of that table? Look up a later article by Cavalli-Sforza (1966) in which more extensive data (more loci and more populations sampled) are presented. The F estimates range from a low of 0.029 for the Kell blood group with little populational differentiation up to a high of 0.382 for the R^0 allele of the Rh blood group. What possible factors might account for differences in this estimate? Should the estimate be expected always to be uniform? Why or why not?
17. DeFries and McClearn (1972) described experiments in which they tested certain genetic inbred and outbred strains of house mice for male social dominance and for siring of litters. They introduced three males and three females into a Y-shaped triad of joined mouse cages between which animals were free to migrate. The males began to fight within a few minutes, and social dominance could be evaluated by counting

wounds on tail and hindquarters; the male with least wounds indicated the most dominance. A summary of results was as follows:

Type Strains Used	Litters Obtained	No. Litters Sired by Dominant ♂
Inbreds both sexes	61	56
{♀ heterogeneous stock {♂ inbred	42	40

(In testing relative outbred-inbred genotypes for social dominance, outbred ♂♂ had significantly fewer wounds than inbred ♂♂.)

(a) Does social dominance have a role in "choice" of mate and progeny production? How would results such as these affect estimates of effective breeding unit size if natural populations of house mice showed mating behavior like that described here
(b) What is the likely population structure of house mice like those described by Selander for the large barn in Texas (Figure 12-3)?

13

MUTATION: SOURCE OF NEW VARIATION

The decay of genetic variability seems inevitable from some of the considerations set forth in the last few chapters. When population size is limited over several generations or when nonrandom-mating systems lead to fixation or loss of alleles, both phenomena being more realistic for many populations than the large random-mating ideal of the Castle-Hardy-Weinberg equilibrium principle, the tendency to lose stored genetic variability would appear to be more pervasive than any trend to increase or to conserve genetic potential. We must recall, however, that in making these considerations, we have not looked at the effects of any counterbalancing forces. Now it is vital to study the main deterministic forces that counteract genetic decay: mutation (supplying new variation), selection (organizing adaptive genetic systems), and migration (gene flow among populations).

It would seem perhaps to the novice geneticist that mutations occur at most genetic loci so infrequently and sporadically that single mutations are unique events or at most so rare that they would have very little effect in the long run on any sizable population of organisms. They would seem therefore not to be worth the effort of consideration. While it is true that spontaneous mutation at single loci is on the average so slow per cell generation as to take a very long time to change gene frequencies noticeably, many features of mutation are of vital concern to population dynamics.

Fundamentally, mutations increase the hereditary variety in a population, providing the resources for the action of natural selection and evolution. From current evidence bearing on the molecular nature of the mutation process, structure of DNA and chromosomes, and the types of mutations being monitored, we can say unequivocally that mutations occur in general without regard to the functional needs of the organism. They occur for reasons determined by the fine structure and biochemical dynamics of cells in which they take place independently of any adaptive needs by the organism. We say "mutation takes place at random" without known directive determination; this is at present unpredictable as far as particular locus mutational events are concerned. (That statement is not to imply that the overall amount of mutation per organism may not be regulated in some way by the adaptation process, a point of present-day inquiry.)

When we realize that the vast store of genic variation in most populations today must have arisen by the mutation process in the past, we must consider whether any of that variation has arisen by unique events or by a recurrent process and, if by both singular and continuing events, by what relative proportions of each. In the case of any unique genetic event (mutation to a new allele or a new chromosomal aberration that acts like a new allele), what is the likelihood that it could be perpetuated via reproduction, be increased in frequency, and perchance become fixed in the population? We need to answer this question apart from any adaptive function the new genetic unit may give to its possessor. Once having described the probability for the fate of unique mutational events, we will be in a position to look at the array of genetic variants that do exist in a population and to attempt an accounting of their origin and likely reasons for their perpetuation—because of adaptive functions, random factors, or more likely *both* to some degree. In this chapter, however, our emphasis is on the initiation process of new variation alone, not on selection. We assume large population size and no selection to be acting on genotypes.

When mutation recurs at known genic loci, we need to know the types of mutational events, their average rate, range of rates, and conditions that affect those rates before we can estimate populational consequences of the process. If genic loci have measurable rates of change, the consistent directional force acting to alter the status quo will constitute a pressure on the total genome and over a long period of time bring about a change in allelic frequencies. How strong a force recurrent mutation can be and how much of total genetic polymorphism can be accounted for by that process needs to be estimated.

Mutation at separate loci is slow, but the eukaryotic organism's genome consists of thousands of loci. If each locus is capable of mutation at some constant rate, a considerable number of mutants might be expected in every gamete each generation. Mutations affecting general functions such as viability, fertility, developmental rate, disease resistance, behavior, and so forth can accumulate, making the total genetic variation affecting these functions greater than individual locus mutants would per generation simply because several loci take part in determining the functions. For example, lethal mutants are a general cumulative group; a chromosome with a recessive lethal effect might be considered by a novice as having a single lethal locus when in fact the chromosome will have an effect of lethality whether it has one or a large number of nonallelic lethals linked together. Alternatively, some nonlethal genes may act synergistically as "synthetic" lethals under particular conditions (see Chapter 17). Consequently, the phenotype "lethality" may have a wide range of genetic determination, and mutations that bring about such a phenotype cannot often be analyzed individually but only as a set, or group, whose total mutation rate may be considerably higher than the mutation rates of its component loci. In this chapter, most of our considerations will be in terms of the single locus, but in the event of possible ambiguity in the discussion, it will be important to specify exactly what level of mutation is meant.

TYPES OF MUTATION

Before embarking on the populational dynamics of the mutation process, it will be important for the student to keep in mind that our concept of that process has become sophisticated since the molecular structure of DNA was worked out in the 1950s. Subsequent complexity of function in types of genic loci (structural protein, regulator, operator, tRNA, and rRNA loci, for example) make us realize that a gene mutation can occur at a variety of sites on a chromosome and still effect the same phenotypic change. Detailed descriptions of these numerous avenues to the same product and the diversity of mutational types are beyond the scope of this book, but a very limited summary is here included for convenience in this discussion. See Drake (1970, 1973) or Stent (1971) for extensive discussions on the molecular nature of mutation.

It is now accepted generally that gene mutation in the narrow sense consists of an alteration in the DNA of chromatin; the alteration may be a change either in the sequence or in the number of nucleotide pairs (purine or pyrimidine bases attached to the sugar-phosphate backbone of the DNA molecule). The following list summarizes the main types that have been demonstrated.

1. *Missense*: a change in the critical "letter" of the triplet codon of a structural gene (cistron) causing an altered transcription of mRNA, which in turn may bring about a substitutional change in one amino acid of the protein product.
2. *Nonsense*: a codon change resulting in a triplet of mRNA not translatable as an amino acid; for example, a change from GAA (glutamic acid) to UAA, or from GAG (glutamic acid) to UAG, or from CGA (arginine) to UGA.
3. *Silent*: when the third letter of the codon is changed with no resulting change in the amino acid product.
4. *Frameshift*: deletion or duplication of code letters causing mRNA to be transcribed in different sequence of codons, usually resulting in defective protein or complete lack of protein product.
5. At the initial point along the DNA adjoining the structural gene, the RNA polymerase control of a cistron probably is located as an *operator* that may mutate and bring about a "constitutive" production of mRNA unaffected by "regulation"; alternatively, a mutation in the operator may shut off or greatly decrease mRNA synthesis in its adjoining structural gene.
6. *Regulator* genes specify a protein to which a particular structural gene or its operator is sensitive. Two general types of regulation are *activation*, if there is positive control over mRNA synthesis, or *repression* if the regulator product reduces or represses mRNA formation at the site of the structural gene or its operator. Missense mutations in a regulator gene may allow constitutive protein production with altered ability to interact with the operator locus it regulates. Loss of regulation for the structural gene then may result, with loss of repression or activation.

In addition to these genes, which control or regulate structural proteins, there are genic loci that control synthesis of tRNA directly. Loci controlling rRNA have been demonstrated with RNA-DNA hybridization techniques (in *Xenopus*, for example—see Brown, et al., 1972). In mRNA, transcription errors would amount to the equivalent of mutations.

The details and proof of site for mutational origin are in the field of molecular genetics and are beyond the scope of this book, but it should be obvious from this brief summary that an observed change in what appears to be a single gene product can be caused by a variety of mutational events (see Auerbach and Kilbey, 1971, for a review of mutation in eukaryotes). In discussing mutational rates, then, it may be technically very difficult to ascertain mutation events at particular gene sites. Beyond single genic changes, larger hereditary changes include chromosomal breaks resulting in rearrangements within or between chromosomes and changes in chromosome number (aneuploidy and polypoloidy). For complete discussions of chromosomal changes, see White (1969, 1973) and Stebbins (1971) in particular.

MUTATION PREADAPTATIONAL

The evidence that mutation occurs without direction regardless of selective agent presence or absence is incontrovertible. Exposure to any selective agent does not induce hereditary change in a Lamarckian, or postadaptive, sense. The exact nature of a mutational event appears to be a matter of chance at the molecular level, so that all organisms over a period of time acquire a supply of genic mutations—some of which may have adaptive value (preadaptive in the Darwinian sense). Critical evidence was established by the following main experimental techniques (see Adelberg, 1966).

1. Luria and Delbruck in 1943 demonstrated that bacterial cultures begun as mutant-free small inocula possessed highly variable numbers of phage-resistant mutants (with variances of mutant incidence about four to eight times greater than the mean incidence) compared with control subcultures derived from a single source culture. In view of the fact that mutational incidence in the control cultures approximately fits a Poisson distribution, the variance excess over the mean among independent cultures indicated a distribution of mutational events quite independent of the exposure to the selective agent (phage); that is, cultures with high incidence had early mutational events to resistance while low incidence cultures had later events.
2. The Lederbergs in 1952 devised a simple technique for testing bacterial cells for mutation without exposing them to a selective agent—the "replica-plating technique." A colony of bacteria uniformly growing on nutrient agar was replicated onto a piece of sterile velvet first pressed gently against it and then pressed against an agar surface containing bacteriophage. Locations of phage-resistant bacterial colonies could then be pinpointed easily on the original master colony, which had not been exposed to the phage. By selecting cells from the resistant areas on the master plate, the Lederbergs greatly increased the number of resistant cells in subsequent generations without exposure of cells directly to the phage, thereby demonstrating that the mutations to resistance had occurred sporadically in the original colony independently of the phage.
3. In diploid biparental organisms, the technique of sib selection has often been used to establish the presence of genetic variation (mutational derivation) independent of selective agents. By separating out a few offspring within a family to be exposed to a selective agent, but using the sibs of the individuals tested for propagation of the selected line, Bennett (1960) found selection to be quite effective in increasing resistance of drosophila to DDT. Many other investigators have employed sib selection effectively (see Falconer, 1960; Lerner, 1958).

FATE OF A SINGLE MUTATION: LARGE POPULATIONS

In most natural populations that reproduce in vast numbers but maintain a constant population size, only a small fraction of offspring survive from birth to parenthood—usually two on the average for each pair of parents. With such low probability for individual survival, a mutant gene newly arisen in one parent runs a high risk of chance elimination in its immediate descendants. Eventually, the descendants of a neutral or unfavorable mutant *in a large population* will be completely eliminated from the population. Only selectively beneficial mutants have small finite probabilities of getting established permanently.

The problem of chance survival or extinction of a single mutant allele was first investigated by Fisher in 1922 (1930) and by Haldane in 1927 (1932). It was solved by using a technique already employed in the sort of situation posed by Galton in the nineteenth century on the extinction of surnames; the formal inheritance of a surname is analogous to that of a gene (Y-linked actually for paternally oriented cultures), and the extinction of a mutant allele can be treated in the same manner as a surname. Watson in 1875 responded to Galton's question by using the method of generating functions and the distribution of family size as a branching process. For the solution it is evident that we must rely on some reasonable distribution of family size (assumed to be stable) and on the Mendelian expectation of a single-locus heterozygote from any parent having the mutant. Fisher made the assumption that family size distribution is approximated by the Poisson* distribution (Appendix A-4C), and the following description is based on that assumption.

When a single mutational event occurs, the probability of its extinction can be described as follows: We make the simplifying assumptions that (1) the population size is constant ($N_0 = N_1 = N_2 \cdots$) so that the average number of surviving offspring for each parental pair is two on the average, (2) all mothers have the same opportunity to produce offspring per generation, and (3) the distribution of offspring among families will be constant over generations. The single mutation event ($A \rightarrow a$) produces an Aa individual who must mate with an AA only. Families with zero offspring, one offspring, two offspring, and so forth will be distributed (Table 13-1A, row 1) with probabilities generally according to the binomial expansion $(p + q)^n$, where

$$p = \text{probability of not surviving to maturity.}$$
$$q = \text{probability of surviving to maturity}$$
$$n = \text{constant number of zygotes produced per parent}$$

If n is large, p close to 1.00, and q close to zero, as is likely for wild species in a stable state, then we use the Poisson approximation as given in row 2 of Table 13-1A for the distribution of survivorship. Probabilities for each class of family (P_0, P_1, P_2, $P_3 \cdots$) will sum to 1.00,

* In human populations, it has been shown that the Poisson distribution is too simple; of course, all women do not have the same probability of bearing children. Some U.S. Census data (Kojima and Kelleher, 1962) and data from other representative populations (see Cavalli-Sforza and Bodmer, 1971, pp. 311–314) indicate that a negative binomial distribution (Appendix A-4E) provides a better fit to observed progeny size distributions than the Poisson. Mean offspring per mother was about three, with variance about seven for U.S. data, while means varied from 2.8–5.4 for other countries with variances ranging from 5–21. Nevertheless, we shall present the solution based on the Poisson distribution because it is likely to be reasonable for stable natural populations and it is simpler to describe. Ultimate survival probabilities for a mutant under the negative binomial assumption are smaller than those under the Poisson by about two-thirds, so that Fisher's calculation of survival probability of single mutants tends to be too large for human populations, according to Kojima and Kelleher.

TABLE 13-1 *Distribution of offspring growing to maturity from parents* AA × Aa *in which* a *is a new mutant; probability of zero* Aa *offspring is the probability of loss for the mutant in a single generation*

A	Distribution of Surviving Offspring in G_1 Families				
	Zero	One	Two	Three	...
1. If distribution binomial	p^n	$+ np^{n-1}q$	$+ \frac{n(n-1)}{2!} p^{n-2}q^2$	$+ \frac{n(n-1)(n-2)}{3!} p^{n-3}q^3$	$\cdots = 1.00$
Probability (symbolically)	P_0	P_1	P_2	P_3	$\cdots \sum P_r = 1.00$
2. If distribution Poisson	$e^{-m}\Big[1$	$+ \frac{m}{1}$	$+ \frac{m^2}{2!}$	$+ \frac{m^3}{3!}$	$\cdots\Big]$
3. If Poisson with mean = 2 surviving offspring	$e^{-2}\Big[1$	$+ 2$	$+ \frac{4}{2!}$	$\frac{8}{3!}$	$\cdots\Big]$
4. Mendelian expectation of zero *Aa*	1	$\frac{1}{2}$	$\left(\frac{1}{2}\right)^2$	$\left(\frac{1}{2}\right)^3$	...
(symbolically)	x^0	x^1	x^2	$x^3 \cdots$	...
5. Product (3 × 4)	$e^{-2}\Big[1$	$+ 1$	$+ \frac{1}{2!}$	$\frac{1 \cdots}{3!}$	$\cdots\Big] = e^{-2}(e) = e^{-1}$
(symbolically)	P_0x^0	$+ P_1x^1$	$+ P_2x^2$	$+ P_3x^3 \cdots$	$\cdots = \sum P_r \cdot x^r = f(x)$

Therefore, when offspring are distributed according to the Poisson, with a mean of two survivors per family, the probability for losing a single neutral mutant *a* in a single generation is $e^{-1} = 0.3679$. The sum of row 5, or total probability, is a function of x—that is, $f(x)$.

Distribution of Aa *offspring from* Aa × AA *parents and the probability of loss* (*zero* Aa) *for mutant* a *in succeeding generations, based on one* Aa *surviving per family*

B	Distribution of Surviving *Aa* Offspring				
	Zero	One	Two	Three	...
1. Distribution probability (P_r)	P_0	$+ P_1$	$+ P_2$	$+ P_3$	$+ \cdots = 1.0$

2. Distribution of one *Aa* offspring per family	$e^{-1}\Big[1$	$+\ 1$	$+\ \frac{1}{2!}$	$+\ \frac{1}{3!}$	$+\ \cdots\Big] = 1.0$
3. Expectation of zero *Aa* in G_1	1	0	0	0	$\cdots$
(x)	x^0	x^1	x^2	x^3	$\cdots$
Probability of loss *Aa* in G_1	$e^{-1} = 0.3679$ (as in Table 13-1A)				
4. Expectation of zero *Aa* in G_2	1	0.3679	$(0.3679)^2$	$(0.3679)^3$	$\cdots$
5. Expectation in terms of G_1	$[P_0x^0P_1x^1P_2x^2\cdots]^0 = [f(x)]^0$	$[P_0x^0P_1x^1P_2x^2\cdots]^1$ $[f(x)]^1$	$[\text{same}\cdots]^2$ $+\ [f(x)]^2$	$[\text{same}\cdots]^3$ $+\ [f(x)]^3$	$\cdots$ $\cdots$
6. Product (rows 1 × 5) constant probability $= f(f(x))$	$P_0[f(x)]^0$	$+\ P_1[f(x)]^1$	$+\ P_2[f(x)]^2$	$+\ P_3[f(x)]^3\cdots$	$= \sum P_r[f(x)]^r$
Value for G_2 (rows 2 × 4)	$e^{-1}\Big[1$	$+\ 0.3679$	$+\ \frac{(0.3679)^2}{2!}$	$+\ \frac{(0.3679)^3}{3!}\cdots\Big]$	
	$e^{-1}[e^{0.3679}] = e^{-0.6321} = 0.5315$ (probability for losing *a* in the second generation)				

Probability for loss of neutral mutant a *in generation following its occurrence (at G_0) in large population with family size distribution according to Poisson*

C	*At Generation* (G)	Probability for loss of $a = e^{x-1}$, where x = probability of loss in previous generation
	1	0.3679
	2	0.5315
	3	0.6259
	5	0.7319
	20	0.9125
	50	0.9624
	100	0.9807
	limit	1.0000

that is, ($\sum P_r = 1.00$), where P_r = probability for class of family with r survivors. In row 4 the expectation of losing the mutant allele is based on Mendelian segregation of having 0, 1, 2, 3, . . . *Aa* individuals from $AA \times Aa$ parents (that is, the single-event probability is $\frac{1}{2}$, which can be designated x from $AA \times Aa$ parents). Total probability, then, for the zero, one, two, three, etc., *Aa* is given by the product of rows 3 and 4 to obtain row 5. Symbolically, these successive terms combining $P_r \cdot x^r$ are a function of x, or $f(x)$. The sum of this product we find to be $e^{-1} = 0.3679$, the probability that the mutant allele will be lost in a single generation. The student should note that the first term is the probability for extinction of the mutant in a single generation; relative to the total sum, the P_0 term is $e^{-2}/e^{-1} = e^{-1}$, which is identical with the sum itself, since all the remaining terms of the product line (row 5) have at least one *Aa* remaining in each family class. Thus, considering the distribution of family sizes, we may visualize that about 37 percent of all new mutants are lost in a single generation after their origin.

Each *Aa* survivor in the G_1 progeny can independently mate to *AA* and give rise to a second generation distribution array equivalent to row 5 in Table 13-1A, with the probability of each type of family assumed to be equal to that in the G_1. Intricate branching of the original array seems complicated. We shall simplify the expectation in the G_2, skipping the algebraic description for the sake of brevity. (For details of probability-generating functions, see Li (1955), pp. 248–249; Crow and Kimura (1970), pp. 419–421; Schaffer (1970).) The reasoning should be intuitively obvious as follows: from any parental mating of $AA \times Aa$ genotypes, the average survival for *Aa* offspring in a stable population should be just *one* per family. Now, instead of utilizing the expectation of zero *Aa* based on two survivors, it is more convenient to consider the expectation based on just one *Aa*. We repeat the first progeny (G_1) distribution for clarity at the top of Table 13-1B, restated for single *Aa* survivorship instead of a mean of two progeny per family. In row 3, the expectation of zero *Aa* in G_1 is obviously certain ($x = 1$) in the first type of family, by definition. In the second type of family, with one *Aa* offspring, obviously the expectation for zero *Aa* is by definition impossible ($x = 0$), as it is for all remaining family types in the array. The first term alone of this $f(x)$ is the probability of extinction, exactly as we saw from consideration of two survivors per family.

In the second generation (G_2), the probability distribution of *Aa* survivors is identical with the top row of Table 13-1B (assumed to be constant), but the expectation for zero *Aa* individuals in each type of family has been determined by the outcome of the first generation. In row 4, the first type of family is always $x = 1$ (no *Aa* survivors), but if there is one survivor, the expectation of not being $Aa = 0.3679$; if two survivors, expectation for both to be not $Aa = (0.3679)^2$, and so forth across the array. This expectation is, in fact, the entire function from the first generation for each *Aa* individual, as indicated in row 5 with $f(x)$ raised to the power of survivor number in each type of family. Final probability for loss of a is the sum of the new products; in the function of x, the x now is the function itself (since every *Aa* survivor can produce the array independently), so that $\sum P_r[f(x)]^r$, expressing the probability generating function as $f(f(x))$, can then be continued to further generations. For this G_2, the probability for loss of a or zero $Aa = 0.5315$. In other words, more than half of the new mutants would be lost within two generations. Note that this probability of loss has the form $[e^{-1}(e^x)] = e^{x-1}$ generally, where x = probability for loss of one a (or zero *Aa*) in the previous generation. After five generations, nearly three-fourths of the mutants would be lost.

As is evident from Table 13-1C, the trend is to lose the mutant completely under these conditions, but the most rapid elimination takes place in the first few generations. It may be noted that Fisher (1930) also calculated the expected rate of loss if the mutant were not neutral to selection but had a slight continuous advantage; with a one percent advantage for the *Aa* genotype compared with *AA*, the limit probability for extinction after many generations would be 0.9803, or approximately a 2 percent probability for persistence of the mutant (fixation) in a large population (see Exercise 3).

As summarized by Schaffer (1970), Lotka first investigated the probability of extinction for male lines of descent in the United States (about 1920), using a tedious numerical iteration method. He found the average number of sons to be 1.175, and the ultimate probability of extinction for a male line of descent, starting from a newborn male, was 0.872. However, a useful approximation formula for the ultimate probability of extinction (*upe*) of a single mutant was worked out by Bartlett (1966, cited by Schaffer) ; it gave the probability in terms of mean and variance of the distribution of children per family:

$$upe = e^{-2(m-1)/V} \tag{13-1}$$

the ultimate probability of extinction, where m = mean surviving offspring and V = variance of survivors per family.

The ultimate probability of survivorship (*ups*), or establishment of the mutant, is $(1 - upe)$. If we are concerned with a particular genotype (*Aa*) carrying a single surviving mutant, and survivorship is distributed as a Poisson, both m and $V = 1$, and $upe = 1$, as expected. Lotka's data for male lines of descent gave $m = 1.175$ and $V = 2.768$. Substitution in (13-1) gives $upe = 0.880$, which is only slightly greater than the exact value worked out by Lotka. Thus, if the distribution of progeny per generation is such that $V \geq m > 1.0$, the probability will be greater than zero for establishment of the mutant (or variable) in a large random-mating population.

FATE OF A SINGLE MUTATION: MORE REALISTIC CONDITIONS

Population sizes are largely determined by environmental factors such as space available, nutrition, predation, and competition that in turn influence family size; expansion and contraction of population size (and corresponding family size) will affect the survival of new mutant genes through changes in the average survivorship ($m > 2$ when the population is expanding or $m < 2$ when contracting). The branching process used in the last section for generating probabilities of mutant extinction (or survival) can be used to explore expansion and contraction cycles, as has been done by Kojima and Kelleher (1962). The importance of the *distribution* of descendants becomes evident (see Exercise 2) when we see that the probability for extinction is less (survival greater) when the population is expanding for a given number of generations. It is in the opposite direction under a contraction phase. With a neutral mutant, expansion only postpones the inevitable loss, but for a slightly advantageous mutant, the ultimate survival probability is improved if that expansion continues for a long time. Nevertheless, assuming that the population does not expand forever, the most influential effect on the ultimate survival probability will be that taking place in the initial stages after the mutational event. As a result, we would expect greater genetic variation from new mutations while populations are expanding and less while populations are contracting.

If a population's effective size is small continuously, as discussed in Chapter 12, random events can bring about fixation of neutral mutants as well as eliminate them, in contradiction to the classic expected outcome described in the section above. One could devise computer simulation experiments with any convenient small-sized population. Assume that a single mutation occurs to put the frequency $q = 1/(2N)$ at the start, further assume a binomial distribution, and obtain an average probability for fixation or for elimination of the mutant after a definite number of generations with constant population size and structure. Mathematically, the general solution for the probability of extinction under such stochastic conditions (random events with time) is difficult, especially when it is desirable to achieve a formula combining selection (Darwinian fitness) properties with any mutant and population size, as we shall need to consider later. Instead of using the branching process of Fisher as discussed in the above section, the general solution has been achieved by treating the process of gene frequency change as a continuous one. Even though changes in gene frequencies are discontinuous, no essential accuracy is lost. In any population larger than the extreme "Noah's Ark" case, it is not far from precision to think of gene frequency changes as if they were continuously changing such that as the time interval becomes smaller the gene frequency change would also be correspondingly smaller. Mathematical treatment of the solution for probability of fixation or extinction and the number of generations involved has been worked out by Kimura (1955, 1962) by means of the diffusion equations (Kolmogorov forward and backward equations), following Wright's (1945) introduction of the forward equation (known as the Fokker-Planck equation in physics). Mathematical description of these equations including their applicability to a wide number of stochastic processes is beyond the scope of this book. We shall simply summarize some of the pertinent principles in the use of these equations and their solutions for the probability and timing of fixation or extinction of neutral mutants. For further mathematical description see Crow and Kimura (1970, pp. 371–382, 423–432), Kimura (1970), and Kimura and Ohta (1971, appendix).

The process of change in gene frequencies as a continuous stochastic one is analogous to the random movement of molecules of a diffusing substance dropped into a liquid—like sugar crystals dropped into water. The liquid itself does not move, only the molecules dissolved in it. It can be shown that the concentration of the molecules, high at the point where they were introduced, tends to diminish, and regions of lower concentration far from that point will increase in concentration. Rate of change in concentration then can be described by differential equations (concentration per unit time as a function of distance the molecules have traveled). In similar manner, a gene frequency can be thought of as a point x on a continuum of probability density, where x is the gene frequency at the tth generation after having changed from a fixed p value. Actually, in a finite population, any change in x (that is, Δx) will have to be finite, but it is approximated by a curve of probability distribution. From one generation to the next, the x gene frequency can be considered a random variable (analogous to the molecules in a liquid), free to move up or down the probability distribution. It can then be shown that the diffusion equation (Kolmogorov forward equation) will estimate the change per unit time in x in terms of the second derivative of the variance (random fluctuation of x up or down) minus the first derivative of the mean x displacement per generation. Solutions of these equations obtained by Kimura give probabilities for fixation or for extinction of neutral alleles arising by mutation, which we shall simply summarize as follows: when a population is subdivided into isolate breeding units of limited size (N), the ultimate probability for fixation over all isolates (ups) $= q_0$, the initial frequency of a neutral allele. For a single mutant, when it occurs just once, $q_0 = 1/(2N)$. Consequently, the vast majority of new mutants are lost: $(1 - q) = 1 - 1/(2N)$, the ultimate probability of extinction.

If we recall the principles of random distribution among isolates of limited size described in Chapter 12, the ultimate limit of the dispersion process is the state for all isolates to be either "fixed" or "lost" in proportion to the p,q frequency of the ancestral random-mating population, because for any species population broken up into subpopulations, or isolates, the gene frequency over all isolates is expected to be constant. Similarly, inbreeding leads to the same final result (Chapter 8).

Knowing the probability of fixation, we may wish to find out how long a time on the average it will take for complete fixation from the generation of a single mutant's origin. Kimura and Ohta (1971) gave the average time to fixation ($\overline{t_1}$) in generations as the following (disregarding the cases of extinction):

$$\overline{t_1} = -\left(\frac{1}{q}\right)\left[4N_e(1 - q)\log_e(1 - q)\right] \tag{13-2}$$

For the average number of generations to extinction of the mutant ($\overline{t_0}$) (disregarding the cases of fixation)

$$t_0 = -\left[\frac{4N_e q}{1 - q}\right]\log_e(q) \tag{13-3}$$

For the special case of a single mutant's occurrence, at the time of origin, $q = 1/(2N)$, so that (13-2) becomes approximately

$$\overline{t_1} = 4N_e \tag{13-2A}$$

and (13-3) becomes approximately

$$\overline{t_0} = 2\left(\frac{N_e}{N}\right)\log_e(2N) \tag{13-3A}$$

It is evident that the length of time for fixation of neutral alleles is very much greater than for loss by a factor of $2N/\log_e(2N)$.

In Example 13-1 we demonstrate the random fixation or loss of a single mutant in populations (isolates) of small size by simulation in a high-speed computer. The average time in generations to fixation or to loss is confirmed according to the Kimura and Ohta formulas given above. It is important to keep in mind that the isolates' size must be constant throughout these procedures for close agreement with expectation. Even so, the average time to loss will be less dependable (have greater variance) than time to fixation because a new mutant can easily increase its frequency by chance and persist for a long time at high frequency yet eventually be lost. The vast majority of isolates lose the mutant within the first five generations, as might be expected from the deterministic result of Fisher. However, a new mutant may become fixed, and the number of generations to pass before a single isolate becomes homozygous for it is about equal to N.

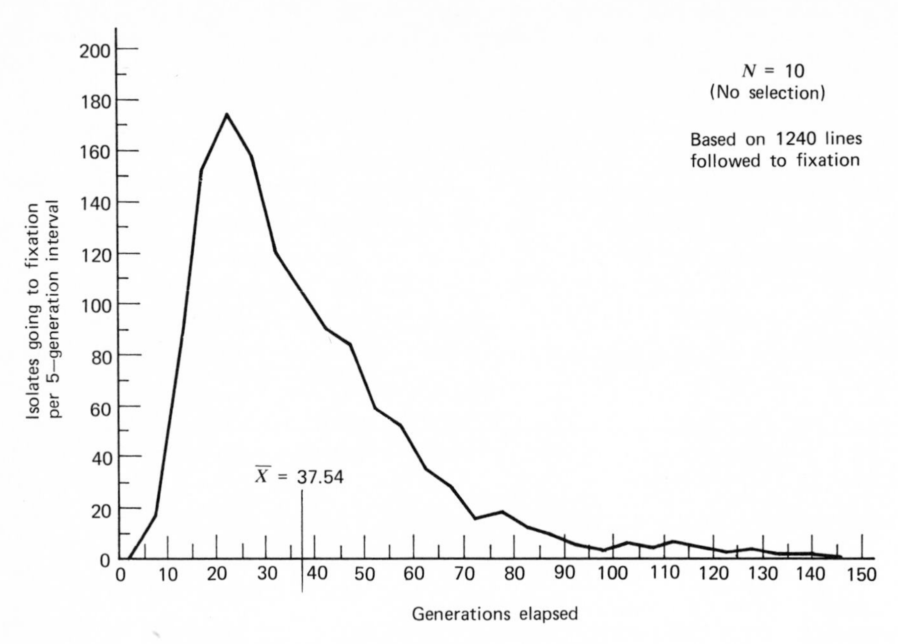

Figure A. Distribution of isolates ($N = 10$) going to fixation of neutral mutants at five-generation intervals from the time of mutation. Based on 1240 lines of IBM computer simulation followed to fixation. Mean time to fixation ($\bar{X}$) = 37.54 generations.

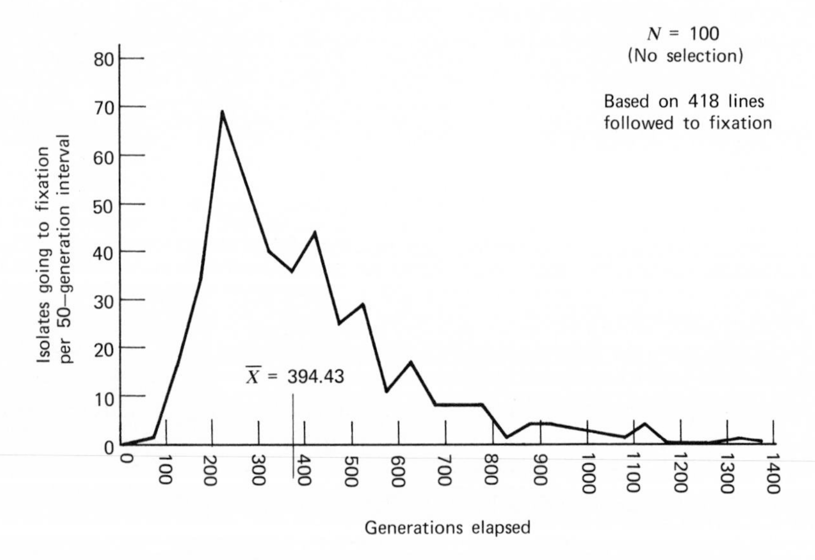

Figure B. Distribution of isolates ($N = 100$) going to fixation of neutral mutants at 50-generation intervals from the time of mutation. Based on 418 lines followed to fixation (IBM computer simulation). Mean time ($\bar{X}$) = 394.43 generations.

EXAMPLE 13-1

COMPUTER SIMULATION OF MUTATION FIXATION AND LOSS

Simulation of fixation and extinction of single mutants ($q_0 = 1/(2N)$) in different sized populations was done with the IBM 370 computer program used in Examples 12-1 and 12-2 set for an initial input of $q = 1/(2N)$, with no selection (all genotypes equal in Darwinian fitness), and various population sizes: $N = 4$, 10, 20, or 100. The distribution of isolates going to fixation of the initial mutant (*a*) allele is illustrated in Figure A and B for populations of $N = 10$ and 100, respectively. In Figure C a typical set of $N = 10$

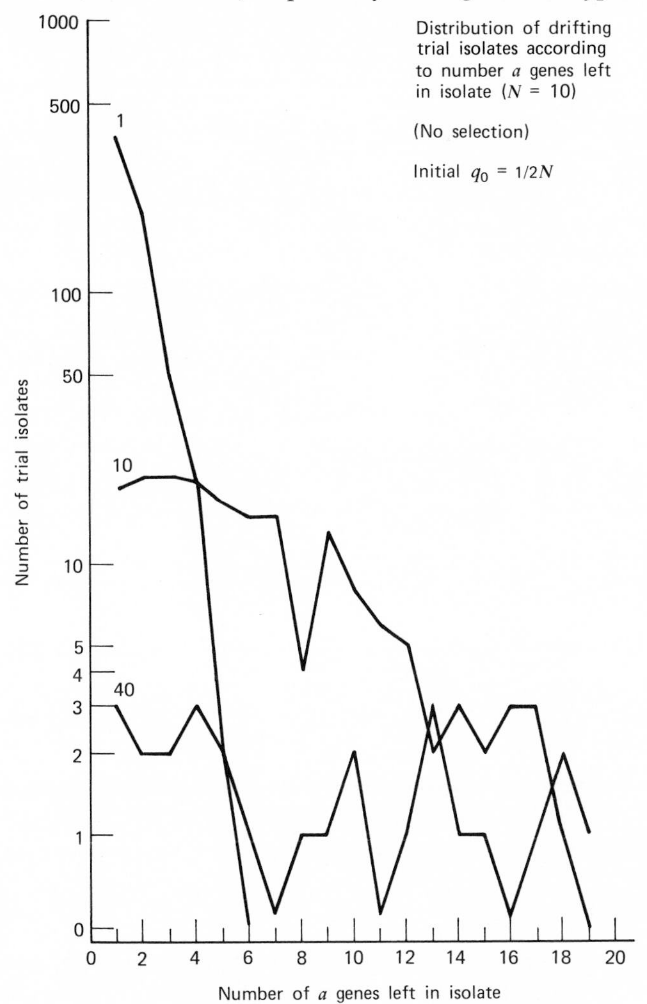

Figure C. Distribution of still-drifting isolates ($N = 10$) according to the number of *a* alleles (mutants) remaining in the isolate at G_1, G_{10}, and G_{40}; vertical scale logarithmic.

TABLE A *Neutrality—Distribution of single mutation among isolates of constant size from simulation on IBM computer*

N	4	10	20	100
Initial q $(1/(2N))$	0.125	0.050	0.025	0.005
Final q	0.1203	0.0516	0.0241	0.0049
Mean fix generation $\bar{t}_1$	13.83	37.54	78.66	394.43
$4N$	16	40	80	400
Total lines run	8000	24000	40000	~100000
No. lines followed to fixation	961	1240	964	418
1st generation with isolate fixed	3	8	20	88

isolates "still drifting," or unfixed, within G_1, G_{10}, and G_{40} is illustrated to show the approach to a flat distribution with all isolates of about equal probability for any number of unfixed alleles after 40 generations.

From the distribution of fixations, it appears that most of the fixations take place before $4N$ generations have elapsed, although they continue to occur at increasingly slowed rate until ultimately all go to fixation. Table A summarizes the outcome confirming expectations of Kimura and Ohta that the mean time for fixation of neutral mutants is a little less than $4N$ generations (13-2A) with isolates' size kept constant. Final q value over all isolates is very close to the initial $1/(2N)$. The number of generations for a new mutation to reach fixation is generally a little less than N, as shown in the last row of Table A.

Extinction of the mutant a was much faster. For example, the first five generations loss of a (fixation of A) and average time to loss (t_0) were as follows for the $N = 4$ and $N = 10$ series:

	G	% Isolates Lost a		G	% Isolates Lost a
($N = 4$ isolates)	1	33.4	($N = 10$ isolates)	1	36.6
	2	15.4		2	16.2
	3	9.6		3	9.4
	4	6.5		4	5.6
	5	4.0		5	4.1
Average G_t to loss $= 4.13$		68.9	Average G_t to loss $= 5.30$		71.9

In five generations, about 70 percent of all isolates have lost the new mutant and the average time to loss is in agreement with formula (13-3A).

RECURRENT MUTATION

While a single mutant occurring just once has its fate in the population determined primarily by the distribution of progenies in families, by population size, and by its selective value, we should note that each cistron (gene locus) is expected to have a specific mutation rate per mutable locus under specified conditions. Repeated genic change from one allele to a mutant allele (DNA nucleotide site change being the most narrow mutational change) will slowly bring a shift in allelic frequencies toward the "mutated" direction. Thus, a vectorial "pressure" exists at a very low level per locus. However, with the enormous number of genic loci mutating per generation, a cumulative effect of total genic change at molecular and chromosomal levels becomes quite sizable.

The technical problem of ascertaining the exact mutation rate for a specific mutable site in a cistron is immense, as we have already implied by summarizing the types of mutation detectable. Dobzhansky (1970) listed (his Table 3.1) mutation rate estimates for a few genes in a variety of organisms. For loci in *Drosophila melanogaster*, *Zea mays*, *Mus musculus*, and *Homo sapiens*, the range is from 0.1 to 25.0×10^{-5} as mutations per cell or gamete per generation with a clustering in the range $0.4\text{–}4.0 \times 10^{-5}$. Extensive data on the mouse from Schlager and Dickie (1967) using five coat color loci and forty other loci for 3.5 million mice counted gave 9×10^{-6} forward mutations per locus per gamete for coat color genes and about 6×10^{-7} on the average for the other loci. It is very difficult to estimate human genic mutation rates. We need to monitor human populations very carefully; sources of error are considerable and pitfalls many. Neel (1962, 1971) discussed the detection of human population mutation rates and the problems involved both from direct methods (dominant traits, codominant and sex-linked) and from indirect methods (based on consideration of populations at selection-mutation equilibrium). The latter method will be postponed to Chapter 18. However, Table 13-2 lists some of the estimates given by Neel (1971). See also Vogel and Rathenberg (1975) for a summary of mutation rates in humans.

In contrast, mutation rates in the microbial organisms listed by Dobzhansky (*Escherichia coli*, *Salmonella typhimurium*, *Diplococcus pneumoniae*, and *Neurospora crassa*) range from 4×10^{-10} to 1×10^{-7} with clustering between 10^{-8} and 10^{-9}. (See also Drake (1970), who gives the rate of 2×10^{-8} per base pair replication in bacteriophage T4.) Undoubtedly, the number of cell generations between parents and offspring in the higher organisms is far greater than in the microbial ones. Thus, there is more opportunity for mutation between zygote and gamete than between successive cell divisions of microbes. Another possible source of difference may lie in the fact that the rates for the higher organisms are based on *any* mutations at most of the loci studied, while the estimates for the procaryotes and *Neurospora* are the reversion rates *to* wild types of specific mutant gene sequences, as pointed out by Watt (1972), or they represent fine structure changes within cistrons. Thus, the microorganism mutation rates are site-specific, while many of the others are overall perlocus rates.

The basic problem of ascertaining rates of mutation for genic loci is even more compounded by the presence or absence of "mutator" systems (see Drake, 1973). Mutable genes and mutator systems that act to increase mutation rates on specific loci point to the fact that ascertainment of mutation rates for intracistronic sites involves so many complex conditions that a constant mutation rate over a long period of time for most organisms is practically unattainable. To put the situation in other words, while we can make the assumption that some constant slow mutation rates may exist for stable systems in some organisms, the diversity of environmental and internal genetic conditions is so great that such an assumption is far too simple. In fact, we might well leave our consideration of mutation in populations simply to the fate of a single mutational event. Nevertheless, it is important for the student to realize that recurrent mutation *can* bring about a long-term slow change in allelic frequencies and that the order of magnitude of time might be very slow in a deterministic large population.

TABLE 13-2 *Some spontaneous mutation rates in humans** (*from Neel*, 1971)

Character	Method of Estimation**	Mutations per Gene per Generation	Remarks
Dominant genes			
Epiloia	Direct	$(0.4–0.8) \times 10^{-4}$	—
Chondrodystrophy	Direct	4.2×10^{-5} 4.9×10^{-5}	Estimates may be spuriously high because of some evidence for occurrences of phenocopies
	Direct	7×10^{-5}	
	Indirect	4.3×10^{-5}	
Pegler's nuclear anomaly	Direct	2.7×10^{-4}	—
Aniridia	Direct	0.5×10^{-5}	—
Retinoblastoma	Direct	1.4×10^{-4} 2.3×10^{-5}	Estimate based on assumption that all sporadic cases are due to mutation
		4.3×10^{-6}	Estimate based on proposition that approximately 75% of all sporadic cases are phenocopies
Waardenburg's syndrome	Direct	3.7×10^{-6}	—
Neurofibromatosis	Direct	$(1.3–2.5) \times 10^{-4}$	—
	Indirect	$(0.8–1.0) \times 10^{-4}$	—
Facio-scapulo-humeral progressive muscular dystrophy	Direct	4.7×10^{-6}	—
	Indirect	4.7×10^{-6}	—
Multiple polyposis of the colon	Indirect	$(1.3) \times 10^{-5}$	—
Sex-linked recessive genes			
Hemophilia	Indirect	3.2×10^{-5}	Estimate may include three distinct types of hemophilia
Childhood progressive muscular dystrophy	Direct	3.2×10^{-5}	—
		1×10^{-4}	Not a true direct estimate but an approximation that overestimates the mutation rate
	Indirect	3.8×10^{-5}	—
		$(4.5–6.5) \times 10^{-6}$	—
		1×10^{-4}	—

* Knowledge of mutation rates in humans, as in other organisms, is still considered provisional, and the traits for which estimates are available are a highly select fraction of all genes. Estimates for chondrodystrophy, aniridia, and neurofibromatosis may be more reliable than those for epiloia and Waardenburg's syndrome.

** Estimates are considered to be direct when based on observed mutations and indirect when not so based. Indirect estimates make use of determinations of the relative fitness and frequency at birth of the trait and assume that the population is in equilibrium.

If we assume a constant slow mutation rate (u) from A to a, measured per generation, we may estimate the time in generations for change in gene frequency as follows. First, we obtain a recurrent series for change in q from one generation to the next

$$q_t = q_{t-1} + up_{t-1} \tag{13-4}$$

where up_{t-1} is the proportion of A allele mutated to a in production of the t generation. Putting (13-4) into the form

$$q_t = q_{t-1} + u(1 - q_{t-1}) = u + q_{t-1}(1 - u)$$

we see that repeated substitutions for previous generations of q, as

$$q_t = u + (1 - u)u + q_{t-2}(1 - u) = u + (1 - u)u + (1 - u)^2 q_{t-2}$$

produce a geometric series in which the last term with q_0 refers to the initial generation of interest.

$$q_t = u + (1 - u)u + (1 - u)^2 u + (1 - u)^3 u + \cdots (1 - u)^t q_0 \tag{13-5}$$

From Appendix A-13, it is apparent that the sum of terms in this geometric series, with u the initial term and $(1 - u)$ common ratio, becomes

$$q_t = 1 - (1 - u)^t + (1 - u)^t q_0 \tag{13-6}$$

so that

$$(1 - u)^t = \frac{1 - q_t}{1 - q_0} = \frac{p_t}{p_0} \tag{13-7}$$

As an example, let us estimate the number of generations for recurrent mutation process to increase an allele's frequency from $q = 0.10$ to 0.20 at a rapid rate of mutation ($u = 1 \times 10^{-4}$). Solving for t in (13-7), we may use logs

$$t \log(1 - u) = \log(p_t/p_0) \qquad \text{or} \qquad t = \frac{\log(p_t/p_0)}{\log(1 - u)}$$

$$t = \frac{\log(0.88889)}{\log(0.99990)} = 1279 \text{ generations}$$

As an approximation (Appendix A-4C footnote) of expression (13-7),

$$(1 - u)^t \approx e^{-tu} \tag{13-7A}$$

We may use natural logs to solve (13-7) so that it becomes $-tu = \log_e(p_t/p_0)$ and the example becomes $t = 0.1178/0.0001 = 1178$ generations. Even though this approximation (13-7A) is about 100 generations less than (13-7), it is clear that the time it takes to change allelic frequencies by mutation alone is much longer than most rates of observed change are likely to be. Of course, with a mutational rate 10 times slower (10^{-5}), the number of generations would be 10 times that in the example, using the approximation.

As t increases indefinitely, the left term of (13-7) reaches the limit of zero and q_t reaches 1.00. Finally, the rate of change in q depends on its value; stating (13-4) in terms of change in q

$$\Delta q = u(1 - q_{t-1}) \tag{13-8}$$

where $\Delta q = q_t - q_{t-1}$, so that as q gets close to the limit of 1.00, the rate of change is even slower.

All these calculations point to the extreme slowness of mutation with the major conclusion that recurrent mutation, in large populations with no selection affecting the mutations concerned, is quite ineffective in achieving any substantial change by itself. If mutation *alone* were responsible for much of the observed variation in populations it is obvious that the present level of genetic polymorphism could be achieved only through stochastic processes in populations of small size. Of course, a vast number of mutational events affect traits important to Darwinian fitness, as we shall see in subsequent chapters.

In a population of N diploid individuals, the expected number of new mutant alleles at every generation will be $2N\bar{u}$ with $\bar{u}$ = average mutation rate over all loci. Under stochastic conditions mentioned previously, the probability of ultimate fixation for a neutral is $1/(2N)$. Then the number of new mutant alleles (M) appearing per generation that are destined to become fixed will be

$$M = \frac{2N\bar{u}}{2N} = \bar{u} \tag{13-9}$$

It has been estimated by Kimura and Ohta (1973) that in human hemoglobin, β = chain variants arise at a rate of 2.5×10^{-5} per generation, or approximately 9×10^{-8} per amino acid site (codon) per generation, higher than the estimate cited by these authors given in 1969 by Motulsky, who concluded that the nucleotide mutation rate lies somewhere between 10^{-8} and 10^{-9}. From these estimates, we might expect that it would take about 250,000 generations for a hemoglobin variant to become established (assuming it to be neutral, which some variants probably are not), provided the human population were very large and panmictic. It is clear that selection must account for many of the observed changes in human hemoglobin evolution, because it is far more likely that most hemoglobin variants established in populations have probably spread rapidly after their origin. It is also possible for neutral variants to be established if human breeding units are small.

ACCUMULATION OF MUTANTS AFFECTING GENERAL FUNCTION

Total genetic variation affecting broad functions such as viability, fertility, developmental rate, disease resistance, and behavior is certainly of selective value to the organism. We have been considering mutation apart from its effects on Darwinian fitness. Here it is important to consider, however, that genic changes arising by mutation at several loci may act apparently as a cumulative group because they may affect particular biotic functions additively or epistatically. Many of these functions are easy to measure, so that much information about the rate of accumulation for the mutants has been compiled.

We discussed in Chapter 7 the problem of whether polygenes exist as a specific class of genetic material, as originally proposed by Mather, or whether the additivity of small units is attributable to pleiotropic "side effects" of oligogenes. There is much evidence that mutations of small effect take place at many loci over time. Mutational changes affecting a trait such as viability may accumulate over a number of generations and can be ascertained by selection techniques; however, with selection it is difficult to indicate frequency or magnitude of individual mutational events. That can more easily be done by sampling chromosomes. Mutant chromosomes will be distinguishable from nonmutant ones only when the magnitude of deviation is great in comparison with the confidence limits of the control (unchanged chromosomes).

Viability is a special trait. It is a fitness property of an individual, but it is defined as survival ability expressed as a proportion compared with a control. It can only be expressed on the individual level as an all-or-none phenomenon—alive or dead at the moment of scoring—but it is measured as the survival rate of a group, a trait that is expressed by the penetrance of a full viability factor and is sensitive to environmental variation as well as to genetic change. The probability for individual survival expressed under carefully defined environmental conditions may change if mutations accumulate to affect that probability.

A technique to disclose new mutations that affect viability or other general biotic functions likely to be affected by cumulative action of numerous loci is that of making all individuals genetically identical by use of dominant markers with crossover suppressors, as outlined in Example 9-5. Once chromosomes in a strain are made homozygous (isogenic), uncovering new mutants can be done by selection or screening techniques. The customary expectation for new mutations is to find them deleterious (lowered fitness) compared with "normal" wild-type genomes. In Example 13-2, considerable evidence was found that mutations beneficial to viability (improving survival) take place at significant rate, while in Example 13-3, a demonstration of the more expected lethals and low viability mutant changes is presented from Wallace's work. In Example 13-4, some long-term experiments by Mukai and his colleagues on spontaneous mutation of viability modifiers indicate that mutation rates for factors of slight phenotypic effect ("polygenes") may be far more rapid than those with major effects (lethals and visible mutants). In all cases where genic changes are either beneficial or slightly detrimental to viability, we must conclude that strains of laboratory organisms like drosophila cannot be considered as unchanging genetically over a short space of time in the laboratory. Therefore, given the expected low rate of mutation per locus measured for structural genes, the fast accumulation of changes observed in these examples must be accounted for by either an enormously great number of mutational events occurring over all the loci contributing to the function being tested or much higher rates of mutation per locus for genes affecting these traits. Note that Mukai (Example 13-4) attributed his observations to the latter. Proof of these alternatives, of course, must await our analysis of individual loci, a tedious task. A third consideration lies in the complexity of intracellular biotic functions along with the expected pleiotropic secondary, tertiary, or further-derived consequences among intracellular functions that may ensue following just a few genic changes at particular loci. If intracellular organelles and conditions are intricately balanced, we have no idea how much variety of function could be the consequence of a small number of focal gene changes.

EXAMPLE 13-2

MUTATIONS BENEFITTING SURVIVAL

Dobzhansky and Spassky (1947) designed experiments to find out whether mutations beneficial to viability occur often enough to be responsible for bringing about improvement through natural selection. The response of populations to "improvement" mutations was examined in a long-term series of experiments begun in 1942 using *Drosophila pseudoobscura* collected in southern California and tested for viability in second and fourth chromosome homozygotes. Seven strains of homozygotes for either of these chromosomes, in which viability was significantly reduced under particular laboratory conditions and thus were carrying detrimental genetic combinations initially on either chromosome 2 or 4, were treated by four parallel techniques:

1. Homozygous controls—20–25 pairs of flies carried on the line each generation with dense cultures of progeny. Natural selection would be expected to encourage the spread of mutations increasing vigor under these conditions.
2. Homozygous but irradiated with 1000 r of X-ray each generation applied to the males. Radiation is expected to increase the spontaneous mutation rate.
3. Balanced heterozygotes over markers dominant (*D*) with crossover suppressor (*Bare* on second or *Curly* on fourth). Five pairs of *D*/+ flies maintained each line each generation. Recessive mutations of all kinds could accumulate without being exposed to natural selection.
4. Balanced heterozygoes (*D*/+), as in 3, except that the five males were given 1000 r of X-ray each generation.

This experiment was continued for five years (1941–1947), covering 50 generations with over 400,000 flies counted. When wishing to test the viability, rate of development, or other fitness trait of the homozygous flies, Dobzhansky and Spassky outcrossed wild-type males from the first or second treatments to marker females, and after three generations

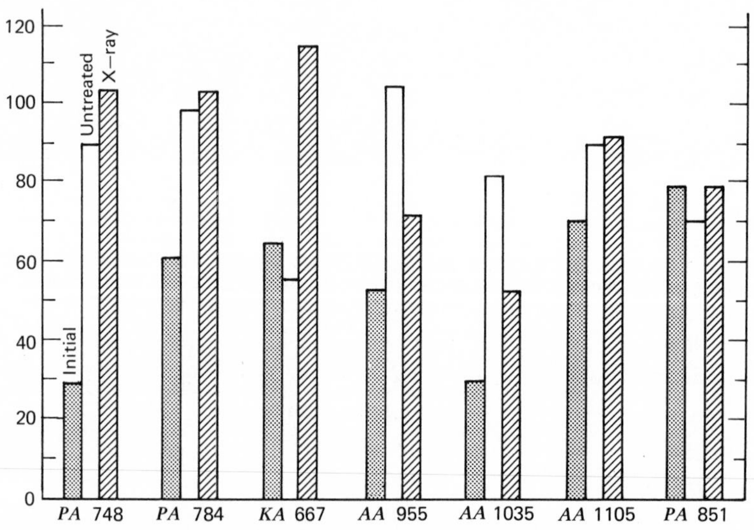

Figure A. (V) Viability of seven homozygous strains before and after 50 generations of maintenance in homozygous condition. Black columns = initial viability, open columns = untreated controls, hatched columns = X-ray treated (from Dobzhansky and Spassky, 1947).

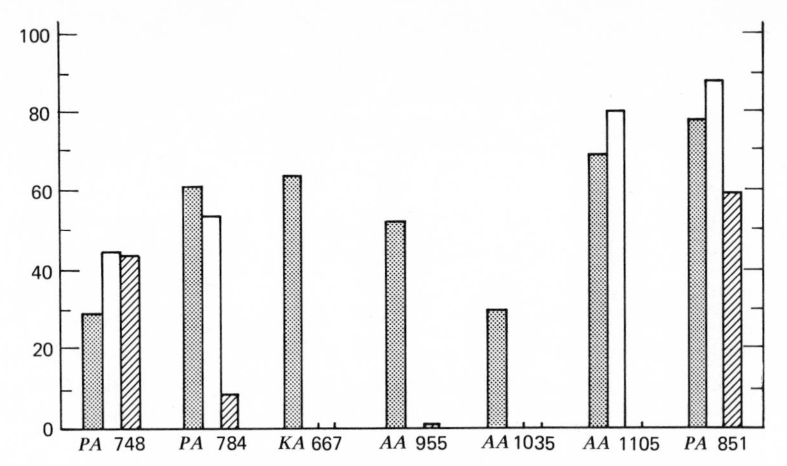

Figure B. Viability of the same seven strains before and after 50 generations of maintenance in balanced condition over dominant markers (D) with crossover suppressors. Columns of the same designations as in Figure A.

of suitable crosses (as in Example 9-5), estimates of these traits could be made by comparing wild-type with marker flies. With the balanced lines, it was a simple matter merely to isolate a few $D/+$ balanced pairs without outcrossing. Progeny from those pairs are expected as $33.3 + : 66.7D$ ratios.

In Figures A (homozygotes) and B (balanced strains), the following conclusions can be seen: improvements of viability occured in five out of seven homozygote control lines and in six out of seven homozygote irradiated lines; the others nor were not significantly changed. Among the balanced lines, however, three of the untreated developed lethals, three were unchanged, and one improved slightly (PA 748); in the X-rayed balanced lines, five had gained lethals and two were not significantly changed.

The contrast is striking: most of the homozygous populations improved, while most of the balanced ones degenerated as far as their fitness effect in tested homozygotes was concerned. Polygenic mutation producing improved fitness (viability) does occur and can be utilized in the population's adaptation to the laboratory conditions. The genetic load of the balanced cultures increased without disturbing the total productivity or the perpetuation of these lines. Therefore, in selection experiments lasting more than a very few generations, we must take into account that new variation can become available through mutation for a significant selection response.

EXAMPLE 13-3

CHRONIC IRRADIATION AND LETHAL MUTATIONS

Adaptation of random-mating populations to such a detrimental environmental agent as chronic irradiation has been investigated for several years by Wallace (summarized in his books, 1968, 1970). While the mutation rates of populations under chronic irradiation are very high, leading mostly to lethal genetic conditions, it is the spontaneous lethal mutation accumulation in Wallace's (1956) control population that is of interest here.

Beginning in 1949, Wallace set up several populations, most of which he exposed to radiation. His unirradiated control (population 3) was maintained as a large random-mating population of Oregon-R *Drosophila melanogaster*, consisting of about 10,000 individuals kept in a lucite and screen cage modified from the population box described in Example 2-4. It was initiated with flies from 16 strains that had been made isogenic for normal viability on chromosome 2, using a *Curly-Lobe/Pm* marker technique. At regular intervals (about every two weeks), Wallace tested a sample of flies by removing eggs from the population, culturing them under near-optimal conditions in separate conditions or low density to produce a number of males to be tested for lethal or semilethal mutations. The males to be tested every two weeks were mated to *Cy-L/Pm* females and run through the system of marker matings in order to test homozygous wild chromosomes. Semilethals (with viability less than half the expected) and lethals (less than 1 percent wild-type homozygotes) were then sampled over a period of about three years.

Wallace's data on lethal + semilethal frequencies in the control (unirradiated) population are given in Table A for the first year of tests. After six months, the frequencies rose to

TABLE A *Observed frequencies of lethals and semilethals on chromosome 2 accumulated in a control population of Drosophila melanogaster (from Wallace, 1956, 1969)*

Weeks from Initiation	Sample Tested Males	Percent Lethals and Semilethals
2	133	0.8
4	52	0.0
6	183	2.2
9	212	1.4
13	263	5.7
17	285	3.5
21	283	5.7
25	386	6.2
29	377	6.1
33	408	8.1
37	409	6.4
41	289	9.0
50	434	12.0
60	278	14.4
68	254	19.7
76	253	20.6
84	179	16.2
92	177	18.6
96	80	27.5
100	78	14.1
104	80	26.3

a plateau of about 30 percent lethals-semilethals, which was maintained until the population was terminated in 1958. Thus spontaneous detrimental mutations must accumulate to a point where presumably a balance is reached between mutation and selection (see Chapter 18).

Following Wallace's maintenance of this population, Spiess and coworkers continued to maintain a derived sample population in their laboratory to study lethals on chromosomes 2 and 3 as well as sterility factors (Spiess and Allen, 1961; Spiess, Helling, and Capenos, 1963; Sweet and Spiess, 1961). By linkage analysis, these workers found many of the lethals to be multilocus; out of 18 second chromosome lethals isolated for study, one was a double locus, one a triple, and one a quadruple, and there was a clustering of lethal mutant sites along the chromosome. Frequencies of 3–5 percent sterility in one or both sexes were also found for the second chromosome. Consequently, the second chromosome initiated as free of detrimental effects (isogenic) by Wallace came to have considerable accumulation of such factors contributing to lowered fitness of homozygotes after ten years of maintenance in the laboratory (see also Temin, 1966, who analyzed the Wallace population for viability and fertility factors).

Wallace (1968, pp. 137–138) estimates the rate of mutation to lethality on chromosome 2 of *D. melanogaster* to be approximately 5 out of 1000 chromosomes per generation. Three sets of experiments intended to measure the rate of mutation to lethality gave the following results:

1. A direct test of previously lethal-free homozygous males gave 23 lethals out of 3778 tested, or 0.61 percent.
2. From accumulation of lethals in balanced heterozygotes maintained for several generations, 33 lethals were recovered after 6764 chromosome generations (chromosomes tested × number of generations of accumulation), or 0.49 percent.
3. Chromosomes recovered from the control population carried 181 lethals after 41,103 chromosome generations, or 0.44 percent.

These last two estimates may be low, since they do not take into account any multilocus lethals that may have accumulated.

EXAMPLE 13-4

SPONTANEOUS POLYGENIC MUTATIONS AFFECTING VIABILITY

Long-term experiments to study the accumulation of spontaneous polygenic mutations (viability) were begun by Mukai in 1961. In Figure A the method of propagation is outlined. The second chromosome of *Drosophila melanogaster* was followed exclusively by comparing viabilities of wild versus dominant marker. A wild-type isogenic stock was initially outcrossed to the marker (abbreviated as *Cy*/*Pm* from *Curly* and *Plum*—both are homozygous lethals on chromosome 2 included within an inversion suppressing crossing over) co-isogenic with the wild-type stock. Single wild-type chromosomes were passed on in balanced condition over the *Pm* marker via single males exclusively in 104 lines established in the second generation. Of the original 104 lines, 3 were lost by accident, so that 101 actually were followed. Each male in a line was allowed to mate with $5Cy/Pm$ females always taken from the original balanced Cy/Pm stock. Any recessive mutation has about equal probability of being established in a line, since it is introduced in heterozygous condition with no chance to become homozygous until the investigator wishes to test for homozygous affects by mating five identical sibs of $Cy/+$ (last two lines in Figure A). Viability was simply measured as the number of wild-type flies occurring in the progeny out of the total flies. According to the *Cy*-method of Wallace (1956), the expected percentage of wild-type is 33.3 for normal viability, zero for lethal mutations, and from 1 to 20 for semilethals. A quasi-normal line was defined by Mukai as any with a viability greater than 20 of this index:

$$v_i = 100x\left[\frac{(+_i/+_i)}{(Cy/+_i) + (+_i/+_i)}\right]$$

where $(+_i/+_i)$ = number of wild-type homozygotes counted within the ith line.

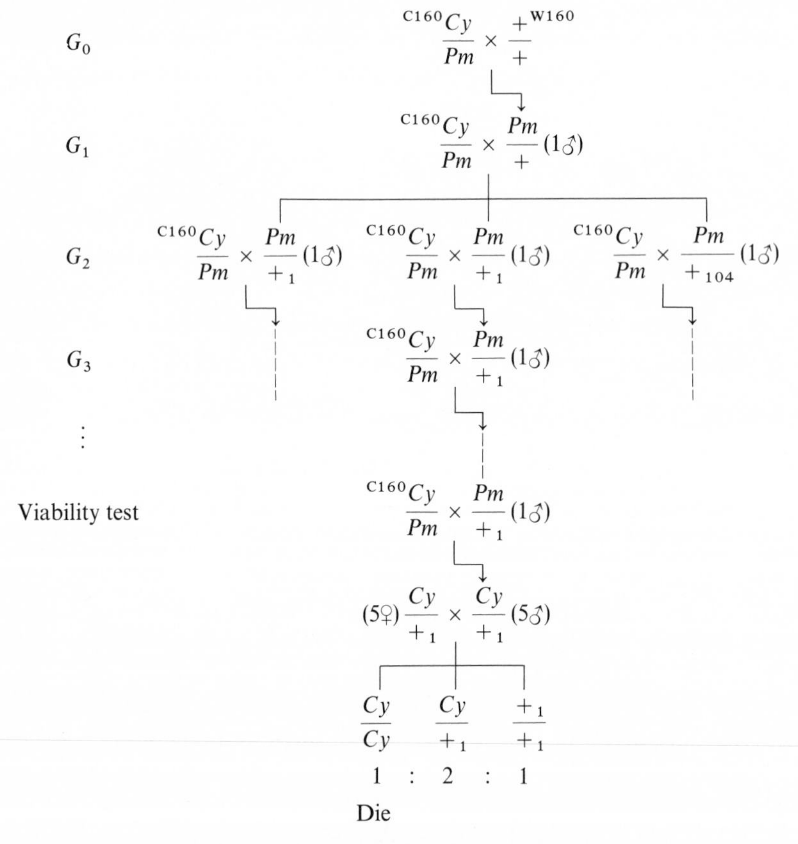

Figure A. Mating scheme to detect the accumulation of mutations. W160 = isogenic wild-type whose second chromosome is to be followed. C160 = Co-isogenic with W160, containing crossover suppressors and Cy/Pm dominant markers (Mukai, 1964).

TABLE A *Average viability indexes of quasi-normal homozygotes excluding controls, their control viabilities, standardized quasi-normal viabilities on the basis of the controls, the expected standardized viabilities (on the basis of $p = 0.1411$ and a quadratic relationship between viability and number of mutant polygenes estimated in later generations), and the phenotypic variance among lines (based on standardized viabilities) (from Mukai, 1969, based on 72 lines out of original 104).*

	Generation					
	10	20	25	32	52	60
Average viability	31.57	30.85	28.53	28.14	21.32	16.42
Control	32.94	32.23	32.92	32.84	33.12	32.41
Standardized viability	0.9584	0.9572	0.8666	0.8569	0.6437	0.5066
Expected viability	0.9751	0.9281	0.8963	0.8425	0.6292	0.5191
Phenotypic variance	0.0028	0.0027	0.0040	0.0062	0.0374	0.0642

Differentiation between lines for polygenic mutation was measured by analysis of variance, excluding from consideration any lethals or semilethals that had accumulated.

At generations 10, 20, 25, 32, 52, and 60 over five years, Mukai measured the viability of homozygotes raised at 25° C under standard conditions. The changes in mean viability of quasi-normal homozygotes are summarized by the data in Table A and the distributions by generations in Figure B. Control lines (unchanged) were determined as follows: for G_{10} and G_{20}, five lines showing best viabilities in G_{15}, G_{20}, and G_{25} were chosen as "unchanged" and their data pooled as a control viability index. (If viability mutations had actually occurred in these control lines, the mutation rate estimates would be *under*- rather than overestimates.) In G_{25} and subsequent generations, controls were chosen on the basis either of replicated lines with greater than 30.0 viability index or of normal viability lines in later generations.

Table A data indicate a progressive lowering of homozygous viability during the accumulation process for 60 generations. We may now make certain simplifying assumptions about the nature of those changes, based on reasoning expressed by Bateman (1959) and developed by Mukai (1964). The average deterioration of viability among the lines is presumed to be the result of accumulation of deleterious small-effect mutations. (Some increases in viability could also occur by mutation, of course, as demonstrated in Example 13-2 by Dobzhansky and Spassky, and Mukai does record one line out of 104, his No. 15, which did increase in viability above the controls; others may have had increasing viability balanced by decreasing without detection.) Estimation of the number of mutations per chromosome per generation was done by assuming that spontaneous mutations occur on chromosomes according to a Poisson distribution. Thereby, the average number of mutations per chromosome (symbolized p) will equal the variance in number. Further, since these mutations are assumed to be more or less equal in their individual effects (x) in reducing viability below the controls, the observed lowering of viability over all quasi-normal lines in a generation (exclusive of the control lines) then equals the average

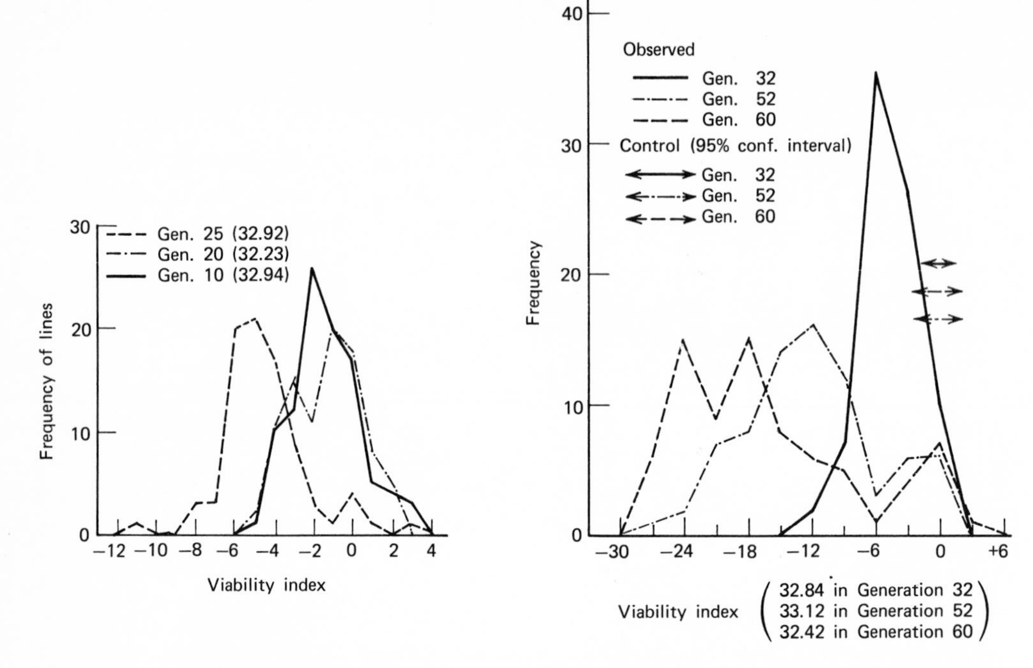

Figure B. Left: Distribution of viability indexes (deviation from control number of wild-type flies counted out of 100) at G_{10}, G_{20}, and G_{25} when cultured at 25°C. Control numbers in parentheses (from Mukai, 1964). Right: Distribution of viability indexes at G_{32}, G_{52}, and G_{60} (from Mukai and Yamazaki, 1968).

effect of a mutant ($\bar{x}$) times the number of mutations: $\bar{x}p = v_c - \bar{v}$, where v_c = control viability average and $\bar{v}$ = quasi-normal viability average.

With increased mutation, the genetic variance must increase among lines. The variance of mutation number will be equal to the average frequency of quasi-normal mutants, or p, but not all mutants will have a unit effect on viability. Therefore, the viability affect (x) measured as a deviation from the control can be squared so that the variance due to mutation would be px^2. Finally, in addition, the magnitude of viability effects will perhaps not be constant over all mutations, so that the variance due to mutational effect must have another component $\sigma_x^2 \cdot p$. Putting these components together then should equal the observed variance among the lines due to real differences (after subtracting variance due to sampling error, determined from variance among replicates within lines), symbolized σ_G^2.

The parameter p can then be estimated by solution of the two simultaneous equations as follows:

$$\begin{cases} \bar{x}p = v_c - \bar{v} \\ (\bar{x}^2 + \sigma_x^2)p = \sigma_G^2 \end{cases}$$

where the right hand portions are obtained from the observed data. For the first 25 generations, Mukai estimated increment in variance and decrement in mean viability to be about linear. Consequently, on a *per generation* basis, the linear regressions $b_v = 0.1127$ and

$b_m = 0.1261$ were the regression coefficients, respectively, for variance increment and the amount of decrease in mean viability. Supplying these estimates into the above formula with the assumption that σ_x^2 = zero gives the following solution:

$$\begin{cases} \bar{x}p = 0.1261 \\ (\bar{x}^2 + \sigma_x^2)p = 0.1127 \end{cases}$$

assuming $\sigma_x^2 = 0$, then

$$\bar{x} = 0.8937$$

(average reduction in viability effect of single mutation)

$$p = 0.1411$$

(number of mutations per chromosome per generation)

It should be evident that if $\sigma_x^2 > 0$, then the estimate of $p > 0.1411$. Therefore, the estimated mutation rate is a minimum on the basis of no variance in magnitude of the viability index from line to line. Undoubtedly, there is some variance so that it is likely that p is really greater than the estimate.

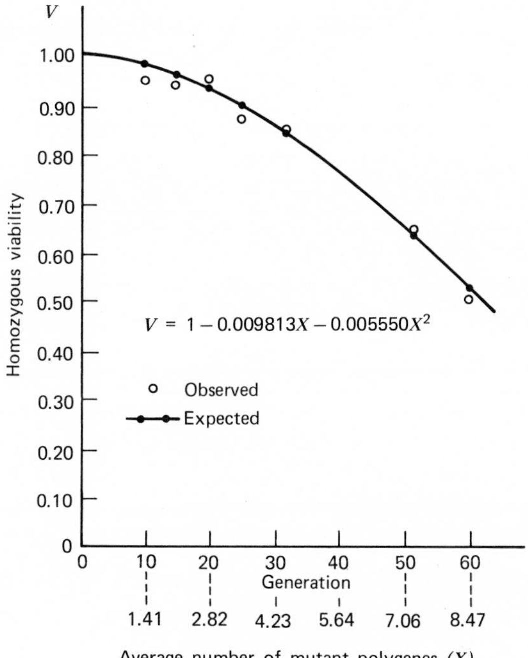

Figure C. Relationship between the standardized average viability and generation number with estimated number of mutant polygenes, assuming the rate of mutation is $\bar{m} = 0.1411$ per generation (from Mukai, 1969).

The spontaneous mutation rate of lethals was estimated by Wallace (Example 13-3) at approximately 0.0049 per chromosome per generation. Mukai's estimate of polygenic mutations with slight depression effect is greater than 28 times that of lethals. In Mukai's experiment, 15 lines had become lethal within 25 generations, or 15/2525 chromosome generations; assuming a Poisson distribution of single locus, double locus, etc., lethals, the estimate was 0.0063 for lethals. His estimated polygenic mutation rate is 22 times greater. About six lines had shown no changes by the 25th generation so that they were either mutation free or balanced by increases in viability.

Crosses between lines at G_{25} produced a genotypic correlation between homozygotes and their heterozygous viabilities of 0.75, not only indicating incomplete dominance of transphase mutants/original polygenes, but also supporting the evidence that the change was a real set of mutational events. The homozygote-heterozygote correlation and its significance to the genetic load is discussed in Chapter 18.

Finally, by the 60th generation, viability had fallen on the average for nonnormal lines to nearly half what had been the original value. Mukai (1969) showed that the relationship between homozygous viability and generations was curvilinear (see Figure C), fitted best by a quadratic line given in the figure. Newly arising mutant polygenes, when building up at orders of magnitude beyond a small number per chromosome, then, are best interpreted as having synergistic action in homozygous condition.

REVERSE MUTATION

While forward mutation (A-a) is slow, reverse mutation is usually slower (see Drake, 1970, 1973). In fact many mutations, or chromosomal changes where specific breakage in chromatin is localized at particular sites, have only the remotest probability of being repeated exactly, let alone being reversed exactly. Back-mutants phenotypically resembling the original alleles, however, can take place from at least three different general genic changes: (1) true restoration of the original base pair at the same site in a missense codon, (2) mutation to a "suppressor" separable by recombination within the cistron, and (3) mutation to an extragenic "suppressor." When attempts at recombination between revertant and original mutant fail, it may be assumed that a true reversion has occurred.

With a directional process going on both forward and reverse, a dynamic equilibrium is possible—that is, a point of gene frequency held in place by opposing forces, gain versus loss to either allele. Because the equilibrium is simple to see, it can illustrate the principle of a dynamic equilibrium perhaps better than any other example.

Let Δp be increment in A allele, u = mutation rate of $A \rightarrow a$, and v = reverse mutation rate $a \rightarrow A$. Then the proportional gain in $A = vq$ and loss in $A = up$ so that net increment in A will be

$$\Delta p = vq - up \qquad (13\text{-}10)$$

or, in terms of q: $\Delta q = up - vq$. If the forward and backward rates are balanced, then Δp comes to zero at a nontrivial equilibrium ($0 < p < 1.0$) such that neither allele is fixed or lost, at the point where $up = vq$, or $p/q = v/u$. That point in terms of gene frequency is easily found by changing q to p in (13-10) with $\Delta p = 0$, so that

$$\hat{p} = \frac{v}{u + v} \qquad (13\text{-}11)$$

where $\hat{p}$ is the frequency of the A allele at equilibrium. As an example, if $u = 0.5 \times 10^{-5}$ and $v = 0.25 \times 10^{-5}$, then $p = 0.333$. It is clear from (13-11) that, under forward and reverse mutation only, the final equilibrium point is independent of starting frequencies, although the rate of approach to the point is exceedingly slow.

Understanding the meaning of this equilibrium can be gained by an example (based on an illustration in Li, 1961, pp. 102–105). Let us assume two gene pools, representing frequencies of the alleles at some initial point. Let $p_A = 0.20$ and $q_a = 0.80$. If each pool gives up a constant fraction of itself to the opposite pool each generation, say half of p is converted to q while one-quarter of q is converted to p by forward and reverse "mutation," respectively, then the balanced state will occur as indicated by setting $\Delta p = 0$ in (13-10), as in Table 13-3. Note that the outcome is entirely determined by the relative fractional exchanges, up or vq, amount of "mutations," but the products of fractional exchange times allelic frequencies are equal at the equilibrium point. In other words, the relative values $\hat{p}/\hat{q} = v/u$ give the equilibrium condition, not the absolute values.

Finally, this equilibrium is a stable one (recall Chapter 1) because a change in p value imposed on the population will be counteracted by the tendency to return to the p value. Of course, mutation will change the p value very slowly (see Exercise 5). Consequently, we cannot realistically ascribe persistent equilibriums in most populations to the action of forward and reverse mutation. Perhaps this particular equilibrium has only theoretical interest basically, but it should serve to illustrate the simplest type of equilibrium. A more realistic and fundamentally similar yet more powerful equilibrium may be the case when gene flow between populations occurs (Chapter 19).

TABLE 13-3 *Changes in gene frequencies leading to equilibrium due to forward* (u) *and reverse* (v) *"mutation"* ($\times$ *10*^{5} *compared with the example in text). Decimals in parentheses are the fractional exchanges, up or vq, each generation (based on Li, 1961)*

	$p_A \xrightarrow[(u)]{\frac{1}{2}}$	$\xleftarrow[(v)]{\frac{1}{4}} q_a$
G_0	0.20 (−0.10) →	← (−0.20) 0.80
G_1	0.30 (−0.15) →	← (−0.175) 0.70
G_2	0.325 (−0.1625) →	← (−0.16875) 0.675
G_3	0.33125 ⋮	0.66875 ⋮
Limit	0.3333 (−0.1667) →	← (−0.1667) 0.6667

Symbolically $up = vq$ at equilibrium

EXERCISES

1. (a) If a gene mutates at a rate of 1×10^{-6} for just one generation in a population of 5 million diploid individuals, how many genes will be passed on in the mutant state to the next generation, assuming constant population size?
   (b) If the gene continues to mutate at the same rate and back mutation is absent, what will be the mutant frequency expected after 10 generations?
   (c) At the same rate of mutation, how many generations would it take to raise the frequency q of the mutant allele from zero to $q = 0.095$?
2. (a) Using Table 12-1A, suppose a single mutation occurs in an expanding population. Assume progeny to be distributed as a Poisson but with a mean number of four per family, what will be the probability of losing the mutation in a single generation?
   (b) How is the probability of losing the mutation affected by a decreasing population? Let the mean number of progeny per family be one.
3. Suppose the survivorship of an *Aa* genotype following initial mutation is 1 percent greater than complete neutrality (no advantage in survival).
   (a) How would the calculation in Table 13-1B be changed? (Assume Poisson distribution and large population size.)
   (b) What will be the ultimate probability of survival for the mutant (likelihood of fixation)? How much greater would that be than for a completely neutral mutant?
4. (a) What would be the expected probability of fixation if a single neutral mutation occurs in a population broken up into a large number of isolates of size $N_e = 20$, within each of which mating is random?
   (b) What is the average time in generations for fixation of a single mutant in isolates of size $N_e = 20$? Average time to loss of the mutant?
5. If in a certain strain of drosophila, the white eye (w) locus mutates from w^+ to w^0 (eosin) at the rate of 1.3×10^{-4}, but the reverse mutation back to w^+ occurs at the rate of 4.2×10^{-5}.
   (a) What would be their frequencies in the next generation if a large panmictic population had exactly 50 percent of these two alleles?
   (b) Graph Δp (13-10) for p values at intervals of 0.05 from $p = 0.05$ to 0.95.
   (c) What would be the eventual frequencies of these two alleles if the population stayed large and there were no advantage to either allele?
   (d) Show that at equilibrium due to forward and reverse mutation, $\hat{p}/\hat{q} = v/u$.
6. How does the stochastic situation for extinction of a mutant (as in Example 13-1) compare with the rate of extinction expected for a large deterministic population? How does the final outcome for survival of the mutant compare between stochastic and deterministic models?
7. In Example 13-3, Table A indicates a gradual increase in number of lethals and semi-lethals. If a generation is estimated at two-week intervals, assume each new lethal occupies a single chromosome and there are no multilocus lethals in the population. What rate of lethal mutation per generation would satisfy this accumulation rate observed by Wallace?
8. In Example 13-4, Mukai estimated the rate of polygenic mutation to subvitality at about 0.1411 mutations per generation on the second chromosome of *Drosophila melanogaster*.

What are the assumptions made by Mukai in arriving at this estimate? Do you agree that the estimate is a minimum? Why?

9. From considerations of viability in Chapter 13, how would you envisage the nature of the genotype (number of loci, phenotypic effects of mutants, and range of effects) controlling viability as an inherited trait?

14

NATURAL SELECTION: DIRECTED GENE CHANGE WITHIN POPULATIONS

There is no doubt that natural selection as conceived by Charles Darwin and Alfred Russel Wallace is the central mechanism bringing about adaptation, the perfecting of a population's or a species' ability to exploit its environment. Darwin realized that his term *natural selection* was a metaphorical expression for the net effect of two or more variants' differing in reproductive capacity (their total expectation of offspring) when living together under the same conditions. Given time, genetic determination of reproductive differences, and the replicating ability of the genetic material, simple arithmetic or logic will indicate that any unselected population would become more "fit"—that is, more efficient at getting through the life cycle and producing offspring—by eliminating "harmful" genetic variants that lower efficiency and by increasing the frequency of "useful" variants until maximum *fitness* (capacity for reproduction) has been attained. The genetic potential for change must exist in the population's store of mutational and chromosomal variation, diversity of linkage blocks, and other heterogeneous genetic elements if selection is to achieve a response to environmental stresses. With the immense genetic diversity available in most species and their innate capacity for reproduction, populations respond to environmental necessities by changes in their gene pools in ways that give them improved chances for survival and reproduction. Genic changes over time within populations are the most significant of directed long-term biological processes; they represent the conversion of inherited variation within a biological entity into permanent differences among groups and thus the evolution of new groups.

Multiformity of environments through space and time on this planet is recognized by all biologists. Organisms respond to this diversity of conditions (climates, physical features, biological interactions, and environmental exigencies) via many strategies that involve becoming more efficiency at living and utilizing the available resources to create their own biomass; thereby, species evolve new gene pools, and, as we say, become "adapted." They gain in fitness as a response of their population in which genotypes vary in fitness; individuals are powerless to respond by changing their genotypes in a directed fashion. Mutations occur "at random"—that is, not being produced to satisfy the needs of organisms—thus, they constitute the essential building blocks for the selective process.

At the risk of sounding repetitive, we must emphatically point out that natural selection is neither a tangible force (like radiant energy) nor is it a process that is an essential property of organisms (like mutation). As Lerner (1958) has clearly stated:

> *. . . natural selection is really not an a priori cause of any phenomenon observed in nature, in the laboratory, or on the farm. Natural selection is a term serving to say that some genotypes leave more offspring than others. Natural selection has no purpose. . . . For*

> *any generation, natural selection is a consequence of the differences between individuals with respect to their capacity to produce progeny. The individuals who have more offspring are fitter in the Darwinian sense. To speak of natural selection as causing one array of individuals . . . to have offspring, and another not to, is a tautology. This fact, however, should not prevent us from attributing a major part of evolutionary change, viewed in retrospect, to natural selection.*

Perhaps these phrases may sound inconsistent with the notion that a selective agent or an environmental condition evokes a selective result, such as one or more genotype producing more offspring than another, following that agent's effect. It is common practice to speak of selection in an active sense favoring or eliminating genotypes; this language is figurative and simply convenient as a shorthand. The student should remember that Darwinian fitness differences between genotypes brought about by countless agencies result in differential propagation of genetic material to the next generation. After the inequality of genetic contribution has taken place, we say that natural selection has occurred.

In this chapter we focus attention on the consequences to the population's genetic system resulting from differences in Darwinian fitness (selective coefficients) between genotypes in the simplest genetic system (one pair of Mendelian factors). We have seen in previous chapters that gene pools may change for several reasons—for example, by sampling of gametes or by mutation—but none of those causes produces directed changes rapidly enough to account for adaptation. For understanding of the fundamentals in the selective process, we describe simple selection models in this chapter without examples from the experimental literature. Identification of selective agents and fitness properties of genotypes under laboratory and field conditions will be discussed later when the student has had some elementary experience thinking about the outcome of the selective process. We do, however, illustrate rates of change under selection from computer simulations in examples, including the effects from reduction in population size. Here we make the following simplifying assumptions, which we shall remove one at a time in a later chapter to approach more realistic conditions: (1) genetic variants with fitness differences affect just one portion of the life cycle, (2) fitness values are constant, (3) a single genic locus affects fitness with one pair of alleles, (4) the mating system is panmictic and general conditions for the square law exist outside the selective condition, and (5) the environment is single and constant.

MODES OF SELECTION

The countless agencies by which phenotypes come to differ in fitness and thus be "favored" or "unfavored" within populations have general features in common so that they may be grouped as modes of selection. Selection usually acts only on phenotypes irrespective of genotype so that any genetic change resulting depends on the relationship between genotype and phenotype. Only when we have a genetic marker to follow from one generation to another can we learn much about the genetic determination of fitness traits and thus the resulting action of selection on gene frequencies. Nevertheless, we can infer much about the amount of genetic variation available for selective action when phenotypes are under polygenic control and response can be quantified (as we saw in Chapters 6 and 7).

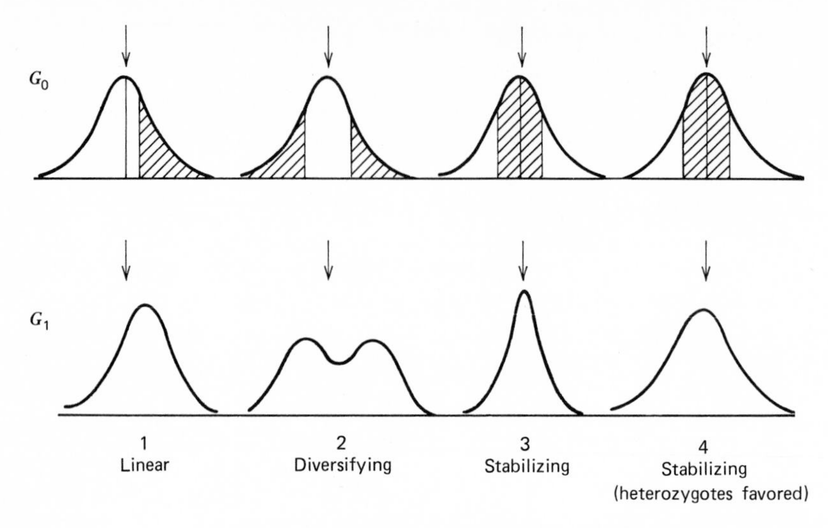

Figure 14-1. Diagram of the major modes of selection. Upper curves represent distributions in the G_0 from which individuals are selected as parents (shaded areas). Arrows indicate the mean for the phenotypic trait. Lower curves represent the expected progeny (G_1) responding to selection applied. Parents are considered as mated at random except in (2 Diversifying) where they tend to mate assortatively. In (3 Stabilizing) polygenic additivity is assumed, while in (4) the trait is assumed to depend on heterozygosity to be favored.

The following are some of the general ways in which we may consider selection to be applied, as illustrated in Figure 14-1:

1. Linear (dynamic or directional). Selection favors one extreme of a phenotypic distribution so that the mean of the offspring tends to be altered in the direction of the favored phenotype compared with the population from which the parents were selected. Response may continue in the same direction either until genetic variation for the trait is exhausted or until a "balanced" state, or plateau, is reached. This classic mode of selection was outlined in Chapter 7.

2. Diversifying (centrifugal or disruptive). Selection favors two or more diverse types at a time, such as the opposite extremes of a distribution with elimination of intermediates but with gene flow between the favored phenotypes as a necessary condition for maintaining the population. This mode is conceptually the equivalent of having the population occupy two or more ecological niches with gene flow between niches and favoring a different genetic complex in each niche. This mode tends to increase genetic variance compared with an unselected population, and the gene pool will tend to retain polymorphism (diversity).

3. Stabilizing (centripetal or unifying). Selection favors individuals near the mean of a distribution, eliminating extremes, so that genetic variance tends to be reduced.

The integrating properties of the population may thus be reinforced, particularly if genotypes (heterozygotes) promoting overdominance in fitness are being selected.

Some details of these selection modes will be examined in Chapters 16–19. However, it is important to realize that fitness values are not likely to remain constant (as in the classic case discussed in this chapter); they may depend on a number of conditions that bring about modification and flexibility to these modes of selection, often making them complex. In this chapter we restrict our attention to simplified selection principles in which we assume constant fitness values throughout; some of the conditions that bring about variable fitness will be examined in Chapter 16.

The levels at which selection may act have been stressed by Wright (1956): selection among genes, cells, gametes, clones, biparental organisms, within populations (demes), between populations, and among species. According to Lewontin (1970), the biparental level "gives the great advantage that the evolution of the entire population, rather than a single lineage, is promoted by selection." Under constant selection conditions, the multiplicity of genotypes resulting from the sexual process in biparental organisms allows for an immense diversity on which selection may act, while with clonal organisms genic diversity is more restricted to acquisition by mutation. Under changing conditions, it is obvious according to Wright "that the process of continual readaptation would be more effective if it could be based on the adaptiveness of genotypes or systems of these than merely on the momentary net effects of the separate genes." Consequently, in the adaptiveness of several gene pools to communities or several demes within a species to the entire species, interdeme selection leading to various adaptive levels including races and species is of vast importance. Of course, linear, diversifying, and stabilizing selection acts at all these levels, not as mutually exclusive forces but as overlapping ones, so that in natural populations the mode of selection may be difficult or nearly impossible to discern. Only when we apply selection experimentally in the laboratory or in breeding can we be more certain of the goal to be attained. The instances in which actual specific selective action has been ascertained in nature are very few indeed, and a considerable amount of attention should be given in future to finding out what selective differences genotypes may have in nature. At this time, more than 100 years after the Darwin-Wallace papers on the subject, we are still quite in the dark when we ask somewhat metaphorical questions: "What is selection aiming at in the population?" "What are the selective agents in our environmental niche?" Nevertheless, the few cases of which we are reasonably certain demonstrate how powerful and how subtle the action of natural selection may be.

While natural selection and artificial selection differ in that the first is directed by relative fitness of hereditary traits under specified conditions and the second is goal-directed by the experimenter, they share the common element that the effectiveness of selection in producing a response depends on (1) modes and intensity of selection, (2) action of genes as manifested in phenotypes being selected, and (3) the amount and architecture of genetic variability in the population. The result of these factors—the adapted population following the action of natural selection or the "improved" population following artificial selection—demonstrates the powerful yet varied property of the entire process. With numerous genetic pathways available to selection, populations with various genotypes initially may respond with parallel outcome; conversely, under various selective pressures, a single population may give diverse responses.

THE SIMPLEST CASE OF SELECTION

For selection to produce a directed change in the gene pool, only two ingredients are essential: a reproductive difference between genotypes and the duplicating ability of the genetic material. In order to establish some fundamental principles and definitions, we begin with a hypothetical example of selection in a haploid organism. We may visualize a population of haploid prokaryotic organisms stressed by a selective agent such as an antibiotic at low concentration (in order not to kill all the cells) or a low dose of radiation to separate sensitive from the more resistant cells. Further, we assume:

1. Two alleles exist in the population initially, A and a, in equal frequencies ($p = q = 0.5$).
2. Reproduction is by simple fission with no sexual phase.
3. The ability of A to get through its life cycle and reproduce, leaving *one* offspring per *one* parent individual, is better than the ability of a allele under the specified environmental conditions; for example, for every 100 new A individuals produced by one life cycle from A parents, only 90 new a individuals are produced in the same life cycle by a parents. In brief, the Darwinian "fitness" of A relative to a is 1:0.9; and we assume fitness remains constant over all generations.
4. The environment can support a constant number of individuals in a generation. If these haploid organisms double their number in their life cycle, the "struggle for existence" will cause all to be lost except that constant number. We call this number 100 percent. For example, if the population consisted of 1000 individuals initially (500 A:500 a), and they all divided by binary fission, making a total of 2000 progeny, the number must be reduced back to 1000 in the "struggle for existence."

Given these assumptions, the changes in frequencies of the two kinds of organisms (or genotypes) will take place as given in Table 14-1A. It is important to have clearly in mind the meaning of the calculations in the first generation. First, we might imagine that the organism only reproduces by making a single copy of itself and that the A individuals are 10 percent better at doing that than the a individuals. But this interpretation is not what is meant by the calculation, because such an implication would result in a *lowering* of the total progeny number. It is more meaningful to recall condition 4, above—that both A and a reproduce themselves by doubling and that the environment will support only a *constant* number. To make the relative numbers add up to a constant total, we "normalize" the relative survival numbers from the row "after selection." These relative "after-selection" numbers are merely the product of fitness × frequency of each genotype in order to calculate the mean fitness, given at the end of that line as 0.950. (The student should recall the formula for arithmetic average $\bar{Y} = \sum f Y_i / n$ for data grouped by frequency.) We now have a total progeny = 100 percent. There would consequently be a probability for any individual in G_1 to be A genotype = 0.500/0.950 = 0.5263; for a = 0.450/0.950 = 0.4737, or a ratio of 1:0.900.

If the fitnesses remain constant (condition 3) each generation, for each round of life cycle the mean fitness and relative proportions of progeny surviving will be occurring as in the calculated illustration. It is evident that there is a rise both in the "favored" allele (A) and in the mean fitness (boxed in Table 14-1). Briefly, the objective of natural selection is to

TABLE 14-1 *Haploid selection*

A Simple haploid selection: assuming initial $p = q = 0.5$, relative fitness values of $1A:0.9a$ per generation, and a constant number of individuals (100 percent) per generation. Mean fitness values ($\sum f W_i$) are in boxes at right.

Generation	Genotypes		Totals
	A	a	
Initial frequency (parents)	$0.500 = p_0$	$0.500 = q_0$	1.000
G_0 fitness	1	0.90	
After selection	0.500	0.450	[0.950]
G_1 (relative)	$0.526 = p_1$	$0.474 = q_1$	1.000
Fitness	1	0.90	
After selection	0.526	0.427	[0.953]
G_2 (relative)	$0.552 = p_2$	$0.448 = q_2$	1.000
Fitness	1	0.90	
After selection	0.552	0.403	[0.955]
G_3 (relative)	$0.578 = p_3$	$0.422 = q_3$	1.000
⋮			⋮
Limit G_∞	$1.000 = p_\infty$	$0 = q_\infty$	[1.000]

B Symbolically, these data may be represented as follows

G_0	p	$+ q$	1.00
Fitness (W_i)	W_1	W_2	
Or	1	$(1 - s)$	
Product (after selection)	pW_1	$+ qW_2$	$fW_i = \bar{W}$
Or	p	$+ q - sq$	$(1 - sq)$
G_1 (relative)	$p/(\bar{W})$	$+ (q - sq)/(\bar{W})$	1.00
Or	$p/(1 - sq)$	$+ (q - sq)/(1 - sq)$	1.00

Change in q per generation: $q = \dfrac{q - sq}{(1 - sq)} - q = \dfrac{-sq(1 - q)}{(1 - sq)}$

or more generally in terms of W_i: $q = \dfrac{qW_2}{pW_1 + qW_2} - q = \dfrac{(W_2 - W_1)pq}{\bar{W}}$

maximize the mean fitness of the population by ridding the population of the less-fit allele. We may say that the population "suffers a genetic load" (using H. J. Muller's term) as long as the less fit allele is present. We merely imply by this metaphorical phrase that the population is not reproducing at its potentially maximum efficiency for the conditions under which these genotypes have the fitness values we have assumed.

The rate at which a is eliminated changes as the limit is approached. We symbolize these changes algebraically to generalize them, using the symbols employed by Wright (1931). For relative fitness (selective) values of the two genotypes: W_1 and W_2 (1 and 0.90, respectively), which may stand for "weight" given by relative selective effects of conditions on the two genotypes. Alternatively, the less fit genotype may be thought of as having a disadvantage (0.10), which may be symbolized as the selection coefficient $s = 1 - W_2$ in this case. More generally, the genotype fitness is $W_i = 1 - s_i$. Mean fitness is $\bar{W} = \sum f W_i$, with frequencies usually p,q values (or genotype frequencies in diploids). Calculations would proceed then as in Table 14-1B. The student should verify the calculations in the A portion of the table by applying the symbols from the B portion.

The change in allele frequency per generation (Δq) is easily calculated by subtracting the original q value from the new q value $[q(1 - s)/(1 - sq)]$ at the bottom of Table 14-1B.

$$\Delta q = \frac{-sq(1 - q)}{1 - sq} = \frac{-(W_1 - W_2)q(1 - q)}{\bar{W}} \tag{14-1}$$

(rate of change in q under haploid selection).
From Exercise 1 the student should note that this rate of change decreases as the less-fit allele becomes rarer. The denominator in (14-1) approaches a value of 1 as $q \to 0$, and the rate approaches the allelic frequency product $(pq) \times$ the selection coefficient s. It is constantly negative (downward in q), so that the limit can only be $q = 0$. If we ask whether there can be a stable equilibrium by setting $\Delta q = 0$, we see that no solution except $q = 0$ or 1 will satisfy the equation. We call this equilibrium "trivial" because it is more or less self-evident.

Finally, we may consider the number of generations it would take for selection to produce a given change in q. From (14-1), if we dismiss the denominator as negligible when the product sq is very small, the rate of change can be envisioned as a continuous function so that (14-1) may be expressed as a differential equation where t refers to time in generations.

$$\frac{dq}{dt} = -sq(1 - q) \qquad \text{or} \qquad \frac{dq}{q(1 - q)} = -s\,dt$$

Integrating both sides gives

$$\int_{q_0}^{q_t} \frac{dq}{q(1 - q)} = -s \int_0^t dt$$

where $t =$ generations for a given change in q value from q_0 to q_t). The solution is

$$t = \log_e \left[\frac{q_0(1 - q_t)}{q_t(1 - q_0)} \right] \bigg/ s \tag{14-2}$$

(where $\log_e = 2.3026 \log_{10}$). For example where $s = 0.10$, if we wish to estimate the number of generations to change from $q_0 = 0.05$ to $q_t = 0.005$, we should have as follows:

$$t = \log_e \left[\frac{0.05(0.995)}{0.005(0.95)} \right] \left(\frac{1}{0.1} \right)$$

$$t = \log_e[10.4737] \left(\frac{1}{0.1} \right) = 23.5 \text{ generations}$$

SELECTION AGAINST RECESSIVE HOMOZYGOTES

With diploidy and the accompanying increase in genotypes at a locus, the relative differences in fitness between genotypes can be described in several different ways, depending on dominance, partial or incomplete dominance, or overdominance in fitness. We first examine the outcome of selection when the less-fit genotype is recessive—that is, $AA = Aa > aa$ in fitness. Simplifying assumptions must be made here to include (1) standard conditions (a large population, random mating to produce genotypes from gametes, Mendelian segregation, and no other forces acting on the population), (2) constant Darwinian fitness throughout, (3) a single pair of alleles at the locus, and (4) selection acting between the zygotic stage and the adult. We shall discuss fitness and the total life cycle in later chapters. At this point it is the simplest model to assume that selection acts after fertilization and before mating.

Recessive Lethals

The most extreme selection against a single genotype would be the case if *aa* is entirely eliminated each generation before it can reproduce—that is, for *aa* to be lethal. If parents are initially $Aa \times Aa$ to provide maximum frequency of the lethal allele, the changes over the first few generations would be as in Table 14-2A. One-fourth of the progeny zygotes

TABLE 14-2 *Diploid selection*

A Selection against a completely recessive lethal (*aa* has a fitness (W) of zero). Parents are $Aa \times Aa$. Lethality is assumed to act between zygote and adult stages

Generation	Genotypes			Totals	q_a
	AA	Aa	aa		
G_0 parent frequency	0	1	0	1	$0.50 = \frac{1}{2}$
G_1 zygotes $(p + q)^2$	0.25	0.50	0.25	1	
Fitness (W)	1	1	0		
After selection	0.25	+ 0.50	+ 0	$0.75 = \bar{W}$	
G_1 (mature)	0.3333	0.6667	0	1.00	$0.3333 = \frac{1}{3}$
Zygotes after random mating $(p + q)^2$	0.4444	0.4444	0.1111	1.00	
Fitness (W)	1	1	0		
After selection	0.4444	+ 0.4444	+ 0	$0.8889 = \bar{W}$	
G_2 (mature)	0.5000	0.5000	0	1.0000	$0.25 = \frac{1}{4}$
⋮	⋮	⋮	⋮	⋮	⋮
G_∞	1.0000	0	0	1.0000	0.0000

B Symbolically

	Genotypes			Totals	q_a
	AA	Aa	aa		
G_1 zygotes	p^2	$+ 2pq$	$+ q^2$	1	q_0
Fitness	1	1	0		
After selection	p^2	$+ 2pq$	$+ 0$	$1 - q^2 = \bar{W}$	
Relative frequency	$p^2/(1 - q^2) = \dfrac{1 - q_0}{1 + q_0}$	$2pq/(1 - q^2) = \dfrac{2q_0}{1 + q_0}$			$\dfrac{q_0}{1 + q_0}$

die before maturity, so that one-third of the remainder is AA and two-thirds Aa adults, which then mate and produce zygotes. One-ninth is lethal ($q^2 = \frac{1}{9}$) in the zygotes following random mating of the G_1 adults. It is easily seen that the lethal allele frequency rapidly diminishes in a simple series: $\frac{1}{2}, \frac{1}{3}, \frac{1}{4}, \ldots$, so that the denominator increases by 1 for each generation. We proceed to symbolize the allele frequencies and fitness values constantly generation by generation in Table 14-2B. The mean fitness ($\overline{W}$) is $(1 - q^2)$, giving the relative frequencies of the two visible genotypes and the new q values as one-half the heterozygote frequency:

$$q_1 = \frac{q_0}{1 + q_0} \tag{14-3}$$

(new q in terms of previous generation q for recessive lethal, where subscripts refer to any two generations), or, in general,

$$q_t = \frac{q_{t-1}}{1 + q_{t-1}}$$

Successive q values from generation to generation form a harmonic series (see also Chapter 10). If we symbolize q_2 (after two generations) in terms of the two previous generations we have

$$q_2 = \frac{q_1}{1 + q_1} = \frac{q_0/(1 + q_0)}{1 + q_0/(1 + q_0)} = \frac{q_0}{(1 + q_0) + q_0} = \frac{q_0}{1 + 2q_0}$$

Therefore, q values over a few generations will be

Generation	q	terms of q_0
0	q_0	$= q_0$
1	q_1	$= q_0/(1 + q_0)$
2	q_2	$= q_0/(1 + 2q_0)$
3	q_3	$= q_0/(1 + 3q_0)$
t	q_t	$= q_0/(1 + tq_0)$

(14-4)

(general value of q after t generations of selection against recessive lethal).

The change in q per generation is calculated as with (14-1) by subtracting the original q_0 from the new q_1 values:

$$\Delta q = \frac{q}{1 + q} - q = \frac{-q^2}{1 + q} \tag{14-5}$$

It should be noted that the elimination of a recessive lethal from a population becomes much less efficient as the lethal becomes rare. Because of property 3 (Chapter 1) of the square law in large random-mating populations—with increasing rarity, a greater proportion of an allele is preserved in the heterozygote—the lethal allele will be exposed to selection less and less, and time for complete eradication will be extended. Intuitively, we see that some limitation in population size with accompanying random sampling of gametes is a condition in real populations that may help to complete the eradication of a rare allele.

Nonlethal Recessives with Low Fitness

If selection is not so extreme as a lethal factor but still decreases the probability of progeny from the homozygote (*aa*), rate of change in q will be less rapid but in the same direction as for the more extremely deleterious genotypes. Semilethals (viability about half of normal), viability depressors ("subvitals"), and those genotypes changing fitness slightly for the worse compared with normal may fit this scheme if they have no depression on heterozygotes. Relative fitness values for the three genotypes then may be represented as follows:

Genotypes:	*AA*	*Aa*	*aa*
Relative fitness	1	1	$1 - s$
Symbols for fitness	W_1	W_2	W_3

where $s = 1 - W$, as before. With a lethal, $s = 1$ (complete disadvantage, or $W = 0$). The more general case for any lowered-fitness homozygote will change as given in Table 14-3A and B, and in Example 14-1, Figure A. If selection acts between the fertilized egg

TABLE 14-3 *Selection against a homozygote*

A Selection against a recessive homozygote (*aa*). Let *aa* fitness be 70 percent of the other genotypes and the initial allele frequency be $q_0 = 0.90$. Selection is assumed to act between zygote and mature adult

	Genotypes				
Generation	*AA*	*Aa*	*aa*	Totals	q_a
G_0 zygotes	0.01	0.18	0.81	1.00	0.9000
Fitness	1	1	0.7		
After selection	0.01	+ 0.18	+ 0.567	$0.757 = \bar{W}$	
G_0 (mature)	0.0132	0.2378	0.7490	1.00	$0.8679 = q_1$
G_1 zygotes $(p_1 + q_1)^2$	0.0174	0.2293	0.7533	1.00	
Fitness	1	1	0.7		
After selection	0.0174	+ 0.2293	+ 0.5273	$0.774 = \bar{W}$	
G_1 (mature)	0.0225	0.2962	0.6813	1.00	$0.8295 = q_2$
B Symbolically					
G_0 zygotes	p_0^2	$+ 2p_0q_0$	$+ q_0^2$	$= 1$	q_0
Fitness	(W_1)	(W_2)	(W_3)		
Or	1	1	$1 - s$		
After selection	p_0^2	$+ 2p_0q_0$	$+ q_0^2 - sq_0^2$	$= 1 - sq^2 = \bar{W}$	
G_0 (mature)	$\left[\frac{p_0^2}{\bar{W}}\right]$	$+ \left[\frac{2p_0q_0}{\bar{W}}\right]$	$+ \left[\frac{q_0^2 - sq_0^2}{\bar{W}}\right]$	$= 1$	$q_1 = \frac{q - sq^2}{1 - sq^2}$
G_1 zygotes	$(p_1)^2$	$+ 2p_1q_1$	$+ (q_1)^2$	$= 1$	

and the adult, the relative frequencies after selection are, of course, not in square-law equilibrium. After random mating, their progeny zygotes will be, however, and selection acts on those zygotes in the progeny generation. For calculating the new q among the G_1 after selection, it should be apparent that we merely need to apply the simple relationship of $q = R + \frac{1}{2}H$ from (1-3) and divide by the net fitness (normalizing the q value in G_1). There is no need to calculate the G_1 adult relative frequencies for the purpose of estimating allelic frequencies; however, to continue to G_2, we must apply the square law to the new allelic frequencies—$(p_1 + q_1)^2$—because we are assuming selection to be acting from the zygote stage to adult. The new q value found in Table 14-3B is $q_1 = (q - sq^2)/(1 - sq^2)$, so that the change in q per generation is

$$\Delta q = \left[\frac{q_0 - sq_0^2}{1 - sq_0^2}\right] - q_0 = \frac{-sq_0^2(1 - q_0)}{1 - sq_0^2} \qquad (14\text{-}6)$$

This relation reduces to (14-5) if the recessive is a lethal with $s = 1$. Its similarity with (14-1) for the haploid condition is noteworthy except that the recessive allele frequency is squared in this diploid case. More generally, (14-6) can be expressed in terms of relative fitness coefficients (W_i) instead of as selection coefficients (s_i); from *net fitness* of $p^2W_1 + 2pqW_2 + q^2W_3 = \bar{W}$ and zygotes *after* selection,

$$\frac{p^2W_1 + 2pqW_2 + q^2W_3}{\bar{W}}$$

Then the new q value after selection is

$$q_1 = \frac{p_0q_0W_2 + q_0^2W_3}{\bar{W}}$$

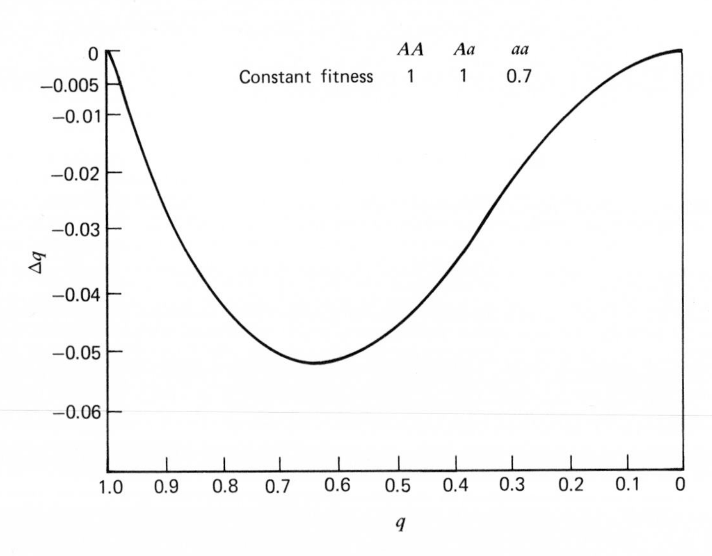

Figure 14-2A. Plot of Δq from formula (14-6) when fitness values from Table 14-3A are held constant.

and change in q is

$$\Delta q = \left[\frac{p_0 q_0 W_2 + q_0^2 W_3}{\bar{W}}\right] - q_0 \tag{14-7}$$

If $W_2 = W_1 = 1$, this statement reduces to

$$\Delta q = \frac{(W_3 - 1)p_0 q_0^2}{\bar{W}} \tag{14-7A}$$

It should be noted that this rate of change then depends on q. If q is high (0.90), as in Table 14-3A, the rate of change will be large ($\Delta q = -0.0321$); but if q is low (0.10), then $\Delta q = -0.0027$. If we plot the (14-6) relationship on a graph as in Figure 14-2, we find the maximum point, when q is changing most rapidly. The first derivative to be calculated and set equal to zero may be simplified by assuming the denominator of (14-6) to be nearly 1 if s is small—at $s \leq 0.10$, for example. Then the solution, if $d\,\Delta q/dq = 0$, becomes $q = \frac{2}{3}$ for maximum value of Δq. As q diminishes, approaching zero, Δq becomes infinitesimally small. Also, $\bar{W}$ is plotted in Figure 14-2, and we note that it is continually increasing in the direction selection goes.

As in the haploid case, the number of generations to change q from any value to a lower value by this kind of selection, assuming selection coefficients to remain constant, can be calculated by using (14-6) as a differential equation with respect to time in generations and simplified by assuming s to be small so that the denominator is almost 1. Then

$$\frac{dq}{dt} = -sq^2(1 - q) \qquad \text{or} \qquad \frac{dq}{q^2(1 - q)} = -s\,dt$$

Integrating both sides gives

$$\int_{q_0}^{q_t} \frac{dq}{q^2(1 - q)} = -s \int_0^t dt = -st$$

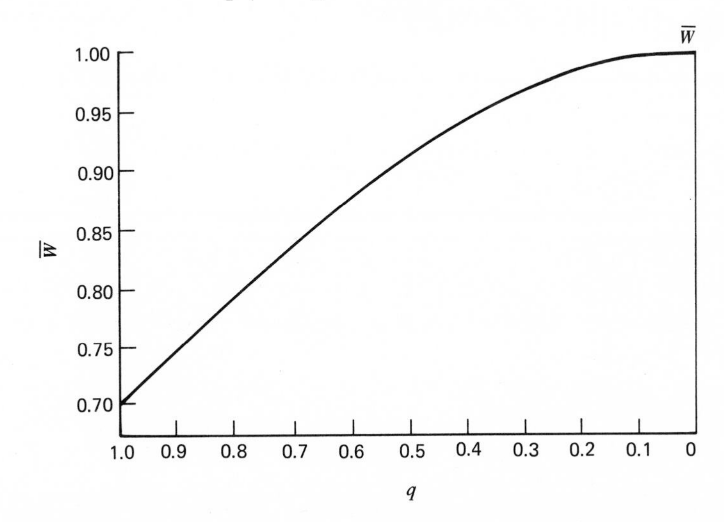

Figure 14-2B. Plot of $\bar{W}$ over the same array of q values with fitness values from Table 14-3.

whose solution is

$$t = \left\{\left[\frac{q_0 - q_t}{q_0 q_t}\right] + \log_e\left[\frac{q_0(1 - q_t)}{q_t(1 - q_0)}\right]\right\} \cdot \frac{1}{s} \tag{14-8}$$

If we assume $s = 0.10$ in the diploid case and compare with the haploid, estimating the number of generations to change $q_0 = 0.05$ to $q_t = 0.005$, we would have

$$t = \left\{\frac{0.05 - 0.005}{(0.05)(0.005)} + \log_e\left[\frac{0.05(0.995)}{(0.005)(0.95)}\right]\right\} \frac{1}{0.1} = 1823.5 \text{ generations}$$

This result is enormously longer than for the haploid case, and it emphasizes the immensely more difficult task for deterministic selection to rid a population of a completely recessive deleterious allele in diploid organisms compared with haploid. It is clear that for elimination of the allele to occur at a faster rate, either limitation of populations size or inbreeding must be invoked to increase the probability of homozygosity when alleles become rare.

In Example 14-1, the deterministic changes in a large panmictic population are illustrated along with the contrasting outcome for subdivided populations with reduced-size isolates as simulated by a computer with fitness values as given in Table 14-3. If the *a* allele starts as a deleterious mutation in $1/2N$ genes, it may spread for a while due to drift but tends to be eliminated from small-sized isolates, while it still exists in a transitory state (polymorphic) in the large population. At the opposite high q of $(2N - 1)/2N$ genes, the *A* allele may be considered a single favorable mutant that displaces *a* rapidly at first and then more slowly in the large population. However, in the drifting small isolates, fixation for the less-fit but common *aa* genotype occurs so rapidly that by G_{100} the outcome of high average $\bar{q}$ is hardly recognizable as similar to the deterministic trend. (See Exercise 4 for the ultimate probability of survival of an advantageous dominant under stochastic conditions.)

SELECTION AGAINST DOMINANTS

In effect, when the fitnesses of three genotypes are in relative values, as follows:

AA	*Aa*	*aa*
$1 - s$	$1 - s$	1

it can easily be shown that if a recessive allele is rare, it will increase more rapidly for a given q value than it would have declined when selection was against it (see Exercise 7). The student should verify that when fitness values favor the recessive as above, the rate of change in q will be as follows:

$$\Delta q = \frac{+sq^2(1 - q)}{1 - s(1 - q^2)} \tag{14-9}$$

This rate of change in q is similar to (14-6), although it is of opposite sign and with a slightly different denominator. If s is small for both cases, (14-6) and (14-9) will become nearly equal, though in opposite direction. Whatever the direction of selection, the student may note that dominance always reduces the efficiency of selection as a limiting q value is being reached or as $\bar{W}$ approaches maximum.

Example 14-2 illustrates both the deterministic and average stochastic results when a rare recessive is advantageous. In many respects, this result appears to be simply an upside-down version of the outcome in Example 14-1.

EXAMPLE 14-1

DELETERIOUS RECESSIVE = ADVANTAGEOUS DOMINANT

Simulation of selection coupled with a constant population size was programmed for the IBM 370 computer (UICC Computer Center), as already outlined in Examples 12-1 and 13-1. Selection against a complete recessive with $W_{aa} = 0.7$ (Table 14-3) is diagrammed in Figure A. Population sizes were either $N = 10$ or 20 (stochastic cases) or very large (deterministic cases). Initially, the lower fitness allele (a) was set at either $(2N - 1)/(2N)$ or $1/(2N)$ (stochastic cases) on the assumption that a single favorable dominant mutant (A)

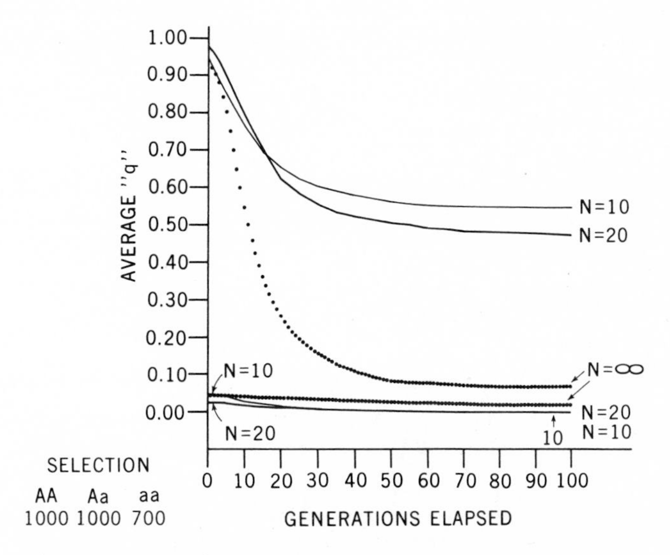

Figure A. Changes in frequency of a (q) with generations when selection is against a complete recessive or favoring a dominant (A). Large dots are successive-generation q values expected in a very large population (deterministic result), initially set at either $q_0 = 0.95$ or 0.05. Solid lines are average q ($\bar{q}$) over all isolates from computer simulations: $N = 20$, thick lines; $N = 10$, thin lines. Initial q values (q_0) represent cases of single (a) mutants occurring in all isolates at low q ($1/2N$); at high q, mutants would be advantageous dominant (A) so that $q_0 = (2N - 1)/2N$.

has occurred (former frequency) or that a single recessive mutant (a) has occurred (latter frequency). In the deterministic case, q_0 was set at 0.95 or at 0.05 to compare with the $N = 10$ case.

When the population size is large, the deleterious allele falls rapidly for about 50 generations to about $q = 0.10$ and after 50 more generations (lower dotted sequence) to $q = 0.06$, and its rate of change is very much slower after its frequency falls below 0.06. Conversely, in the stochastic cases, the changes in average frequency ($\bar{q}$ over all isolates of constant N size) are much slower when at high frequencies because several isolates become fixed for the more favorable allele (A), and an equilibrium is achieved by eventual fixation of most isolates. In this simulation experiment, the a allele was ultimately elim-

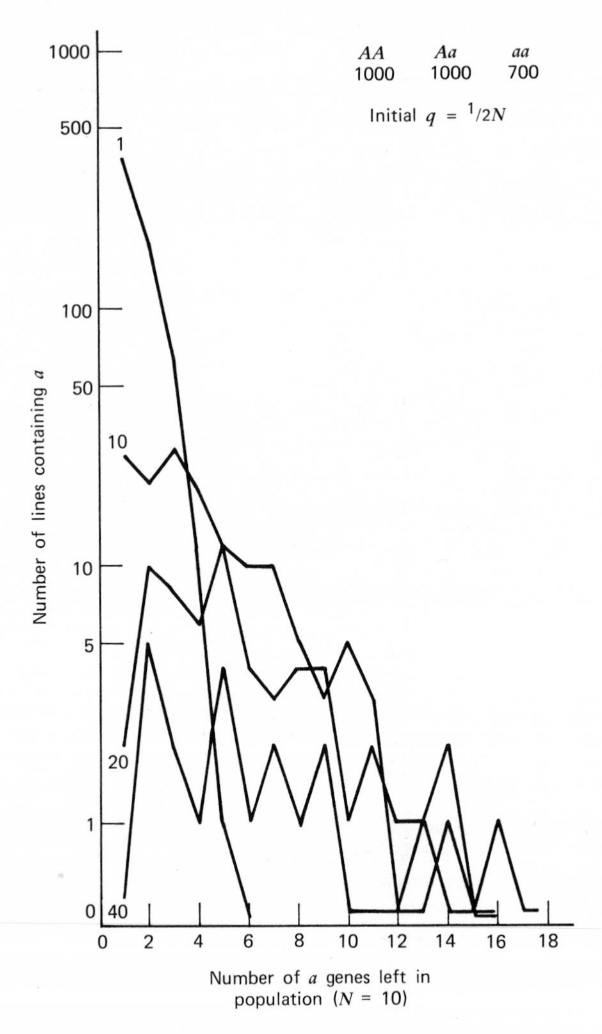

Figure B. The distribution of 1000 isolates of size $N = 10$ for generations G_1, G_{10}, G_{20}, G_{40} after starting at $q_0 = 0.05$.

inated by G_{94} from all isolates of $N = 10$ in which it had begun as a mutant $(1/2N)$. In isolates of $N = 20$, the allele still existed in a single isolate at G_{100}. In contrast, for the deterministic case, q was 0.02 at G_{100}. Thus, elimination of a detrimental recessive mutant is aided by reducing population size.

In Figure B, the distribution of isolates (for 1000 isolates run at $N = 10$) is diagrammed for certain generations (G_1, G_{10}, G_{20}, G_{40}) after a single mutational start $(1/(2N))$. The vertical scale for numbers of isolates is logarithmic for convenience. By the 40th generation, the *a* allele had spread so that, while it would be expected at $q = 0.03$ in the large population, it occurred at higher frequency in $\frac{19}{1000}$ isolates and in a single isolate had achieved $q = 0.70$! No doubt with a larger sample of isolates, some would have become fixed for that allele. However, $\bar{q}$ was only 0.005 over all isolates at G_{40}. But the spreading tendency of these few isolate frequencies was reversed, and the *a* allele was eliminated from all 1000 isolates by G_{100}.

At the opposite starting frequency of $q_0 = (2N - 1)/(2N)$, the new mutant is *A*, the favorable allele. At G_{100} with small isolate sizes, the computer outcomes were as follows:

Isolate Size N	No. Isolates Fixed *AA*	Fixed *aa*	Unfixed
10	447	552	1
20	478	514	8

Final $\bar{q}$ over all isolates was still sizable then, and very far from the deterministic expectation:

$$\text{for } N = 10, \quad \bar{q}_{100} = 0.5525$$
$$\text{for } N = 20, \quad \bar{q}_{100} = 0.5180$$
$$\text{for } N = \infty, \quad \bar{q}_{100} = 0.0060$$

EXAMPLE 14-2

ADVANTAGEOUS RECESSIVE

If populational genetic structure is regarded as mostly homozygous (see "classic" interpretation in Chapter 18 under genetic loads), adaptive genetic changes would be rare and would come about mostly from occasional mutants that improve fitness. If such a new beneficial mutant (*a*) is completely recessive to the previously "normal" allele (*A*), it will increase as shown in Figure A, where $N = \infty$ (deterministic case) with fitness values of *A*- genotypes at 0.7 of the *aa* genotype, or $W_{AA} = W_{Aa} = 1.00$, $W_{aa} = 1.43$.

When small population size ($N = 10$ or 20) is simulated by the IBM computer (as used in Example 14-1), it is evident that the initial average rise in $\bar{q}$ value (average q over all isolates) can be faster than by deterministic selection. After a few generations, the $\bar{q}$ rise was slowed as some isolates became fixed for the old, less-adapted allele. By G_{70} the more advantageous allele became fixed in 153 out of 1000 isolates of size $N = 10$;

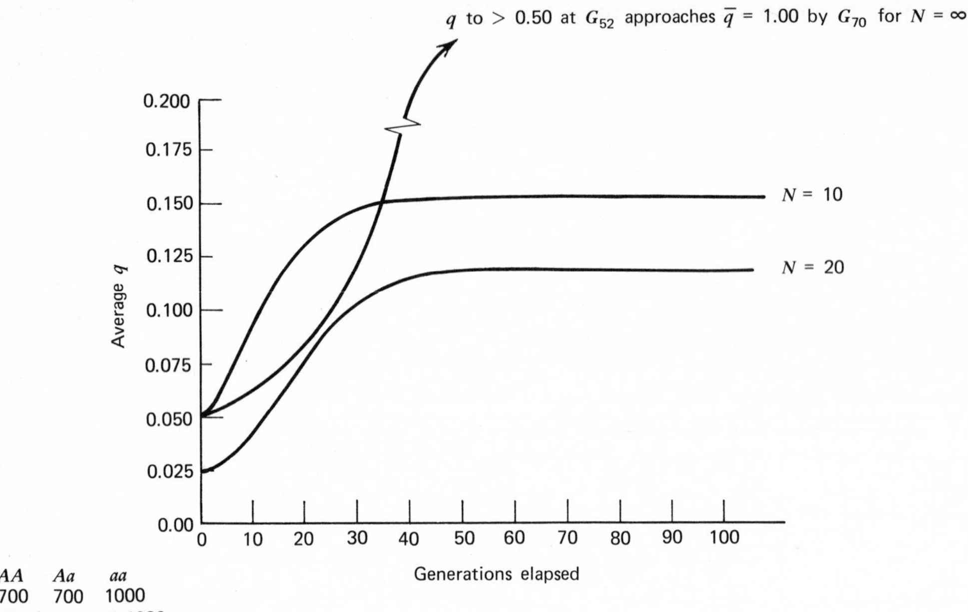

Figure A. Changes in frequency of a (q) with generations when selection favors a complete recessive. Curve with arrow is q value in a large population (as in Example 14-1, Figure A). Solid lines are average q ($\bar{q}$) over all isolates from computer simulations; $N = 20$, lower line and $N = 10$, upper line.

while in those of size $N = 20$, it became fixed in 119 out of 1000 isolates by G_{56}. These are, of course, far short of the expected complete fixation of an advantageous allele!

SELECTION AGAINST INTERMEDIATE HETEROZYGOTE AND RECESSIVE, OR INCOMPLETE DOMINANCE IN FITNESS

When the heterozygote is slightly lower in fitness than the favored homozygote such that relative fitnesses for the three genotypes are as follows:

$$\begin{array}{ccccc} AA & & Aa & & aa \\ W_1 & > & W_2 & > & W_3 \end{array}$$

selection can be more efficient at eliminating the a allele because there will much less protection in the heterozygote. The general condition (14-7) will always hold; in terms of selection coefficients, the Δq statement becomes algebraically more complex. For a meaningful notation, we may express the W_2 fitness value as some proportion of the selection coefficient in W_3, or the other way around. Let us suppose the fitnesses of the genotypes are as follows: $W_1 = 1.0$, $W_2 = 0.94$, $W_3 = 0.70$. We may choose to use the selection coefficient ($s = 0.30$) as before for the recessive homozygote, but the selection coefficient of the heterozygote is 0.06. Instead of using a different coefficient symbol (say, t for the heterozygote), it is somewhat easier and more comparable with what has already been described to express the heterozygote's selection coefficient as a proportion of the recessive's coefficient; because $0.06 = 0.20\ (0.30)$, we may say that $W_2 = 1 - hs$, where h is the fraction of the recessive's coefficient found in Aa—that is, the heterozygote's selective disadvantage is only 20 percent of the recessive's. Expressions for the change in q under this type of selection are given in Table 14-4 with the solution for $W_1 : W_2 : W_3 = 1 : 1 - hs : 1 - s$.

$$\Delta q = \frac{-spq[q + h(p - q)]}{1 - sq[q + h(p - q) + h]} \tag{14-10}$$

In comparing the rate of change in q with the case of the completely recessive allele with fitness of 0.7 (Table 14-3A), it is evident that the rate is slightly slower here when the allele is common; when $q = 0.9$, $s = 0.3$, and $h = 0.2$, then $\Delta q = -0.0268$ compared with -0.0321 in the pure recessive case. As the allele becomes rare, the rate of elimination is increased nearly three times. In this example, $\Delta q = -0.071$ for the depressed heterozygote case compared with -0.0027 in the pure recessive when $q = 0.10$. Selection is thus more efficient at eliminating the incompletely recessive as it becomes rare than in the complete recessive case; compare Example 14-3, Figure A (the dotted curve), for the large population with the corresponding curve of Figure A in Example 14-1. With reduced population sizes, the fixation of aa isolates is nearly the same as that of a complete recessive (Example 14-1) when an advantageous dominant (A) mutation occurs initially; however, when recessive a is rare initially, it is about completely eliminated within less than 100 generations.

When the heterozygote is exactly intermediate between homozygotes in fitness so that $h = 0.5$ (for example, if $W_2 = 0.85$ in the numerical example, or $(0.5)(0.3) = 0.15 = hs$),

TABLE 14-4 *Selection against intermediate heterozygotes*

Generation	Genotypes			Totals
	AA	Aa	aa	
G_0 zygotes	p^2	$2pq$	q^2	1
Fitness	W_1	W_2	W_3	
Or	1	$1 - hs$	$1 - s$	
After selection	$p^2 + 2pq - 2pqhs + q^2 - sq^2 = 1 - sq[q + 2ph] = \bar{W}$			

$$\Delta q = \frac{q - pqhs - sq^2}{\bar{W}} - q = \frac{-pqhs - sq^2 - sq^2[q + 2ph]}{\bar{W}}$$

$$= \frac{-sq^2(1 - q - 2ph) - pqhs}{\bar{W}}$$

$$= \frac{-spq(q - 2qh + h)}{\bar{W}}$$

$$= \frac{-spq[q + h(p - q)]}{1 - sq(q + 2ph)} = \frac{-spq[q + h(p - q)]}{1 - sq[q + h(p - q) + h]}$$

then (14-10) reduces to the simpler form

$$\Delta q = \frac{-(0.5)spq}{1 - sq} \tag{14-11}$$

This expression is very similar to formula (14-1). Thus, the rate of change in q for an exactly intermediate heterozygote is similar to the rate for a haploid population. In other words, the recessive allele has a complete effect on fitness, or in the terminology of Chapter 6, it is additive in fitness.

Finally, note that (14-10) reduces to (14-6) when $h = 0$. There is no equilibrium in the change of q except the trivial one.

EXAMPLE 14-3

DELETERIOUS INCOMPLETE RECESSIVE

In similar design to the two previous examples, the outcome of selection for genotypes is illustrated in Figure A for the case of intermediate, slightly deleterious heterozygotes with the following fitness values: $AA = 1$, $Aa = 0.94$, $aa = 0.70$. Thus, $h = 0.2$ and $s = 0.3$. Note that in the deterministic case, the fall in q value from an initial high of 0.95 is slightly slower than in Example 14-1, where the a allele was a complete recessive, but elimination after about the 20th generation is much faster with some deleterious dominance. These differences in Δq are expected from (14-10).

In the stochastic cases ($N = 10$ or 20), when assuming a single original mutation to the A allele (advantageous), the a recessive became fixed in this set of trials about 55 percent

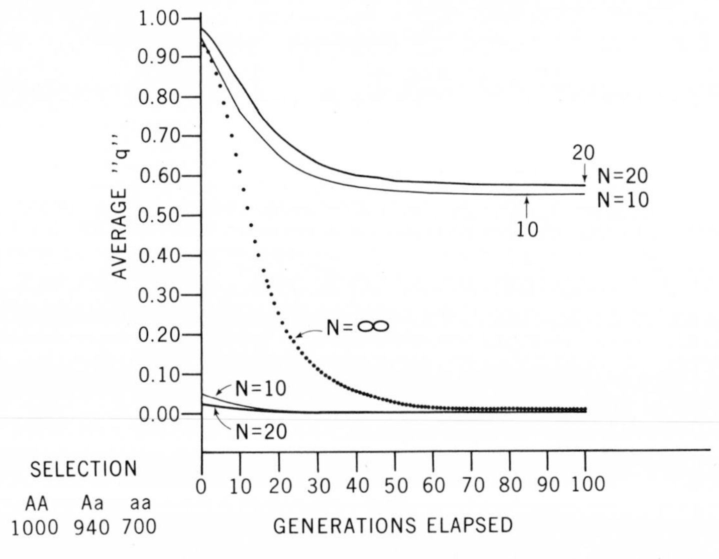

Figure A. Changes in frequency of a (q) with generations when the deleterious allele is incompletely recessive. Labels as in Example 14-1.

of the time for isolates of $N = 10$ by G_{54} or about 57 percent of the size $N = 20$ isolates by G_{22}. At opposite starting frequencies ($q_0 = 1/(2N)$), however, the *a* allele was eliminated very rapidly, as seen in the lowest curves in Figure A. Spreading of the allele did produce an array of unfixed isolates throughout early generations, but complete loss of *a* from all 1000 isolates of size $N = 20$ was accomplished by G_{89}; for isolates of size $N = 10$, only $\frac{1}{1000}$ was still segregating by G_{100}.

SELECTION FAVORING HETEROZYGOTES

When the *Aa* genotype has highest fitness, selection will tend to preserve both alleles, as can be seen intuitively, by recalling the most extreme case known as a "balanced lethal"; if both homozygotes are lethal, the only surviving genotype is the heterozygote. If homozygotes are simply less fit than the heterozygote so that relative fitnesses are as follows:

$$\begin{matrix} AA & & Aa & & AA \\ W_1 & < & W_2 & > & W_3 \end{matrix}$$

then allelic frequencies will come to a nontrivial equilibrium maintained by selection as a stable point. Suppose a population exists with $p = 0.6$ and $q = 0.4$ and fitnesses of 0.80: 1:0.70, respectively as given in Table 14-5A. After selection, the mean fitness ($\bar{W}$) will be 0.88, but the adjusted genotype frequencies will be unchanged. Even with unequal fitnesses of homozygotes, we have started at the stable equilibrium point, as we now proceed to show. If the population is initiated at $q = 0.2$ (lower *a*), q will rise (Table 14-5B); if the population is initiated with $q = 0.6$, *a* will fall (Table 14-5C). These two initial alternatives demonstrate that change will take place in the direction of some stable point between the two.

Let us examine the nature of this equilibrium by proceeding as before. If we use selection coefficients s_1 against the *AA* genotype and s_2 against the *aa* genotypes, then Table 14-5D presents the essential steps toward the Δq statement

$$\Delta q = \frac{pq(s_1 p - s_2 q)}{1 - s_1 p^2 - s_2 q^2} \tag{14-12}$$

The student should verify that this change in q takes place when applying the q values from the examples in Table 14-5A, B, and C. Note that $s_1 = 0.20$ and $s_2 = 0.30$. When taking the population I, which is stable with $q = 0.40$, we note that the products of selection coefficient times frequency—within the parentheses of the numerator in (14-12)—are equal so that the numerator is zero in that case and $\Delta q = 0$. In population II, Δq is positive, while it is negative in population III. To find the point of equilibrium ($\hat{q}$), we may set (14-12) equal to zero, as given in Table 14-5D, which reduces to the simple relation based on a balance between homozygote selective coefficients

$$\hat{q} = \frac{s_1}{s_1 + s_2} \tag{14-13}$$

(equilibrium point for q defined by selection coefficients of homozygotes).
As can be seen from the example, $\hat{q} = 0.2/(0.2 + 0.3) = 0.4$, a result independent of starting frequencies. This mode of selection is often referred to as leading to balanced

TABLE 14-5 *Selection against both homozygotes (= favoring heterozygotes)*

A Population I (at equilibrium)

Generation	Genotypes AA	Aa	aa	Totals	q_a
G_0 parent frequency	0.36	0.48	0.16	1.00	$q_0 = 0.40$
Fitness	0.8	1.0	0.7		
After selection	0.288	0.480	0.112	$0.88 = \bar{W}$	
G_1 (mature)	0.32727	0.54545	0.12727	1.00	$q_1 = 0.4000$
		[q does not change!]			

B Population II (q below equilibrium)

Generation	AA	Aa	aa	Totals	q_a
G_0 parent frequency	0.64	0.32	0.04	1.00	$q_0 = 0.20$
Fitness	0.8	1.0	0.7		
After selection	0.512	0.320	0.028	$0.86 = \bar{W}$	
G_1 (mature)	0.5953	0.3721	0.0326	1.00	$q_1 = 0.2186$
		[q increases!]			

C Population III (q above equilibrium)

Generation	AA	Aa	aa	Totals	q_a
G_0 parent frequency	0.16	0.48	0.36	1.00	$q_0 = 0.60$
Fitness	0.8	1.0	0.7		
After selection	0.128	0.48	0.252	$0.86 = \bar{W}$	
G_1 (mature)	0.1488	0.5581	0.2930	1.00	$q_1 = 0.5721$
		[q decreases!]			

D

	AA	Aa	aa	Totals
G_0	p^2	$2pq$	q^2	1.00
Fitness	$1 - s_1$	1	$1 - s_2$	

After selection $(p^2 - s_1p^2) + 2pq + (q^2 - s_2q^2) = 1 - s_1p^2 - s_2q^2 = \bar{W}$

$$\Delta q = \frac{q - s_2q^2}{\bar{W}} - q = \frac{-s_2q^2 + s_1p^2q + s_2q^3}{\bar{W}}$$

$$\Delta q = \frac{-s_2q^2p + s_1p^2q}{\bar{W}} = \frac{pq(s_1p - s_2q)}{1 - s_1p^2 - s_2q^2}$$

If $\Delta q = 0$ (no change in q at equilibrium)

$$0 = pq(s_1p - s_2q)$$

$$s_1p = s_2q$$

$$\hat{q} = \frac{s_1}{s_1 + s_2}$$

where $\hat{q}$ = equilibrium a frequency

polymorphism due to "heterosis" in fitness of heterozygotes. For a single locus, heterozygote superiority is referred to as overdominance (Chapter 6).

In Example 14-4, the outcome of selection is established in relatively few generations at the "balanced" point in a large population. In contrast, for a population comprised of small-sized isolates, many isolates will become fixed at $q = 1$ or 0, depending on their initial q values. An initially high p_A population in the computer simulation had greater A fixation among isolates than the opposite initially high q_a population. However, in contrast with the previous examples, after many generations (G_{100}) there were still a significant number of isolates unfixed, or polymorphic (8 to 10 percent), within which the average $\hat{p},\hat{q}$ was approximated. Drift will then take much longer to fix or to eliminate alleles when selection is strongly "balanced." Over all the isolates, however, average gene frequency ($\bar{q}$) will be far above or far below the deterministic expected $\hat{q}$ value, depending on initial q.

Before exploring the equilibrium point for a more general definition, let us examine the deterministic numerical example in Table 14-5 further. We should note that the mean fitness ($\bar{W}$) is at highest value for the three cases when the allelic frequencies are at equilibrium. We find this $\bar{W}$ value to be a maximum. Second, when the G_1 values have been normalized (mature G_1) in all cases, the homozygotes have essentially failed to replace themselves (*AA* decreases from 0.36 to 0.327, *aa* from 0.16 to 0.127), while the heterozygotes have increased compared with their starting frequencies (*Aa* increases from 0.48 to 0.545). This result may be interpreted to mean that in the struggle for existence heterozygotes have had to survive in greater numbers than mere replacement to make up for the losses in the homozygotes. We may visualize, in parallel with the simplest case of haploid prokaryotes (Table 14-1), that each genotype actually has been produced by random mating in greater numbers than the "100 percent" (constant N) carrying power of the environment. When selection has finished its action, homozygotes are deficient in frequency, while heterozygotes are in excess compared with their initial frequencies. Under these circumstances, the population will "carry a genetic load" (*AA* and *aa*, genotypes with lower fitness than the heterozygote) permanently. Elimination of these less-fit genotypes is not possible with the conditions set forth by the selection regimen.

Just how much of a biological burden the genetic load may be on the reproductive potential of the population is examined in Chapter 18. We should think more about the meaning of Darwinian fitness before we can appreciate fully the significance (or lack of significance) of the genetic load. At this point, it is important to realize what is *not* true. The student should avoid anthropomorphic feelings that the genetic load necessarily implies suffering on the part of any individuals in the population, recalling that we have defined Darwinian fitness as a function of reproductive ability *only*. Many other factors may affect lowered reproduction than simply the death or illness of an individual, as we shall discuss in the following chapter.

EXAMPLE 14-4

SUPERIOR HETEROZYGOTE ("HETEROSIS")

The outcome of selection for genotypes with heterozygotes of highest fitness both in deterministic and stochastic cases as in the IBM simulation of the previous examples is

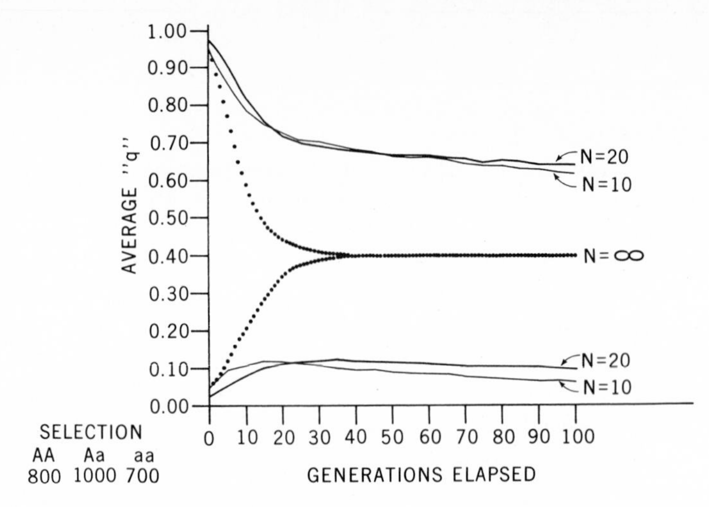

Figure A. Changes in frequency of a (q) with generations when selection favors the heterozygote. Labels as in previous examples in this chapter.

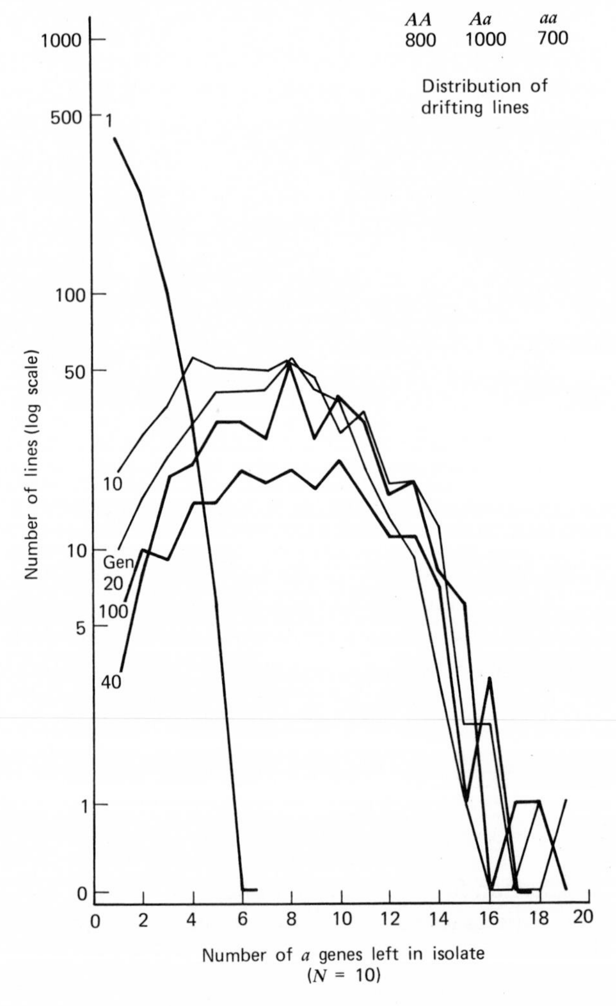

Figure B. Distribution of 1000 isolates of size 10 for G_1, G_{10}, G_{20}, G_{40}, G_{100} after starting at $q_0 = 0.05$.

illustrated in Figure A with fitness values as in Table 14-5. In the large population, equilibrium at $\hat{q} = 0.40$ is established by G_{40} while the $\bar{q}$ over all the small populations has a different outcome depending on the initial q value. With initial low $\bar{q}(1/(2N))$, an increase at first occurred slowly until G_{15}, when $\bar{q}$ over all isolates began to fall again when fixation of A took place in many isolates. A distribution of unfixed isolates ($N = 10$) over generations G_1, G_{10}, G_{20}, G_{40}, G_{100} (Figure B) indicates that their modal q tended to approach the expected equilibrium (deterministic) $\hat{q}$ value; but by the 100th generation, fixation had overwhelmed most isolates: 888 were AA, 30 aa, with only 82 still unfixed. From the opposite initial $\bar{q} = (2N - 1)/2N$, many more aa became fixed (Figure A), so that the final average over all isolates ($\bar{q}$) was far above the deterministic result.

THE STATIONARY STATE: SELECTION EQUILIBRIUM

If selection alone is acting on the genotypes in populations, or in less metaphorical language, if genotypes differ in fitness value, change in allelic frequencies will occur directed by that selection pressure (fitness differences) until limiting frequencies are achieved either at trivial equilibriums (complete loss or fixation or an allele) or at nontrivial points. These latter points, which can be stationary in p,q values over many generations, represent the dynamic equilibrium points of balanced forces acting on the genotypes. The zero rate of change in $q(\Delta q = 0)$ is graphically shown in Figure 14-3A at the point of selection balance.

We may express the equilibrium condition due to balancing selection as follows: $q_0 = q_1 = q_2 = \cdots$ (allele frequency before selection equals that after selection for several generations). We use a more general expression for q_1 in terms of q_0 based on letting $\Delta q = 0$ in (14-7)

$$q_1 = \frac{p_0 q_0 W_2 + q_0^2 W_3}{\bar{W}} = \frac{q_0(p_0 W_2 + q_0 W_3)}{\bar{W}}$$

When in the stationary state, the subscripts may be discarded so that $\bar{W}q = pqW_2 + q^2W_3$, and because q is not zero, it can be cancelled on both sides to yield the equilibrium condition

$$\bar{W} = pW_2 + qW_3 \qquad (14\text{-}14)$$

(equilibrium condition in terms of q) or

$$\bar{W} = pW_1 + qW_2 \qquad (14\text{-}14A)$$

(equilibrium condition in terms of p).

The student should verify these relationships using the numerical example in Table 14-5A. Also note that net fitness is greater than the fitness of either homozygote. Algebraic solution of (14-14) in terms of q (equilibrium) is given in Table 14-6, where it is a fundamental step with the solution

$$\hat{q} = \frac{(W_1 - W_2)}{(W_1 - W_2) + (W_3 - W_2)} \qquad (14\text{-}15)$$

(equilibrium q expressed in terms of relative fitness for diploid genotypes with single pair

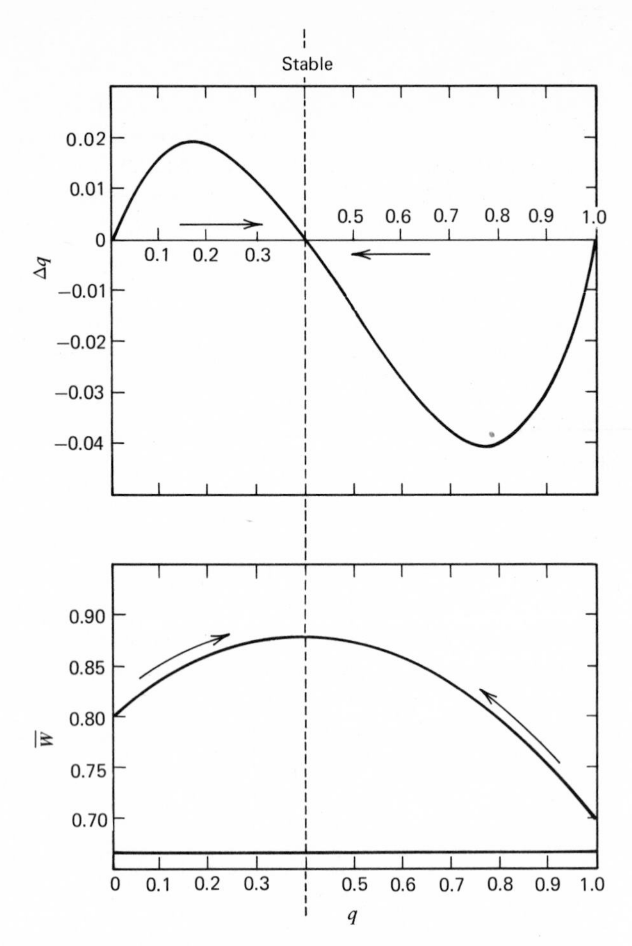

Figure 14-3. (A) Rate of change in q (Δq) and (B) $\overline{W}$ as a function of q in the stable heterotic case in which constant fitness values are $W_1 = 0.80$, $W_2 = 1.00$, $W_3 = 0.70$, so that the stable point is at $\hat{q} = 0.40$ toward which selection tends to move allelic frequencies (from Li, 1955).

of alleles). Note that if selection coefficients are substituted as $s_i = 1 - W_i$ and W_2 is expressed as equal to 1.00, then (14-15) reduces to:

$$\hat{q} = \frac{-s_1}{-s_1 - s_2} \tag{14-13A}$$

We shall see that this form of the equilibrium has an important distinction from (14-13).

Another significant point is that in the approach to equilibrium, $\overline{W}$ increases each generation (Figure 14-3B). The change in $\overline{W}$ can be a continuous function similar to that of Δq, except that the latter is more complex algebraically because it includes $\overline{W}$; thus, we

TABLE 14-6 *Algebraic solution for the stationary-state condition of formula (14-14).*

$\bar{W} = pW_2 + qW_3$. Changing $p \rightarrow q$ gives $\bar{W} = W_2 + q(W_3 - W_2)$, and with $\bar{W}$ expressed completely, we have $p^2W_1 + 2pqW_2 + q^2W_3 = W_2 + q(W_3 - W_2)$

1. Changing $p \rightarrow q$ and putting right terms on the left

$$(1 - 2q + q^2)W_1 + (2q - 2q^2)W_2 + q^2W_3 - W_2 - qW_3 + qW_2 = 0$$

2. Rearranging and grouping

$$(W_1 - W_2) - q(W_1 - W_2) - q(1 - q)(W_1 - W_2) + q(1 - q)W_2 - q(1 - q)W_3 = 0$$

3. Reduces by factoring and dividing out $(1 - q)$

$$(W_1 - W_2) - q(W_1 - W_2) - q(W_3 - W_2) = 0$$
$$q[(W_1 - W_2) + (W_3 - W_2)] = W_1 - W_2$$

4. So that equilibrium

$$\hat{q} = \frac{W_1 - W_2}{(W_1 - W_2) + (W_3 - W_2)}$$

may find it easier to take a first derivative with respect to q in the expression for $\bar{W}$ and set it equal to zero to find the stable point, instead of taking the derivative of the Δq (14-12) expression. First, defining $\bar{W}$,

$$\bar{W} = \sum fW_i = p^2W_1 + 2pqW_2 + q^2W_3 \tag{14-16}$$

(and changing p to q)

$$= W_1 - 2qW_1 + q^2W_1 + 2qW_2 - 2q^2W_2 + q^2W_3$$

Then, taking the derivative of $\bar{W}$ with respect to q,

$$\frac{d\bar{W}}{dq} = -2W_1 + 2qW_1 + 2W_2 - 4qW_2 + 2qW_3$$

$$= 2[q(W_3 - W_2) + (1 - q)(W_2 - W_1)]$$

$$= 2[q(W_3 - W_2) + p(W_2 - W_1)] \tag{14-17}$$

Setting this derivative equal to zero and changing p to q, we have $0 = W_1(1 - q) + W_2(1 - 2q) + qW_3$, from which we arrive at (14-15) exactly. Thus, we find the stable point by a more sophisticated method. However, having taken the first derivative, we can find out whether it is at a maximum or minimum point by taking the second derivative of (14-17):

$$\frac{d^2\bar{W}}{dq^2} = 2[(W_1 - W_2) + (W_3 - W_2)] \tag{14-18}$$

If this is negative, we are at a maximum point; if it is positive, we are at a minimum in the $\bar{W}$ curve. In the case given above for selection favoring the heterozygote, this second derivative (14-18) is negative and the equilibrium point is stable. Now we shall explore the case where all signs are reversed, (14-18) is positive, and the equilibrium point is *unstable*.

SELECTION AGAINST THE HETEROZYGOTE

A case with fitness values favoring both homozygotes and not favoring the heterozygote, which has lowest fitness, as follows:

$$\begin{array}{ccc} AA & Aa & aa \\ W_1 > & W_2 < & W_3 \end{array}$$

will have an equilibrium that is not at the trivial $p = 0$, or 1 point. For example, suppose a large population exists with $p = 0.6$, $q = 0.4$ with fitnesses of genotypes 0.9:0.7:1.0, respectively, as given in Table 14-7A. After selection, the allelic frequencies do not change (right-hand column). Here the *aa* genotype is best, so the lack of change in q is surprising with *aa* in the minority! By starting with p, q values other than the equilibrium point, however, we soon see the sort of equilibrium it is. In the B and C portions of the table, we take values for q away from the equilibrium point; once disturbed in that way, the allele frequencies tend to move even farther from the equilibrium point. We know, then, that the equilibrium is an *unstable* one; as long as the population is poised on that point exactly, it is unchanged, but putting it off the point in one direction makes it move away in the same direction (Figure 14-4).

By using (14-15), we find that the selection coefficients are identical with those used in the stable case (Table 14-5), but now the signs are all positive and the second derivative

TABLE 14-7 *Selection against the heterozygote*

Generation	*AA*	*Aa*	*aa*	Totals	q
A Population I					
G_0	0.36	0.48	0.16		$q_0 = 0.4$
W	0.9	0.7	1.0		
After selection	0.324	0.336	0.16	$0.82 = \bar{W}$	
G_1	0.395122	0.409756	0.195122	1.00	$q_1 = 0.4000$
	(q does not change!)				
B Population II					
G_0	0.64	0.32	0.04		$q_0 = 0.20$
W	0.9	0.7	1.0		
	0.576	0.224	0.040	$0.84 = \bar{W}$	
	(q decreases!)				$q_1 = 0.1810$
C Population III					
G_0	0.16	0.48	0.36		$q_0 = 0.6$
W	0.9	0.7	1.0		
	0.144	0.336	0.360	$0.84 = \bar{W}$	
	(q increases!)				$q_1 = 0.6286$

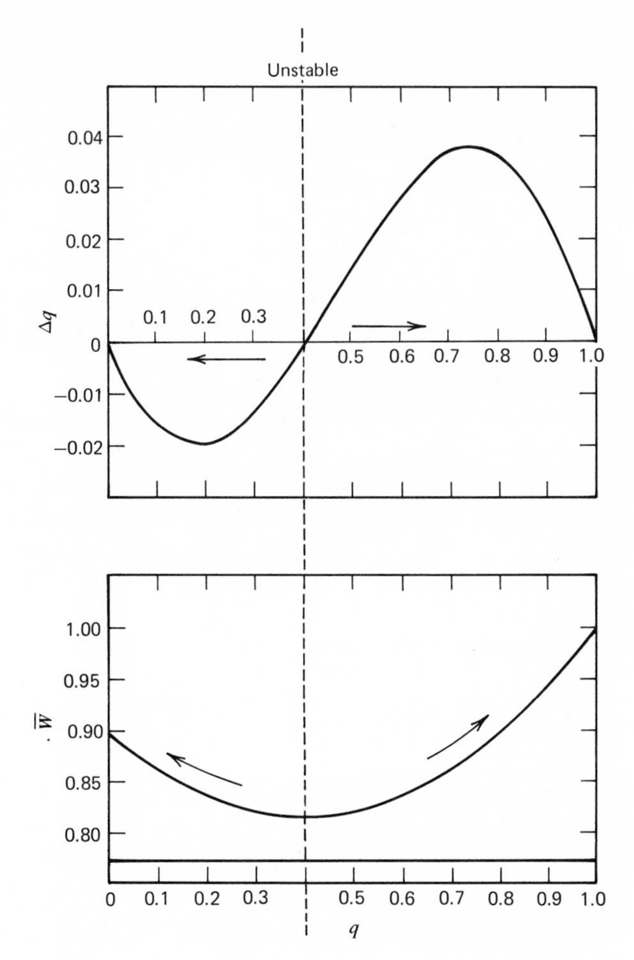

Figure 14-4. (A) Rate of change in q (Δq) and (B) $\bar{W}$ as a function of q in the unstable equilibrium case in which heterozygote fitness is least among three genotypes: $W_1 = 0.90$, $W_2 = 0.70$, $W_3 = 1.00$. Selection tends to move allelic frequencies away from $\hat{q} = 0.40$ when $q \neq 0.40$ (from Li, 1955).

(14-18) will be positive, indicating that the equilibrium point for $\bar{W}$ is a minimum, as is evident from Table 14-7. In that table note that $\bar{W}$ is greater when gene frequencies are away from the unstable equilibrium point.

CAN WE ASSUME NO SELECTION WHEN ZYGOTES FIT THE SQUARE LAW?

When fitness values of genotypes can be ascertained, we can predict the outcome of selection; but we *cannot easily work in reverse.* If genotype frequencies are known for a single generation, we cannot infer whether selection is acting or not simply by examining their distribution, even if we assume genotypes to be at equilibrium. (See the next chapter for estimation of fitness values from genotype frequencies.) Selection may be favoring a homozygote (directional) or favoring a heterozygote (balanced), and yet adult frequencies can approximate the binomial distribution, or square law. This problem of unreliability in attempting to attribute genotype frequencies either to selection or lack of it has been examined by Li (1959, 1967*c*), Wallace (1958), Lewontin and Cockerham (1959), and reviewed by Workman (1969). Many experimental population geneticists early in their experience—including the present author (Spiess and Langer, 1961)—"fell into the trap" of concluding no selection to be occurring merely because of confirmation by zygotic data of the square law at some stage in the life cycle.

Let us examine a few conditions in which selection is acting but zygotes are binomially distributed. Usually, after selection has acted, genotype frequencies are not in binomial proportions, but one of the important conditions under which they *can* be is as follows (Workman, 1969). The "fixation index" (F) may be used generally to describe any deviation from square-law proportions (Chapters 8–11); that deviation may be caused by any one or a combination of a large number of forces including breeding structure, sampling of gametes, or selection. Recalling formula (9-9) that $F_{est} = (1 - H_t/H_0)$, we transform F into a measure dependent on relative fitness values by writing H_s as the normalized heterozygote frequency under selection in place of H_t.

$$H_s = \frac{2pqW_2}{\bar{W}}$$

(normalized heterozygote frequency after selection), and with $H_0 = 2pq$, estimated F becomes

$$F_{est} = \frac{\bar{W} - W_2}{\bar{W}} = \frac{p^2(W_1 - W_2) + q^2(W_3 - W_2)}{\bar{W}} \tag{14-19}$$

If genotypes were neutral to selection, all W values would be equal, and the numerator would reduce to zero (panmixis and no selection). If F is positive, heterozygotes are deficient (see Table 14-7 at G_1, for example); if F is negative, heterozygotes are in excess (Tables 14-2 and 14-5 when comparing G_1 mature after selection with their G_2 zygotes produced by random mating).

Further, if there is a nontrivial equilibrium (dynamic stable point not at $p, q = 0, 1$), we may utilize the stationary-state definition of $\bar{W}$ from (14-14) supplied into (14-19). Multiplying both numerator and denominator by $\bar{W}$, then simplifying, gives the following.

$$F = \frac{(\bar{W})^2 - (W_2)(\bar{W})}{(\bar{W})^2} = \frac{pq(W_1W_3 - W_2^2)}{(pW_1 + qW_2)(qW_3 + pW_2)} \tag{14-20}$$

This relationship indicates that when $W_1W_3 = (W_2)^2$, the value of F will be zero in the adult population following selection. Second, for this product of homozygote fitnesses to equal the square of heterozygote fitness value, the fitness values must form a geometric progression—for example,

$$W_1:W_2:W_3 = 1:r:r^2 \quad \text{or} \quad x^2:xy:y^2 \tag{14-21}$$

Let us suppose that we observe genotype frequencies following the action of selection:

AA	Aa	aa
$\frac{9}{16}$	$\frac{6}{16}$	$\frac{1}{16}$

These genotypes fit the expansion of $(\frac{3}{4} + \frac{1}{4})^2$. Are we justified in concluding that no selection is acting? Of course not, because any fitness values that form a geometric progression as given in (14-21) could exist and could change genotype frequencies to produce these observed values. For example, if the genotypes at birth were

	$\frac{9}{25}$	$\frac{12}{25}$	$\frac{4}{25}$	or $(3/5 + 2/5)^2$ at fertilization
With fitnesses	1	$\frac{1}{2}$	$\frac{1}{4}$	
After selection	$\frac{9}{25}$	$\frac{6}{25}$	$\frac{1}{25}$	$= \frac{16}{25}(\bar{W})$
Relative frequency	$\frac{9}{16}$	$\frac{6}{16}$	$\frac{1}{16}$.	

The fitness values indicate that the a allele will be eliminated by selection after a few generations, but the zygotes occur in frequencies that fit the square law because of the (14-21) relationship of their fitness values. Thus, the mere fit to the square law does not in itself rule out the action of natural selection in determining the observed genotype frequencies.

In Exercise 14, the student should verify that for the relationship (14-21) the multiplier (λ) in the geometric progression of fitnesses may be estimated from the solution of the simple selection model used in deriving formula (14-7), provided that gene frequencies are known before and after selection, arriving at the following:

$$\lambda = \frac{q_1}{p_1} \cdot \frac{p_0}{q_0} \tag{14-22}$$

where the subscript (1) indicates the frequency after selection and (0) the frequency before selection. Note that λ is r or y/x in (14-21).

CAN WE ASSUME SELECTION WHEN ZYGOTES DO NOT FIT THE SQUARE LAW?

It is quite generally true that it is hazardous to draw specific conclusions about the action of selection merely from observing particular genotype frequencies in a single generation. One example (illustrated by Wallace, 1958) should suffice to indicate the nature of the hazard involved:

	AA	Aa	aa	
Given fitness values	0.2	0.6	1.0	
Given frequency before selection	0.49	0.42	0.09	$p = 0.7000$
After selection (relative)	0.22273	0.57273	0.20454	$p_1 = 0.5095$

An observer recording the frequencies after selection might conclude in view of the high heterozygote frequency that the population is stable in balanced equilibrium with heterozygote superiority in fitness. Formulas (14-19) and (14-14) might be employed under those assumptions with quite erroneous conclusions (see Exercise 15 to verify these results). Using (14-19), $\bar{W} = 0.8727$, with $W_2 = 1.00$; using (14-14), $W_1 = 0.75$ and $W_3 = 0.74$, far off from the given fitness values above.

Obviously, the observer's main mistake was to make the assumption of stability in the population, and there is no shortcut to the realization that more than one generation is needed to prove the genotypes to be stabilized. Another point to notice is that F_{est} (14-19) may well be negative (excess heterozygotes) without implying heterozygote superiority in fitness.

GENERAL EXPRESSION FOR GENE FREQUENCY CHANGE

This chapter has illustrated the fundamental nature of the Darwinian process with the simplest of genetic systems. We have seen that what we call natural selection ensues because of differential reproductive "fitness" among genotypes. Allelic frequencies change as a result of weighting various hereditary combinations differentially in their relative contributions to successive generations.

It is important to symbolize these changes as generally as possible, and then to take note of the general outcome for this genetic system under random mating—the maximization of fitness for the population. These ideas in the past have led to Wright's concept of "adaptive peaks" ($\bar{W}_{max}$) and to Fisher's "fundamental theorem of natural selection." There are important exceptions to these principles under complex genetic systems and nonrandom mating where their exact meaning must be qualified, but they are nevertheless generalizations of solid importance.

First, let us put into general form the statement for allelic frequency change: taking (14-7), changing p to q, and simplifying, we have (see Exercise 12):

$$\Delta q = \frac{pq[q(W_3 - W_2) + p(W_2 - W_1)}{\bar{W}} \tag{14-23}$$

We note that the bracketed portion of the numerator (14-23) is one-half of the first derivative (14-17) of the $\bar{W}$ curve, so that the Δq curve is related to the $\bar{W}$ curve by this relationship (Wright, 1937)

$$\Delta q = \frac{pq}{2\bar{W}} \cdot \left[\frac{d\bar{W}}{dq}\right] \tag{14-24}$$

(generalized change in q with the following conditions: W is constant, single genic locus, and random mating).

Thus, in Figures 14-3 and 14-4 maximum or minimum of the $\bar{W}$ curve coincides with the zero point in the Δq curve. The stationary condition, setting (14-24) equal to zero, is equivalent to setting the derivative $d\bar{W}/dq = 0$, or $[q(W_3 - W_2) + p(W_2 - W_1)] = 0$, by which we have obtained (14-15).

In all the illustrative cases mentioned in this chapter, $\bar{W}$ *increased* following selection whenever there was a change in allelic frequency. Most biologists will intuitively understand that the differential set of fitness values leads to a change in p,q such that a higher average fitness will always be achieved until a stationary point is reached; that is the outcome of *natural* selection. We shall not attempt to prove mathematically that average fitness always tends to increase toward a stationary point, an adaptive peak, under these specified conditions. (Consult Li, 1967*c*, for references on mathematical proofs for maximization of average fitness under multiple alleles.)

In Chapter 7, we saw that artificial selection applied to particular phenotypes (the breeder choosing a selection differential) will produce a response in terms of progeny average changed in the direction of the selection applied. So also will the natural fitness values of genotypes in a particular set of conditions evoke changes as a response in similar fashion; but the difference between the artificial and natural situations lies in the fact that for the latter the selection differential is an intrinsic value of genotypes and cannot be ascertained easily. The process is what we might term "automatic" because it "just happens" given the initial conditions, and it is not goal directed.

In 1930, Fisher following a complex mathematical argument based on a continuous-generations model and the "intrinsic rate of natural increase," stated somewhat abstrusely his fundamental theorem of natural selection: "The rate of increase in fitness of any organism at any time is equal to its genetic variance in fitness at that time." As pointed out by Li (1967*b*,*c*), Fisher improved the clarity of his statement in 1941: "The rate of increase in the average fitness of a population is equal to the genetic variance of fitness of that population"; this modified and less ambiguous statement points to the "average fitness of a population" instead of "any organism."

The fundamental theorem of natural selection for the discrete-generation model was derived by Li (1967*b*,*c*), who showed that Fisher was referring to the *additive component* of variance in fitness. In Chapter 6, the additive (or linear) component of genetic variance was defined as $2pq\alpha^2$, where $\alpha = a + d(q - p) = p(Y_2 - Y_1) + q(Y_1 - Y_0)$; that is, α is the average phenotypic increment per allelic substitution, or regression slope of linear ("breeding") values with increasing doses of the positive quantitative allele to the genotype. If we let fitness (W_i) be our phenotypic (Y_i) value, we may use formula (6-9) as the regression slope for fitness values among the three genotypes as follows:

$$\alpha = p(W_1 - W_2) + q(W_2 - W_3) \tag{14-25}$$

(linear slope of fitness values for genotypes with two alleles). Thus, the additive component of variance among the fitness values is given by (as formula 6-8)

$$\sigma_A^2 = 2pq\alpha^2 \tag{14-26}$$

The effect of natural selection (differential fitness values) is to change the gene frequency according to (14-23) and (14-24). Taking note of the fact that $\Delta p = -\Delta q$, we may use (14-23) in terms of Δp since the bracketed numerator of (14-23) will then be positive and equal to the slope α.

$$\Delta p = \frac{pq[p(W_1 - W_2) + q(W_2 - W_3)]}{\bar{W}} = \frac{pq\alpha}{\bar{W}} \tag{14-27}$$

With the new gene frequency, however, the average fitness has also changed. The estimation of that change is made by the mathematical statement for the value of any variable with a small increment (positive or negative) in the neighborhood of a point on the curve of that variable (such as the $\bar{W}$ curve); this is known as Taylor's expansion of a function of one variable.* We have then, for $\bar{W}_{t+1}$ (generation following selection) according to Taylor's expansion,

$$\bar{W}_{t+1} = \bar{W}_t + \frac{d\bar{W}}{dp}(\Delta p) + \frac{d^2\bar{W}}{dp^2} \cdot \left[\frac{(\Delta p)^2}{2!}\right] \tag{14-28}$$

because all higher derivatives are zero. The derivatives of $\bar{W}$ are from (14-17) and (14-18). Making the first derivative with respect to p the negative of that with respect to q, we note that $d\bar{W}/dp = 2\alpha$ and $d^2\bar{W}/dp^2 = 2(W_1 - 2W_2 + W_3)$ so that we may write (14-28) after transposing the term $\bar{W}_t$ to the left side of the equation and supplying (14-27) for Δp.

$$\Delta\bar{W} = \bar{W}_{t+1} - \bar{W}_t = \frac{2pq\alpha^2}{\bar{W}_t} + (W_1 - 2W_2 + W_3)\left(\frac{pq\alpha}{\bar{W}_t}\right)^2 \tag{14-29}$$

This result expresses the Fisher "fundamental theorem" in exact terms. The increment in fitness equals the additive component of variance (14-26) in units of $\bar{W}$ plus a fitness difference factor $(W_1 - 2W_2 + W_3)$ multiplied by $(\Delta p)^2$. Note the parallel between this coefficient and the slope of differences between phenotypic values β given in formula (6-10). The second term in (14-29) will usually be nearly zero, so that for most purposes the first term (or $\sigma_W^2/\bar{W}_t$) constitutes the essential portion of the increment. This formula must always be positive for the genetic system we have been discussing, and fitness must be always increasing until a stationary state is reached (when the variance equals zero).

For illustration of this principle, let us use data from different sorts of selection schemes. In Table 14-8, (1) the recessive allele is decreasing (first row data from Table 14-3); (2) a heterotic balanced state (data from Table 14-5B); (3) an unstable condition (data from Table 14-7B); (4) an assumption of fitnesses with an intermediate heterozygote ($W_1 = 1$, $W_2 = 0.75$, $W_3 = 0.50$). In each case, $\Delta\bar{W}$ is positive, as it should be, and is equal to the sum of the last two columns, fulfilling the relationship (14-29). In the two equilibrium conditions (14-5A and 14-7A), $\alpha = 0$ and the stationary state is achieved with zero increment in $\bar{W}$. When, in the case of the stable equilibrium (Table 14-5B), p initially is higher than the stationary point ($\hat{p}$), the regression of fitnesses (α) is negative and the additive component of fitness variance (normalized) is a little greater than the amount of increment in $\bar{W}$; while with the unstable equilibrium (Table 14-7B), when initial p is higher than the unstable equilibrium point, the variance estimate is slightly less than the increment in $\bar{W}$.

* Brook Taylor, British mathematician in 1715, showed the following: let $y = f(x)$ be a continuous function of x in the neighborhood of a certain point x_o. Let $x - x_o = \Delta x$, a small increment in the neighborhood of x_o that may be positive or negative. Suppose all $(n + 1)$ derivatives of x exist and all are continuous in the neighborhood of x_o. Taylor proved that $f(x)$ can be expressed in the form

$$f(x) = f(x_o + \Delta x)$$

$$= f(x_o) + f'(x_o)\,\Delta x + \frac{1}{2!}f''(x_o)(\Delta x)^2 + \cdots \frac{1}{(n+1)!}f^{(n+1)}x_1(\Delta x)^{n+1}$$

where $f', f'' \ldots f^{(n+1)}$ represent successive derivative functions of x_o. Since $\bar{W}$ is a function of p, we let $\bar{W}_t$ = a function of p_t and $\bar{W}_{t+1}$ = the new $\bar{W}$ determined by genotype frequencies after selection.

TABLE 14-8 *Components of variance in fitness and increment in fitness*

Data from Table						Additive Variance in Fitness			
	$\bar{W}_t$	p_t	p_{t+1}	$\bar{W}_{t+1}$	$\Delta\bar{W}$	α	$2pq\alpha^2 = \sigma_A^2$	$\sigma_A^2/\bar{W}_t$	$(W_1 - 2W_2 + W_3)\left(\frac{pq\alpha}{\bar{W}_t}\right)^2$
14-3	0.7570	0.10	0.1321	0.7740	0.0170	+0.2700	0.013122	0.0173	−0.0003
14-5A	0.88	0.60	0.6000	0.8800	0	0	0	0	0
14-5B	0.86	0.80	0.7814	0.86355	0.00355	−0.10	0.0032	0.00372	−0.00017
14-7A	0.82	0.60	0.6000	0.8200	0	0	0	0	0
14-7B	0.84	0.80	0.8190	0.84398	0.00398	+0.10	0.0032	0.00381	+0.00017
Assume $\begin{cases} W_1 = 1 \\ W_2 = 0.75 \\ W_3 = 0.50 \end{cases}$	0.80	0.60	0.675	0.8375	0.0375	+0.25	0.0300	0.0375	0

Finally, in the bottom row of Table 14-8 where the genotypes' fitness values are in exact arithmetic progression (W_2 exactly intermediate between W_1 and W_3), the small-coefficient-factor last term of (14-29) is zero so that the increment in $\bar{W}$ is entirely determined by the additive component first term. In a sense of fitting Fisher's fundamental theorem, additivity of fitness values provides the simplest form of selection.

Under nonrandom mating (inbreeding), we shall demonstrate in Chapter 18 that $\bar{W}$ may decrease under selection. Also, with multiple genic locus situations, the average fitness is not necessarily increasing at all times or under all conditions. At such times neither Fisher's fundamental theorem nor Wright's adaptive peak concept holds exactly; yet those very exceptions are vital to a more general natural selection theory.

In the next chapter, we turn our attention to a fuller meaning of the Darwinian fitness and its estimation from real populations.

EXERCISES

1. (a) Calculate q and graph the q values for 10 generations in the haploid case as starting at $q_0 = 0.60$ where the a allele is at a 10 percent disadvantage compared with A, as in Table 14-1A.
   (b) Graph the q values for 10 generations in a diploid case starting at $q_0 = 0.6$ with a complete recessive whose fitness in aa is 0.90 compared with the A- phenotype. Assume panmixis.
   (c) In which case is the a allele eliminated faster, haploid or diploid? Calculate initial Δq and at the G_9 to G_{10} sequence.
   (d) At what generation (t) will q be approximately one-half q_0 in these haploid and diploid cases?
2. A compltely recessive lethal occurs in all members of a population (all $+/l$), as in Table 14-2A.
   (a) Graph the frequency of the lethal allele for 10 generations.
   (b) How many generations will it take to diminish the lethal from $q = 0.05$ to 0.025?
   (c) What can you say about the efficiency of selection in eliminating a deleterious allele?
   (d) At what frequency is the rate of change (Δq) maximum?
3. Use the fitness values of genotypes as given in Table 14-3A for a deleterious recessive (aa).
   (a) Calculate the rate of change in $q(\Delta q)$ over the range of q from 0.60 to 0.70 and narrow down the q value at maximum Δq to two decimal places.
   (b) What will be the mean fitness ($\bar{W}$) for this q value? At what q value will $\bar{W}$ be maximum?
4. Kimura and Ohta (1971) described the stochastic outcome with ultimate probability of survival (*u.p.s.*) for an advantageous dominant mutant. The authors used the symbol s to be the *advantage*, rather than disadvantage, of the heterozygote so that $W_2 = 1 + s$. Furthermore, their model assumed the heterozygote to be intermediate in fitness with each A allele adding s to total fitness so that $W_1 = 1 + 2s$. Applying Kolmogorov backward diffusion equations, they derived the following solution:

$$u.p.s. = \frac{1 - e^{-4Nsp}}{1 - e^{-4Ns}} \qquad (N = \text{effective size})$$

In the case of an initial dominant mutant (as in Example 14-1), with $W_{Aa} = 1$, and $W_{aa} = 0.7$, initial $p_A = 1/(2N)$.

(a) Verify that $s = 0.4286$ as the Aa advantage over aa in this use of symbols.

(b) For $N = 20$, verify that $u.p.s. = 0.576$. We may note that this amount of fixation is somewhat greater than observed (0.478). Why should that be an expected discrepancy?

(c) For an advantageous recessive mutant, Kimura gives

$$u.p.s. = \sqrt{\frac{2s}{\pi N}} \quad \text{(where } s = \text{advantage of } aa\text{, approximately if } s \text{ is small).}$$

In Example 14-2, s is not "small", but with $N = 20$, the observed fixation proportion is 0.119. What is the expected $u.p.s.$ according to the given formula?

5. Visualize a selective value difference between gametes segregating in any eukaryotic diploid heterozygote. Would the outcome of selection be parallel with that of a bacterial culture (haploid) such as the "simplest-case" model of Table 14-1A? Where does such a parallel break down? How would you incorporate gametic selection into a total life-cycle model for a eukaryotic diploid?

6. (a) Show that Δq is one-half for the diploid heterozygous intermediate (14-11) case as for the haploid (14-1) when $s = 0.1$, $p = q = 0.5$, and $h = 0.5$.

   (b) At what q value is the haploid selection rate of change greatest (Δq maximum)? You may take the derivative of formula (14-1), set it equal to zero for the maximum, and verify that

$$\max \Delta q = \frac{1 \pm \sqrt{1 - s}}{s}$$

   (c) Go back to Exercise 1(b) and find that the maximum rate of change in q occurs between $q = 0.5$ and 0.6 when $s = 0.10$.

   (d) In the diploid heterozygous intermediate case, where would you expect to find the maximum Δq. Check to verify that by using formula (14-11) and trying q values between 0.5 and 0.6 with $s = 0.10$.

7. (a) In Example 14-1, a large population is initiated with $q_0 = 0.95$, assuming a favorable dominant (A) is 30 percent greater in fitness than aa. What would be Δp, the rate of increase for A at that initial frequency?

   (b) Suppose the opposite were true in fitness values—that the dominant phenotype (A-) had a 30 percent disadvantage to the aa phenotype. If initial $q_0 = 0.05$, what would be the Δq value?

   (c) Which increases faster in a large panmictic population, a rare advantageous dominant or a rare advantageous recessive?

8. How would you summarize in general terms the effect of population size limitation on the probability of fixation in the case of: (a) Detrimental recessives? (b) Detrimental dominants? (c) Advantageous recessives? (d) Advantageous dominants?

9. Given the fitness values of AA, Aa, and aa, respectively, as 0.60, 1.00, and 0.20.

   (a) What will be allelic frequencies one generation after selection in a large population if original frequencies are equal ($p = q$)? What is Δq?

(b) What will be the final outcome if these fitness values remain constant?
(c) Plot $\bar{W}$ on a graph as a function of q for enough points to establish the maximum point.
(d) Plot Δq similarly and note its value at the equilibrium $\hat{q}$ value.

10. (a) Suppose fitness values of *AA*, *Aa*, and *aa* are 0.50, 0.90, and 0.10, respectively; how will the equilibrium value compare with that of the previous question?
(b) What would be the Δq if $p = q$? Hint: Use formula (14-23). How does that value compare with Exercise 9(a)?
(c) What is the relationship between these fitness values and those of Exercise 9? How do you explain the differences and similarities between the outcome ($\hat{q}$) and rate of change in Exercise 9 and this case? In your answer, determine the values of formulas (14-15), (14-18), (14-23), and (14-24) for these fitnesses and for those in Exercise 9.

11. (a) If a panmictic population has $p = 0.6$, $q = 0.4$, but the heterozygotes are lethal (*Aa* $W_2 = 0$), what would be the frequencies of alleles in the next generation? (Assume $W_1 = W_3 = 1$.) If the initial $p = 0.3$, $q = 0.7$, what would be the next generation's allele frequencies?
(b) What would be the allelic frequencies after one generation if the original frequencies had been equal ($p = q$)? What kind of equilibrium is this?
(c) If heterozygotes are inferior to all homozygotes and the population size is reduced into small isolates, what would you suppose to be the eventual allelic frequencies over all the isolates?

12. Derive (14-23) from (14-7) by rearranging and simplifying to verify as follows:

(a) $$\Delta q = \frac{pqW_2 + q^2W_3 - \bar{W}q}{\bar{W}}$$ Complete $\bar{W}$ in the numerator last term.

(b) Grouping terms in the numerator only and putting W_3 first, we have (numerator):

$$q^2W_3 - q^3W_3 + pqW_2 - 2pq^3W_2 - p^2qW_1$$
$$= q^2W_3(1 - q) + pqW_2(1 - 2q) - p^2qW_1$$

Because $1 - 2q = p - q$, we have simplified the numerator to

$$= pq^2W_3 + p^2qW_2 - pq^2W_2 - p^2qW_1$$

(c) Rearrange the two middle terms of this numerator and supply into (a) above to obtain formula (14-23).

13. Show algebraically that under a balanced equilibrium, $\bar{W} > W_1, W_3$; that is, net fitness exceeds each of the homozygote fitnesses. Use formula (14-14) or (14-14A).

$$\bar{W} = pW_2 + qW_3 > W_3$$

(or, correspondingly, $(pW_1 + qW_2) > W_1$). Let $W_2 = 1$, $W_3 = 1 - s$, for example, and arrive at $(1 - sq) > (1 - s)$ so that $s(1 - q) > 0$ as long as q is not 1.

14. (a) When fitness values occur in a geometric progression so that $W_1 = 1$, $W_2 = \lambda$, and $W_3 = \lambda^2$, then net fitness will be $\bar{W} = p^2 + 2pq\lambda + q^2\lambda^2$. Thus q_1 (after selection) $= [(pq\lambda + q^2\lambda^2)/(p + q\lambda)^2]$. Factor, simplify, and verify formula (14-22) as a solution for λ.

(b) Genotypes occur in these frequencies after selection: $AA = \frac{4}{9}$, $Aa = \frac{4}{9}$, $aa = \frac{1}{9}$. It is known that these genotypes occurred in reverse frequencies before selection, with $AA = \frac{1}{9}$ and $aa = \frac{4}{9}$. Find the relative fitness values of these three genotypes.

(c) After selection, genotypes occur as follows: $AA = \frac{1}{4}$, $Aa = \frac{2}{4}$, $aa = \frac{1}{4}$. Before selection, genotypes produced by random mating came from $p = \frac{5}{6}$, $q = \frac{1}{6}$ allelic frequencies. What are the fitness values of these genotypes?

15. (a) Verify the $\bar{W}$ values from using (14-19) and (14-14) from Wallace's example of drawing conclusions about the balanced nature of an equilibrium merely from examining a single generation. First calculate F_{est} by the departure from the square-law expectation and supplying $W_2 = 1$.

(b) The following genotype frequencies are found after selection: $AA = 0.3934$, $Aa = 0.5246$, $aa = 0.0820$. Suppose it is known that the allelic frequencies were $p_A = 0.8$, $q_a = 0.2$ before selection. Assuming that mating had been at random to produce genotypes and that selection acted between the zygote and adult stages, find the relative fitness values (W) for the genotypes. Does the high frequency of heterozygotes correspond to a superior fitness of that genotype?

16. In Fisher's fundamental theorem of natural selection, $\bar{W}$ depends on the linear (additive) component of normalized variance in fitness between genotypes, $\sigma_A^2/\bar{W}$.

(a) Show that this component will be zero at equilibrium if fitness values are $W_1 = 0.75$, $W_2 = 1.00$, $W_3 = 0.50$.

(b) For the fitness values of (a), let genotype frequencies be $AA = 0.36$, $Aa = 0.48$, $aa = 0.16$. Then calculate $\Delta\bar{W}$ and $\sigma_A^2/\bar{W}$, the additive component of variance in fitness, plus the final term of formula (14-29), one-half the product of the second derivative times $(\Delta p)^2$.

(c) If fitnesses of AA and Aa are interchanged, so that $W_1 = 1$, $W_2 = 0.75$, with W_3 remaining at 0.50, show that the additive component of variance in fitness equals exactly the change in $\bar{W}$ for $p_0 = 0.3, 0.4, 0.5$.

17. If a deleterious condition can be "cured" or alleviated by medical treatment, it has often been asked what the consequences would be for change in the gene frequency. Assume a condition so deleterious that no one with the condition reproduces. Then a "cure" (treatment) is found to alleviate the condition permitting normal reproduction and not affecting the mutant allele.

(a) If the condition is dominant (Aa not reproducing until "cured"), and if the mutation rate is 1×10^{-4} per generation, what would be the incidence after one generation? after two generations? (Assume all who have the condition were to be "cured.") How fast would this dominant increase?

(b) If the condition is recessive (aa not reproducing until "cured"), and if $q^2 = 1 \times 10^{-4}$ when all aa individuals become "cured," how would the gene frequency be altered after one generation? after two generations? If the mutation rate from A to a is also $u = 1 \times 10^{-4}$, what would be the incidence of aa in two generations? How fast would this recessive increase with mutation at that rate?

(c) What are the general consequences of improving conditions or relieving symptoms (eliminating selection pressure) on rare dominants versus rare recessives? Consider whether mutation is occurring or not.

18. With improved techniques for detecting heterozygotes, carriers of deleterious recessive traits could be advised about their likelihood of having defective children. Assume

$q_a = 0.01$ and that the *Aa* frequency is approximately $2q$ while *AA*'s frequency is $(1 - 2q)$.

(a) What would be the frequency of q_a in the next generation if all married couples who are $Aa \times Aa$ avoided having any children whatsoever?

(b) What would be the frequency of q_a in the next generation if all $Aa \times Aa$ couples ascertained the phenotype of their children by amniocentesis early enough to have defective *aa* children aborted in every case so that only *AA* and *Aa* children would be produced?

(c) Do you feel that society would be benefited by any eugenic program that would advocate either of these procedures? Why?

15

DARWINIAN FITNESS

When gene frequencies change in a population as a result of an improvement in reproductive efficiency for the population, we say that natural selection has occurred. The rate and direction of genetic changes are limited by at least the following factors: (1) the amount of variation in "reproductive value," or fitness, among the phenotypes determined by genotypes in the population (indicated in Chapter 14), (2) the heritability of fitness properties of the phenotypes, and (3) the genetic architecture contributing to the variation in fitness (amounts of dominance, linkage relationships, nonallelic interactions, recombination frequencies, and so on).

We cannot tell whether selection has or has not acted on a population merely by observing frequencies of genotypes at a single point of time (one generation). If we are to understand and analyze the process of natural selection, we cannot avoid examining populations over several generations, primarily to discover any changes in genetic composition that have occurred. However, the discovery of changes does not reveal what mechanisms may have caused the changes. Even today we have discovered few cases of selective agents definitely known to bring about genetic change. That is, we know little of what specific contributions genotypes make to traits affecting fitness in populations. Some genotypes may have greater fertility than others; some may contribute indirectly to fertility by increasing longevity or speed of development with reduced time to maturity, mating, or fertilization; some may be resistant to diseases or increase the probability for survival through competitive ability or escape from predators. These fitness differences may be manifested under "ordinary" environmental conditions or under a particular stress (environmental selective agent), depending on genic action through development and how phenotypes are expressed under particular conditions.

In this chapter we discuss the meaning of Darwinian fitness in terms applicable to the genetic outcome of intrapopulational selection. We need to know how selection is able to modify genetic systems in the process of achieving a state of adaptedness and how diversity among biological entities may arise during that process. It is important not only to estimate the order of magnitude for the rate of natural selection, but also to discern the likely outcome of the process, how subtle its action is for the creation of genotypes, and how powerful it can be in doing so.

HOW TO MEASURE DARWINIAN FITNESS

When we used the relative fitness values of 1 for A and 0.9 for a in the simplest case of selection (haploid in Table 14-1), the fundamental implication was to quantify the relative contributions of the two genotypes to the total hereditary dosage of the next generation. We could as easily have expressed the relative contributions of those genotypes by calling the fitness of $a = 1.0$ and that of $A = 1/0.9 = 1.111\ldots$. In fact, there is an infinite number of pairs of values with identical relative proportionality between the two. These fitness values imply a summation, or net effect, of all biological components in the life cycle that brings about the differential contribution of heredity to the progeny generation. We might find that A reproduces faster (more progeny per unit time), has more total progeny (greater fertility over a long time), or has higher-quality progeny that in turn survive through their life cycle better than a. Conversely, we might find a has greater mortality so that its probability for survival before it can reproduce is less than for A. In other words, a multitude of biological functions, or any one of them, could affect the relative contributions of genetic material to the progeny or future generations of progeny.

Relative fitness of an individual cannot usually be predicted, although we may compare individuals for fitness traits. As pointed out by Li (1967*c*), the fitness of a genotype is a "retrospective" quality generally and "determined only after its performance has been completed. Therefore, the term 'fitness' is very much like a title such as 'champion.' We do not know who the champion is until after the contest."

The biological "contest" that ultimately tells us what the relative fitness (or selective) values of genotypes may be is an immensely inclusive set of processes. When we reflect on the life cycle of a bisexually reproducing organism, what components of that life cycle contribute to the production of offspring and the features of the offspring that can result from any differences among the hereditary contributions of their parents? What counts in the "struggle for existence" toward getting genetic material of a particular sort across the generation gap? A summary of major life-cycle components contributing to Darwinian fitness—total number and quality of progeny over several generations as applicable to sexually reproducing species—is outlined in Table 15-1. Each of the traits mentioned is measurable, but each depends on an enormous number of physiological and biochemical features, especially advanced animal types having behavioral features with various regulatory mechanisms consisting of internal and external feedbacks (sensory and motor nerve pathways, effectors, endocrines, and interconnecting functions in growth and development). The student is no doubt aware of the complex organization of living things, and it should be evident that with the interdependency of these components for optimal functioning, nearly any genetic change is likely to have some consequences on fitness. It is necessary to consider the magnitude of fitness differences between genotypes that can result in significant change for producing adaptedness and diversity in the biological world. We find that organisms may utilize particular and often unique strategies by capitalizing on certain fitness components to the exclusion or lesser emphasis of others, depending on the conditions under which they attempt to make a living. For example, weedy types may combine high fecundity with fast development to make rapid invasions of wide areas, while other species may rely on efficient utilization of narrow resources.

Of the various measures of fitness, no doubt the most all-inclusive is the "innate capacity for increase"—the "intrinsic rate of natural increase," or what Fisher (1930) called "the Malthusian parameter of population increase." That parameter measures the *net reproductive rate of a population* living under a certain set of conditions; specifically, it is the ultimate rate of increase in population size "when the quantity of food, space, and animals of the same kind are kept at an optimum and other organisms of different kinds are excluded from the environment" (Andrewartha and Birch, 1954)—that is, when competition is eliminated, along with effects of crowding, nutritional shortage, predators, diseases, and other troubles and stresses. (Detailed description of this parameter r and its utility are beyond the scope of this book; however, a clear and concise summary of the concept is given by Mertz, in Connell, Mertz, and Murdoch, 1970; also see Wilson

TABLE 15-1 *Some major life-cycle components contributing to Darwinian fitness in sexually reproducing diploid organisms*

Generation	Traits (Stages) in Fitness
t: zygotes	*At Fertilization*: Genotypes into Population
	E-1. Embryonic development ability (prehatching or prebirth growth)
E = Early in life cycle	E-2. Viability, survival to maturity (resistance to disease, avoidance of risks, competitive ability, adolescent survival ability)
	E-3. Rate of development (reciprocal of time to maturation)
	Maturation: Adults
	L-4. Viability, survival probability from maturity
	L-5. Competitive ability, use of resources in short supply with other organisms
	L-6. Longevity, average age at death
L = Later in life cycle	L-7. Mating ability ("drive," dissemination of gametes, fertilization ability)
	L-8. Mating associations ("preferences" or grouping by phenotypes)
	L-9. Fecundity, gametic production
	L-10. Age-specific fertility (family or litter size, frequency of litters, care and protection of young)
	L-11. Total progeny = fertility
Gametes	*At Meiosis*: Segregation of Homologous Chromosomes
	G-1. Gametic formation (segregation "distortion")
	At Prefertilization:
	G-2. Gametic "competition" (differential penetration or elimination of male gametes, compatibility of ova and sperm)
$t + 1$: zygotes	*At Fertilization*: Genotypes into Population
	E-1. Repeat of features above and any features influenced by maternal or special parental effects

and Bossert, 1971.) Fisher pointed out that the net gain in offspring over deaths, which can be expressed in continuously breeding populations by the net reproductive rate when a stable age distribution has been reached (constant proportions of individuals in each age group), may be the widest-scope definition of Darwinian fitness for a population. Even though this measure is a theoretical ideal, it has many drawbacks: (1) it cannot predict competitive ability since the reproductive potential of a population when measured under optimal conditions may be greatly modified when the population is put under stress conditions or under competition; (2) it is time-consuming to construct life tables giving rates of deaths and births in a population; (3) while it is theoretically desirable to estimate the innate capacity for increase (r) for particular genotypes, if we applied such measures to those genotypes in a population, we would have insufficient information because interactions between genotypes in real populations would undoubtedly be so great as to make predictability of outcome virtually impossible. Dobzhansky (1968) has expressed this difficulty of measuring what he termed *adaptedness*; for individuals, it comprises the probability of survival, reaching the stage of reproduction in the life cycle, and, for humans, at least, a sense of well-being; for populations, the Malthusian parameter is a measure of adaptedness. This measure of net reproductive rate for a population might well give us predictive information for *inter*populational fitness (see Chapter 19). Competitive ability and feedback recognition ability for the organism when an upper limit to the carrying capacity of an environment (K) is being reached are vital fitness traits for a population as an integrated unit. The various strategies open to populations for improvement of such fitness qualities (r and K selection) will be described briefly in Chapter 16. (See also Levins, 1968; Lewontin, 1961, 1965, for general discussions of fitness strategies.)

Operationally, we are forced to estimate relative fitness of individual genotypes from a few main methods.

1. We can monitor genotypes either in experimental or wild populations over several generations to estimate overall fitness values sufficiently to fit some or all of the genotype changes observed.
2. Alternatively, after observing the genotype changes, we may infer what relative fitness components the genotypes tend to control from experimental examination of the life cycle. Each stage (Table 15-1) may be evaluated by rating the genotypes cultured separately.
3. The "early" and "late" stages may be evaluated by working out expected changes for genotype frequencies before and after selection if those genotypes can be monitored at those stages in the population.
4. After finding that genotypes differ in some relative fitness values for certain components of the life cycle, we may attempt selecting under population conditions for those components and see whether we can control the frequencies of the genotypes; when changes in genotype frequencies can be altered in a predictable way, we could be confident that the selective influence relevant to the population's adaptedness has been determined.

ESTIMATING FITNESS VALUES FROM GENOTYPE FREQUENCY CHANGES IN POPULATIONS

Haploid Selection

In the simplest case—haploid selection—there is relatively little problem in estimating relative fitness values. From formula (14-1), we may solve the rate of change in q for s, the selection coefficient, either for each generation or over several generations by use of (14-2). Problems of life-cycle components may not apply, although the microbial geneticist may have the more fundamental problem of deciding whether changes observed are due to selection or to mutation, because generation time may be very fast and care must be taken to rule out all forces except selection before trying to estimate selective values. Example 15-1 illustrates a clear case of selective change in bacteria (*E. coli*) following beneficial genetic changes in regularly transferred cultures. The fitness superiority of the h_1^+ type was about 6 percent greater than its allele the h_0^-, although it should be clear from the experiments described that the adaptive superiority did not lie with the simple genic mutational change of h^- to h^+ (histidine-requiring to wild-type). Without recombination occurring in the culture, the *h* locus could be considered only as a marker for some genetic adaptive change acquired during the particular transfer technique used. Once the adaptive genotype was produced, the changes in frequency of the competing markers were quite straightforward. These changes were sufficiently constant to indicate fitness values for the two genotypes over the range of frequencies recorded in the experiment.

Diploid Selection

Does the stage of life cycle make a difference? Before attempting to estimate overall fitness values, we ought to be reassured that the model we have assumed for action of selection in Chapter 14 is not necessarily restricted to that between zygote and adult (development viability). That model was given for simplicity of conceptualization in numerical examples and understanding of rate of change in p and q values. Differences in fitness at any stage of the life cycle (Table 15-1) could be applied to achieve the same change in gene frequency (with the single exception of trait L-8—preferential mating—that accounts for the special interactions seen in Chapter 10). For example, we may consider L-7, genetic differences in mating ability ("drive"), as an alternative to differences in development viability (E-2) used in Tables 14-2, 14-3, 14-5, and 14-7. Using the simplest model for selection since there are fewest coefficients to consider, let us illustrate the case from Table 14-3 where recessive homozygotes had a disadvantage in viability. Instead of a viability effect, suppose the recessives have a mating activity of only 0.7 in both sexes compared with the dominant phenotype. We may assume zygotes had been formed at proportions that were the square of previous-generation genotype frequencies. When they reach the adult stage they mate, with recessives at proportion m mating ability relative to the dominant forms, so that the mating proportions can be symbolized.

$$\begin{aligned}[p^2 + 2pq + mq^2] \times [p^2 + 2pq + mq^2] &= (p^2 + 2pq + mq^2)^2 \\ &= p^2 + 2pq(p + mq) + q^2(p + mq)^2 \qquad (15\text{-}1)\end{aligned}$$

where m is relative mating ability of homozygous recessives. Note that expression (15-1) fits the square law with fitness values in a geometric progression as in (14-21) and (14-22), with $\lambda = (p + mq)$. If we apply the same coefficients and gene frequencies used in Table 14-3, we would obtain frequencies for those which have mated:

G_0 adults: $0.01AA + 0.1314Aa + 0.431649aa = 0.573049$ for $\bar{W}$.

Normalizing produces: $0.0174 + 0.2293 + 0.7533$, identical with the frequencies *after random mating* in G_1 of Table 14-3. As far as the net contribution of parental genotypes to the G_1 genotypes is concerned it made no difference whether the relative fitness contributions were applied before mating or as a factor in the mating itself. The $\bar{W}$(mean fitness) values in the two cases are very different, but the genotype frequencies resulting from selection are the same irrespective of the point in the life cycle where the selection was applied.

That is *not* to say, however, that it does not matter where in the life cycle selective differencies arise. First of all, it has just been noted that the net fitness will be affected more by applying the selection during mating than during development. In Exercises 2 and 3 it is demonstrated that application of selective differences at the adult stage in mating or fertility incorporates W values when both sexes have the same values so that mean fitness ($\bar{W}$) becomes the square of what it would be from application of selection during development. We cannot then be unconcerned about the point in the life cycle where selection acts.

Second, and more important, when fitness values are to be estimated from population changes we must realize that any genotype may have a particular selective value at a specific stage or stages of the life cycle. Components of total fitness may be positive at one stage and negative at another or any combination of positives and negatives may follow in sequence of life-cycle stages. Problems of estimating fitness values become complex as genotypes are analyzed for relative fitness at various stages. We now proceed to look at the problems of estimation in diploids. First, in an "overall naive" sense, we may estimate W values from populational changes in genotype frequencies ascertained from sampling done at the same stage of the life cycle each generation. Second, we may estimate W from components of the life cycle by considering the genotype frequency changes from one stage of the life cycle to the next. It will then be necessary to try measuring fitness values from several techniques.

EXAMPLE 15-1

MEASURING FITNESS VALUES IN BACTERIA

Atwood, Schneider, and Ryan (1951) analyzed pure cultures of an *Escherichia coli* strain requiring histidine (h^- grown on a nutrient medium supplemented with L-histidine) looking for h^+ mutants (found by plating known quantities of organisms on nutrient agar lacking histidine). Cultures of h^- were maintained by 1/100th serial transfer regularly, allowing the bacteria to be in continuous growth phase for about six hours after each transfer. Numbers reached about 2.5×10^{10} bacteria held in stationary phase until the end of 12 hours when about 1/100th volume was transferred again. Any h^+ bacteria that had accumulated due to mutation and growth, counted immediately before transfer, tended to reach a more or less constant frequency equilibrium within 100 generations at about $1.3h^+/10^6h^-$, held for about 2760 generations when the experiment was terminated.

Mutation rate determinations were calculated from separate tests as follows:

$$h^- \rightarrow h^+ = 2.80 \times 10^{-8} = v$$

$$h^+ \rightarrow h^- = 1.17 \times 10^{-6} = u$$

Using (13-11), we find $\hat{p}/\hat{q} = v/u$ at mutational equilibrium, so that $\hat{p}(h^+) = 2.4 \times 10^{-2}$. However, h^+ never increased beyond 1/10,000th of that frequency. The authors concluded that a selective difference between mutant revertants must exist, with the h^- strain having a strong advantage over h^+. To be more certain of selective differences, the serial transfer technique was reinitiated with original h^- mixed with h^+ in 1–100 times the amount at equilibrium. In Figures A and B, initial h^+/h^- ratios (in $\log_{10}$) are given for cells bearing an additional genetic marker (lactose fermentation) either as a different allele (in A) or the same allele (in B) from that in the h^- cells. Thus, h^+ arising as mutants from h^- could be distinguished from h^+ rising due to other mechanisms. For about 200 generations the h^+ frequency remained at the initial level, but then it suddenly dropped to either the equilibrium point ($1.3h^+/10^6h^-$), as shown in Figure B, or close to zero h^+ cells in Figure A.

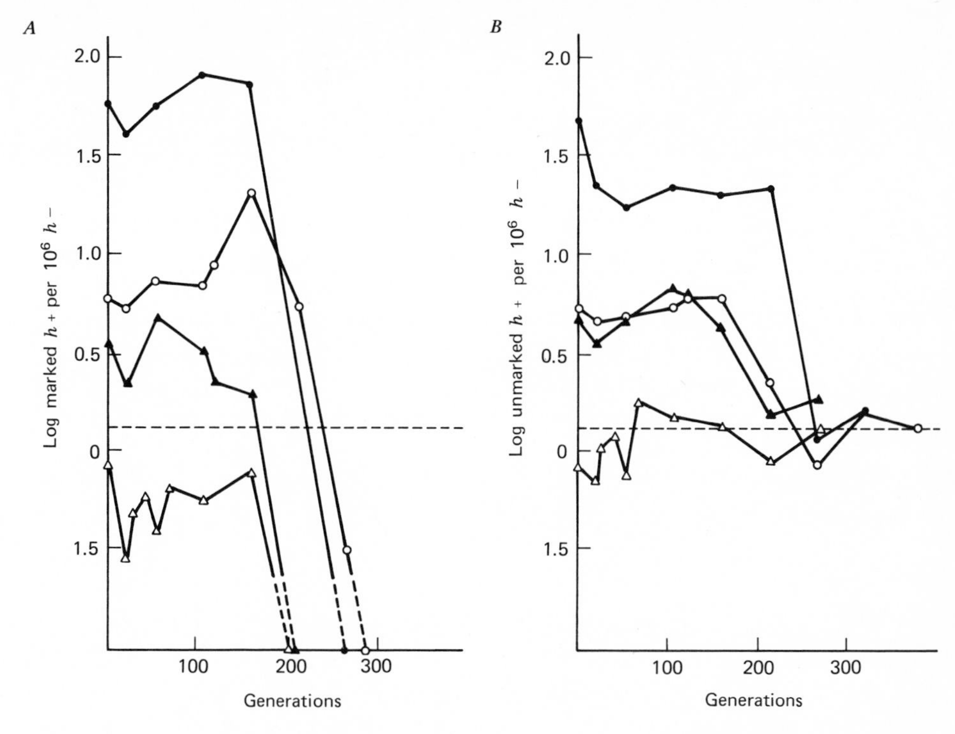

Figure A. Simultaneous behavior of h^+ bacteria in serial transfers started at various h^+/h^- ratios and bearing a genetic marker, lactose fermentation, different allele from the h^- strain.

Figure B. Various starting ratios of h^+ having the same lactose fermentation marker allele as the h^-. Solid circles, open circles, and triangles are samples from the same cultures in A and in B. Broken line indicates equilibrium. Both figures from Atwood, Schneider, and Ryan (1951).

Apparently, selection against h^+ was negligible for a time and then suddenly became manifested.

Accordingly, h^+ and h^- bacteria were isolated before and after the period of selection and tested together in various combinations to see whether the two types of cells had changed in relative fitness. The results of many experiments are summarized in Figure C. First (left) the original strains, designated h_0^+ and h_0^-, were put in competition at a ratio above equilibrium as in the upper sample curve of Figure B. At time Y, h_Y^+ and h_Y^- were extracted from the population and put together in three new combinations: h_Y^+ with original h_0^-, h_Y^+ with new h_Y^-, and original h_0^+ with new h_Y^-. Note that the new h_Y^- strain (middle diagram) clearly displaced both h^+ strains by selection immediately while the new h_Y^+ held its former temporary plateau against original h^-. Thus, the h^- had acquired a selective advantage at time Y which it did not have at time 0, but the h^+ had remained unchanged. A second isolation was done at time 1 (left diagram following the equilibrium point), and isolated strains were put into competition with originals again (right diagram results). Now h_1^+ (new form) was no longer the equivalent of h_0^- but was clearly superior in fitness (right diagram), although it was equivalent to h_1^-. A succession of extractions of strains and isolations at later times following the adaptive changes was made continually with similar results

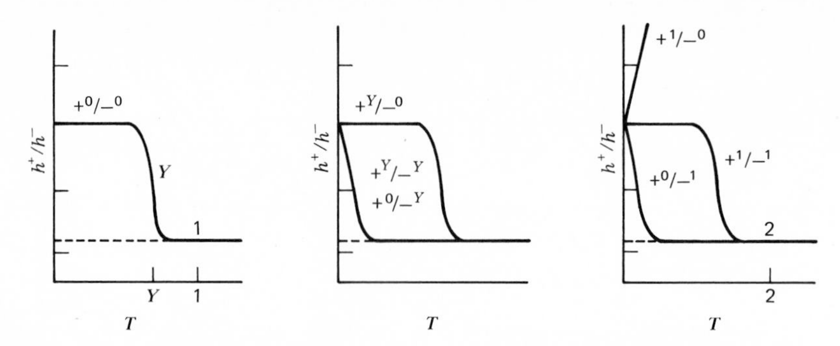

Figure C. Diagrams of changes in h^+/h^- ratios. *Left*: original h_0^+ and h_0^- put in competition (as in Figures A and B). At time Y both + and – forms are extracted from the population and put together in combinations. *Center*: new h_Y^+ versus original $_0^-$ (top), new h_Y^+ versus new h_Y^-, and original h_0^+ versus new h_Y^-. New h_Y^- selects against both h^+ strains. At time 1 (left), second isolation is made. *Right*: new h_1^+ is now superior to original h_0^-. Other strains as marked similar to center.

so that the authors could conclude that during the serial transfer technique, the more common h^- form acquires certain fitness improvement, and because h^- is about 10^6 times more frequent than h^+, the h^+ mutants, when they did occur, had in their genome whatever improved genetic change was acquired during the transfers over about 200 generations. A periodic selection process may be diagrammed briefly as follows:

$$\begin{array}{ccccccc} h_0^- & \rightarrow & h_1^- & \rightarrow & h_2^- & \rightarrow & h_3^- \quad \cdots \\ \downarrow & & \downarrow & & \downarrow & & \downarrow \\ h_0^+ & & h_1^+ & & h_2^+ & & h_3^+ \end{array}$$

By testing growth rates of the isolated strains in various nutrient media, Atwood and colleagues demonstrated that some of the adaptive changes involved increases in growth rate in the h^+ medium but decreased ability to grow in other media. The adaptative genic changes were then specifically geared to the environmental components of the serial transfer stages. These changes must have occurred as progressive remodeling of the genotype in periods of selection.

The h^+ mutants detected in these experiments did not have improved selective value in themselves, since in competition with contemporary h^- they were not adaptively superior. However, h_1^+ was clearly superior to h_0^-. If no recombination occurred between these strains when put in competition, simple selection displacement would occur. By diluting the h_1^+ strain, it was put in competition with h_0^- at a frequency of about 2.5×10^{-8} and serially transferred. The proportion of h_1^+ increased rapidly, as shown in Figure D, and continued to increase until the entire population was essentially h_1^+. The time required for this replacement was about the same as the time for acquiring the new adaptive genotypes in the main experiments (300–350 generations). Thus, when the population fluctuated between growth and stationary phases, natural selection for adaptation to the culture medium was in continuous operation. Selective changes took place in about the most simplistic way we can suppose.

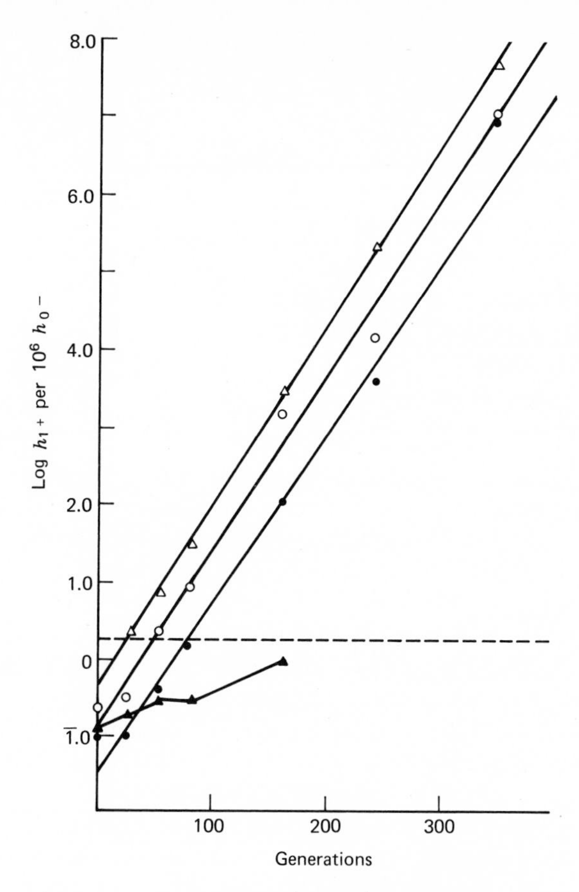

Figure D. Increase in h_1^+ on H^+ medium following the addition of one or few h_1^+ cells to a population predominantly h_0^-. Solid triangles are h_0^+/h_0^- as determined in the same experiment. Solid circles, open circles, and triangles are from replicate experiments giving the log of h_1^+/h_0^- over generations.

From the rate of increase in h_1^+ diagrammed in Figure D, we may compute the relative fitness values of the two genetic strains growing differentially in the serial transfers as follows. All three repeats of the experiment gave parallel results. Approximately the increase in h_1^+ goes from $1 \times 10^{-6} = p_0$ up to equality with h_0^- ($0.5 = p_t$) within about 250 generations. To compute the selection coefficient s, we may use formula (14-2).

Letting	h_1^+	h_0^-
frequency	$p_0 = 1 \times 10^{-6}$	$q_0 = 0.999999$
fitness	1	$1 - s$
at 250 generations	$p_t = 0.5$	$q_t = 0.5$

then

$$st = ln\left[\frac{q_0(p_t)}{q_t(p_0)}\right] = ln\left[\frac{0.999999(0.5)}{(0.5)(0.000001)}\right]$$

$$s(250) = 2.30259 \log_{10}[999999] = 16.118123$$

Therefore, $s = 0.064$, so that if h_1^+ fitness $= 1$, h_0^- has the approximate fitness $= 0.936$.

ESTIMATING FITNESS VALUES FROM OVERALL POPULATIONAL GENOTYPE FREQUENCY CHANGES

In principle, if q and Δq values are known accurately from generation to generation, with q being estimated from individuals *before* selection acts each time, solving for the selection coefficient s in (14-6) can be done: $s = -\Delta q/q^2(1 - q - \Delta q)$. Use of that expression must be limited to cases where the recessive homozygote is truly recessive in fitness ($AA = Aa > aa$). Error in estimation of q can be troublesome, and one must obtain data from large and reliable samples or from the averages of several replicate populations in order to ascertain three genotype frequencies carefully.

The demonstration by Wright and Dobzhansky (1946) that some of the cytogenetic changes in natural populations could be reproduced in laboratory experimental populations opened a way to explore the magnitude of the selective process in evolution. Accordingly, the need for more general methods of estimation of fitness values arose, particularly for populations maintained in relatively constant conditions of the laboratory where organisms have short adult life spans, specific breeding times, little or no overlap of generations, and basic Mendelian processes in reproduction. Populations could be started with frequencies of genotypes presumably far from points of equilibrium so that rate of change in q could be explored over a large range of q values. Fitness values (W_i) for all genotypes being monitored were parameters for estimation. With k alleles at a locus, there would be $k(k + 1)/2$ genotypes, each with W. The task becomes formidable operationally for more than a single pair of alleles, and our treatment here will be confined to the estimation problem with just two alleles (or segregating units such as chromosomal arrangements) because the fundamental problems to be solved are complex enough in the biological sense at our present levels of understanding. As we shall see, the pitfalls and problems of estimating just a few parameters are quite substantial, and until they are clarified, there is little point in extending methods of estimation further, at least for the present.

What we term the "classic" model of the selection process in populations whose genotype frequencies are to be monitored over several generations includes the following assumptions, which may well be too simple and thus need to be modified as biological information is added and the model improved: (1) W is assumed to be constant for genotypes over the range of frequencies, (2) sexes have the same fitness values and have no preferential mating interactions, and (3) environmental conditions are constant. Continued use of this classic model, however, brings considerable difficulty in interpretation, so that in many cases the classic model has to be abandoned.

The usual procedure is to estimate two W's relative to a third to reduce the number of parameters. With two unknown quantities, we must obtain estimates of Δq from a succession of at least three generations. Unfortunately, small sampling errors in q make large differences in estimates of W, although accuracy can be increased with adequate numbers of replicate populations run. In 1946, Wright devised a method of least squares based on formula (14-12) with which to fit Dobzhansky's experimental population data for *Drosophila pseudoobscura* chromosome arrangement frequencies. In Example 2-4, one of the experimental populations containing ST and CH chromosomal arrangements derived from a natural population (Piñon Flats) displayed increase in the ST sequence relative to the CH over about six generations. In all, seven populations were run by the same technique in Dobzhansky's laboratory under constant warm temperature for the same arrangements from Piñon Flats. Some were run in light, some in the dark, and some with different foundation stocks. It was significant that all populations tended to reach equilibrium at levels about 75 percent ST: 25 percent CH after about 10 generations. In other words, the result was determined and predictable with the heterokaryotype (ST/CH) seeming to exhibit superior net fitness throughout. In order to provide a more precise test for uniformity of changes taking place and reproducibility of the equilibrium point (and also to estimate fitness values for comparison with other populations), Dobzhansky & Pavlovsky (1953) repeated these experiments with carefully controlled conditions and freshly caught flies from the same natural population. Their results on four replicated populations are summarized in Example 15-2. An improved method for estimating fitness values (over Wright's method of fitting data to formula 14-12) was worked out by Levene (Dobzhansky and Levene, 1951) based on Haldane's selection principles in which the frequency ratio p/q is used to simplify the algebra. While the method improves on Wright's least-squares method by *providing for any change in fitness values*, it is subject to large differences in fitness estimates when small departures are made in the observed sample points from a perfectly smooth selection curve. The method is illustrated here for these reasons: (1) it has been used to produce first-approximation fitness values, which may then be used in fitting expected to observed population frequency changes, (2) comparisons between results from different populations can be made, (3) detection of fitness changes during the course of selection in possible (thus, one of the major assumptions of the "classic" model may need to be relaxed), and (4) the student may acquire a more real sense of the problems involved in detection of selection effects on populations' genotypes.

The basic theory for estimation of W is given in Table 15-2. Haldane found the algebra of selection simpler using the ratio p/q for which he used the symbol u. Change in the ratio becomes a simple regression equation with two constant W values to be estimated relative to a third fixed W. We do not need to know the frequencies of genotypes immediately following the action of selection—only the overall change in p and q. Frequencies should be based on incidence as close as possible to zygote formation (fertilization) so that the point of selective action can be assumed to occur between the egg stage in one generation and the same stage in the next. With these limitations, we may employ this formula of estimation, as illustrated in Example 15-2.

$$W_3 = \left[\left(\frac{u_0}{u_1}\right) u_0\right] W_1 + \left[\left(\frac{u_0}{u_1}\right) - u_0\right] \tag{15-2}$$

From Table 15-2, W_3 is in terms of change in the p/q ratio and W_1. A solution necessitates having a minimum of three generations in sequence for two simultaneous equations with

TABLE 15-2 *Estimating fitness values (W) using a method of Haldane modified by Levene (Dobzhansky and Levene, 1951)*

Let $u = p/q$ (It is easier for computation if p = frequency of the *more common* allele at the start of selection—the allele expected to decrease)

After random mating, genotype frequencies are expressed relative to q^2

$$\frac{p^2}{q^2} + \frac{2pq}{q^2} + \frac{q^2}{q^2} \qquad \text{or in terms of } u \rightarrow u^2 + 2u + 1 = (u + 1)^2$$

After selection, genotypes will be summing to net fitness.

A
$$W_1 u_0^2 + 2u_0 + W_3 = \overline{W}$$

where W_1 and W_3 are fitness values for the more common and rarer homozygotes, respectively, at the start of selection assuming the $W_2 = 1.00$ for the heterozygote and $u_0 = p_0/q_0$ before selection. Then, if $u_1 = p_1/q_1$ after selection

B
$$\frac{p_1}{q_1} = \frac{W_1(u_0)^2 + u_0}{u_0 + W_3} = u_1$$

Dividing by u_0

C
$$\frac{W_1(u_0) + 1}{u_0 + W_3} = \frac{u_1}{u_0}$$

Cross-multiplying and solving for W_3

D
$$W_3 = u_0 \frac{(W_1 u_0 + 1 - u_1)}{u_1}$$

Taking the component terms on the right

E
$$W_3 = \left(\frac{u_0}{u_1}\right) u_0 W_1 + \left(\frac{u_0}{u_1}\right) - u_0$$

Letting

$$\left[\frac{u_0}{u_1} - u_0\right] = a$$

and

$$\left(\frac{u_0}{u_1}\right) \cdot u_0 = b$$

this equation becomes a simple regression:

$$W_3 = bW_1 + a$$

for which a solution for W_3 in terms of W_1 necessitates two simultaneous generation sequences for $u_0 \rightarrow u$ and $u_1 \rightarrow u_2$.

In Levene's modification for Dobzhansky's continuously producing populations, where samples were not taken at discrete generation times, the ratio $(u_1/u_0)^{1/t}$ adjusts in **C** for time longer than one t-generation on the basis that the rate of change is approximately linear between observed points. Then **E** becomes

$$W_3 = \left(\frac{u_0}{u_1}\right)^{1/t} \cdot u_0 W_1 + \left(\frac{u_0}{u_1}\right)^{1/t} - u_0$$

Levene uses symbol x for Haldane's u and r for the ratio $(u_1/u_0)^{1/t}$.

u_0/u_1 followed by u_1/u_2. (For a maximum-likelihood method of computing W from observed allelic frequency changes in populations, see DuMouchel and Anderson, 1968. Their method necessitates solutions by iteration of the likelihood equations.)

EXAMPLE 15-2

FITNESS ESTIMATION UNDER CLASSIC MODEL ASSUMPTIONS

Dobzhansky and Pavlovsky (1953) needed a reliable test for the selection outcome in experimental populations of *Drosophila pseudoobscura* with strains derived in 1949 from Piñon Flats, California, as controls for those populations with chromosomes of mixed origin (to be described in Example 15-3). They initiated four replicate populations with flies raised as homokaryotype G_1 progenies ST/ST and CH/CH from intercrossing one dozen strains within each karyotype. Initial frequencies in the experimental replicate populations were 20 percent ST/ST: 80 percent CH/CH.

Figure A diagrams the changes in frequencies of CH observed in the four replicates averaged (solid line). However, Dobzhansky and Pavlovsky took egg samples from these populations at intervals that did not exactly correspond to generation time (they had estimated a generation to be 25 days under the population cage conditions at 25°C). While

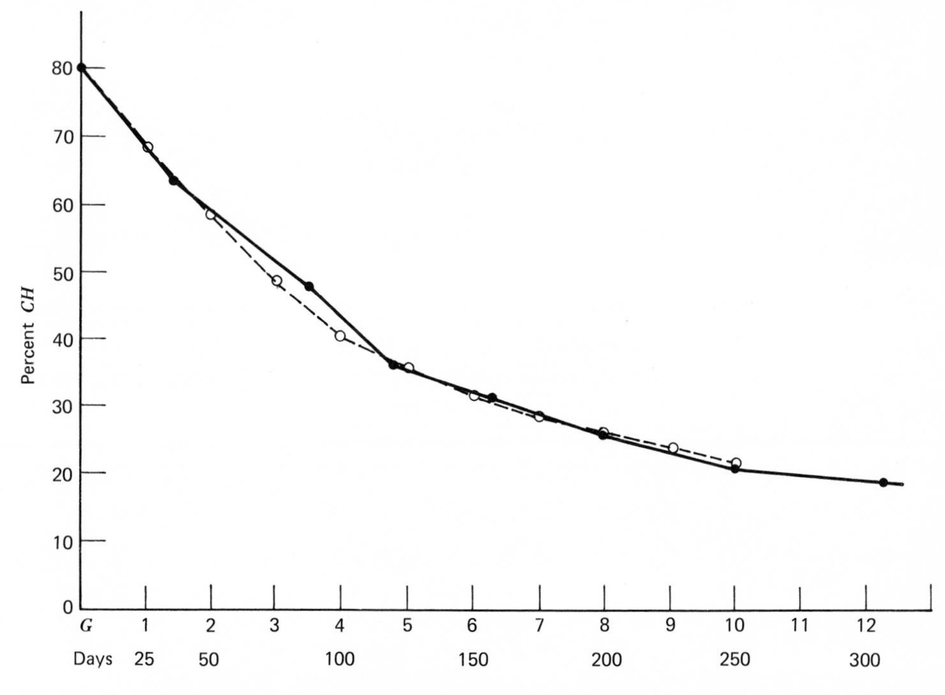

Figure A. Frequencies of CH observed in four replicate experimental populations averaged (solid line) over 12 generations (*Drosophila pseudoobscura* from Piñon Flats). ST frequencies = 1 − CH frequencies. Dashed line adjusted by smoothing out small irregularities between generations. From Dobzhansky and Pavlovsky, 1953.

the analysis for fitness coefficients was adjusted for generation time (as given at the bottom of Table 15-2), we here simplify the analysis by estimating a smooth selection curve in Figure A, drawn by eye (dashed line) to fit the observed values with as close to what would be the curve if selective differences between karyotypes were producing regular changes in the frequencies each generation. The frequencies of *CH* observed by the investigators are given in the left column of Table A, while the fitted values by generation are given to the right of the vertical line in that table. These changes have occurred on a "microevolutionary level" determined as to selective outcome by the genetic contents of the chromosomal complexes represented in this single population.

TABLE A *Percentages of CH (Chiricahua) arrangement chromosomes (versus ST) averaged over four replicate experimental populations initiated in late fall of 1951 by Dobzhansky and Pavlovsky (1953), containing flies derived from Piñon Flats, California. Analytical ratios needed for estimation of W_i values by method of Levene (Dobzhansky and Levene, 1951). Observed data on left, adjusted on the right of the vertical line for single generations.*

Days	G	% CH Observed	G	% CH	u	u_0/u_1	$(u_0/u_1) - u_0 = a$	$(u_0/u_1)u_0 = b$
0	0	80	0	80	4.0000	—	—	—
35	1.44	63.7	1	68.7	2.19488	1.82242	−2.17758	7.28968
80	3.20	47.6	2	58.5	1.40964	1.55705	−0.63783	3.41754
120	4.8	36.2	3	48.5	0.94175	1.49683	0.08719	2.10999
160	6.4	31.2	4	40.5	0.68067	1.38356	0.44181	1.30297
200	8.0	26.0	5	35.5	0.55039	1.23670	0.55603	0.84178
250	10.0	20.6	6	31.5	0.45985	1.19689	0.64650	0.65876
310	12.4	18.5	7	28.5	0.39860	1.15366	0.69381	0.53051
365	14.6	18.0	8	26.0	0.35135	1.13448	0.73588	0.45220
			9	23.5	0.30719	1.14375	0.79240	0.40185
			10	21.0	0.26582	1.15563	0.84844	0.35500

Solutions (using data on right adjusted for discrete generations)

G Intersects	W_1 (*CH*/*CH*)	W_3 (*ST*/*ST*)	(assuming $W_2 = 1$)
1–2	0.40	0.73	
2–3	0.42	0.835	
3–4	0.76	1.42	4th sequence clearly "not good"—see graph of solutions
4–5	0.27	0.78	
3–5	0.52	0.99	
5–6	0.46	0.94	
6–7	0.38	0.89	
7–8	0.58	1.00	
Average solutions (omitting #4)	0.46	0.89	
Dobzhansky and Pavlovsky's solutions	0.413	0.895	

[Note that before the fourth sequence, W averages: $W_1 = 0.41$, $W_3 = 0.80$. After that sequence W averages approximately $W_1 = 0.50$, $W_3 = 0.95$.]

We may now proceed to analyze this sequence of changes as a model for illustrating the methods and problems in estimating fitness values under what we have termed the "classic" model assumptions for the selection process. We develop a table of frequency ratios u as needed from Table 15-2 in the solution of formula (15-2). To estimate W_3 and W_1 (assuming $W_2 = 1.00$), we take the ratio relations of the two right-hand columns of Table A. It is easy to plot linear regressions (Figure B) of successive generations with $W_1(CH/CH)$ as the X axis and $W_3(ST/ST)$ the Y axis. Solutions for W take a sequence of

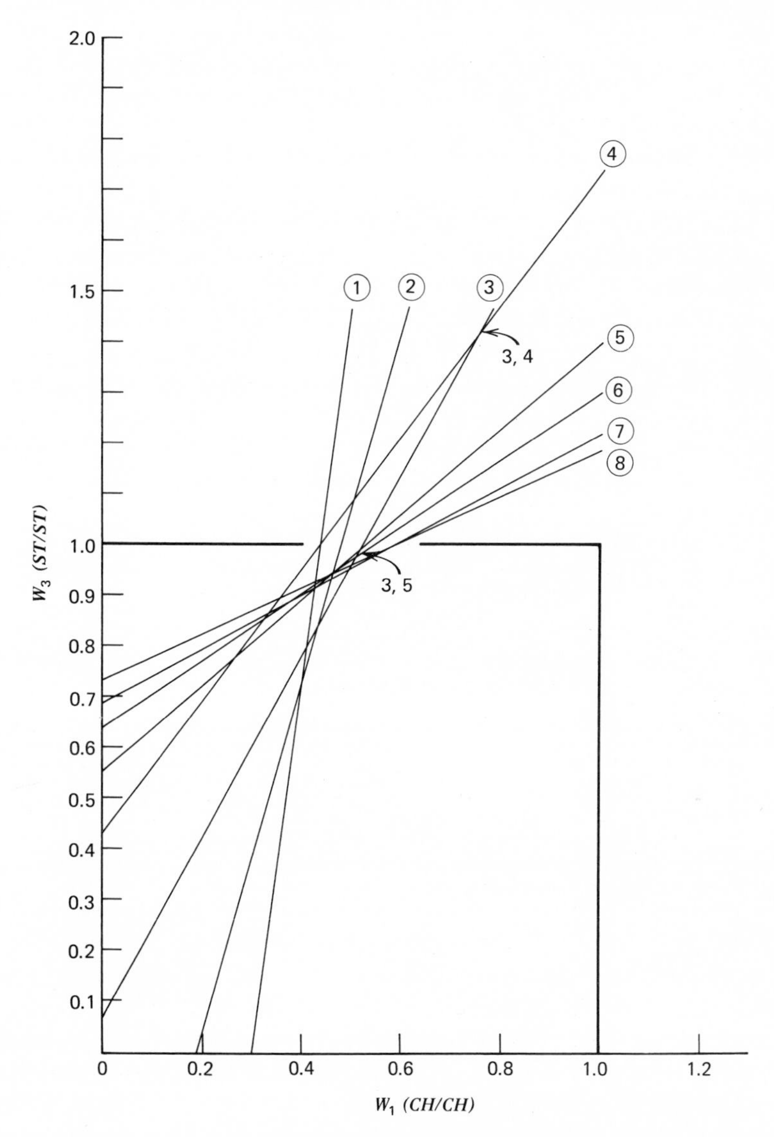

Figure B. Linear regressions of successive generation pairs of frequency data taken from Table A. W_1 (CH/CH fitness) on horizontal axis and W_3 (ST/ST fitness) on vertical.

at least three generations for an estimate. Intersects of successive lines are shown for generations 1–8 in the lower part of Table A.

If under the "classic" model, a single set of W's is assumed to be constant throughout all frequencies, the average of intersects (average solution) is given as $W_1 = 0.46$, $W_3 = 0.89$, which is quite close to the authors' estimates made from their observations not conforming exactly to generation times. It should be noted that the fourth sequence clearly is not good for a solution because it intersects with the third line far above the others and with the fifth far below. Such a discrepancy can easily occur if a very slight sampling departure from the smooth curve is made in the original data, and it demonstrates how difficult it is to achieve precision in this method. The student should try plotting the selection curve starting with 80 percent *CH* and using these W values to discover that the expected curve closely approximates the observed.

If we relax the assumption that a single set of fitness values satisfies the observed curve, we should note that (as mentioned at the bottom of Table A) the first three generations as a group have slightly different values from the last few generations. By using those values for the specific corresponding sections of the observed curve, the student should demonstrate that the separate two segments of the observed curve fit much better than any single set of constant values for the entire curve. We may tentatively make the judgment that as the *ST-CH* frequencies approach equilibrium, a change in fitness values might be more likely than constant values, with the *ST/ST* karyotype approaching equality with the heterokaryotype (*ST/CH*), while *CH/CH* also improves slightly.

COADAPTATION, MICROEVOLUTION, AND MESOEVOLUTION

As long as experimental populations contained two variants derived from the same natural population, a balanced equilibrium in variant frequencies was achieved, implying a superior fitness of structural heterozygotes (heterokaryotypes) and a deterministic predictable outcome to selection. That predictability was tested by Dobzhansky and coworkers when they asked the question of whether these chromosomal variants, which are widely distributed over much of the species' area, possess about the same fitness values irrespective of their source (wild population); that is, whether the genic contents of any chromosomal arrangement (*ST*, *AR*, *CH*, *TL*, *PP*, and others mentioned in Example 3-1) would be constant from one locale to another. The test of chromosome arrangement constancy was made by initiating experimental populations with the same chromososal variants as before (*ST* and *CH*, for example), but deriving one variant from a locality over 1000 miles away from the derived locality of the second variant. Population cage experiments utilizing *ST* chromosomes from Piñon Flats and *CH* chromosomes from Chihuahua, Mexico, are described in Example 15-3. Levene's method of estimating fitness values had the advantage of utilizing generation-by-generation segments of the selection curve in each population so that by inspection one could see whether the coefficients appeared to be constant over the range of q values. The fitness values of the karyotypes (*ST/ST*, *ST/CH*, and *CH/CH*) proved to be diverse from one population to another; balanced equilibriums appeared to evolve in some populations, while *ST* eliminated *CH* in others; a single set of selection coefficients rarely fit an entire selection curve because of abrupt changes in frequencies within a few generations. It was established by these experiments that the sequence of chromosomal banding (gene arrangement) did not by itself determine a fixed and predictable selective function or imply any intrinsic adaptive property independent of the population from which the chromosomal variant came Dobzhansky's (1954) major conclusion was stated: "Within the population of each geographic region, the polygene complexes in the chromosomes have become mutually adjusted, or 'coadapted,' by a process of natural selection"—as demonstrated in Example 15-3. These coadapted complexes do not exist when chromosomes of different geographic origin are mixed in laboratory populations, and consequently adaptive novelties may be created, as in Example 15-3. New "races" appeared in these laboratory populations that were uniquely adapted to their own genotypic background and physical environment.

Evolutionists have used the terms *microevolution* and *macroevolution* (Simpson, 1953) as if they were completely unrelated phenomena, and Simpson suggested that "clarity might now be improved by abandoning them." However, with new definitions for these terms and insertion of a new one, *mesoevolution*, Dobzhansky (1954) revived them to describe phenomena that are either creative and unique in evolution (meso- or macroevolution) or repeatable, reversible, and predictable (microevolution). Specifically, he pointed out that "Mesoevolution involves alteration in more or less numerous genetic units, emergence of new adaptively integrated genotypes, and appearance of at least new races, or of new species and genera. . . . [It] may be rapid enough to be observed by human observers." He classified the attainment of new equilibriums in laboratory populations containing chromosomal arrangements from diverse localities as a mesoevolutionary process. The coadaptation that brings about heterozygote superiority in fitness is not simply a property of inversion heterozygosity, intrinsic to the zygotic combination *per se*, but must have been acquired and perfected under specific local ecological conditions through natural selection. Each complex from a local population differs in genic contents from every other so that the response from mixing complexes is indeterminate and unpredictable. However, a balanced system can evolve and often does, apparently when genetic combinations are available to promote it. The principle of coadaptation established by these experiments of Dobzhansky recalls the principle of relational balance between polygenic systems proposed by Mather in 1953 (see Chapter 7). Together these principles helped form the concept of balanced genetic structure in natural populations of outcrossing species—high levels of heterozygosity and balanced linkages (see Chapter 18).

EXAMPLE 15-3

MESOEVOLUTION IN POPULATION HYBRIDS

A fundamental test of the coadaptation principle was made by Dobzhansky and Levene (1951) and by Dobzhansky and Pavlovsky (1953), when they initiated six experimental populations of *Drosophila pseudoobscura* with strains homokaryotypic for *ST* from Piñon Flats, California, and an equal number of strains for *CH* from Chihuahua, Mexico. (Experiments of *AR* from Piñon and *CH* from Chihuahua also were done, but we give details only of the *ST-CH* experiments because the *AR-CH* experiments illustrate the same principles). All populations were thus provided with approximately equal amounts of genetic heterogeneity in a set of interracial hybrid genotypes.

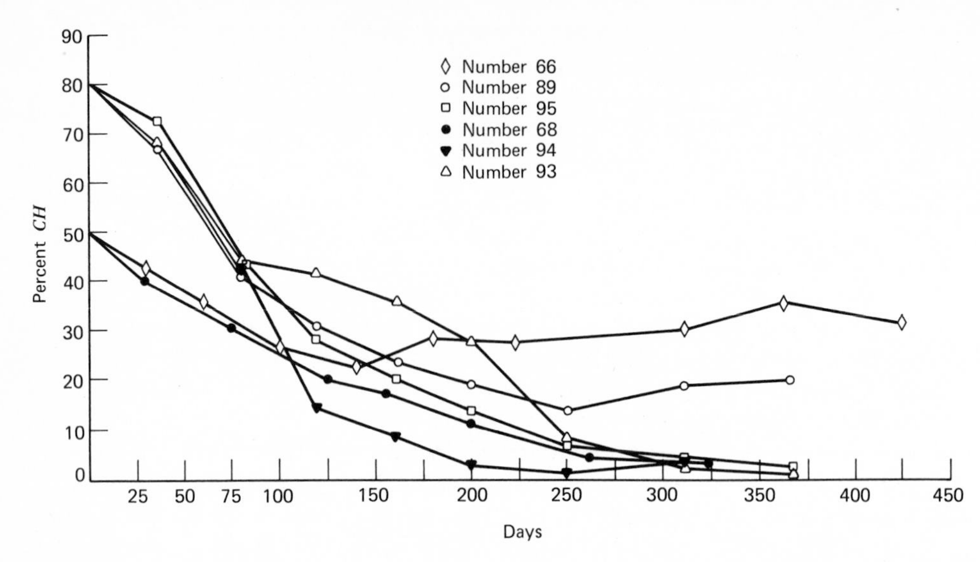

Figure A. Frequencies of CH observed in six experimental populations of *Drosophila pseudoobsura* containing ST from Piñon Flats, and CH from Mexico. From Dobzhansky and Pavlovsky, 1953.

Sets of populations were initiated as follows where superscript C = California, M = Mexico; 12 strains from each locality were included:

Date Initiated	Population No.	Initial Flies Each Population
November 1949	66	2000 ST^C/CH^M
October 1951	68	
November 1951	89,93, 94	900 CH^M/CH^M + 600 ST^C/CH^M
December 1951	95	

Results of the six population-cage experiments are presented in Figure A. These populations with interpopulational hybrids produced diverse results quite in contrast with the uniformity expressed in Example 15-2, where flies from a single locality were tested. Two populations came to a stable equilibrium (66 and 89), although their equilibrium frequencies were different, while the other four populations' selection curves indicate that CH would have been eliminated in time.

From the selection curve of population 66, it is evident that an abrupt change in fitness values of karyotypes must have taken place at about the sixth generation. Before then, that population, along with No. 68, proceeded in parallel with no significant difference between them. Approximate W was calculated for these segments of the selection curve (66) as follows:

	CH/CH	ST/CH	ST/ST
W up to G_6	0.50	1.00	1.00
W after G_6	0.30	1.00	0.65

In short, heterokaryotype superiority had developed, particularly with a lowering of *ST*/*ST* relative fitness and a lesser change in that of *CH*/*CH*.

In population 68, as well as 94 and 95, there was fairly continuous rise in *ST* with probable eventual elimination of *CH*. In 93 at about the third generation, an abrupt change took place as if a temporary equilibrium were going to be established; within another four generations, it was clear that *ST* would probably eliminate *CH* in that population also.

Some evidence indicating that the change in population 66 developed from an improvement in preadult viability in the *ST*/*CH* karyotype is presented in Example 15-8.

The main result is the diverse outcome for selection in contrast with the uniform and predictable outcome for populations with flies from a single natural population. Thus, under laboratory conditions when individuals were hybrids with genotypes of racially mixed origin (Californian versus Mexican), entirely new and unpredictable genetic complexes must have evolved uniquely, via different routes in at least three of the six populations.

FITNESS VALUE CHANGES AT SPECIFIC FREQUENCIES IN POPULATION CAGES

In testing for the selection outcome from chromosomal arrangements taken from a single population, Spiess found (1957) that while chromosomal frequencies tended to come to a selective equilibrium in *Drosophila persimilis*, the approach to that equilibrium was not interpretable based on a single set of fitness values, and therefore those values must be frequency dependent, as described in Example 15-4. The Levene-Haldane method of fitness estimation was used, but it is clear simply from the selection curves of the populations involving *WT* arrangement that no single set of values would fit as well as two sets of values in sequence. This work led Spiess to examine the karyotypes from the natural population in terms of fitness properties with entirely different techniques (see section on measuring fitness from analysis of life cycle, below).

Further extensive tests (Pavlovsky and Dobzhansky, 1966) for selective outcome among laboratory populations of *D. pseudoobscura* chromosomal variants taken from the Yosemite region (Mather, California) revealed that coadaptive relationships between the variants are far more complex than had been evident from the earlier studies. Populations containing more than two variants were analyzed, and relative adaptive values, despite low precision of measurement estimates, indicated that karyotype fitness in multichromosomal populations was often unpredictable from fitness estimates derived from the two-chromosomal populations. But more important, these workers found when analyzing some of the data (*ST* and *AR* populations raised at cool temperatures) that the outcome of selection depended on the *frequencies* of those variants when the populations were initiated. Replicate populations started at 70 percent *ST*, 30 percent *AR* maintained or slightly increased in *ST* to 75 percent over about 8 generations, while replicate populations initiated at 30 percent *ST* rose slowly over about 18 generations to 50 percent *ST*. The latter populations did not appear to be converging on the equilibrium of 75 percent *ST*. The different outcome may have depended on the starting frequencies, although differences in initial genetic contents cannot be ruled out.

The principle that fitness of chromosomal variants might often depend on their frequency in the population was reinforced by observations made by Tobari and Kojima (1967) using inversions on second and third chromosomes of *Drosophila ananassae*. Some of their data are examined in Example 15-5. While a balanced equilibrium outcome for selection seems obvious from the frequency changes observed, there was considerable discrepancy between W estimates from replicate populations containing the same arrangements but opposite initial frequencies ($q_0 = 0.10$ versus $q_0 = 0.90$). Consequently, these authors proposed that the model of constant fitness over the range of frequencies was undoubtedly in error, and they proceeded to examine the estimated W values generation by generation by directly counting chromosomal variant (karyotypes) numbers taken out of the populations. This analysis (described in the section below on measuring fitness by life-cycle components and in Examples 15-6 and 15-7) revealed that when an inversion was rare in the population, it had an advantage that it did not have when it was common. Such minority selective advantage could, of course, lead to a balanced equilibrium with or without essential heterokaryotype superiority throughout the range of frequencies (see Chapter 16).

EXAMPLE 15-4

FITNESS CHANGE AT SPECIFIC VARIANT FREQUENCY

Specimens of *Drosophila persimilis* were collected by Spiess (1957) at an 8000-foot elevation in the Yosemite Park region where third-chromosome arrangements occur in the following frequencies: *WT* (Whitney) 87 percent, *KL* (Klamath) 7 percent, *MD* (Mendocino) 3 percent, *ST* (Standard) 2 percent, and *SE* (Sequoia) 1 percent. The first three were tested by population-cage technique for overall fitness with two at a time put into competition, using constant laboratory conditions with cool temperatures (15°C). Figure A indicates the selection curve in percent *KL* when interbreeding with *WT*, averaged for

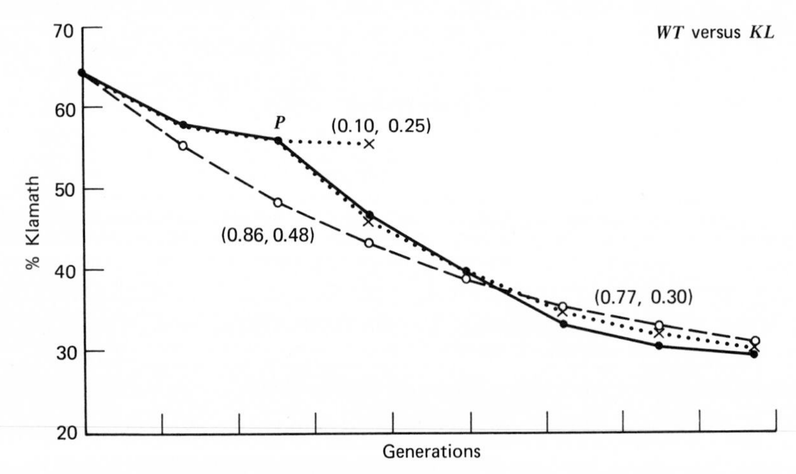

Figure A. Frequencies of *KL* observed in three experimental populations containing *WT* and *KL* of chromosome 3 from *Drosophila persimilis* from Yosemite; averaged (solid line) over nine generations. Single constant set of W (WT/WT first, KL/KL second, assuming $WT/KL = 1$) dashed line. Set of W values changing at point P, dotted line (from Spiess, 1957).

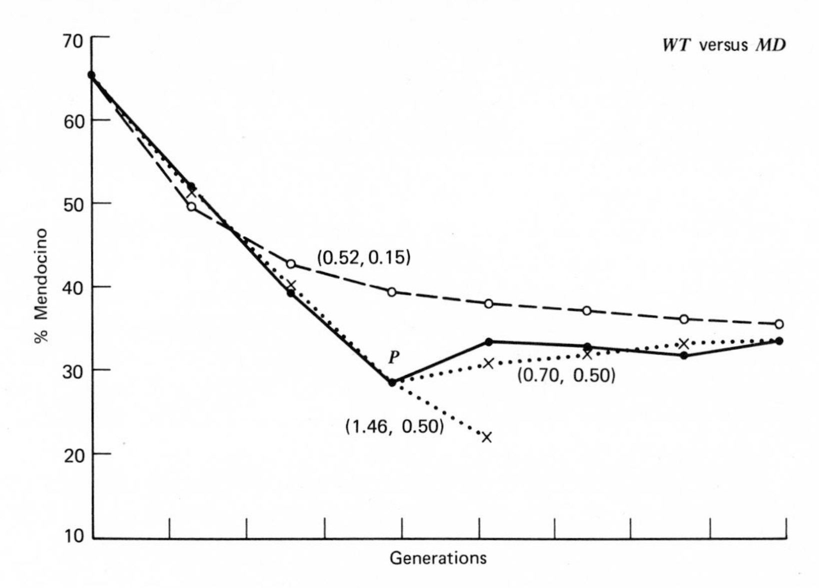

Figure B. Frequencies of *MD* observed in two experimental populations with *WT* and *MD*, averaged over nine generations (solid line). Single set of *W* (*WT/WT* first, *MD/MD* second) dashed line. Set of *W* changing at point *P*, dotted line.

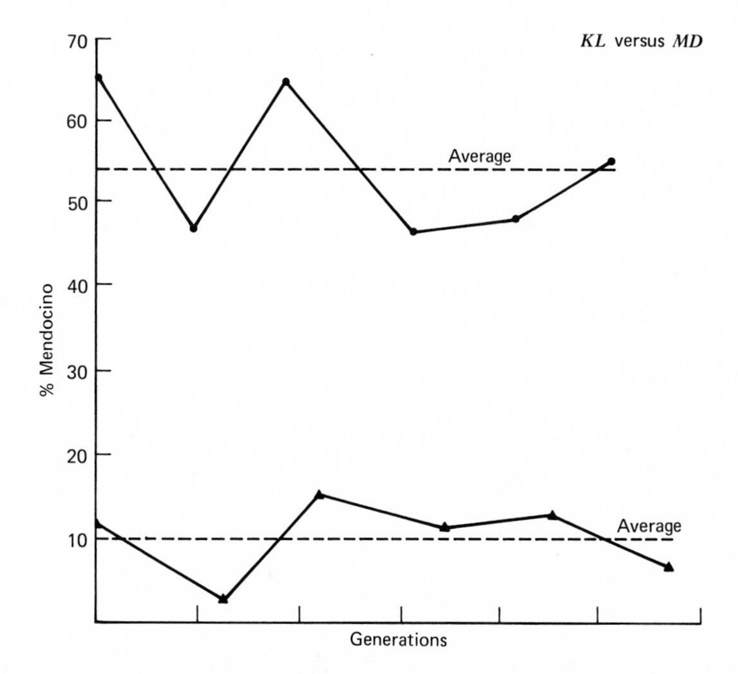

Figure C. Frequencies of *MD* observed in two populations containing *KL* and *MD*, over five to six generations.

three replicate populations; Figure B gives the average for two replicate populations in percent *MD* interbreeding with *WT*; Figure C indicates observed changes in two populations of the rarer arrangements *KL* and *MD* (without *WT*).

Fitness values for the three karyotypes in populations of *WT* versus *KL* and in populations of *WT* versus *MD* were calculated using the method of Levene and Haldane outlined in the text. Single-solution selection curves are given as dashed lines in both Figures A and B; in both cases, while a stable state is reached, the approach to that state does not follow the sequence fitted by the single constant fitness set.

The most likely explanation for the abrupt changes at points marked *P* in each case was that the fitness value of a karyotype was dependent on its frequency. In Figures A and B, the best-fitted values (with selection curves given as segments of dotted lines) are those involving changes at point *P*; *W* is given in order of *WT*/*WT*: *KL*/*KL* (or *MD*/*MD*), with heterokaryotype fitness assumed to be 1.00. In Figure A *KL*/*KL* seemed superior to *WT*/*WT* above 55 percent *KL* frequency, but *WT*/*WT* fitness increased sevenfold at that point. In Figure B, *MD* declined very rapidly, as if it were being eliminated, until it reached about 28 percent *MD* when it "recovered." In both cases, the significant shift in *W* value occurred probably in the *WT*/*WT* karyotype relative to the heterokaryotype since the homokarytype of rarer arrangements did not change in fitness very much. It was expected that *WT*/*WT* would be superior to the other homokaryotypes because it is most common in the wild population, but possibly selection may have perfected a phenotype for that karyotype only when it occurs in large enough numbers. When *WT* was less common in the laboratory populations, its karyotypes indicated neither superiority of *WT* (interbreeding with *KL*) nor balanced polymorphism (interbreeding with *MD*). To achieve fitness related to its adaptedness in the natural population, it was necessary to bring the frequencies of karyotypes close to those found in that population.

In contrast, *KL* and *MD*, the rarer arrangements, probably seldom occur in heterokaryotypes in that natural population and thus have little opportunity to become coadapted. When interbreeding without *WT* (Figure C), no selective changes took place within the six generations sampled; from the cages in which they separately were breeding with *WT*, they obviously did not appear to be equal in fitness. Consequently, relative to each other, the three karyotypes *KL*/*KL*, *KL*/*MD*, and *MD*/*MD* were dependent on the presence of *WT* to effect fitness differences.

EXAMPLE 15-5

FREQUENCY-DEPENDENT FITNESS

A useful technique employed to test for dependency of fitness values on frequency of genotypes in experimental populations is simply to initiate populations with contrasting frequencies. If fitness values are always constant and independent of frequency, respective selection curves should be predictable, based on the fitness values estimated from the opposite starting frequency. A large number of laboratory populations with many replicates were studied in this way by Tobari and Kojima (1967) and Kojima and Tobari (1969*a*) using two sets of segregating inversions in *Drosophila ananassae*, a set on the left arm of chromosome 2 and the second set on chromosome 3. Changes in frequencies of inversions on each chromosome were followed with the nonhomologous chromosome's arrangement held constant and some of the genetic background controlled as well. Here we illustrate only the changes in frequencies of $A:B$ inversions of chromosome 2(L) holding the chromosome-3 arrangement constant, from the earlier report of these authors, because the changes in frequency $A:B$ inversions of chromosome 3 were similar. (We shall mention this again in Example 15-7 and under frequency dependency in Chapter 16.)

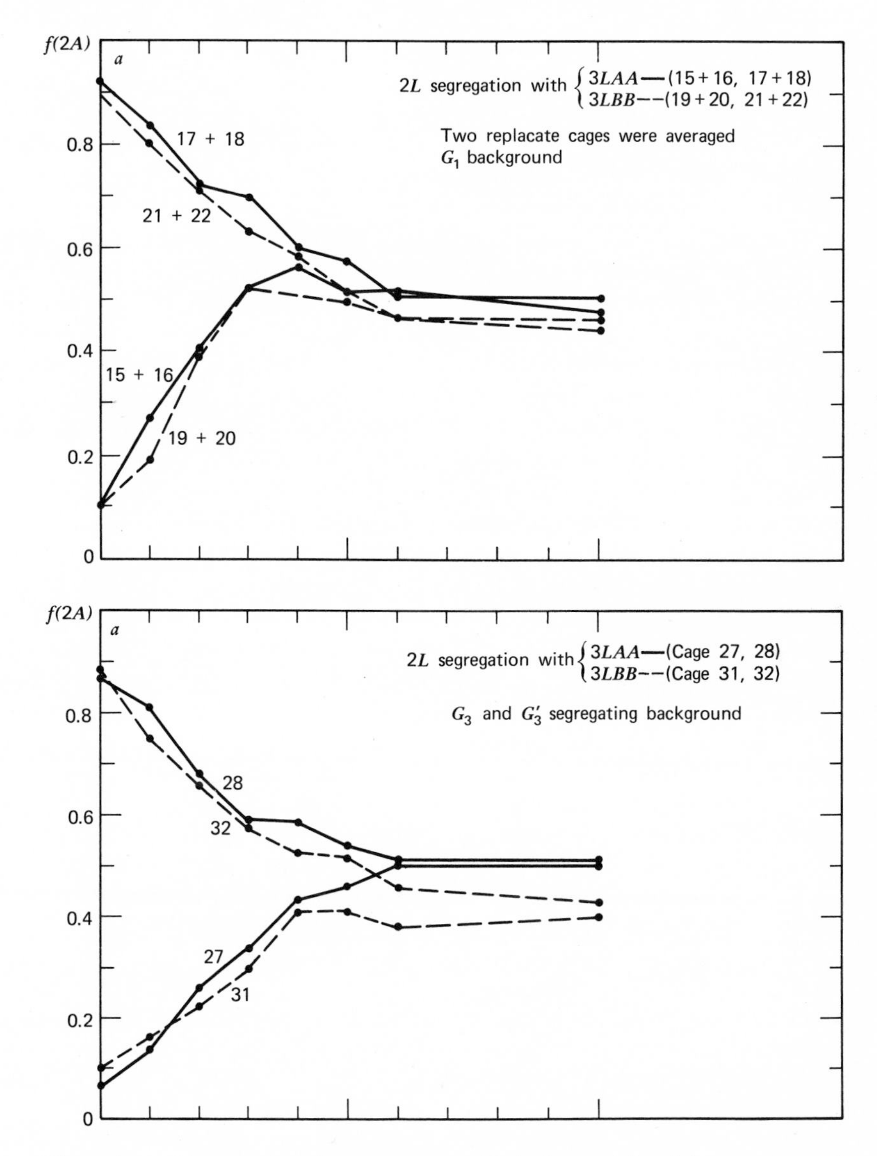

Figure A. Frequencies of arrangements $2LA$ observed in eight experimental populations containing $2LA$ versus $2LB$ of *Drosophila ananassae* over 10 generations (third chromosomal background was controlled) (from Kojima and Tobari, 1969*b*).

Eight populations were initiated as follows. First, all populations started with 1000 flies, 900 homokaryotypic for one inversion and 100 for the other. Second, the four different "treatments" were essentially a 2 × 2 design where the chromosome-3 background was either A or B inversion (constant) and the starting frequency of chromosome-2 inversion was either 90 percent A or 10 percent A. Third instar larvae were sampled directly from the population cages for salivary gland chromosome analysis. The changes in frequencies of these eight populations are presented in Figure A. Estimates of fitness were made first from the rate of change in each population by using Wright's least-squares method to get a survey of selection coefficients (s_1 and s_2), assuming constant fitness values in the "classic" model: $W_{AA} = 1 - s_1$, $W_{AB} = 1$, $W_{BB} = 1 - s_2$. From initial W values, chi-square tests comparing expected selection curves to those observed were run using a computer program for minimizing the deviations between observed-expected. Successive new estimates of W (= iteration) gave a better fit of observed data until chi-squares were minimized.

As was evident from the frequency changes, a stable equilibrium was attained in all eight populations with the A inversion ($2L$) at about 50 percent with $3L$-AA background and slightly below 50 percent with $3L$-BB background. In Table A, the fitness estimates are given relative to $W_{AB} = 1$. We note that out of the eight pairs of replicate fitness estimates (rep 1 versus rep 2), five are reasonably close; one pair is very different (0.72 versus 0.32 estimating W_{BB} from 0.9 initial q_A with $3L$-AA background), while the other two pairs (W_{AA} at $q_A = 0.1$) are different by about 0.20. This lack of agreement may cause some lack of confidence in the use of those estimates, but it can be seen that the frequency changes in the two replicates in the upper part of Figure A show the largest discrepancy. If the uppermost curve (17 and 18) is considered out of agreement because of slight sampling or laboratory error (making a little plateau at the third generation), the curve could be smoother and would agree better with its replicate. The W_{BB} would be put at about 0.40 instead of 0.72. Assuming the other slight discrepancies also would fit better by smoothing the observed curves, we may look at the fitness estimates in Table A as four blocks. The two blocks of fitness for W_{AA} do not change much between starting frequencies $q_A = 0.10$ and 0.90. However, in contrast, the W_{BB} fitness estimates do change; all of them are greater when their starting frequency is less—when $q_A = 0.90$. Thus, there appears to be an advantage to B when it is rare compared to its fitness when common (see Example 15-7).

TABLE A *Estimates of fitness values computed by minimum chi-square procedures under the assumption of constant fitness over generations. $W_{AB} = 1$. Data from Tobari and Kojima (1967): inversions A and B on chromosome 2L, only. Rep = replicate populations at same starting frequencies*

		W_{AA}		W_{BB}	
Input Frequency q_A	Fixed on Chromosome $3L$	Rep 1	Rep 2	Rep 1	Rep 2
0.10	A	0.43	0.26	0.24	0.17
	B	0.20	0.43	0.22	0.25
0.90	A	0.49	0.40	0.72	0.32
	B	0.42	0.43	0.53	0.62

ESTIMATING FITNESS VALUES FROM COUNTING GENOTYPES BEFORE AND AFTER SELECTION

The problem of relating a genic unit (allele at one locus) or a cytological variant with some fitness value relevant to its adaptedness for a particular population need not be solved only by measuring its rate of change over several generations, although that measure may give us an overview of its adaptive function for the conditions under which we find the population. Any attempt to measure relative fitness under conditions different from those the population has been experiencing (laboratory environment) runs the risk of simply finding a unique result not relevant to the organisms' adaptedness in nature. One available technique merits some exploration—the one by which the main life-cycle components itemized briefly in Table 15-1 may be surveyed for their contributions to fitness. For example, we may culture the genotypes (morphotypes or karyotypes) under specified conditions that may be experimentally varied. Starting at the zygotic stage, with two or more genotypes suitably marked, we may monitor the "early" and "late" components of the life cycle for developmental ability, rate of development, viability, competitive ability, mating, fecundity, and so forth. Perhaps one or a few of the components will show marked differences among the genotypes, and we may acquire some clue as to the critical component (or components) the genotype may control. If that component is relevant to the adaptedness of the genotypes in the population, we may attempt to manipulate the system by selecting in the following way and predict the outcome based on the life-cycle component differences we have discovered. Under population conditions where the monitored genotypes are interbreeding and segregating with genetic background recombining independently, the fitness property can be selected in diverging directions with the usual linear techniques to see whether the monitored genotypes respond. If they do in the expected direction, we can be more confident that those genotypes do in fact control the fitness property they seemed to have when tested separately. This method of analysis is illustrated in Example 15-6 to pinpoint the fitness components that may be controlled by genic contents of certain chromosomal variants in *Drosophila persimilis*. After testing of major life-cycle components, Spiess found karyotypes differing mostly in rate of development in preadult stages and mating propensity of adults, but with slight heterosis in viability and female fecundity. Confirmation of control over development rate was made by selecting for that component and obtaining a resultant change in frequencies of karyotypes in the expected direction. (See also B. Clarke, 1975, for experimental verification that segregating variants of alcohol dehydrogenase in *D. melanogaster* can be selected at will by varying concentrations of alcohol in food medium.)

A useful technique that ought to assess fitness values efficiently employs the counting of genotypes each generation. Unfortunately, unless one knows when in the life cycle selection has acted, this procedure may produce erroneous or misleading results, as we shall consider in the next section. Example 15-7 illustrates the method and the difficulties involved. If it is surmised that selection has acted before a sample is taken each generation, we may estimate fitness values based on the $\bar{W}$ relation of Chapter 14:

$$p^2W_1 : 2pqW_2 : q^2W_3 = D:H:R \quad (15\text{-}3)$$

where D, H, and R are frequencies of genotypes (or karyotypes) sampled *after* selection has acted so that $[(D + H + R)/\bar{W}] = 1.00$, and p^2, $2pq$, and q^2 are the frequencies expected *before* selection. A set of fitness values for the transition from generation t to $t + 1$ can then be estimated by

$$W_1 = \frac{D}{p^2}, \qquad W_2 = \frac{H}{2pq}, \qquad W_3 = \frac{R}{q^2} \tag{15-4}$$

The details of selective action would be known better if the frequencies before selection were determined with some confidence. Genotypes should be monitored at least two or three times during the life cycle. If selection is known to be acting in a certain interval, then the most efficient estimates of W are to be obtained from the maximum-likelihood method as given in (15-4) (see Appendix A-9 and Exercise 12). If three genotypes enter a population in which they are then exposed to selective action so that they emerge in differential frequencies as given in (15-3) normalized (dividing through by $\bar{W}$), we may estimate the relative fitness values in this way: (15-4) estimates W in such a way that $D + H + R = 1$. We may readjust the W values relative to W_2 to reduce the number of parameters to be estimated from three to two. Then we have the maximum-likelihood estimates

$$\frac{W_1}{W_2} = \left(\frac{D}{p^2}\right)\left(\frac{2pq}{H}\right) = \left(\frac{2D}{H}\right)\left(\frac{q}{p}\right) \quad \text{and} \quad \frac{W_3}{W_2} = \left(\frac{R}{q^2}\right)\left(\frac{2pq}{H}\right) = \left(\frac{2R}{H}\right)\left(\frac{p}{q}\right)$$

so that

$$W_1 = \left(\frac{2D}{H}\right)\left(\frac{1}{u_0}\right) \quad \text{and} \quad W_3 = \left(\frac{2R}{H}\right)(\mu_0) \tag{15-4A*}$$

where each W is relative to $W_2 = 1.00$ and $u_0 = p_0/q_0$.

Ideally, if a genotype could be ascertained at all stages of the life cycle, it could be monitored for these fitness values from one stage to the next so that the exact level at which selective differences are operating could be pinpointed. For the student to be reassured that these W values could be obtained from observed data, Exercises 10, 11, and 13 are provided. Examples 15-8 and 15-9 illustrate the utility of these formulas, giving cases where the time of selective action is fairly certain.

EXAMPLE 15-6

SELECTION AT STAGES OF THE LIFE CYCLE

Spiess and colleagues tested *WT* and *KL* karyotypes of *Drosophila persimilis* from a Yosemite locality (8000 feet) by breeding each karyotype separately and measuring fecundity, viability, rate of development, and mating activity in various amounts of detail (1958; Spiess and Langer, 1961, 1964*b*; Spiess and Schuellein, 1956; Spiess and Spiess, 1964, 1966;

* From maximum-likelihood estimates variances are the negative reciprocals of the second derivative of the likelihood function (see Appendix A-9). Variances of fitness estimates are as follows:

$$V_1 = \frac{\bar{W}(W_1 p + 2q)W_1}{2p^2qN} \quad \text{and} \quad V_3 = \frac{\bar{W}(W_3 q + 2p)W_3}{2pq^2N}$$

where N = observed sample size, V_1 = variance of W_1, and V_3 = variance of W_3. Covariance of W_1 and $W_3 = \bar{W} W_1 W_3/2pqN$. Standard error of W then is the square root of the variance.

and Appendix to Ehrman, 1966). In all these studies, the same strains of *WT* and *KL* were intercrossed to simulate wild-type flies from the natural population with high genic heterozygosity but constant for karyotype in order to discover consistent fitness differences between the karyotypes as units, especially when grown under the conditions of laboratory experimental populations (as in Example 15-4).

Heterokaryotype superiority was found to a significant degree in female egg-laying capacity and in preadult viability at high larval density, as given below in the average data from several replicates (Spiess, 1958):

Karyotype	Mean Eggs/Day/♀	Average Preadult Survival (%) At Low Density	At High Density
WT/WT	18.9	37.3	41.0
WT/KL	21.1	48.3	47.8
KL/KL	19.7	50.0	38.4

In time of development (days from egg to adult), the predominant homokaryotype (*WT/WT*) took less time (24.9 days) by one and one-half days than the *WT/KL* (26.4 days), which was in turn nearly two days shorter than *KL/KL* (28.2 days) on the average, so that the rate of development control by these chromosomal arrangements was "additive."

To evaluate the extent to which the rate of development may contribute to the adaptedness of these chromosomal arrangements (genic complexes) in the laboratory populations, experimental populations were designed in which rapid or slow development was favored by selection. Populations with best results were maintained by collecting all emerging adults each generation and selecting the first or last fraction from the emerging distribution. Figure A shows the frequencies of *KL* (versus *WT*). Earlier populations (*F* and *S*) only selected about the first half of emergents and did not show any differences in elimination of *KL* chromosomes (500 adults selected per generation). With greater intensity (*i*) of selection, at 100 adults per generation, however, the slow development population (*iS*) did increase in the *KL* arrangement, in contrast with the fast (*iF*). Then, at third and fifth generations, respectively, the *iS* and *iF* populations were back-selected (in reverse of their original selection regime) with significant though asymmetrical responses. *WT* increased faster by selecting early-emergence flies from the slow population (*iS*) than it decreased by selecting last emergences in the fast population (*iF*). Those results were interpreted as evidence (1) that a real difference in developmental rate did exist under population-cage conditions between these karyotypes even though the trait must have had low heritablity, (2) in spite of recombination in the genetic background the control was maintained even though genetic change probably occurred to make the back-selection responses asymmetrical, and (3) in the less intense selection populations, uniformity of *KL* decrease indicated that other fitness properties may have been of more critical value to total fitness.

One additional fitness property turned out to be the mating activity of adults. In initiating all these populations for development rate selection, only homokaryotype flies had been used for parents. In every case, a discrepancy was found between introduced proportions of *WT* and *KL* arrangements and their proportions among first-generation larvae removed for salivary gland analysis. Frequencies among introduced adults (from

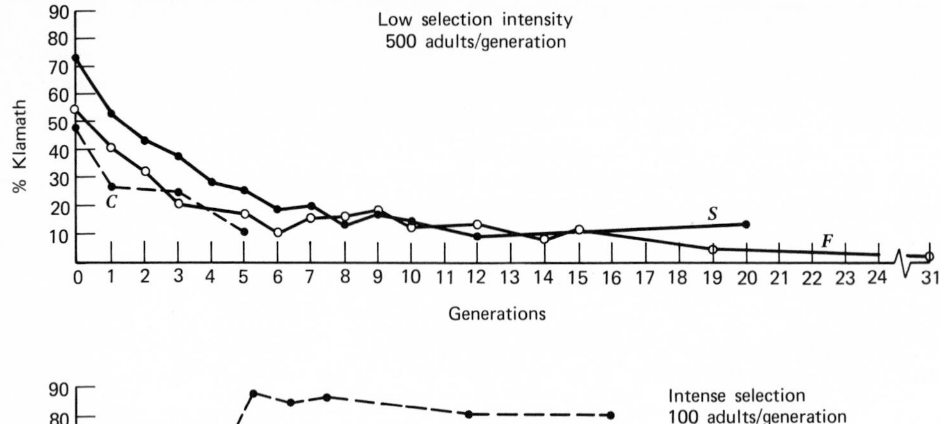

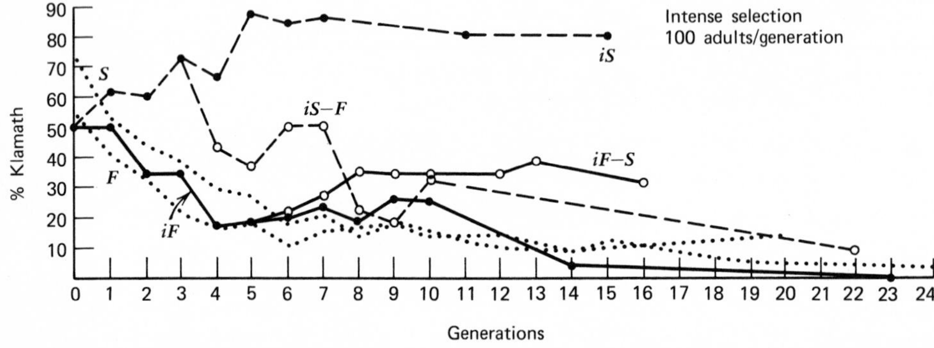

Figure A. Above: observed frequencies of *KL* in two populations of *Drosophila persimilis* from Yosemite, selected for fast (*F*) or slow (*S*) development with low intensity by utilizing just 500 adults per generation selected out of about 2000 emerging flies. Below: *KL* frequencies in four populations selected with greater intensity (100 adults per generation). *iF* (intense fast, solid line with black dots), *iS* (intense slow, dashed line with black dots), *iF-S* (*iF* back-selected at G_5, solid line with open circles), *iS-F* (*iS* back-selected at G_3, dashed line with open circles). Dotted lines are *F* and *S* populations from above for comparison (from Spiess and Spiess, 1966).

600 to 1000 per population) and the resulting G_1 larvae from egg samples (100 larvae per population cultured separately from the population cages in low-density food bottles) were as follows.

Population Number	Introduced Parents		G_1 Larvae from Egg Sample		
	WT/WT	*KL/KL*	*WT/WT*	*WT/KL*	*KL/KL*
FF	0.204	0.796	0.14	0.39	0.47
SS	0.208	0.792	0.11	0.53	0.36
FF – 2	0.251	0.749	0.25	0.58	0.17
SS – 2	0.248	0.752	0.28	0.48	0.24
DF	0.244	0.756	0.21	0.48	0.31
DS	0.250	0.750	0.03	0.47	0.50

The zygotic frequencies of F_1 larvae had a consistent deficiency of *KL/KL* karyotype. Differential viability and rate of development in the sample culture bottles before salivary analysis were ruled out, and therefore a mating activity difference in one or both of the sexes among parents was suspected of being the main cause of the discrepancy.

The student should note that the zygotic data agree fairly well with the expectation of the square law for their own (larvae) haploid frequencies. For example, in *SS* – 2 $(q_{WT})' = 0.52$ $(p_{KL})' = 0.48$ (where $'$ indicates G_1 frequencies). On squaring, "expected" values would be $WT/WT = 0.270$, $WT/KL = 0.499$, $KL/KL = 0.230$, which are not significantly different from the values observed. Except for *FF*-2 and *DS* zygotic data, which do not fit as well, the populations fit their own squared haploid frequencies. The square law can only fit if gametic frequencies are effectively the same in both sexes (Chapter 1). If we assume that a mating activity (m) increment has occurred in both sexes of WT/WT with respect to KL/KL, similar to the expression (15-1) in the text, we may estimate m easily since only two parent karyotypes are involved (also formula (14-22).

Let p and q denote proportions of KL/KL and WT/WT parent flies introduced into the population cages, and let the mating activity be $1:m$, respectively. Expected proportions of G_1 larvae will be the terms of

$$(p + mq)^2 : \begin{array}{ccccc} KL/KL & & WT/KL & & WT/WT \\ p^2 & + & 2mpq & + & m^2q^2 \end{array}$$

(p is chosen for the KL symbol to conform with changes in Example 15-4). In the larval generation (G_1), the new p value is

$$p_1 = \frac{p^2 + mpq}{(p + mq)^2} = \frac{p}{p + mq}$$

which we solve for m

$$m = \frac{p}{p_1 q} - \frac{p}{q} = \frac{q_1}{p_1}\frac{p}{q} = \frac{u_0}{u_1}$$

where $u_0 = p/q$ and $u_1 = p_1/q_1$

Variance of m, from the maximum-likelihood estimate, is given by Li in the Appendix to Spiess and Langer, 1961.

$$V(m) = \frac{u_1(1 + u_1)^2}{2Nu^2}$$

with the square root = standard error of m.

In the first two populations (*FF* and *SS*), the initial chromosome frequencies are nearly equal, and the G_1 larvae are very nearly so also (so that averaging will not involve weighting). Therefore, $p_{KL} = 0.794$, $q_{WT} = 0.206$, and $u_0 = 3.854$. Pooled G_1 gives (for 200 larvae):

KL/KL	WT/KL	WT/WT
83	92	25

and $u_1 = 1.817$, $m = 2.12$, $V(m) = 0.0491$, standard error $(m) = 0.222$. Therefore, WT/WT flies had effectively achieved parentage by a factor greater than twice that of KL/KL flies. Further tests for mating activity (Spiess and Langer, 1964) confirmed these results and further indicated that heterokaryotype WT/KL mated at intermediate speeds.

EXAMPLE 15-7

ESTIMATING FITNESS BEFORE AND AFTER SELECTION

When Tobari and Kojima (Example 15-5) found that fitness values (W) calculated from populations with opposite initial frequencies of *D. ananassae* inversions were not in agreement, they treated their observed data from generation counts in a new way. Instead of assuming constant W, they proposed estimating W from observed proportions of the three karyotypes each generation, using formula (15-4). The expected frequencies each generation were those expected from the squaring of haploid frequencies of the A and B inversions of the previous generation. A technical problem was the sampling error fluctuation around a smooth selection curve, and they chose first to generate the smooth curve (assuming constant selection coefficients as mentioned in Example 15-5) so that p_A, q_B was read from the point for generation t on the smooth curve. They then applied formula (15-4). Fitness values were estimated for six transitions between generations. For example, from $t = 0$ to $t = 1$, with chromosome $2L$ segregating in $3LAA$ constant background, the data from opposite starting frequencies were as follows (two replicate populations combined):

$p_A = 0.10$	W_{AA}	W_{AB}	W_{BB}
Observed No. / Expected No.	19 / 3.5	125 / 58	156 / 238.5
W estimate	5.70	2.17	0.66
σ_W	± 0.53	± 0.12	± 0.03
$p_A = 0.90$			
Observed No. / Expected No.	206 / 247.7	87 / 49.7	7 / 2.6
W estimate	0.83	1.79	2.63
σ_W	± 0.03	± 0.13	± 0.67

Most of the other estimates showed this trend for higher fitness to characterize the less frequent karyotype—that is, for the minority type to be advantageous. As equilibrium was approached, however, these fitness differences were erased with a slight though nonsignificant superiority of heterokaryotypes in many cases. For the two populations given above, the W estimates at the sixth transition (G_5 to G_6) were as follows

From initial $p_A = 0.1$:

	W_{AA}	W_{AB}	W_{BB}
Observed No. / Expected No.	65 / 86.5	183 / 149.1	52 / 64.3
W estimate	0.75	1.25	0.81
σ_W	± 0.09	$+0.06$	$+0.11$

From initial $p_A = 0.9$:

	W_{AA}	W_{AB}	W_{BB}
Observed No. / Expected No.	71 / 94.9	169 / 147.7	60 / 57.4
W estimate	0.75	1.14	1.05
σ_W	±0.09	±0.06	±0.12

Tobari and Kojima stressed that there were indeed changes in W values, so that the "classic" model was not holding true; the equilibrium was achieved by the minority advantage, which was reduced on approach to the equilibrium frequency.

Their observation that there was no significant heterosis proved to be erroneous, however. Prout (1969, 1971*a*, *b*) pointed out the error of counting genotypes before selection had been completed (see next section). These investigators then extended their work on the problem of estimating fitness (Kojima and Tobari, 1969*a*) to clarify and improve their methods when they discovered an allozyme marker locus (acid phosphatase, ACPH) within the chromosome $2L$ inversion system, so that they could monitor at least two stages—larvae and adults. With the three genotypes being distinguished by electrophoretic banding, the adults of both sexes in the populations displayed an excess of heterozygotes in every generation when compared with square-law expectations.

Egg-to-adult viability differences were estimated from separate experiments done outside the population cages, using the ACPH identification markers. Females of known inversion types (AA or BB extracted from population cage 1) were premated so that input proportions were controlled for frequency of fertilized eggs. All culture bottles were maintained under crowded conditions similar to those of the population cages. All tests used 100 AB virgin females outcrossed to the three different karyotypes of males. Total adult numbers of progeny tested for ACPH allozymes emerging from three of these crosses were as follows:

	Authors' ♀♀ × ♂♂		Karyotypes AA	AB	BB	N (sampled)
III	$AB \times BB$ (10) $AB \times AA$ (90)	Input	0.45	0.50	0.05	
		Obs. total	254	306	40	600
		Viability	0.92	1.00	1.11	
IV	$AB \times AB$ (100)	Input	0.25	0.50	0.25	
		Obs. total	112	280	105	497
		Viability	0.80	1.00	0.75	
V	$AB \times BB$ (90) $AB \times AA$ (10)	Input	0.05	0.50	0.45	
		Obs. total	32	285	258	575
		Viability	1.12	1.00	1.01	

It is clear that minority genotypes were favored under the two opposite (III and V) input conditions. (The student may verify the significance of differences in W estimates by using the footnote to (15-4A), taking the square root of the variance as the standard error.) That result means that survival rate depended on the frequency of the inversions in the population. However, when nearing equality (IV, which is about the equilibrium point in the population cages), the heterokaryotype was most fit. Heterosis then was superimposed on the minority advantage effect. Fitness was not constant over the array of frequencies in which the population was found, and equilibrium was not achieved by a single set of fitness values.

EXAMPLE 15-8

PREADULT VIABILITY MESOEVOLUTION IN POPULATION HYBRIDS

Along with the population-cage sequence of changes in frequencies of *ST* and *CH* arrangements in *D. pseudoobscura* illustrated in Example 15-3, Dobzhansky and Levene (1951) also tested preadult viability in their population with mixed origin (California and Mexico). We utilize formula (15-4A) with the data given below. The three segregating karyotypes from the interpopulational cross were obtained by sampling G_1 offspring from 2300 ST^C/CH^M parents introduced into a population cage. Emerging virgin females (120) and males (140) were crossed to a known CH/CH stock to ascertain the adult G_1 karyotypes by salivary chromosome analysis from the outcross progeny larvae. If there were no selective differences in preadult components of fitness, the familiar 1:2:1 ratio would be expected among progeny karyotypes. The following data were obtained:

	CH^M/CH^M	ST^C/CH^M	ST^C/ST^C	Total
Observed ♀♀	25	48	47	120
Relative W (using 4′)	1.04	1.00	1.96	
Observed ♂♂	32	74	34	140
Relative W	0.86	1.00	0.92	

It is clear that the sexes differ; among females the *ST*/*CH* karyotype is inferior to *ST*/*ST*, while among males there is no significant difference between those karyotypes. In both sexes, however, there is no significant difference between *CH*/*CH* and *ST*/*CH*. With such superiority in *ST*/*ST* females, if the preadult viability were the only fitness component of relevance to the population, it would be expected that CH would eventually be eliminated.

After nearly one year from starting, when the 66 population had reached and maintained equilibrium for about 10 generations, a test was again made for viability of preadult stages similar to the test made at the start. Flies hatching in the cage were isolated and inbred to produce 10 *ST*/*ST* strains and 10 *CH*/*CH* strains. Those strains were intercrossed to produce *ST*/*CH* heterokaryotypes, 2500 of which were put in a population cage to test for offspring conformity with the 1:2:1 ratio in similar manner to that done initially (outcrossing to *CH*/*CH* known flies in single pairs and checking salivary chromosomes of larvae within pairs). Results were as follows:

	CH^M/CH^M	ST^C/CH^M	ST^C/ST^C	Total
Observed ♀♀	34	117.	49	200
Relative W	0.58	1.00	0.84	
Observed ♂♂	35	117.	48	200
Relative W	0.60	1.00	0.82	

Here the sexes were impressively alike. But more important, the contrast with the relative viabilities at the start of the population cage was highly significant. Heterokaryotypes now had superior viability to both homokaryotypes in both sexes. That feature of relative fitness must have been acquired during the year's running of the population under the laboratory conditions.

EXAMPLE 15-9

FITNESS ESTIMATES FROM FIELD OBSERVATIONS

In nature, populations can be monitored for incidence of genotypes with good ascertainment of selection action when initial frequencies before selection can be made certain. Sheppard and Cook (1962) recorded field studies of the moth *Panaxia dominula* (see also Example 10-4). In regions where the moth does not generally occur, artificial colonies were founded. The one of interest at the Botanical Garden at Ness, Cheshire, England was segregating for the three genetic "units" *dominula* (typical homozygote), *medionigra* (heterozygote), and *bimacula* (rare homozygote). The founder population was started only with heterozygous eggs in 1959, supplemented with eggs from *medionigra* × *medionigra* matings the following year. Consequently, the expected frequency of the three genetic types was 1:2:1. The population at Ness was sampled about two hours each day in late June and July 1961, with the following captured: 36 *dominula*, 42 *medionigra*, 21 *bimacula*. This ratio does not conform to square law. The investigators ruled out differences in rate of emergence between these types. Differential survival between egg and captured adult could be measured by using formula (15-4A):

Letting

$$W_1 = \textit{dominula} \text{ viability} = \frac{2D}{H}, \qquad \text{since initial } p = q$$

$$W_2 = \textit{medionigra} \text{ viability}, \qquad \text{set} = 1.00$$

$$W_3 = \textit{bimacula} \text{ viability} = \frac{2R}{H}$$

We estimate

$$W_1 = \frac{2(36)}{42} = 1.71$$

$$W_3 = \frac{2(21)}{42} = 1.00$$

The *dominula* homozygote is clearly a better survivor than the other two types, which are equal to each other at about 58 percent in survivorship compared to *dominula*. In all localities where this moth was colonized, similar selection was found against *medionigra* and *bimacula* in egg to adult survivorship.

The stable polymorphism for these genetic units in a natural colony observed by these authors was ascribed to the counterbalancing effects of disassortative mating (Example 10-4) when the *bimacula-medionigra* unit is low in frequency and the survival disadvantage observed in these artificial colonies.

PITFALLS IN ESTIMATION OF FITNESS FROM GENOTYPE FREQUENCIES

A glance at Table 15-1 will convey the complexity of the problem in estimating the total fitness value of a genotype over the life-cycle components that may be affected in diverse ways. Let us consider that fitness may have two major components, first during the preadult stages of growth ("early" part of the life cycle) and then as a result of differential mating ability of adults ("late" part of the life cycle). These two components may contribute to total fitness of the genotypes as the product of their successive values because they represent proportions of selective change occurring between the fertilized egg stage in G_0 and the same stage in G_1.

It has been emphasized by Prout (1965) that conventional techniques for ascertaining fitness values by taking a census of adult genotypes in an experimental population each generation run the risk of counting the genotypes before selection is completed, thereby calculating by the usual methods values that may be in considerable error. We may see the principles involved and how the erroneous fitness values emerge by carrying out calculations when the population is being counted before selection is completed. We designate the relative fitness values from the early stage E_1, E_2, and E_3; those of the late stage are L_1, L_2, and L_3. The net fitnesses obtained by the products E_1L_1, E_2L_2, E_3L_3 will then correspond to W_1, W_2, and W_3, respectively, as the net contributions of the three genotypes to their progeny. These principles are illustrated in Table 15-3. In Table 15-3A, the frequencies of zygote genotypes at G_0 are given in line 1; if early stage selection occurs we would observe these frequencies weighted by E_i symbolized as D', H', R' in line 2; after later stage selection has acted, we observe frequencies weighted by L_i in line 3 symbolized as D, H, R since selection has been completed. In Table 15-3B, we use a numerical illustration for fitness components; assume that these net fitness values combine the action of early stage (E) viabilities: $E_1 = 0.72$, $E_2 = 1.00$, $E_3 = 0.80$ (heterozygotes being favored) with a late stage (L), let us say differences in mating ability: $L_1 = 1.00$, $L_2 = 0.72$, $L_3 = 0.63$ (heterozygotes about intermediate). If we further assume that these two components of total fitness have equal weight over the life cycle, then the amount of contribution from the three respective genotypes ought to be the products. These overall fitness (W_i) values correspond to the case of simple selection against the homozygous recessive in Table 14-3.

In Tables 15-3C and D, the magnitude of disparity is illustrated between the true fitness (W_i) and spurious fitness values (K_i) if a stage of the life cycle (young adults before mating) is counted before completion of the selection process. We begin with zygote frequencies at G_0 with $q = 0.90$ as in Table 14-3. Early stage viability differences (E_i) (line 2)

TABLE 15-3 *The model (modified after Prout, 1965) for two components (E and L) to each fitness value. If the census of genotypes is made in young adults before mating, it will be made at the time of "partial selection." Postselection = adults that have survived E and L stages of selection (they have mated). D', H', R' = frequencies among young adults in G_0 after E stages; D, H, R = frequencies after selection is completed; D'', H'', R'' = frequencies among young adults after a second round of partial selection (E stages) in G_1.*

A	G_0 eggs to G_1 eggs	AA	Aa	aa
	G_0: zygotes	p^2	$2pq$	q^2
	Partial selection	$p^2E_1 = D'$	$2pqE_2 = H'$	$q^2E_3 = R'$
	Postselection (after mating)	$p^2E_1L_1 = p^2W_1 = D$	$2pqE_2L_2 = 2pqW_2 = H$	$q^2E_3L_3 = q^2W_3 = R$

B	Fitness components	AA	Aa	aa
	Early (E_i)	$E_1 = 0.72$	$E_2 = 1.00$	$E_3 = 0.80$
	Late (L_i)	$L_1 = 1.00$	$L_2 = 0.72$	$L_3 = 0.63$
	Overall fitness (W_i)	$W_1 = 0.72$	$W_2 = 0.72$	$W_3 = 0.504$
	Relative to $W_2 = 1.00$	$W_1 = 1.00$	$W_2 = 1.00$	$W_3 = 0.700$
	(Corresponding to Table 14-3)			

C	Numerical example: G_0 eggs to G_1 young adults	AA	Aa	aa	Totals	q_a
	G_0 zygotes	0.01	+ 0.18	+ 0.81	= 1.00	0.90
	Early fitness (E_i)	0.72	1.00	0.80		
	Young adults	0.0072	+ 0.1800	+ 0.6480	= 0.8352 $\bar{E}$	
	Relative frequency	0.00862	+ 0.21552	+ 0.77586	= 1.00	0.8836*
		(D')	(H')	(R')		
	Mating, late fitness (L_i)	1.00	0.72	0.63		
	Postselection	0.00862	+ 0.15517	+ 0.48879	= 0.65258 = $\overline{EL} = \bar{W}$	
	Relative frequency of adults that have mated	0.0132	+ 0.2378	+ 0.7490	= 1.0000	0.8679**
		(D)	(H)	(R)		

TABLE 15-3 (*Continued*)

Fertilization (random)					
G_1 zygotes	0.0174	+ 0.2293	+ 0.7533	= 1.00	
Early fitness (E_i)	0.72	1.00	0.80		
Young adults	0.01253	+ 0.22930	+ 0.60264	= 0.84447 = $\bar{E}$	
Relative frequency	0.01484	+ 0.27153	+ 0.71363	= 1.00	0.87282
	(D'')	(H'')	(R'')		

* Δq from young G_0 adults to young G_1 adults = −0.01078, but Δq from G_0 zygotes to G_1 zygotes = −0.0321.
** Same as Table 14-3 (G_0 mature after selection).

D Spurious fitness estimates (K_i) when selection is not complete at the stage sampled
Let $E_2L_2 = W_2 = 1.00$, and omit normalizing of frequencies for brevity

		AA	*Aa*	*aa*
(1)	G_1 zygotes	$(p^2W_1 + pq)^2$	$2(p^2W_1 + pq)(q^2W_3 + pq)$	$(q^2W_3 + pq)^2$
(2)	Partial selection	$(p^2W_1 + pq)^2E_1 = D''$	$2(p^2W_1 + pq)(q^2W_3 + pq)E_2 = H''$	$(q^2W_3 + pq)^2E_3 = R''$

Then using (15-4A), but calling the fitness K_i when selection is incomplete, we have

$$K_1 = \frac{2D''}{H''} \cdot \frac{q'}{p'}$$

Subsituting from line (2) gives

(3)
$$K_1 = \left(\frac{2(p^2W_1 + pq)^2E_1}{2(p^2W_1 + pq)(q^2W_3 + pq)}\right)\left(\frac{q(qE_3 + p)}{p(pE_1 + q)}\right) = E_1\left[\frac{(pW_1 + q)(qE_3 + p)}{(qW_3 + p)(pE_1 + q)}\right]$$

Similarly,

(4)
$$K_3 = \frac{2R''}{H''} \cdot \frac{p'}{q'} = E_3\left[\frac{(qW_3 + p)(pE_1 + q)}{(pW_1 + q)(qE_3 + p)}\right]$$

In the numerical example from **C** above, we have spurious estimates (K_i)

$$K_1 = 0.72\left[\frac{0.82}{0.70956}\right] = 0.832$$

$$K_2 = 1.000$$

$$K_3 = 0.80\left[\frac{0.70956}{0.82}\right] = 0.692$$

This result indicates heterosis (!), a misleading estimate when selection is presumed to have acted before the genotypes are counted in the population. Complete fitness values are $W_1 = 1$, $W_2 = 1$, $W_3 = 0.7$, as in Table 14-3.

produce frequencies D', H', and R' in which q_a is diminished slightly ($q_a' = 0.8836$). However, if the adults mate in the frequencies given in line 3, the net change in q for one life cycle is to $q_a = 0.8679$ ($\Delta q = 0.0321$). If formula (15-4A) is applied, we obtain the net fitnesses $W_1 = 1.0$, $W_2 = 1.00$, and $W_3 = 0.70$, using frequencies from line 4. With random union of gametes, G_1 zygotes occur in the frequencies indicated in line 5. Partial selection again is presumed to act (line 6) during preadult stages, so that we may observe frequencies D'', H'', and R'' among young adults (line 7).

The hazard of estimating fitness values from counts of stages in the life cycle before selection is completed can be illustrated by deriving spurious fitnesses (K_i), using the principle of formula (15-4A), as indicated in Table 15-3D. In G_1, if young adults are sampled after partial selection only (early stage), fitness calculated on the change from the young adults in G_0 frequencies (line 2) will be incomplete and spurious. We call this spurious fitness estimate K_i. Lines 1 and 2 in Table 15-3D represent the three genotype frequencies among the young adults in G_1 in terms of the G_0 frequencies. The double prime (″) implies a second round of partial selection. If we use (15-4A) to estimate the fitness (K) at this point, we obtain lines 3 and 4 for K_1 and K_3 (assuming $K_2 = 1$)—the fitness values obtained when calculating the change from young adults (partially selected) in one generation to the next. When the data are supplied from the numerical illustration of Table 15-3C, the result falsely indicates not only superiority of heterozygotes but also inequality between *AA* and *Aa* genotypes. Furthermore, as emphasized by Prout, K_i is erroneous in that its value depends on allelic frequencies! Thus, the estimate based on constant fitness produces a gene frequency component that is misleading in this context.

EVALUATION OF FITNESS COMPONENTS AND MAKING PREDICTIONS OF CHANGE IN *q*

In order to comprehend the precision and subtlety of natural selection, we need to have improved techniques and models for making estimates of fitness values more inclusive of the total life cycle and at the same time more precise as to the points in the life cycle where fitness differences are due to genotype. Thus, we ought to arrive at the technical ideal of being able to pinpoint the action of selection and also to evaluate the points of action in terms of their relative importance to the organism's population and its ultimate adaptedness. Population geneticists who have attempted to estimate fitness from the alternative methods of either measuring Δq in experimental populations or from testing for viability and other single life-cycle components may reach a limited sense of selective action involved in the populations being analyzed, but the test of their estimates lies with the predictability of their models, which can be applied to the outcome of selection. Few cases of comprehensive estimates (overall W_i) or pinpointing stages in the life cycle where fitness differences arise have been worked out. The more difficult attempt to evaluate the relative importance of the fitness components in terms of any population's adaptedness has been made even less often.

The general problem of making predictions of selection outcome (magnitude and direction of Δq) and the estimation of fitness components from populational data has been investigated by Prout (1969, 1971*a*,*b*), who reviewed the methods used in experimental population cage studies, particularly with drosophila, by surveying life-cycle components for fitness differences. First, he applied a comprehensive model including general components of the life cycle in any transition from parents to offspring, in the manner of Table 15-1, and specified each component for its contribution to total fitness. Then, assuming that an experimental population is started with allelic frequencies far from equilibrium, he pointed out that

> *. . . some distance from the final equilibrium, there will be a series of transitions linked in sequence, with the parent generation of each transition having different genotype frequencies. It is evident that the more transitions available before equilibrium, the more information there is concerning the fitnesses governing the system. The basic concern . . . is the specification of the minimum number of transitions which are necessary for the estimation of fitnesses under the assumption of certain models.*

Specifically, the "classic" model (constant W, random mating, sex independence) becomes too restrictive, so that fitness values estimated by using it are so far from precise that they have very doubtful value. If genotype frequencies are known only at "partially selected" stages each generation, at least *three transitions* are necessary for estimation of W values, but when the classic model is assumed, the result is disappointing in that values are extremely various (large standard errors, sometimes negative W that is biologically meaningless, and lack of agreement between estimates based on different segments of the selection curve).

In view of these difficulties made from the restrictive "classic" model, Prout devised a set of experiments to test a more general model including at least both early and late components of fitness with sufficient precision to allow prediction of selection outcome for real populations. Some of his experimental work, the reasoning, and the accuracy of prediction are illustrated in Example 15-10. We conclude by pointing out, first, that when the entire life cycle is included (at least for the mutants tested) the adult components far outweighed the preadult in their magnitude and effect on the selective outcome. Second, partial analysis of the life cycle gives biased estimates; if viability of preadult stages alone had been studied (as has been characteristic of work in the field of polygenic analysis using dominant marker techniques such as illustrated in Examples 9-3), we might conclude erroneous net fitness values and selective outcome (mild heterosis in viability and a weak equilibrium at about 50 percent of the two mutants tested). Alternatively, if we studied fecundity alone, the prediction would be a rapid elimination of one mutant (*sv*) by the other mutant (*ev*). Neither conclusion would be correct, because when relative male mating activity was taken into account, the system showed a strong sex-dependent heterozygote superiority slightly favoring *ey*. Third, although the entire life cycle ought to be included, it does not seem necessary to study every detail of it; for population prediction, it is sufficient to lump segments of the life cycle together into a small number of fitness components. The problem then becomes identifying "sufficient parameters" of fitness (Levins, 1966) as a minimum for prediction and utility. Both the number and kind of net components that must be measured depend on the particular levels at which selection is operating. In general, we know little of the specifics in that process, so "it is safest to assume a model of selection of maximal complexity, limited only by biological plausibility, for the population regime under study" (Prout, 1971*a*). It is recommended that as a matter of procedure, adult components be evaluated first in any study. Doubtless it would be revealed that all components of the life cycle are affected by most genetic variation, but we cannot avoid dissecting the life cycle for evaluation of the selection process, and in the past most studies have concentrated on the easier viability of preadult stages than on the more complicated later stages.

EXAMPLE 15-10

ESTIMATING FITNESS VALUES IN A MODEL SYSTEM

Prout (1971*a*,*b*), following some methods used by Spofford in 1956 and Wright and Kerr in 1954, devised an experimental system by which he intended to estimate a sufficient number of life-cycle components, especially in connection with the adult phase, by contriving numerical mixtures of parents so that they could be evaluated in terms of those most important to the outcome of selection in laboratory populations. He hoped to predict the selective changes in populations with discrete generations.

Prout chose a simple genetic system to monitor. As he points out, it had no intrinsic value in itself because it was not a "natural system," but it was easy to monitor, and the events that could affect gene frequency changes could involve much complexity. Given the simple genetic system, the effort was to estimate as much of the life cycle as possible without details and then make predictions of how selection would make the breeding populations tend to come out. He chose in *Drosophila melanogaster* two mutants located on the dot fourth chromosome—eyeless (ey^2) and shaven-naked (sv^n)—that recombine so rarely that they can be considered segregating "alleles" with their "heterozygote" (repulsion) wild-type due to complementation.

Five replicated populations were maintained starting with $ey+/+sn$ heterozygotes simply by allowing egg-laying in half-pint bottles for 24 hours each generation. All populations quickly came to equilibrium at about 60 percent *ey*: 40 percent *sn*. Perturbations were artificially introduced twice (at G_9 and G_{21}) by etherizing the entire population and discarding all except females with one of the homozygous phenotypes (eyeless or shaven). Then only those females, most of which had mated previous to the etherization, were used as founders of the next generation by 24 hours of egg-laying. The sequence of changes over 25 generations in average p_{ey} and the perturbations of the separate populations are illustrated in Figure A. The perturbations indicate clearly the strength of the stable selection equilibrium.

In a separate set of experiments, Prout tested the "early" and "late" components of the life cycle. The larval and pupal viability measure was straightforward; in the mating of $ey/sn \times ey/sn$, each replicate culture ($N = 47$ *reps*) would yield an estimate of viability by using formula (15-4A). For estimating the "late," or adult, components, the matter was far more complex. Prout chose to mix the three mutant phenotypes in different combinations and allow mating at random; mixtures contained 60 flies of each sex, with one sex as a single phenotype but the opposite sex varied in ratios of the three phenotypes such as 30 *ey*: 20 *ey*/*sn*: 10 *sn*, and the reverse, for each of the constant parent types. Varying the ratio among females and keeping the male phenotype constant produced data (six sets) testing for female receptivity and egg-laying ability ("fecundity" symbolized F), while varying ratios among males produced six sets testing for male courtship and rivalry sperm volume ("virility" symbolized V). Two day-old flies were used for these adult tests; they were allowed to mate, and females that laid eggs for three days were then transferred to

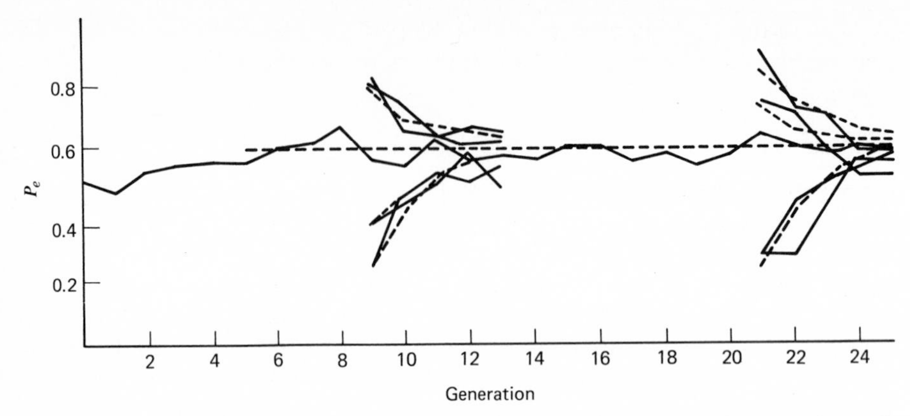

Figure A. Frequencies of the *eyeless* (*ey*) allele = P_e observed in five experimental populations containing *ey* and *sn* (chromosome 4 of *Drosophila melanogaster*) averaged over 25 generations. Perturbed as described in text at G_9–G_{13} and G_{21}–G_{25}. Dashed lines are theoretical predictions from the fitness model (from Prout, 1971*b*).

fresh food to get a 12-hour egg sample. From this last sample in each test about 100 progeny flies of each sex were classified and counted.

From these original parent mixtures, knowing the relative viability of each phenotype, it was possible to estimate what the frequencies must have been among the fertilized eggs and thus algebraically solve for the adult components F and V (for the explicit algebra of the model, see Prout, 1971*a*). In summary, while there was considerable heterogeneity between replicates, it became clear from analyses of variance that female (F) adult fitnesses were not affected by the constant male phenotype, but the male (V) fitnesses were affected by the type of constant female present. These estimates from the 12 sets were combined in order to estimate all the components. For F and V, the ratios of *ey* genes to *sn* genes *after* adult selection were calculated from the observed data and compared with original mixture combination frequencies. Their solutions (from Prout's Table 10, 1971*a*) were as follows (assuming the heterozygote *ey*+/+*sn* fitness = 1.00):

Fecundity (F)	$F(ey/ey)$	$F(sn/sn)$	
	1.037 ± 0.122	0.458 ± 0.068	(from pooling over all constant male types)

Virility (V)		$V(ey/ey)$	$V(sn/sn)$
Constant female genotype	*ey/ey*	0.363 ± 0.074	0.039 ± 0.033
	ey/sn	0.243 ± 0.042	0.122 ± 0.037
	sn/sn	0.135 ± 0.036	−0.018 ± 0.030

In contrast, the larval-pupal early component was

Viability	*ey/ey*	*sn/sn*
Females	0.865 ± 0.039	0.934 ± 0.039
Males	0.839 ± 0.036	0.777 ± 0.038

In larval-pupal viability, a mild heterosis occurs with a suggestion of sex difference in the *sn*/*sn* phenotype. However, a much larger effect arises from the adult components; in females, the fecundity-recessive deleterious effect of *sn*/*sn* is independent of the male partner, while the male component (V) depends significantly on the female, with *sn*/*sn* very nearly being discriminated against completely in mating, except with heterozygous females, while *ey*/*ey* males do best with *ey*/*ey* females. Heterosis among the males is very strong throughtout ($V = 1.0$). It is clear that the overwhelming fitness differences arise in the adult component, and if only the preadult early stages are analyzed, there will be no hint of the large total amount of selection at work in these populations.

Finally, with these estimated fitness values, generation-to-generation transitions were examined by substituting the estimated values, assuming in an overall model that heterosis in the virility component would hold the equilibrium at a stable point. Predicted frequencies based on theoretical equilibrium values (*ey*/*ey* = 0.35, *ey*/*sn* = 0.51, *sn*/*sn* = 0.14) and rate of change were compared with the five populations maintained for 25 generations. Only one of the five indicated a significant difference between results expected from the fitness tests and the population frequency changes; all the other four populations showed good agreement.

Thus, the method of estimating fitness components was capable of predicting population changes by selection reasonably well. Undoubtedly, the virility components found to vary so much with constant females might be expected to vary even more under multiple-choice conditions of any total population. Nevertheless, the degree of confidence in the method can be quite good for estimating total fitness components. In contrast, how little information would have been obtained from studying only the viability component! Also, the kind of selection taking place for all practical purposes would be inaccessible through analysis of population data frequency changes alone. The method used did not allow for the further complication of a change in fitness values with frequency change (minority advantage or frequency dependency).

HUMAN GENOTYPES AND FITNESS

There is considerable evidence for the operation of selection on human phenotypes (see Bajema, 1971; Post, 1971, for summaries), but it is rare that particular genic loci can be evaluated for genotype fitness values. The most widely known and best authenticated case for fitness differences in maintaining a genetic polymorphism comes from data on sickle cell and other hemoglobin abnormalities among African peoples particularly (Examples 1-1 and 3-2). On the basis that geographic distribution of malaria is positively correlated with that of the thalassemia condition, Haldane suggested in 1949 that the maintenance of that polymorphism could be due to increased resistance to malarial infection by heterozygotes compared with normal hemoglobin homozygotes. In 1950 Neel presented data on the frequencies of both thalassemia and sickle cell trait, but for the latter it seemed that the incidence was higher among children than among adults—the opposite of what would be expected if heterozygotes were more resistant than homozygous normals to malaria in their early years. However, Neel pointed out that nearly all of his data were from hospital records, not from populations at large, so "the apparent decrease in the incidence of sickling in older age groups might mean that the sicklers are enjoying better health and so are not represented. . . ." Finally, in 1954, Allison brought out considerable evidence from African populations that sickle cell trait (Hb^A/Hb^S) individuals did exhibit protection from malarial infection. With the severe anemia and near lethality of sickle cell homozygotes, a case of balancing selection in malarial areas was evident; numerous lines of evidence support that view (see Allison, 1964; Motulsky, 1964; Motulsky, Vandepitte, and Fraser, 1966; Livingstone, 1967).

Allison emphasized that the time from birth to age two to three years is the stage of greatest susceptibility to infection from *Plasmodium falciparum*, although mothers give some in utero immunological protection to their fetuses. For the heterozygous individual, those years are the most critical for the malarial resistance benefit. Nearly all children are bitten by infected mosquitoes in hyperendemic areas at least once every night, and therefore they all become infected. But the incidence of mortality due to severe malaria is less among A/S than among A/A children. Motulsky (1964) summarized five independent studies by stating that of 100 children dying of malaria, only one was A/S genotype, and the level of severe infection was twice as high among AA as among AS children. Example 15-11 presents some data on the superior fitness of heterozygous children under these conditions, but it makes clear the difficulty of getting accurate estimates from human populations. Without knowledge of the genotype incidence at birth, it is not easy to estimate the parameters needed for accuracy.

Along with sickle cell trait, G6PD deficiency in red cells also is likely to express resistance to malarial infection (*P. falciparum*). Polymorphism at this sex-linked locus is quite widespread and is as well correlated with the geographic distribution of malaria along with the hemoglobin variants (see Example 4-4 and summary by Kirkman, 1971). The most convincing evidence for G6PD deficiency advantage against malarial infection comes from a study by Luzzato, et al. (1969) of red blood cells from female children heterozygous for G6PD (Gd^{A-}/Gd^B). According to the X chromosome inactivation theory of Lyon (1961, 1962), only one X chromosome is activated in human females' cells to produce primary proteins so that the female heterozygote is effectively a mosaic for X chromosome products, with chance determining which X chromosome is active in any embryonic cell line. With a simple stain test for the enzyme activity of G6PD (methemoglobin elution test), red cells can be distinguished for normal versus deficient and then stained for presence of the parasite. Luzzato and coworkers found that nearly all parasites found in infected females were within normal enzyme cells, while parasite rates in deficient cells ran from zero to at most about half the rate for normal enzyme cells. Because mortality from malaria is correlated with the parasite concentration in the blood, it is likely that only one-half of the erythrocyte population of any heterozygous female is available for parasitization, and such an individual enjoys a direct advantage when malaria is endemic. Parasites do invade deficient cells, and how much advantage a hemizygous Gd^{A-} male or homozygous deficient female may have is not known. They do run the risk of hemolysis, although that risk is not usually great without administration of drugs that put stress on the red cell pentose-shunt pathway. Kirkman (1971) mentioned that acute bacterial infections and infectious hepatitis may cause suffering to the African G6PD-deficient male. With the rare homozygous-deficient female, as well as the hemizygous male, parasite counts are lower in mature red cells than in younger red cells. At present, we do not know what the fitnesses of all genotypes may be, but we can consider what they might be by making the simple assumption that the deficient males and females have approximately the same fitness.

We did not consider the outcome of selection for sex-linked genes in Chapter 14, but the procedure of analysis is outlined in Example 15-12 with data on G6PD alleles in Nigerian populations. (Solutions for sex-linked selection have been published by Haldane and Jayakar, 1964; Bennett, 1958; Mandel, 1959; Li, 1967*c*.) Selection for sex-linked alleles is the equivalent of haploid selection in males and diploid females. Thus, for a constant selection coefficient against one allele (say, a recessive), the rate of change in q will be faster for sex-linked than for an autosomal equivalent, especially as the allele nears elimination; only one-half as many recessive alleles will be in heterozygotes. When heterosis occurs in the female, however (or any equivalent balancing state), conditions necessary for a nontrivial stable equilibrium are a bit more complex than for the autosomal case (see Exercise 16). The stable-state condition given in Example 15-12 leads to the relationship for $\hat{q}$ in females.

$$\hat{q} = \frac{\frac{1}{2}(V_3 + V_1) - W_1 V_1}{(V_3 + V_1) - W_3 V_3 - W_1 V_1} \tag{15-5}$$

where $W_2 = 1.00$ in females and V_1 and V_3 are the fitnesses of hemizygous males AY and aY, respectively. This formula corresponds to (14-15) for autosomal genes. It is important to note that this sex-linked equilibrium contains product terms from the fitnesses in the two sexes (W and V). The stable state then can be reached in at least two main ways: (1) there may be heterozygote advantage in females with homozygous females and hemizygous males varying in the same or in opposite ways, or (2) no heterozygous advantage is necessary if the recessive female homozygote and corresponding male hemizygote (aY) differ from the heterozygote in opposite directions. The case of G6PD variants being advantageous to females in malarial areas is probably a case of the first alternative, although it is not at all clear what the fitnesses actually may be of the remaining four genotypes.

For the other hemoglobin variant polymorphisms and thalassemias, we are in a less secure position with regard to evidence indicating specific resistance to malaria, although there are many convincing arguments in the same direction (Allison, 1964; Livingstone, 1967). Genetic variants that may well have been subject to strong selection in the past cannot be demonstrated as being so today under modern living conditions, so we must assume relaxation of selection for them. Post (1971) summarized the evidence for likely cases such as colorblindness, myopia, and visual acuity in general, abnormalities of the nasal septum, and abnormally small opening of the tear duct. It is quite likely that many blood antigen and protein polymorphisms could be similar cases that in the past had adaptive functions but have them no longer under modern conditions.

Doubtless certain antigen-antibody variants have a special form of selection resulting from the compatibility of mother and fetus. The A-B-O, Rh, and histocompatibility genic loci (*HL-A* system) have important clinical significance, especially in obstetrics, blood transfusion, and transplantation. Hemolytic disease due to erythroblastosis in the fetus or newborn child resulting from anti-Rh (anti-*D*) formed in an Rh-negative mother when the father is Rh-positive (*D*—) is well known (Giblett, 1969; Race and Sanger, 1968). Only from matings in which the mother is *dd* and father *Dd* or *DD* and thus capable of producing a child with the *D* antigen is there likely to be any buildup of anti-Rh in the mother, provided some leakage from fetus to mother of red cells carrying the antigen occurs in the placental circulation. The first *D*— child and *dd* children would not be affected by the mother's antibodies, but only for the second Rh-positive offspring is there a potential risk. While the frequency of hemolytic newborn from Rh incompatibility is quite low in Caucasian populations (about 1/150 until after 1950), the heterozygote *Dd* is the genotype affected, so that if no other influences were operating, the selection would lead to an unstable condition (Table 14-7 and Figure 14-4).

In addition, a few cases of hemolysis are brought about by incompatibility between type O mothers and A, B, or AB children (about 1 in 500 cases), as well as cases involving the Kell, Duffy, and Kidd blood type systems, among others (Levene and Rosenfield, 1961). More important, perhaps, may be the interactions between blood group systems; fetuses incompatible with their mothers in the A-B system almost never suffer from Rh hemolytic disease (Levine, 1958; Reepmaker, et al., 1962) because it is presumed that red cells entering the mother from fetus will encounter either anti-A or anti-B or both and thus be destroyed before they are able to stimulate the mother's anti-Rh production. Thus, some A-B-O heterozygotes would have an advantage because they would be protected from the more severe Rh hemolysis. All these observations lead to the realization that specific mechanisms of selective functioning as a result of blood type interactions may be exceedingly complex, although the frequencies of blood type incompatibilities are low in most populations and the occasions for interactions seem rare. They may have been more important in early human history.

Besides these well-known mother-child interactions, there is suggestive evidence that some naturally occurring antibody substances may have differential effects on diseases that are no longer common in the world. Several microorganisms possess antigens similar to human blood substances, and in fact antigenic A and B substances may be commonly found in plants and animals. It is especially likely that in times past, when smallpox and the plague were common, the distribution of A-B-O groups may well have been selectively important. The vaccinia virus and (by inference) the smallpox virus were found to possess an antigen similar to blood group A antigen. Persons of blood groups B and O who have have natural anti-A might be expected to neutralize the virus more easily during the infectious stage and thus have a milder case than those lacking anti-A. After it was found that the A-like substance was present in the egg culture material in which the vaccinia virus was grown and apparently not in the virus, the hypothesis fell into disrepute. However, Vogel and Chakravartti (1966; reprinted in Bajema, 1971) obtained positive results in field experiments where smallpox was endemic in rural areas of India (West Bengal and Bihar), where few people had been vaccinated and little medical treatment was available. They attempted to ascertain all cases of smallpox in the area. To avoid bias from any association between blood group and susceptibility, these investigators used as controls the exposed but unaffected siblings of the smallpox patients. It was found that individuals with A and AB blood groups much more commonly had smallpox than B or O individuals, and the course of the disease was more severe in the former. Survival for the B + O persons was about 1.6 times that of the A + AB. Thus, fitness to survive diseases that were more common in the past may be conferred by certain genetic polymorphisms.

EXAMPLE 15-11

FITNESS ESTIMATES OF HEMOGLOBIN VARIANTS IN MALARIA AREAS

Motulsky, Vendepitte, and Fraser (1966) reported hemoglobin and G6PD abnormality data collected on nearly 2000 persons from the Congo. In Kwango Province (south-western Congo), six villages among the Yaka (Bantu) tribe were chosen because of high malarial endemism. No abnormal hemoglobin other than Hb^S was found. Out of 587 children in those villages, 116 were heterozygotes (Hb^A/Hb^S), or 19.76 percent. The investigators estimated the frequency of Hb^S allele *before* selection (against the homozygous sickle cell anemics) as follows:

Using the change in q for a lethal (14-3), we have

$$q_1 = \frac{q_0}{1 + q_0} = q_0(1 - q_1)$$

Then $q_0 = q_1/p_1$. If $q_1 = H/2$, where H = frequency of heterozygotes after selection, and $p_1 = 1 - H/2$, then $q_0 = H/(2 - H)$. For the six Yaka villages, $q_0 = 0.1976/1.8024 = 0.1096$, Hb^S frequency estimate before loss of sickle cell anemic children. Unfortunately, this estimate is slightly in error because of selection against homozygous normals from malarial infections, but we would need to know the incidence of the alleles *at birth* to avoid bias due to both selections. This estimate of q, then, may be somewhat higher than the true incidence *before* selection.

The separation of children by age groups—infants (from birth to age 3) and older (from 3 to 15)—gave the following frequencies:

Age Group	No. Tested	*Hb A/S* Frequency
Birth to 3 years	107	0.103
3 to 15 years	456	0.226
Total	563	
(Age of 24 children not recorded)		

This rapid rise in frequency of the sickle cell trait was presumably caused by high mortality among nonsicklers before age 3–4 when active acquired immunity to malaria develops. If the younger children's frequency is used to estimate the q_0 parameter, it comes out $q = 0.054 \pm 0.022$, which is probably too low for the q_0 value. No doubt, there is also some inbreeding, which will tend to lower the estimate of Hb^S. The upper 95 percent confidence limit for q of 0.098 is not significantly different from the estimate above. For the purpose of the illustration, we shall use the first estimate to calculate the amount of selection going on—that is, the W values for the two viable genotypes. Using (15-4A), we estimate W_1 for Hb^A/Hb^A fitness relative to W_2 for Hb^A/Hb^S as follows:

$$W_1 = \frac{2D}{H} \cdot \frac{q}{p} = \frac{2(0.774)(0.1096)}{0.226(0.8904)} = 0.843$$

There is then about 16 percent greater mortality for the normal hemoglobin homozygote child from birth to age 3 than for the sickle-cell trait child, when malaria is hyperendemic. This mortality estimate is likely to be a minimum since the q_0 estimate was probably slightly high.

EXAMPLE 15-12

SELECTION FOR SEX-LINKED MALARIAL RESISTANCE

In Nigeria, the Yoruba tribes' incidence of G6PD variants Gd^A and Gd^{A-} averages about 20 percent or greater for both alleles, with the normal Gd^B at about 60 percent (Kirkman, 1971). With a definite advantage to heterozygous females protecting them against malarial severe infection, we may postulate relative fitness values that could account for the stable equilibrium for the alleles in that part of the world.

We first consider the equilibrium conditions for a single pair of sex-linked alleles as follows (using Li's 1967 terminology and our own): we list gene frequencies separately for the two sexes (p, q for females and r, s for males as used in Chapter 1) as well as genotype fitness values (W for females and V for males). Then we adopt the gene frequency ratios $u = p/q$ and $v = r/s$ for simplicity of algebra. Thus, frequencies and fitnesses may be symbolized as follows (u without subscript is p/q *before* selection):

	Females			Males		
Genotypes	AA	Aa	aa	AY	aY	(since males' genotypes are maternally determined)
Zygote frequency	uv	$+ (u + v)$	$+ 1$	u	$+ 1$	
Fitness	W_1	W_2	W_3	V_1	V_3	
After selection	$uvW_1 + (u + v)W_2 + W_3$			$uV_1 + V_3$		

Normalizing fitnesses to $W_2 = 1.00$ gives, after selection,

$$\text{Females} \quad u_1 = \frac{uvW_1 + (u + v)/2}{W_3 + (u + v)/2} \qquad \text{Males} \quad \frac{V_1 u}{V_3} = v_1$$

$$u_1 = \frac{2uvW_1 + u + v}{2W_3 + u + v}$$

Because males have their mothers' p,q values at zygote formation, we need only arrive at the equilibrium condition for female gene frequencies. The stable state is defined as $u_0 = u_1 = u_2, \ldots,$ so that the subscript may be dropped and the expression above for females solved for u

$$\hat{u} = \frac{\frac{1}{2}(V_1 + V_3) - W_3 V_3}{\frac{1}{2}(V_1 + V_3) - W_1 V_1}$$

(female gene frequency ratio at equilibrium with fitnesses relative to $W_2 = 1.00$). Li (1967*c*) stated that if both numerator and denominator were positive, the equilibrium would be stable and if negative, unstable; if the signs were different, there would be no nontrivial stationary state.

Livingstone (1964) proposed some sets of fitness values in malarial areas assuming that heterozygous females benefit from malarial resistance and the homozygous *Gd* variant *A*− females have as low fitness as the hemizygous variant males due to their risk of bacterial infection and hepatitis. From the little that is known about these genotypes' fitness values in West Africa, we could assume a reasonable set of fitness values to give the observed equilibrium as follows:

	Females			Males	
Gd	B/B	$B/A-$	$A-/A-$	B/Y	$A-/Y$
Fitness	0.90	1.00	0.85	1.00	0.90

Using the u expression above, we have u = (0.950 − 0.765)/(0.950 − 0.900) = 3.70. Thus, $\hat{p} = 0.787$, $\hat{q} = 0.213$ (female frequencies) and $\hat{r} = 0.804$, $\hat{s} = 0.196$ (male frequencies), which are reasonably close to the observed frequencies in Nigeria. Note that the sexes in general will have different frequencies when selection affects sex-linked traits.

A SEARCH FOR SELECTION EFFECTS OF GENETIC POLYMORPHS IN A HUMAN POPULATION

If any genic locus displays polymorphism in a human population, it ought ideally to be possible to ascertain whether the genotypes occurring have any differential fitness values under current conditions, or whether they are selectively neutral, by collecting incidence data extensively throughout the various age groups in the population, the parental combinations, and their offspring ratios, in order to apply the principles we are discussing. As outlined by Neel and Schull (1968), several methods could be tried if a reasonably homogeneous population could be ascertained for incidence of several genic systems. Such a well-defined human population was studied by a research team from the University of Michigan, Department of Human Genetics (Shreffler, et al., 1971; Sing, et al., 1971). A population of about 10,000 persons, nearly all Caucasians of West European ancestry inhabiting the town of Tecumseh, Michigan, had been systematically studied from 1959 with the aim of evaluating health and disease in that community. By 1965, data had been collected to include biochemical analyses of blood samples, various anthropometric measurements, and physiological tests so that several genetic systems could be classified for nearly the entire population. Evidence for selection was looked for from the data using the following general methods: (1) searching for age trends or differences between sexes in genotype or phenotype frequencies, (2) searching for nonrandom associations between two or more polymorphic systems (see Chapter 17), (3) searching for distortions of genetic ratios among progenies segregating for polymorphs, and (4) searching for associations of genotypes or phenotypes with life-cycle component traits such as fertility and disease resistance.

In all, 12 genetic systems were recorded on all individuals: A-B-O, MNSs, Rh, Kell, Duffy, Kidd, P, Lewis, ABH secretion (blood groups), and haptoglobin, transferrin, and group specific component (Gc) (protein variants). Overall population frequency data did not differ significantly from the published values on comparable populations in Western Europe, and tests for agreement with expectation from the random-mating square law revealed no significant deviations. Analyses for the effects of age and sex on gene and phenotype frequencies yielded nine significant effects among 120 χ^2 tests for heterogeneity; four of the nine effects involved the P+ blood system, where it is known that the degree of expression of the antigen is not reliable in young girls (birth to nine years age group). The remaining tests were significant only at the five percent level, and because there were just five such tests, that would be the number of significant tests expected by chance. The investigators (Shreffler, et al.) then discussed the likelihood of real biological significance for these age and sex effects and concluded that for several reasons they do not indicate more than chance deviations. Some associations between systems were more significant than chance would allow, and these will be discussed in Chapter 17.

Family data were analyzed from 2507 families in four ways (for each genetic system):

1. Mating-type frequencies to determine whether mating was at random. Expected mating frequencies were derived according to principles set forth in Chapter 1, Table 1-1, with gene frequencies estimated from parents by the maximum-likelihood method.
2. Number of children per mating to find out whether fertility differences would be apparent.
3. Parent and offspring phenotype distributions. Children's expected distributions were calculated from gene frequencies estimated from children's and parents' phenotypes.
4. Genetic ratios from matings to ascertain segregation effects, checking on distortion effects of normal Mendelian segregation.

From the detailed analysis, there was no evidence that parental mating-type frequencies deviated from square-law expectations. Also, there was no evidence that the number of children in a family was a function of parental mating type for any system.

In contrast, significant departures from expectation were found in 8 of 17 analyses of children's phenotype frequencies either because of deviations from the square law (Rh-E showed an excess of heterozygotes (*Ee*) based on children's frequencies), or because of gene frequency difference between parents and offspring (MN, haptoglobin, and Gc), or a combination of both deviation from the square law and a difference in gene frequencies between parents and offspring (Duffy, Kidd-b, Lewis-secretion, and ABH-secretion).

Finally, when genetic ratios were tested for departure from expectation, seven tests were significant when offspring data were pooled within mating types (χ^2 values exceed the 5 and 1 percent critical levels about two to three times more often than expected by chance alone). Two of the tests, however, were in the P system, and they must be disregarded because of misclassification of younger children. Of the remaining five, four (MN, Rh-E, Kell, and ABH-secretion) had χ^2 with probabilities of less than 1 percent to be due to chance alone. If all ratios were averaged, the segregation ratio of 0.50 would be realized; if one looked only at that average, one might conclude that there was thus no evidence for selection. However, Sing, et al. (1971) points out:

> *. . . an alternative explanation might be that the observed variation in genetic ratios is indicative of nonrandom, deterministic, forces operating in this population on at least a portion of these polymorphic loci . . .* [*and*] *we suggest it is improper to pool numbers across systems. . . . Given the existence of some polymorphic loci at which selection is not occurring, pooling across loci can only obscure the findings at those loci at which selection*

is occurring . . . . If (as Dempster, 1955, states so clearly) diversity is being maintained to some degree by variation of selection coefficients in time and space, we must treat the data in a way so as to recognize the systems of interaction which have been emphasized by the work of Wright.

It seems clear, viewing past efforts to detect selection (including our own), that the studies of the "next round" in the effort to understand the forces maintaining the polymorphisms must be both far more extensive and far more detailed than those to date . . . there is also a need for studies in diverse environments . . . man in his numbers and his technology may have so altered the environment that the original selective pressures can no longer be detected. We must also recognize the composition of the study population. If genetic coadaptation exists, then estimates of selection based on populations where major hybridization has occurred recently may have no relevance to populations where only minor hybridization has occurred in the ancestry.

. . . the only spur to investigators at this point is the fundamental nature of the question concerned. One is accordingly led to wonder whether future studies can be combined with other large scale genetic investigations to which society finds itself committed. Should, for instance, systems of continuous monitoring for increased mutation rates be established in designated areas, involving studies of a large series of proteins derived from cord bloods from newborn infants with corresponding studies of maternal and paternal specimens, here might well be the basis for a prospective study of selection on the requisite scale.

In similar vein, Dobzhansky (1972) points out:

The present understanding of how natural selection operates, especially in man, is far from satisfactory. One can infer from circumstantial evidence that various forms of natural selection act on human populations, but only in a few exceptional instances has conclusive direct evidence become available. . . . The view that the biological evolution of mankind became arrested when the cultural evolution began is uncritically accepted by many nonbiologists. The invalidity of this view is demonstrable on the ground of theoretical inferences but not, it must be admitted, on the basis of concrete observations. This is a really shocking state of affairs: scientifically and technologically advanced countries have seen fit to expend huge amounts of effort and money to perfect means for self-destruction and to fly to the moon, but not to learn the most basic facts about the state and the possibilities of mankind's own biological endowment.

EXERCISES

1. It is sometimes difficult to decide how to assess relative fitness. Li (1967*c*, 1970*a*) pointed out the difficulty in human genetics by the following example:

	No. at Birth	No. Surviving and Reproducing	Total Living Children	Average No. Children per Parent
Hemophiliacs	100	40	70	1.75
Normal (males)	100	96	240	2.50

(a) If you were measuring the relative fitness of hemophiliacs compared to normal males, which set of relative values above would you choose and what would the fitness value be? Why is your choice the best one?

(b) If there were 50 hemophiliacs who produced an average of 1.4 children per father and 80 normals who produced an average of 3 children per father, what would be your estimate of relative fitness for hemophiliacs compared with normals? (Assume 100 for each at birth as before.)

(c) Suppose three genotypes *AA*, *Aa*, and *aa* have relative fitnesses of 2:3:1, respectively. Also suppose *Aa* parents have an average of three children per family. What kinds of life-cycle events could bring about the lower fitnesses in the other genotypes (for example, suppose all genotypes have equal average family sizes but mortality differences among children)?

(d) What information would you need for better estimates of fitness than those implied above?

2. How might a completely recessive allele that causes sterility (lack of germ cells) in both sexes differ from a recessive lethal (death of zygote) in its effect on successive generations? Assume the population's mating is completely monogamous and that dominant genotypes have high fertility. First consider the population size to be large, then consider it to be small. Apply the principles to the diploid selection model such as that in Table 14-2 but consider sterility as a factor in adults instead of in growth and development of the zygote.

3. Go back to Chapter 1, Table 1-2. Assume m is a proportion of mating for *aa* in both sexes as a fraction less than that of the dominant genotypes so that the relative amounts of mating are D, H, and mR.

(a) Show algebraically that the proportions of genotypes after random mating with the lower mating by *aa* is as follows:

$$AA = (D + \tfrac{1}{2}H)^2,\ Aa = 2(D + \tfrac{1}{2}H)(mR + \tfrac{1}{2}H),\ aa = (mR + \tfrac{1}{2}H)^2.$$

(b) If $D = 0.01$, $H = 0.18$, $R = 0.81$ at the adult stage, and recessive homozygotes mate at 70 percent of the amount of mating displayed by the dominant genotypes, what will be the relative proportions of genotypes among the mated adults?

(c) In reference to Table 14-3, does the frequency of mated genotypes among adults correspond to mature G_0 adults or to G_1 progeny zygotes?

4. We may see the correspondence between the outcome of selection during early (E) stages and that of late (L) stages by the following relationships with a simple deleterious recessive (as from Table 14-3).

(a) First assume the recessive is deleterious during developmental stages. If zygotes D, H, and R develop such that $D + H + W_3R = \bar{W}$, then in relative numbers they occur as $D/\bar{W} + H/\bar{W} + W_3R/\bar{W} = 1.00$, and after random mating progeny genotypes will be

$$\left[\frac{D + \frac{1}{2}H}{\bar{W}}\right]^2 + 2\left[\frac{(D + \frac{1}{2}H)(W_3R + \frac{1}{2}H)}{(\bar{W})^2}\right] + \left[\frac{W_3R + \frac{1}{2}H}{\bar{W}}\right]^2$$

Verify this statement and use numerical values from Table 14-3 for G_1 zygotes.

(b) If the recessive is deleterious to mating activity of G_0 adults, let m be the amount of mating relative to the dominant genotypes, and then the adults having mated would occur in frequencies similar to formula (15-1) but expressed as arbitrary genotype frequencies:

$$(D + H + mR)^2 = D^2 + 2DH + H^2 + 2HmR + 2DmR + (mR)^2 = \bar{K}$$

where we let $\bar{K}$ be the normalizing total after mating analogous to $\bar{W}$ above. Progeny zygotes arising after the matings then will have the same frequencies as follows:

$$\begin{array}{ccc} AA & Aa & aa \\ \hline (D + \frac{1}{2}H)^2 + & 2(D + \frac{1}{2}H)(mR + \frac{1}{2}H) + & (mR + \frac{1}{2}H)^2 = \bar{K} \end{array}$$

This is the same result as in Exercise 2 (above).

(c) If $m = W_3$, then show that $\bar{K} = (\bar{W})^2$, or that the normalizing total after selective mating equals the square of the normalizing total ("net fitness") from development.

(d) Verify that G_1 zygotes will be in identical frequencies whether selection occurs during development (early stages) or during mating (late stages).

5. If we "normalize" the fitness values weighted by frequencies after selection by $\bar{W}$, thus making $\bar{W}$ equal to 1.00, we discover the relative contributions of each genotype to the next generation. Thus, $\bar{W}$ may be considered as 100 percent, or some constant population size (as in Table 14-1). Use the data of Table 14-3.
   (a) What are the relative contributions of genotypes to progeny after completion of development? after random mating? (Assume selection to act during development.)
   (b) Assume selection to act in mating proportions, as in Exercise 3. What would be the relative contributions of genotypes to progeny?
   (c) Continue for a second generation of selection. Assume first selection to act during development, then in mating proportions. How do the relative contributions of genotypes change? Remember to make $\bar{W} = 1.00$.
   (d) Given selection during development or selection in mating proportions as alternatives, how does the random combining of males × females as parents affect the changes in genotype frequencies?
6. Knight and Robertson (1957) devised an index of overall ability to "compete" for both mates and preadult growth stages in drosophila cultures. Their method was to allow a stock (a strain or a population to be tested for "competitive ability") to be introduced as virgins in equal numbers with a standard "tester" strain (virgins) characterized by having dominant marker mutations on homologous chromosomes (*Curly*/*Plum* second chromosome dominants in *D. melanogaster*). After mating and growth through one life cycle, the competitive index was simply the relative proportion of wild-type flies emerging out of the total (wild type + *Curly*/*Plum*) flies. Crossmated progenies (*Curly*/+ and +/*Plum*) were disregarded.
   (a) Do you think any accuracy is lost by disregarding the crossmated progenies?
   (b) A wild outbred stock ("Kaduna") gave a competitive index = 3.74 ± 0.64 for 10 replicates. A series of highly inbred lines from the Kaduna stock (full sib mated for 12 generations) averaged an index of 0.14 with a range from 0.65 down to less than 0.01. How inbred was the series of inbred lines?
   (c) To measure the effect of nonmarker (wild-type) chromosomes from the inbred lines with the worst competitive index on preadult viability (early stages), *Cy*/+ × *Cy*/+ flies were mated and viability of segregant adults measured as the ratio of

+/+ : Cy/+ flies. Since Cy/Cy is lethal, the expected ratio is 1:2, so the +/+ would be doubled for an index of viability (+/+) × 2 out of the total (wild-type × 2 + Cy). Comparisons of competitive index and viability were as follows:

Inbred Line	Competitive Index	Viability
4	0.03	0.83
7	0.01	0.69
14	0.01	0.76

(d) Why should the competitive index and viability be different? Comment on where in the life cycle the lowering of fitness had been most affected by inbreeding?

7. The following are two sequences of successive frequencies for an allele with a fitness effect. In sequence (A), the allele is in a haploid organism without a Mendelian process of reproduction. In sequence (B), the allele is in a diploid organism with regular sexual reproduction and it is completely recessive in fitness to a dominant "normal" allele. Estimate the selection coefficient (s) over these sequences and tell whether selection is constant or changing. What would you predict for the outcome of selection in each sequence?

Sequence A		Sequence B	
G	q	G	q
0	0.9	0	0.9
1	0.8710	1	0.8746
2	0.8351	2	0.8450
3	0.7915	3	0.8113
4	0.7401	4	0.7741
5	0.6811	5	0.7343

8. The following are two sequences of successive frequencies for an allele with a different effect on fitness in each sequence. Estimate the fitness values of genotypes (W_i) over each sequence and tell whether the values seem to be constant or changing. How would you predict the outcome of selection for each sequence?

Sequence A		Sequence B	
G	q	G	q
0	0.9	0	0.9
1	0.8591	1	0.8354
2	0.8060	2	0.7551
3	0.7406	3	0.6715
4	0.6643	4	0.5967
5	0.5808	5	0.5358
6	0.4950	6	0.4888
7	0.4120	7	0.4530

9. In Examples 11-1 and 12-3, two brown eye color alleles (*bw* and bw^{75}) of drosophila were used because of their easy detection in the heterozygote when scarlet was also homozygous, but in small populations there was no positive evidence of any fitness difference between genotypes for those alleles. A series of large-population experiments was done in the Spiess lab in 1968 by T. Yacher (unpublished). Homozygous strains were crossed to produce heterozygotes (bw/bw^{75}; *st*/*st*), a few of which were introduced into one of two population boxes along with a large number of homozygotes for one allele or the other so that starting frequencies would contain about 90 percent of one allele and 10 percent of the other. Each generation, food containers were plugged after a week of egg-laying by parent females. After two weeks, emerging progeny were entirely counted for genotype and a new generation initiated. Frequencies of the bw^{75} allele (p values) are listed below for two populations ($p = bw^{75}$, $q = bw$ frequencies):

Generation	0	1	2	3	4	5	6	7
Population I	0.10	0.205	0.310	0.415	0.510	0.615	0.690	0.730
Population II	0.90	0.845	0.800	0.770	0.750	0.740	0.739	0.738

After G_7, both populations maintained approximately equilibrium at $p = 0.73$.
(a) Using formula (15-2) and the method outlined in Table 15-2 and Example 15-2, estimate fitness values for sequences of two generations of changes in p in each population. Verify, for example, that G_4–G_5 estimates $W_1 = 1.31$, $W_2 = 1.00$, $W_3 = 0.60$ in population I.
(b) Does it appear that W values are similar for these two populations or are they very different? Would you say that a single set of fitness values could explain the changes in frequencies as well as the maintenance of an equilibrium at G_7 on? Explain why you think the fitness values are constant or not over the range of frequencies recorded.

10. Efficient estimates of W values are obtained from maximum-likelihood estimates as given in formulas (15-4) and (15-4A). Use data from tables in Chapter 14 as follows to confirm W values from frequencies of genotypes before and after selection.
(a) Using Table 14-3, if $p_0 = 0.1$, $q_0 = 0.9$, and genotypes observed after selection are $AA = 132$, $Aa = 2378$, $aa = 7490$. Find W_1 and W_3, assuming $W_2 = 1.00$. Two-decimal accuracy is sufficient.
(b) Using Table 14-5B, if $p_0 = 0.8$, $q_0 = 0.2$, and genotypes observed after selection are $AA = 5953$, $Aa = 3721$, $aa = 326$. Find W_1 and W_3, assuming $W_2 = 1.00$. Two-decimal accuracy is sufficient.
(c) Would it be correct to apply the same maximum-likelihood estimate formulas to genotype frequencies following random mating of these selected observed numbers? Why?

11. Battaglia recorded (*Evolution* 12: 358–364, 1958) populations of a marine copepod (*Tisbe reticulata*) from Roscoff, France, in which certain color variants are controlled by a pair of codominant alleles: *violacea* (V^v) and *maculata* (V^M). Homozygous color forms violacea (V^vV^v) mated × maculata (V^MV^M) produce heterozygotes (V^vV^M), which then can be inbred for a segregating generation in which the expected ratio is 1 violacea:2 violacea-maculata:1 maculata. In seawater culture dishes, larval mortality is usually high and it is more so with increasing density. Experiments were done using

three density regimens: high crowding (I), medium (II), and low density (III). Observed numbers of adults were as follows:

	Phenotypes			
Density Condition	V	VM	M	Total
I (high)	353	1069	329	1751
II (medium)	343	1015	385	1743
III (low)	904	2023	912	3839

(a) What are the expected numbers among the phenotypes at each density?
(b) What are the relative viabilities of the three types in each density?
(c) What can you say about the effect of density on expression of heterosis?

12. The likelihood function (see Appendix) for selection of three genotypes, given initial frequencies before selection at square law $(p + q)^2$ values, would be based as follows:

$$\left(\frac{W_1p^2}{\overline{W}}\right)^A \left(\frac{2pq}{\overline{W}}\right)^B \left(\frac{W_3q^2}{\overline{W}}\right)^C$$

where A, B, C = observed numbers of three genotypes after selection, with $W_2 = 1.00$.

$$\log L = A \log\left(\frac{W_1p^2}{\overline{W}}\right) + B \log\left(\frac{2pq}{\overline{W}}\right) + C \log\left(\frac{W_3q^2}{\overline{W}}\right)$$

(a) $$\frac{\delta L}{\delta W_i} = \frac{Ap^2[\overline{W} - W_1p^2]}{\overline{W}^2} \cdot \frac{\overline{W}}{W_1p^2} - \frac{Bp^2}{\overline{W}} - \frac{Cp^2}{\overline{W}}$$

(b) $$= \frac{A}{W_1} - \frac{Np^2}{\overline{W}} = \frac{A\overline{W} - NW_1p^2}{W_1\overline{W}}$$

Similarly,

$$\frac{\delta L}{\delta W_3} = \frac{C\overline{W} - NW_3q^2}{W_3\overline{W}}$$

and setting these partial derivations equal to zero gives

$$A\overline{W} - NW_1p^2 = 0$$
$$C\overline{W} - NW_3q^2 = 0$$

$\overline{W}(A + C) = N(W_1p^2 + W_3q^2)$. Because $N = A + B + C$ we have the basic statement $(A + C)(W_1p^2 + 2pq + W_3q^2) = (A + B + C)(W_1p^2 + W_3q^2)$.

(c) Multiplying terms and solving for fitnesses we have

$$W_1 = \frac{2A}{B} \cdot \frac{q}{p}$$

$$W_3 = \frac{2C}{B} \cdot \frac{p}{q}$$

The student should verify derivatives of logs at (a) and simplifications at (b) and (c).

(d) Variances of W's from maximum-likelihood estimates are derived from the reciprocal of the second derivative of the log function and are as follows (Vw_i = variance of W_i):

$$Vw_1 = \frac{\bar{W}(W_1 p + 2q)W_1}{2p^2 qN}$$

$$Vw_3 = \frac{\bar{W}(W_3 q + 2p)W_3}{2pq^2 N}$$

13. The relative amounts of selective predation can be estimated from release-recapture experiments with prey. Kettlewell (1956, cited in 1973 ref.) described the industrial melanic moth *Biston betularia*, carbonaria phase (C) versus typical phase (T), when released in a heavily industrialized woodland near Birmingham, England. Most C forms appeared inconspicuous to the human eye, while T forms were conspicuous. Thus, melanic forms had an advantage against blackened tree trunks when birds were feeding. The following data were obtained:

Released		Recaptured	
C_0	T_0	C_1	T_1
154	64	82	16

(a) If C and T forms have abilities W_1 and W_2 to avoid capture by predators, respectively, the following relation should hold: $T_1/C_1 = W_2 T_0/C_0$, letting $W_1 = 1.00$. Why?
(b) Find the value of the W_2 estimate.

14. In Example 15-8, flies tested for preadult stages with chromosomes from Piñon Flats and from Mexico. Earlier Dobzhansky (1947) had sampled flies from population cages containing only Piñon Flats chromosomes at three stages in the life cycle: (1) egg samples cultured separately from the population cage at optimal conditions for growth, (2) freshly emerging adults (males) from the crowded population cage, outcrossed separately to virgin known ST/ST females to ascertain karyotypes, and (3) older adult males from the population cage also outcrossed for karyotype. The three sets of data are as follows:

Stage Observed	CH/CH	ST/CH	ST/ST	Total
Egg sample	42	88	20	150
Young adults (cage)	16	83	31	130
Old adults (cage)	13	86	26	125

(a) If we use the egg sample to represent frequencies before selection, find the W_1 and W_3 fitness values among the young adults and older adults. (Assume $W_2 = 1.00$.)
(b) Calculate variances of W_1 and W_3 for each type of adults using the variance of W_i from Exercise 12. Take the square root of these variances to obtain standard errors for the W_i estimates. What W_i's appear to be significantly different from 1.00?

15. (a) Verify the algebraic succession of statements on estimation of spurious fitness (K_i) from Table 15-3D. Show the steps in derivation line by line. Then supply data from Table 15-3C for the numerical example given.
(b) What precautions must be taken by an investigator who wishes to estimate fitness differences between genotypes?
(c) Suppose early stage fitness (E) and late stage fitness (L) are as follows:

	AA	Aa	aa
E	1	1	0.7
L	0.63	0.875	1

Assuming these fitness values act multiplicatively as in Table 15-3, what will be the overall fitness (W) for these genotypes through the life cycle?
(d) Is the example in (c) a case of genuine heterosis or overdominance? This result is called "marginal overdominance" (Wallace, 1959). Why?

16. In Example 15-12, derive the equilibrium condition for sex-linked alleles with balanced heterozygosity in females by working out the frequency statements given.
(a) If $u = p/q$ and $v = r/s$ before selection, show that the zygote frequencies of females are $AA = uv$, $Aa = (u + v)$, $aa = 1$, and the zygote frequencies of males are $AY = u$, $aY = 1$. Assume selection acts during early stages of development. Then verify the u_1 (allelic ratio after selection with $W_2 = 1.00$) as equal to $(2uvW_1 + u + v)/(2W_3 + u + v)$ for females and $v_1 = V_1u/V_3$ for males (where u or v without subscripts refers to allelic ratio before selection).
(b) At equilibrium $u_0 = u_1 = u_2 \cdots$ and $v_0 = v_1 = v_2 \cdots$. Then supply V_1u/V_3 for v in the females' allelic ratio to verify the equilibrium

$$\hat{u} = (\tfrac{1}{2}(V_1 + V_3) - W_3V_3)/(\tfrac{1}{2}(V_1 + V_3) - W_1V_1).$$

First factor out u from the ratio, then cross-multiply with u and simplify to obtain the equilibrium $\hat{u}$.
(c) Determine equilibrium $\hat{u}$, $\hat{v}$, $\hat{p}$, $\hat{q}$, $\hat{r}$, $\hat{s}$ values for the following three sets of W_i and V_i values:

W_1	W_2	W_3	V_1	V_3
0.8	1.0	1.2	0.9	0.6
0.9	1.0	0.9	0.9	0.6
0.9	1.0	0.7	1.0	1.0

Tell whether each equilibrium is stable, unstable, or trivial.

16

EXTENSIONS OF SELECTION PRINCIPLES: (1) VARIABLE FITNESS

Natural selection has been intentionally introduced with a focus on the simplest aspects to concentrate on the fundamental outcome of genetic changes when just a single pair of alleles at one locus accounts for constant differences in fitness. This selection specification is highly restrictive compared with real populations, and the "classic" model of constant fitness (and implied constant ecological conditions) can be misleading at times when looking for evidence of selective differences between genotypes. In this chapter and the next we shall explore lifting these restrictions one at a time so that we may discern at an introductory level what may be the primary selection outcome, attainment of nontrivial equilibrium, and the likely consequent population genetic structure under these complexities. After understanding the consequences of lifting each restriction in an elementary way, the student may wish to explore combining two or more factors and thus attempt to simulate natural populations. It is not our intention here to go beyond the primary principles to any degree, partly because such extensions are outside the scope of this book and partly because many have not yet been modeled in the literature. With high-speed computers, simulations are feasible, and the student ought to be able to continue many of these multiple extensions once fundamentals are understood—undoubtedly discovering unexpected and exciting outcomes for selection when combining changes in fitness with complex genotypes.

Some of the extensions implied in the last chapter that we shall now consider are variable fitness values as functions of (1) genotype frequencies, (2) sex of genotype, and (3) more than one constant set of ecological conditions ("niches") or multiple fitness optima within the population's area (space or time conditions). These single extensions might be called ecological, or behavioral, extensions since they do not involve lifting the genetic restriction of just two alleles at one locus, as held in the last two chapters.

In the following chapter, we shall hold the set of fitness values constant and explore the extensions for genetic complexities: (1) multiple alleles at a genic locus and (2) two loci, each with a single pair of alleles. Diploidy and sexual reproduction with Mendelian processes will be assumed throughout.

RANDOM FLUCTUATIONS IN FITNESS

If fitness values fluctuate randomly and independently of genotype frequencies or environmental forces—that is, without predictable relationship between fitness and frequency over long periods of time—the distributions of genotypes would take a form similar to those of purely neutral (without selective value) genotypes under small population sizes (Chapter 12). Wright (1948) considered the distribution of gene frequencies with random fluctuations in selection pressure (σ_s^2) in large populations; selection outcome would be deterministic in short periods of time—say, for any generation t to $t + 1$ transition—but fluctuating so that over the long run the average selection coefficient would be zero, $\bar{s} = 0$. Assume a set of large-sized populations ($N > 10^6$) with an input of a small proportion of migrants each generation whose gene pool has $p = q = 0.5$, which tends to bring the population to equilibrium at that $p = q$. Also assume the genotypes concerned have intermediate heterozygote fitness (as in Table 14-4) with, say, the W_2 ranging 95 percent of the time from 0.99 to 1.01, but a mean $\bar{W}_2$ $(Aa) = 1$, relative to $\bar{W}_1$ $(AA) = 1.00$, and $\bar{W}_3$ $(aa) = 1$ but W_3 ranging twice that of W_2. If the fluctuations in W_2 and W_3 are completely at random, the W_2 standard error (σ_{W_2}) would be 0.005 in that case. The immigration pressure would restrict the gene frequencies to range from $q = 0.25$–0.75 over the set of populations. However, if the fluctuations were 10 times greater ($\sigma_{W_2} = 0.05$ so that the range would be 0.90–1.10 for W_2 and twice as great with 95 percent confidence

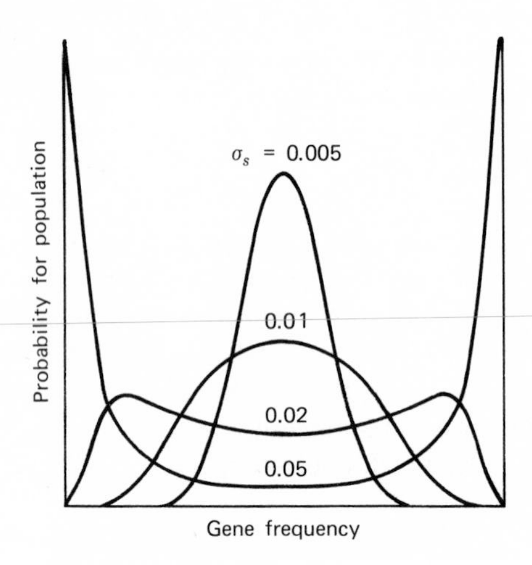

Figure 16-1. The distribution of gene frequencies among populations with different amounts of random fluctuation in selection pressure ($\sigma_s = 0.005$, 0.01, 0.02, 0.05), with the only systematic pressure due to immigration from outside where $q = 0.50$. Average selection ($\bar{s}$) is zero within each population and migration rate (m) is 0.0001. Size of each population is assumed to be so great ($N > 10^6$) that fluctuations in gene frequencies due to sampling are negligible. From Wright, 1948.

for W_3), then in spite of the immigration tendency to bring the population frequency to 0.50, most of the time gene frequencies would be close to fixation or loss. In short, according to Wright, random fluctuation in fitness values have the same effect on distribution of gene frequencies in large populations as random genetic drift does in small populations. The outcome for random fluctuations in W_2 to give a standard error of $\sigma_{W_2} = 0.05$ is roughly equivalent in a large population ($N > 10^6$) to the accidents of sampling in a population of 1000. These distributions are shown in Figure 16-1. These effects would take place over the long run, and they are mentioned here to emphasize that to be effective, selection must have some relationship to the conditions under which the population exists. Randomness of selection leads to depletion of genetic variability just as reduction in population size does with no selection acting at all.

FITNESS DEPENDENT ON FREQUENCY OF GENETIC UNITS

Some evidence was mentioned in Chapter 15 (Examples 15-2, 15-4, 15-5, 15-7) that genotypes (or karyotypes) could be interpreted to have fitness values differing according to genotype frequencies and in fact likely to be determined by some prescribed relationship to those frequencies. It was implied in those examples that differences in fitness brought about by change in frequencies of genetic units were independent of genetic background recombinations (making new phenotypic interactions possible with the monitored genetic units) or of any extrinsic environmental changes that could account for the fitness differences. Considerable evidence now accumulating shows not only that fitness values are often variable instead of constant over the range of gene frequencies observed in populations, but also that there is often a selective advantage for rare genotypes ("minority advantage") that they do not enjoy as they become common.

For a long time, evidence pointing to frequency-dependent selection has been accumulating in the literature. As pointed out by Kojima (1971), the founders of population genetics theory (Wright, Haldane, and Fisher), together with experimentalists (Castle, Dobzhansky, Lerner, and Lush) have indicated in different ways that the fitness values of genotypes may not be generally constant (see Haldane, 1949; Wright, 1955; Dempster, 1955; Li, 1955*a*). Earlier documented evidence for minority advantage came from two sources: (1) Batesian mimicry, in which mimetic resemblance of a prey species to a distasteful model species helps protect the mimic while it is rare because predators learn about the distasteful quality only while the model is more common than the mimic (Carpenter and Ford, 1933; Sheppard, 1958; Ford, 1964, 1975); and (2) finding that certain drosophila mutants persisted in experimental populations (*Bar* and *sepia*) because their relative fitness increased with their decreasing frequency in the population, as discovered by L'Heritier and Teissier (1937) and Teissier (1943, 1954). Both of these general types of selection are now documented with much additional light thrown on the mechanisms that promote the rare form advantage in each case. Mimicry can be considered a special case of the general predator-prey relationship summarized in the term *apostatic selection* (Clarke, 1969), which implies that the rare phenotype gains advantage because it is so noticeably different from the common "normal" form. The second case of mutant phenotypes having minority advantage often results from adult interactions in mating behavior (Petit, 1972; Petit and Ehrman, 1969; Ehrman, 1972*a*,*b*; Spiess, 1970), particularly when males of two or more genotypes are rivals at different frequencies for acceptance by females. Finally, a third general body of evidence for frequency-dependent fitness values comes from interactions between two or more genotypes competing during preadult stages of growth (Lewontin, 1955; Sokal and Huber, 1963; Sokal and Karten 1964; Kojima, 1971).

We focus on this minority advantage mode of selection for at least these reasons: (1) a stable equilibrium is possible, (2) some of the principles that apply to constant fitness values (Chapter 14) do not hold, and (3) combinations of "minority advantage" with constant fitness can produce mixtures of stable and unstable equilibriums at more than a single point in the gene frequency array of a population. First, we treat these principles by following the rationale and discussions of frequency-dependent models given by Li (1955*a*, 1962, 1967*c*), Lewontin (1958), and Clarke and O'Donald (1964). Further, an extensive summary of mathematical treatment can be found in Wright (1969, Vol. II, Chapter 15). Examples giving evidence for minority advantage will follow this elementary theoretical summary.

Minority Advantage: the Stable Point

If selection pressure is regulated by frequencies of genetic units (that is, by a "frequency-stat" analogous to a temperature-controlling mechanism known as a thermostat, as pointed out by Li, 1962), the W values will vary as selection proceeds to change those frequencies until a stationary state is achieved. We proceed in the same manner as in Chapter 14 to arrive at formula (14-15), with new symbols for W to relate fitness as a function of q.

Let $W_1(q)$, $W_2(q)$, $W_3(q)$ be the fitnesses of AA, Aa, and aa genotypes, respectively, as functions of the allelic or genotypic frequencies. If we supply these fitness symbols in place of W_1, W_2, and W_3 in the *net fitness* statement preceding (14-7) and use the general stationary state expressed in (14-14), we obtain as a solution the equilibrium condition (Lewontin, 1958)

$$\hat{q} = \frac{W_1(q) - W_2(q)}{[W_1(q) - W_2(q)] + [W_3(q) - W_2(q)]} \tag{16-1}$$

This is analogous to (14-15) with the constant fitnesses replaced by fitness values as functions of q.

After brief contemplation of this formula, the student will no doubt realize that the equilibrium (or equilibriums) here described could be exceedingly complex, since the fitness functions of q could be simple proportionality, quadratic, cubic, or logarithmic; they could be directly or indirectly related to q; they might not be the same for all $W(q)$—that is, $W_1(q)$ might be linear while $W_2(q)$ might be quadratic, etc.—in fact, there are innumerable possibilities one could explore. We illustrate the simplest model of minority advantage with a linear proportional relationship between frequency (q) and fitness values. In more complex cases, transformation of the data into a linear system can often be made, no matter how complex the function of q may be. The equilibrium condition above (16-1) will have to be solved ultimately by supplying fitness values as functions of q.

In Table 16-1, some simple models of minority advantage leading to stable equilibrium are given. In Table 16-1A (from Clarke and O'Donald, 1964, and from Li, 1967*c*), let t = proportional relationship between selection coefficients and genotype frequencies such

TABLE 16-1 *Minority Advantage Selection*

A Fitness values proportional to genotype frequencies

	Genotypes			
Fitness	AA	Aa	aa	
$W_i(q)$	$1 - tp^2$	$1 - 2tpq$	$1 - tq^2$	
If $q = 0.1$	$1 - 0.81t$	$1 - 0.18t$	$1 - 0.01t$	(if t is a positive ratio)
If $q = 0.9$	$1 - 0.01t$	$1 - 0.18t$	$1 - 0.81t$	

Substituting fitness values $W_i(q)$ into (16-1) gives

$$\hat{q} = \frac{1 - tp^2 - 1 + 2tpq}{1 - tp^2 - 1 + 2tpq + 1 - tq^2 - 1 + 2tpq} = \frac{2pq - p^2}{4pq - p^2 - q^2}$$

Then $2pq - p^2 = 4pq^2 - p^2q - q^3$
Changing $p \to q$ and grouping, $6q^3 - 9q^2 + 5q - 1 = 0$
Factoring, $(2q - 1)(3q^2 - 3q + 1) = 0$
Whose only real root is $q = \frac{1}{2}$

B Frequency-dependent fitness values of *ST-CH* karyotypes proposed by Wright and Dobzhansky (1946) for experimental populations of *Drosophila pseudoobscura* from Piñon Flats

	CH/CH	ST/CH	ST/ST
$W_i(q)$ fitness	$[1 - a + bq]$	1	$[1 + a - bq]$

Substituting these values $W_i(q)$ into (16-1) gives (letting q = frequency of ST)

Equilibrium

$$\hat{q} = \frac{(1 - a + bq) - 1}{(1 - a + bq) - 1 + (1 + a - bq) - 1}$$

but the denominator is zero!

We may avoid the problem of dividing by zero in this way. At equilibrium, according to formula (14-14), $\bar{W} = pW_2 + qW_3$, which is true whether W is frequency dependent or not. Therefore, the equilibrium condition is $\bar{W} = p + q + aq - bq^2 = 1 + aq - bq^2$. If we write out $\sum fW_i(q)$ in any generation, we have $\sum fW_i(q) = p^2(1 - a + bq) + 2pq + q^2(1 + a - bq)$. Changing p to q and simplifying, and equating to the equilibrium condition, we have $1 - a + bq + 2aq - 2bq^2 = 1 + aq - bq^2$, so that $-a + bq + aq - bq^2 = 0$. Factoring gives $(a - bq)(1 - q) = 0$. Solving for q at equilibrium, we have $\hat{q} = a/b$ (stable equilibrium) or $\hat{q} = 1$ (trivial).

that

$$t = \frac{1 - W_1(q)}{p^2} = \frac{1 - W_2(q)}{2pq} = \frac{1 - W_3(q)}{q^2}$$

Then fitness values will be as given in the top row of the table. If q is small, a will increase; if q is large, a will decrease. When we substitute the fitness values from the table into (16-1), the equilibrium condition, we obtain a result independent of t (as shown in the table): $(2q - 1)(3q^2 - 3q + 1) = 0$. The only real root gives the equilibrium gene frequency: $q = p = \frac{1}{2}$.

In this particular model, when the three fitness values are noted at the equilibrium point, it may come as a surprise that the heterozygote's fitness is lowest among the genotypes ($W_2 = 1 - 0.5t$, compared with $W_1 = W_3 = 1 - 0.25t$). Under constant fitness values, such a result was an unstable point. That this minority advantage equilibrium is a stable one can easily be demonstrated; the student should try p,q values away from the equilibrium on either side to find the direction of change. For example, if we let $p = 0.520$, $q = 0.480$, let the deviation (x) from equilibrium (p) be $x = 0.020$. Then, after selection (if we let $t = 1$), $p' = 0.516$, moving closer to the equilibrium value of $p = 0.500$. The new deviation is $x' = 0.016$. It should be clear that continuous applications of this model in sequences will generate smaller and smaller deviations leading toward the equilibrium point. The fractional change in x becomes $x' = (8 - 4t)/(8 - 3t)\,(x) = p' - \frac{1}{2}$ (see Exercise 3). Thus, x' is approximately $(\frac{4}{5})(x)$, which is about (0.8)(0.02) in this case.

This example with heterozygotes having the least fitness value stresses that we cannot draw conclusions about the outcome of selection simply by looking at genotypes separately from the population or at a single generation in a population. When fitness values depend on frequencies of genotypes, we would not be able to tell what the outcome of selection would be unless we compared fitness values over at least a few different frequencies of genotypes. We might easily look at relative fitnesses, find the heterozygote to be lowest in fitness, presume the array of fitness values to be constant, and come to the completely erroneous conclusion that the equilibrium point was unstable, with changes in p,q expected to go in the opposite direction from that actually taking place. There is no shortcut to learning about the nature of the selection pressure and the stability of the population.

Of course, it should occur to the student that this case of heterozygote inferiority is not necessarily characteristic of the frequency-dependent model. It could easily be that the heterozygote is less affected by its frequency than the homozygotes. For example, if t for the *Aa* genotypes is just half of what it is for the homozygotes given in the previous case, its fitness would be $1 - tpq$ compared with $1 - tp^2$ and $1 - tq^2$ for the two homozygotes. In such a case, the equilibrium gene frequency would be $p,q = \frac{1}{2}$, as before, but all three genotypes would be equal in fitness at that point ($W_1 = W_2 = W_3 = 1 - 0.25t$). (This case is easily extended to multiple alleles, as shown by Li, 1962.)

Both of these examples illustrate the same stable point, and we may look at another example. Wright (Wright and Dobzhansky, 1946), when estimating the fitness values of *ST* and *CH* chromosomal arrangements in *Drosophila pseudoobscura* (see also Example 15-2), pointed out that the homokaryotype W could easily involve a function of q that would be linear, with heterokaryotype (ST/CH) exactly intermediate in fitness and thus give a good fit to the observed changes in the experimental populations. He proposed the fitness values given in Table 16-1B with the fitness of each homokaryotype expressed as a linear regression of opposite sign to the other. The equilibrium *ST* frequency was $q_{ST} = 0.70$ approximately. At equilibrium $\hat{q} = a/b$, therefore, the linear regression coefficients (a and b) relating fitness and arrangement frequency would be $a = 0.70b$. (To obtain an estimate of these two unknowns, we need another equation; Wright uses the equivalent of our (14-7) Δq formula, supplying an average $\overline{\Delta q}$ from the observed changes in the populations and then calculates that $a = 0.902$, $b = 1.288$.) The actual values of a and b are not as important here as their relationship in giving the equilibrium point. Note that in this case, as in the previous one above, at equilibrium all karyotypes have equal fitness values.

Minority Advantage: Maximum $\bar{W}$

One of the the principles we noted when fitness values were constant could be epitomized in Wright's concept of the "adaptive peak," or Fisher's "fundamental theorem," that selection tended to maximize $\bar{W}$ (Chapter 14). With frequency-dependent fitnesses, however, this general principle does not necessarily hold. In a numerical example, we may look at the case above where Wright used the a and b linear regression coefficients as a possible model for *ST-CH* karyotypes. It is easy to see that at equilibrium in that case $\bar{W} = 1.00$ when $q = 0.70$. However, if we take the chromosomal frequencies as $p_{CH} = q_{ST} = 0.50$, $\bar{W}$ will be also 1.00, irrespective of a and b values. Between $q = 0.50$ and equilibrium q, $\bar{W}$ will increase—for example, when $q = 0.6$, $\bar{W}$ becomes 1.026.

When fitnesses include a function of q, the derivative of $\bar{W}$ includes at least two main components: (1) the same component as in (14-17)—namely, $\sum W\, df/dq$ in which W is constant and frequencies (f) are derived with respect to q—and (2) the component in which frequencies are constants and W is derived with respect to q.

$$\frac{d\bar{W}}{dq} = \sum W \frac{df}{dq} + \sum f \frac{dW}{dq} \tag{16-2}$$

As pointed out by Wright (1949) and Li (1955*a*), the new term on the right is the *average* of dW/dq for the genotypes in the population at a particular q and may be written as $\overline{dW/dq}$. It measures the average effect of allele frequencies on genotypic fitness values over all genotypes. This term is zero when the W is constant and (16-2) reduces to (14-17).

At the equilibrium point of the $\bar{W}$ curve under frequency-dependent selection, $(d\bar{W}/dq)$ may not equal zero. Instead, at equilibrium, while the component with W as constant becomes zero (left component of (16-2)), the right component may not be zero. At equilibrium, then,

$$\frac{d\bar{W}}{dq} = \sum f \frac{dW}{dq} = \frac{\overline{dW}}{dq} \tag{16-3}$$

while $\sum W\, df/dq = 0$. The derivative of $\bar{W}$ (net fitness) with respect to q equals the average derivative of fitnesses with respect to q.

We may easily demonstrate this principle by using the model from Table 16-1B. Taking $\bar{W} = 1 - a + bq + 2aq - 2bq^2$, its derivative with respect to q is, first, $d\bar{W}/dq = b + 2a - 4bq$. At equilibrium, $\hat{q} = a/b$ so that this expression $= b - 2a$. Second, $\sum f\, dW/dq = q^2(-b) + 2q - 2q^2 + (1 - 2q + q^2)b$, and since $\hat{q} = a/b = b - 2a$. Finally, for the q value at the point where $\bar{W}$ is maximum, we may set the $\bar{W}$ derivative above equal to zero, whereupon $q_{\bar{W}_{\max}} = (b + 2a)/4b = 0.600$ for the case where $a = 0.7b$, and $\bar{W}_{\max} = 1.020$. This is a point just halfway from $q = 0.5$ ($\bar{W} = 1.00$) to $\hat{q} = 0.7$ ($\bar{W} = 1.00$). Thus, the stationary population point is not at the adaptive peak, and Fisher's fundamental theorem will not apply to this case.

Why does q continue to rise above the $\bar{W}_{\max}$ point? We see that the *aa* genotype has the highest fitness until the equilibrium point is reached. Consequently, under frequency-dependent selection, there is no "homozygous genetic load" at equilibrium; all genotypes will be as equal as they can get in fitness under the conditions imposed by selection. (Note that in the first case, we considered *Aa* was less fit than the two homozygotes at the stable equilibrium point, but the two homozygotes were equal to each other.)

Comparison between the outcomes for frequency-dependent selection and for constant fitness selection with heterosis may be made to show that rate of change is very much faster in the frequency-dependent model at extreme frequencies than in the heterotic constant model, but as equilibrium is approached it is quite impossible to distinguish the two models purely by looking at the rate of change (Δq) (see Exercise 5).

Minority Advantage: Combination of Constant Fitness and Frequency-Dependent Fitness

There was some evidence from Example 15-7 that both heterozygote superiority and frequency-dependent selection were combined to bring about a stable equilibrium. The same may also be true for a large number of cases, including the classic *ST-CH* of *D. pseudoobscura* (Examples 2-4, 15-2), so that we ought to examine the nature of such equilibriums briefly. As illustrated by Li (1967*c*) and Clarke and O'Donald (1964), if the three genotypes are different in fitness due to constant W (perhaps from one stage of the life cycle) and frequency dependent $W(q)$ (from some other stage) so that their respective net fitnesses over their entire life cycle can be the product, symbolically, using the case given in Table 16-1B,

Types	AA	Aa	aa
	$W_1(1 - a + bq)$	W_2	$W_3(1 + a - bq)$

Substituting these net fitnesses in (16-1) for the equilibrium q value we obtain,

$$\hat{q} = \frac{W_1(1 - a + bq) - W_2}{W_1(1 - a + bq) - 2W_2 + W_3(1 + a - bq)}$$

which on rearrangement gives the equilibrium statement with frequency-dependent fitnesses on the left and constant fitnesses on the right of the equation.

$$\begin{aligned} W_1(a - bq) - q(a - bq)(W_1 - W_3) &= (W_1 - W_2) - q(W_1 - 2W_2 + W_3) \\ (a - bq)(W_1 p + qW_3) &= (W_1 - W_2) - q(W_1 - 2W_2 + W_3) \end{aligned} \quad (16\text{-}4)$$

When all constant W's are equal ($W_1 = W_2 = W_3 = 1.00$), the right-hand portion equals zero and on the left we have only the first term, which then reduces to $\hat{q} = a/b$, as before with just minority advantage. If, on the other hand, the left portion is zero because there is no frequency-regulating component (with $a - bq = 0$), the right-hand portion reduces to (14-15). Consequently, formula (16-4) represents a balance between the minority advantage component and the constant fitness component. These forces may work together in the same direction, as they presumably do for the *ST-CH* case (Examples 2-4, 15-2) and the *2L* inversions of *D. ananassae* (Example 15-7). They could be conflicting in direction to give two equilibriums, or one stable and the other unstable, or they could produce a single equilibrium depending on magnitude of opposing forces. Li and Clarke and O'Donald give examples with three equilibriums. In such cases, there may be two stable equilibriums and one unstable between the stable points or two unstable with a single stable point.

It is possible to test the stability of such equilibriums by supplying the known coefficients into (16-4) and arranging terms into the familiar quadratic (or cubic or higher form, as the case may be) as follows for the numerical example we have been considering:

Letting $a = 0.7$, $b = 1$, $W_1 = 0.8$, $W_2 = 1.0$, $W_3 = 0.7$ (for the stable case), so that using (16-4), $(0.7 - q)(0.8p + 0.7q) = (-0.2) - q(-0.5)$. Solving to get the quadratic, we obtain: $0.76 - 1.37q + 0.1q^2 = 0$, whose only root between 0 and 1 is $\hat{q} = 0.58$. (Compare with the $\hat{q} = 0.7$ for minority advantage or $\hat{q} = 0.40$ for constant fitness values as in Table 14-5.)

When the q value is substituted in the quadratic above, and the left side is differentiated, the value is found to be negative, indicating a stable equilibrium (Lewontin, 1958). On differentiating such an expression and obtaining a positive value, the equilibrium would be unstable (see Exercise 6).

EXPERIMENTAL EVIDENCE FOR FREQUENCY-DEPENDENT SELECTION

Experimental geneticists, it must be admitted, did not put much effort into demonstrating variable fitness values among genotypes until the 1950s and afterwards. There is good reason for the lack of information in view of the technical difficulties involved in attempting to discriminate between variable and constant fitness when measurement of fitness is far from an exact operation. Now, owing to the immensely increased genetic resolving power we have available to detect genotypes via their allozyme or protein products (Chapter 3), and our motivation to discover general explanations for the maintenance of the high-order genetic variation observed in most sexually reproducing populations, we have the inducement and curiosity to examine frequency-dependent selection as a likely mechanism for providing a substantial portion of genetic polymorphism.

It is fitting that the student's attention be drawn to the significance of minority advantage selection by quoting from Haldane (1949):

> *Probably a very small biochemical change will give a host species a substantial degree of resistance to a highly adapted microorganism. This has an important evolutionary effect. It means that it is an advantage to the individual to possess a rare biochemical phenotype. For just because of its rarity it will be resistant to diseases which attack the majority of its fellows. And it means that it is an advantage to a species to be biochemically diverse.*

There is no reason to doubt that this concept may well be fundamental for a large number of polymorphisms of functional significance to organisms. Another fundamental instance of frequency dependency has been expressed in a general model article by Cockerham, et al. (1972):

> *The idea that mixtures contribute more than can be accounted for by the separate components has been around a long time.... It is obvious that a society which requires many talents benefits from an ensemblage of individuals providing these talents and that, moreover, the individual is in a relatively better position when his talent is in short supply.*

There is considerable evidence that a genotype's performance is affected by its neighbors. From self-fertilizing plant populations, Allard and Adams (1969) have shown that fitness increases with rarity. In studies of drosophila viability, the following have shown similar interactions between genotypes occupying a single culture, which can lead to frequency-dependent selection: Lewontin (1955), Lewontin and Matsuo (1963), Dawood and Strickberger (1969), Beardmore (1963), and Young (1970). All these experiments indicate that genetic heterogeneity often leads to environmental heterogeneity and thus provides for a diversity of "niches" (see page 522) within which genotypes may find out their optimal niche and thus maximize the population's total fitness. Later in this chapter we discuss how the corollary situation of diverse niches with different fitnesses in each niche may lead to maintenance of polymorphism.

Minority Advantage in Predation

"Apostatic selection" (Clarke, 1962*a,b*, 1969) illustrates the principle that protective coloration among a prey species—snails under predation by birds, for example—can give a minority advantage when the minority stands out from the common type. Predators apparently concentrate on one or a few common varieties of prey and tend to overlook rarer forms even if they seem obvious to human observers. By a learning process, the predators develop a specific "searching image" for the most frequently encountered types of prey. Thus, Clarke pointed to observations of mixed populations of the polymorphic snail species *Cepaea nemoralis* and *C. hortensis*, where one species was common and the other more rare; if the two prey species shared the same predator, then that predator, no doubt treating both species as a single one, would take the species and variant of that species. The rarer species would then tend to lose the variant that most closely resembled the common species' variant. Thus, a negative correlation would be brought about between the frequencies of the same phenotype in the two species because the variant would decrease proportionally faster in the smaller species than in the larger. Murray (1972) summarized much of the evidence that predators that depend on visual cues for prey take more of a common type than a rare form, because once they acquire an image of the prey they continue to search for it preferentially to any other. Experiments by Allen and Clarke (1968), where artificially colored baits, dyed green or brown, were displayed on a green lawn for birds to take, demonstrated clearly the birds' "searching-image" behavior. Most birds took the more common bait exclusively, but two exceptional blackbirds (*Turdus merula*) out of fourteen took the rare form exclusively, thus demonstrating that a bird continues to search for a particular bait even when it is scarce.

Polymorphism is advantageous to Batesian mimics in which the model is distasteful to predators but the mimic edible. Such mimicry, unlike the Müllerian type, which depends on both species being unpalatable, depends on the deception of predators through learning about the more common model. The Browers and their colleagues have described many such cases and have established the behavioral basis for this type of mimicry (Brower, Cook, and Groze, 1967; Brower, Alcock, and Brower, 1971, summarize much experimental evidence). Brower (1960) points out that the mimic must not be too common relative to the model lest the predator encounter it so often that the deception is discovered or, alternatively, fail to form the initial association of model appearance with unpalatability.

In view of the theoretical selection favoring *scarcity* among Batesian mimics, then, it becomes necessary for the mimic species to avoid the constant threat of extinction, so that a strategy for increased mimic forms in a species has taken place in many instances. The emergence of models in the spring to ensure predators' learning has also been documented (see Rothschild, 1971; Ford, 1964, 1975).

Murdoch (1969) presented a general experimental and theoretical study on predation, using marine snails preying on mussels and barnacles. He pointed out that when a predator has strong preference for a particular prey, there is no minority advantage, but when preferences are less discriminating, predators take the most abundant prey in greatest numbers until the prey becomes difficult for the predators to find. At this time, they switch to the next most abundant prey, provided that they can become trained to whichever prey is abundant. A patchy prey distribution might provide the right opportunity for this sort of predation to occur in nature. The response necessary for minority advantage was a function of prey density. Murdoch pointed out: "One possible consequence of switching a predator feeding on two potentially competing prey species is that the two prey might be able to coexist indefinitely owing to heavy predation on whichever species was winning in competition." Thus, diversity would be likely to increase, as predicted by minority advantage models (Holling, 1965), whether diverse forms are separate species or genetic polymorphs within a species population.

Minority Advantage in Mating Activity

EXPERIMENTAL EVIDENCE. Fitness differences expressed at the adult stage are probably the most critical in the life cycle. A diversity of genotypes acting on the efficiency of mating activity will no doubt exert considerable influence on the efficiency of the sexual process. Intrasexual rivalry, in which minority genotypes may participate in mating in greater numbers proportionally to their frequency than majority genotypes, has been found to occur especially in various drosophila species. Petit (1951) first observed in multiple-choice matings between *Bar* and wild-type *D. melanogaster* that *Bar* males mated with relatively greater success when they were rare than when common. Later (1958) she described in detail, using a "coefficient of sexual selection," which is similar to our m in formula (15-1), for *Bar* versus wild and for *white* eye versus wild, that for both mutants there was a minority advantage plus a high frequency advantage for *white* (see Petit and Ehrman, 1969, for summary). In 1964, Ehrman, working with *D. pseudoobscura AR* and *CH* chromosome arrangements, and Spiess, using *D. persimilis WT* and *KL* arrangements, independently made tests on mating activity with changing frequencies of competing karyotypes and found positive evidence for minority advantage. Techniques of these two investigators differed in certain details, but the results were clearly similar and indicative of the same phenomenon.

Ehrman's experiments were begun with *AR* and *CH* flies, which had been selected by Dobzhansky and Spassky for positive or negative geotaxis (see Example 7-4 and Ehrman, et al., 1965). In particular, pairs of experimental populations had been maintained by allowing a small migration of opposite selected flies each generation between members of the pair. Each pair started with one population completely *AR* (monomorphic) and the other completely *CH*, but within just four generations the opposite arrangement introduced had risen in frequency far faster than simple heterosis would have allowed. By the twelfth generation, the populations had become reversed with respect to their original arrangement; the original *AR* population became high in *CH* and the original *CH* high in *AR*. The explanation for that reversal was found when matings between *AR* and *CH* homokaryotypes were tested at unequal frequencies, as described in Example 16-1. When equal numbers of both sexes were present in mating chambers, all individuals had about equal probability of mating, but when one karyotype was more frequent than the other (80 versus 20 percent), the rare karyotype was favored in mating success.

In *D. pseudoobscura*, Ehrman found the mating advantage of rare genotypes not only for mutants but also for chromosomal arrangements from different geographic origin and even for flies of the same strain grown at different temperatures (Ehrman, 1966, 1967, 1968, 1972*a,b*). Also, in the *D. willistoni* group, minority males from geographic strains—some of which exhibit a tendency for ethological isolation—displayed greater mating success when rare (Ehrman and Petit, 1968). In all these studies, Ehrman and Petit used an observation chamber within which matings could be recorded without disturbing the flies (Elens-Wattiaux chamber). No flies were withdrawn from the mating chamber, but males could, of course, mate more than once, and it is technically difficult to distinguish males that had mated from those that had not. Consequently, some males may have mated two or more times and given their genotype even more advantage than if they had had only a single chance to mate. Spiess (1968*b*) and L. D. Spiess and E. B. Spiess (1969) used a technique of aspirating out mating pairs to prevent males from mating more than once. The technique allowed recording only the rarer males that had mated just once. Positive results have indicated that the advantage to rare males is expressed often in first matings (Ehrman and Spiess, 1969).

From Ehrman's earlier experiments where both sexes were varied for input ratios simultaneously, it was not possible to distinguish which sex (or both) might be the one that "recognized" the cue of frequency among the opposite sex or among its own sex to alter its mating activity to favor the minority. Spiess had done all experiments by varying the ratio in one sex at a time and had found a minority advantage only when male ratios were varied. When female ratios were varied, frequency of mating followed from the simple proportionality of input. Meanwhile, Ehrman worked out a technique employing a double chamber (1967), one chamber on top of the other with a partition of cheesecloth between chambers. By introducing two types of flies in the upper chamber and a single type in the lower, it was possible first to use a "minority" ratio above (say, 20 *AR*:5 *CH*); but by introducing an excess of *CH* in the lower chambers, if transmission of information between chambers either by odors (chemical pheromones) or sound (male wing vibrations) was necessary to provide the cue of frequency, the flies in the upper chamber would mate as if they existed at equal proportions. Later, Ehrman (1970) was able to induce a minority situation by putting equal numbers in the upper chamber and introducing a single type in the lower. When Ehrman changed frequencies among males only (or both sexes) in the lower chamber, the cue for altered frequency was received in the upper and the mating response of females was altered. If females alone were varied in the lower chamber, no cue for frequency was communicated to the upper chamber, and the flies there mated as if there had been no change below. Some data from these experiments are given in Example 16-2. Thus, it was demonstrated that the mating result depended on relative frequencies of genotypes (or karyotypes) among males, which are then selected by the females. We may suppose (Ehrman and Spiess, 1969)

> *. . . that with the two kinds of males present, one being rare with respect to the other, when a female is courted she is stimulated by some cue from the male (either chemical or auditory). Her first reaction is most often reluctance, and consequently she receives more than one stimulus, since she usually encounters more than one male. If the majority produce a set of stimuli which bring about a "sensory adaptation," it may be the difference in stimulus brought about by the minority which induces her to accept the courting male, simply because his cue is different.*

An interesting alternative mechanism for the minority mating advantage was proposed by Dews (1970). Components of mating cues are partitioned so that one genotype emphasizes a certain component while another genotype emphasizes another, both components being important for female acceptance of the male. When a majority of males provides a preponderance of one component, the minority may provide the other component sufficiently for female acceptance. This kind of mechanism may well occur in some cases, but it would be more economical to expect that the mechanism would involve single behavioral factors than compound ones. In addition, two genotypes (or environmentally produced variants) would have to differ in just the right components to make the minority advantage occur as frequently as has been observed.

A MODEL FOR MATING ADVANTAGE OF RARE MALES. In these instances of rare male mating advantage, not only will there be an equilibrium because of frequency-dependent selection, but the problem of different fitness values in the two sexes also arises. Anderson (1969*a*) proposed a model that fits the observed mating success results for Ehrman's *AR-CH* experiments quite well, illustrating the use of our considerations in Table 16-1. The advantage of rare *AR* or *CH* males is diagrammed in Figure 16-2A; for various frequencies of *AR* or *CH* homokaryotypes, the m value for male mating success ranges from low values of about 0.6 when those flies are common to a high value of 3.6 (*AR* geotactically positive males when rare).* This rapid increase in mating success with decrease in gene frequency does not fit a linear relation between fitness value and frequency; instead, it suggests an inverse proportionality of the form $W_1 = 1 + x/G_i$, where x is some parameter of selection and G_i is the frequency of the ith male genotype. Anderson presented a series of hyperbolas for several values of x to see which would be appropriate for these observed curves, illustrated here in Figure 16-2B. Anderson gave a general case in which both sexes might have the same sort of selection, but we shall abbreviate the model simply for the male advantage since the algebra is simpler and the outcome is appropriate to the observations of Ehrman (1967).

* Ayala (1972) pointed out that mating success can be expressed as a single statistic by regression, the amount of mating on the frequencies at which the two competing types are present: the logarithm of the ratio of A:B that have mated regressed on the log of the input ratio. Intersects of these regressions with the "equality" ratio (45-degree slope) indicate the relative ratio of type A:type B at which mating sucess is equal.

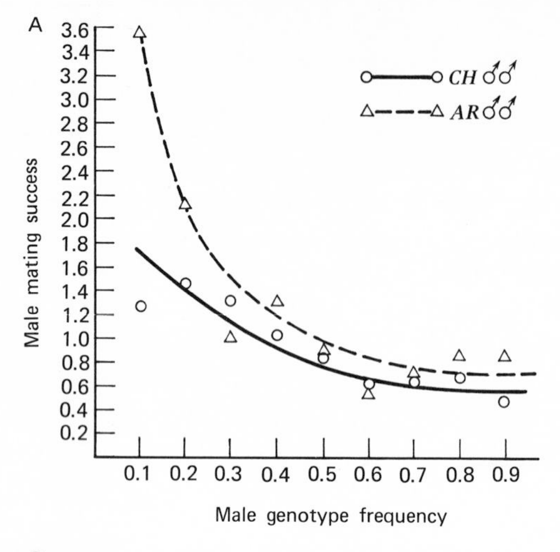

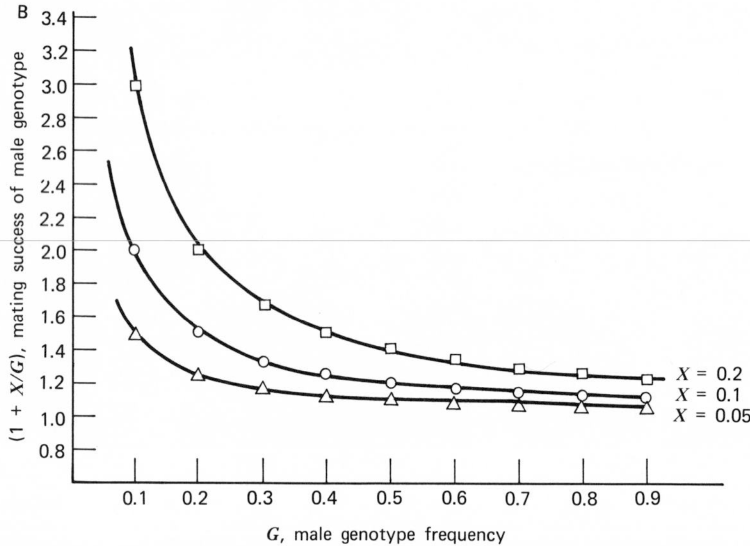

Figure 16-2A. Male mating success as a function of male genotype frequency in mating tests between *AR/AR* and *CH/CH* karyotypes of *Drosophila pseudoobscura.* Data from Ehrman's Table 1A, 1967; graph from Anderson, 1969.

Figure 16-2B. Male mating success as a function inversely proportional to male genotype frequency. From Anderson's model, 1969.

We shall assume all genotypes are equal in fitness in females. Furthermore, the frequencies of genotypes (karyotypes) at the zygotic stage we shall assume to be the same in both sexes, and only in the adult stage of males will the fitness difference become expressed. At the start, however, we should recall that the sexes could be different in allelic frequencies, so we shall use p,q symbols for females, but r,s for the corresponding alleles in males, as in formula (1-6); however, at the end of selection, we shall find that both sexes will have the same allelic frequencies for this case. At the zygotic stage of G_0, then,

	Females			Males		
	AA	Aa	aa	AA	Aa	aa
Frequencies	p^2	$2pq$	q^2	r^2	$2rs$	s^2
Fitness	1 :	1 :	1	$1 + x/r^2$:	$1 + y/rs$:	$1 + z/s^2$

Females	Males
Gametic frequencies	Gametic frequencies after selection (r'_A and s'_a):

A	a
p,	q

$\underline{A}$

$$r' = \frac{r^2\left(1 + \frac{x}{r^2}\right) + rs\left(1 + \frac{y}{rs}\right)}{r^2\left(1 + \frac{x}{r^2}\right) + 2rs\left(1 + \frac{y}{rs}\right) + s^2\left(1 + \frac{z}{s^2}\right)}$$

$$r' = \frac{r^2 + rs + x + y}{r^2 + 2rs + s^2 + x + 2y + z} = \frac{r + x + y}{1 + x + 2y + z}$$

$\underline{a}$

$$s' = \frac{s + z + y}{1 + x + 2y + z} = 1 - r'$$

Frequency A after selection (recalling 1-6)

$$f'_A = p(r') + \frac{ps' + qr'}{2} = \frac{p + r'}{2}$$

then, substituting

$$f'_A = \frac{1}{2}\left[p + \frac{r + x + y}{1 + x + 2y + z}\right]$$

At equilibrium, $(p + r)/2$ (before selection) $= (p + r')/2$ (after selection). Therefore,

$$\frac{p + r}{2} = \frac{1}{2}\left[p + \frac{r + z + y}{1 + x + 2y + z}\right]$$

$$0 = p(x + 2y + z) + x + y$$

$$\hat{p} = \frac{x + y}{x + 2y + z} \qquad (16\text{-}5)$$

(frequency of A at equilibrium).

From Anderson's series of hyperbolas in Figure 16-2B, the appropriate value for $x = 0.10$. In addition, he estimated values for the other coefficients to be $y = 0.05$ and $z = 0.3$, which when substituted into (16-5) give $\hat{p} = 0.30$, $\hat{q} = 0.70$ at equilibrium. (What will be the relative fitness values for males at equilibrium? And how do those relative values differ from the case cited above for *ST-CH* in Table 16-1B?)

How does Molecular Variation Assert Its Influence on Fitness?

While it is probably partly erroneous to think that allozyme variants exert selective effects independently for each genic locus (no doubt cumulative fitness and epistatic interactions are of great significance, as discussed in Chapter 17), we nevertheless may expect to find that each allozyme or protein locus contributes in some way and in a significant amount to total fitness by its polymorphism if that is stable in the population. A particularly lucid case in which there was apparent evidence for selective neutrality of allozyme alleles until a test of frequency dependency was made with positive results is that of the esterase-6 locus in *Drosophila melanogaster* (chromosome 3 at 36.8 Morgan units). The fast (F) and slow (S) allelic forms of this esterase, distinguishable by relative electrophoretic mobility of bands in starch gel, were studied extensively by population-cage technique by MacIntyre and Wright (1966), who found that Est-6 alleles established equilibrium at intermediate frequencies in a few generations when the genetic background was heterozygous. They concluded, however, that these alleles could be interpreted to be selectively neutral if they were removed by crossing over from linked blocks of fitness modifier loci. In Sweden, Rasmuson and coworkers (see Example 5-6) examined populations segregating for Est-6 and the LAP locus, and they found stable equilibrium levels for both sets of alleles with linkage equilibrium established for 35 generations. While the approach to equilibrium was slow for each locus, probably reflecting small differences in selective value between genotypes, a general excess of heterozygotes was demonstrated for the Est-6 locus when tested at equilibrium. These two groups of workers then agreed that slight selective differences with some heterozygote superiority were indicated but not with the magnitude expected for a locus whose polymorphism must be determined by strong selective forces.

Considerable light was shed on the problem by Kojima and his colleagues by studies of the same locus, illustrated in Example 16-3. After developing a selection model based on frequency-dependent fitness values for the Est-6^F and Est-6^S genotypes, their population results and viability estimates showed a much better fit than with constant fitness values. Similar results were obtained with the alcohol dehydrogenase (ADH) locus of *D. melanogaster* (Kojima and Tobari, 1969*a*) and with the 2L inversion of *D. ananassae* marked with an allozyme variant, acid phosphatase (ACPH) (Kojima and Tobari, 1969*b*). All large-scale experiments in Kojima's laboratory indicated almost equal fitness values for genotypes near the equilibrium point in populations, but with reversal of fitnesses from one extreme frequency to the other. With such genetic variants, there is no substantial segregational genetic load at equilibrium. Kojima (1971) was of the opinion, now shared by many, that a large proportion of the observed protein polymorphism in populations may be kept at stable equilibrium by frequency-dependent selection.

At this stage, we cannot yet rule out the possibility that closely linked fitness modifiers not recombinable with the protein variant loci were actually responsible for the positive selection results Kojima described. Certainly, in *D. ananassae*, the ACPH locus was linked to the 2L chromosome inversion system, although in the other cases no cytologically obvious variation was found. Yamazaki (1971) tested the sex-linked Est-5 variants in *D. pseudoobscura* populations using many techniques similar to those employed by Kojima without finding evidence for significant selection of any kind, including frequency dependency (although mating propensity was not tested). While F and S alleles of that locus are found at nearly equal frequencies in California populations of that species, and the locus is highly polymorphic in all populations so far examined (Example 3-4), no experimental laboratory evidence has yet been found attributable to selection differences among those variants. Prakash, Lewontin, and Hubby (1969) reported that there were gradual changes in allelic frequencies from western to eastern populations of that species, so that from geographic distribution it is logical to conclude that some selective influences under natural conditions determine polymorphism at the Est-5 locus. (However, see also Powell, 1973, for evidence of selection between esterase alleles in *Drosophila willistoni*.)

EXAMPLE 16-1

MINORITY MATING ADVANTAGE IN DROSOPHILA

Strains of *Drosophila pseudoobscura* selected for geotactic scores (Example 7-4) by Dobzhansky and Spassky were made homokaryotypic for *AR* or *CH* arrangements in four populations and run as two pairs: one pair as (*AR*) – (*CH*) selected for positive geotaxis and the other pair (*AR*) – (*CH*) selected for negative geotaxis. Between members of the pair, a certain small fraction of migration was made each generation; in the *AR* populations (with 25 parents of each sex), one of five flies was a migrant from the *CH* member of the pair, while in the *CH* populations (250 parents of each sex), one of fifty flies was a migrant from the *AR* population. Surprisingly, within four generations, the original frequencies of *AR* and *CH* in these populations became reversed so that *CH* was predominant in the original *AR* populations and likewise *AR* was predominant in the original *CH*. As the amount of migration alone could not explain the suddenness of the alteration in frequencies, a likely mechanism was proposed and tested by Ehrman, et al. (1965), where flies that were the minority immigrants into each population were at some advantage—in mating, presumably—each generation.

Observations of matings were made using a chamber described by Elens and Wattiaux (1964)—essentially a wood block about 2 cm thick with a circular central depression about 10 cm in diameter, a glass top, and a sectional grid on the floor for reference when mating pairs are observed (see illustration in Ehrman, 1965, or in Parsons, 1973). Two types of male and two types of female were distinguished by wing clipping so that when a pair mated, four types of mating could be recorded. (Other ways of marking by colored dyes or poking small holes in wings have been used in the Spiess laboratory.) In *D. pseudoobscura*, copulation averages were 5–7 minutes under room temperature conditions, and mated pairs remainded stationary long enough for making observations easily.

Tests for mating frequency (in an hour) were made on virgin flies matured for three days in isolation on yeasted food. The *AR*:*CH* ratio was set either at equal numbers of the two homokaryotypes (10 *AR*:10 *CH* of both sexes) or set at 20 of one type to 5 of the other (both sexes) in a mating chamber test. Replicate tests were run to produce the data in Table A using strains first selected for geotaxis (left) and strains unselected (right) but from the same populations.

It is apparent that when the ratio of *AR*:*CH* is 1:1, both sexes mate equally (although there is some evidence of positive assortative mating from these karyotypes), as given in the top row of the table. When one type is in the minority, however, it has the advantage in

TABLE A *Number of matings recorded in observation chambers with flies selected (left column) or unselected (right column) by running through geotactic maze. Input ratios same in both sexes. Positive and negative geotactic score types are pooled here because there was no significant difference between + and − in the original data (from Ehrman, et al., 1965)*

Selected for Geotaxis				Unselected for Geotaxis			
Input 10 *AR*:10 *CH* (32 tests)				Input 10 *AR*:10 *CH* (15 tests)			
	AR♀♀	*CH*♀♀	Σ		*AR*♀♀	*CH*♀♀	Σ
AR♂♂	86	51	137	*AR*♂♂	81	50	131
CH♂♂	45	83	128	*CH*♂♂	56	78	134
Σ	131	134	265	Σ	137	128	265
Input 20 *AR*:5 *CH* (8 tests)				Input 20 *AR*:5 *CH* (17 tests)			
	AR♀♀	*CH*♀♀	Σ		*AR*♀♀	*CH*♀♀	Σ
AR♂♂	66	18	84	*AR*♂♂	98	38	136
CH♂♂	6	16	22	*CH*♂♂	40	31	71**
Σ	72	34*	106	Σ	138	69**	207
Input 5 *AR*:20 *CH* (8 tests)				Input 5 *AR*:20 *CH* (19 tests)			
	AR♀♀	*CH*♀♀	Σ		*AR*♀♀	*CH*♀♀	Σ
AR♂♂	16	22	38**	*AR*♂♂	43	62	105***
CH♂♂	12	50	62	*CH*♂♂	26	78	104
Σ	28*	72	100	Σ	69**	140	209

* $0.05 > P > 0.01$ for null hypothesis based on input numbers.
** $0.01 > P > 0.001$.
*** $0.001 > P$.

mating, especially among the unselected flies (both sexes in right column) and in the case of *AR* selected males, although not for *CH* selected males. Slight significance among selected *AR* and *CH* females was not important after further work showed that females were usually not significantly affected in their total receptivity in favor of the minority. Significance among unselected *CH* females was apparently unique to this experiment.

EXAMPLE 16-2

RECOGNITION OF MALE GENOTYPE FREQUENCY BY FEMALE DROSOPHILA

When both sexes were varied in mating chambers using the techniques of Ehrman, Spiess (1968*b*) proposed four possible models that might be invoked to account for the advantage to mating: (1) females "recognize" that two types of males are present and change their receptivity in favor of the minority males; (2) females "recognize" that their sex occurs in two kinds, and the minority female type increases its receptivity; (3) males "recognize" that they occur in two kinds, and the rarer increases its courtship activity; and (4) males "recognize" that there are two kinds of females present and court the rare type of female. Models 2 and 4 were eliminated by the experimental evidence; when two types of drosophila female occur together, the rarer type is only seldom observed to mate in excess of expectation, and males do not court rare females more than common ones. Therefore, it was the ratio of males that made the difference either to other males (3) or to the females (1).

Ehrman helped resolve these last two possibilities into likelihood for the first model by utilizing a double chamber for mating tests; one chamber was fixed above the other with a space between chambers separated by a cheesecloth partition. Flies may certainly become aware of each other between the chambers, but there is no possibility of one type competing or interfering through the partition. Experiments were done first by introducing a minority ratio in the upper chamber (say, 20 *AR*:5 *CH*), but sufficient flies of the minority type (15 *CH*) were introduced below to equalize the total of each type and possibly to nullify the minority advantage in the upper chamber if recognition of frequency was occurring through the partition. Direct contact between flies in upper and lower chamber was prevented. Only matings in the upper half were scored.

In her first experiments (1967), Ehrman continued to vary the ratio of *AR* to *CH* in both sexes simultaneously. However, owing to results of Spiess in which varying the ratio among females made no difference, Ehrman was persuaded to do tests with just one sex varied at a time in the lower chamber. Numbers mating in these double chambers with minority ratios in the upper portion are given in Table A. Rows 1 and 2 give results of experiments in which only females or males were varied: the third row is a repeat of previous experiments with both sexes varied (positive and negative original geotactic trend strains are pooled in this table from Ehrman and Spiess, 1969). With the addition of the rare type

TABLE A *Numbers of* D. pseudoobscura *males mating in double chambers with the rarer type in the upper chamber and the same type added below to equalize the two types. Positive and negative geotactic strain results pooled from Ehrman and Spiess, 1969*

Input 5 *AR*:20 *CH* (per run)	Runs	Observed Males Mating		Expected Males Mating	
		AR	*CH*	*AR*	*CH*
AR females added below	10	88*	124	42.4	169.6
AR males added below	10	40	168	41.6	166.4
AR both sexes added below	9	43	159	40.4	161.6

Female mating results were not significantly different from expected proportions.

* Highly significant departure from expected: minority increased and majority decreased.

TABLE B *Numbers of* D. pseudoobscura *males mating in double chambers with AR = CH (12 pairs each) in the upper chamber, but one type added (both sexes) to the lower in large numbers. From Ehrman, 1970*

Type in Lower Chamber	Runs	Observed Males Mating		Observed Females Mating	
		AR	*CH*	*AR*	*CH*
CH	6	63*	39	49	53
AR	6	37*	64	52	49

* Highly significant departure from equality: minority increased and majority decreased.

in the lower chamber, there was no longer a rare type in the upper in totality of both chambers, but a change in frequency was "recognized" only when males (or both sexes) were added below. When females were added below, there was no nullifying of the input minority advantage effect.

In a later set of experiments, Ehrman was able to induce a minority effect with double chambers by introducing equal numbers above but adding extra flies of one sort in the lower (both sexes were added). Data from those experiments are given in Table B. Males that would be in the minority if both chambers were considered together clearly had an advantage in mating, while the females' frequency of mating was not affected.

EXAMPLE 16-3

FREQUENCY DEPENDENT SELECTION FOR ALLOZYME VARIANTS

Kojima and his colleagues (with Yarbrough, 1967; with Huang and Singh, 1971, 1972) began working on the esterase-6 polymorphism by population-cage technique in 1964. Two wild-type inbred lines of *Drosophila melanogaster* were crossed and progeny maintained in a large random-mating population for 30 generations. Then it was found to be segregating for the Est-6^F and Est-6^S alleles at a frequency of about 30 percent *F* allele. At G_{31}, single-pair matings from this base population were made and subsequently inbred to produce 14 isolated lines of homozygous *SS* and 6 lines of *FF*. Many of the separate viability tests to be described below were started from all 20 lines. Eight populations of about 1000 flies each were initiated, four cultured on cornmeal medium and four on banana, with initial frequencies of alleles at 90 percent *F*, 10 percent *S*, or vice versa. Generations were kept discrete by anesthetizing all adults each generation, sacrificing about 200 for electrophoretic determination of *F-S* genotypes and using the remainder for starting the next generation. Changes in the frequencies of the *F* allele are illustrated in Figure A for the four populations on cornmeal medium (Yarbrough and Kojima, 1967). The changes in banana medium were similar in most respects, so our example will be confined to the results on cornmeal.

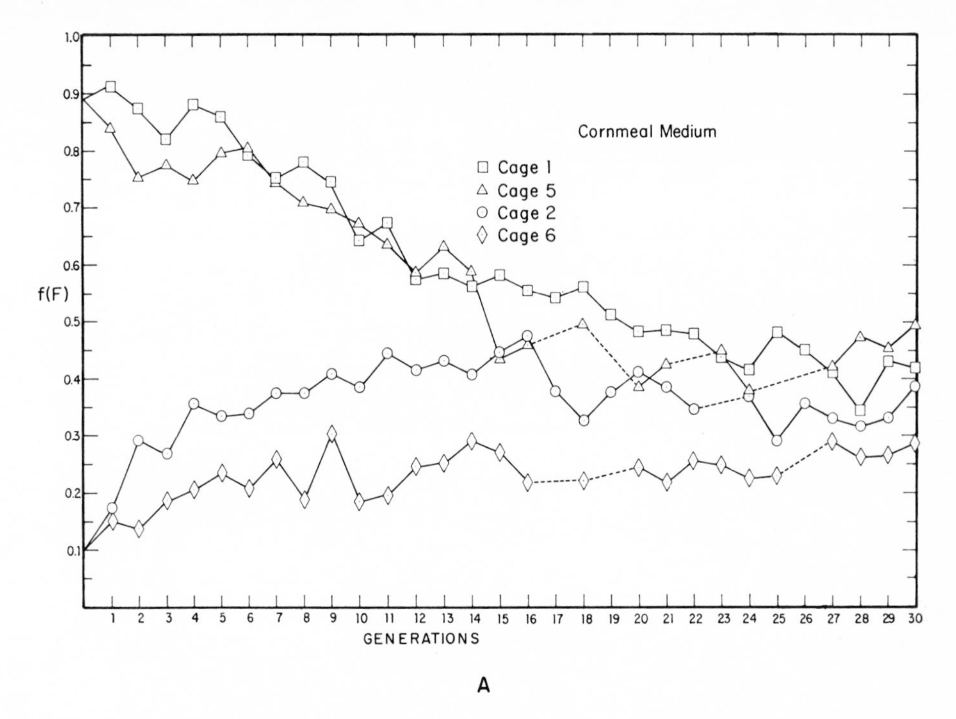

Figure A. Fast allele (F) frequencies at the Esterase-6 locus in two high and two low F frequency population cages maintained on cornmeal medium. Generations omitted from counting are indicated with dashed lines. From Yarbrough and Kojima, 1967.

All populations showed convergence over about 30 generations on about the frequency of the base population (30 percent F, 70 percent S). At first, these results could be interpreted as a response due to constant fitness values with FS (heterozygote) superiority. W value constants were then estimated from "smoothed" selection curves in the manner described in Chapter 15. The estimates are as follows (assuming $W_{FS} = 1.00$):

Frequency	Population	W_{FF}	W_{SS}	Chi-square(G_1–G_{15})
High F	1	0.80	0.84	45.0
	5	0.80	0.90	56.4
Low F	2	0.38	0.50	16.5
	6	0.56	0.80	43.2

Chi-squares for the best fit of the observed generation frequencies over 15 generations to selection curves based on these constant W values are given in the right column. These fitness estimates were unsatisfactory for at least two reasons. First, chi-square at the 0.05 level of significance for the 15 generations tested is about 25, so that three of these selection curves significantly departed from the selection curve prescribed by constant W. Second, the W values estimated from high F populations were not in agreement with those from low F populations.

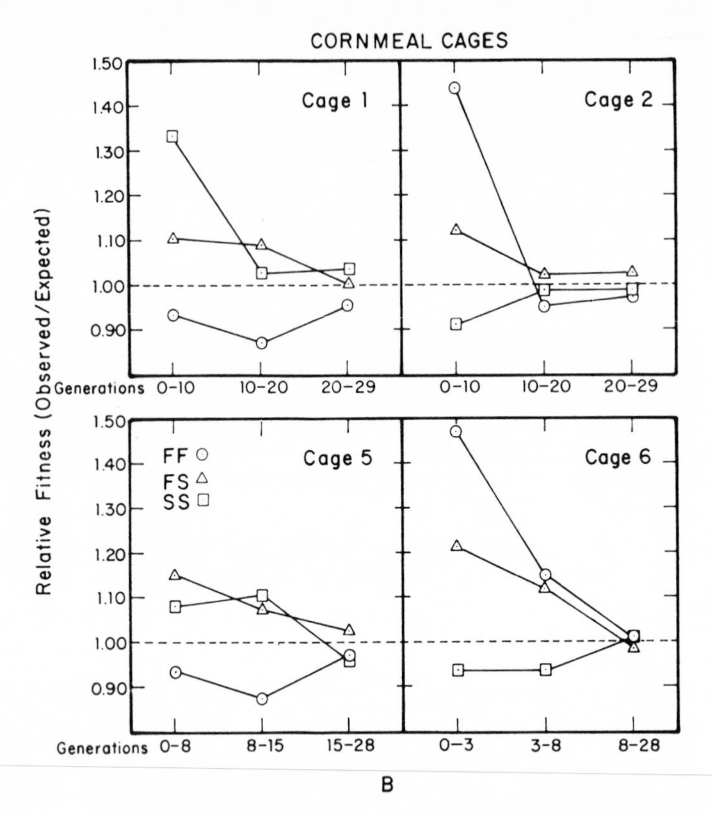

Figure B. Relative fitness for population cages with *FF*, *FS*, and *SS* genotypes of Esterase-6 locus, pooled over various intervals of generations specified below each graph. From Yarbrough and Kojima, 1967.

It was then decided that fitness values should be estimated based on generation-by-generation sequences in the manner of formula (15-4). Fitness estimates are with respect to survival from egg to adult (early). They were pooled over groups of generation sequences in these four populations, as shown graphically in Figure B. It is evident that during the early generations, there was considerable difference between fitness in three out of four populations with the rare homozygote superior, while during the later generations all genotype fitnesses became nearly equal to each other, as expected from a frequency-dependent model. The *FS* genotype tended to be intermediate between the two homozygotes in early generations. Further elucidation of fitness components followed.

By crossing the 14 *SS* and 6 *FF* lines in a randomized scheme, known *FF*, *FS*, and *SS* females were produced and tested for egg-laying capacity (mating to known male genotypes).

	FF	*FS*	*SS*
Mean daily egg counts (110 ♀♀/genotype)	49.8	53.0	46.7 (st. error ± 1.6)

The slight superiority of *FS* females was significant, with *FF* second and *SS* lowest. Obviously, female fecundity alone could not maintain the stable equilibrium with 30 percent *F* allele in the population.

Tests were made using known genotypes by putting 50 mated females into population-cage food cups so that eggs would be laid about in the same density as in the total population. Quasi-random mating was achieved first by mating only homozygotes ($FF \times FF$, $FF \times SS$, $SS \times FF$, and $SS \times SS$), discarding the males, and introducing the females into food cups so that their eggs would be expected to occur according to the square law following random mating, at frequencies of $p_F = 0.7, 0.5, 0.3$, and 0.15.

Expected proportions of emerging adults were calculated by adjusting the expected square-law frequencies to the known female fecundity rates. For example, at $p_F = 0.70$,

	FF	*FS*	*SS*
Frequencies expected	0.511	0.393	0.096
Frequencies observed (pooled rep's.)	0.387	0.448	0.164
Relative viability	0.759 ± 0.048	1.140 ± 0.061	1.713 ± 0.151

In the same way, frequencies expected and observed gave the following relative viabilities:

	FF	*FS*	*SS*
$p_F = 0.50$	0.842 ± 0.078	1.034 ± 0.049	1.099 ± 0.084
$p_F = 0.30$	1.148 ± 0.151	1.016 ± 0.062	0.954 ± 0.052
$p_F = 0.15$	1.677 ± 0.195	1.726 ± 0.075	0.697 ± 0.028

For each genotype, the viability was highest when it was rare. Also, at the equilibrium frequency ($p_F = 0.3$), the three genotypes showed the least difference in fitness. Therefore, it would appear that *F* and *S* allozymes must differ physiologically. There could be a competition for specific resources among the majority that lessens as their excess frequency is reduced, or alternatively, there could be mutualism between genotypes in that the culture medium may be affected in such a way as to benefit the opposite genotype (as larval conditioning of medium had been found by Weisbrot, 1966, and by Dawood and Strickberger, 1969).

When larval viability was tested in culture media conditioned by each of the three genotypes (Huang, Singh, and Kojima, 1971; Kojima and Huang, 1972), it became apparent that larval turnover of culture medium, particularly at high densities of larvae, changed the condition of the medium in such a way as to benefit the rarer forms in the same medium. Conditioning was done by allowing larvae of a single genotype to grow through their life cycle to pupation, when the medium was frozen on dry ice for one hour to kill any larvae still alive. After thawing, to provide three types of conditioned media (reciprocal *FS* and *SF* pooled into a single type), live first-instar larvae of each genotype were planted on the conditioned surface and subsequently grown at optimal temperature. Emerging adults were counted until all had emerged. Nine treatments were used (three genotypes × three conditionings). Viability estimates based on the emergence frequency of *FS* larvae grown on *FS*-conditioned food are given in Table A. Note that there was no *FS* superiority, although *FS* was least affected by the media conditioning. Most important is the fact that the viability of homozygotes was lowest when conditioned by their own genotype. Thus, the viability

TABLE A *Viability coefficients for combinations of Est-6 genotypes on culture media conditioned by larvae of these genotypes (from Huang, et al., 1971)*

	Conditioning Genotypes (st. errors from 0.053–0.059)		
Genotypes	*FF*	*FS*	*SS*
FF	0.923	1.068	1.130
FS	1.090	1.000	1.087
SS	1.146	1.078	0.928

of a given genotype was frequency dependent, since the more common type would be most effective in conditioning its medium. Relative viability values based on the relative amounts of conditioning at different allelic frequencies were then estimated by using marginal mean values. Illustrated below is the method for determining relative viabilities when the medium is conditioned at a frequency of $p_F = 0.9$, for example.

	Media Conditioned by			
Larvae	*FF*	*FS*	*SS*	$\sum fx_{ij}$
FF	0.81(0.923)	0.18(1.068)	0.01(1.130)	0.9512
FS	0.81(1.090)	0.18(1.000)	0.01(1.087)	1.0738
SS	0.81(1.146)	0.18(1.078)	0.01(0.928)	1.1316

$\sum fx_{ij}$ becomes the average viability for the genotype of larvae (i) in a culture where the relative amounts of conditioned (j) food depend on the frequencies of genotypes in the cul-

TABLE B *Relative viability values estimated from using the coefficients in Table A and genotype frequencies from the random-mating expectation*

f(*F*)	*FF*	*FS*	*SS*
0.90	0.886	1.000	1.054
0.80	0.904	1.000	1.054
0.70	0.953	1.000	1.044
0.60	0.981	1.000	1.031
0.50	1.003	1.000	1.013
0.40	1.021	1.000	0.990
0.30	1.033	1.000	0.962
0.20	1.040	1.000	0.930
0.10	1.042	1.000	0.894

ture. Thus, at $p_F = 0.9$, the viability of FF relative to FS is $0.9512/1.0738 = 0.886$, while SS relative to FS is $1.1316/1.0738 = 1.054$. An array for these genotypes' viabilities then can be given over all frequencies encountered in the populations in similar manner. Some of these relative viabilities are given in Table B. It is clear that in spite of sampling errors, the estimates of viability are sufficiently precise to show minority advantage at the top and bottom of the range as well as near equality of the three genotypes near the stable point ($p_F = 0.35$). This reversal of viability from one end of the range to the other is most significant for frequency-dependent selection.

Kojima, in a lecture at the University of Chicago in 1972, said that experiments had been done with *FF* and *SS* larvae introduced into the center of a dish culture divided into four quarters of conditioned medium (two opposite quarters conditioned by *FF* and two by *SS*). After larvae had fed and wandered throughout the medium, they pupated and emerged as adults with *FF* adults tending to emerge in greater numbers from the medium conditioned by *SS* and vice versa. Thus, a mutualism was found, dependent on heterogeneity and conditioning of the medium.

DIFFERENTIAL SELECTION IN THE SEXES

The sexes may differ in their fitness values, a situation probably more common than is usually recognized. Our examples of minority advantage clearly demonstrate, for drosophila species at any rate, that among mature adults males have advantages while rare, but females do not seem to display minority advantage in respect to their mating receptivity. Female drosophila express much variation in developmental time to switch-on of receptivity, however (Manning, 1967; Spiess and Stankevych, 1973). No doubt as fitness differences are looked for, many sex-influenced traits will be found, but traits unrelated to sex may well have different fitness values for males and females.

On examining tumorous-head (*tu-h* on chromosome 3) strains of *Drosophila melanogaster* for fertility (number of progeny produced) Woolf and Church (1963) found that nearly half the *tu-h* males were sterile while females were nearly all quite fertile. The *tu-h* effect is recessive, but strains in laboratory populations were mostly heterozygous for a third chromosome containing the Payne inversion (In(3L)P), which is lethal in homozygous condition. Stable frequencies of lethal and nonlethal chromosomes were accounted for by female heterokaryotypes, which were far more productive than *tu-h* homokaryotypes with standard third chromosomes, while male heterokaryotypes did not show such superiority over homokaryotypes. These differences between the sexes were predicted by Li (1963*b*) to produce stable selection equilibrium with different frequencies of alleles in the two sexes. Li's analysis and selection models (1967*c*, par. 84) follow procedures similar to what we have already illustrated in the Anderson model for minority mating advantage and in the human sex-linked trait G6PD (Example 15-12), so that we shall not examine the details of the models proposed. We simply point out the major conclusions that when fitness values differ between the sexes in some degree, stable and unstable equilibriums are possible depending on whether the low fitnesses are balanced (say, against *AA* in one sex and against *aa* in the other, which leads to a stable state) or whether there might be a superior heterozygote in one sex and inferior heterozygote in the other (*Aa* superior in one sex and inferior in the other leads to two stationary equilibriums and one unstable equilibrium). (See also Haldane, 1962, and Mandel, 1971, for selection differing in males and females.)

SELECTION PROVIDING MORE THAN ONE SET OF CONDITIONS: MULTIPLE FITNESS OPTIMA

Selection may favor two or more phenotypes (say, opposite extremes in a normally distributed phenotypic trait). From our consideration of selection against heterozygotes in Chapter 14, we know that if the genetic control of the phenotype were a simple additive pair of alleles with *AA* and *aa* both more fit than *Aa*, with random mating throughout and fitness constant, no stable polymorphism would be possible because any equilibrium would be unstable (Table 14-17), and the population would soon become either all *AA* or all *aa*.

But we may further consider when selecting for two extremes that these simple genotypes may be environmentally affected sufficiently so that under selection for some "high" or "low" quantitative trait, we could envisage the three genotypes to be available (although presumably in different frequencies) and selectable symbolically as if they occupied a niche* for high or a niche for low selection as follows:

$$\text{Selection for high} = \text{Niche I} \qquad AA > Aa > aa$$

$$\text{Selection for low} = \text{Niche II} \qquad AA < Aa < aa$$

In niche I, *AA* might have highest fitness because it would tend to have a high value, while in niche II, *aa* would tend to be selected. If high selected and low selected were allowed to reproduce in such a way that survival of the population depended on their exchanging genetic material, then a stable polymorphism could exist without heterozygote superiority or any other stable selection mechanism operating (such as minority advantage).

The favoring of opposite phenotypes in this way could take place in alternate generations by switching the direction of selection (as with seasonal changes or longer cycles affecting populations in nature where generation time may be geared to the cycles). Alternatively, at a single point of time, favoring of such phenotypes could take place in subenvironments into which the population is extended. One necessary condition for maintaining the polymorphism with selection aimed at two or more optima is gene flow between the subdivisions of the population.

One might postulate all sorts of ecological subdivisions into which portions of a species population could be extended in time or space—various density conditions, competitive conditions, or physical conditions, for example. Any variation in fitness a genotype might experience from one set of conditions to another could easily lead to the effective mode of selection called "disruptive" (Mather, 1953, 1955), or "diversifying" (Dobzhansky, 1970). The main ingredients are: (1) that a genotype will have different fitnesses from one set of environmental conditions to another, and (2) the sets of environments subdivide the

* The term *niche* is used here to denote any subenvironment into which a population may be extended and within which natural selection will favor genotypes differently from other subenvironments. Such niches may be characterized by substantially diverse conditions or microdifferences. The terms *niche*, *habitat*, and *ecotope* have been reexamined by Whittaker, Levin, and Root, 1973.

population, but gene exchange between the subpopulations must occur for the population's perpetuation. As an alternative to ecological subdivision, one may visualize diverse genotypes controlling opposite phenotypes, such as high and low bristle-number flies, which could be favored simultaneously (ingredient 1), and gene exchange between these opposites must occur to a significant extent (ingredient 2).

Mather (1953, 1955) proposed that the response to such diversity of selection optima may be

> *. . . an adjusted discontinuity at the phenotypic level which is a feature of both isolation and polymorphism. In the case of isolation, however, this is accompanied by disruption of the erstwhile common gene pool, whereas in polymorphism the morphic types continue to share a common gene pool (apart from any switch genes that may be involved), on whose adjustment the polymorphism indeed depends for its efficient working.*

While the "disruption of gene pools" is of considerable significance to the origin-of-species problem, it is the polymorphic outcome for a single population that is pertinent to this discussion. For that outcome, an essential ingredient is continuous gene exchange within the total gene pool. Mather (1955) mentioned many illustrations of selection for two or more optima, the most ancient being the sex difference itself.

> *The males and the females in dioecious populations, for example, will be subjected to different forces of selection and will represent different optimal phenotypes. These phenotypes are nevertheless tied together, for neither has any meaning except in relation to the other: each is an integral part of the other's effective environment. Thus there can be no tendency towards isolation because fertility and fitness depend on the cooperation of the two. . . . Wherever we can discern a sufficiently strong tie, or cooperation, between structures or entities, we might expect to find all the features of polymorphism.*

Experimental verification of Mather's predictions has been achieved many times, but the first successful experiments were those done by Thoday (illustrated in Example 16-4). These experiments combined the effects of mating phenotypic extremes for a morphological trait with either negative or positive assortative mating. An increase in genetic variance occurred over the controls (both the original unselected and the stabilizing selected lines) whether gene flow was at 50, 25, or 0 percent between extreme phenotypes. Heterogeneity of phenotypic optima then not only can maintain but also can promote genetic diversity. In effect, the diversifying selection applied uncovered a few specific chromosome linkages with high and low modifier combinations that maintained the polymorphism on both major autosomes. Thoday summarized (1967, 1972) the general likely consequences of this mode of selection as the following: (1) an increase in the phenotypic and genetic variance of a population, (2) maintenance of linkage disequilibrium, (3) establishment of polymorphisms for switching supergenes and modifying (enhancing) genetic backgrounds, (4) producing and maintaining of divergence between two subpopulations between which there may be considerable gene flow, and (5) splitting a population's gene pool into two parts between which there can be reproductive isolation (see Chapter 19). Thoday warned that these are "restricted generalizations," not incompatible with the finding that other experiments may fail to produce these results if genetic structure of the foundation population lacks the necessary genetic variability.

Attempts to repeat diversifying selection experiments based on Thoday's design have been made by Scharloo and his colleaques (1967) and by Barker and Cummins (1969). In these cases, selection for bristle numbers led to large increases in additive genetic variance and bimodality of the phenotypic distribution so that the establishment of polymorphism by diversifying selection was supported. However, in none of those experiments was there any evidence for development of reproductive isolation between phenotypic extremes.

Similar achievements of diversity within a single gene pool have been accomplished by Dobzhansky and Spassky (1967*a,b*; with Sved, 1969; and with Levene, 1972), who utilized two pairs of *D. pseudoobscura* populations already selected for geotaxis in a maze and then selected for tolerance to concentrations of NaCl. Each of the four populations was treated differently in a two-by-two design with a constant rate of migration between the two pairs of populations. Despite the migration, the four populations became genetically distinct; all differed in chromosomal constitution (*AR-CH*), geotactic behavior, and salt tolerance.

EXAMPLE 16-4

DISRUPTIVE SELECTION IN DROSOPHILA

Thoday and his colleagues carried out a series of experiments on disruptive selection using *Drosophila melanogaster* beginning in 1959 (Thoday, 1959, 1960, 1965, 1967; Thoday and Boam, 1959, 1961*a,b*; Gibson and Thoday, 1962, 1963, 1964; Millicent and Thoday, 1961; Wolstenholme and Thoday, 1963). Much of the work was done on subpopulations descended from a single inseminated wild female fly caught in Sheffield in 1954! With that presumed limitation of genetic variation, it is quite astonishing that consistent response to disruptive, stabilizing, and directional selection should occur, as has been described by these authors.

After culturing the progenies of the single wild pair ("Dronfield" stock), Thoday designed three lines, two of which were "disruptive" and the third "stabilizing." In all three lines, the design was the same insofar as the matings were always *between* four sublines (A, B, C, and D) to avoid close inbreeding; within each subline, 20 individuals of each sex were measured for sternopleural bristle number (total sum of bristles on both sides). For high or low selection, four of each sex were chosen, one of which was used for propagating the subline and three of which were used as backups for insurance against loss. The mating scheme of the two disruptive lines is outlined in Table A, as used in the first experiments. Sublines were designated according to the female parent with A and B = high bristle number and C and D = low.

The line called $D-$ (negative assortative disruptive) consisted of all parent pairs with the low selected half of the line mated to the high selected half each generation; thus, the scheme was "disruptive" since extremes were mated. Negative assortative matings only were allowed each generation with all pairs of high × low. High and low designations were made by female subline, so that maternal effects would be selected for if bristle number had any "cytoplasmic" determination.

The line called $D+$ (positive assortative disruptive) consisted of parent pairs chosen so so that all matings were positive assortative ($H \times H$ and $L \times L$), but the high individuals

TABLE A *The mating and selection systems for disruptive selection experiments (from Thoday, 1959). The entries designate the parents used to produce the culture in the generation shown in the first column. H indicates the highest and L the lowest chaeta-number fly found in the appropriate culture. A, B, C, and D indicate the culture from which the fly was selected*

		Culture (i.e., female subline)			
		A	B	C	D
Parents of Generation		"High"		"Low"	
$D-$	t	$HA \times LC$	$HB \times LD$	$LC \times HA$	$LD \times HB$
	$t+1$	$HA \times LD$	$HB \times LC$	$LC \times HB$	$LD \times HA$
	$t+2$	$HA \times LC$	$HB \times LD$	$LC \times HA$	$LD \times HB$
	$t+3$	$HA \times LD$	$HB \times LC$	$LC \times HB$	$LD \times HA$
	etc.				
$D+$	t	$HA \times HC$	$HB \times HD$	$LC \times LA$	$LD \times LB$
	$t+1$	$LA \times LD$	$LB \times LC$	$HC \times HB$	$HD \times HA$
	$t+2$	$HA \times HC$	$HB \times HD$	$LC \times LA$	$LD \times LB$
	$t+3$	$LA \times LD$	$LB \times LC$	$HC \times HB$	$HD \times HA$
	etc.				

chosen each generation came from a low line and low individuals from a high, a scheme that would avoid selecting for any maternal effects on bristle number. Because exchange was always made each generation between high and low halves of the line, and switched from high to low within sublines, the selection was again disruptive.

The line called *S* (stabilizing) was maintained similarly with four sublines, except that only flies nearest the mean bristle number were chosen within sublines each generation. The order of subline matings ($A \times C$, etc.) was identical with that in the $D-$ and $D+$ lines.

Over several generations, the $D-$ and $D+$ lines rose in phenotypic variance to about double their initial variance. However, in the $D-$ line, the mean bristle number also increased so that the coefficient of variation (Appendix A-7) was essentially unchanged during that time ($CV = 8$ percent throughout); thus, increased variance was completely associated with increase in mean phenotype in that line and could not be ascribed to the selection. The $D+$ line rose gradually in variance and maintained its original mean bristle number (CV increased from 7 percent to more than 10 percent). In contrast, the phenotypic variance diminished in the *S* line while the mean was unchanged.

These lines, including the original unselected Dronfield stock, were subjected to directional selection to check on realized heritability as evidence of genetic variation. Results are given in Figure A in terms of differences between high and low selection for three generations. Both *D* lines were more responsive than the *S* line, but the $D+$ line became more responsive with generations of selection, so that by its 34th generation of disruptive

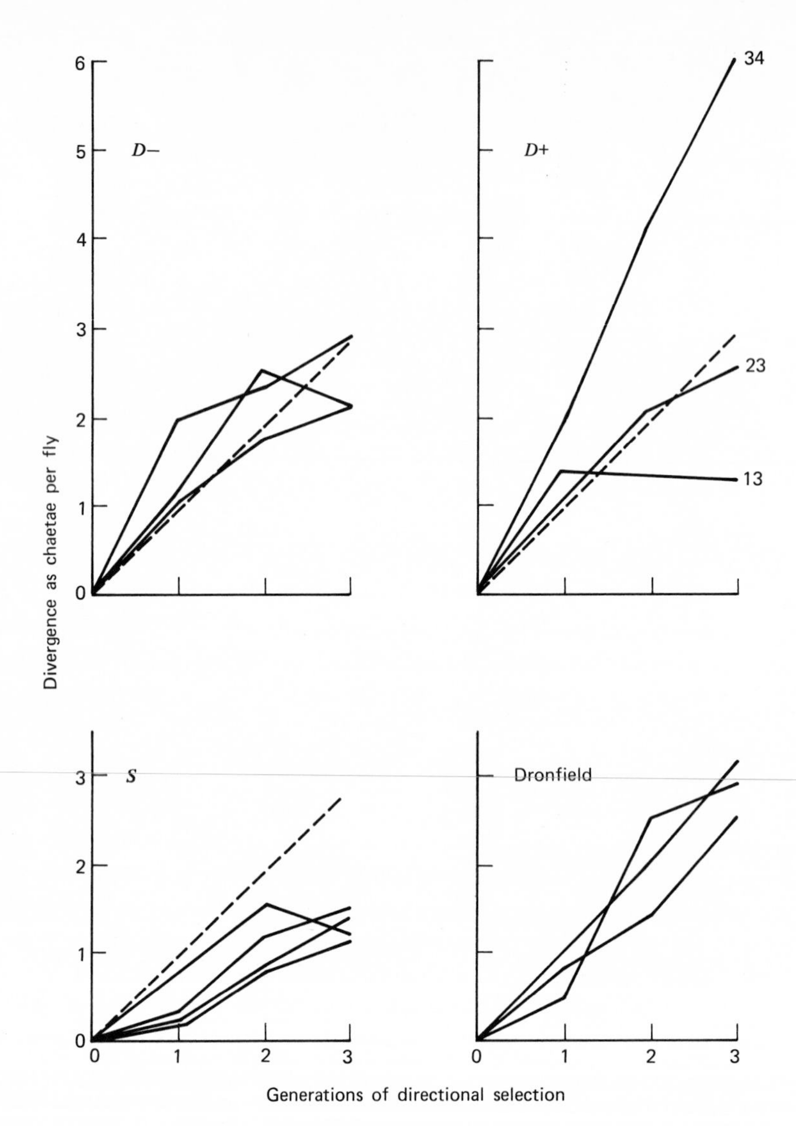

Figure A. Results of divergent-directional selection tests on the disruptive and stabilizing selection lines and on the base line (unselected Dronfield) from which the selected lines were derived. Each solid curve represents the difference in bristles per fly between high and low selected test lines. The broken line represents the mean of three tests on the base stock. Generations in which D^+ was tested are indicated at the ends of their curves, while those plotted for D^- and S lines are G_{22}, G_{32}, and G_{43}. From Thoday, 1959.

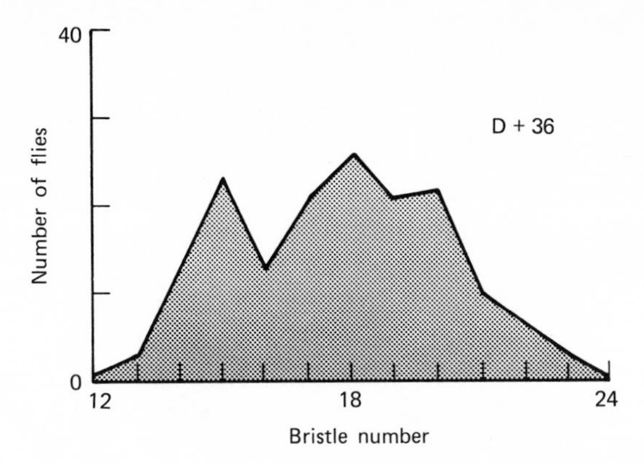

Figure B. Distribution of sternopleural bristle numbers for 160 flies of the D^+ line at G_{36}. From Thoday and Boam, 1959.

selection, the test for directional change was far more responsive in $D+$ than in any other line.

This $D+$ line was carried on further, including a chromosome analysis to elucidate the nature of the genetic polymorphism being maintained by the disruptive technique. Thoday and Boam (1959), restructuring their designated sublines in terms of high and low males instead of females, found that by the 13th generation of disruptive selection, a change occurred that permitted the line to maintain a consistently positive difference between its high and low components. By the 36th generation, the distribution of flies was clearly multimodal for bristle number, as evidenced in Figure B. In fact, two particular sublines (their $H1$ and $L4$) stood out as clearly separated from then on with increasing deviation in opposite directions until termination of the experiment. Sublines $H2$ and $L3$ were more intermediate, although they tended to differ after about 18 generations. Presumably, some recombination of repulsion linkages into coupling must have been involved in the abrupt change that led to maintenance of the polymorphism. X chromosome linkages were ruled out because the factors that distinguish males would be lost in a single generation (sons were chosen on the basis of a father's high or low bristle number). Also, the differences between $H1$ and $L4$ were equal in both sexes.

Test crosses were then made from G_{33} using $H1$ and $L4$ sublines: outcrosses of females from the disruptive sublines were made to males homozygous for mutants brown (*bw* on chromosome 2) and scarlet (*st* on chromosome 3). The mutant males came from a stock with intermediate bristle number. Assuming that a female from the $H1$ subline was being tested,

$$H♀ \times \left[\frac{bw}{bw}; \frac{st}{st}\right] ♂$$

$$\text{Backcross } F_1 ♂[H/bw; H/st] \times [bw/bw; st/st]♀$$

Disregarding the X chromosome, then, there were four phenotypes:

1. Wild-type = H/bw; H/st (high factors on both autosomes).

2. Brown eye = bw/bw; H/st (high factors on three).

3. Scarlet eye = H/bw; st/st (high factors on two).
4. White eye = bw/bw; st/st (no high factors).

Similarly test crosses with $L4$♀♀ were made.

Results of these tests, illustrated in Figure C, indicated that chromosome 2 was segregating for low and intermediate factors, while chromosome 3 had high, intermediate, and low factors compared to the marker stock.

In further detailed analyses of chromosome 2 (Gibson and Thoday, 1962), the $D+$ population was found to have just three viable linkages for two "plus" and "minus" loci: all H flies had either $(+\,-)$ or $(-\,+)$ phases, in a ratio of 3:1, while all L flies had about equal numbers of $(+\,-)$, $(-\,+)$, and $(-\,-)$ linkages, with the latter a recessive "synthetic" lethal—that is, lethal because of detrimental epistatic interaction of both $(-)$'s even though either $(-)$ alone was viable in homozygous condition. The $(+\,+)$ was never found and

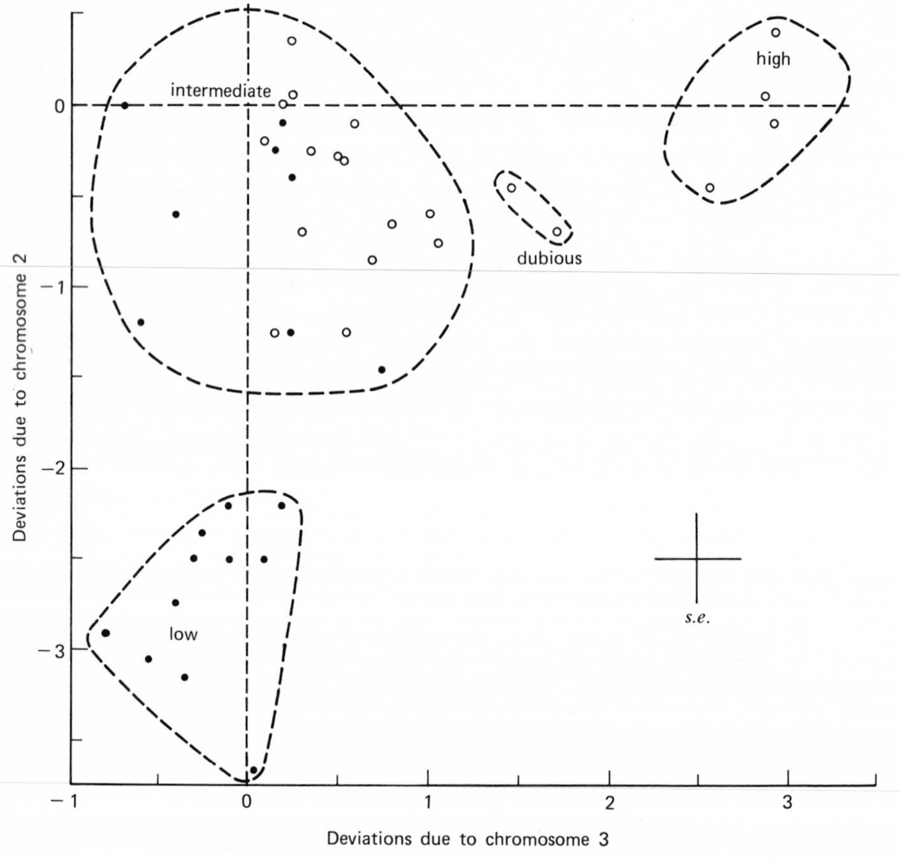

Figure C. Deviations—in bristles per backcross progeny fly from homozygous *y bw st* standard flies—produced by chromosomes extracted from extreme H_1 and L_4 subline females. Solid points are L_4 genomes. Circles are H_1 genomes. The mean standard error of a single entry is indicated with a cross. From Thoday and Boam, 1959.

was considered a dominant lethal. Consequently, for chromosome 2, the selection scheme was maintaining five distinct genotypes with two major loci located at about 27.5 and 47.5 centimorgans. Most of the low flies were $- -/- +$ or $- -/+ -$, with a few repulsion heterozygotes $- +/+ -$, while the high flies were either the latter genotype (commonly) or homozygous (rarely) for one locus with + and the other $-$ ($+ -/+ -$ or $- +/- +$). The presence of repulsion heterozygote females on the high side, which could produce low bristle number males by crossing over in the 20-unit region between loci, must have been responsible for providing a portion of the polymorphism.

The high bristle numbers were caused by two fully dominant loci on the chromosome 3 (analyzed by Wolstenholme and Thoday, 1963), found to be located at 49 and 51 centimorgans. Each locus added about 1.2 bristles per fly compared with the standard number. Disruptive selection preserved these two loci in the coupling phase, in spite of 2 percent recombination from crossing over.

Later, with the discovery by Gibson (1962) of the proximity of two markers (*dachs* and *black*) to the ($- -$) loci on chromosome 2, a screening technique was developed to test other populations including those from nature. Several natural populations were found to be polymorphic for the same loci.

Gibson and Thoday (1963, 1964) carried out disruptive selection lines with quasi-random mating (50 percent interchanged between H and L) with different base populations. Even after lowering the intensity of selection and allowing for the reduction in gene flow, variance in the genetic component increased in much the same way as it had in the earlier experiments. In fact, the extreme phenotype cultures diverged so much and the hybrid (interchanged) cultures produced so few extreme bristle numbers that they ceased to produce any flies with bristle numbers extreme enough to be selected, and an isolation was then effectively achieved between extreme phenotypes but imposed by the selection scheme. In still other experiments, reproductive isolation was achieved (Thoday and Gibson, 1962), a result we shall discuss briefly in Chapter 19.

SIMPLE MODELS FOR MULTIPLE-NICHE SELECTION

A simple model for genetic disruptive selection based on subdivision of a population over two subenvironments was examined by Levene (1953), Li (1955*a*,*b*; 1967*c*), and Maynard Smith (1962, 1966). These authors emphasized slightly different aspects of the conditions leading to a stable polymorphism, but their solutions are based on similar models and their algebra is equivalent—a subdivided population that occupies two or more niches within which conditions would perhaps not promote a stable polymorphism if either set of selective agencies alone were acting on the population as a whole. In Example 16-4 we illustrated some of the experiments of Thoday in disruptive selection featuring two quantitative phenotypic optima each generation, but these optima could as well be two subenvironments (niches) into which a population is extended. As mentioned by Maynard Smith (1962), "In phytophagous insects the niches might be different food plants, in tits they might be nesting boxes and holes in trees, and in man they might be slums and palaces." In a drosophila culture bottle, there are microhabitats of, for example, varying nutrition, density of larvae, moisture, and yeast activity.

As an important condition in the multiple-niche model, we assume that at the time of reproduction, all individuals mate randomly with respect to their original niche. While this condition may not be realistic for natural populations, for many organisms it could be achieved in some degree; for example, in marine organisms that shed their gametes into the sea, insects that mate in flight or at a common food source (as in drosophila) even though their younger stages may occupy separate localities, and flowering plants whose pollen is windblown or carried by nondiscriminating insects. When we contemplate the variety of microhabitats available in nature, it is quite conceivable that this model could apply to organisms whose populations are adjusted to a diversity of environmental conditions.

In the experiments illustrated in Example 16-4, genotypes from opposite-extreme phenotypes were crossmated each generation in a "forced gene flow" scheme or later by "quasi-random mating." If we assume a model with no conditions that would lead to stable polymorphism based on what we have considered previously (superior heterozygotes, minority advantage, etc.), the question we wish to resolve is whether merely by allowing for different selective conditions in two (or more) niches a stable polymorphism is possible. Thus, we envisage what Levene called "the worst case" for promoting a stable polymorphism. If subpopulations happen to have greater isolation from each other so that only a fraction (m) of migrants is exchanged to allow some gene flow, that factor alone would allow much greater independence for selective action in diverse niches and would promote even more the totality of polymorphism throughout the species population.

Intermediate Heterozygotes Within Each Niche

If heterozygotes are intermediate within each niche, then let relative fitnesses (W_i) of three genotypes in two niches be (from Maynard Smith, 1962, 1966)

	AA	Aa	aa
Niche I	$(W_1) = (1 + K)$	$(W_2) = (1 + K/2)$	$(W_3) = 1$
Niche II	$(W_4) = 1$	$(W_5) = (1 + J/2)$	$(W_6) = (1 + J)$

where K is an additive effect in fitness in niche I and J is a similar effect in niche II. If mating takes place at random among adults irrespective of the niche occupied and if there is no preference for particular niches, we let p,q be allelic frequencies at the time of reproduction among all adults over the niches. After selection, proportions of adults in each niche will be

	AA	Aa	aa
Niche I	$p^2(1 + K)$	$2pq(1 + K/2)$	q^2
Niche II	p^2	$2pq(1 + J/2)$	$q^2(1 + J)$

If n_1 is the number of individuals in niche I, then in that niche the number of AA after selection becomes

$$\frac{n_1 p^2(1 + K)}{1 + Kp^2 + Kpq} = \frac{n_1 p^2(1 + K)}{(1 + Kp)}$$

and so on for the remaining genotypes in that niche; similarly, for niche II, with n_2 as the number of individuals there, AA would be $p^2n_2/(1 + Jq)$. Now a stable equilibrium exists when $\Delta q = 0$ (or when $d\bar{W}/dq = 0$, if W is constant). Let $n_1/(n_1 + n_2) = c_1$ (proportion of total population in niche I) and $n_2/(n_1 + n_2) = c_2$ (proportion of total population in niche II); then overall change in q value for the entire population will be

$$\overline{\Delta q} = c_1\,\Delta q_1 + c_2\,\Delta q_2 \tag{16-6}$$

or the weighted average change in q over the niches, where subscripts refer to values within respective niches. Formula (14-23) can be used for gene frequency change between generations in any general case (irrespective of whether W is variable or constant). We first set formula (16-6) in terms of our given fitness values and then set $\overline{\Delta q} = 0$ to find the stationary state value of q at nontrivial equilibrium for the population as a whole:

$$\overline{\Delta q} = pq\left[c_1\left(\frac{(W_3 - W_2)q + (W_2 - W_1)p}{\bar{W}_1}\right) + c_2\left(\frac{(W_6 - W_5)q + (W_5 - W_4)p}{\bar{W}_2}\right)\right] \tag{16-7}$$

Substituting our given W_i values and setting $\overline{\Delta q} = 0$, we have

$$0 = c_1\left[\frac{(-K/2)q + (K/2 - K)(1 - q)}{\bar{W}_1}\right] + c_2\left[\frac{(J - J/2)q + (J/2)(1 - q)}{\bar{W}_2}\right]$$

$$0 = c_1\left[\frac{K/2 - K}{\bar{W}_1}\right] + c_2\left[\frac{J/2}{\bar{W}_2}\right]$$

then, expressing $\bar{W}_i$ and eliminating denominators, we obtain

$$0 = c_1(-K - JKq) + c_2(J + JK - JKq)$$

Solving for q (equilibrium), we obtain

$$\hat{q} = c_2 + \frac{c_2J - c_1K}{JK} \tag{16-7A}$$

If an equilibrium is nontrivial, q must lie between 0 and 1. That would be the case if the numerator $(c_2J - c_1K)$ is zero, or if the second term is positive and not as great as $(1 - c_2)$, or negative and not as large as c_2.

For a numerical example, assume the population is divided equally between two niches so that $c_1 = c_2 = 0.50$; then, if we let $K = 1$, $J = 0.80$, fitness values will be as follows:

Niche I	2:1.5:1
Niche II	1:1.4:1.8

On supplying these values in (16-7A) we obtain $\hat{q} = 0.375$.

While the restrictions in (16-7A) are perhaps somewhat narrow, in that the relative sizes (c_1 and c_2) must be adjusted to J and K for nontrivial equilibrium, we must remember that the example was chosen to be the "worst case" in terms of maintaining polymorphism at all. It was chosen to help explain the disruptive selection results illustrated in Example 16-4. If there is more dominance, so that AA nearly equals Aa in fitness, all positive values of K and J would allow a stationary nontrivial state. If the heterozygote has an overall advantage because each homozygote has some niche where it would be less fit, that situation would be the equivalent of Wallace's concept of "marginal overdominance" (Wallace, 1959; 1968, p. 213) for a set of conditions into which a population may be subdivided in time or space but between which heterozygotes show less variation in fitness than homozygotes. An overall heterozygote advantage would be analogous to a more constant phenotype throughout development and the life cycle (Table 15-1) for heterozygotes than for homozygotes. In terms of selection coefficients, if AA has fitness in some environments (or at some stage of its life cycle) amounting to the form $1 - s_1$, where s_1 is a positive disadvantage compared with the heterozygote (Aa), and also if aa has fitness of $1 - s_2$ in any environment (or stage of life cycle), then over all such conditions the net fitness of the heterozygote would be superior (see Exercise 11 and Chapter 15, Exercise 15).

Fitness Values Constant Within But Variable Among Niches

More generally, if fitness values within each niche are known, we may use (16-6) as before to find stable or unstable equilibrium points. If desired, the method can be extended to any number of niches (Levene, 1953; Li, 1955*a*, 1967*c*). If W is constant within each niche, the $\bar{W}$ distribution can be calculated based on components ($\bar{W}_1, \bar{W}_2, \ldots, \bar{W}_k$) for k niches with the relation (14-17) in each niche instead of (14-23). When fitnesses are constant, then overall fitness for the population is

$$\bar{W} = (\bar{W}_1)^{c_1}(\bar{W}_2)^{c_2} \cdots (\bar{W}_k)^{c_k} \tag{16-8}$$

(geometric mean of fitnesses over k niches). Stationary points ($\hat{q}$) may be found first by taking logs of (16-8).

$$\log_{W} = c_1 \log_{\bar{W}_1} + c_2 \log_{\bar{W}_2} + \cdots c_k \log_{\bar{W}_k} \tag{16-8A}$$

Setting the derivative of (16-8A) equal to zero (because setting (14-24) equal to zero is the equivalent of setting the derivative of $\bar{W}$ equal to zero), $d(\log_{\bar{W}})/dq = 0$, we may solve for the roots to obtain $\hat{q}$.

Using (14-17), and recalling that $d(\log_{\bar{W}})/dq = (d\bar{W}/dq)/\bar{W}$, we set this expression equal to zero for the stationary condition and let $s_1 = 1 - W_1$, $s_2 = 1 - W_3$ within each niche so that we have

$$0 = \frac{c_1(ps_1 - qs_2)_1}{\bar{W}_1} + \frac{c_2(ps_1 - qs_2)_2}{\bar{W}_2} \tag{16-9}$$

(where subscripts outside parentheses refer to *within-niche* values and $s_1 = 1 - W_1$, $s_2 = 1 - W_3$ for each niche). In general, then, the nontrivial equilibrium points give, for i niches,

$$0 = \sum c_i \frac{(ps_1 - qs_2)_i}{\bar{W}_i} \tag{16-9A}$$

For example, let the AA genotype vary in fitness from one niche to another as follows, while keeping the other two genotypes constant (from Levene, 1953):

	AA	*Aa*	*aa*
Niche I	2	1	1.1
Niche II	0.5	1	1.1

so that in no niche is the *Aa* genotype superior—in fact, it is inferior to both homozygotes in one of the niches. Further, let equal numbers of individuals occur in each niche so that $c_1 = c_2 = 0.5$. Using (16-9), we would have

$$0 = c_1 \frac{[q(0.1) + p(-1)]}{\bar{W}_1} + c_2 \frac{[q(0.1) + p(0.5)]}{\bar{W}_2}$$

Then, because $\bar{W}_1 = 2 - 2q + 1.1q^2$ and $\bar{W}_2 = 0.5 + q - 0.4q^2$, the expression becomes a cubic equation: $0 = 0.5 - 2.25q + 2.85q^2 - 0.88q^3$. The biologically meaningful roots give $\hat{q} = 0.3992$ (approximately 0.40) and $\hat{q} = 0.6501$ (approximately 0.65).

The first stationary point ($\hat{q} = 0.4$) is stable ($\bar{W}_{\text{peak}} = 1.073$), while the second is unstable ($\bar{W}_{\text{valley}} = 1.069$) (see Figure 16-3). Note that the first point has a peak in the $\bar{W}$ curve, while the second is at a "saddle," or valley. Note that the peak at $q = 0.4$ is *not* a maximum point; it is simply the point of higher fitness while retaining both alleles in the population as a whole, given the reproductive conditions imposed by the model.

In general, with selection occurring in two niches in this way, stable and unstable points must occur alternately. It is impossible to have two stable equilibriums adjacent in q value, and it is impossible to have two successive unstable ones without a stable point in between. A case of three equilibrium points is given in Exercise 12. These peaks and valleys of the adaptive value surface ($\bar{W}$ curve with q) represent graphically Wright's concept that gene frequencies under the deterministic directed process of natural selection tend toward their adaptive peaks provided that the population sizes are large and conditions of selective pressure constant.

As we discussed in the beginning of this chapter, fitness values may not be constant over the distribution of q values (frequency dependency) so that much of this model is too simple. Of course, constancy of conditions for many generations is probably far more restrictive than is ever the case, even allowing—as we have in the present discussion—that the niches represented could be considered as alternations in conditions of a cyclical nature. Nevertheless, stable polymorphism is thus shown to be possible without heterozygote superiority when the population occupies more than one set of conditions.

Given that environments in nature are usually heterogeneous in time and space, generalization of conditions for stable polymorphisms and extension of principles set out in this chapter to include the "grain" structure of populations, with and without inbreeding, can be examined (Levins and MacArthur, 1966). The model of two (or more) niches we have been considering is called "coarse-grained" in the sense that each individual spends the part of its life relevant to the mode of selection in a single niche (or "patch"). In a fine-grained environment, the individual is exposed to niches (patches) of several kinds and may have sampled each in proportion to experiencing each. In terms of our expression (16-8) for mean fitness over niches, a graph will be curvilinear in coarse-grained conditions, but it will become more a straight-line function with increasingly fine-grained population structure. In other words, the population will have an adaptive value of $c_1 W_1 + c_2 W_2$, instead of the logs of adaptive values as in (16-8A). These authors point out that by finer-grained they refer either to patches becoming smaller or to the members of the population becoming more mobile. Such fine-grained structure would make polymorphism less likely to be achieved or would reduce the stability of polymorphism already attained. With coarseness of grain, however, the population can be thought of as using a mixed strategy in the face of environmental uncertainty to maintain some choice of genotypes over the types of conditions available.

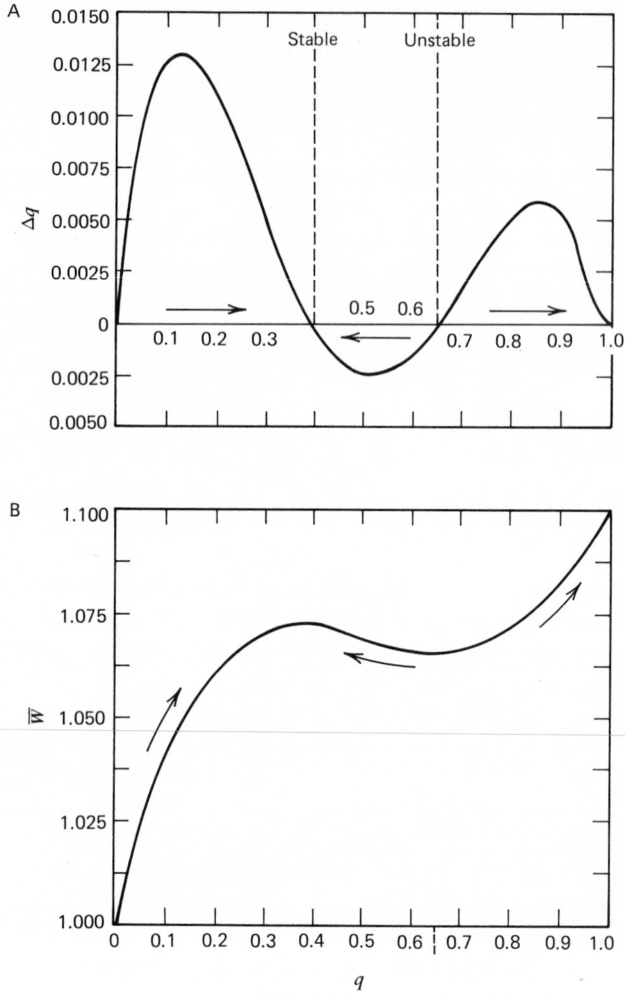

Figure 16-3. (A) Rate of change in $q(\Delta q)$ and (B) $\bar{W}$ as a function of q in the entire population distributed equally between two ecological niches. In niche I fitness values are 2:1:1.1, and in niche II the same genotypes have fitnesses of 0.5:1:1.1. The stable nontrivial equilibrium point is at $q = 0.40$ with a peak for and an unstable point at $q = 0.65$ with a valley of $\bar{W}$. From Li, 1955.

Evidence that heterogenity of habitats will influence the amount of genic variation in populations experiencing them, both within single life cycles and over many generations or spatial extensions of the population, was summarized by Beardmore (1970). He pointed out how markedly the fitness differences of morphs in a population may change over time or space, so that heterogeneity of habitat may produce an "ecological load" depending on the adaptedness of genotypes for habitats into which they are existing. Polymorphism could be considered a strategy for the population to gain flexibility in its future survival and continued adaptedness to the variety of niches that may potentially come to exist and be exploitable (sec Dobzhansky, 1968). Example 16-5 presents evidence that environmental heterogeneity promotes retention of protein-variant polymorphism even under fine-grained conditions.

These ideas were extended, mostly on a theoretical standpoint, by Levins (1968), who examined the strategies of populations over environmental conditions—whether organisms might respond to changing environments via maintaining genetic polymorphisms, a greater degree of individual developmental homeostasis, spatial differentiation, an amount of inbreeding, flexibility, or other strategies. Details of the methods are beyond the scope of this book. Rationale for the principles developed by Levins has grown out of the considerations we have made in this chapter involving multiple-niche selection analysis. General conditions fostering genetic polymorphism in temporally and spatially varying environments have been explored by Gillespie and Langley (1974). Also see Taylor (1975) for a model of equilibrium determined by habitat choice in heterogeneous environments.

It may have occurred to the reader that these principles are further extended by consideration of population density selection (r and K selection) or of age-specific selection (Anderson and King, 1970). MacArthur and Wilson (1967) proposed that selection in an uncrowded expanding population would be aimed at achieving a high intrinsic rate of increase (per capita rate of net increase), while under crowded conditions selection will favor "genotypes which can at least replace themselves with a small family at the lowest food level," maximizing the efficiency of conversion of food into offspring so that the carrying capacity of the environment (K) will not be exceeded. Genotypes sensitive to density then would be selected for their feedback to lower reproductive output total but increase their efficiency to use resources. We have little specific genetic information on control of such density sensitivity. The main ecological concepts were summarized by Pianka (1972) following fundamental discussions of Roughgarden (1971), Charlesworth (1971), and Clarke (1972). When a set of genotypes exists at diverse densities such that their fitness values depend on those densities, a useful strategy for the population would be to maintain sufficient polymorphism so that those genotypes best suited to each density would persist at stable frequencies.

A classic case of density-dependent polymorphism was found by Birch (1955) when testing *D. pseudoobscura* larvae of *ST*/*ST* and *CH*/*CH* karyotypes under varying conditions of crowding. In the natural population at Piñon Flats, *CH* increases in spring while it is replaced by *ST* in summer and fall. The *CH*/*CH* larvae in Birch's experiment had superior viability to *ST*/*ST* when density among larvae was low, while at high density *ST* larvae were superior. This reversal of viabilities at low and high densities may well account for much of the cycle of those arrangement frequencies in that locality through the seasons.

The observations of Wallace (1968*a*, 1975), which he called "soft selection," are cases where the fitness scale is not invariant but is dependent on conditions, such as when density or microenvironmental nutritional differences affect changes in fitness values. (see Exercise 14). It is found, for example, that heterozygote superiority (heterosis) is observed more often when organisms develop under stress than under more optimal conditions (see Parsons, 1971, 1973; Young, 1971; and Chapter 19). Certainly, the innumerable cases of genotype-environment interactions in the literature attest to the delicate balance between adaptive genetic systems and environmental conditions. Fitness of genotypes is far more likely to be flexible in time and space and over various conditions met by organisms than it is to be invariant.

EXAMPLE 16-5

GENETIC DIVERSITY IN DROSOPHILA

Powell (1971) exposed experimental populations of *Drosophila willistoni* to three levels of environmental heterogenity; genetic heterozygosity was monitored for 22 allozyme loci. Environmental conditions varied: (1) yeast, either Fleischmann's bakers (y_1) or Budweiser's brewers (y_2); (2) food, either Carolina instant drosophila medium (f_1) or Spassky's cream-of-wheat-molasses medium (f_2); and (3) temperature, either 25°C or 19°C, in alternate weeks (19/25). Five populations were maintained under uniform and constant conditions, six populations had a single environmental condition varied (for example, with yeast, y_1/y_2 refers to seeding of half the food cups in the population cage with one yeast and half seeded with the other yeast, and similarly for foods half with f_1, half with f_2, symbolized as f_1/f_2), and two populations had all three conditions varied.

Allozymes were monitored under these regimens.* Populations were initiated with 500 flies each, derived from a recently collected Brazilian locality. After 45 weeks of random mating (about 15 generations), samples of adult flies were assayed for these allozymes by starch-gel electrophoresis with 50 flies from each population assayed for each enzyme.

Two measures of genetic diversity are given in Table A: (1) average heterozygosity observed per individual, and (2) average number of alleles per locus. Since the sample size was the same for all populations, an allele had to be present at least once to be included in this measurement.

As indicated in Table A, populations from more variable environments have on the whole retained more genetic variability, both for heterozygosity and for numbers of alleles. Thus, the allozymic variation is maintained by environmental heterogeneity. Natural populations of *D. willistoni* from Brazil average about 19 percent heterozygosity, but neotropical rain forests undoubtedly have more diversified and complex environments than laboratory population cages.

Powell's populations were pointed out to be probably polymorphic for multiple chromosomal inversions—it is well known that *D. willistoni* is chromosomally one of the most variable species—and thus it was argued that the allozymes may not have been the targets of selection but instead blocks of loci in chromosomal arrangements. To counter that argument, McDonald and Ayala (1974) performed similar experiments with *D. pseudoobscura* freshly derived from natural populations in Napa County, California, since that species has chromosomal variants only on the third chromosome and the allozymes assayed were from 17 loci located on the remaining (not third) chromosomes. These investigators varied environmental conditions as Powell had, and they used the variable of light dark. From Table B, it is apparent that the effect of selection in more variable conditions

* Acph-1 (acid phosphatase), Aph-1 (alkaline phosphatase), Est-1, -3, -4, -5, -7 (esterases), Lap-5 (leucine aminopeptidase), α-Gpdh (α-glycerophosphate dehydrogenase), Mdh-2 (malic dehydrogenase), Adh (alcohol dehydrogenase), Odh-1 (octanol dehydrogenase), To (tetrazolium oxidase), Idh (isocitrate dehydrogenase), Tpi-2 (triosephosphate isomerase), Pgm-1, -2 (phosphoglucomutase), Adk-1, -2 (adenylate kinase), Hk-2, -3 (hexokinase), Me-1 (malic enzyme, TPN-dependent malate dehydrogenase).

TABLE A *Proportions of heterozygosity per individual* D. willistoni *for 22 loci (percent ± standard error) and alleles per genic locus in either constant environments, where a single condition was varied or where three factors were varied (from Powell, 1971)*

Cage	Conditions	Heterozygosity per Individual (% ± *s. e.*)	Alleles per Locus
		Constant environment	
1	$y_1 f_1 25$	7.29 ± 0.83	1.68
2	$y_1 f_1 25$	7.80 ± 0.82	1.68
3	$y_2 f_1 25$	8.04 ± 0.81	1.68
6	$y_1 f_2 25$	7.96 ± 0.85	1.68
9	$y_1 f_1 19$	7.98 ± 0.84	1.64
		One variable	
4	$y_1/y_2 f_1 25$	11.09 ± 0.89	1.91
5	$y_1/y_2 f_1 25$	9.06 ± 0.84	2.00
7	$y_1 f_1/f_2 25$	10.16 ± 0.88	2.05
8	$y_1 f_1/f_2 25$	9.33 ± 0.88	1.91
10	$y_1 f_1 19/25$	8.86 ± 0.80	1.95
11	$y_1 f_1 19/25$	9.21 ± 0.88	1.68
		Three variables	
12	$y_1/y_2 f_1/f_2 19/25$	13.90 ± 0.98	2.00
13	$y_1/y_2 f_1/f_2 19/25$	12.81 ± 0.92	2.23

TABLE B *Proportions of heterozygosity per individual* D. pseudo-obscura *for 17 loci in environments where either no factor or up to four factors was varied (from McDonald and Ayala, 1974)*

Number of Environmental Conditions	Heterozygosity per Individual (% ± *s. e.*)
0	15.9 ± 3.2
1	19.3 ± 4.4
2	21.7 ± 4.6
3	22.4 ± 5.2
4	20.4 ± 5.1

parallels the trend in Powell's experiments up to two environmental conditions. These authors found temperature to be the most effective variable, followed by yeast, while varying food was barely significant and light had no significant effect.

ENVIRONMENTAL FACTORS AND SELECTIVE PRECISION

The influence of environmental variables both on developmental expression of genotypes and on their fitness values brings about genotype-environment interactions. Selection has been demonstrated to be very precise in bringing about genetic changes that gear a population toward optimal fitness for specific conditions. Thus, experimental work of this nature is needed to give us further insight into the ways in which populations may exploit environmental diversity with its genetic potential. Some examples of particular environmental factors that elicit genetic response were listed by Spiess (1968*a*): humidity and developmental rate in *Tribolium*, temperature, wing venation and wing size in drosophila, larval density and body weight and mating propensity in drosophila, and nutrition and body weight in mice.

Under complex conditions to which populations are exposed in nature, few cases are known where specific environmental factors can be proved to be selective agents providing sufficient selective pressures to be measurable over limited distances. Antonovics (1971) summarized some basic cases. In particular, Bradshaw and his colleagues (Antonovics, Bradshaw, and Turner, 1971) described several plant populations (grasses of genera *Agrostis*, *Anthoxanthum*, *Holcus*, and *Festuca*; other flowering plants such as *Silene* and *Plantago*) growing on soils contaminated by heavy metals (copper, zinc, or lead) adjacent to populations where the soil was more free of those elements. The populations growing on contaminated soil displayed genetic complexes permitting tolerance for those metals ("tolerant ecotype") in contrast with their progenitor populations ("nontolerant ecotype"). In spite of considerable gene flow between these adaptively contrasting populations, their precision of genetically controlled tolerance to those conditions has evolved and been maintained over short distances, probably since the industrial revolution. The physiological mechanism by which these flowering plants achieve metal tolerance is designed to keep metal ions away from active sites of metabolism by chelation in the cell wall; in fact, in *Agrostis* there are independent systems for zinc and for copper tolerance. However, along with the metabolic alterations that occur in the tolerant races, their genotypes seem to have lost competitive ability, because, in competition with rye grass grown under experimental greenhouse conditions, metal-tolerant ecotypes were uniformly lower in plant size and weight than nontolerant types (Cook, Lefébvre, and McNeilly, 1972).

A similar correlation between genotype and environment occurs in populations of wild oats (*Avena barbata*) in California (Allard and Kahler, 1972; Clegg and Allard, 1972; Hamrick and Allard, 1972, 1975). In the portion of its range characterized by hot dry conditions (xeric), seven enzyme loci are monomorphic in all populations (see Example 8-1). In more moist conditions (mesic), as in the Coast Range, the populations are polymorphic for these loci; especially loci Est-4, Est-9, and APX-5 have alleles that vary as a group from more moist to dry conditions over short distances (about 400 feet up a hillside, for instance). In this highly selfed species of grass, with little chance for recombination, linkage is tight and a complex of genic alleles has been selected as a group for specific environmental conditions, although the relationship between the physiological basis of the adaptation and moisture in the habitat has not been elucidated, as in the case for metal-tolerant plant populations mentioned above.

The general interpretation of these cases is that highly specific polymorphisms may evolve with genotypes (polygenes or oligogenes) that influence the penetrance or expressivity of major genes or, alternatively, that have greater sensitivity to environmental factors critical to the population being selected in specific niches adjacent to each other.

An extensive series of studies shedding light on the precision of selective action in terms of construction of genotypes sensitive to specific conditions was carried out by F. W.

Robertson (1963, 1964, 1966; Church and Robertson, 1966) on ecological genetics of growth in drosophila. The properties of polygenic variation and the action of selection in nutritive, chemically defined environments on two traits were investigated: (1) a metric but homeostatically regulated trait, body size, and (2) a fitness trait, larval growth rate. These two traits are often positively correlated in insects. The main analysis consisted of measuring adult size and growth rate under diets limited in protein, fructose, or RNA (which drosophila does not synthesize fast enough for maximum growth rate), and selecting for genotypes controlling body size and growth rate under the suboptimal diets. After selection had been effective, the performance of selected and unselected controls on the same or other diets was compared.

Selection for larger body size on a low-protein diet tended to separate out two component sets of genotypes: those with special ability to increase size on the deficient diet and those for large size on nutritious food. Selection for speed of growth on a low-RNA diet was effective, but size reduction occurred in the fast-growing selected line as a correlated response, while on the same diet selection for larger body size lengthened the time of growth as a correlated response. When these selected lines were grown on well-yeasted food, their sizes were retained, but the small strain's development time was extended; thus, the fast growth rate was restricted to the narrow range of conditions under which selection had been effective. Selection on RNA-deficient media accumulated genotypes that controlled changing the duration of larval growth up to a critical stage in early third instar when adult structures are determined. The special low-RNA diet uncovered normally undetectable genetic variation, a result similar to the phenomenon of "genetic assimilation" (Waddington, 1957, 1961), or "genetic recruitment" (Milkman, 1970), to be discussed more fully in Chapter 19 on developmental homeostasis. Continued selection in an environment that favors gene expression for the character selected will often raise the level of expression to a stabilized point by fixing stabilizing genetic modifiers so that the expression remains even when the particular environmental stimulus (in this case, low RNA) is removed. Normal wild-type growth is controlled by homeostatic influences and mechanisms that buffer the developing system against the potentially disturbing effects of stresses in the environment.

These results demonstrate how misleading it is to refer to body size, or in fact any metric (polygenic) character, as ultimately a unimodal category with single causation; several physiological processes are contributory to the character, and selection under specified conditions can be aimed at a particular process with considerable precision.

Finally, when we consider selection in diverse niches into which a population is extended, we begin to realize the precision of conditions that can evoke change in genotypes via specificity of genotype-environment interactions. Environmental stress may lower the sensitivity threshold of the developmentally buffered wild-type. Alternatively, introduction of a mutant genotype into the genome to which developmental pathways are not adapted may often bring about increased phenotypic diversity in an analogous disturbance of normal ontogenetic processes. Techniques for studying genotypes with sensitivity to particular environments of stress will be discussed in Chapter 19. In the present context of multiple-niche selection, we wish to emphasize that the subtlety of microniche selection may be extended to include a wide variety of circumstances within which fitness can be tightly geared to very precise conditions. Both ecologists and developmental biologists need to appreciate how much valuable information is yet to be found from investigations of this sort.

EXERCISES

1. (a) Using the model of frequency-dependent fitnesses $W_i(q)$ from the top row of Table 16-1A, graph the fitness values $W_1(q)$, $W_2(q)$, and $W_3(q)$ for q values from $q = 1$ to $q = 0$ at 0.10 intervals. Assume $t = 1$.
   (b) Can you tell from estimating fitness values at any single allelic frequencies what the outcome of selection might be? Explain.
   (c) Start with $q_0 = 0.10$ and calculate successive allelic frequencies and mean fitness ($\bar{W}$) until G_{15}. Assume $t = 1$.
   (d) Let $t = 0.5$ and calculate q successive values from $q_0 = 0.10$ for five generations. Do the same with $t = 2$. What is the effect of proportionality between $[1 - W_i(q)]$ and genotype frequency?
2. (a) Using the fitness values dependent on allelic frequencies as given in Table 16-1A, let $t = 1$ and calculate the q values for five successive generations starting at $q_0 = 0.90$.
   (b) Suppose W values are constants (as in Chapter 14). Let $W_1 = W_3 = 0.768$ with $W_2 = 1.000$ (heterotic heterozygote). Let $q_0 = 0.90$ and calculate q values for five generations.
   (c) In which case does q change faster, with frequency dependent W or with constant W? What would you expect for rate of approach to the stable equilibrium point of $p = q$?
   (d) At the equilibrium point, what will be the fitness values of the genotypes in the first (frequency-dependent) and in the second (constant-fitness) cases?
3. When fitness depends on simple proportionality (t) between $1 - W_i(q)$ and genotype frequency, as given in Table 16-1A, we may prove the equilibrium $\hat{q} = \hat{p} = \frac{1}{2}$ to be stable by trying p values on either side of $\hat{p}$. Let x = a small deviation above $\hat{p}$ so that x^2 is negligible:

$$p^2 = (\tfrac{1}{2} + x)^2 \approx \tfrac{1}{4} + x, \qquad 2pq \approx \tfrac{1}{2}, \qquad q^2 = (\tfrac{1}{2} - x)^2 \approx \tfrac{1}{4} - x$$

   Then $\sum f W_i$ will be as follows:

AA	$(\frac{1}{4} + x)[1 - (\frac{1}{4} + x)t]$
Aa	$\frac{1}{2}(1 - \frac{1}{2}t)$
aa	$(\frac{1}{4} + x)[1 - (\frac{1}{4} - x)t]$

   (a) Verify that $fW_i = 1 - \frac{3}{8}t$, and show that the new p value will be

$$p_1 = \frac{4 + 8x - (\frac{3}{2})t - 4xt}{8 - 3t}$$

   (b) Find the general value of x and show that each new x will be smaller than x of any generation. If $t = 1$, what is the value of x when $p = 0.525$? after another generation?
   (c) What would be the effect on Δq of changing the proportionality t.
4. Suppose the heterozygote is less influenced by frequency than the homozygotes. For example, let $W_2 = 1 - tpq$ instead of the value given in Table 16-1A.

(a) What will be the rate of approach to equilibrium ($\hat{p} = \hat{q}$) following a deviation as in the previous exercise? Is it faster or slower than the previous case?
(b) At the equilibrium state, what can be said about any differences between fitnesses of the genotypes? What happened to $\bar{W}$?

5. Formula (14-7) can always be employed for calculating new q values from one generation to the next irrespective of the basis for fitness. Given the W values from Table 16-1B (Wright and Dobzhansky, 1946), supply them into formula (14-7).

$$\Delta q = \frac{pqW_2 - q^2W_3 - q\bar{W}}{\bar{W}}$$

Let $a = 0.70$, $b = 1.00$.
(a) Work out the $W_i(q)$ values for $q = 0.10, 0.50, 0.70, 0.90$.
(b) Find the Δq for each of those cases. What is the equilibrium $\hat{q}$ value?
(c) What will be the $\bar{W}$ values for each case? What happens to $W_i(q)$ values and $\bar{W}$ as equilibrium is approached? Where is $\bar{W}$ maximum? Where does it equal 1.00?

6. The following observed proportions of three genotypes are found after selection, given that zygotes before selection occur according to the square law:

	Genotypes After Selection		
Before Selection	AA	Aa	aa
0.5	0.20	0.50	0.30
0.6	0.1412	0.4706	0.3882
0.7	0.0900	0.4200	0.4900
0.8	0.0468	0.3404	0.6128

(a) Estimate fitness values for these genotypes according to the maximum-likelihood estimates from formula (15-4A). Are they constant throughout? If not, what relationship do they have?
(b) What do you think the outcome of selection might be for this case?
(c) If $q = 0.1$ before selection, what would you estimate the genotype frequencies to be after selection?

7. The changes in frequency of ST and CH chromosomes in Example 15-2 led Dobzhansky and Pavlovsky (1953) to estimate constant adaptive values of $W_{CH/CH} = 0.413$, $W_{ST/CH} = 1.000$, $W_{ST/ST} = 0.895$. Formerly, Wright and Dobzhansky had estimated that if fitness was not constant but dependent on frequency for these karyotypes, the observed changes might be explained by the fitnesses given in Table 16-1B. Let $a = 0.85$, $b = 1.00$.
(a) Calculate single generation changes in q value (with the frequency of $ST = q$) starting with $q_{ST} = 0.20, 0.30, 0.40, 0.50, 0.60, 0.70, 0.80$ first by using the constant W values and then by applying the frequency-dependent fitness model from Table 16-1B. How does Δq compare over this range of frequencies between the constant and the frequency dependent model?
(b) Are the two models distinguishable by the Δq method? What is needed to predict the outcome of selection?

8. Suppose fitness values are complex, having a constant component as well as a frequency-dependent component. Suppose constant W_i as follows: $W_1 = 1.0$, $W_2 = 0.8$, $W_3 = 1.0$,

while the frequency-dependent components are the same as those in Table 16-1A, top row with $t = 0.30$.

(a) First work out the $\hat{q}$ statement and equation similar to formula (16-4) with frequency-dependent fitnesses on the left and constant fitnesses on the right.

(b) Supply the parameters given to verify the following cubic equation: $0 = 1 - 9.8q + 23.4q^2 - 15.6q^3$.

(c) Factor the right side of that equation into $(1 - 2q)(1 - 7.8q + 7.8q^2)$. Find three stationary states for q.

(d) For each $\hat{q}$ value, test its stability first by taking deviations on either side of $\hat{q}$ and then by taking the derivative of the cubic equation above with respect to q and finding either a negative (stable equilibrium) or positive (unstable) value.

9. If a genotype has an advantage when it is rare because of mating success in males, the model proposed by Anderson for some data of Ehrman's (Example 16-2) could fit the selection process. Use formula (16-5) to estimate frequencies of A and a gametes after males have mated and then the frequencies of progeny zygotes. Let initial $q_0 = s_0 =$ frequency of the ST chromosome arrangement among females and males before mating, while $p_0 = r_0 =$ frequency of CH, and let $x = 0.10$, $y = 0.05$, and $z = 0.30$, as estimated by Anderson.

(a) Calculate single generation changes in q value starting at $q_{ST} = 0.20, 0.30, 0.40, 0.50, 0.60, 0.70, 0.80$.

(b) How does Δq compare over this range of frequencies with the changes you obtained in Exercise 7 from the constant and from the frequency dependent models?

(c) What happens to $\bar{W}$ throughout these frequencies?

10. In multiple-niche selection, using Maynard Smith's model for two niches, the $\overline{\Delta q}$ statement (16-7) leading to formula (16-7A) is the weighted average change over niches, using formula (14-23) for each niche component.

(a) Verify the algebra leading to formula (16-7A) by completing the steps after substituting the niche I and niche II fitness symbols from page 530.

(b) Why must the relative sizes of niche populations be adjusted to J and K for non-trivial equilibrium?

11. (a) From the numerical example given after formula (16-7A), $c_1 = c_2 = 0.5$, $K = 1$, $J = 0.8$, calculate the $\overline{\Delta q}$ and $\bar{W}$ for the following q values (remembering q is over all the population, which occupies two niches): $q = 0.20, 0.3, 0.4, 0.5, 0.6$. Does the equilibrium point of $\hat{q} = 0.375$ appear to be stable? How do you know?

(b) Let $c_1 = 0.6$, $c_2 = 0.4$, $J = 1$, $K = 0.5$. Will there be a stable equilibrium point? What is the $\hat{q}$ value? Test it by calculating $\overline{\Delta q}$ from taking deviations in q above and below the equilibrium point.

(c) In each of the equilibrium points in (a) and (b) above, what would be the fitness of each genotype averaged over both niches?

(d) Is the heterozygote ever superior to both homozygotes (overdominant) in any niche? What about the heterozygote's average over both niches? This example illustrates "marginal overdominance." Why?

(e) Suppose the niches were stages in the life cycle, as if one were an early stage (E) and the other a late stage (L), as in Table 15-1. Would the model be the same as separate niches for a single stage in the life cycle? Why or why not?

12. Assume two niches with equal numbers of individuals selected differently within each niche in early stages; they all mate together in the adult stage. Let the fitness values of genotypes be as follows:

Niche I			Niche II		
AA	Aa	aa	AA	Aa	aa
0.5	1	0.5	3	1	3

  (a) Using formula (16-9), find the equilibriums ($\hat{q}_1$, $\hat{q}_2$, $\hat{q}_3$).
  (b) Calculate $\bar{W}$ at each equilibrium point and above and below each equilibrium point to test the equilibrium for stability. Which points are stable? which unstable?
  (c) Find the value of the $d(\log \bar{W})/dq$ statement for these values, then the second derivative, supplying the equilibrium q values from (a) to test for stability or unstability of the equilibrium points.

13. Refer to Thoday's $D-$ and $D+$ lines in Example 16-4.
  (a) What does each type of assortative mating tend to do for genotypes and gene frequencies?
  (b) Which type of mating would be expected to maintain polymorphism more efficiently if no disruptive selection were being practiced?
  (c) With an additive genetic model, what would be the maximum divergence expected between H and L lines under disruptive selection with quasi-random mating (as in Thoday's experiment) compared with directional selection lines?

14. Wallace (1970) illustrated his concept of "hard" and "soft" selection as follows: (1) hard selection is a "complete correspondence of genotype and its fitness" (as with a completely penetrant lethal) under all growth conditions; (2) soft selection is variable fitness depending on specific conditions; for example, visualize a fixed number of sites to be occupied, (say, 100) for survival of some stage in the life cycle (as a puparium for an insect or a nest for a bird). If genotype B is alone, it will occupy all 100 sites. Genotype A has better competitive ability than B and will occupy any site available more efficiently than B, so that if just one A is present, it will occupy a site. Visualize 200 individuals in a population but only 100 sites are available for occupation. The following will be relative fitness of B:

No. Type		Frequency of B		Relative Fitness of B
		Before	After	
A	B	A Competes		(After/Before)
0	200	1.00	1.00	1.00
1	199	0.995	0.99	0.99497
5	195	0.975	0.95	0.9744
10	190	0.950	0.90	0.9474
20	180	0.90	0.80	0.8889
50	150	0.75	0.50	0.6667
75	125	0.625	0.25	0.4000
100	100	0.500	0.00	0.0000

(a) How would you characterize the fitness of B?
(b) How does this change in fitness differ from frequency-dependent minority advantage selection?
(c) Suppose homozygotes are like type B above, while heterozygotes are like type A and occupy niches more easily and efficiently. Is that kind of selection "heterotic"? Explain why you think it is or is not.

17

EXTENSIONS OF SELECTION PRINCIPLES: (2) GENETIC SYSTEMS

Genetic systems include variation in at least two dimensions: "vertical" (between homologous chromosomes in diploids) and "horizontal" (between linked or unlinked loci). To be more specific, the first includes variation produced by mutation in the broad sense, creating new "segregating units," or intralocus variants, as well as chromosomal segments that act as units because of low internal recombination, illustrated in Chapters 3 and 13. The second includes variation produced by recombination between nonallelic units, illustrated in Chapter 5. Obviously, a consideration of selection based simply on two alleles at a single locus is far from realistic. Polymorphisms are often those involving extension in one or both of these dimensions: multiple alleles or multiple loci. The latter dimension, which had been largely disregarded in most selection models because of complexity in mathematical expressions, is experimentally demonstrable to be of first-order significance; nonallelic gametic combinations are known to persist or undergo cyclic changes in natural or experimental populations, and thus genetic system stability is implied with much phenotypic interaction, or epistatic control ("supergenes") of fitness traits.

Some of the fundamental features of selection which we considered with single loci in the previous chapters will introduce the student to methods for dealing with increasingly more complex genetic systems. The student should realize that proceeding from two alleles to three alleles can be relatively easily extended to more than three alleles once the general features of the three-allele system are understood. In a similar sense, consideration of two loci introduces the general concept of gametic combination selection as the next order of magnitude above the single genic locus; however, mathematical models of more than two loci rapidly become ponderous and intractable for analysis of selection outcome or conditions for equilibrium. With two loci we get sufficient appreciation of the influences of epistasis, genic interaction, and linkage on selection outcome to grasp intuitively their significance for the development of higher-order genetic systems.

THREE ALLELES AT ONE LOCUS UNDER SELECTION

With the high-order genetic polymorphism observed in natural populations of sexually reproducing species implied from Chapters 3 and 13, we realize that at least a considerable fraction of total variability must have adaptive significance for those species, because much of the polymorphism is maintained in a stable or cyclically stable state. Although we must later ascertain just what the relationship is between observed genetic polymorphism and fitness, we now consider to a limited extent whether several alleles can be maintained at stable or unstable equilibrium under selection.

For simplicity, we must assume fitness values to be constants. If the conditions of the last chapter are taken into consideration—fitness as functions of allelic frequencies or subdivisions of the environment—models can be devised to incorporate the features thus imposed, or restrictions of the constant-fitness models can be lifted as needed. Conditions for stability of equilibrium of multiple allelic systems have been investigated when W is constant, and maximum $\bar{W}$ can be found from the continuous function of $\bar{W}$ with respect to q (allelic frequency). With variable fitness values—that is, $W(q)$—it becomes more feasible to use the general continuous function for change in $q(\Delta q)$ and solve for the zero point (no change in q), as outlined in the previous chapter, owing to the fact that the point of maximum $\bar{W}$ may not be the equilibrium point when W is not constant. The following description of expressions for change of multiple-allele frequencies is based on Li (1967*c*), who derived his discussion from Mandel (1959).

With three alleles, each allele can exist in three of the six genotypes. For k alleles, each can exist in k genotypes, of which one is homozygous for that allele and the remainder are heterozygotes. We arrive at a more general concept of average fitness when genotypes are arranged as in Table 17-1 according to the gametic parental dosage. W is symbolized according to row and column so that W_{ii} = homozygous fitness values and W_{ij} and W_{ji} = heterozygote fitness values. On the assumption that maternal effects are zero, we can let $W_{ij} = W_{ji}$. When each genotype frequency is weighted by its fitness value ($p^2W_{11} \cdots$), then we obtain the usual *frequencies after selection*, whose sum is the net fitness: $\sum fW_{ij} = \bar{W}$. If $i = j$, the genotype is homozygous; if $i \neq j$, the genotype is heterozygous. It should be evident that our notation symbols here for W are useful in expressing a more general case than that employed in previous chapters with a single pair of alleles. Here, W_{11}, $[W_{12} + W_{21}]$, W_{22} correspond to our previous W_1, W_2, and W_3, respectively, for example. (The student ought to realize that notation symbols cannot be considered rigid entities, although for

TABLE 17-1 *Notation for selection with three alleles (A, a, a') determining six genotypes and random union of gametes (see Figure 3-1)*

	Frequency of Gametic Combinations (f)			Relative Fitness of Genotypes (W_{ij})			Frequencies After Selection (fW_{ij})		
	$p(A)$	$q(a)$	$r(a')$						
$p(A)$	p^2	pq	pr	W_{11}	W_{12}	W_{13}	p^2W_{11}	pqW_{12}	prW_{13}
$q(a)$	pq	q^2	qr	W_{21}	W_{22}	W_{23}	pqW_{21}	q^2W_{22}	qrW_{23}
$r(a')$	pr	qr	r^2	W_{31}	W_{32}	W_{33}	prW_{31}	qrW_{32}	r^2W_{33}
Totals	$\sum f = 1.00$						$\sum fW_{ij} = \bar{W}$		

Row (or column) totals in $\bar{W}$		Numerical Notation	Gene Symbol Notation
$p(pW_{11} + qW_{12} + rW_{13})$	=	$p\bar{W}_{1.}$	$p_{(A)}\bar{W}_A$
$q(qW_{21} + qW_{22} + rW_{23})$	=	$q\bar{W}_{2.}$	$q_{(a)}\bar{W}_a$
$r(pW_{31} + qW_{32} + rW_{33})$	=	$r\bar{W}_{3.}$	$r_{(a')}\bar{W}_{a'}$

clarity it is better to avoid changing symbols indiscriminately. Notation is most useful when it is flexible for incorporation into increasingly complex models.)

Each row (or column) of the $\bar{W}$ array in Table 17-1 may be factored to see that each gives the *net fitness due to each allele* ($\bar{W}_A$ or $\bar{W}_a$ or $\bar{W}_{a'}$), weighted by that allele's frequency. Also, it is evident that when each of the nine factors in the $\bar{W}$ array is normalized (divided by $\bar{W}$), *allelic frequencies after selection* (p_1, q_1, and r_1) will be each of these row (or column) totals normalized.

$$p_1 = \frac{p_0 \bar{W}_1.}{\bar{W}} \qquad q_1 = \frac{q_0 \bar{W}_2.}{\bar{W}} \qquad r_1 = \frac{r_0 \bar{W}_3.}{\bar{W}} \tag{17-1}$$

($\bar{W}$ is the normalizing total that makes $p_1 + q_1 + r_1 = 1.00$). These expressions are analogous to the expression for a single pair of alleles in Chapter 14, preceding (14-7). Subscripts for the allelic frequencies refer to *before* (0) and *after* (1) selection, as we have used in previous chapters, while the subscripts of W refer to alleles A, a, and a', respectively.

Change in any allelic frequency is then expressed as follows:

$$\Delta p = \frac{p_0 \bar{W}_1.}{\bar{W}} - p_0 = \frac{p_0(\bar{W}_1. - \bar{W})}{\bar{W}} \tag{17-2}$$

(similarly for Δq and Δr), which is analogous to (14-7) for a single pair of alleles.

STATIONARY STATE FOR THREE ALLELES

At a stationary state, each of the expressions in (17-1) will have $p_1 = p_0 \ldots$, or in (17-2) $\Delta p = 0 \ldots$, so that all W's will be equal.

$$\bar{W}_1. = \bar{W}_2. = \bar{W}_3. = \bar{W} \tag{17-3}$$

We may now set the row (or column) totals from Table 17-1 all equal to $\bar{W}$ as a constant.

$$\begin{aligned} pW_{11} + qW_{12} + rW_{13} &= \bar{W} \\ pW_{21} + qW_{22} + rW_{23} &= \bar{W} \\ pW_{31} + qW_{32} + rW_{33} &= \bar{W} \end{aligned} \tag{17-3A}$$

Li (1967*c*) showed mathematical proof that this expression will also be true for the $\bar{W}$ continuous curve with allele frequencies by differentiating the $\bar{W}$ with respect to q_i, assuming constant W and employing Lagrange's multiplier. We omit this proof for brevity. The basic assumption of constant W is an ingredient of the continuous $d\bar{W}/dq$ curve.

The stationary allelic frequencies can be obtained by solving the equations in (17-3A) by using determinants (third degree) and "Cramer's Rule" (see Appendix A-12 and Exercise 3). (An algebraically equivalent alternative solution using (17-2) in two simultaneous equations is found in Exercise 4.) Thus, $\hat{p}$ at equilibrium is given as follows:

$$\hat{p} = \frac{\begin{vmatrix} \bar{W} & W_{12} & W_{13} \\ \bar{W} & W_{22} & W_{23} \\ \bar{W} & W_{32} & W_{33} \end{vmatrix}}{\begin{vmatrix} W_{11} & W_{12} & W_{13} \\ W_{21} & W_{22} & W_{23} \\ W_{31} & W_{32} & W_{33} \end{vmatrix}} = \frac{|\bar{W} \quad W_{\cdot 2} \quad W_{\cdot 3}|}{|W|} \quad \text{[determinants in brief notation]} \tag{17-4}$$

Similarly,

$$\hat{q} = \frac{|W_{\cdot 1} \quad \bar{W} \quad W_{\cdot 3}|}{|W|} \qquad \text{and} \qquad \hat{r} = \frac{|W_{\cdot 1} \quad W_{\cdot 2} \quad \bar{W}|}{|W|} \qquad [\text{determinants in brief notation}]$$

At this stage, we cannot estimate $\bar{W}$ without knowing allelic frequencies; because it is a constant at equilibrium, we may substitute unity for it so that the numerators in (17-4) are *proportional*, respectively, to

$$D_1 = \begin{vmatrix} 1 & W_{12} & W_{13} \\ 1 & W_{22} & W_{23} \\ 1 & W_{32} & W_{33} \end{vmatrix} \qquad D_2 = \begin{vmatrix} W_{11} & 1 & W_{13} \\ W_{21} & 1 & W_{23} \\ W_{31} & 1 & W_{33} \end{vmatrix} \qquad D_3 = \begin{vmatrix} W_{11} & W_{12} & 1 \\ W_{21} & W_{22} & 1 \\ W_{31} & W_{32} & 1 \end{vmatrix} \tag{17-5}$$

where D_i is the determinant of the W matrix with the ith column replaced by 1's. Algebraically, this modification is easily accomplished by dividing each of the expressions in (17-4) by $\bar{W}$, so that for example the expression for $\hat{p}$ is

$$\frac{\hat{p}}{\bar{W}} = \frac{|\bar{W} \quad W_{\cdot 2} \quad W_{\cdot 3}|}{|W|}\left(\frac{1}{\bar{W}}\right) = \frac{|1 \quad W_{\cdot 2} \quad W_{\cdot 3}|}{|W|}$$

Then, letting $D_1 = |1 \quad W_{\cdot 2} \quad W_{\cdot 3}|$, which then equals $\hat{p}|W|/\bar{W}$, each D_i in (17-5) becomes

$$D_1 = \frac{\hat{p}|W|}{\bar{W}} \qquad D_2 = \frac{\hat{q}|W|}{\bar{W}} \qquad D_3 = \frac{\hat{r}|W|}{\bar{W}} \tag{17-5A}$$

Then, summing, we have

$$\sum D_i = \frac{\hat{p}|W| + \hat{q}|W| + \hat{r}|W|}{\bar{W}} = \frac{|W|}{\bar{W}}$$

and

$$\bar{W} = \frac{|W|}{\sum D_i} \tag{17-6}$$

Substituting (17-6) into (17-5A) gives the equilibrium allele frequencies

$$\hat{p} = \frac{D_1}{\sum D_i} \qquad \hat{q} = \frac{D_2}{\sum D_i} \qquad \hat{r} = \frac{D_3}{\sum D_i} \tag{17-7}$$

For an equilibrium in which all three alleles coexist, all three determinants *must be of the same sign.* Their sum ($\sum D_i$) becomes a normalizing total. Any $D \leqq 0$ obviously implies that allele to have been eliminated if the other D's are positive.

NUMERICAL EXAMPLES OF THREE-ALLELE FITNESS

Before proceeding to examine the conditions for a stationary state and stability of equilibriums of multiple alleles, let us examine a few numerical examples. First, we may ask what relative fitness values would be necessary for homozygotes if all heterozygotes had equal heterosis. The simplest case (equality of fitness for homozygotes with equal superiority of heterozygotes, so that $0 = W_{ii} < W_{ij}$) would obviously produce $p = q = r$ at a stationary state. For a case with homozygote fitness unequal but with equal heterosis among heterozygotes, we might have the following for fitness values:

$$\begin{array}{c} \\ |W| = \end{array} \begin{array}{c|ccc|} & A & a & a' \\ \hline A & W_{11} & W_{12} & W_{13} \\ a & W_{21} & W_{22} & W_{23} \\ a' & W_{31} & W_{32} & W_{33} \end{array} = \begin{vmatrix} \frac{2}{3} & 1 & 1 \\ 1 & \frac{1}{2} & 1 \\ 1 & 1 & \frac{1}{2} \end{vmatrix} = \tfrac{1}{2}$$

According to (17-5), we would have

$$D_1 = \begin{vmatrix} 1 & 1 & 1 \\ 1 & \frac{1}{2} & 1 \\ 1 & 1 & \frac{1}{2} \end{vmatrix} = \tfrac{1}{4} \qquad D_2 = \begin{vmatrix} \frac{2}{3} & 1 & 1 \\ 1 & 1 & 1 \\ 1 & 1 & \frac{1}{2} \end{vmatrix} = \tfrac{1}{6} \qquad D_3 = \begin{vmatrix} \frac{2}{3} & 1 & 1 \\ 1 & \frac{1}{2} & 1 \\ 1 & 1 & 1 \end{vmatrix} = \tfrac{1}{6}$$

All D's are positive and their sum is $\sum D_i = \frac{7}{12}$. Then, from (17-6) and (17-7), $\bar{W} = \frac{1}{2}/\frac{7}{12} = \frac{6}{7}$ and equilibrium frequencies are $\hat{p} = \frac{3}{7}$, $\hat{q} = \frac{2}{7}$, $\hat{r} = \frac{2}{7}$. This equilibrium is stable (see Exercise 1), and the allele frequency values are the only solution for the set of W's given.

Wright (1949, p. 372) showed in this simplified case, where all heterozygotes have equal heterosis, the following relationship for k alleles at a locus using selection coefficients (s_i) instead of fitness values: if any homozygote A_iA_i is selectively inferior by a fraction s_i to the heterozygote A_iA_j, equilibrium is reached when

$$\hat{q}_i = \frac{1}{s_i \sum (1/s_i)} \tag{17-7A}$$

In this numerical example, we would have

$$\hat{p}_A = \frac{1}{\frac{1}{3}\left[\frac{1}{\frac{1}{3}} + \frac{1}{\frac{1}{2}} + \frac{1}{\frac{1}{2}}\right]} = \tfrac{3}{7}$$

and the remaining alleles, similarly. We shall refer to this expression in a discussion of the segregation load in Chapter 18.

Second, using an example from Li (1967*c*), for a case of heterozygotes with unequal fitness and not all with heterosis, we assume two heterozygotes (Aa' and aa') to display heterosis, but a third (Aa) to have intermediate fitness, with aa homozygote to be superior to all other homozygotes.

$$\begin{array}{c} \\ |W| = \end{array} \begin{array}{c|ccc|} & A & a & a' \\ \hline A & 1 & 2 & 4 \\ a & 2 & 2.2 & 3 \\ a' & 4 & 3 & 0 \end{array} = 3.8$$

$D_1 = 0.2$, $D_2 = 1.0$, $D_3 = 0.4$, $\sum D_i = 1.6$, and $\bar{W} = 2.375$. Then $\hat{p} = \frac{1}{8}$, $\hat{q} = \frac{5}{8}$, $\hat{r} = \frac{2}{8}$, which is a stable equilibrium. The delicate balance of this equilibrium can be seen if just one fitness value is altered. If the W_{22} value is increased to any value greater than 2.25, these three alleles would no longer be maintained by this selection scheme. The student should try a value of 3, for example, in place of 2.2 given above and find that only the a allele would be preserved since D_1 and D_3 then become negative while D_2 remains positive. The high heterosis in the W_{13} (W_{31}) position would not be sufficient to hold those alleles in the population in that case.

Third, we may look at relative amounts of heterosis. If we lower the W_{22} value to 1 in the above example, we would have

$$\begin{array}{c} \\ |W| = \end{array}\begin{array}{c|ccc|} & A & a & a' \\ \hline A & 1 & 2 & 4 \\ a & 2 & 1 & 3 \\ a' & 4 & 3 & 0 \end{array}$$

so that $\hat{p} = \frac{1}{2}$, $\hat{q} = \frac{1}{10}$, $\hat{r} = \frac{2}{5}$. Suppose the Aa' genotype becomes more heterotic with W_{13} (W_{31}) increasing from 4 to 5; then $D_1 = +7$, $D_2 = -9$, $D_3 = +5$; the second determinant is negative, and the a allele would be eliminated.

THE STATIONARY STATE WITH THREE ALLELES

By examining these examples, we find that a balance of a rather restricted sort must be made between homozygote and heterozygote fitness values. Some heterozygote fitnesses must be high enough to raise the net fitness above any homozygote fitness, in analogous fashion with the case of a single pair of alleles (formula (14-14) and Exercise 8). However, if heterosis gets too great for one pair of alleles, the third allele may get displaced as in the last example where the polymorphism is reduced from three to two alleles.

So far in these examples with three-allele equilibriums, where there is a stable state, all D's were positive. If all D's are negative in (17-5), the stationary state will be unstable, as in the following example (from Li, 1967*c*):

$$|W| = \begin{vmatrix} 1 & 4 & 3 \\ 4 & 2 & 0 \\ 3 & 0 & 2 \end{vmatrix} = -46$$

$D_1 = -10$, $D_2 = -3$, $D_3 = -8$; $\bar{W} = -46/-21 = 2.19$; $\hat{p} = \frac{10}{21}$, $\hat{q} = \frac{1}{7}$, $\hat{r} = \frac{8}{21}$. In this example, two of the three heterozygotes display heterosis, but the remaining one is less fit than its homozygotes. Consequently, this equilibrium set of frequencies will tend to depart from this value with the slightest perturbation. Yet the net fitness is higher than any homozygote fitness. It would be higher still without the a' allele (if the system were reduced to the upper left set with A and a only).

We may now summarize all the necessary and sufficient conditions for a nontrivial equilibrium with three alleles:

1. All net fitnesses must be equal—conditions (17-3) and (17-3A).
2. All determinants of fitness values (D_i, or numerators of (17-4) normalized) must be of the same sign. If they are all positive, the equilibrium is stable; if all negative, the equilibrium is unstable.

3. No homozygote can have a greater fitness than $\bar{W}$.

4. To be stable, the determinant of fitnesses must be greater than 0: $|W| > 0$, or the equivalent that all D_i's are positive (or that the determinant cofactors ("minors") of the homozygotes are all negative, as illustrated in Exercise 10).

RELATION OF MULTIPLE ALLELES TO A SINGLE PAIR OF ALLELES WITH FITNESS AS A FUNCTION OF FREQUENCY

The equilibrium based on heterosis and constant fitness values may in some cases be shown to determine real polymorphisms, as in Example 17-1. Allison found the African tribes possessing both sickle-cell and C hemoglobins (West Africa from the region of Ghana) to be similar to the second example in the previous section where one heterozygote does not display heterosis (Hb^S/Hb^C), but the other two compensate so that a stable state may result incorporating Hb^A, Hb^S, and Hb^C (as in Example 3-2).

With the delicate balance necessary between constant fitnesses for such equilibriums to be maintained, it seems reasonable either that not many cases of multiple-allele steady states can exist or, a more likely but less investigated situation, that fitness values could well be functions of allelic frequencies so that equilibrium fitness values between zygotic types would tend toward equality (as exemplified in Chapter 16 for one pair of alleles). In order to extend the features of this discussion on multiple alleles to the "minority-advantage" case, we need to see the relationship between expressions (14-15), (16-1), and (17-7). Since (16-1) has already been pointed out as the general steady-state solution from consideration of (14-15) with $W(q)$, we must now point out how (14-15) can be considered a special case of the multiple-allele expression (17-7). In effect (as pointed out by Li, 1967*c*), all that is necessary is to eliminate the third allele from (17-3A) and (17-4) to obtain for one pair of alleles

$$|W| = \begin{vmatrix} W_{11} & W_{12} \\ W_{21} & W_{22} \end{vmatrix}$$

so that similarly to (17-5) and (17-7)

$$\hat{p} = \frac{\begin{vmatrix} 1 & W_{12} \\ 1 & W_{22} \end{vmatrix}}{\begin{vmatrix} 1 & W_{12} \\ 1 & W_{22} \end{vmatrix} + \begin{vmatrix} W_{11} & 1 \\ W_{21} & 1 \end{vmatrix}} \quad \text{and} \quad \hat{q} = \frac{\begin{vmatrix} W_{11} & 1 \\ W_{21} & 1 \end{vmatrix}}{\begin{vmatrix} 1 & W_{12} \\ 1 & W_{22} \end{vmatrix} + \begin{vmatrix} W_{11} & 1 \\ W_{21} & 1 \end{vmatrix}} \qquad (17\text{-}8)$$

These are the equivalent of (14-15) with the new subscript notation for fitness. Formula (17-7) then could be applied with W's as functions of q_i in like manner.

While positive evidence is scanty, some negative evidence is very impressive. Example 17-2 describes what happens to the well-known *Drosophila pseudoobscura* chromosomal arrangements taken from a highly polymorphic group of natural populations into the laboratory. Reduction of population size down to that maintained in population cages is not a sufficient explanation for the consistent loss of most of those arrangements in the laboratory environment, since population sizes were quite large, and, more importantly, the selection outcome for the arrangements' frequencies was very much the same from a wide variety of donor populations. Consequently, the implication is that the natural populations containing many of these segregating units must retain their steady state of frequencies by fitnesses as functions of frequencies, as functions of diversity among niches (micro- or macro-), or as some combination of both those modes of selection. We might expect to find some heterosis, some constancy of fitness values for a certain range of frequencies, and some variety of fitness values dependent both on niche and on chromosomal type frequency. The expectation from nature should be that we will not find a simple explanation to inform us as to *the* mode of selection acting on these genetic units. Yet some possibility for a combination of these modes merits our efforts, so that we may recognize them and learn what to expect from their separate or concerted action.

EXAMPLE 17-1

FITNESS ESTIMATES FOR HEMOGLOBIN VARIANT POLYMORPHISM

For the tribes of the West African region (Ghana, for example), the β-chain hemoglobin alleles, Hb^A, Hb^S, Hb^C were shown to be maintained at high frequencies (see Example 3-2) by Allison (1956*a*,*b*). However, the Hb^S and Hb^C frequencies were negatively correlated because the heterozygote Hb^S/Hb^C has more anemia and lower fitness than Hb^C/Hb^C persons. Allison made calculations of relative fitness for the six genotypes occurring among the 1042 adult Ghanians mentioned in Example 3-2 based on observed viability between birth and reproductive age as follows:

$$|W| = \begin{array}{c|ccc|} & Hb^A & Hb^S & Hb^C \\ \hline Hb^A & 0.976 & 1.138 & 1.103 \\ Hb^S & 1.138 & 0.192 & 0.407 \\ Hb^C & 1.103 & 0.407 & 0.550 \end{array}$$

This leads to D_i from (17-5): $D_1 = 0.01436$, $D_2 = 0.00119$, $D_3 = 0.00172$ so that from (17-7), $p_{(Hb^A)} = 0.83$, $q_{(Hb^S)} = 0.07$, $r_{(Hb^C)} = 0.10$. These values are close to those observed (Example 3-2), except that the relative values of Hb^S and Hb^C are reversed. However, the remarkable proximity of values expected to those observed serves to reinforce reliance on these fitness values as very nearly the mode of selection actually existing in the population.

Cavalli-Sforza and Bodmer (1971) tabulate (in their Table 4.8) the observed incidence of these six genotypes from 72 West African populations (after the summary data of Livingstone, 1967). Most of those data were taken on adults, and the authors calculate fitness values based on expected frequencies using the reasoning that selection occurs before maturity, and if the allelic frequencies are stable in the population, application of our (15-4) modified for multiple alleles could be used. If we overlook the hazards in the assumption that selection is completed when the hemoglobins are observed in adults, we may agree with those estimates. However, even more hazardous is the pooling of data from several populations with diverse frequencies over the region of West Africa. It is highly likely that conditions in localities (tribes) could be sufficiently different to produce widely heterogeneous selection effects. For comparison with Allison's earlier estimates, the Cavalli-Sforza and Bodmer estimates are as follows:

$$|W| = \begin{array}{c|ccc|} & Hb^A & Hb^S & Hb^C \\ \hline Hb^A & 0.99 & 1.10 & 0.98 \\ Hb^S & 1.10 & 0.22 & 0.79 \\ Hb^C & 0.98 & 0.79 & 1.45 \end{array}$$

The major differences between this estimate and the earlier are the high value for Hb^C homozygotes, higher Hb^S/Hb^C, and lower Hb^A/Hb^C. This is curious, because on the basis of condition 3 for nontrivial equilibrium, we see that because Hb^C homozygotes have higher fitness than $\bar{W}$, that allele would eventually eliminate the others including Hb^A! With the obvious error involved in such an estimate, it hardly seems worthwhile pursuing the matter. Nevertheless, the student may come to realize how to pinpoint obvious errors in such estimates. For a stable equilibrium, the Hb^C/Hb^C fitness must be lower than 0.91, assuming all other fitness values are accurate and constant. Cavalli-Sforza and Bodmer indicate that the largest error in the estimate would be in that particular homozygote. One must also note the possibility that fitness values for that genotype (as well as the others involving the Hb^C allele) may not be constant. Livingstone (1967) points out that the advantage to Hb^A/Hb^S may occur only with falciparum malaria; when that parasite is replaced by quartan malaria (*Plasmodium malariae*), the fitness of Hb^C may increase relative to Hb^S. Possibly the majority of samples was taken among adults when the Hb^C allele was on the rise. On the average, however, fitness in those populations of Hb^C homozygotes probably is less than 0.90 compared with the remainder given in the above estimate.

EXAMPLE 17-2

CHROMOSOMAL POLYMORPHISM IN NATURAL POPULATIONS REDUCED IN LABORATORY POPULATIONS

When experimental populations of *Drosophila pseudoobscura* have been maintained in laboratory cages in order to test for a selective outcome between two or more chromosomal arrangements (Examples 2-4, 3-1, 15-2), they have usually been started with frequencies of arrangements far different from their natural population progenitors in order to take advantage of maximum change before equilibrium is established. In this species, chromosomal adaptive polymorphism is "flexible" (Dobzhansky, 1962), referring to the microevolutionary and predictable changes in frequencies that usually take place (in contrast with "rigid" polymorphism found in species like *D. subobscura* in which relatively little change takes place over different environments). Under laboratory conditions, which obviously cannot duplicate natural conditions in many details, the arrangement frequencies often contrast to some degree with the observed frequencies of the natural populations from which they originated. Most natural populations of *D. pseudoobscura* are polymorphic for three or more arrangements of the third chromosome. If this high-order polymorphism were retained under laboratory conditions, which are far more constant and restricted than natural conditions, we might have a clue toward understanding the ecological and selective pressures maintaining that polymorphism in nature.

Anderson, Dobzhansky, and colleagues (Anderson, Dobzhansky, and Kastritsis, 1967; Anderson, Dobzhansky, and Pavlovsky, 1972) initiated several laboratory populations of *D. pseudoobscura* with the same chromosomal constitution as samples from the natural population from which the flies were derived. Two populations (1972) included three other species of the *obscura* group (*D. persimilis*, *D. azteca*, and *D. miranda*) proportional to their frequencies in nature (Mather, California), while eleven populations (1967) consisted only of *D. pseudoobscura*, taken from samples ranging from Canada to Mexico and the Pacific coast to eastern Texas.

In all instances, populations monitored for chromosomal arrangement frequencies over a year and a half in the laboratory showed reduction in the number of arrangements by loss of one or more arrangements along with sharp decrease in the frequency of most arrangements and emergence of usually one in the majority, with that one dominating arrangement being characteristic of the geographic region from which it came. Thus, the genetic differentiation of chromosomal races was demonstrated by the different outcome from each geographic region represented. The general reduction in total polymorphism and emergence of a single arrangement demonstrated that the laboratory conditions are not suitable for retention of the high level of polymorphism, possibly because the natural environment provides a greater diversity of niches into which the numerous karyotypes of the wild population are particularly adapted. Alternatively, initial frequencies could determine diverse selection outcome if frequency dependency is important.

Three representative results are given below (N = sample of larvae for salivary analysis)

Population Origin	Date Lab Population Examined	Third Chromosome Arrangements (%)							
		ST	*AR*	*CH*	*PP*	*TL*	*EP*	*SC**	*N*
Berkeley (1967)	May 1964	45.0	2.0	24.0	4.0	17.0	6.0	2.0	200
	May 1965	85.0	1.0	4.0	1.0	5.0	4.0	—	200
	July 1966	99.3	—	0.7	—	—	—	—	300
Austin (1967)	April 1964	2.6	16.4	—	72.0	3.5	1.3	4.3	232
	May 1965	13.0	61.3	—	17.3	8.0	0.3	—	300
	July 1966	48.7	45.7	—	0.3	5.3	0.0	—	300
Mather (1972)	July 1969	37.8	44.6	3.2	2.2	11.9	—	0.3	312
	Oct. 1969	50.0	38.7	1.7	3.4	6.3	—	—	300
	Dec. 1969	59.3	33.3	1.3	1.3	4.7	—	—	300
	Feb. 1970	61.7	34.3	0.3	0.3	3.3	—	—	300
	Sept. 1970	72.7	26.0	0.3	0.3	0.7	—	—	300

* Includes *SC* at Berkeley and *OL* at Austin and Mather localities.

In both California populations, *ST* predominates, although from Mather *AR* tends to be maintained at about 20 percent, while from Berkeley *AR* will be eliminated as well as the other arrangements. From Austin, ST and AR may be maintained at about equality, with a small frequency of TL (Treeline).

Anderson calculated by a maximum-likelihood method (DuMouchel and Anderson, 1968) the fitness values of the predominant arrangements for the Mather populations and found by chi-square tests that the following fitness values could be considered as constants:

Population I	W Values
ST/ST	0.94
ST/AR	1.00
ST/TL^*	0.83
AR/AR	0.73
AR/TL^*	0.60
TL/TL^*	0.76

where $TL^* = TL + CH + PP + OL$. These fitness values lead to a stable equilibrium for ST and AR with elimination of the remaining arrangements.

TWO GENIC LOCI: RECOMBINATION AND EPISTASIS IN FITNESS

Unquestionably, the components of the phenotype related to fitness are not only the consequence of complex interactions of genotypes with environmental factors but also the product of multitudes of genic interactions within the genome. A genotype is not a collection of independent loci but a set of intricately organized, spatially related, interdependent, and mutually adjusted units of primary cellular activity. The ultrasimplicity of studying the outcome of selection for single loci is a first step in our understanding of the selection process, but we cannot continue to look only through the simplistic aspect of that process. As pointed out by Lewontin (1973):

> *Many of the difficulties of understanding how selection can maintain so many polymorphisms arise because the models of population genetics deal chiefly with single loci, so that the genome is regarded as a collection of individual independent loci each undergoing its separate evolution. But genes are organized on chromosomes. . . .*

There are two general ways in which we may analyze the relationship of genetic variation to a population's adaptedness: we may start with fitness properties (Table 15-1) and search for genotypes controlling those properties (as discussed in Chapter 15) or we may start with observed genetic variation in a population and attempt to discover what relationship to fitness the variants may have. The latter method is widely used in present surveys of protein variation, following the techniques of Lewontin and Hubby and Harris in 1966. This procedure has led to the "paradox of variation" (Lewontin's 1974 term)—an immensity of genetic diversity *seeming* to be selectively neutral (see the controversy of neutralists versus selectionists in Chapter 18). For a sizable portion of this protein variation, however, consideration of two or more loci together in nonrandom association has revealed that the magnitude of genic variation is less paradoxical when viewed in the light of interlocus effects on fitness.

The importance of genic interaction (epistasis) and linkage to genotypes under natural selection was first recognized by Fisher in *The Genetical Theory of Natural Selection* (1930):

> Two factors, the alternative genes in which may be represented by *A*, *a* and *B*, *b* may maintain each other mutually in genetic equilibrium if the selective advantage of *A* or *a* is reversed when *B* is substituted for *b*, or vice versa. . . . *A* is advantageous in the presence of *B* but disadvantageous in the presence of *b*, and that *B* is advantageous in the presence of *A* but disadvantageous in the presence of *a*.
> Equilibrium in such a system evidently implies that the increase in the frequency of *A* which takes place in the presence of *B* shall be exactly counterbalanced by its decrease in the presence of *b*; and that the increase in *B* which takes place in the presence of *A* shall be exactly counterbalanced by its decrease in the presence of *a*. But it is important to notice that the equilibrium of the frequencies of the gametic combinations *AB*, *Ab*, *aB*, *ab* requires a third condition of equilibrium. . . . *AB* and *ab* are favored by Natural Selection, and increase in their zygotic stages, while the opposite pair *Ab* and *aB* decrease. The adjustment of the ratio between the frequencies of these two pairs of gametic types must take place by recombination in those individuals which are heterozygotes for both factors. Of these so-called double heterozygotes some arise by union of the gametic types *AB* and *ab*, and in these the effect of recombination is to diminish the frequencies of these two types. This effect will be partially counteracted by recombination in heterozygotes of the second kind, arising from the union of *Ab* and *aB*; and, if the net effect of recombination is to decrease the frequencies of *AB* and *ab*, it is obvious that double heterozygotes derived from gametes of these kinds must be the more numerous.

Thus, a balance between recombination and epistasis in fitness leading to nonrandom association between combinations of alleles from two or more loci was postulated. During the intervening years between Fisher's statement and the present, it was much easier to assume that each genic locus acted independently on fitness. Tacitly, most evolutionary geneticists assumed that they could consider gametic combinations to be at "linkage equilibrium." Only since about the 1950s has evidence begun to accumulate pointing to the importance of genic interaction, linkage, and recombination in our total description of evolutionary dynamics. It has become apparent that it is unrealistic to consider genic variation one locus at a time apart from its aspect as an ingredient of the structural genome.

In an important review, Bodmer and Parsons (1962) summarized the evidence pointing to the essential roles played by linkage, recombination, and genic interaction in Darwinian fitness. These authors stressed the linkage aspect, especially in connection with Mather's "relational balance model" of polygenic complexes (Chapter 7) without documenting the full importance of genic interaction (epistasis) as a vital ingredient in maintaining a multilocus equilibrium. Mather's classic polygenic structure model was based on the proposition that intermediate phenotypes tend to have optimal fitness in nature (extremes being selected against in the stabilizing mode of selection); if "plus" and "minus" modifier polygenes affecting the fitness trait were linked in coupling (*AB*/*ab*), more gametes potentially forming low-fitness genotypes (double homozygotes with extreme phenotypes such as *AABB* and *aabb*) would be produced than if the modifiers were linked in repulsion (*Ab*/*aB*) where homozygous segregants (*AAbb* and *aaBB*) would be intermediate. In the short run, then, close linkage of "balanced gametes" *Ab* and *aB* (repulsion) would be favored by selection. Too-close linkage would be a disadvantage in the long run, however, if conditions were to change in favor of coupling linkages, so that recombination was viewed as a necessary compromise in selection between short-term and long-term needs for future adaptation. With three loci, the heterozygote *AbC*/*aBc* would be the most balanced genotype and would give the least amount of double or triple "unbalanced gametes" leading to extreme homozygotes. With four loci, the *AbCd*/*aBcD* would be most balanced, and so on. Mather (1943) summarized his model by stating, ". . . a balanced polygenic combination is characterized by the twin properties of having a phenotypic effect near to the optimum for the constituent polygenes, and of releasing its variability slowly by recombination with other homologous combinations of the same chromosome." Bodmer and Parsons emphasized the "poise between the complementary needs for preserving balanced polygenic combinations and releasing adequate variability by recombination" through control of crossing over within chromosomes. Most of their summary points to the evidence for relationally balanced genetic systems in selection experiments and in natural populations as well as various controlling mechanisms for crossing over. We have already noted some of the evidence for balanced linkage complexes from artificial selection experiments (Examples 7-1, 7-2), when sudden increased selection response takes place after reaching a plateau or sudden acceleration occurs after slow-steady response (Thoday and Boam, 1961*b*). A corollary phenomenon is the tendency for a genetic system under artificial selection to return to its original state when selection regimen is relaxed (the concept of populational homeostasis discussed under stabilizing selection in Chapter 19).

Analysis of *Drosophila melanogaster* chromosome segments for polygenic activity in determining bristle number and viability traits by Breese and Mather (referred to in Chapter 7) revealed a major difference in the organization of the genetic systems affecting these two traits. Bristle number is mostly controlled by additive and simple dominant genes, while viability is mostly controlled by dominant and epistatic systems. Viability is certainly a trait influenced by an enormous number of biochemical processes, so that the number of loci affecting that trait must be not only vast but also highly interdependent; it is a trait that natural selection tends to maximize. Bristle number, on the other hand, is a trait with intermediate optimum, probably influenced by a much smaller number of loci, and more balanced for plus and minus modifiers since natural selection would tend to eliminate extreme forms (stabilizing). For viability, then, selection will have been directional, and high viability would be expected to display dominance and epistasis; for the morphological trait of bristle number, a balance of modifiers would be expected, and dominance might be ambidirectional, or lacking altogether.

Evidence is overwhelming that genotype coadaptation depends to a considerable extent on specific linkage relations and epistatic interactions selected for high fitness (see Prakash and Levitan, 1973; Prakash and Lewontin, 1968, 1971; Prakash and Merritt, 1972). Crossing over might be expected to break up linked combinations on chromosomes lacking inversions or other crossover suppressor mechanisms. When there is no obvious cytogenetic machinery for maintaining certain linkage phases intact at the expense of others, we would like to know the nature of genetic systems that allow recombination at normal rates but can preserve linked combinations against that constant drain. How much selection is necessary to counteract the effect of recombination?

Viability, being a trait dependent on "good" epistatic interactions, can be dissected genetically by recombination (crossing over) of chromosomal units taken from natural populations and tested for viability control as homozygotes. Dobzhansky and his colleagues carried out extensive experiments on the release of variability from recombination between "good-viability" chromosomes, as summarized in Example 17-3. While homozygous recombinants such as those observed in these studies are probably seldom or never produced under natural population conditions of drosophila owing to high-order heterozygosity and balanced genetic structure, the array of the homozygotes measures the relative diversity engendered by new linked combinations and the resulting nonallelic interactions (epistasis) of those combinations. No doubt, a substantial portion of the genetic variation controlling viability (and by inference many other fitness properties) in wild populations arises from intrachromosomal recombination even when the chromosomes before recombination are selected for their effects in heterozygous condition. In this extensive set of experiments for all the species tested, average viabilities tended to be lower after recombination, approaching the average of all larger samples of chromosomes from natural populations. The nonadditive component of variance was highly significant, indicating that crossing over between "normal-viability" homologues allows segregation of viabilities in a wide spectrum from lethality to supernormal approaching the distribution of wild chromosomes (as in Figure B, Example 9-5). Even as small a sample of normal viability chromosomes as 10 in most of these populations must contain several isoallelic differences affecting viability, and the populations must have a high order of relational balanced heterozygosity and epistatic interaction. It is important to realize that selection has perfected these balanced genic complexes for their fitness *while heterozygous and balanced*—that is, in both "vertical" and "horizontal" dimensions. While the experiments described in Example 17-3 did not test for viability of heterozygotes, there is much evidence to indicate that extremely deleterious recessives show no significant correlation between their homozygous and heterozygous viabilities. To the crossmating organism under natural selection, the phenotypic effect of completely homozygous chromosomes is probably quite irrelevant to their adaptedness as heterozygotes (see Wallace and Dobzhansky, 1962; Dobzhansky and Spassky, 1963).

At a higher level of organization, second-order interactions between polygenic complexes (inversions on right and left arms of the same chromosome) have been extensively described in *Drosophila robusta* by Levitan in a series of papers entitled "Studies of Linkage in Populations" (1955–1973), partially summarized by Bodmer and Parsons (1962). Particularly on the *X* chromosome (see Carson (1958) review), associations between linked arrangements (inversions) may be widespread over many populations: (1) the *XL*-2 arrangement is almost exclusively linked to *XR*-2 despite considerable linkage distance between them available for crossing over; (2) *XL*-1 is often linked to *XR*-1, but in some northeastern United States localities, the *XL* (standard arrangement) is linked exclusively to *XR*-1; (3) some *cis* or *trans* linkages predominate in particular populations. *XL-XR* (standard) with *XL*-1.*XR*-2 are in excess, while XL.XR-2 and XL-1.XR are deficient in the northeast; the opposite frequencies are found at high elevations of the Great Smoky Mountains of the southeast. Some of these associations appear to be constant and others cyclic with the seasons. In many other species of drosophila, similar phenomena of nonrandom associations between linked gene arrangements have been described, including studies by Stalker (1960, 1964) in *D. paramelanica* and *D. euronotus*; by W. B. Mather (1963) in *D. rubida*; by Krimbas (1964) in *D. subobscura*; by Brncic (1961) in *D. pavani*; by Carson and Stalker (1968) in the Hawaiian species group of *D. grimshawi*; by Carson, et al. (1970) in other Hawaiian species groups; and by Yamaguchi and Moriwaki (1971) in *D. bifasciata*. In the genera *Simulium* and *Chironomus*, associations have also been found by Pasternak (1964) and Martin (1965).

Interactions between polymorphic inversion systems occurring on nonhomologous chromosomes have been described by White and his colleagues in the grasshopper *Keyacris* (formerly *Moraba*) *scurra*, as mentioned in Examples 5-1 and 5-2. The several populations in which Lewontin and White studied the dynamics of these polymorphisms showed the nonhomologous inversions (on the *CD* and *EF* chromosomes) to be out of gametic equilibrium by a significant amount. By assuming that their observed frequencies of nine karyotypes for these two polymorphic systems were following the action of natural selection, they could apply the reasoning we described in Chapter 15 by comparing observed with frequencies expected on the basis of random mating with no selection in order to estimate fitness values. Turner (1972) pointed out, however, that Lewontin and White had calculated expected values based on single chromosome frequencies and not on the expected gametic (two-chromosome system) frequencies. In Example 17-4, we summarize Turner's modification of the Lewontin and White data for one locality in Australia to illustrate the working out of the problem and the likely conclusion that the population of grasshoppers occupies a stable equilibrium in frequencies of the "adaptive topography." These observations lead to the conclusion that these two polymorphic systems must interact in an epistatic way on fitness.

Further evidence that the total genome is adapted as a unit (or high-order organized interacting entity) comes from analyses of genetic backgrounds such as the extensive work of Mukai and his colleagues (cited in Example 13-4) on the viability of newly arising mutant polygenes in homozygous and heterozygous genetic backgrounds. Mukai, Chigusa, and Yoshikawa (1964, 1965) compared homozygotes for chromosomes that had accumulated spontaneous mutations for 32 generations to their heterozygous combinations; on a homozygous background the spontaneous mutant chromosomes displayed heterosis when outcrossed (103.02 relative viability) compared with their average as homozygotes (84.12). On a heterozygous background, the apparent heterosis disappeared. Also, for radiation-induced mutations, Mukai, Yoshikawa, and Sano (1966) found heterosis to be about 3 percent on a homozygous background but no heterosis when the background was heterozygous. Thus, Mukai argued that there is an optimum level for heterozygosity, a concept introduced by Wallace (1958).

Similar background effects of neutrality, heterosis, or lowering of viability were found by Anderson (1969*b*) when he tested 45 lethal bearing chromosomes from natural populations of *D. pseudoobscura* in genetic backgrounds from the natural population. Five out of the 45 were heterotic and none were deleterious, while on unrelated genetic backgrounds 10 out of 45 were deleterious. The same lethal chromosomes on different backgrounds were not significantly correlated. Thus, the interactions of naturally occurring recessive deleterious effects with their coadapted backgrounds probably played a significant role in the maintenance of the lethal component of the genetic load in the natural populations sampled.

EXAMPLE 17-3

RELEASE OF GENETIC VARIABILITY THROUGH RECOMBINATION

In 1946, Dobzhansky analyzed the viabilities of recombination products from three second chromosomes lacking structural rearrangements and derived from a natural population of *Drosophila pseudoobscura*. Recombination between these chromosomes generated a diversity of viabilities extending from lethality to slightly supervital as well as varying degrees of retardation in developmental rate. Of much interest was the production of lethality by recombination between "normal" viability chromosomes. These recombination products must have arisen as deleterious interactions between two or more loci whose alleles did not disturb viability as long as they did not occur together in linked homozygous combination. They were called "synthetic lethals."

Wallace, et al., in 1953 tested recombinants of chromosome 2 from *D. melanogaster* populations including viability classes and abdominal bristle numbers. Five nonlethal homozygous chromosomes chosen from two different experimental populations were crossed in all possible combinations. Crossovers collected were made homozygous by a dominant marker technique. Homozygous recombinants displayed lethality plus a wide distribution of viabilities. The results were less striking for bristle numbers, but phenotypic variance increased significantly following recombination. Some chromosome combinations produced little variation, while others produced high variation.

To make potential variability observable, we may employ techniques to reveal the effect of homozygosity for whole chromosomes (unrecombined) or allow recombination to release linked complexes first and make the recombined chromosomes homozygous (see Figure A).

Early in 1954, Dobzhansky and his colleagues agreed to use a common plan to study the effects of recombination on viability in four species of drosophila: *D. pseudoobscura*, *D. willistoni*, *D. prosaltans*, and *D. persimilis*. The first two species are widely distributed, abundant, and ecologically versatile, and the second two are more narrowly adapted and more limited in distribution. In each species, populations were sampled from two localities separated by some distance, by ecological conditions, or by both. It was decided that chromosome 2 in all these species should be studied for the effect of homozygosity and recombination on viability. Homology had been established between loci of the sibling species of the obscura group (*pseudoobscura* and *persimilis*) and between loci of the two tropical species (*willistoni* and *prosaltans*).

In each species, 10 strains were chosen for normal or quasi-normal viability in homozygous condition, with each strain descended from a single wild female inseminated in nature. Strains were free of inversion differences (except in *D. willistoni*, where some of the strains differed by two small inversions in the left arm, and we omit the results for that species here). Detailed techniques are to be found in the following references: Spassky, et al. (1958); Spiess (1958); Dobzhansky, et al. (1958); Levene (1959); Krimbas (1961). An extension of the study to an experimental population of *D. melanogaster* (Wallace's "control population") included both chromosomes 2 and 3 (Spiess and Allen, 1961). To produce recombinant chromosomes, the 10 strains carrying normal or quasi-normal second chromosomes from each locality were intercrossed in all possible combinations—that is, 45 intra-

Generation	Line 1 × Line 2 (and all other interline crosses)
0	$♂\frac{+_A}{+_B} \times ♀\frac{+_C}{+_D}$
1	$♀♀\left[\frac{+_A}{+_C}, \frac{+_A}{+_D}, \frac{+_B}{+_C} \text{ or } \frac{+_B}{+_D}\right] \times \frac{r}{r}♂♂$
2	$\left[\text{single } ♂\left(\frac{+_A \cdots +_C}{r} \text{ or other } \frac{+_A \cdots +_D}{r}, \text{etc.}\right) \times ♀♀\frac{rD}{r+}\right]$ × 10 replicates within line intercross
3	$\frac{+_A \cdots +_C}{rD} \times \frac{+_A \cdots +_C}{rD}$ sibs mated
4	$\frac{+_A \cdots +_C}{+_A \cdots +_C} : \frac{+_A \cdots +_C}{rD} : \frac{rD}{rD}$
	33 : 67 (expected) dies (exactly 100 flies counted)

FIGURE A Technique for revealing recombinant homozygous chromosomes following erossovers in G_1 females heterozygous for two wild population chromosomes. Chromosome 2 in *D. pseudoobscura*, *D. persimilis*, *D. prosaltans*, and *D. willistoni.* See Example 9-5, Figure A for symbols.

locality (or 90 from the two localities) and 100 interlocality crosses for a total of 190 crosses. G_1 females heterozygous for two wild chromosomes were mated to recessive marker males (see Figure A). Their G_2 progeny then contained chromosomes, most of which must have resulted from crossovers in G_1 females, now heterozygous with the recessive marker chromosome. Ten of these G_2 males as a sample in each cross were mated *singly* to females of the dominant marker also containing the recessive marker on the same chromosome. Flies from any G_2 sire showing the dominant but not the recessive in the G_3 generation all contained in identical wild recombinant chromosome since there is no crossing over in male drosophila. Brother-sister matings produced a G_4 generation in which exactly 100 flies were counted in each culture. If viability was normal, 33 wild-type flies would be expected and 67 dominant marker flies in G_4.

The data comprising 1900 progenies for each species were ponderous. They may be summarized first by the average effects of recombination on the mean viability of normal chromosomes and then by the variances engendered through recombination. For intralocality crosses, each chromosome tested by mating to nine others produced 90 derived, or recombinant, chromosomes; the mean viability of those might be termed the "recombinability" of that chromosome. Between localities, the average recombinability is the mean viability of 100 cultures per chromosome. Table A presents the general results; the control "normal" viabilities in each species are the means of 20 homozygous chromosomes selected for quasi-normal expectation before recombination; recombinabilities are less than controls in all species by about 3–5 wild type (out of 33 expected). This average decrease occurred in about 78 percent of all crosses in all three species, so that it could be considered

TABLE A *General results of recombination in terms of mean viability (expected $\frac{33}{100}$ wild type) and variance ("real" within crosses) before and after recombination. Natural population chromosomes sampled previously are given for comparison, from Dobzhansky, et al. (1958); Spiess (1958); Dobzhansky (1970, p. 91). Natural population variance in viability includes all chromosomes from lethals to supervitals*

Species—Origin of Chromosomes	Mean Viability	Mean Variance	Recombination Variance / Natural Variance (=%)
D. pseudoobscura			
Natural populations	20.29	140	$\frac{64}{140}$ =
Control "normal"	28.46	5.10	
After recombination	23.02	64.00	45.7
D. persimilis			
Natural populations	23.54	110	$\frac{27.2}{110}$ =
Control "normal"	31.90	2.23	
After recombination	28.06	37.24	24.8
D. prosaltans			
Natural populations	21.17	200	$\frac{52.75}{200}$ =
Control "normal"	31.82	1.00	
After recombination	28.13	52.75	26.4

a consistent effect of recombination. For the remaining 22 percent of crosses, recombinabilities were equal or greater than the corresponding control viability, but the parental chromosomes had below-average viability.

Do these results show that crossing over lowers viability? Not at all. Stocks maintained over the course of the experiment in wild-type homozygous condition had free crossing over continually, yet in retests of those original chromosomes, there was no lowering of viability with respect to their viabilities when initially isolated as homozygotes.

The average viability of naturally occurring second chromosomes in these species, given in Table A as the top value under each species, amounts to subvitality bordering on semilethality. The drop in viability following recombination of "good" chromosomes is explained by Dobzhansky: "The distribution of viabilities found in the natural population tends to be restored by recombination." The average of recombinants regresses toward the mean of the population from which the normal chromosomes were derived.

For each pair of chromosomes crossed, 10 recombinant chromosomes were tested. If the 10 derived viabilities were distributed randomly around the mean for the cross, the observed variance $S^2 = \sum d^2/9$ should equal the binomial sampling variance (Appendix A-4 and A-6E) obtained by counting 100 individuals in the average sample—$S^2_{\text{bin}} = pqn = (0.33)(0.67)100$, expected $= \bar{Y}(100 - \bar{Y})/100$. If the recombinant chromosomes were so different in viability that their variance was greater than the expectation from sampling, a positive difference would be obtained by subtraction of the binomial from observed variance. These differences were termed "real" variance, and their means are presented in Table A. None of the control chromosome variances are significantly greater than zero; they measure a component of total variance due to environmental factors, residual genotype segregations, and possible mutations. In contrast, recombination variances are immensely larger than their controls in every case, and the consistency of that increased variance occurred 65–92 percent of the time over all the species tested. Finally, when these variances were partitioned into "additive" or "nonadditive" components of parent chromosome effects (see Chapter 6), additive effects were small and generally not significant, while all the nonadditive components were large and highly significant. In other words, particular crosses released widely differing amounts of variability. Correlation between the means of parent chromosomes with their recombinabilities was about zero.

There was some correlation between the viability after recombination of intra- and interlocality crosses ($r = +0.57$ with confidence limits from 0.37 to 0.73). Therefore, one cannot predict the recombinability of a chromosome from measuring it in a control before recombination; when viability is raised or lowered by recombination with one chromosome, there is a slight tendency for the same raising or lowering when that chromosome recombines with other chromosomes of a larger sample (other locality).

That correlation is significant only between recombination products, and raising or lowering viability after recombination must be dependent on epistatic interactions after new linkages are formed. Genes determining viability must act in epistatic ways to satisfy these results. Minus modifiers cannot be added by recombination in excess of those already present in the two parents to bring about consistent lowering of viability following recombination. Some of the crosses produced lethals either by mutation or by recombination. From the distributions arising through recombination and from further testing (Dobzhansky and Spassky, 1960), many of the lethals proved to be "synthetic," representing the extreme lowest value in the distribution engendered by intrachromosomal recombination.

Table A presents in the right-hand column the proportion of total natural population variance in viability represented by these recombined chromosomes—a substantial portion of the total observed variation that a random sample of wild chromosomes would display, from about 46 percent in *D. pseudoobscura* down to about 25 percent in *D. persimilis*. Thus, a very large proportion of homozygote variation observed might result from constant recombination in nature. New linkage combinations that interact epistatically, not new mutation, become the chief immediate source of genetic diversity in natural populations of these species.

EXAMPLE 17-4

GAMETIC ASSOCIATIONS IN GRASSHOPPER POPULATIONS

The frequencies of inversions in chromosomes *CD* and *EF* of the Australian grasshopper, *Keyacris* (formerly *Moraba*) *scurra*, indicate an equilibrium condition with the Blundell (*Bl* arrangement of *CD*) and the standard (*St* of *EF*) predominating in natural populations. Estimates of fitness were made by Lewontin and White (1960) based on deviations of observed adult frequencies from proportions expected from squaring the chromosomal arrangement (single chromosome independence frequencies in the usual manner, as in Example 5-2, Assumption 2: *no* gametic association ($D = 0$) and random mating (see Table A(1) here). However, we could base expectation of karyotype frequencies on an assumption that there *is* gametic association at equilibrium (constant nonzero D) followed by random mating. Turner (1972) pointed out that for all of 16 population samples of this grasshopper collected by White from 1955 to 1961, there was a positive D value; therefore, it was highly likely that an overall gametic association in all populations did exist (heterogeneity between samples was very small). Undoubtedly, those populations were out of gametic equilibrium. Relative fitness estimation must then be based on a comparison of observed adult values with those expected from *gametic* frequencies, not individual chromosomal arrangement frequencies. Gametic frequencies (Example 5-2, Table C) for the Royalla B locality, 1956–1958, may be squared for expected frequencies based on karyotypes *before selection*, as shown in Table A(2) for comparison with the values taken from Example 5-2, calculated by Lewontin and White (Table A(1) for convenience). Expected values from gametic frequencies indicate that the double heterokaryotype indeed

TABLE A *Calculation of fitness values in the Royalla B (1956–1958) population of grasshoppers by the methods (1) of chromosome frequency and (2) of gametic combination frequency expectations (modified from Turner, 1972).*

Observed Karyotype Numbers

EF Chromosome	CD Chromosome *St/St*	*St/Bl*	*Bl/Bl*
St/St	59	282	231
St/Td	24	152	150
Td/Td	2	14	19

Total = 933

Gametic Frequencies

	St	*Bl*	$\sum$
St	0.268	0.520	0.788
Td	0.063	0.149	0.212
$\sum$	0.331	0.669	

(1) Expected based on chromosome frequencies (Lewontin and White, 1960) (see Example 5-2)

63.52	256.53	259.00
34.22	139.19	139.53
4.61	18.61	18.79

Fitness estimates (standardized relative to *St/Bl, St/St*)

0.85	1.00	0.81
0.64	1.00	0.98
0.39	0.68	0.92

(2) Expected based on gametic combination frequencies (Turner, 1972)

66.99	259.91	252.12
31.62	135.83	144.51
3.73	17.58	20.71

Fitness estimates

0.81	1.00	0.84
0.70	1.03	0.96
0.49	0.73	0.85

has highest fitness in addition to contrasts in other karyotypes that in fact lead to considerable alteration in the interpretation of adaptive state for the population and the stability of the genetic system.

Lewontin and White invented a graphic method of indicating net fitness ($\bar{W}$) in two-locus systems as a function of joint allelic (chromosome) frequencies. By assuming the W values to be constants and calculated as given in Table A(1), they described the contours of a topographical $\bar{W}$ map, as shown in Figure A-1. Frequencies of Bl:St and Td:St for *CD* and *EF* chromosomes respectively were varied at intervals of 0.05 in all possible 441 combinations, and $\bar{W}$ was computed according to the fitness values given in Table A(1). The dot represents the observed population, and it rests in a shallow saddle of net fitness, slightly lower than two peaks to upper left and lower right. Thus, this equilibrium would be unstable, and frequencies could move either way to an adaptive peak. In contrast, recalculation of fitness values for this population according to Turner (1972) given in Table A(2) would make the observed point a *stable* equilibrium, as in Figure A-2. While there is no compelling reason for feeling that a stable state for two loci must be at the point of maximum $\bar{W}$ (as indicated in the text, pp. 577–578), it can be verified with (17-17) that the two estimates of fitness values give unstable or stable points as shown. Indeed, there is no strong justification for the use of the adaptive topography, because we do not know enough about the restrictions of the two locus $\bar{W}_{max}$ as a stable point (see Moran, 1964). Turner stated that computer simulation using the fitness values as given does agree with the adaptive topography, however, and he pointed out that "although it lacks a rigorous foundation, the adaptive topography is comparatively robust." That is, small departures from specifications will not materially affect the reliability of the results. Turner showed by using gametic combination frequencies to estimate fitness values that several of the grasshopper population samples lie at stable equilibrium. However, with the fitness values reevaluated in this way, some populations were still shown by Turner to be at unstable "saddle" points on the

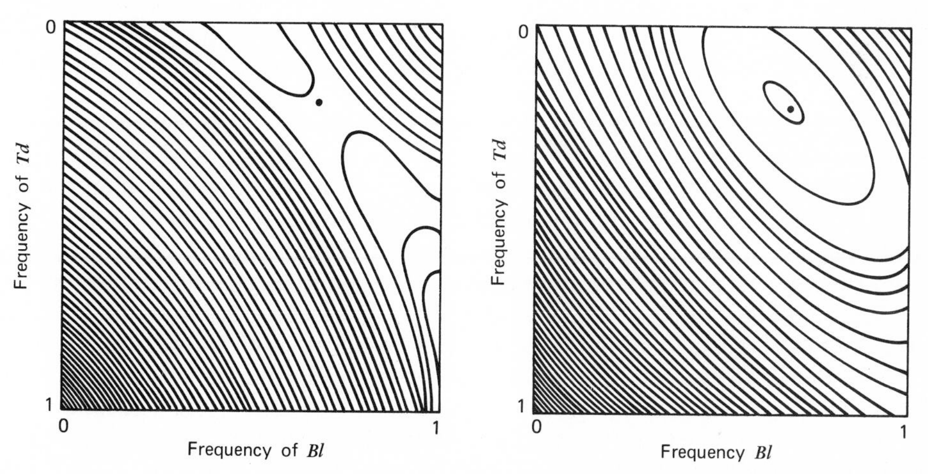

Figures A-1 and A-2. Adaptive topographies for the population of *Keyacris scurra* at Royalla B locality calculated from the viabilities in A-1 by the method of Lewontin and White and in A-2 by Turner (1972). The black dot is the observed population point. Contour intervals at 0.01 in chromosomal frequency. From Turner, 1972.

adaptive topography. It is likely that variable fitness, some inbreeding, or other less restrictive selection conditions than those in our classic model may play a part in maintaining stable points for these inversions. Therefore, there is little doubt that the inversions of the two nonhomologous chromosomes have direct effects on fitness and that the two chromosomes interact epistatically, often tending to have the double heterozygote superior in fitness and single heterozygotes dependent on the karyotype of the nonhomologous chromosome for superiority or intermediate fitness values.

CHANGE IN GAMETE FREQUENCIES UNDER SELECTION

With two alleles at each locus, the genotypes produced by random combination of gametes can be symbolized in frequencies ($g_{AB}, g_{Ab}, \ldots$), as arranged in Table 17-2A and B. W is symbolized with subscripts according to genotypes produced at each locus (1 = dominant homozygote, 2 = heterozygote, 3 = recessive homozygote as in Chapter 14, while the left position ($W_{i\cdot}$) is for the A locus and the right position ($W_{\cdot j}$) is for the B locus). When each genotype, given in terms of parent gamete frequencies (g), is weighted by its fitness value, we obtain the usual *frequencies after selection*, whose sum is the net fitness, or the total of Table 17-2B: $\sum gW_{ij} = \bar{W}$. Each row (or column) may be factored, as with multiple alleles in Table 17-1, to obtain the *net fitness due to each gametic combination* ($\bar{W}_{AB}$, $\bar{W}_{Ab}$, etc.) as given in Table 17-2C. When there is no "linkage disequilibrium" ($D = 0$ from Chapter 5), the gamete frequencies after selection would be simply as follows:

$$g'_{AB} = \frac{g_{AB}\bar{W}_{AB}}{\bar{W}} \qquad g'_{Ab} = \frac{g_{Ab}\bar{W}_{Ab}}{\bar{W}} \tag{17-9}$$

and so forth (where prime (′) indicates the following generation). Then the change in gametic frequencies is

$$\Delta g_{AB} = \frac{g_{AB}(\bar{W}_{AB} - \bar{W})}{\bar{W}} \qquad \Delta g_{Ab} = \frac{g_{Ab}(\bar{W}_{Ab} - \bar{W})}{\bar{W}} \tag{17-10}$$

etc. Genotypes are arranged by locus zygosity in the usual 3 × 3 manner in Table 17-2D for convenience in consideration of the nine genotypes and their fitness symbols.

We recall from Chapter 5 that gametic frequencies may not be at equilibrium after random mating. It is not sufficient to ignore the gametic product determinant. With linkage or with independent assortment, the amount of recombination must be taken into account when considering the gametic output. With any deviation from gametic equilibrium (nonzero "disequilibrium," or $|D| > 0$), gametes will be formed in meiosis from double heterozygotes at rates of recombination that will modify formulas (17-9) and (17-10). If we write the expression for any gametic frequency (g_{AB}, for example) from parents' output and incorporate the recombination value, we would do so as in Table 17-2E. The modification to incorporate recombination then is found as an extra term containing the product of percent recombination (c), fitness of the double heterozygote, and the gametic determinant (D). When normalized, these expressions give us the general changes in gametic frequencies per generation (Δg) as follows:

$$g'_{AB} = \frac{1}{\bar{W}}[g_{AB}\bar{W}_{AB} - cW_{22}D] \qquad g'_{Ab} = \frac{1}{\bar{W}}[g_{Ab}\bar{W}_{Ab} + cW_{22}D] \tag{17-9A}$$

TABLE 17-2 *Notation for selection with two loci ($A - a$ and $B - b$) determining nine genotypes if independently assorting or ten genotypes if A and B are linked (modified from Lewontin and Kojima, 1960, and Kojima and Lewontin, 1970)*

A Symbols

Frequencies

g_{AB} = gametic frequency of $\overline{AB}$ combination
g_{Ab} = gametic frequency of $\overline{Ab}$ combination
g_{aB} = gametic frequency of $\overline{aB}$ combination
g_{ab} = gametic frequency of $\overline{ab}$ combination

Fitnesses (W_{ij})

W_{11} = fitness of $AABB$ genotype
W_{12} = fitness of $AABb$ genotype
W_{13} = fitness of $AAbb$ genotype, etc., with subscript 1 denoting homozygote AA or BB, 2 heterozygote, and 3 homozygote aa or bb. Left position ($W_{i.}$) = A locus; right position ($W_{.j}$) = B locus

B Frequencies of genotypes with random mating and selection in terms of gametic contributions from parents

	g_{AB}	g_{Ab}	g_{aB}	g_{ab}
g_{AB}	$(g_{AB})^2$ W_{11}	$(g_{AB})(g_{Ab})$ W_{12}	$(g_{AB})(g_{aB})$ W_{21}	$(g_{AB})(g_{ab})$ W_{22}
g_{Ab}	$(g_{Ab})(g_{AB})$ W_{12}	$(g_{Ab})^2$ W_{13}	$(g_{Ab})(g_{aB})$ W_{22}	$(g_{Ab})(g_{ab})$ W_{23}
g_{aB}	$(g_{aB})(g_{AB})$ W_{21}	$(g_{aB})(g_{Ab})$ W_{22}	$(g_{aB})^2$ W_{31}	$(g_{aB})(g_{ab})$ W_{32}
g_{ab}	$(g_{ab})(g_{AB})$ W_{22}	$(g_{ab})(g_{Ab})$ W_{23}	$(g_{ab})(g_{aB})$ W_{32}	$(g_{ab})^2$ W_{33}

$$\sum g W_{ij} = \bar{W}$$

C Row (or column) totals after selection

Row 1 $g_{AB}[(g_{AB})W_{11} + (g_{Ab})W_{12} + (g_{aB})W_{21} + (g_{ab})W_{22}] = (g_{AB})\bar{W}_{AB}$
where $\bar{W}_{AB}$ is net fitness due to the $\overline{AB}$ gamete

Similarly,

Row 2 $(g_{Ab})\bar{W}_{Ab}$ where $\bar{W}_{Ab}$ = net fitness due to the $\overline{Ab}$ gamete
Row 3 $(g_{aB})\bar{W}_{aB}$ where $\bar{W}_{aB}$ = net fitness due to the $\overline{aB}$ gamete
Row 4 $(g_{ab})\bar{W}_{ab}$ where $\bar{W}_{ab}$ = net fitness due to the $\overline{ab}$ gamete

Sum = $\bar{W}$

TABLE 17-2 *(Continued)*

D Frequencies of genotypes after random mating and selection arranged according to their state at the two loci, omitting maternal differences

	BB	Bb	bb
AA	$(g_{AB})^2$ W_{11}	$2(g_{AB})(g_{Ab})$ W_{12}	$(g_{Ab})^2$ W_{13}
Aa	$2(g_{AB})(g_{aB})$ W_{21}	$\begin{bmatrix} 2(g_{Ab})(g_{ab})+ \\ 2(g_{Ab})(g_{aB}) \end{bmatrix}$ W_{22}	$2(g_{Ab})(g_{ab})$ W_{23}
aa	$(g_{aB})^2$ W_{31}	$2(g_{aB})(g_{ab})$ W_{32}	$(g_{ab})^2$ W_{33}

E Gametic output of parents after selection with recombination (c) between the two loci (A and B) (note $\bar{W}$, normalizing total on left for convenience)

$$\begin{aligned}\bar{W}(g_{AB}) &= (g_{AB})^2 W_{11} + \tfrac{1}{2}[2(g_{AB})(g_{Ab})W_{12} + 2(g_{AB})(g_{aB})W_{21}] \\ &\quad + \tfrac{1}{2}[2(1-c)(g_{AB})(g_{ab})W_{22} + 2c(g_{Ab})(g_{aB})W_{22}] \\ &= (g_{AB})^2 W_{11} + (g_{AB})(g_{Ab})W_{12} + (g_{AB})(g_{aB})W_{21} + (1-c)(g_{AB})(g_{ab})W_{22} + c(g_{Ab})(g_{aB})W_{22} \\ &= g_{AB}(\bar{W}_{AB}) - cW_{22}[(g_{AB})(g_{ab}) - (g_{Ab})(g_{aB}) \\ &= g_{AB}(\bar{W}_{AB}) - cW_{22}D, \text{ where } D = \text{gametic determinant of parent generation}\end{aligned}$$

So that $(g'_{AB}) = [g_{AB}(\bar{W}_{AB}) - cW_{22}D]\dfrac{1}{\bar{W}}$

Similarly,

$$(g'_{Ab}) = [g_{Ab}(\bar{W}_{Ab}) + cW_{22}D]\frac{1}{\bar{W}}$$

$$(g'_{aB}) = [g_{aB}(\bar{W}_{aB}) + cW_{22}D]\frac{1}{\bar{W}}$$

$$(g'_{ab}) = [g_{ab}(\bar{W}_{ab}) - cW_{22}D]\frac{1}{\bar{W}}$$

etc. Then changes in gametic frequencies are

$$\Delta g_{AB} = \frac{1}{\bar{W}}[g_{AB}(\bar{W}_{AB} - \bar{W}) - cW_{22}D)]$$

$$\Delta g_{Ab} = \frac{1}{\bar{W}}[g_{Ab}(\bar{W}_{Ab} - \bar{W}) + cW_{22}D] \qquad (17\text{-}10A)$$

and so forth. These equations are modified from Lewontin and Kojima (1960) and were originally derived by Kimura in 1956.

An illustration will help the student see the important contrast between change in allelic frequencies with a single locus and those changes occurring when two loci are considered. Before proceeding to describe the conditions for a stationary state and the stability of gametic equilibriums, we shall illustrate the use of these formulas to examine the major features that come into play when fitness depends on two loci.

Numerical Illustration 1

Both loci are equal in fitness with heterosis and no linkage ($c = \frac{1}{2}$) as given in Table 17-3A. It appears that all of the double homozygotes are lethal; this fact may bother some students, but it will become apparent that ease in calculation justifies the bother of the lethality. (In Exercise 12 we add just 1 to each fitness value—that is, linear transformation—to avoid the lethality problem.) This illustration is called "additive" because the amount of heterosis derived from the two loci together ($W_{22} = 2$) is the simple sum of the amounts of heterosis of each locus (1 for each locus taken separately). Also on a phenotypic scale, the increment in fitness as the A locus is varied (increments of $+1$ from AA to Aa and -1 from Aa to aa) is true no matter what the B genotype, and vice versa for the B locus with constant A genotype. We call this an *additive system* of two loci, to be contrasted in the next section with an interaction, or *epistatic*, system. (Do not be confused by the obvious fact that lethality appears as an interaction of two loci; that could be called a special-threshold condition in this case because of the *finality* of lethality.) The student ought to consider this illustration as if the fitness values were simply points on a linear scale that conceivably could go into "minus" values. The effects of each locus are additive in the sense given in Chapter 6, Table 6-8.

If gametes originally enter a population with the frequencies (g^0) given in Table 17-3B at "linkage equilibrium," zygotes will be formed as in Z_0 (Table 17-3C). Then if selection acts during a period before gametes are produced, the g' output (after selection) will be according to (17-9), since $D = 0$. The student should verify the gametic output using the usual method given in Chapter 5. If employing (17-9), we must obtain the respective net fitnesses due to each gametic combination as follows (using Table 17-2C or successive rows factored from 17-2B):

$$\bar{W}_{AB} = 0 + 0.08 + 0.18 + 2(0.02) = 0.30$$
$$\bar{W}_{Ab} = 0.72 + 0 + 2(0.18) + 0.02 = 1.10$$
$$\bar{W}_{aB} = 0.72 + 2(0.08) + 0 + 0.02 = 0.90$$
$$\bar{W}_{ab} = 2(0.72) + 0.08 + 0.18 + 0 = 1.70$$

Thus, $g'_{AB} = 0.72(0.30)/0.5 = 0.432$, as given in Table 17-3D, etc.

Note the sum:

$$\begin{array}{ccccccccc} g_{AB}\bar{W}_{AB} & + & g_{Ab}\bar{W}_{Ab} & + & g_{aB}\bar{W}_{aB} & + & g_{ab}\bar{W}_{ab} & = & \bar{W} \\ 0.216 & + & 0.088 & + & 0.162 & + & 0.034 & = & 0.50 \end{array}$$

One major result of selection here is to produce a gametic output that is no longer at "linkage equilibrium"; that is, there is now a significant departure from gametic independence and $D \neq 0$! The next generation zygotes after selection then will have a gametic output (g'') according to the formulas of Table 17-2E. Thus,

$$g''_{AB} = [g'_{AB}(\bar{W}_{AB}) - cW_{22}D]\frac{1}{\bar{W}}$$

TABLE 17-3 *Numerical illustration 1 of nine genotypes, with fitness of A locus and B locus equal in having heterosis and additive in double heterozygote; no linkage:* $c = \frac{1}{2}$

A Fitness

	BB	Bb	bb
AA	W_{11}	W_{12}	W_{13}
Aa	W_{21}	W_{22}	W_{23}
aa	W_{31}	W_{32}	W_{33}

W Values

$$\begin{vmatrix} 0 & 1 & 0 \\ 1 & 2 & 1 \\ 0 & 1 & 0 \end{vmatrix}$$

B Original gametic input

$$g^0 = \begin{vmatrix} g^0{}_{AB} & g^0{}_{Ab} \\ g^0{}_{aB} & g^0{}_{ab} \end{vmatrix}$$

	B	b	
A	0.72	0.08	$p_A = 0.80$
a	0.18	0.02	$q_a = 0.20$
	$u_B = 0.90$	$v_b = 0.10$	

$D = 0$

C Original zygote frequencies

Before Selection

$$Z_0 = \begin{vmatrix} 0.5184 & 0.1152 & 0.0064 \\ 0.2592 & 0.0576 & 0.0032 \\ 0.0324 & 0.0072 & 0.0004 \end{vmatrix} \begin{matrix} 0.64 \\ 0.32 \\ 0.04 \end{matrix}$$

$\sum$ 0.8100 0.1800 0.0100

After Selection

$$\begin{vmatrix} 0 & 0.1152 & 0 \\ 0.2592 & 0.1152 & 0.0032 \\ 0 & 0.0072 & 0 \end{vmatrix}$$

$\sum = \bar{W} = 0.5000$

D Gametic output from surviving zygotes

$g' =$

	B	b	
A	0.432	0.176	$p_A = 0.608$
a	0.324	0.068	$q_c = 0.392$
	$u_B = 0.756$	$v_b = 0.244$	

$D = -0.027648$

E Successive generations of gametic output

From	$\bar{W}$	Gametic Output			p_A	D
Z_1	0.8456		B	b		
		$g'' =$ A	0.3576	0.2062	0.5638	−0.009667
		a	0.2938	0.1424		
			$u_B = 0.6514$			
Z_2	0.9460		B	b		
		$g''' =$ A	0.3069	0.2268	0.5337	−0.002646
		a	0.2731	0.1932		
			$u_B = 0.5800$			
Z_∞	1.00		0.25	0.25	0.50	0
			0.25	0.25		
			0.50			

so that in the example of Table 17-3E,

$$g''_{AB} = [0.432(0.636) - \tfrac{1}{2}(2)(-0.027648)]\frac{1}{0.8456} = 0.3576$$

The remaining gamete frequencies from the Z_1 are given in Table 17-3E. The student should verify those frequencies by performing the random product for Z_1 and calculating the gametic output in the usual way (Table 5-1).

As selection proceeds, it is important to note the sudden lack of independence between loci; even though the initial gametic combinations (g^0) are at exact linkage equilibrium and loci are unlinked ($c = \frac{1}{2}$), selection brings about a deviation from independence (significant D) immediately in the gametic output. This will be true in all instances of selection for gametic combinations (unless the two loci are exactly multiplicative in fitness, when initial zero D may remain but is unstable, according to Lewontin, 1974, p. 290). The observed D value for the initial gametic output (-0.027648) in Table 17-3D is 28.9 percent of maximum D (5-9) for the allelic frequencies observed. That D value indicates a sizable association between loci. Gradually, D will approach zero, but for this particular set of fitness values it has gone through an increase to a maximum value and then a decline.

Ultimately, the changes produced by selection will lead to the symmetrical result of equality between all gametic combinations in this case ($g_{AB} = g_{Ab} = g_{aB} = g_{ab} = 0.25$) with net fitness at maximum ($\overline{W} = 1.00$) and $D = 0$, which is a stable equilibrium.

It is better for understanding how two loci affect the rate of change in frequencies to examine each locus alone. For example, if only the A locus is considered in Table 17-3C, the rate of change depends on what state the B locus is in; if that is either BB or bb genotype, then in a single generation all zygotes would be Aa and frequencies would be immediately established at $p_A = q_a = \frac{1}{2}$. If the B locus has Bb genotype, however, the A genotypes would have fitness values of 0.5:1.00:0.5, so that for those progeny zygotes, p_A would be 0.7278. (Alternatively, the same relationships would hold for the B locus genotypes in A background, since the two are exactly equivalent in the illustration.) For the population as a whole, the genotypes of A can be weighted by the relative frequencies of B locus genotype surviving after selection to obtain the p_A value identical with that in Table 17-3D:

p'_A	*B* Genotype	*B* Genotype Frequency Surviving	*Wgt'd* $p'_A \left[\frac{1}{\overline{W}}\right]$
0.5	*BB*	0.2592	0.2592
0.5	*bb*	0.0032	0.0032
0.7278	*Bb*	0.2376	0.3459
		Total = 0.5000 = $\overline{W}$	0.6083

Thus, the rate of change for a single locus in this case cannot be considered realistically without reference to the second locus.

Introducing linkage (letting $c < 0.50$) would not change the final equilibrium in this case, but the rate of attainment of gametic equilibrium will be slowed down. For each locus, allelic frequencies will approach the selection equilibrium at about the same rate with or without linkage, however. If we assume close linkage (for this example, assume A and B

loci to be linked with a crossover value of $c = 0.05$), the g'' output from second-generation survivors will be as follows (compared with those in Table 17-3E):

$$g'' = \begin{array}{c|cc|l} & B & b & \\ \hline A & 0.3282 & 0.2357 & p_A = 0.5639 \\ a & 0.3231 & 0.1130 & q_a = 0.4361 \\ \hline & u_B = 0.6513 & v_b = 0.3487 & \end{array} \qquad D = -0.039068$$

Both gametes with A allele are closer to equilibrium ($g_{AB} = g_{Ab} = 0.25$) than for the unlinked case, while the gametes with a are either almost unchanged from the previous generation (g_{aB}) or approaching equilibrium more slowly (g_{ab}). More important, the degree of association between the loci measured by the gametic determinant (D) has increased in absolute value. This D value is 25.7 percent of maximum D, just a little less than in the previous generation gametic output. We may conclude, therefore, that linkage will tend to keep any association between loci in gametic combinations longer than under independent assortment, but with fitness values that are "additive," there will be *no permanent disequilibrium.*

THE STATIONARY STATE OF GAMETIC FREQUENCIES

When we consider two (or more) genic loci, our attention is drawn to the question of whether selection can hold *gametic combinations* of alleles in a stationary state. Without selection we know that random combination of genes will bring about "linkage equilibrium," and the question is resolved into whether selection can hold gametic combinations at nonzero D, or a constant deviation from random combination. Specifically, we wish to know the conditions of selection and linkage that bring about such a nonrandom association between loci.

If we set every Δg in formula (17-10A) equal to zero (no change in gametic frequencies), as we have done for other genetic systems,

$$\Delta g_{AB} = \Delta g_{Ab} = \Delta g_{aB} = \Delta g_{ab} = 0$$

the frequencies of gametes and zygotes as well as the value of D must be a constant. When this set of conditions is substituted into (17-10A), the four equations resulting are the statement of the stationary state:

$$\begin{aligned} g_{AB}(\overline{W}_{AB} - \overline{W}) - cW_{22}D &= 0 \\ g_{Ab}(\overline{W}_{Ab} - \overline{W}) + cW_{22}D &= 0 \\ g_{aB}(\overline{W}_{aB} - \overline{W}) + cW_{22}D &= 0 \\ g_{ab}(\overline{W}_{ab} - \overline{W}) - cW_{22}D &= 0 \end{aligned} \qquad (17\text{-}11)$$

Unfortunately, these equations are deceptively simple looking. They have not been explicitly solved for a general case, although it is always possible to use numerical cases to explore the full range of linkage and selection. Even in cases of "symmetrical" fitness values, solutions require finding cubic roots. It will be sufficient for us to explore two numerical examples to illustrate how a permanent "linkage disequilibrium," or association of alleles between loci, can be maintained by selection and how it is influenced by linkage. To go beyond these cases, the student is referred to summaries by Lewontin (1973, 1974).

It is essential first to define more explicitly what is meant by additive and nonadditive (epistatic) fitness values. In Table 17-3A, change in fitness due to allelic substitution will be the same irrespective of the genetic background. The corner difference ($W_{11} - W_{12}$) equals ($W_{21} - W_{22}$), or $W_{11} + W_{22} = W_{21} + W_{12}$. The same would be true of any of the other three corner values because each of the corner values is equal ($W_{11} = W_{31} = W_{13} = W_{33}$) and the single heterozygotes are all equal ($W_{12} = W_{21} = W_{23} = W_{32}$). These relationships mean that the effect of the *B* locus is the same at various levels of the *A* locus, and the same is true of the *A* locus at levels of the *B* locus. We say there is no interaction between loci if (symmetrically true for all corners)

$$E = W_{11} + W_{22} - W_{21} - W_{12} = 0 \qquad (\textit{additive}\text{ fitness}) \qquad (17\text{-}12)$$

If $E \neq 0$, there is nonadditivity, or *epistasis* (E) in fitness. Suppose fitness values are as follows for the nine genotypes:

$$\begin{vmatrix} W_{11} & W_{12} & W_{13} \\ W_{21} & W_{22} & W_{23} \\ W_{31} & W_{32} & W_{33} \end{vmatrix} \qquad \begin{vmatrix} 1 & 1 & 0 \\ 1 & 2 & 1 \\ 0 & 1 & 1 \end{vmatrix}$$

The fitness values are symmetrical along both diagonals, but it is clear that there is dominance and no heterosis in *AA* and *aa* when in *BB* and *bb* backgrounds (outer columns) and similarly for *BB* and *bb* in *AA* and *aa* (top and bottom rows), while there is overdominance (heterosis) in each of the heterozygote backgrounds (midcolumn and midrow). For the upper left corner and the lower right, $E = 1$, but for the other two corners $E = 0$. (Compared with Table 17-3, the latter are the two corners unchanged.) we shall designate these different epistatic effects as E' and E'', respectively, for later use.

One can see at a glance that selection will produce single-locus equilibrium with $p = q$, $u = v$, since the marginal totals will be identical for homozygotes with heterosis at both loci. It is important to realize, however, that the gametic combinations will not be independent at their equilibrium state. Instead the two loci will be associated, as we now proceed to show (using reasoning worked out by Lewontin and Kojima, 1960, and summarized by Li, 1967*c*).

Numerical Illustration 2

At first we shall assume that *A* and *B* loci are unlinked so that $c = \frac{1}{2}$. Linkage will be added after we see the outcome for the simpler case. Using (17-11) as the stationary condition, we find that if we divide the expression for one coupling gamete (g_{AB}) by the expression of the other (g_{ab}) (then similarly divide the two expressions for repulsion gametes), we obtain a simplifying result that allows an easy solution. The fitness set given above with $E' = 1$, $E'' = 0$, and $W_{22} = 2$ produces for the top of (17-11): $g_{AB}(\overline{W}_{AB} - \overline{W}) = D$ and the fourth equation of (17-11): $g_{ab}(\overline{W}_{ab} - \overline{W}) = D$, or on rearranging each and dividing the top by the fourth, we obtain

$$\frac{\overline{W}g_{AB}}{\overline{W}g_{ab}} = \frac{g_{AB}\overline{W}_{AB} - D}{g_{ab}\overline{W}_{ab} - D}$$

Canceling $\overline{W}$ and cross-multiplying gives $g_{AB}(g_{ab}\overline{W}_{ab} - D) = g_{ab}(g_{AB}\overline{W}_{AB} - D)$. With the fitness values given above, $\overline{W}_{AB} = g_{AB} + g_{Ab} + g_{aB} + 2g_{ab}$ and $\overline{W}_{ab} = 2g_{AB} + g_{Ab} + g_{aB} + g_{ab}$. Performing the algebra and simplifying gives

$$g_{AB} = g_{ab} \tag{17-13A}$$

Similarly, using the repulsion gamete expressions in (17-11),

$$\frac{\overline{W}g_{Ab}}{\overline{W}g_{aB}} = \frac{g_{Ab}\overline{W}_{Ab} + D}{g_{aB}\overline{W}_{aB} + D}$$

With $\overline{W}_{Ab} = g_{AB} + 0g_{Ab} + 2g_{aB} + g_{ab}$ and $\overline{W}_{aB} = g_{AB} + 2g_{Ab} + 0g_{aB} + g_{ab}$, we find

$$g_{Ab} = g_{aB} \tag{17-13B}$$

Therefore, at equilibrium, the symmetry of fitness values allows the following relationships:

$$g_{AB} + g_{Ab} = g_{aB} + g_{ab} = \tfrac{1}{2} \tag{17-13C}$$

$$g_{AB} + g_{aB} = g_{Ab} + g_{ab} = \tfrac{1}{2} \tag{17-13D}$$

so that $p_A = u_B = \frac{1}{2}$ (but, of course, $g_{AB} + g_{ab} \neq g_{Ab} + g_{aB}$ necessarily!). We may then let $g_{AB} = g_{ab} = x$ (either coupling gamete) and $g_{Ab} = g_{aB} = \frac{1}{2} - x$ (either repulsion gamete).

Owing to these simplified relationships, we only need solve the equilibrium statement for *one* gametic combination. Using Table 17-2D, we substitute x for any coupling gamete and $(\frac{1}{2} - x)$ for repulsion and supply genotype fitness values from the numerical set above to obtain

$$\overline{W} = 2x^2 + 8x(\tfrac{1}{2} - x) + 2[2x^2 + 2(\tfrac{1}{2} - x)^2] = 2x^2 + 1 \tag{17-14A}$$

Substituting for $\overline{W}$ in the first equation of (17-11), we obtain

$$x[(x + 1) - (2x^2 + 1)] - [x^2 - (\tfrac{1}{2} - x)^2] = 0,$$

simplifying to the cubic equation

$$2x^3 - x^2 + x - \tfrac{1}{4} = 0 \tag{17-14B}$$

$x = 0.28492$ (the other roots are complex). Therefore, the gametic output at equilibrium will be (as obtained by Li, 1967*c*):

$$\begin{vmatrix} \hat{g}_{AB} & \hat{g}_{Ab} \\ \hat{g}_{aB} & \hat{g}_{ab} \end{vmatrix} = \begin{vmatrix} 0.28492 & 0.21508 \\ 0.21508 & 0.28492 \end{vmatrix} \quad \text{with } \hat{D} = +0.03492 \tag{17-14C}$$

There is then a stable equilibrium with a permanent association between loci, at about 14 percent of the maximum gametic determinant (D_{max}). The selection, based on epistasis but no linkage, has prevented random allelic combinations in gametes!

Successive changes over generations that would take place from an arbitrary set of initial gametic frequencies are most important for comparison with our additive first numerical illustration. A graphic method based on gametic frequencies, represented as perpendicular distances from four sides of an equilateral tetrahedron, was shown by Bodmer and Felsenstein, 1967; with all the graphic elegance of such a device, it is not as informative as a table of data, given in Table 17-4 (modified from Li, 1967*c*). In the first illustration of changes, initial gametes are at "random" combination ($g = \frac{1}{4}$) and will change so that within about four generations they reach the nonrandom equilibrium point. $\overline{W}$

TABLE 17-4 *Approach to equilibrium under selection pattern (numerical illustration 2) for two unlinked loci (from Li, 1967c), t = generation*

A

t	g_{AB}	g_{Ab}	g_{aB}	g_{ab}	$\bar{W}$
0	0.2500	0.2500	0.2500	0.2500	1.125
1	0.2778	0.2222	0.2222	0.2778	1.154
2	0.2834	0.2166	0.2166	0.2834	1.161
3	0.2846	0.2154	0.2154	0.2846	1.162
4	0.2849	0.2151	0.2151	0.2849	1.162
5	0.2849	0.2151	0.2151	0.2849	1.162
					(increasing)

B

t	g_{AB}	g_{Ab}	g_{aB}	g_{ab}	$\bar{W}$
0	0.5000	0	0	0.5000	1.500
1	0.3333	0.1667	0.1667	0.3333	1.222
2	0.2955	0.2045	0.2045	0.2955	1.175
3	0.2872	0.2128	0.2128	0.2872	1.165
4	0.2854	0.2146	0.2146	0.2854	1.163
5	0.2850	0.2150	0.2150	0.2850	1.162
					(decreasing)

C

t	g_{AB}	g_{Ab}	g_{aB}	g_{ab}	$p = u$	D	$\bar{W}$
0	0.4500	0.1000	0.1000	0.3500	0.5500	0.1475	1.3150
1	0.3498	0.1882	0.1882	0.2738	0.5380	0.0603	1.1915
2	0.3233	0.2086	0.2086	0.2595	0.5319	0.0404	1.1678
3	0.3141	0.2132	0.2132	0.2595	0.5273	0.0360	1.1630
4	0.3092	0.2143	0.2143	0.2622	0.5235	0.0351	1.1621
5	0.3056	0.2146	0.2146	0.2651	0.5202	0.0349	1.1620
10	0.2945	0.2150	0.2150	0.2755	0.5095	0.0349	1.1623
25	0.2859	0.2151	0.2151	0.2839	0.5010	0.0349	1.1624
70	0.2849	0.2151	0.2151	0.2849	0.5000	0.0349	1.1624
							(decrease – increase)

increases in this sequence. In Table 17-4B, starting from the extreme case of coupling gametes only at $g_{AB} = g_{ab} = \frac{1}{2}$, we find rapid approach to equilibrium, but the net fitness $\bar{W}$ *decreases* constantly. Third, initiation of a gametic set as given in Table 17-4C takes much longer to reach equilibrium, during which time $\bar{W}$ first decreases and then increases! Here both gene and gametic frequencies change, while the other cases involved only gametic changes.

This illustration will serve to point out that for two pairs (or more) of loci under epistatic selective pressure, the "adaptive peak" concept, or $\bar{W}_{\max}$ at the point of equilibrium, does not necessarily hold (Chapter 14), as was originally pointed out by Kojima and Kelleher (1961) and later elucidated by Moran (1964), although it does hold approximately for small selection coefficients. It is better to say that selection leads to a balanced state of genotypes instead of to maximization of fitness.

As for linkage, we may return to our illustration, but consider the A and B loci as linked with a small crossover fraction; assume $c = \frac{1}{4}$. All that is altered from our previous consideration is the incorporation of crosses into formula (17-11) so that formula (17-14B) must be recalculated. Gametic relationships (17-13A, B, C, D) and the $\bar{W}$ statement (17-14A) will be identical irrespective of crossing over. We may then supply into (17-11) the top equation as follows:

$$x[(x + 1) - (2x^2 + 1)] - \tfrac{1}{4}(2)[x^2 - (\tfrac{1}{2} - x)^2] = 0 \qquad (17\text{-}15)$$

This simplifies to the cubic equation

$$2x^3 - x^2 + \tfrac{1}{2}x - \tfrac{1}{8} = 0 \qquad (17\text{-}15A)$$

$x = 0.324$, so that the gametic output at equilibrium will be as follows, and the gametic determinant will be greater than when loci were unlinked:

$$\begin{vmatrix} 0.324 & 0.176 \\ 0.176 & 0.324 \end{vmatrix} \qquad (17\text{-}15B)$$

$D = +0.0740$ or 29.6 percent of D_{max}.

SOME GENERALIZATIONS FROM NUMERICAL ILLUSTRATIONS OF GAMETIC SELECTION

From these numerical illustrations, there are several points for our attention, even though we avoid the complexity of general solutions for the stable state of gametic associations.

1. Linkage is *not* a necessary condition for a stable gametic association or constant disequilibrium.
2. It *is* necessary to have some epistatic interaction in fitness between loci ($E \neq 0$) in order to maintain a stable gametic association or constant disequilibrium.
3. Linkage enhances the magnitude of the association at equilibrium when epistasis in fitness characterizes genotypes; it thus plays a supporting role rather than a primary one.
4. When fitness values are additive between loci ($E = 0$), changes in gametic frequencies with selection involve temporary associations—that is, nonrandom combinations of alleles at two loci. Linkage tends to preserve such temporary associations longer than for unlinked loci; however, the outcome of selection eventually will be $D = 0$, and linkage will make no difference to that final state.

These equilibrium results (D constant) based only on the numerical illustrations may be summarized as follows:

	Fitness Values—9 Genotypes		
Illustration	Additive (1)	Epistatic (2)	
	D	D	Percent of D_{max}
Loci independently assorting	0	0.035	14
Loci linked	0	0.074	30

CONDITIONS FOR STABILITY OF GAMETIC EQUILIBRIUM

Further summary of the symmetrical fitness case is given in Table 17-5. Numerical illustration 2, just discussed, is of the type in Table 17-5B, where there are two different corner fitness values leading to two different epistatic values, E' and E''. In Table 17-5A is the simpler form with all double homozygous fitness values equal ($W_{11} = W_{33} = W_{13} = W_{31}$, called i) so that there is just one epistatic value E. This latter symmetrical set of fitnesses was investigated by Lewontin and Kojima (1960) for the existence of stability of equilibrium. A real solution of the cubic equation in Table 17-5A requires either that $D = 0$ or that the quantity under the radical sign have a value between 0 and 1 inclusive; that is, the conditions for a solution are

$$E > 0 \qquad \text{and} \qquad 4\,ch < E \qquad \text{that is, } c < E/4h \tag{17-16}$$

The crossover percentage must be less than the ratio of epistasis to four times the double heterozygote fitness (W_{22}). When these conditions are fulfilled, there will be three separate

TABLE 17-5 *Epistatic fitness—two loci. Two types of symmetrical fitness pattern with equilibrium statements, $\bar{W}$ and cubic equation for equilibrium condition (modified from Lewontin and Kojima, 1960; and from Li, 1967c)*

	BB	Bb	bb
AA	W_{11}	W_{12}	W_{13}
Aa	W_{21}	W_{22}	W_{23}
aa	W_{31}	W_{32}	W_{33}

x = either coupling gamete

$(\frac{1}{2} - x)$ = either repulsion gamete

A All corners identical (see Exercises 15 and 16)

$$\begin{vmatrix} i & k & i \\ j & h & j \\ i & k & i \end{vmatrix}$$

$$E = i + h - j - k$$

$$\bar{W} = 4Ex^2 - 2Ex + \tfrac{1}{2}(i + h)$$

Cubic equation

$$Ex(4x^2 - 3x + \tfrac{1}{2}) + cDh = 0$$

Equilibrium solutions

1. $g_{AB} = g_{Ab} = g_{aB} = g_{ab} = \frac{1}{4}, D = 0$
2. $g_{AB} = g_{ab} = \frac{1}{4} \pm \frac{1}{4}D$
3. where $D = \sqrt{1 - \dfrac{4ch}{E}}$

(2 and 3 bracketed together)

and $g_{Ab} = g_{aB} = (\frac{1}{2} - g_{AB})$

B Two corners (upper left and lower right) different from other two corners

$$\begin{vmatrix} i & k & l \\ j & h & j \\ l & k & i \end{vmatrix}$$

$$E' = i + h - j - k$$
$$E'' = l + h - j - k$$
$$\bar{W} = 2x^2(E' + E'') - 2xE'' + \tfrac{1}{2}(l + h)$$

Cubic equation

$$2x^3(E' + E'') - x^2(E' + 2E'') + (\tfrac{1}{2})xE'' + cDh = 0$$

(In numerical illustration 2 given in the text, $E'' = 0$ so that this cubic equation reduces to (17-14B) when $c = \frac{1}{2}$, or 15A if $c = \frac{1}{4}$)

gametic equilibrium sets: (1) the "central" gametic set ($g_{AB} = 0.25$, equal to all others), and (2) and (3) are the two "side points" expressed by $g_{AB} = 0.25 \pm 0.25D$. The central point always exists due to the symmetry of the fitnesses, but the side points may or may not exist. Three conditions for stability of the central point are as follows:

$$h > i, \qquad (h - i) > (k - j), \qquad \text{and} \qquad c > E/4h \tag{17-17}$$

Note: the last is the opposite of (17-16).

Therefore when the central point is stable, there can be no side points; conversely, when side point equilibriums exist (stable or unstable), the central point is necessarily unstable. The important generalization here is that in order to have stable side equilibriums, linkage has to be close, relative to the amount of epistasis.

As a numerical illustration of the stability of three gametic equilibriums, assume the following fitness values with A and B linked at $c = 0.0525$:

$$\begin{vmatrix} 2 & 3 & 2 \\ 5 & 8 & 5 \\ 2 & 3 & 2 \end{vmatrix} \quad h = 8, \qquad E = 2 + 8 - 5 - 3 = 2$$

then using Table 17-5A, the equilibriums would be (1) $g_{AB} = g_{Ab} \cdots = \frac{1}{4}$; (2) and (3) $g_{AB} = g_{ab} = \frac{1}{4} + \frac{1}{4}D$, where $D = \sqrt{1 - 4(0.0525)8/2} = \sqrt{1 - 0.84} = 0.4$ so that the three equilibriums for g_{AB} are

$$\underset{\text{(stable)}}{g_{AB} = 0.25 + 0.10 = 0.35} \qquad \underset{\text{(unstable)}}{g_{AB} = 0.25} \qquad \underset{\text{(stable)}}{g_{AB} = 0.25 - 0.10 = 0.15}$$

The two side points are stable and the central point unstable. If $c > 0.0625$, there would be no side equilibrium points and the central one would be stable. Thus, linkage values must be small (loci must be tightly linked) for the association between loci to remain stable. (See Exercise 16 for a case where all three equilibriums are unstable.) Do you find the $\overline{W}$ value to be maximum at the stable point?

The relationship between linkage and epistasis given in (17-16), then, defines a critical value of c (crossover percentage) above which linkage has no effect on the final selection equilibrium points because loose linkage allows D to become zero. Lewontin (1974) describes several cases of complexities that may arise when linkage and gametic selection interact. In general, the greater the epistasis in fitness, the less tight linkage has to be for association between loci to be permanent; conversely, the less the epistasis, the tighter the linkage must be to play its subsidiary role in bringing about stable association of loci.

LINKAGE AND NET FITNESS

While the conditions for a stable gametic association indicate that linkage plays a subsidiary role, one might conclude that linkage is relatively unimportant, but nothing could be more erroneous. We have only to call attention to two cases to point out how the role of linkage should be judged.

1. If nonallelic recessive lethals occur in *trans* phase of linkage with no recombination between them (balanced lethals), the population will have higher net fitness than if

recombination is allowed. Only the double heterozygote (which has highest fitness of all genotypes possible) can exist; if the potential production of the six out of the nine genotypes that would have recombinant chromosomes were suppressed by prevention of crossing over, it is obvious that total loss of genotypes by lethality would be least. The same would be true for the case illustrated in Table 17-3.

2. If crossing over is zero in the case illustrated in Table 17-5, all linkages at equilibrium would be only *cis* or *trans*, depending on starting conditions, and $\bar{W}$ would be maximized compared with any case involving crossing over. The student should verify that statement.

We conclude that there is no higher mean fitness at equilibrium than for the case when $c = 0$. Lewontin (1974) points out that there are cases where there may be more than one local fitness maximum, and for intermediate values of c a slight tightening of linkage may cause a decrease in net fitness, but in the long run the tightest linkage gives the highest net fitness. If that is the only consideration to be made in the dynamics of genotypes under selection, why has not recombination been reduced to zero more frequently? As Turner (1967) expressed it: "Why does the genome not congeal?" A partial answer comes from what we have already noted in considering two loci—a stable state does not necessarily occupy an adaptive peak for multiple loci. But we must also remember that these considerations have been premised on the assumption of constant fitness values for the expediency of working out simple selection models. From our last chapter, we realize full well that fitness values are likely to be variable over space and time, and that the genotype must evolve in a population as a compromise with best strategy for the conditions, which are more likely to be complex than simple.

In effect, a general attitude toward recombination was expressed by Carson (1957), who emphasized the significance of genetic flexibility in populations:

> *The amount of recombination permitted . . . is best looked upon as an adaptive property which is under the control of natural selection. Too much recombination is inadaptive because it tends to break up adaptive complexes of genes which have been welded together by natural selection. This process reduces the immediate fitness of the organism. In order to survive, a species must maintain high immediate fitness, and it is not surprising that devices limiting or preventing recombination are common in species populations. . . . It appears that too little recombination jeopardizes the ability of the species to meet drastic changes in environmental conditions over very long periods of time. Most organisms which have survived to the present day display a balance between these two forces.*

OBSERVED LINKAGE DISEQUILIBRIUM

When associations of nonallelic genes in gametes are found in any population, one of the following four conclusions might be drawn: (1) selection has been acting on genotypes determined by the loci concerned or loci tightly linked with the observed ones, and the population is in approximate selection equilibrium; (2) selection has been acting as in (1), and the population is in a transient phase (not at equilibrium); (3) the disequilibrium observed is due to chance, but owing to the fact that we do not yet have a clear description of the D random distribution, we cannot distinguish between a significant and nonsignificant D value; in this case, several ancillary statistical tests may be needed to lend support to one view or another (see Example 17-6 for the use of two such tests in a human population); (4) the population size may be small enough to cause random linkage disequilibrium via inbreeding, drift, or stratification (Wahlund effect if both loci show significant allelic frequency differences; see Prout's appendix to Mitton and Koehn, 1973, and Kimura and Ohta, 1971).

In addition to the cases of gametic association mentioned previously in this chapter, the reader may be reminded of two examples in Chapter 5 (Example 5-3 from the locus studies of Cannon in *Drosophila melanogaster* and Example 5-4 from the shell color polymorphism studies of Cain, Sheppard, and others in *Cepaea nemoralis*) in which there was evidence for linkage disequilibrium. Another example (Example 5-6 from two-locus allozyme studies of Rasmuson, et al., in *D. melanogaster*) did not show any disequilibrium. Cannon's observations do not have any obvious explanation from selection theory since we have no idea what the relationship to fitness may be for the linkages she studied. The snail color polymorphism, on the other hand, is one of the better-known cases where at least some selective agent is known and fitness values for phenotypes can be estimated (see Chapter 16).

In certain populations of grasses (*Avena* and *Hordeum*), Allard and his colleagues have monitored several linked allozyme loci and have found very significant gametic associations dependent on moisture in the natural environment (*Avena* referred to in Chapter 16) and in experimental populations of *Hordeum* (Weir, Allard, and Kahler, 1972). In the latter case, a composite cross of 30 barley varieties initiated in 1937 and allowed to pollinate naturally (mostly by selfing) was maintained at Davis, California, for the intervening years until these authors had the seed made available to them for electrophoretic study of young seedlings. Four esterase loci monitored over 25 generations showed gametic phase disequilibrium for each of the six paired combinations that had developed from about the first six generations following the original cross.

In addition to these instances, Lewontin (1974, p. 316) lists other cases of allozyme linkage disequilibrium. Three other instances include the following:

1. The nonrandom association of gametic combinations of closely linked *XDH* and *AO* loci in *Drosophila subobscura* in two localities in Greece, where the D values were substantially alike, as reported by Zouras and Krimbas (1973). This case is complicated by the fact that several alleles are present in each population, and the disequilibrium was analyzed by grouping several possible gametic combinations into four.

2. Mitton, Koehn, and Prout (1973) describe assocations between two allozyme loci (*Lap* and *Ap*) in the blue mussel *Mytilus edulis* taken from different tidal zones over a sand bar on Long Island, New York. Unfortunately, in their case, one of the loci (*Lap*) shows geographic variation in allelic frequencies along the Atlantic coast so that some heterogeneity may exist between the tidal zones for that locus; also, their data do not indicate a very large D value (only 7.42 percent of D_{max} as calculated from their data), so that the case may not be highly significant.

3. Roberts and Baker (1973) indicate a highly significant case in α-esterase allozymes in natural populations of *Drosophila montana*; their data are summarized

in Example 17-5, illustrating a case in which there is close linkage between apparently functional tandem duplications that have come to control two essential component esterases as beneficial for the organism.

Finally, in a certain human population (Example 17-6) sampled for the amount of gametic association among eleven different genetic systems, there is considerable evidence for association between at least nine pairs of loci. While two of the eleven systems are known to be linked (*Duffy* and *Rh* on chromosome 1, according to McKusick and Chase's review, 1973), those two show no evidence of linkage disequilibrium. For the systems that do, some may later prove to be linked and thus may be considered worthy of further detailed investigation for linkage on the basis of this finding. If these associations are confirmed by later study, it will be concluded that epistatic selection must be a significant factor in maintaining the disequilibrium.

EXAMPLE 17-5

ASSOCIATION BETWEEN LINKED ALLOZYME VARIANTS IN A DROSOPHILA POPULATION

Populations of *Drosophila montana* (belonging to the *D. virilis* group and nearctic, particularly at high elevations) were sampled at 10,000-foot elevation in Gunnison County, Colorado, from 1968 to 1970 by Roberts and Baker (1973). The mode of inheritance of 15 electrophoretically distinguishable α-naphthyl acetate-specific esterases was worked out by mating a strain with no esterase activity ("null" allele) to wild-caught flies and analyzing the progeny. These 15 specific bands, lettered from *A* through *P* (except *O* to avoid confusion with the null allele) in order from slowest migrating band to the fastest, were accounted for by four closely linked loci on the second chromosome. At each of the four loci there was a null allele as well as either three or four active allozyme sites. Allocation of allozyme to a specific locus was reasoned on the basis that no chromosome would contain two allozymes determined by the same locus.

By means of outcrossing a sample of wild-caught flies ($N = 237$) to the completely null strain (all four loci with a null allele), the authors were able to examine the G_1 progeny sufficiently to establish the exact genotype of each parent wild fly with respect to these four loci. Each wild chromosome provided the linkage form found in the parent, and after examination of about eight progeny the likelihood of obtaining both wild parent chromosomes was sufficient to establish the wild genotype. Over 90 percent of the flies had an active esterase at three or four of the possible eight sites on homologous chromosomes in a sample of 474 ($=2N$).

The α-esterase loci arranged in linkage order and with their within-locus observed frequencies, separated into active and null allelic portions are diagrammed in Table A. If there were random association between alleles at the four loci, the frequencies of chromosome types could be predicted by the products of active and null alleles at each locus. For example, the predicted number of all-null chromosomes would be: (0.726)(0.363)(0.452)(0.654)474 = 35.3, but the observed number of such chromosomes was just three. There was a highly significant linkage association between active and null alleles at these four

TABLE A *Within-locus observed frequencies of alleles at four alpha-esterase loci arranged in linkage order from* Drosophila montana *(237 flies = 474 chromosomes sampled) from Roberts and Baker, 1973). Letters A through P = active esterases, Null = no activity allele.*

Locus	#1		#3		#2		#4	
Allele	*A B C D*	Null	*H I J P*	Null	*E F N*	Null	*G K L M*	Null
Active-Null observed frequencies	0.2743	0.7257	0.5675	0.4325	0.6371	0.3629	0.3460	0.6540

loci, as shown in the Figure A histogram, where observed and expected (random association) of active and null alleles over the four loci are compared. Clearly, there was a much higher than expected number of chromosomes with two active allozymes. When the detailed data were examined further for association between loci, active versus null alleles pooled into a 2×2 table between pairs of loci displayed a result of linkage disequilibrium; for example, locus 1-locus 2. Using *A B C D* and *E F N* active alleles, the authors obtained

	Locus 1	
Locus 2	Active	Null
Active	31	273
Null	97	73

or in frequencies $\begin{vmatrix} 0.0654 & 0.5760 \\ 0.2046 & 0.1540 \end{vmatrix}$

$D = -0.1078$

Similarly, it could be shown that a large D value occurs for loci 3 and 4. However, the remaining four pairs of loci (1 and 3, 1 and 4, 2 and 3, 2 and 4) were much closer to linkage equilibrium; for example, locus 1-locus 3 data give

	Locus 1	
Locus 3	Active	Null
Active	74	195
Null	53	152

or in frequencies $\begin{vmatrix} 0.1561 & 0.4114 \\ 0.1118 & 0.3207 \end{vmatrix}$

$D = +0.004$

The student should verify that a contingency chi-square on each of these two sets of data gives highly significant nonrandom association for the first case of 1 and 2, while the second case of 1 and 3 is random.

Roberts and Baker presented evidence for considering the left two loci (1 and 3) a tandem duplication of the right two loci (2 and 4) and that the first locus of each pair specifies an essential component esterase (E_X from either 1 or 2), while the second locus specifies another component esterase (E_Y from either 3 or 4). Then the linkage disequilib-

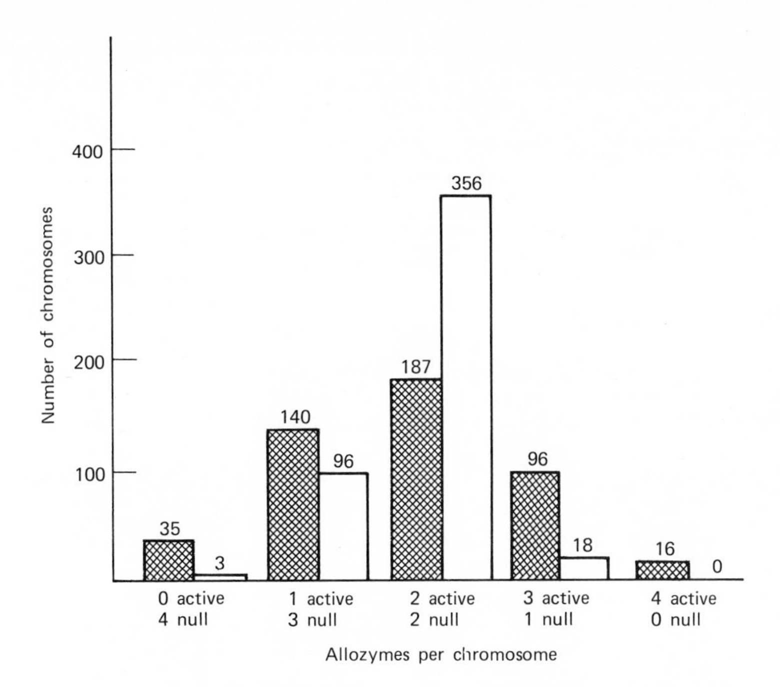

Figure A. Comparison of the expected (shaded) with the observed (open) number of chromosomes with one, two, three, or four active loci. From Roberts and Baker, 1973.

rium occurs between the two first members or the two second members of the duplication with the generalization that the optimum linkage has to have either one or the other duplicate active allozyme but not both or neither. For each chromosome, 75 percent had two active alleles linked, but they represented the E_X and E_Y substances together predominantly, and chromosomes lacking both substances or having duplicates of one or the other were in lower than expected frequency. Therefore, the polymorphism between active and null esterase alleles appears to be the result of a balance between having an optimum number of active alleles and the presence of both component substances.

EXAMPLE 17-6

EVIDENCE FOR GAMETIC ASSOCIATION IN A HUMAN POPULATION

The human population from Tecumseh, Michigan, was described in Chapter 15, pp. 487–489. Sinnock and Sing (1972*a*,*b*) utilized the data collected on more the 9000 individuals for evidence of gametic associations between 11 different genetic systems, considering two loci at a time to find at least sufficient first approximations to the existence of any associations. Of the 9000 persons who furnished blood samples for this project, 6756 were completely typed for the 11 genetic systems (representing 14 loci) and thus constituted the data sample: Hp, Gc, MN-Ss (two loci), Rh (three loci), Duffy, P, ABH-secretion, Lewis-secretion, Kell, Kidd, A-B-O. There were thus 91 possible two-locus combinations. For each combination, a maximum-likelihood procedure was used to estimate gametic frequencies. Five groupings of the population were analyzed: total, kin (consisting of parent and child subgroups), and no kin (all persons who were neither parent, child, nor sibling of any other in the sample).

Owing to the fact that we know little of the gametic determinant distribution, estimates of gametic nonrandom association and its significance were made by using three statistical techniques:

1. Calculation of D and its percent of $D_{\max}$ by formula (5-9).
2. Chi-square from a 2 × 2 contingency based on gametic frequencies:

$$\chi^2 = \frac{(g_{AB}g_{ab} - g_{Ab}g_{aB})^2 N}{p_A q_a u_B v_b} \qquad \text{(Lewontin, 1964)}$$

   However, the correlation (r) of gene frequencies between loci

$$r = \frac{D}{\sqrt{p_A q_a u_B v_b}}$$

   (according to Hill and Robertson) is related to the χ^2 in that upon squaring the correlation and multiplying by the total sample size N, we have the same value distributed as a chi-square with one degree of freedom. We specify this chi-square as Nr^2.

3. A chi-square may be partitioned based on the fact that one may compute expected values for phenotypes by using either gametic frequencies or gene frequencies. Each two-locus gametic frequency is a function of gene frequency and D, that is, $AB = pu + D$, $Ab = pv - D$, etc. There are usually nine phenotypes in the case of these codominant two-locus sets. Thus, if expected values for phenotypes are calculated from gene frequencies and D, there will be $(9 - 3 - 1) = 5$ degrees of freedom, because of the four gametic frequencies only three are independent; that chi-square is called χ_G^2. A second chi-square can be calculated by disregarding D and using only gene frequencies to compute expected phenotype frequencies; in this case the number of independent gene frequencies will be two (since we only need to know p and u, for example), so that there will be $(9 - 2 - 1) = 6$ degrees of freedom. This second chi-square is called χ_T^2. The difference in these chi-squares is attributable to fitting the estimate of the additional parameter D. Consequently, Sinnock and Sing used the partitioned difference $\chi_T^2 - \chi_G^2 = \chi_D^2$ with the remaining degrees of freedom as a test of gametic association.

Of the 132 values of D estimated for the total, kin, and no kin groups, the largest values were as expected for the MN-Ss and Rh combinations. Excluding those from further consideration because they are presumably very closely linked, there were 114 independent combinations for analysis (note that Duffy and Rh are now known to be located on chromosome 1). In each combination, the value of D_{max} and the percentage of D_{max} that gives an order of magnitude to the amount of gametic association were computed. The average percentages of D_{max} were 3.6, 8.7, 5.2, and 6.0 for total, no kin, parent, and child groups, respectively. Although there was no test available of the hypothesis that percent of $D_{max} = 0$, it is informative to note that 17 of the paired loci combinations gave a percent of D_{max} as large or larger than triple the mean for at least one of the groups in upper Table A, and nine combinations exceeded the overall average for each of the groups. In addition, when D had the same sign in kin, parent, and child groups, it would be more likely to be nonzero. Thus, in lower Table A are presented those combinations with consistent high D or D of the same sign in parent and child.

TABLE A *Combinations with proportion of D_{max} (sign of D included) values that are triple the absolute mean proportion for at least one of the groups or exceed the absolute mean for all groups studied (from Sinnock and Sing, 1972b) (values in parentheses are not triple the mean of all proportion of D_{max} values)*

	Group				
Combination	Total	No Kin	Kin	Parent	Child
Lewis secretion—ABH secretion	−0.157	(−0.185)	−0.153	−0.424	(−0.035)
Kell-Gc*	+0.163	+0.372	+0.125	(+0.072)	(+0.154)
Kell-MN	(+0.107)	(−0.074)	+0.145	+0.166	+0.187
Kell-Rh-E*	−0.164	(+0.141)	−0.307	(−0.079)	−0.565
Kell-Rh-D	(+0.007)	+0.381	(+0.042)	(−0.057)	(−0.058)
Kell-P	(−0.023)	−0.307	(+0.032)	(−0.111)	(+0.118)
Kell-Lewis secretion*	+0.247	+0.308	+0.236	+0.194	+0.342
Kidd a-Lewis secretion	(+0.083)	(−0.020)	(+0.098)	+0.176	(+0.066)
Kidd a-Kell	(+0.075)	+0.401	(+0.013)	(+0.146)	(−0.092)
Kidd b-Ss*	(+0.091)	(−0.231)	+0.140	+0.171	(+0.082)
Kidd b-Rh-E*	−0.114	(+0.148)	−0.145	(−0.057)	−0.276
Kidd b-Lewis secretion*	−0.136	(+0.215)	(−0.197)	(−0.133)	−0.291
Kidd b-Kell	+0.183	+0.570	+0.122	+0.217	(−0.004)
ABO B-MN	(−0.014)	+0.284	(−0.061)	(−0.048)	(−0.074)
ABO B-Rh-E	(−0.054)	(+0.054)	(−0.109)	(+0.024)	−0.230
ABO B-Lewis secretion	(−0.028)	(−0.183)	(−0.003)	(−0.144)	+0.226
ABO B-Kidd b	(+0.084)	(+0.025)	(+0.093)	+0.179	(+0.077)
Average proportion of D_{max} for this subset	0.102	0.229	0.119	0.141	0.169
Duffy-ABH secretion*	+0.056	+0.098	+0.050	+0.058	+0.096
Kell-Duffy*	+0.087	+0.102	+0.085	+0.080	+0.090
Kidd A-ABH secretion*	+0.059	+0.123	+0.048	+0.076	+0.062
Average proportion of D_{max} from all independent paired loci	0.036	0.087	0.040	0.052	0.060

* Proportion of D_{max} exceeds its average in all groups.

For parent-child groups, there were correlations of D between parent-offspring amounting to $r = 0.415$ for both D's positive and 0.294 for both D's negative, while correlations were about zero when parent-offspring had opposite-sign D's.

The second and third statistical measures Nr^2 and χ_D^2 were expected to be no greater than their respective degrees of freedom if there was random association of nonalleles in gametes. However, these measures were significant, and the hypothesis for $D = 0$ was rejected at least at the 5 percent level for the Nr^2 in 64 combinations and for the χ_D^2 in 22 combinations. Mean values for these measures computed from the 114 independent two-locus combinations were according to group as follows:

Measure	Group			
	Total	No Kin	Parent	Child
Nr^2 (1 *d.f.*)	2.59	1.95	2.89	2.52
x_D^2 (1 *d.f.*)	1.53	1.12	1.32	1.74

It seems unlikely that these measures would be preponderantly above 1.5 times their expected values unless there was a significant degree of nonrandom association in gametes. The high values in the two kin groups may be a reflection of the effect of the mating system on organization of this set of polymorphisms.

EXERCISES

1. In the first numerical example (page 549) of three alleles with heterosis among heterozygotes and $W_{11} = \frac{2}{3}$, $W_{22} = W_{33} = \frac{1}{2}$, test the stability of the equilibrium frequencies ($p = \frac{3}{7}$, $q = r = \frac{2}{7}$) by calculating allelic frequencies after one generation of selection starting with each of the following in G_0:

	p_0	q_0	r_0
(a)	0.5	0.25	0.25
(b)	0.3	0.4	0.3
(c)	0.2	0.4	0.4

   (d) In which case is total amount of gene frequency change the greatest?
   (e) Obtain tricoordinate graph paper; represent heights of the triangle as p,q, and r, and position the three cases above (a), (b), and (c) showing them before and after selection.

2. Example 17-1 gives fitness values for six genotypes of hemoglobin variants, Hb^A, Hb^S, and Hb^C in Ghana as calculated by Allison (1956*a*).
   (a) Show that for these values at $\hat{p} = 0.83$, $\hat{q} = 0.07$, $\hat{r} = 0.10$, the fitness relationships of formula (17-3) hold—that net fitness equals the net fitness due to each allele.
   (b) Use the fitnesses calculated by Cavalli-Sforza and Bodmer (1971) with $W_{11} = 0.99$, etc., except change the fitness of $Hb^C/Hb^C(W_{33})$ from 1.45 to 0.90 in the fitness set. Calculate the determinants from formula (17-5). Is a stable equilibrium possible? What are the allelic frequencies with that change in W_{33}? How does the result compare with Allison's estimate?

3. In Appendix A-12, Cramer's rule for solving third-order simultaneous equations gives

$$x = \frac{|kbc|}{|abc|}, \; y = \frac{|akc|}{|abc|}, \; z = \frac{|abk|}{|abc|},$$

provided that $|abc| \neq 0$. Let $p = x$, $q = y$, and $r = z$. Then derive formula (17-4) from (17-3A) by this method of determinants.

4. It may be more satisfying to find the equilibrium allelic frequencies by utilizing formula (17-2) for change in two out of the three frequencies as a set of simultaneous equations that may be put equal to zero for the equilibrium condition. However, we need two sets of simultaneous equations to cover three alleles; the first set might be as follows:

$$\begin{cases} \Delta p_i = p_1 - p_0 = 0 \\ \Delta q_i = q_1 - q_0 = 0 \end{cases} \quad \text{becomes} \quad \begin{cases} p_0 \bar{W}_{1\cdot} - p_0 \bar{W} = 0 \\ q_0 \bar{W}_{2\cdot} - q_0 \bar{W} = 0 \end{cases}$$

(a) Show that the solution will be $p(W_{11} - W_{21}) + q(W_{12} - W_{22}) + r(W_{13} - W_{23}) = 0$.
(b) A second pair of equations might be as follows:

$$\begin{cases} \Delta p_i = p_1 - p_0 = 0 \\ \Delta r_i = r_1 - r_0 = 0 \end{cases} \quad \text{becomes} \quad \begin{cases} p_1 \bar{W}_{1\cdot} - p_0 \bar{W} = 0 \\ r_0 \bar{W}_{3\cdot} - r_0 \bar{W} = 0 \end{cases}$$

Show that the solution will be $p(W_{11} - W_{31}) + q(W_{12} - W_{32}) + r(W_{13} - W_{33}) = 0$.
(c) What would be a third possible pair of equations? Why is this pair not necessary?
(d) Supply into each of the two solutions above the values for W_i from the second numerical example on page 549 and verify that the solution gives $\hat{p} = \frac{1}{8}$, $\hat{q} = \frac{5}{8}$, $\hat{r} = \frac{2}{8}$.

5. In the fitness set below, at what value of $W_{13}(W_{31})$ would the outcome of selection switch from a stable equilibrium for three alleles to an elimination of the a allele? (page 550, third numerical example).

	A	a	a'
A	1	2	?
a	2	1	3
a'	?	3	0

6. Show that the fitness set below gives an unstable equilibrium at $\hat{p} = \frac{10}{21}$, $\hat{q} = \frac{1}{7}$, $\hat{r} = \frac{8}{21}$ by calculating p, q, r values after one generation of selection for the starting (G_0) frequencies given below.

	A	a	a'
A	1	4	3
a	4	2	0
a'	3	0	2

(a) G_0: $p = 0.5$, $q = 0.1$, $r = 0.4$.
(b) G_0: $p = 0.4$, $q = 0.2$, $r = 0.4$.
(c) G_0: $p = 0.4$, $q = 0.3$, $r = 0.3$.
(d) On tricoordinate graph paper, position these three cases showing them before and after selection.

7. Show that formula (17-8) for two alleles is the equivalent of formula (14-15). Also show that for the two-allele case, $\bar{W}_{1\cdot} = \bar{W}_{2\cdot} = \bar{W}$. Use the numerical example from Table 14-5 and calculate these net fitnesses due to each allele, showing that they equal the net fitness.
8. What would happen to a three-allelic locus if one homozygote (AA) had greater fitness than $\bar{W}$? For a nontrivial equilibrium with three alleles, the following condition is a necessary one:

$$\bar{W} = \bar{W}_{1\cdot} > W_{11}$$

The last 2 terms (inequality) can be represented as follows:

$$pW_{11} + qW_{12} + rW_{13} > pW_{11} + qW_{11} + rW_{11}$$

Thus,

$$\frac{qW_{12} + rW_{13}}{q + r} > W_{11}$$

Show that this condition holds in each of the equilibrium examples given in pages 549–550 including the unstable equilibrium on page 550.
9. If all heterozygotes for three alleles have equal fitness ($W_{ij} = 1$) and two homozygous fitness values are just one-half that ($W_{22} = W_{33} = \frac{1}{2}$), what would be the limits for W_{11} to produce a stable three-allele equilibrium?
10. In a third-order determinant (Appendix A-12), cofactors ("minors") are those determinants of order one less (that is, second order) obtained by consideration of any of the nine elements and eliminating the row and column in which that element stands. For example, in the following fitness set

$$\begin{vmatrix} 1 & 2 & 4 \\ 2 & 1 & 3 \\ 4 & 3 & 0 \end{vmatrix}$$

the cofactor C_{11} would be obtained by crossing out row 1 and column 1 to obtain

$$\begin{vmatrix} 1 & 3 \\ 3 & 0 \end{vmatrix}$$

whose value is -9. According to Li (1967c), the equivalent of saying "all D_i's are positive," as in formula (17-5) for a nontrivial stable state of three alleles, is to state that all cofactors for homozygotes must be negative.
(a) What are the other two homozygote cofactors in the above set (C_{22}, C_{33})? Do they indicate a stable state? What would be the homozygous cofactors if $W_{13} = W_{31} = 5$?
(b) Find all the remaining six cofactors in the above fitness set. Arrange them in a 3×3 set corresponding to the fitnesses; then total the rows and obtain the total cofactors.

$$\begin{aligned} C_{11} + C_{12} + C_{13} &= \ ? \\ C_{21} + C_{22} + C_{23} &= \ ? \\ C_{31} + C_{32} + C_{33} &= \ \underline{?} \\ \text{Total} &= \end{aligned}$$

Divide each row total by the cofactor total. Obtain the allelic frequencies at equilibrium p, q, r. Note also that each cofactor row total $= D_1, D_2, D_3$ in turn. What relationships do these have to $\bar{W}$ and $|W|$?

11. Example 17-3 presents evidence that homozygous chromosomes have lower viability following recombination between natural population chromosomes selected for good viability. In the following three cases, which simulate crossing strains of good viability homozygotes, determine the average phenotypic value before and after recombination assuming equal crossover and noncrossover classes for simplicity.
    (a) Case I. Additive genic action. Let each + or − have a value of $1/a$ increment.

$$G_0: \frac{+-+-+-}{+-+-+-} \times \frac{-+-+-+}{-+-+-+} \qquad G_1: \frac{+-+-+-}{-+-+-+}$$

    (Net increment is 0 in each). Let a crossover occur anywhere between loci. Will the average of homozygous recombinants be greater, less, or equal to the parent phenotype? Average the crossovers and noncrossovers.
    (b) Case II. Geometric mean of loci effects. Assume linked loci contribute to viability as the product of their individual viability effects, as follows. Let $A/A = B/B$ make an increment of 2, and $a/a = b/b$ make an increment of $\frac{1}{2}$. Then let loci $A(a)$ and $B(b)$ determine viability as the geometric mean of their two-locus increments.

$$G_0: \frac{Ab}{Ab} \times \frac{aB}{aB} \qquad \text{(Each parent's increment will be } (2)(\tfrac{1}{2}) = 1.00\text{)}$$

$$G_1: \frac{Ab}{aB} \qquad \text{Let a crossover occur between } A \text{ and } B \text{ loci and average equal crossover}$$

    and noncrossover homozygotes. Verify that the average progeny would have value $= 1.125$ (increased viability compared with parents!). What would be the outcome if original parents had been in "coupling" phase of linkage? Since the experimenters in Example 17-3 would have no way of knowing which linkage form existed in the parents, could it be concluded that geometric gene action could explain consistent lower viability following recombination?
    (c) Case III. Epistatic interaction dependent on the sequence of modifiers in linkage. Let any x-y combination in either order $= 1$ increment in viability. Let any x-x or y-y combination $= 0$ increment in viability.

$$G_0: \frac{x\text{-}y\text{-}x\text{-}y}{x\text{-}y\text{-}x\text{-}y} \times \frac{y\text{-}x\text{-}y\text{-}x}{y\text{-}x\text{-}y\text{-}x} \qquad G_1: \frac{x\text{-}y\text{-}x\text{-}y}{y\text{-}x\text{-}y\text{-}x}$$

    (Each parent gamete has 3 x-y combinations $= 3$). Let a crossover occur anywhere between loci, and average homozygotes with equal numbers of crossovers and noncrossovers. Note whether there is a lowering of average viability. Will the same effect occur irrespective of the crossover that has taken place? What can you conclude about the type of genic action necessary to explain the results of Example 17-3?

12. If all fitness values in Table 17-3A are increased by 1 so that $W_{11}, W_{13}, W_{31}, W_{33} = 1$, $W_{22} = 3$, and the remaining W's $= 2$, then we say the values are linearly transformed.

(a) What will be the expected gametic frequencies at equilibrium ultimately? Does linear transformation of fitness values change the final equilibrium?
(b) What will be the gametic output from surviving genotypes of G_0? What will be the allelic frequencies of each locus? Compare with Table 17-3D. Also, find $\overline{W}_{AB}$, $\overline{W}_{Ab}$ $\cdots$ and $\overline{W}$. How do they differ from the values you would get from Table 17-3C?
(c) What is the trend in gametic output for G_1 survivors with linearly transformed fitness values? How does transformation affect the outcome of selection?

13. Consider each locus (A-a, B-b) separately with allelic frequencies as given in Table 17-3B. Assume fitness values for genotypes as $W_{AA} = 1$, $W_{Aa} = 2$, $W_{aa} = 1$ (and, similarly, the B genotype fitnesses).
(a) Using formula (14-7), calculate expected changes in u_B, v_b from 0.9, 0.1 and p_A, q_a from 0.8, 0.2 for two generations.
(b) How do the allelic frequencies of single loci differ from those you calculated in Exercise 12 above? To what might you attribute the outcome difference?
(c) What happens to the average fitness of *Aa* (or *Bb*) with change in gene frequencies in the case of Exercise 12 where the *AaBb* genotype has a fitness value of three? Do you consider this a case of frequency dependency as defined in Chapter 16? Why or why not?

14. (a) From formula (5-8), calculate the maximum D (departure from gametic independence) for the gene frequencies in Table 17-3D. Is D_{max} negative or positive?
(b) For D_{max} to be opposite in sign, how would the gametic frequencies have to be arranged and what would be their values?
(c) What is D_{max} for the following generation's gametic output? Is the ratio D/D_{max} increasing or decreasing in absolute value?
(d) In Table 17-4, there is permanent D at the point of gametic equilibrium. What is the D_{max} for the gametic frequencies at the stable point?
(e) What is D_{max} for the gametic frequencies in the case of linkage as given in (17-15B)?

15. Given the following fitness set for two loci,

$$\begin{vmatrix} 1 & 1.5 & 1 \\ 2.5 & 4 & 2.5 \\ 1 & 1.5 & 1 \end{vmatrix}$$

(a) How much epistasis in fitness is there? If A-a and B-b loci are independent, what will be the gametic equilibriums possible?
(b) If the loci are linked with $c = 0.20$, what will be the gametic equilibriums possible?
(c) Calculate gametic frequencies D and D_{max} after one generation of selection starting with each of the following allelic frequencies and assuming $D = 0$ initially (G_0). First assume the loci to be independent and then assume them to be linked with $c = 0.05$.

$$p_A = u_B = 0.5$$
$$p_A = 0.6,\ u_B = 0.7$$

(d) At which frequencies is a greater gametic disequilibrium generated by selection? What is the effect of linkage?

16. Given the following set of fitness values, and assuming linkage with $c = 0.3150$,

$$\begin{vmatrix} 9 & 2 & 9 \\ 2 & 10 & 2 \\ 9 & 2 & 9 \end{vmatrix}$$

   (a) How many equilibriums exist for gametic frequencies? What are they?
   (b) For the equilibriums occuring, which are stable and which unstable?
   (c) How many equilibriums would be possible if $c = 0.10$? Would any be stable?
17. Verify statement 2 under *Linkage and Net Fitness*—that if crossing over is zero, $\bar{W}$ would be maximum at gametic equilibrium compared with any case involving crossing over. Assume the fitness set

$$\begin{vmatrix} 1 & 1 & 0 \\ 1 & 2 & 1 \\ 0 & 1 & 1 \end{vmatrix}$$

   (a) If only linkage is in repulsion so that *Ab* and *aB* alone are in the population compare with the independent case given in numerical example 2 on page 574.
   (b) If linkage were only in coupling (*AB* and *ab* only), what would be the gametic equilibrium? Does the phase of linkage make a difference?
18. In Example 17-4, assume the observed gametic frequencies are at equilibrium for the *CD* and *EF* chromosomal arrangements.
   (a) Use the fitness estimates of Turner. Assuming random mating, calculate the gametic frequencies after selection. Verify that these frequencies are at equilibrium.
   (b) Let the gametic frequencies be as follows:

	St	*Bl*
St	0.12	0.38
Td	0.08	0.42

   Assuming random mating, show that with Turner's fitness estimates the following generation after selection will have frequencies approaching the stable point.
   (c) Use the Lewontin and White fitness estimates and show that gametic and chromosome frequencies will be going away from the unstable point observed in the natural population.
19. In Example 17-5, show that loci 1 and 2 are significantly associated by using a contingency chi-square test on the observed gametic frequencies of active and null alleles. Show that loci 1 and 3 are not significantly associated. Use the active versus null allelic data.

18

GENETIC POLYMORPHISM: A LOAD OR NOT A LOAD? NEUTRAL OR ADAPTIVE?

In Chapters 14 through 17, as the fundamentals of selection theory have been presented, our interpretation for the likely functioning of the genetic structure in diploid populations has often been implied. The student may have gained a broader scope of basic theory, and now should see more perspective of this field and begin to realize that along with new techniques and discoveries have come problems of interpretation. Disagreement is a natural consequence among workers in the field brought about by differences in their personalities, bias in concepts and ideologies, their training, and their main research interests. Contrary to popular belief about scientists, their decisions are not always free of their attitudes toward themselves or the world in general. The functional significance of what is observed is often fuzzily mixed with opinion and subjectivity. However, scientific endeavor thrives on difference of opinion. Proof and establishment of principle require rigorous attempts to disprove and to correct hypotheses. If some of this book impresses the reader as biased toward one concept or another, the author earnestly hopes the student will utilize the more objective contributions toward final assessment, one way or the other.

At first, population genetics was established as an extension of Mendelian facts in the earlier decades of this century by theoreticians who generalized the gene concept to populations of genotypes as outlined in the introductory chapter. Considering genetic treatment of organisms as essential to evolutionary theory, they combined the Darwin-Wallace natural selection concept with genetic processes of breeding, mutation, hybridization, and random consequences of limited population size into the synthesis that is still the main cement of ideas in the field and that is presented throughout this book. Today their mathematically explicit statements describing expected changes in genotypes under these forces are agreed on by most workers in the field.

The problems of interpreting the evolutionary function of genetic variation have accompanied population genetics since Tshetverikov pointed out in 1926 that mutations were the source of raw materials in evolution, that populations absorb mutations as a sponge takes up water, accumulating a store of potential variation in heterozygotes, and that the mechanism of maintenance of the genetic reservoir was a central problem for evolutionary genetics. However, the visible mutants recorded during the decades preceding the 1960s, necessary for assignment of genotype and genetic ascertainment, were seldom analyzed for their relationship to Darwinian fitness, while the genetic conditions such as lethals, steriles, semilethals, and subvital modifiers—certainly of significance to Darwinian expectations—were difficult to elucidate as simple genic loci; more commonly, such fitness conditions were found to be controlled by numerous modifiers, each of small effect (polygenically). Although genetic precision was more or less limited to the "visible" mutants or to chromosomally distinguishable karyotypes, experimental work was nevertheless highly successful in enlarging the concepts begun by the synthesizer theoreticians. A wealth of empirical information from laboratory, field, and human populations gave us sophistication in our opinions about the genetic architecture as it is molded by natural selection, the breeding system, and other evolutionary forces. By Mendelian methods and reasoning with sampling of natural populations, selective differences between composite genotypes could be measured but not often traced to individual loci. The subtlety of natural selection in its dynamic control of genetic potential under particular environmental conditions had to be surmised from making assumptions about genotypes without access to specific loci. Thus, an immense field of experimental population genetics and related observations from human populations were described, for example, by Dobzhansky (1937, 1962, 1970); Lewontin (1967, 1973, 1974); Spiess (1962, 1968*a*); Wallace (1968*b*); Cavalli-Sforza and Bodmer (1971). The exact nature of genotypes in a population could only be inferred, although often to a considerable degree of understanding (for example, the effects of recombination on viability in Example 17-3). Depending on the viewpoints of the investigators and the problems they were seeking to analyze, opinions varied with regard to the specifications of naturally occurring genotypes.

For a geneticist working on the mutation process and particularly the artificial induction of mutations by radiation or other factors (hazardous chemicals used commonly in modern technological societies) as H. J. Muller (1950) had been doing for most of his carrer, it appeared that nearly all new mutants have some lowering effect on fitness, commonly in homozygotes and often also with slight detriment in heterozygotes (partial dominance) (see also Wallace and King, 1952; Wallace, 1962, 1965). Any unusual new mutant with higher than average fitness arising spontaneously would be expected eventually to replace the ancestral alleles of its locus in randomly mating populations, so that the "wild type" was presumed to be a composite of all the high-fitness alleles acquired over geologic time. For Muller, then, new mutants in general ought to have slightly detrimental effects in heterozygotes, and the aggregate of the mutants would add up to a substantial "load of mutations" with slight incapacitations in heterozygous individuals compared with those homozygous for the time-honored high-fitness alleles. (See Chapter 9 for examples of the genetic load expressed by inbreeding from outcrossed populations.)

In contrast, Dobzhansky and his coworkers, consistently impressed with the high order of heterozygosity and maintenance of polymorphisms in natural populations, realized that naturally occurring mutants and "supergenes" often displayed superior fitness in heterozygotes (due either to complementation of nonalleles or to overdominance of alleles) compared with homozygotes. Dobzhansky, in contrast with Muller, characterized genetic variants that are established in natural populations as selectively advantageous, at least in some genotypes (often heterozygotes) and under some environmental conditions in which the organism's population is adapted. These two interpretations of genetic architecture to be found in crossmating populations were formally set forth by Dobzhansky (1955) in terms of genotypes to be expected from sampling such populations: "the classical and balanced hypotheses of the adaptive norm."

CLASSIC AND BALANCE HYPOTHESES OF GENETIC STRUCTURE

Darwin consistently confessed ignorance with respect to the causes of hereditary variation; with no realization that segregation of alleles from hybrids and their combination in sexual reproduction was a fundamental mechanism promoting genetic equilibrium, he conceived of natural selection as a cleansing process where more advantageous forms in the struggle for existence replaced the less fit. Darwin saw that hereditary variation was necessary as a resource in populations for this cleansing action with adaptations as the result, but he could not visualize natural selection as leading to the *preservation* of hereditary variation; he saw only its elimination. To Darwin, adaptation meant becoming, as we say now in Mendelian terminology, homozygous for high-fitness alleles. If we could sample genotypes from populations in which natural selection was preserving an "adaptive norm" in this way, we might expect to find most genic loci homozygous and most individuals alike. Occasional heterozygotes would presumably represent alleles unfixed but in a state of transition on the way toward fixation in homozygotes of highest adaptive value; alternatively, such alleles might be entirely neutral with regard to fitness and thus fluctuate at random according to gametic sampling. As summarized by Dobzhansky, a classic genetic structure would feature an adaptive norm consisting of "a fairly small number of genotypes; the genetic diversity would be either neutral or transient, or morbid." This concept of adaptive structure is pre-Mendelian in that diploidy, segregation, and genic interactions such as complementation and heterosis are entirely unnecessary to its operation. In Chapter 14 (p. 406) we saw that the only necessary ingredients for simple classic selection are a difference in fitness between two haploid genotypes and the ability to reproduce. Diploidy was a complicating and superfluous factor.

In contrast, when diploidy and Mendelian processes are considered, the "vertical" and "horizontal" dimensions within the genome take on considerable possible significance for production of fitness differences, and preservation of genetic equilibrium with nonzero allelic frequencies becomes most likely. Natural selection can be conceived as not only preserving heritable variation but also increasing it within populations when alleles are favored for their fitness values as heterozygotes or as epistatic combinations with other components of the genome. Thus, the adaptive genetic structure would be expected as an array of genotypes heterozygous for more or less numerous gene alleles, gene complexes, and chromosomal structures. Homozygotes for these genetic units would occur in outbred populations only in a minority of individuals more or less inferior to the norm of fitness. This concept of a "balanced" genetic structure also encompassed the views of Mather (1953) on the relational balance of polygenes and of Lerner (1954) on genetic homeostasis at an obligate level of heterozygosity. Dobzhansky summarized the balanced structure as follows:

> *The fate of an allele or a gene complex under selection will then depend chiefly on its effects in heterozygous combinations with the other alleles and gene complexes present in the gene pool of the same population. A genetic good mixer becomes superior to a genetic rugged individualist. The gene pool . . . becomes an organized system. . . . Evolutionary changes will not be limited to simple allele substitutions; they will tend to alter the whole genetic system and to produce a re-patterning of the gene pool of the population.*

Undoubtedly, these types of population genetic structure are not mutually exclusive. A spectrum of genetic structures from one extreme to the other might exist. Some species that are more widely distributed and in larger population sizes may tend toward the balance extreme, while smaller, more narrowly confined species may tend toward the classic. It may well be that the purifying action of selection (eliminating extremes in a stabilizing way by increasing intermediate phenotype homozygotes or replacing a less fit old allele with a more advantageous new one) predominates at certain times and in many species, but purification is relatively less important when alleles may be selected for their "combinability" in heterozygotes and epistatic complexes. Thus, the classic and balanced views of genetic structure are derived from contrasting opinions about the action of natural selection.

It is very important to realize that the *consequences* of these alternative views are far reaching and of great significance to our world view of individual genotypes, racial differences, typological versus populational thinking about species, and attitudes in general about human as well as nonhuman biological entities. As pointed out by Lewontin (1974):

> *If the classical hypothesis were correct, the difference between populations would be of far more profound significance than under the balance theory. Since there would be so little genetic variation between individuals within populations, most of the genetic diversity within a polymorphic species would be interpopulational. In man, the manifest genetic differences between geographical races would represent a much greater proportion of total human variation than occurs within races, giving to race a considerable biological importance. A basis for racism may also flow from the concept of wild type, since if there is a genetic type of the species, those who fail to correspond to it must be less than perfect. Platonic notions of type are likely to intrude themselves from one domain into another, and Dobzhansky (1955) was clearly conscious of this problem when he attacked the concept of wild type. "The 'norm' is, thus, neither a single genotype nor a single phenotype. It is not a transcendental constant standing above or beyond the multiform reality. The 'Norm' of* Drosophila melanogaster *has as little reality as the 'Type' of* Homo sapiens.*" The balance hypothesis, conversely, presumes that a vast amount of hidden genetic diversity exists within any population, so that interpopulational differences are less significant.*

With the increased facility of surveying genotypes through protein variation techniques, many (but by no means all!) genotypes can be ascertained without going through the more lengthy Mendelian tests. Random samples of the genome taken from a wide number of species populations seemed to favor the balanced-structure hypothesis (Chapter 3; Lewontin, 1973; Selander and Johnson, 1973; Hopkinson and Harris, 1971), simply from the high magnitude of observed heterozygosity levels. However, those who favor the classic hypothesis can interpret the vast bulk of this variation as entirely neutral to selection! What is needed, of course, is to work out the *relationship of these variants to fitness* of organisms, populations, and species. We first need to examine some of the reasoning that has brought us to this state of difficulty with interpretation, a stage in which we cannot yet agree on the significance of the high level of genetic variation we observe. After going through some of the logical statements, we shall find that a dilemma of interpretation is reached so that we shall need to reevaluate basic assumptions.

SELECTION INTENSITY AND GENETIC LOADS

Given the condition that fitness values are constant over the range of allelic frequencies and genetic conditions, we have noted (Chapter 14) that variation in fitness leads to an equilibrium determined by maximum fitness, an adaptive peak. Haldane in 1937 introduced the concept that if deleterious alleles are segregating there would be some loss in fitness to the population compared with the maximum attainable fitness when only the most fit genotype(s) would exist in the population. Under those "classic" conditions, the less fit genotypes were considered a "cost to the population," or, to use Muller's later term, a "genetic load"—a sacrifice to the struggle for existence and to improved adaptation for the population. Haldane then proposed that we could measure the *intensity of selection* via an expression of the proportional loss to the population by having the genetic load present.

We noted in Chapter 7 that breeders selecting artificially for a desirable trait define selection intensity (i) as the number of standard deviations in the selection differential; that is, since proportionality is related to the normal distribution, the smaller the proportion selected (Figure 7-1), the greater the selection differential and *intensity* of selection. In our natural selection terminology, the phenotypes selected at the tail end of the population's distribution have highest W value: by selecting smaller proportions for propagation, a greater premium is given to those phenotypes while the proportion culled can be represented by the s (selection) coefficient. However, the selection differential, while of importance in measuring gains under artificial selection where mean phenotypes are expected to progress toward predetermined goals, is not always relevant to natural selection. Haldane reminded us (1954*b*) that natural selection is often directed toward stabilization—the maintenance of the mean value and removal of extremes from either end of the distribution. He pointed out that if the mean value of a quantitative trait is the same in survivors as in the original population before selection, the breeder would put the selection intensity at zero. A more general terminology was then proposed by Haldane, a definition with utility both for genotypic selection and for phenotypic. Derivation of the concept for genotypic selection is important for further arguments, so that we shall go through it step by step.

In classic models (Chapter 14), relative fitness values were symbolized as follows:

Selection Against	Genotypes			$\bar{W}$
	AA	Aa	aa	
Deleterious recessive (Table 14-3)	1	1	$1 - s$	$1 - sq^2$
Deleterious semidominant (Table 14-4)	1	$1 - hs$	$1 - s$	$1 - sq(q + 2ph)$
Deleterious homozygotes (Table 14-5)	$1 - s_1$	1	$1 - s_2$	$1 - s_1p^2 - s_2q^2$

Since the maximum fitness has been defined as $W_{max} = 1$ in each case, the proportion lost each generation because of a "load" of deleterious alleles could be considered as the component of $\bar{W}$ following the minus sign at the right, above, for each type of genotypic selection; that is, loss $= 1 - \bar{W}$ as the intensity of selection. Such a measure, however, is *not invariant*; it depends entirely on our choice of values for W (as pointed out by Li, 1963). The outcome of selection depends on the *relative* fitness of genotypes, not on their absolute fitness (which no one has yet succeeded in defining, let alone ascertaining). Let us assume a superior heterozygote stable equilibrium and arrange relative fitness values as in Table 18-1, using three different sets. Each set of fitness values is equal to the other when considered as proportional, so that the equilibrium allelic frequencies are all alike.

TABLE 18-1 *Relative fitness values and the invariant fitness ratio illustrated by three sets of values, using the stable balanced model from Table 14-5D*

All cases $\hat{q} = \dfrac{(W_1 - W_2)}{(W_1 - W_2) + (W_3 - W_2)} = \dfrac{1}{3}$ (from formula 14-15)

	Genotypes AA	Aa	aa	$\bar{W}$	Fitness Difference $(W_{max} - \bar{W})$	"Loss" $(1 - \bar{W})$	Fitness Ratio $(\bar{W}/W_{max})$
1. Fitness	$\frac{2}{3}$	1	$\frac{1}{3}$				
Equilibrium frequency	$\frac{4}{9}$	$\frac{4}{9}$	$\frac{1}{9}$				
After selection	$\frac{8}{27}$ +	$\frac{12}{27}$ +	$\frac{1}{27} = \frac{21}{27}$ =	$\frac{7}{9}$	$\frac{2}{9}$	$\frac{2}{9}$	$\frac{\frac{7}{9}}{1} = \frac{7}{9}$
2. Fitness	2	3	1				
Equilibrium frequency	$\frac{4}{9}$	$\frac{4}{9}$	$\frac{1}{9}$				
After selection	$\frac{8}{9}$ +	$\frac{12}{9}$ +	$\frac{1}{9} = \frac{21}{9}$ =	$\frac{7}{3}$	$\frac{2}{3}$	$-\frac{4}{3}$	$\frac{\frac{7}{3}}{3} = \frac{7}{9}$
3. Fitness	$\frac{6}{7}$	$\frac{9}{7}$	$\frac{3}{7}$				
Equilibrium frequency	$\frac{4}{9}$	$\frac{4}{9}$	$\frac{1}{9}$				
After selection	$\frac{24}{63}$ +	$\frac{36}{63}$ +	$\frac{3}{63} = \frac{63}{63}$ =	1	$\frac{2}{7}$	0	$\frac{1}{\frac{9}{7}} = \frac{7}{9}$

All selection outcomes would be identical if one were to start a population with (nonzero) p,q values above or below equilibrium. Nevertheless, the difference in fitness $(W_{max} - \bar{W})$ as well as the "loss" $(1 - \bar{W})$ varies from one case to another. It is obvious that neither of those two measures would serve as a general estimate of selection intensity. While it has been convenient in much of our discussion to express the highest fitness as $W = 1$ and all the other genotype fitness values as less than 1, that choice has been arbitrary and has no *general* significance. In case 2 of Table 18-1, all fitness values are expressed relative to the lowest fitness genotype. There may be a good reason for using the values in case 3 of the table; for example, if $\bar{W}$ might be assumed to be a constant, such as relating to population size from one generation to the next under equilibrium conditions, it would be useful to consider it as 1, and relative fitnesses could be considered the equivalent of relative contributions of genotypes to the progeny population. Heterozygotes in this scheme would have to do more than replace themselves (9/7) in numbers of progeny to make up for losses among homozygotes. The "loss" of zero in this third case and the negative in the second demonstrate the arbitrary nature of each choice of specific fitness values.

We realize that simply by putting all three fitness values *relative* to one of them (dividing by W_{max} is convenient), we convert the $\bar{W}$ statement into an *invariant* expression as follows:

$$\frac{p^2W_1 + 2pqW_2 + q^2W_3}{W_2} = p^2\left(\frac{W_1}{W_2}\right) + 2pq + q^2\left(\frac{W_3}{W_2}\right) = \frac{\bar{W}}{W_{max}} \tag{18-1}$$

($W_2 = W_{max}$ for the illustration in Table 18-1). Thus, mean fitness expressed as a proportion of maximum fitness is an invariant relationship. This is true irrespective of any W notation.

All three cases are the same (right column) in fitness ratio. Finally, if the fitness difference ($W_{max} - \bar{W}$) is likewise divided by W_{max}, then the "loss" becomes invariant also, as $\frac{2}{9}$ in Table 18-1, $1 - (W/W_{max})$. Either of these two ratios may be used as a measure of selection intensity, because the latter expression is simply 1 minus the ratio of (18-1). It is referred to below (18-3) as the genetic load.

Haldane (1954) chose the natural logarithm of the $\bar{W}/W_{max}$ ratio to be more logical and closer to reality than the ratio itself. (See the discussion of lethal equivalents in Chapter 9.) Since natural logs of values below the value of 1.00 will be negative, he then defined the selection intensity (I) as follows:

$$I = -ln\left[\frac{\bar{W}}{W_{max}}\right] = -[ln\ \bar{W} - ln\ W_{max}] = ln\ W_{max} - ln\ \bar{W} \tag{18-2}$$

where ln is $\log_e$ (Haldane's symbols were s_o for optimal fitness and S for average fitness). In the illustration of Table 18-1, we would have, using case 2, $I = ln(1) - ln(\frac{7}{9}) = ln(10) - ln(7.7778) = 2.30259 - 2.05127 = 0.25132$, or about 25 percent selection intensity. Thus, the difference $ln\ W_{max} - ln_{\bar{W}}$ becomes the equivalent of formula (18-1) in $\log_e$ form. Note that if no selection is acting, the fitness ratio is 1, and $I = 0$ (see Exercises 1 and 2). The student also should note the similarity of (18-2) to the measurement of total genetic load in *lethal equivalents* (9-15).

The population's genetic load (L) is defined (Crow, 1958; Crow and Kimura, 1965) as the "proportion by which the fitness of the average genotype in the population is reduced in comparison with the best genotype."

$$L = \frac{W_{max} - \bar{W}}{W_{max}} = 1 - \frac{\bar{W}}{W_{max}} \tag{18-3}$$

Thus, the genetic load and intensity of selection are related measures of the amount of differential fitness the population is experiencing. In fact, $L = 0.2222$ in Table 18-1, which is slightly less than the difference measured in $\log_e$, or Haldane's I. The student should verify that with a balanced lethal system where both homozygotes (*AA* and *aa*) are lethals so that only *Aa* genotype survives, L becomes 0.50 but I becomes 0.693, maximum for a single pair of alleles.

Depending on genetic origin or mechanism of maintenance, different genetic loads have been distinguished. Muller had in mind the *mutational load* caused by recurrent deleterious mutants. Crow (1958) defined and illustrated mutational load and two additional loads, which he termed *segregational* (due to constant segregation of inferior homozygotes from heterozygotes of highest fitness) and *incompatibility* (genotypes of reduced fitness with certain parental genotypes as an Rh+ child with an Rh− mother). Dobzhansky had also discussed what he termed *mutational* and *balanced* loads (1955, 1957). Crow proposed the term *segregational* because he had conceived of Dobzhansky's term *balanced load* as including polymorphism maintained not only by heterosis but also by disruptive selection or by adaptation to multiple niches without overall heterozygote superiority. However, polymorphism may be the net outcome of having multiple and quite different loads under the various niches, or optima, into which the population penetrates. Obviously, a population can experience more than one kind of genetic load, and the student should realize that each genic locus or genetic unit may have a load characteristic of that unit, although the population may be considered to possess a net load if all loci and genic complexes are taken into account. It will be most difficult in practice to distinguish between diverse components of genetic loads, but it is the net load of preponderantly similar components around which opinions vary and interpretations of populational genetic structure have been proposed.

SELECTION-MUTATION EQUILIBRIUM AND MUTATIONAL LOAD

If mutation is recurrent from a "normal" allele to an allele that is detrimental in fitness, selection working to eliminate the mutant will balance the input so as to produce a constant low proportion of the detrimental allele. From formulas (14-6) and (13-8), we express the increment in q due to both selection and mutation in diploids as follows.

Complete Recessives

Assume a is a completely recessive detrimental allele produced by mutation recurrently from A at rate u, and with selection coefficient s in homozygous condition aa; then net change in q is

$$\Delta q = \frac{up - sq^2(p)}{\bar{W}} \tag{18-4}$$

where $\bar{W} = 1 - sq^2$ (complete recessive) and $p = 1 - q$. Setting $\Delta q = 0$ for the equilibrium condition, we have $0 = up - sq^2p$, from which the simple balance of "gain" = "loss" becomes evident.

$$u = sq^2 \tag{18-5}$$

and the equilibrium frequency of a is

$$\hat{q} = \sqrt{\frac{u}{s}} \tag{18-5A}$$

For example, with $s = 10^{-2}$ and $u = 10^{-6}$, $\hat{q} = \sqrt{10^{-4}} = 0.01$.

If a is a lethal allele, $s = 1.00$ (complete disadvantage) and

$$\hat{q}_{\text{lethal}} = \sqrt{u} \tag{18-5B}$$

The equilibrium frequency of a lethal is the square root of its mutation rate. At $u = 10^{-6}$, a lethal would be expected to reach equilibrium at $\hat{q} = 0.001$. Thus, it would take a very high mutation rate to account for a polymorphism of 1 percent if the recessive had a severely deleterious effect.

The genetic load from complete recessives is estimated not to be of much consequence. In the example of a single mutant, the load under random mating would be

$$L = sq^2 = u \tag{18-6}$$

(completely recessive mutant load). That portion of the genetic load due to complete recessives at two or more loci would be accounted for as the following cumulative net load:

$$L = \sum s_i q_i^2 = \sum u_i \tag{18-6A}$$

(total load due to two or more recessives). Even at a high average mutation rate for a large number of loci, the total load would not be great unless the number of loci were very large. Thus, the load is entirely dependent on mutation rate. It would be $L = 10^{-6}$ in the case of the deleterious mutant (nonlethal) above as well as the lethal because the mutation rates are the same ($u = 10^{-6}$).

Partial Dominants

With a as a partially dominant detrimental allele so that Aa genotype has a fitness of $(1 - hs)$ with h as the heterozygote effect as in Table 14-4, selection will be more effective and the load greater than for a complete excessive. If a is formed by mutation it will be very rare so that the aa genotype may be disregarded in the "gain" balancing "loss" statement paralleling formula (18-5) (see Exercise 5). Since a heterozygote has two chromosomes, either one of which could have mutated, we state the balance as follows with double the mutation rate equal to the "loss" by selection:

$$2u = 2pqhs \tag{18-7}$$

from which the equilibrium frequency of a is approximately, since $2pq \approx 2q$ when a is rare,

$$\hat{q} = \frac{u}{hs} \tag{18-7A}$$

For example, with $s = 10^{-2}$, $u = 10^{-6}$ as above, and $h = 0.5$ (exact intermediate heterozygote), then $\hat{q} = 10^{-6}/(0.5)(10^{-2}) = 2 \times 10^{-4} = 0.0002$.

While this equilibrium frequency is considerably lower than that for the completely recessive example above, the load is greater. As proposed by Crow (1958), the load due to mutation will be mostly from such partially dominant detrimentals. Reduction in fitness will be according to (18-3); using the conventional $W_{max} = 1$ and (18-7) right side ("loss"), we find the load becomes under random mating:

$$L = 2pqhs = 2u \approx 2qhs \tag{18-8}$$

In the above example, the load would be twice as great as the completely recessive case above, or 2×10^{-6}. In 1937 Haldane showed that the fitness of a population would be lowered as a result of recurrent harmful mutation at a locus independently of the selection coefficients of those mutants.

SEGREGATIONAL LOAD

When selection favors the heterozygote ($W_2 = W_{max} = 1$ as in Chapter 14), at equilibrium under random mating the "loss" due to lowered fitness of homozygotes will be according to (18-3): $L = s_1p^2 + s_2q^2$. Since we know $\hat{q} = s_1/(s_1 + s_2)$ according to (14-13), substitution of that expression in the load statement reduces to

$$L = \frac{(s_1)(s_2)}{(s_1) + (s_2)} \tag{18-9}$$

For example, if we use data from Table 14-15, the load would be under random mating

$$L = \frac{(0.2)(0.3)}{(0.2) + (0.3)} = \frac{0.06}{0.5} = 0.12$$

Thus, the load from superior heterozygotes will be considerably greater than that from elimination of partial or complete detrimental recessives. If we choose alleles with homozygous selection coefficients equal to our recessive cases above, say with $s_1 = s_2 = 10^{-2}$, we would have $L = 10^{-4}/2(10^{-2}) = 0.005$, which is 2500 times greater than the case of partially dominant detrimental genes. This load is entirely maintained by natural selection because mutation has little to do with it.

In Chapter 17, with k alleles at one locus, it was shown by Wright that if all heterozygotes are equal in being superior to all homozygotes in fitness, then each allele has a simple stable equilibrium state: (17-7A). The load in a randomly breeding population would then be, paralleling (18-3),

$$L = \sum s_i q_i^2 = \frac{1}{\sum(1/s_i)}$$

$$L = \frac{s_i}{k} \tag{18-9A}$$

(load with k multiple alleles and equally heterotic heterozygotes). Note that (18-9A) reduces to (18-3) and (18-9) if only two alleles exist.

SEPARATING MUTATIONAL AND SEGREGATIONAL LOADS

Theoretically, if selection coefficients were to remain constant over genetic and environmental conditions for sufficient time (and that restriction may be very difficult to achieve in reality), we should be able to compare different genotypes in populations, measure their fitness values, and unambiguously decide whether the classic or balanced genetic structure of the adaptive norm is the more predominant. Particularly important would be a decision for human genotypes and populations. Measurement and interpretation of genetic loads in human populations needs to be estimated carefully and decisively. We need to know how our genetic polymorphism has come about and how it is maintained. A method of deciding whether the human genetic load was predominantly mutational or segregational was first proposed by Morton, Crow, and Muller (1956); it was apparently simple based on the concept of lethal equivalents (Chapter 9) and estimating the amounts of survivorship at various levels of inbreeding. From considerations of load theory (above), those authors pointed out that one could decide the likelihood for type of load from a comparison between the amount of genetic damage following inbreeding and that under random mating. We shall explain and illustrate the method briefly, not so much because it leads to decisive answers about genetic structure of a population (although it does give us a kind of first-approximation basis for further evaluation), but rather because it illustrates the persuasiveness inherent in a simple mathematical device that has not proved to be such a shortcut to reaching our needed answer to a tough biological question. There are too many objections, both theoretical and biological, to merit gathering future data and doing the calculations, but the relationships illustrated by the method have many points of interest. Besides, the student should be aware of the major objections to a potentially useful method.

If a randomly mating population were to be completely inbred without changing gene frequencies (q), what would be the load? Taking the two major types of loads under complete inbreeding so that only AA and aa genotypes would exist (assuming further that selection coefficients are identical whether under random mating or under inbreeding) we would have

Genotype	AA	Aa	aa
Frequency	p	0	q
Mutational fitness	1		$1 - s$
Segregational fitness	$1 - s_1$		$1 - s_2$

Therefore, the loads would be under complete inbreeding

Mutational load (L_i) sq
Segregational load (L_i) $s_1 p + s_2 q$

Then the ratio of the inbred load (L_i) to the outbred load (L_o) from (18-8) may be expressed as follows depending on the type of selection contributing to the load.

Mutational Load Ratio

For a partially dominant detrimental mutant

$$\frac{L_i}{L_o} = \frac{sq}{2pqsh} = \frac{1}{2ph} \approx \frac{1}{2h} \tag{18-10}$$

when the mutant is rare (Crow, 1958). This expression for partially dominant detrimental mutants is commonly used owing to the fact that the classic structure hypothesis claims most new mutants to have slight deleterious effects in heterozygous condition. If a new mutant is completely recessive, the load ratio would be

$$\frac{L_i}{L_o} = \frac{sq}{sq^2} = \frac{1}{q} \tag{18-10A}$$

(load ratio for complete recessive).

Segregational Load Ratio

On supplying equilibrium p and q from formula (14-13) into the inbred load statement above, we obtain

$$s_1\left(\frac{s_2}{s_1 + s_2}\right) + s_2\left(\frac{s_1}{s_1 + s_2}\right) = \frac{2s_1s_2}{s_1 + s_2}$$

We note that the outbred load from (18-9) is just one-half of this value so that the load ratio is

$$\frac{L_i}{L_o} = 2 \tag{18-11}$$

This load ratio is illustrated in Table 18-2A. It should be noted that net fitness decreases when a balanced equilibrium exists and the population is inbred, because the load is thereby expressed in greater proportion than under random mating. In contrast, Table 18-2B indicates that inbreeding a population with the same allelic frequencies but now with homozygote *AA* having maximal fitness (simply interchanging fitness values between *AA* and *Aa* genotypes) improves the population's net fitness. The *increase* in fitness with inbreeding of the mutational case will always be true (see Exercise 6 and Table 18-4). Why?

The load ratio for the mutational case here is less than 1 because the heterozygote effect (h) has been chosen to be fairly large for the sake of the illustration. More commonly, if new mutants are partially dominant, the h value is quite low. If $h < 0.05$, for example, then, using formula (18-10), the load ratio becomes greater than 10. Thus, it is tempting to try measuring such load ratios in order to distinguish the type of load and genetic structure of the population. But the situation is not so simple, unfortunately. Just for a

TABLE 18-2 *Some relationships between inbred and outbred loads and net fitness under inbreeding and outbreeding. Gene frequencies identical with interchange of* AA-Aa *fitness values (from Li, 1963a,c)*

A Segregational fitness
Population at equilibrium (as in Table 14-5)

Outbreeding	*AA*	*Aa*	*aa*	$\bar{W}$
Frequency	0.36	0.48	0.16	
Fitness	0.8	1	0.7	
After selection	0.288 +	0.480 +	0.112 =	0.880
Complete inbreeding				
Frequency	0.60	0	0.40	
Fitness	0.8		0.7	
After selection	0.48	+	0.28 =	0.760

Net fitness ratio
$\bar{W}_i/\bar{W}_o = 0.76/0.88 = 0.864$
Load ratio
$L_i/L_o = 0.24/0.12 = 2.00$

B Mutational fitness: partially dominant detrimental. Population in transition toward elimination of *a*

Outbreeding	*AA*	*Aa*	*aa*	$\bar{W}$
Frequency	0.36	0.48	0.16	
Fitness	1	0.8	0.7	
After selection	0.360 +	0.384 +	0.112 =	0.856
Complete inbreeding				
Frequency	0.60	0	0.40	
Fitness	1		0.7	
After selection	0.60	+	0.28 =	0.880

Net fitness ratio
$\bar{W}_i/\bar{W}_o = 0.88/0.856 = 1.03$
Load ratio
$L_i/L_o = 0.12/0.144 = 0.833$*

* Note: formula (18-10) not applicable here because population is not at equilibrium.

start, we can draw attention to the fact that the segregational load ratio will be greater than 2 with multiple alleles; from (18-9A) for the L_o and having for an inbred load (L_i), $L_i = \sum s_i q_i = k/\sum(1/s_i)$ so that

$$\frac{L_i}{L_o} = k \tag{18-11A}$$

(segregational load ratio for k multiple alleles).

Therefore, the load in a homozygous population would be k times as great for heterotic alleles as it would be under random mating. (If heterozygotes are not all alike in fitness, however, the load ratio would be less than k.) We may foresee a problem in separating loads if there are multiple alleles at some loci with heterotic heterozygotes because load ratios between 2 and 10 could mean either that several alleles are producing a load from segregation or that single mutants have partially dominant effects with $h > 0.05$. In fact, we shall see that such ambiguity is often the case, but first we shall describe the method suggested by Morton, Crow, and Muller for human populations—their "B/A" ratio.

"*B/A*" RATIO

By measuring genetic loads in *lethal equivalents* (Chapter 9) under increasing levels of inbreeding, Morton, Crow, and Muller proposed extrapolating the amounts of mortality from low levels recorded in many human populations to the complete inbreeding ($F = 1$) level for estimation of the numerator in the load ratio (Example 9-4). Their "B/A" ratio estimated the load ratio and was derived and defined as follows.

If a single genic locus is considered, the probability for survival of a particular zygote is 1 minus the detrimental effect of a mutant at that locus:

$$1 - \underset{(a)}{qFs} - \underset{(b)}{q^2(1 - F)s} - \underset{(c)}{2pq(1 - F)sh}$$

where (a) is the probability of death due to homozygosity from consanguineous parents.
(b) is the probability of death due to homozygosity from unrelated parents.
(c) is the probability of death due to heterozygosity.

If all loci that have lethals, or lethal equivalents, are independent in their action on viability and if the environmental causes of death (letting x be the probability of death due to non-genetic causes) are also independent, then the total probability of survival (S) from egg to adult will be the grand product:

$$S = \Pi(1 - x)[1 - qFs - q^2(1 - F)s - 2pq(1 - F)sh]$$

(Π signifies the grand product over all loci). Since the number of causes is large and the separate probabilities small, the survivorship (S) can be expressed as the Poisson term

$$S = e^{-m} \tag{18-12}$$

where $m = \sum x + F\sum qs + (1 - F)\sum q^2 s + 2(1 - F)\sum pqsh$. Under random mating, the component A will be $A = \sum x + \sum q^2 s + 2\sum pqsh$. Under inbreeding, the component B will be the terms factored from F: $B = \sum qs - \sum q^2 s - 2\sum pqsh$. Thus, survivorship can be stated in terms of these two components, A and B:

$$S = e^{-(A+BF)} \tag{18-12A}$$

Taking the log of both sides of this expression, we have

$$-\ln S = A + BF \tag{18-12B}$$

(total lethality due to both nongenetic and genetic causes).

Under complete inbreeding ($F = 1$), the B value will consist mostly of the total genetic damage per gamete ($\sum qs$) since the remaining two terms will be very small. The A value will be mostly a measure of nongenetic effects ($\sum x$). If there is no interaction between loci (no epistasis or multiplicative effects on viability), there will be "additivity" on the log scale, and the $-\ln S$ (18-12B) expression will be a linear regression function of the inbreeding coefficient F. The intercept of this straight line at $F = 0$ will measure A, while B will be the slope of the regression or the added mortality as gametes are converted into complete homozygotes in increasing proportions. Thus, when the linear regression is extrapolated to $F = 1$, the intercept at that point would measure $A + B = \sum x + \sum qs$, the total load from all causes. B is a lower limit to the genetic damage measure, and the needed value $\sum qs$ (lethal equivalents per gamete) must lie somewhere between B and $A + B$. Our load ratio (L_i/L_o) then becomes

$$\frac{L_i}{L_o} = \frac{A + B}{A} = \frac{B}{A} + 1 \tag{18-13}$$

Morton, Crow, and Muller weighted the human data regression (Example 9-4) depending on the theoretical binomial variance for observing particular proportions of survivors out of totals; however, Dobzhansky, Spassky, and Tidwell used an unweighted regression, while Malogolowkin-Cohen, et al. (1964) described in detail a method for using empirical variances of S as weights as well as standard errors for A and B estimates; the latter method is recommended for students wishing to employ the B/A load statistics method.

Table 18-3 lists some of the A and B estimates made for human and insect populations by the method outlined. (Recall for comparison the regression of actual survivors on F in Example 9-3). Morton, Crow, and Muller found B values high (1.5 to 2.5 lethal equivalents were proposed) and the B/A ratios also high in the range of 15–20. In view of these values being too high for the simplest segregational load expectation of 2, they concluded that the evidence strongly favored a preponderance of partially dominant deleterious mutants with about $h = 0.05$ or less, rather than balanced heterozygotes. Other human data collected by Neel and Schull from Japanese populations showed much lower estimated lethal equivalent values and B/A ratios, at about one-quarter those found by Morton, et al. A number of experimental geneticists using drosophila and tribolium also found estimates to be in the lower range (see also Mettler, et al., 1966; Stone, et al., 1963; Mourad, 1964; Torroja, 1964; Sankaranarayanan, 1965*a*,*b*). Obviously, all these estimates are a little too high for the simple case of one pair of heterotic alleles but too low for the prediction expected from partial dominance. In fact, they unhappily fall in the range of high ambiguity. Thus, the empirical load ratios obtained over many different populations bring us back to the difficult problem of interpretation of populational genetic structure we posed at the beginning of this chapter. We realize the B/A ratios are not as useful as they once seemed, and we list objections to using them as follows.

TABLE 18-3 *Some estimates of A (intercept at $F = 0$) and B (regression coefficient estimated from fitting ($-\ln S$) up to $F = 1$)*

Population	A	B	B/A	Reference
1. Human				
France				
Morbihan Dep.	0.141	2.555	18.12	Sutter & Tabah (1953) cited in Morton, Crow, and Muller (1956)
Loir et Cher	0.089	1.482	16.60	
Japan				
Hiroshima	0.172	0.798	4.63	Neel and Schull (1962)
Nagasaki	0.189	0.846	4.48	
2. Insects				
Drosophila pseudoobscura				
Mather	0.161	0.742	4.60	
Arizona (B)	0.119	0.695	5.82	Dobzhansky, Spassky, and Tidwell (1963)
(C)	0.132	0.472	3.57	
D. willistoni				
Venezuela	0.197 ± 0.01	1.052 ± 0.09	5.33 ± 0.69	Malogolowkin-Cohen, et al. (1964)
British Guiana	0.149 ± 0.01	1.100 ± 0.09	7.34 ± 0.96	
Trinidad	0.150 ± 0.02	1.148 ± 0.15	7.67 ± 1.69	
Tribolium castaneum				
Wild, egg-adult, dry, High B/A value	0.229 ± 0.03	0.830 ± 0.26	3.47 ± 1.53	Levene, et al. (1965)
T. confusum				
Wild, egg-adult, dry, High B/A value	0.216 ± 0.03	1.151 ± 0.30	5.09 ± 1.94	

OBJECTIONS TO *B/A* RATIO'S UTILITY

1. As just pointed out, the observed B/A values lie often in an ambiguous range and so cannot discriminate between classic and balanced population genetic structures.
2. Extrapolation over ranges of inbreeding from low F to complete homozygosis would be correct only if all genic loci acted additively on viability. If deleterious effects are synergistic, as with "synthetic lethals" or other epistatic effects on fitness, the expected function with F will no longer be linear. At least there will be no unique slope B, but for each level of inbreeding there will be probably a different slope. Malogolowkin-Cohen, et al., found evidence that indicated the slope B was not uniform at all levels of F. Separate experiments using complete homozygotes from dominant marker techniques (as in Examples 9-5 and 9-6) for two nonhomologous chromosomes simultaneously were done to test for interactions between chromosomes on depression of viability. Spassky, Dobzhansky, and Anderson (1965) using *D. pseudoobscura* found more evidence for interaction than did Temin, et al. (1969) using *D. melanogaster*, but the error variance for interaction values is very great so that no clear decision for or against a substantial amount of interaction could be made.

3. There is probably a mixture of heterotic and partially dominant deleterious alleles contributing to the B/A ratio. A large number of the latter might have little effect averaged with a small number of the former and make the total load ratio indistinguishable from a low heterotic expectation (see Exercise 11).

4. In Chapter 15 we found there was much evidence that viability amounted to a minor fraction of total fitness. Loss of fitness due to viability is probably much less than the load due to other parts of the life cycle, especially the components from fertility (adult components). Owing to the fact that the B/A ratio has been estimated only for viability effects, we have an enormous gap in our estimate of the total genetic load. It has not been computed from overall fitness—that is, consideration of selection at all parts of the life cycle. Some of that load may be segregational and some may be mutational. Unfortunately, at our present level of ignorance, the B/A ratio does not help us to decide.

5. Assumption of genetic equilibrium in the population may lead to serious error, as has been pointed out by Levene (1963) and Haldane and Jayakar (1965). Values of B/A much higher than 2 can be obtained for heterotic balanced loci with two alleles if the allelic frequency is far from equilibrium. It is likely that modern western populations have a large number of genetic variants in the process of being eliminated or increased (see Exercise 10). Conversely, when partially dominant detrimental alleles are high in frequency, their load ratio can be quite low, even less than 2, as illustrated in Table 18-2B.

6. The major objection is the last. The artifactual nature of the load ratio and its poor discriminating power is displayed by the paradox (Li, 1963*c*) "the more similar, the more different" in Table 18-4. Since the mutational load ratio (18-10) has a denominator with an expected small value (h), the smaller h gets, the greater the load ratio. A contrast between segregational and mutational load ratios makes the the point clear. In Table 18-4A, there is a 10 percent difference in fitness between

TABLE 18-4 *Paradox from load ratios and net fitness ratios: "the more similar, the more different" (from Li, 1963c)*

	A Ten percent difference in fitness ($AA - Aa$)		**B** One tenth of one percent difference in fitness ($AA - Aa$)	
Let $q = 0.10$				
	Heterotic	Mutational	Heterotic	Mutational
W_{AA}	0.90	1.00	0.999	1.000
W_{Aa}	1.00	0.90	1.000	0.999
W_{aa}	0.10	0.10	0.001	0.001
$\overline{W}_o$	0.910	0.973	0.999001	0.999997
$\overline{W}_i$	0.820	0.910	0.998002	0.999001
$\overline{W}_i/\overline{W}_o$	0.901	0.935	0.999000	0.999004
L_i/L_o	2	3.33	2	333.3

AA and *Aa* (leaving the *aa* fitness unchanged and assuming equilibrium under segregational selection and with the genotype frequencies identical under the mutational), so that there would be enough of a fitness difference to be biologically detectable. Yet the L_i/L_o ratio differs very little with the mutational well within what might be construed as a segregational ratio with three alleles. In Table 18-4B, however, there is only a very slight difference in fitness between *AA* and *Aa*, $\frac{1}{100}$ the difference of that found in Table 18-4A and hardly detectable biologically. The mutational load ratio becomes enormous (333.3), while the segregational ratio is fixed at 2. With such a jump in load ratio when only a negligible change in fitness exists between *Aa* and *AA* genotypes, the absurdity of its use is demonstrated. Nothing biologically important has occurred in the slight fitness difference. A much better ratio for correlation with the real difference in fitness would be the ratio in net fitness values ($\bar{W}_i/\bar{W}_o$), but the amount of change in that ratio is not great enough to make measuring it worth the effort.

In effect, we have found that the load ratio is an idealized quantity that seemed simple at first with a possibility for discriminating power in the interpretation of genetic polymorphisms, especially in human populations where the need for decision and the experimental limitations are so great. But the load ratio seems at best to be only capable of giving us a first-approximation kind of information; at worst, it is an example of a mathematical model that cannot be measured accurately and has little or no biological significance to the population. There is no shortcut at present to learning what the functional meaning of our genetic polymorphism is outside of differentiating the fitness properties of specific genotypes.

DOES THE LOAD MATTER TO THE POPULATION?

At the risk of repetition—but there are so many opportunities for misconception about selection into which the student as well as the seasoned investigator easily can slip—it is erroneous to suppose the genetic load to be manifest as a loss in reproductive capacity or size of the population unless selection is actually causing loss in numbers through an unconditional mortality independent of density or other environmental conditions (see Wallace's (1968*a*) concept of "hard selection"). In the last section, we were concerned with preadult viability; the elimination of genotypes ("genetic death") was taking place via mortality. As emphasized in Chapter 15 under life cycle components of fitness, it should be clear that actual mortality is only one way out of countless ways in which the life cycle can be affected by genotype frequencies. A load affecting behavior of adults and fertility is doubtless as great or greater than that of preadult viability.

Differential fitness is often conditioned by flexible ecological circumstances or gene frequencies so that at equilibrium in which two or more segregating units are being maintained by selection there may be no load at all (see Taylor, 1975). The student should recall in Chapter 16, for example, that with a minority advantage there may well be no selective differences, no selection intensity, and no genetic load at the equilibrium allelic frequency point. The load would be "experienced" by the population only during the transitional process of change in approach to equilibrium.

Even if a genetic load exclusively from mortality is produced by artificially inducing high lethal frequencies using radiation techniques in an experimental population (Chapter 13), it appears that considerable amount of lethality can be tolerated by the population without suffering substantially in numbers under experimental conditions. In fact, random mating populations of drosophila adapt to such a chronic irradiation rapidly in spite of acquisition of genetic loads up to over 80 percent of their chromosomes carrying lethal factors in heterozygous condition. In Example 13-3, Wallace's irradiated and unirradiated populations indicate that the lethal chromosomes persisting in these experimental populations at high frequencies contain either completely recessive factors or slightly heterotic ones. Random heterozygotes from chronically irradiated populations showed no less viability than those from the unirradiated control population and may have been slightly superior. Also, it is evident that detrimental chromosomes had no lowering effect on viability in heterozygotes with normal viability chromosomes though they may have had about a 2 percent reductional effect when combined in repulsion. In irradiation populations analyzed for egg-adult survival rate by Sankaranarayanan (1965, 1966), frequencies of lethals were very high at equilibrium levels when populations were removed from the radiation and selection was intensified by maximizing larval crowding. The genetic detrimental chromosomes induced by the irradiation were not only retained but also multiplied in the recovery process. Since population sizes were never small, the conclusion was justified that a significant proportion of the induced mutant heterozygotes retained was selectively advantageous.

With an experimental laboratory population, there is no problem of competition with other populations as there would be for natural populations. For such an isolated population, the amount of selective elimination may not matter very much as long as it has a high reproductive rate. But it is quite conceivable that interpopulational competition could be an important factor in nature in putting a population with greater genetic load at a disadvantage relative to one with a lesser load. Unfortunately, attempts to compare populations for total fitness have been few and are limited by the fact that selection dynamics has been entirely modeled on intrapopulational considerations, and comparisons between populations cannot be made on the basis of net fitness values of genotypes competing within populations. However, in some experiments done by Dobzhansky's group (Beardmore, Dobzhansky, and Pavlovsky, 1960; Dobzhansky, Lewontin, and Pavlovsky, 1964; Ohba, 1967; Ayala, 1969), when populations of *D. pseudoobscura* containing single gene arrangements (monomorphic) or two gene arrangements (polymorphic) are compared for their ability to produce numbers of flies, total biomass as well as the parameter r_m (the innate capacity for increase under crowded conditions or stress produced by competition with other species), the polymorphic populations had superior capability in being able to utilize the available resources to increase their numbers compared with the monomorphic populations. In nature, such qualities may well have fitness value. Under more optimal uncrowded conditions, there was no significant difference between polymorphic and monomorphic conditions. From these results, Dobzhansky concluded that adaptedness of a population to stress conditions in nature and the Darwinian fitness of included genotypes measure the capability of greater or lesser levels of genetic organization; reproductive success of a given genotype in relation to other genotypes within a single gene pool divorced from other gene pools (other populations or other species) may or may not predict the adaptedness of the population as a whole in competition with other populations. The innate capacity for increase (r_m) and the Darwinian fitness of polymorphic gene arrangements tend to be correlated, yet they measure different qualities or expressions of adaptedness. From the consideration of genetic load we have been emphasizing, a polymorphic population at balanced equilibrium certainly has a greater genetic load than a monomorphic one. But there is greater diversity of genetic units with known opportunity for more fitness differences than in the monomorphic populations, and thus more selection, and we might therefore have predicted from a priori reasoning that the population with greater genetic load might have been less able to compete with other populations than a population with a smaller load and thus less selection occurring within it. Thus, we must conclude that the amount of genetic load is not necessarily an important criterion for making interpopulation predictions.

THE PARADOX OF HIGH GENETIC POLYMORPHISM

As a consequence of the classic hypothesis of population genetic structure in an extreme form, the deduction would follow that the genetic load should consist mostly of rare and detrimental mutants that are being eliminated by a "purifying" natural selection. In the 1960s, biochemical techniques developed for separation of protein gene products (described in Chapter 3), and the ensuing description of genetic polymorphisms in numerous organisms led to estimates of high heterozygosity and seemed to settle once and for all that the classic hypothesis could not be the correct interpretation of natural population genetic structure, since it had predicted a very low level of genic variation based on mutational-selectional equilibrium. But that prediction was based on the presumption that genetic variants had some selective value; new mutants were generally expected to be deleterious. What if all those variants giving such high levels of heterozygosity were *selectively neutral* with no fitness function whatsoever? As stated concisely by Lewontin (1973): "The 'classical' hypothesis is conserved by the addition of the not unreasonable hypothesis that the variation revealed by electrophoretic studies is, in fact, irrelevant to the physiology and morphology of the organisms and that therefore all of the allelic variants in natural populations are effectively wild type."

Interpreting high-order genetic polymorphism as neutral in fitness is the equivalent of saying that the extensive protein variation detectable as a random sample of the genome in a wide number of organisms is essentially worthless to the organisms possessing it and that it only serves to mark the occurrence of gene change. Such an interpretation has often been called the "neutralist concept" or "neutral mutation theory of evolutionary change" or "non-Darwinian evolutionary theory." However, the proponents of this doctrine (Kimura and Crow, 1964) from the start have emphasized that natural selection is almost always purifying and that the balancing or segregational load would be too great to expect any sizable fraction of observed genetic polymorphism to be based on balancing selection. These proponents, by interpreting what is observed as essentially neutral, are actually holding to the classic hypothesis of population genetic structure, and their arguments ought to be referred to as "neoclassic" (Lewontin, 1974).

Neoclassic arguments center mainly around three lines of evidence and the reasoning derived from them: (1) segregational load is too great, (2) fixation of neutral alleles is highly probable, and (3) rate of evolution by selection is too costly. Their main features are summarized in the following paragraphs along with some counter arguments.

Segregational Load Is Too Great

The segregational genetic load from balanced heterozygotes would be too great for any population to support at the level suggested by the immensity of genetic polymorphism observed. Originally, the experiments of Wallace (1958*a*, 1959, 1963) led him to postulate that, "on the average an individual member of the Drosophila population studied is heterozygous at 50 percent or more of all loci . . ." (see also Example 19-1). Lewontin and Hubby (Chapter 3) similarly were led to postulate high heterozygosity. Kimura and Crow (1964) reasoned that such enormously high levels of heterozygosity could not be supported by the heterotic load. Their calculations and assumptions were as follows. *Assume* drosophila has 10,000 loci, half of which (5,000) are segregating with heterotic equilibrium. Let each homozygote have a slight selective disadvantage ($s_1 = s_2 \cdots s_k = 0.01$) compared with all heterozygotes considered equal in fitness, so that W would be 1.00 for Aa, Aa', Aa'' . . . , and AA, aa, $a'a'$, $a''a''$ . . . would be 0.99. Let there be eight alleles at each of the 5000 loci all in equal frequencies so that $p_A = q_a = r'_a = s_{a''} \cdots = \frac{1}{8}$. Using (18-9A), the load would be

$$L = \frac{1}{\sum(1/s_i)} = \frac{1}{8(1/0.01)} = 1.25 \times 10^{-3}$$

Then *if* all 5000 loci act on fitness additively and independently (without epistatic interaction), the total load (L) would be $L = 5000 \times 1.25 \times 10^{-3} = 6$ and the net fitness would be, using expression (18-12) with $\bar{W}$ representing survivorship S, $\bar{W} = e^{-6} = 0.0025$.

Thus, females would necessarily need to lay 10,000 eggs to get 25 to survive! Wallace (1969) pointed out that if there were just two instead of eight alleles at each of these 5000 loci, the number of heterozygotes would be much fewer, of course, and the load would be much greater—$L = 5000 \times 5 \times 10^{-3} = 25$. Then the survivorship would be immensely smaller—$\bar{W} = e^{-25} = 10^{-11}$. Similar calculations were given by Lewontin and Hubby (1966) to emphasize the paradox (see Exercise 13).

There is an upper limit to the number of selective deaths that can occur in a generation. If W values of genotypes are represented as survivorship relative to a constant population size, as we mentioned earlier under genetic loads, let $\bar{W}$ be considered as 1, a constant replacement of one offspring per parent, and let W_{max} be >1 for "optimal genotype" reproduction in order to offset the losses due to inferior genotypes. In order to perpetuate the population, the fraction of survival must be at least $1/W_{max}$ on the average; alternatively, genetic deaths (mortality in this case) cannot exceed 1 minus that amount, or $(W_{max} - 1)/W_{max}$. Obviously, with the simplifying assumptions of the calculations above, fertility would have to be absurdly immense to maintain the population. A *D. melanogaster* female can lay 1000–5000 eggs in her lifetime; while it is conceivable that the lighter load for multiple alleles calculated above *could* allow a drosophila population to persist, we know that a survivorship of only 25/10,000 ($= \frac{1}{400}$) is far from realistic, and with two alleles the survivorship of 10^{-11} is out of the question for the W_{max} of the complete heterozygote. The neoclassic conclusion is that superior fitness heterozygotes cannot be common enough to account for the high level of protein variation observed in natural populations.

COUNTERARGUMENTS. If some kind of balancing selection *is* maintaining a sizable proportion of observed polymorphism, logic demands that we review the assumptions in these calculations to test their reasonability based on our current experience with fitness traits and genotypes. As pointed out by several authors, the assumptions are fallacious in at least two major ways: (1) independence of genic loci with regard to their action on fitness and their linkage, and (2) the constancy of fitness values.

To emphasize the last first, the student has only to reconsider the main ideas from Chapter 16 to realize that selection is probably seldom "hard" (Wallace's term)—that is, "unconditional" and constant throughout the range of allelic frequencies and environmental circumstances a population goes through during selective change. It is much more likely to be "soft," or "conditional," and dependent on an enormous variety of other circumstances such as the presence of other genotypes, occupation of the niches available, population density, and so forth. In that type of selection, a population may make adjustment for a certain size by eliminating the less fit homozygotes but retaining heterozygotes (or whatever genotypes have best fitness for the conditions) since it will be certain that the most fit genotypes for the conditions will occupy the sites available for the population's survival and reproduction. Thus, whatever genetic load exists is likely to shift considerably in magnitude and direction.

For the independence of genic loci on fitness, the implication of Kimura and Crow's calculation is that for a single locus with two alleles maintained by a superior heterozygote and 1 percent disadvantage for each homozygote, the load would be 0.005, which in the narrow sense of mortality terms could mean $\frac{5}{1000}$ die because they are homozygotes for that locus. A second locus with the same effect on fitness would cause 0.005 of the remainder to die, or $\frac{5}{1000} \times 995$, and so forth, for the remaining 4998 loci. Even allowing that fitness measured in strict mortality terms could be reasonable, we may note that the effects of individual loci probably do not combine as if they were separate probabilities following the product rule of independence. Heterozygous loci can affect fitness in countless ways depending on their grouping and epistatic interaction. Gametic association (Chapter 17) demonstrated that selective forces act on blocks of genic loci making large sections of the genome the units of selection rather than individual loci. A constant disequilibrium or nonrandom association between linked or even unlinked loci while in heterozygous condition is the deduced result of a balanced genetic structure in populations; significant disequilibrium has been demonstrated and strongly indicates that single-locus selection is an oversimplification and far from reality.

Three groups of authors have criticized as an artifact the multiplicative model of fitness (J. L. King, 1967; Milkman, 1967, 1970; and Sved, Reed, and Bodmer, 1967) and have proposed that it is more reasonable to suppose that if the amount of heterozygosity (with heterotic action on fitness per locus) is increased in a genome by increasing the number of heterozygous loci, a plateau of high fitness would be reached asymptotically at a level not far above the mean heterozygosity for the population; increasing heterozygosity beyond the threshold amount that produced the plateau in fitness would not increase fitness. This truncated model briefly can be summarized as follows. Assume that owing to limited resources a certain proportion ($\bar{W}$) of zygotes produced in a generation can survive and reproduce. Suppose that survival depends on an adaptive trait and that increasing the numbers of loci that are heterozygous makes the adaptive trait of greater fitness value. After selection, individuals with greater than some threshold amount of heterozygosity will be surviving in such proportions that $\bar{W}$ of the population remains. An example of this truncation model of calculating the amount of heterozygosity above a minimum level to produce a given amount of mortality is given in Exercises 15 and 16 (derived from Milkman, 1970). A model with 10 percent selection disadvantage to homozygotes could be achieved under a model of truncated selection where a modest increment in heterozygosity does not introduce more than a modest amount of mortality.

While this sort of model helps us see how a population might maintain a particular net fitness with a given amount of heterozygosity, it does not help us in solving the paradox of high-order genetic polymorphism with expectation of considerable inbreeding depression, which has been brought out by the neoclassic argument. Thus, the model of simple heterotic fitness and balanced equilibrium at several independent loci cannot logically be involved as a *general* mechanism to explain the high-order genetic polymorphism that is observed in its entirety. Without doubt, it can explain particular and perhaps a sizable fraction of polymorphisms, but we must be aware that restrictions such as independence of fitness values among loci and constancy of fitness values over all allelic frequencies and environments are naïve and unrealistic. The student should contemplate the numerous other ways in which a genetic balanced state can be achieved and maintained by selection without heterosis.

The Fixation of Neutral Alleles Depends on Population Effective Size and Mutation Rate

According to Kimura and Ohta (1971, 1972), the basic concept in their neutral, random-drift hypothesis on variation within populations is as follows: "At each cistron, a large fraction of mutations are harmful and they will be eliminated by natural selection. A small but significant fraction is selectively neutral and their fate is controlled by random frequency drift. . . . Favorable mutations may occur, and although they are extremely important in adaptive evolution, they are so rare that they influence very little the estimates of the rate of amino acid substitution [over long evolutionary time]."

A proportion of mutation at the molecular level has little or no effect on the phenotype since the genetic code has much redundancy, and changes that do not alter the amino acid sequence of a polypeptide product are obviously neutral. The "synonymous" mutations, however, may alter future mutational possibilities for the DNA molecules. Also, mutations to codons for amino acids with similar properties (for example, GUU for valine to GCU for alanine) are unlikely to affect protein phenotypic properties and thus would tend to be neutral. For these mutations, there is no question about their fate in populations (see Example 13-1). The problem is aimed at all the other mutations that *could* have adaptive value since they bring about amino acid substitutions that are detectable in altered protein. Kimura and Ohta emphasize a "significant fraction" whose fate is controlled by random forces—by the effects of sampling in populations of limited size.

How much genetic variation can be expected in a population of limited size at an equilibrium determined simply by gain from recurrent mutation and loss due to random drift? The equilibrium condition, in terms of heterozygosity (H_t), can be specified by using the probability of identical uniting gametes from inbreeding—from formula (11-2) for a random-mating population of effective size N_e; according to Kimura and Crow (1964),

$$F_t = F_{t-1} + \frac{1 - F_{t-1}}{2N_e}$$

The two alleles will be identical in state only if neither has mutated since the previous generation. The probability that neither has mutated is $(1 - u)^2$. Thus, the formula can be generalized to include mutation by the product

$$F_t = \left[F_{t-1} + \frac{1 - F_{t-1}}{2N_e}\right](1 - u)^2$$

At equilibrium the subscripts can be eliminated since F will be constant. Then, because terms containing u^2 can be ignored, the solution is

$$F = \frac{1 - 2u}{4N_e u - 2u + 1} \approx \frac{1}{4N_e u + 1} \qquad (18\text{-}14)$$

(probability for an individual to be homozygous under mutation and drift—neutral alleles).

If all the k alleles were equally frequent for any locus, the proportion of homozygotes would be the reciprocal of the number of alleles maintained in the population (as in Chapter 3); if the allele frequencies differed, the homozygote proportion would be greater. Therefore, k alleles $= 1/F$ may be used as a measure of the *effective number* of alleles that could be maintained in the population (while the actual number may be greater). For example, let $N_e = 3 \times 10^4$ and $u = 10^{-6}$, then $F = 0.893$ and $1/F = 1.12$, but since that is a minimum estimate, there could be two alleles or more if frequencies were unequal. From Chapter 11, we recall that $1 - F = H_t/H_0$, the panmictic index. Thus,

$$P = 1 - \frac{1}{4N_e u + 1} \qquad (18\text{-}14\text{A})$$

We may consider H_0 to be the ultimate limit of heterozygosity if mutation ran its course (producing an infinite number of alleles at a locus), and thus H_0 can be considered as 1. Therefore, we can let P measure the amount of heterozygosity (H) at equilibrium between mutation and drift, and H may be substituted for P in (18-14A). Thus, in the numerical example, $H = 1 - 1/1.12 = 0.107$.

COUNTERARGUMENTS. This H estimate is close to the observed proportion of the genome heterozygous per individual for a wide variety of organisms sampled according to the methods of Lewontin and Hubby (Example 3-4). Lewontin (1974, Table 22) lists a number of organisms surveyed for protein variation by electrophoretic methods including humans, mice, drosophila, and horseshoe crabs (see also Selander and Johnson, 1973; Prakash, 1973; Powell, 1975). Heterozygosity per locus varies from 0.056 in the house mouse (*Mus musculus*) through 0.067 for *Homo sapiens* up to a high of 0.184 for *D. willistoni.* What we *can* say about this agreement between observed range of H and the expectation from formula (18-14A) is that *if* these populations are in mutational-drift equilibrium and *if* their population sizes and average mutation rates are in the range given in the numerical example, their level of heterozygosity could be accounted for by the neutral (neoclassic) concept. It has been pointed out, however, that equation (18-14A) contains two parameters—a large but virtually unknown N_e and a very small and also more or less unknown u; their product determines the value of the expression H. Any value of Nu that is reasonable can be chosen so as to make agreement with either the classic or the balance theories (see Exercise 17). Thus, applying this formula to every case of high or low heterozygosity leads to the absurdity that one could adjust population sizes and rates of mutation to make any H value fit, and there is no discriminating power to the formula making us favor or disfavor the neutral hypothesis. Finally, it is important to be reminded that populations under this model would have to be at equilibrium. That necessitates the restrictive condition that such a population would have to be isolated with an effective population size of N_e for at least N_e generations. Just how realistic this restrictive condition might be has yet to be determined.

In Chapter 12 we saw that random drift *can* bring about fixation of any allele, although if the allele is counteradaptive, it runs the greater risk of extinction. If neutral, a new mutant has a probability of fixation of $q = 1/(2N)$ in time of approximately $4N$ generations on on the average, as we discussed in Chapter 13. With selection and drift (population size) interacting, we illustrated in the examples of Chapter 14 distributions from a simulation experiment on a computer and saw that probability for fixation (ultimate probability of survivorship) of a favorable allele is given in Exercise 4, Chapter 14. However, the steady-state distribution of all these effects cannot be easily evaluated from observed populations into the respective parameters (N, s, u, m) that determine the distribution. Lewontin and Krakauer (1973) pointed out that "any observed distributed of gene frequencies over space or time, if considered to be in a steady state, can be explained by a suitable choice on N, m, and u with s being made arbitrarily small. . . ." One would expect under limited population size to find the distribution of q to be random, but certainly near-equality or strong similarity in allelic composition of populations at great distances from each other would *not* be expected from neutralist theory. The original Lewontin and Hubby data (Example 3-4) have been extended (see Lewontin, 1974, Tables 26, 27, 30) to encompass observed protein polymorphisms in 12 populations widely spaced over the distribution area of *D. pseudoobscura.* Allelic frequencies at 10 of the genic loci showed remarkable uniformity over the main continental area with no evidence of geographic races (except for the tropical isolated groups in Guatemala and Bogotá, Colombia).

One explanation for similarity instead of a random dispersion of allele frequencies would seem to be that the uniformity is due to some common selective forces in all populations; thus, the evidence would seem to favor the balanced-structure interpretation. However, an equally good fit to the data could be made by an alternative hypothesis: perhaps these protein variants are all neutral and effective migration over the species range is of the order of one individual hybridizing from a neighboring population per generation. It would be impossible to rule out migration rates of that order (see Chapter 19).

Alternatively, Lewontin and Krakauer noted that one feature of genetic drift, inbreeding, and migration contrasts with selection: "While natural selection will operate differently for each locus and each allele at a locus, the effect of breeding structure is uniform over all all loci and all alleles" for neutral variation. In the absence of selection, the steady state of all nonfixed genic loci will be a function of the breeding structure. Variation in frequencies between populations can be used to estimate the parameter F_{est}—the estimated effective fixation coefficient derived from the Wahlund expression (12-7A)

$$F_{\text{est}} = \frac{\sigma_q^2}{pq}$$

where σ_q^2 is the observed variance in q from one population to another and $\bar{p} = (1 - \bar{q})$. This parameter should be identical for all genic loci with neutral alleles among populations that are polymorphic for those alleles. If we calculate F_{est} for each locus separately, each value of F_{est} will estimate the parameter for the group of populations sampled, and those estimates should be homogeneous. On the other hand, if some loci are selected differentially, F_{est} from different loci should not be homogeneous. For any one locus the magnitude of F_{est} would be expected to be low if alleles were selected identically in all populations (numerator close to zero); it would be high if there were considerable local effect of selection making strong differences among populations, while neutral alleles would be expected to give intermediate values, unless the population were structured of small isolates or breeding units.

Table 18-5 shows F_{est} values from a study of human populations by Cavalli-Sforza (1966), reproduced by Lewontin and Krakauer, encompassing nine major loci, with 15 gene frequencies. The range of values from 0.029 for Kell blood group up to 0.382 for the R_o allele of the Rh blood group seemed too great for Cavalli-Sforaza to explain from random factors and neutrality alone, but Lewontin and Krakauer studied the sampling error distribution of F_{est} in order to test the homogeneity of these values. By using Monte Carlo simulation to produce a number of cases, the authors worked out the variance of F_{est} for the data in Table 18-5 to be as follows:

$$\sigma_F^2 = \frac{2(\bar{F})^2}{(n - 1)} \tag{18-15}$$

variance of F, where $\bar{F}$ is the average F_{est} among the genic loci, among n populations. The number of populational groups varies from 25 to 125 in the table, and n is chosen as

TABLE 18-5 *F values for worldwide distribution of human polymorphisms. N = number of groups sampled (from Cavalli-Sforza, 1966)*

Genetic System	Allele	N	F_{est}
ABO	A	125	0.070
	B	125	0.055
	C	125	0.081
MN	MS	45	0.071
	Ns	45	0.094
Rhesus	R_0	75	0.382
	R_1	75	0.297
	R_2	75	0.141
	r	75	0.172
Duffy	Fy	62	0.358
Diego	Di	64	0.093
Kell	k	64	0.029
Haptoglobin	Hp^1	60	0.096
Gm	Gm^a	25	0.226
Gc	Gc^1	42	0.051
	Harmonic Mean	60	$\bar{F}$ 0.148

the harmonic mean number = 60 so that for the human data the theoretical variance of F_{est} is

$$\sigma_F^2 = \frac{2(0.148)}{59} = 0.000742$$

The observed variance is 0.00741, so we have the ratio of

$$\frac{\text{observed variance}}{\text{theoretical variance}} = \frac{0.00741}{0.000742} = 10$$

This variance, ten times greater observed than expected, is highly significant,* and thus it is not possible to explain all the allelic variation among populations as the result of random factors with neutrality. For some genes, there are large differences between populations (Rh), and for others, allele frequencies are similar between populations (Kell); it would seem that selection could account for the result far better than any other mechanism.

In a second study of human data on 10 villages of Yanomama Indians in the Orinoco Basin of South America (data collected by Arends, 1966), Lewontin and Krakauer found the ratio of observed F variance/theoretical only to be 1.380. More complete data on these tribes with more populations sampled (37) and one more locus (Diego blood system) gave a ratio as just slightly larger—1.55, with 14 *d.f.* and $P = 0.10$. Apparently, if selection is operating differentially among these Yanomama tribe villages, it is not significantly detectable or there may be enough migration between villages to tend to equalize them and counteract slight selection effects.

Finally, an experimental technique has been used effectively by Sing, Brewer, and Thirtle (1973) on drosophila populations to test for selective neutrality of seven allozyme polymorphisms. Each generation, exactly five pairs of flies were used to establish progeny in a subline with several (49) sublines established from one large population that was also maintained. Under the neoclassic hypothesis, a predictable dispersion of allelic frequencies should occur among sublines above and below the reference frequency maintained in the large random-mating population. Genic loci varied as to the degree of decay of heterozygosity over 21 generations. At least four of the seven loci showed excess heterozygosity and proportion of lines segregating over that expected on the basis of complete neutrality, and thus selection was implied either in terms of the alleles at the loci themselves or in other very tightly linked but not observed genes controlling the actual fitness properties.

* Significance is found from the fact that the variance in F ratio is distributed as chi-square ÷ *d.f.* (where *d.f.* = number of alleles used in the estimates of F minus 1 *d.f.* for the total) so that *d.f.* = 15 − 3 − 1 = 11 for the data of Table 18-5. Then $X^2/11 = 10$ and $X^2 = 110$ with $P \ll 0.001$. Serious inaccuracies may arise in this test. Alan Robertson, M. Nei, and T. Maruyama (*Genetics* 80:395–396, 1975) warn that Lewontin and Krakauer's variance of F in formula (18-15) may be a serious underestimate because population relationships within a species could arise from migrations or common origins and thus may have correlated gene frequencies so that the expected variance between loci would have an extra component due to correlation coefficients. Lewontin and Krakauer (*Genetics* 80:397–398, 1975) clarify the issue by listing "dos" and "don'ts" for testing the heterogeneity of F. Random sampling of populations is vital for each locus studied unless it is known that there are no migration relationships between the populations involved, in which case the same population could be sampled over and over again for all the loci of interest.

This latter alternative is, of course, difficult to eliminate, but one would have to assume that such juxtaposition of a protein variant locus and a "fitness" locus would have to occur in each of the four cases.

Rate of Evolution by Selection Is Too Costly and Too Fast for the Average Observed Rate of Amino Acid Substitution in Polypeptides

Basing his reasoning on the genetic load concept, Kimura (1968) estimated that evolution of polypeptides in vertebrates proceeds at the rate of about one amino substitution per species every two years. That rate, he argued, would put such a heavy genetic load on any population that it could not be afforded. Therefore, he concluded that most amino acid substitutions in the evolution of proteins must be due to passive fixation of selectively neutral mutations.

Recall from Chapter 13 that the ultimate probability for fixation (*ups*) of a neutral mutant allele is $1/(2N_e)$. In a population of effective size N_e, the total mutations in the population per generation would be $2N_eu$, so that the total fixation rate of new mutants would be the product $(2N_eu/2N_e)$—a constant average mutation rate for neutral new alleles. In 1969, King and Jukes recalled this constant expected rate of fixation when they surveyed nine polypeptides in mammalian evolution for the number of amino acid differences between representative mammals and thus the number of substitutions in their respective molecular evolutionary sequences. These polypeptides and the number of their amino acid differences between mammalian orders are presented in Table 18-6A. On the assumption that most of the mammalian orders diverged about 75 million years ago, therefore making the total elapsed time between any two orders of mammals twice that time, the right-hand column of the table gives the estimated rates in amino acid substitutions per codon per species per year. Table 18-6B gives the ranges of observed differences in amino acids between species sampled. It is apparent that there is a tendency for uniformity *within* polypeptides. This uniformity was confirmed by Ohta and Kimura (1971), who reported on wider phylogenetic comparisons and more samples of hemoglobins (α and β) and cytochrome c; their rates of amino acid substitution for differences between mammals and fishes, reptiles and birds, and amphibians and fishes gave rate averages close to the King and Jukes rate estimates—for hemoglobin $\alpha = 0.97 \times 10^{-9}$, for hemoglobin $\beta = 1.53 \times 10^{-9}$, and for cytochrome $c = 0.28 \times 10^{-9}$.

King and Jukes reasoned that owing to the apparently uniform rate of amino acid substitution per year for each polypeptide, the first seven in the table might be neutral to selection with the rate being approximated by their mutation rates. The two last proteins, although among the fastest evolving of all, were proposed to be evolving by natural selection. From the weighted average, they estimated that the average rate of amino acid substitution would be 1.6×10^{-9} per codon per species per year. [The uniformity *per year* instead of *per generation* may be accounted for by the likelihood that mutation rates in various species are proportional to generation time (Maynard Smith, 1970).] Total mutation rate for an entire individual genome then could be estimated as follows. Total haploid DNA has 4×10^9 nucleotide pairs in mammals (based on total weight of DNA

TABLE 18-6A *Rates of amino acid substitutions in mammalian evolution (from King and Jukes, 1969) based on time elapsed since divergence of euplacental mammalian orders of 75 × 10^6 years*

Polypeptide	Observed Total Number Amino Acid Differences	Per Codon*	Substitutions × 10^{-9} per Codon per Year**
Insulin *A* and *B* (except guinea pig line of descent)	24	0.049	0.33
Cytochrome *c*	63	0.063	0.42
Hemoglobin α chain	58	0.149	0.99
Hemoglobin β chain	63	0.155	1.03
Ribonuclease	40	0.390	2.53
Immunoglobulin light chain, constant half	40	0.498	3.32
Fibrinopeptide *A*	76	0.644	4.29
Bovine hemoglobin fetal chain	97	0.250	2.29
Guinea pig insulin	86	0.411	5.31

* Estimated from actual number observed per codon by assuming that mutants will be distributed at random according to a Poisson distribution. The frequency of unchanged sites then would be $[e^{-m}]$, where m is the true frequency of substitutions per site. For example, fibrinopeptide *A* has 76 amino acid differences among the species sampled. With 16 amino acid's in the peptide, 10 pairs of species were compared, or 160 comparisons. Then $\frac{76}{160} = 0.475$ observed differences per codon. Those unchanged would be $1 - 0.475 = 0.525$, the first term of the Poisson expression so that $m = 0.644$ as given.
** Codon rate over the 150-million-year period.

TABLE 18-6B *Range in number of amino acid differences between pairs of mammalian groups by polypeptide (from King and Jukes, 1969)*

Polypeptide (no. amino acids)	Number of Mammalian Group Pairs*	Amino Acid Differences (largest number/ smallest number)
Insulin *A* and *B* (except guinea pig) (51)	10	3/1
Cytochrome *c* (104)	10	12/2
Hemoglobin α (141)	3	23/17
Hemoglobin β (146)	3	25/14
Ribonuclease (124)	1	40/—
IgG: light chain C half (102)	1	40/—
Fibrinopeptide *A* (16)	10	10/4
Bovine fetal hemoglobin (146)	3	33/31
Guinea pig insulin (51)	5	18/16

* Includes samples from humans, horses, rabbits, sei whales, bovines, pigs, gray whales, mice, rats, donkeys, dogs, and guinea pigs.

in sperm) of which probably about 1 percent codes for proteins in structural locus codons. Thus, there would be about 1.3×10^7 codons in the haploid genome. At the estimated rate of amino acid substitution per year, the total rate of polypeptide change per year would be $(1.3 \times 10^7)(1.6 \times 10^{-9}) = 2.8 \times 10^{-2}$, or about one amino acid substitution in 36 years per species, far slower than Kimura's earlier estimate of one substitution in two years.

If an amino acid substitution occurs by mutation and an advantageous polypeptide is formed, according to simple selection principles (Chapter 14) the new allele will replace the old one and bring about a "substitutional load," a concept proposed by Haldane (1957, 1960). He conceived of the situation where one or more genes (alleles) were rare because of a balance between mutation and selection. After a sudden environmental change, the rare form may become beneficial and the common type deleterious. A number of "genetic deaths" (losses of a gene due to lower fitness of genotypes containing that gene) must occur until the beneficial allele is substituted for the less fit allele. The amount of genetic death needed (total load L) during the process of fixation will be expressed as the exponent m in the Poisson survivorship statement of formula (18-12): $S = e^{-m}$.

From Haldane (1957), the substitutional load (L) is approximately

$$L = -2\, ln\, q_0 \qquad (18\text{-}16)$$

where q_0 is the initial frequency of an advantageous allele that becomes fixed, and ln is $\log_e$. If q_0 is determined by the mutation rate u, and we let K = number of substitutions per generation, then $L = -2K\, ln(u)$. For the estimates given above, assume three years to be average generation time for mammals $L = -2(2.8 \times 3 \times 10^{-2})(ln\, 10^{-5})$ if we let $u = 10^{-5}$ as average mutation rate. Then $L = (-0.17)(-11.5) = 1.95$. Therefore, $S = e^{-1.95} = 0.1416$; so that there would have to be about seven offspring per parent to keep the population in constant numbers. This load does not seem intolerable. On the other hand, Kimura's earlier estimate of one substitution in two years (25 times faster) would be out of the question; $L = -2(70 \times 3 \times 10^{-2})(ln\, 10^{-5}) = 48.3$, and $S = e^{-48.3}$—an absurdly small survivorship. Nevertheless, King and Jukes (1969) followed the earlier Kimura conclusion when they stated: "One amino acid substitution every 50 years [or 36 years, as we calculated above] is still too rapid a rate to be accounted for by classical genetic theory unless most substitutions are selectively neutral." The student ought to realize how easily one can be swayed by preconceptions and what enormously diverse loads can be calculated if one postulated a high or a low probable value either in number of active sites (codons) in DNA or in rate of substitution per year.

As additional evidence for neutrality of polypeptide changes, King and Jukes cited data to indicate that the proportion of amino acid sites that have 0, 1, 2, . . . , n substitutions in variants of numerous proteins follow the Poisson distribution. Before they tabulated the distributions for proteins, they first eliminated from consideration the few sites they believed to be almost certainly of adaptive (functional) significance. For example, in cytochrome c, about 29 of the amino acid residues are invariant in all organisms: these sites undoubtedly must be screened by natural selection, since they are vital for combining with heme group and for interacting with cytochrome oxidase. The remaining 81 residues, which are variable among different species, follow the Poisson distribution for a mean (m) of 2.6 as shown for cytochrome c in Table 18-7. Similarly, for globins and the light chains

TABLE 18-7 *Distribution of numbers of amino acid substitutions compared for 110 sites in cytochrome c chains (from King and Jukes, 1969)*

No. Changes per Site	No. Sites with This No. of Changes per Site	Minus 29 Invariable Sites	Poisson Distribution for $m = 2.6$
0	35	6	6
1	17	17	16
2	18	18	20
3	19	19	18
4	10	10	12
5	6	6	6
6	3	3	3
7	1	1	1.0
8	1	1	0.3
9	0	0	0.1

of IgG immunoglobulins, the *variable* amino acid site substitutions seemed to fit Poisson distributions with means of $m = 3.5$ and 2.4, respectively.

Finally, a third argument was that the amino acid frequencies among 53 vertebrate polypeptides (mostly mammalian) agreed well with those expected from random permutations of nucleic acid bases. From 5492 amino acid residues of these polypeptides, the base composition percentages of relevant mRNA were

1. Uracil = 22.0 percent.
2. Adenine = 30.3 percent.
3. Cytosine = 21.7 percent.
4. Guanine = 26.1 percent.

For each amino acid, the expected frequency was calculated based on the codons' total probability for each; for example, the expected for *tyrosine* (assuming independence of bases) was

Codons	Probability
UAU	(0.220)(0.303)(0.220)
UAC	+ (0.220)(0.303)(0.217)
Total	0.0292

However, because 61 codons specified amino acids out of 64 combinations of nucleotides, this probability must be corrected by $\frac{64}{61} = 1.049$. Thus, the expected frequency of tyrosine on the basis of random permutations of bases in the DNA was $0.0292(1.049) = 0.0306$. The observed frequency of tyrosine was 0.033, which is reasonably close. Similarly, the remaining 19 amino acids gave close agreement between expected and observed (coefficient of correlation $r = 0.89$), except for arginine whose observed frequency was 4.2 percent but the expected was 10.7 percent.

COUNTERARGUMENTS. Those who stress the adaptive nature of molecular change tend to regard those molecules in evolution that are conservative, highly stable, and unvarying (such as chlorophyll *a* and cytochrome *c*) as examples of "evolutionary homeostasis" (Stebbins and Lewontin, 1972). The fundamental principle of a steady state can be applied to the maintenance of certain functions essential for life processes or processes that must be kept constant over a wide variety of conditions. Thus, chlorophyll *a* has been unchanged throughout the evolution of plants. Cytochrome *c* catalyzes one of the most basic processes of cellular metabolism. Hemoglobin, which displays much more variation from one organism to another, is really a molecule within which more change can take place in structure without upsetting basic life processes. Finally, a great deal of molecular variation is found in the fibrinopeptides, which function in blood clotting and are definitely accessory because variation in their molecule does not disturb life functions and they are unlikely to interact with external conditions encountered by an animal. Thus, adaptive variability in molecular structure is likely to occur when a substance is accessory rather than fundamental to metabolic processes. One cannot regard the constancy of a molecule or the variability of its components in an evolutionary line as evidence either for or against its adaptiveness or relevancy to selective forces. We cannot conclude that the more "useless" the protein, the less constraints on its variation and thus the more rapidly it may evolve by random drift. Actually, King and Jukes drew the opposite conclusion for guinea pig insulin; its rapid rate was probably due more to natural selection *because* it was fastest of all in substitution rate. But their concept of selection was that of a purifying-eliminating process, in neoclassic manner; the concept that protein variation might be useful in itself to adaptive processes had not been considered. Fibrinopeptide *A* variants are certainly not neutral because the amino acid substitutions fixed in diverse groups are not randomly distributed along the sequence of residues but tend to be concentrated in certain positions. Stebbins and Lewontin, who analyzed data on fibrinopeptides of artiodactyls for the relative distributions of amino acid substitutions with little effect on the properties of the peptide (conservative) versus those with potentially large effects (radical), found that high concentrations of radical substitutions occurred at certain positions (residues 12-14), where the rate at which thrombin splits the molecule is influenced, while the conservative substitutions were scattered over the molecule where few or no substitutions had become fixed in phylogeny.

How a protein can be presumed to be useful or useless seems entirely to lie within what Stebbins and Lewontin termed "the fallacy of omniscience," which they described as follows:

> *This fallacy, which has often been committed in the past by evolutionists and taxonomists who are comparing macroscopic characters of organisms, runs about as follows: "I can't see what adaptive value this character difference could have; therefore it is inadaptive and was not influenced by natural selection." The fallacy here lies in the implication that the author of the statement knows everything that can be known about the adaptiveness of the organisms concerned. . . . Evolutionists are, therefore, unable to prove, by process of elimination, the null hypothesis that amino acid substitutions in certain portions of enzyme molecules are neutral and do not affect the function of the whole molecule.*

At the same time, interpretations of enzyme structure and function must guard against the reverse fallacy, which would necessarily ascribe an adaptive significance to amino acid substitutions, without definite evidence in favor of it.

We do not know the relative fitness functions of proteins, and it is highly subjective to categorize them into their likely fitness values *after* their rates of evolution have been estimated.

A functional classification of enzymes into whether they are intracellular-metabolic (glycolytic-Krebs cycle) or whether they are "regulatory" (determining pathway rates), nonregulatory (allowing equilibrium between substrate and product), or simply active on variable substrates might be informative. The amount of polymorphism or biochemical diversity available in each class of enzyme has been postulated to be related to function. Enzymes directly involved in energy metabolism would be less variable because their functions are more "essential" than those of other peripheral enzymes. Gillespie and Kojima (1968) found that enzymes that catalyze steps in glycolysis are less polymorphic than the more miscellaneous enzymes, most of which act on external substrates and probably reflect environmental variation in those substrates by their greater polymorphism. When a line of argument points to levels of enzyme polymorphism as a response to physiological function with certain classes of enzymes more variable than others, one cannot regard such enzyme variants as completely neutral. Johnson (1973, 1974, 1975) compiled the available data on the degree of heterozygosity at enzyme loci and classified the enzymes as to their likely role as (1) regulatory, (2) nonregulatory in metabolism, or (3) active on variable substrates. The latter (such as esterases and peptidases) had highest polymorphism; regulatory (such as G-6PD and PGM) had somewhat less; nonregulatory (such as LDH and 6-PGD) had very much less in drosophila, small mammals, and humans. The neutrality hypothesis cannot account for such correlation between enzyme variation and function.

Another similar line of evidence was pointed out by Sing and Brewer (1971), who concentrated on monomorphic (nonsegregating) enzyme systems with isozymes among 22 plant genera; the multiplicity of gene product forms (isozymes) was their measure of the biochemical diversity individual members of a species would have available at the gene product level (for example, the familiar lactic dehydrogenase). They compared enzymes of the glycolytic and Krebs cycle subset versus miscellaneous enzymes for the number of isozymes characterizing each and found the former to be much more closely correlated with one another within its subset than the latter. They concluded that distributions of the number of molecular forms in the metabolic subset cannot be considered independent. Instead, they indicated evidence for selection of concomitant multiplicity in functionally related enzymes. Isozyme diversity was likely to be a reflection of organized subsystems. This analysis remains an area for exploration and for establishing whether enzyme groupings exist as functional units with adaptive value.

Uniformity of the rate of change for each polypeptide sequence has been stressed by the neoclassic argument. These rates are derived from correlations between numbers of amino acid substitutions and the presumed age of divergence in a phylogenetic sequence. A clue to a likely artifact in this uniformity of rate becomes apparent when we compare the rates for fairly closely related mammalian orders compared with those between groups which diverged very much farther back. Figure 18-1 presents a phylogenetic tree for carp and four mammalian groups with the number of amino acid substitutions in the α-chain of

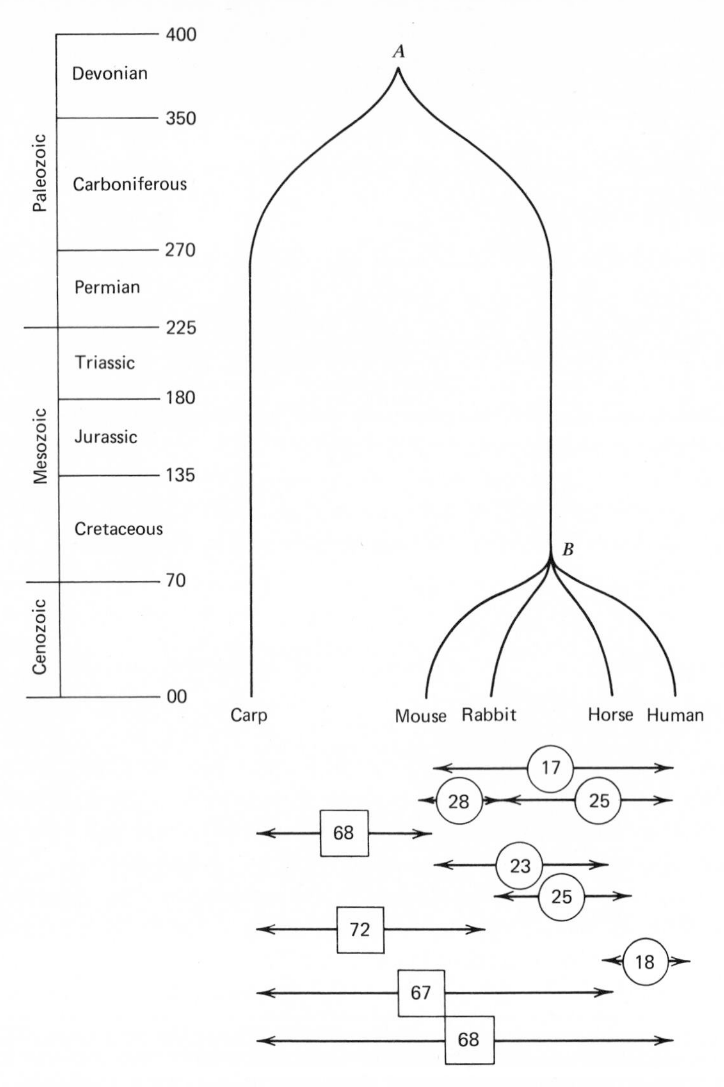

Figure 18-1. A phylogenetic tree of a bony fish (carp) and four mammalian species together with the geologic time scale. Numbers with arrows refer to differences in amino acid sites in the α-chain of hemoglobin between respective animals. From Kimura and Ohta, 1972.

hemoglobin between the groups. There is a difference of 17 amino acids between mouse and human, while there is a difference of 28 amino acids between mouse and rabbit. Each of these mammals has the *same* amino acid difference number (68) as the fish (carp). Thus, the divergence between any two recently separated mammal groups shows greater inequality (11 amino acid residues) for rate change than it shows when compared with the more remote form. A striking exception to the constancy of rate is the guinea pig insulin (Table 18-6A) compared with other mammal groups. Perhaps if more protein molecules were surveyed, the inequality of rates for recently diverged groups would be even more in evidence. However, the constancy of amino acid substitution in all phylogenetic lines can hardly be considered as established fact. Stebbins and Lewontin (1972) pointed out that

> *... there have been many speedings-up and slowings-down of the rate of amino acid substitution differentially in different lines as within the mammals. On the average over vast stretches of time, however, we expect that different phylogenetic lines will have similar average numbers of substitutions in molecules having similar functions. The entire argument is based on a confusion between an average and a constant.*

Certainly, all would agree that phyletic lines have fast or slow periods of evolution; adaptive radiations were periods of fast evolution, but some lines failed to radiate. Simpson (1953) documented the wide differences in rates of evolution within the phylogenetic history of most groups; for example, lungfishes, which may seem to have stagnated in terms of morphological change and speciation for the last 150 million years, had very appreciable change going on 350 million years ago with a rapid explosion of forms and changes in late Devonian and Mississippian times (260–280 million years ago). It would be very misleading to *average* the rate of evolution for such forms or to think of their rate as a *constant*. In calculating an average rate, we would completely lose the details of selection history in the radiation of the group.

King and Jukes stated that the number of amino acid substitutions in polypeptides had followed a Poisson distribution, but they first removed from consideration the few sites they believed to be adaptive. Again, according to Stebbins and Lewontin, classifying molecular regions into "adaptive" (invariant because essential for life) or "nonadaptive" (variant) makes assumptions that include the "fallacy of omniscience," as pointed out by B. Clarke (1969*a*), who showed that the distribution of residues and substitutions in the variant regions is not random. Also, King and Jukes assume that enzyme action is an all-or-none operation without relationship to environmental conditions or that it cannot be modified in adaptation to fit those conditions.

Finally, the neoclassic argument pointed out that the amino acid frequencies among 53 vertebrate polypeptides agreed with random permutations of nucleic acid bases. However, the proteins chosen cannot be considered a random sample of proteins (Clarke, 1969*a*). When structural proteins such as collagen, muscle actin and myosin, keratin and chitin are analyzed, amino acid composition does not agree with the frequencies expected on the basis of the code. Furthermore, when pooling together a great diversity of proteins and species, it would be very strange indeed for any particular amino acids to stand out in view of the fact that when averaging very diverse groups, plus and minus variations tend to cancel each other out. This conclusion by the neoclassic group seems to be using an *average* as a *constant*.

In conclusion, we must ask whether the observed changes in amino acid sequences of proteins represent adaptive changes in response to environmental conditions or not. Perhaps from previous chapters in this book, the student may come to appreciate the following generalization made by Stebbins and Lewontin:

> *To the extent that selection guides the course of evolution, it does so not by accepting or rejecting individual mutations, but through the effects of an altered environment upon the adaptive values of constellations of genes. The adaptive value of a particular allele*

or mutant gene cannot be treated . . . as a constant. Adaptive values of genes vary widely in relation to both the external environment to which the population is exposed, and the internal environment of the cell, in which the primary product of each gene must perforce interact with the products of many other genes.

The issue at hand . . . is . . . whether alterations of entire genotypes represent adaptive shifts of interacting genes that form integrated systems or whether genotypes are altered by random changes in the frequencies of genes that act independently of each other.

The evidence concerning the significance of protein variation has been summarized by Powell (1975) in the following statement:

No conclusion can be drawn whether drift or selection is the predominant force in the evolutionary change of this material. . . . The view that natural populations are highly polymorphic for adaptively significant genetic variation is not necessarily incompatible with the view that most evolutionary changes on the nucleotide level are a result of random fixation of neutral mutations.

EXERCISES

1. (a) For the heterotic equilibrium in Table 14-5, what is the invariant $\bar{W}/W_{max}$ ratio and the invariant fitness difference?
   (b) Let the fitnesses of genotypes be $W_{AA} = 0.9091$, $W_{Aa} = 1.1364$, $W_{aa} = 0.7955$. What is $\bar{W}$ at equilibrium? Show this ratio and fitness difference to be invariant (same as for the fitness values of Table 14-5).
   (c) What is Haldane's selection intensity (I) for these two cases?
   (d) What is the genetic load for these two cases? Is L always less than I? Is L the same as the invariant loss? What is the justification for using L as a measure of genetic load?
2. Selection acts on phenotypes, and Haldane's selection intensity concept can be employed without knowledge of the genetic basis for the phenotypes. Haldane (1954) applied the concept to human birth weight data from a study by Karn and Penrose on survivorship for birth and postnatal period up to 28 days of life. For female babies, Haldane cited the following:

Birth weight	Survival % ± *s.e.*
Under 4.5 lb	41.4 ± 3.7
From 7.5–8.5 lb	98.5 ± 0.3
Over 10 lb	90.5 ± 6.4
Overall survival	95.9 ± 0.2

   (a) Verify that selection intensity I will be for birth weight about 0.027.
   (b) For further data and comparisons with other populations, see Van Valen and Mellin (1967).

3. Relative fitness values may be manipulated in several ways without disturbing the selection sequence or final outcome. Li (1963*a*) illustrated this principle by examining three systems of fitness notation as follows for a stable heterotic equilibrium.

		Systems of W's		
Genotype	W	I	II	III
AA	W_1	1	$1 - t$	$1 - b$
Aa	W_2	$1 + Hs$	1	$1 + \sqrt{bc}$
aa	W_3	$1 - s$	$1 - r$	$1 - c$
Equilibrium $\hat{q}$		$\frac{H}{1 + 2H}$	$\frac{t}{r + t}$	$\frac{\sqrt{b}}{\sqrt{b} + \sqrt{c}}$
$\bar{W}$		$1 + Hsq$	$1 - rq$	1.00

(a) Using these systems and the data from Table 14-5, find H, s, t, r, b, and c as positive proportions.
(b) In systems I and II, recall formula (14-15) for equilibrium $\hat{q}$. Verify the $\bar{W}$ statement by substituting into the net fitness $(p^2W_1 + 2pqW_2 + q^2W_3)$ the appropriate $\hat{q}$ and $\hat{p}$ values above.
(c) In system III, find W_2 given the $W_1 = (1 - b)$ and $W_3 = (1 - c)$ in order to make $\bar{W} = 1.00$. Hint: Verify first that $W_2 = 1 + [(p^2b + q^2c)/2pq]$ and let the bracketed portion be X. Then use the equilibrium condition (14-14) and simplify algebraically.

4. If three gene loci (A, B, and C) all have constant mutation rates (u_i) to recessive deleterious alleles so that homozygous aa, bb, and cc each has a low fitness (W_{ii}), let them have these values:

Locus	W_{ii}	u_i
A	0.75	10^{-4}
B	0.64	10^{-6}
C	0.84	10^{-8}

(a) Assume independence of fitness for each locus recessive. What would be the equilibrium frequency ($\hat{q}$) of each locus recessive allele?
(b) Find the load (L) of each locus at equilibrium and the net cumulative load.
(c) Show that the load formula (18-3) converts into formula (18-5) for these values.

5. With partially detrimental heterozygotes as in Table 14-4, if a is produced by mutation from A, it will be rare enough for the frequency of aa genotype to be disregarded. Then if $W_2 = 1 - hs$, and gain in a due to mutation $= up$, in G_1 following selection and mutation, $q_1 = (pq - pqhs + up)/\bar{W}$ and $\Delta q = (pq - pqhs + up)/\bar{W} - q$. Set $\Delta q = 0$ and let $\bar{W} = 1 - 2pqhs$, approximately. Discard all terms with q^2 as negligible as well as (uq). Solve for $\hat{q}$ as formula (18-7A).

6. Assume fitness values as in Table 18-2B and a mutation rate of A to a of $u = 1 \times 10^{-4}$. Verify that $h = \frac{2}{3}$ to be used in formula (18-7A).

(a) Find $\hat{q}$.
(b) Assume equilibrium conditions and contrast $\overline{W}_i/\overline{W}_0$ from this mutational equilibrium with that ratio at the same allelic frequencies but with a balanced (segregational) set of fitness values from Table 18-2A.
(c) In which case does inbreeding create a lower net fitness than outbreeding? Is this always true when contrasting a mutational with a balanced load?
(d) What is the load ratio (L_i/L_0) for the mutational (Table 18-2B) versus balanced (Table 18-2A) equilibrium conditions?

7. In human populations, particularly, mutation rates have been estimated by an indirect method (Haldane, 1949*b*). With a deleterious allele, if the frequency incidence is known and a fitness value can be estimated for the genotype containing the allele, formula (18-7A) for rare dominants can be used.
(a) Mørch (1941, cited in Neel, 1962) in Denmark studied the incidence of chondrodystrophy (dwarfism due to achondroplasia) as well as the fertility (offspring) per dwarf individual. He found 10 dwarfs were born over a 30-year period out of 94,075 births. By investigating older records, he found 108 known chondrodystrophic dwarfs had produced just 27 offspring compared with their 457 normal siblings who had produced 582 offspring. What is the fitness (fertility component) of the dwarf condition relative to normal? Let this fitness value be W_2.
(b) Using formula (18-7A), let $hs = 1 - W_2$ and let $q = \frac{1}{2}H$—that is, one-half the incidence of dwarfism at birth—what is the estimate of mutation rate for the chondrodystrophy gene from normal?
(c) Cotter (*Jour. Heredity* 58:59–63, 1967) studied the incidence of this kind of dwarfism in descendants of a Mormon family in Utah distributed among 9 sibships with a total of 76 sibs with 34 dwarfs: 42 normals. These 76 persons produced 209 offspring, 70 from dwarf × normal matings and 139 from normal × normal. What is the relative fitness (fertility component) of dwarf: normal? Average number of children per family was 2.06 for dwarfs and 3.31 for normals. Of the 34 dwarf parents, 20 produced no children, while of the 42 normals, 18 produced no children. Average children per producing parent were 5.00 for dwarf parents and 5.80 for normal parents. Which of these data are most relevant to calculation of the fitness (fertility component)? For the choice you make, use that value for W_2 and estimate the mutation rate for the incidence in Denmark. What sorts of errors would be likely in such an estimate?

8. Under inbreeding, a mutational-selection equilibrium for a deleterious recessive (a) will have a balance of the following kind, using formulas (18-5) and (8-3):

$$R = q^2 + Fpq = (1 - F)q^2 + Fq = u/s$$

(a) Verify that there is a quadratic solution for equilibrium $\hat{q}$:

$$\hat{q} = \frac{-F \pm \sqrt{F^2 + 4(1 - F)(u/s)}}{2(1 - F)} \qquad \text{(positive solution only)}$$

(b) Let $u = 10^{-6}$, $s = 10^{-2}$, and $F = 10^{-1}$. What will be the $\hat{q}$? How does that value compare with the expected equilibrium under random mating?

9. Under inbreeding, a heterotic equilibrium can be derived as follows: Partition $\bar{W}$ into a random-breeding portion and an inbred portion, thus:

$$\bar{W} = (1 - F)\bar{W}_0 + F\bar{W}_i$$

where $\bar{W}_i = pW_1 + qW_3$ and $\bar{W}_0 = p^2W_1 + 2pqW_2 + q^2W_3$. After selection $q_1 = [(q^2 + Fpq)W_3 + pq(1 - F)W_2]/\bar{W}$. At equilibrium, $q_1 = q_0$ so that q cancels on both sides. Thus

$$\bar{W} = (q + Fp)W_3 + p(1 - F)W_2.$$

Now we substitute from above as follows:

$$(1 - F)\bar{W}_0 + FpW_1 + FqW_3 = qW_3 + FpW_3 + pW_2 - FpW_2$$

Grouping all terms (except the first) on the right side of the equation gives

$$(1 - F)\bar{W}_0 = (1 - F)qW_3 + (1 - F)pW_2 + Fp(W_3 - W_1)$$

Putting all $(1 - F)$ terms together and factoring on the left side, we have

$$(1 - F)[\bar{W}_0 - qW_3 - pW_2] = Fp(W_3 - W_1)$$

where random mating terms balance the inbreeding terms on opposite sides of the equation.

(a) Simplify and verify the equilibrium solution for $\hat{q}$ at heterotic nontrivial stable point under inbreeding and random mating:

$$\hat{q} = \frac{(1 - F)(W_1 - W_2) - F(W_3 - W_1)}{(1 - F)[(W_1 - W_2) + (W_3 - W_2)]}$$

(b) Assume $F = 0.50$ and W values are successively 0.8, 1.0, and 0.7 (as in Table 14-5), find $\hat{q}$. How does that equilibrium point compare with that under random mating? What is $\bar{W}$ in this case with inbreeding? What is $\bar{W}$ when $q = 0.30$ with inbreeding?

(c) Show that $\bar{W}$ is maximum at $q = 0.30$, which is the halfway point between stable q values under inbreeding and random mating. Thus, is the stationary point under inbreeding at the simple "adaptive peak" as defined earlier? What is the inbreeding effect of the population's $\bar{W}_{max}$? (see Li, 1967*c*).

10. Find the L_i/L_0 ratio $(B/A + 1)$ for the following q values: 0.5, 0.3, 0.1, 0.05, 0.01, 0.001.

(a) First assume a segregational load and let the ratio $s_1/s_2 = 0.5, 0.3, 0.01, 0.001$. Remember that $L_i = s_1p + s_2q$ and $L_0 = s_1p^2 + s_2q^2$. Graph these results with q as the independent variate and the $B/A + 1$ ratio as the dependent variate.

(b) Then assume a mutational load; use formula (18-10). Let $h = 0.20, 0.10, 0.01, 0.001$ as a minimum.

(c) What can you say about the B/A ratio when gene frequencies are not at equilibrium? What is the possible error involved in making an assumption of equilibrium?

11. Suppose there are 90 loci with an average mutation rate of 10^{-5} and all are slightly detrimental in heterozygous condition with $h = 0.025$. Also suppose 10 loci are heterotic with two alleles per locus and with homozygotes all having a disadvantage of $s_i = 0.05$.

(a) What would be the load ratio?

(b) Would it be distinguishable from a segregational load at equilibrium?

12. Discuss whether the load ratio concept is useful in the light of the following points:
    (a) Constant fitness versus variable fitness.
    (b) The size of the load versus the population's survival ability.
    (c) With a segregational load, a more "normal" allele with a fitness of 0.99 for *AA*, 1.00 for *Aa*, and 0 for *aa* contributes 98 times as much to the load as the recessive *a* allele.
    (d) The change in load when a single favorable mutation occurs (see Li, 1963*c*; Brues, (1964).
13. Allowing that net fitness is the equivalent of survivorship so that we may use formula (18-12), assume there are 3000 loci segregating for two alleles, each with an average heterozygosity of $\overline{2pq} = 0.33$ and and all selection coefficients of homozygotes $s_i = 0.10$ compared with heterotic heterozygotes,
    (a) What will be the total load and "net fitness," or survivorship?
    (b) List the main assumptions we need to make in order to make this calculation.
14. Suppose a haploid population exists with the frequency of $A = 0.9999$ and the frequency of $a = 0.0001$. Assume the *a* allele replacement occurs over the course of time to become fixed eventually at frequency of $a = 1.00$.
    (a) From this simple fact, is it possible to distinguish between the change being due to selection or by random genetic drift? If the environment were known to be completely constant, could you distinguish selection from drift?
    (b) Could one calculate a load a posteriori; that is, merely from the fact that a change in frequency has taken place?
    (c) How, then, can we distinguish between a selection effect and one of random drift? If we postulate a "neutral" model when confronted with a certain rate of allelic substitution over long time periods, have we thereby eliminated the problem of genetic load? (see Stebbins and Lewontin, 1972).
15. Milkman (1967, 1970) pointed out that the fallacious reasoning in the dilemma of a segregational load lies in the argument that "each locus exercises its lethal equivalent effect independently." For example, assume there are four nonlinked loci with a recessive lethal each and each of which affects a successive part of an insect's life cycle: *aa* lethal at hatching, *bb* lethal at the first molt, *cc* lethal at the second molt, and *dd* lethal at puparium formation. In a population containing only *AaBbCcDd* individuals, there would be 75 percent hatching *A*-, of which 75 percent would survive the first molt (*A-B-*) or $0.5625 = (0.75)^2$. Thus, by puparium formation $(0.75)^4 = 0.3164$ would be surviving of eggs laid. For these completely penetrant lethals, mortality becomes $[1 - (0.75)^4] = 0.6836$. If the recessive alleles at each locus had only 10 percent penetrance of lethality ($s = 0.10$), verify that the mortality would be $[1 - (0.975)^4]$. If there were 100 loci with $p = q = 0.5$ at each locus and with $s_i = 0.10$ with all heterotic heterozygotes ($W_2 = 1.00$), if these loci all acted independently on fitness, verify that the mortality would be $[1 - (0.95)^{100}] = 0.9941$. Why would you think it unlikely for lethal equivalent effects to be truly independent in their action on fitness?
16. Milkman suggested that the basic assumption of the segregational load would be more realistic if it started with a minimum threshold level of heterozygosity necessary for survival (complementation, overdominance, or some combination of epistatic interaction). Then, to go beyond that level of heterozygosity for mean survivorship would not take much of a load. Thus, a natural population having a mean number of 50

heterozygous loci per individual might "need" about 48 to be heterozygous for survival as a minimum threshold. Individuals with less heterozygosity would die. Let there be $100 = n$ loci in the genome of a population with an average $p = q = 0.5$ and thus maximally heterozygous. The distribution of these heterozygous loci could be binomial or nearly normal with standard error of $\sqrt{npq} = \sqrt{(100)(0.5)(0.5)} = 5.00$. Refer to Figure 7-2 as you visualize a normal distribution of n heterozygous loci. Thus, mean $\bar{Y} = 50 \pm 5.00$. If 48 heterozygous loci is the minimum number for survival in any individual, that threshold is 2/5 of a standard error below the mean. Thus, if we refer to a table giving areas under a normal curve at 0.40 *s.e.* units, the area of 65.5 percent includes the point of truncation to the right. In other words, to maintain that threshold, about 34.5 percent of the population dies because the heterozygosity is too low. What will be the mean number of heterozygous loci among those selected (survivors)? *Answer*: We obtain this number by utilizing a table relating the proportion selected to the height of the normal curve ordinate at the point of truncation (z), as referred to in Chapter 7, p 189. Thus, we find $z/0.655$ becomes 0.5615 *s.e.* units to the right of the mean; that is, $50 + (0.5615)(5) = 52.8$ as the average number of heterozygous loci needed among the selected individuals to support the mean of 50 loci for the population. Thus, 100 loci with heterosis could be maintained with a "load" of 34.5 percent mortality compared with 99.4 percent mortality if all 100 loci were acting independently (as in Exercise 15).

17. With formula (18-14A), graph values of the panmictic index P for $N_e u$ from 0.001 up to 10. Use a log scale for $N_e u$. Place into the graph points represented by known average heterozygosity for at least five organisms.
    (a) What values of N_e and u might be reasonable for these organisms?
    (b) What range of values could just as well fit the observed heterozygosity?
    (c) Does this formula have much discriminatory power?

19

INTEGRATION AND SEPARATION OF GENE POOLS

In the opening paragraph in Chapter 1 and again at the start of Section D, we stressed the continuity of the genetic material in sexually reproducing biparental species within which individuals are interconnected by a fundamental network of relationship. They share a common "gene pool"—our concept of a Mendelian population (Wright, 1931; Dobzhansky, 1950, 1955). Diploidy or higher ploidy in the life cycle provides opportunities for closer integration of genotypes than can be achieved within a simpler haploid system, because special interactions among genic loci can exist in "vertical" as well as "horizontal" dimensions. Selective advantages can be gained via the interactions of (1) overdominance, (2) complementation, and (3) epistasis combined with various levels of heterozygosity. Components of the genetic system (individual as well as populational genomes) would be expected to become integrated, or unified, under the influence of stabilizing selection, which would promote phenotypic stability. *Stability* is meant to imply a steady state for a populational genome including any dynamic cycles or adaptive processes involved and any variation or components within the system that are necessary for maintenance of that integral unity by which the population may be characterized, either in an adaptive sense ("fitting a niche") or in a phenotypic sense (a set of morphological and physiological features). It is *not* meant to imply monomorphism of genotype or phenotype, although monomorphism could easily be a stable state for a population.

Along with integration of a population genome, we must realize that gene flow takes place between populations and that local units differentiate under local forces of effective size and selection into genetically diverse units: semiisolates or, if modified by selective local forces, ecotypes or races. From such diversity, partially or completely reproductively isolated populations may arise. Balance between gene flow and differentiation of local units is expected to produce a complex genetic structure characterizing an entire species—an assemblage of populations capable of interbreeding.

In this chapter we review first those stabilizing selective mechanisms and genetic interactions that are likely to promote unity and stability for populational genomes. Second, while the general problem of origin of species is beyond the scope of this book, we consider at least factors promoting divergence, the measurement of genetic differences between populations ("genetic distance"), and how gene flow (migration) may counteract divergence.

STABILIZING GENETIC SYSTEMS

Heterosis

Elsewhere in this book we have discussed particular genetic models where heterozygosity was related to phenotypic features outside the limits of homozygotes: overdominance (Chapter 6), relational balance of polygenes (Chapters 7 and 17), "luxuriance" among progeny after crossing of inbred lines that had been depressed (Chapter 9), and heterozygotes with superior fitness including single loci as well as "coadapted complexes" (Chapters 14–18). All of these models have at one time or another been lumped or confused together under the loose title of *heterosis*, although Shull's term was meant only for luxuriance. It is a convenient shorthand often used by geneticists as applied to heterozygotes that display some increased vigor or excess of a trait beyong the expectation from pure-line parents or derived homozygous lines. All these heterotic phenomena, though possibly diverse in their genetic details, have in common phenotypic values associated with heterozygosity that can lead toward integration of the total genotype—more genetic differences (alleles plus components of the genotype) to be accommodated per individual genome.

Our aim here is not to review the immense literature on heterotic phenomena, but to pool together evidence and concepts of its nature as an integration mechanism. See Gowen (1952), Grant (1975, pp. 107–119), and Parsons (1973, pp. 152–155) for examples, in addition to such cases as those in Examples 2-4, 3-1, 3-2, 15-2, 15-3, 15-11, and 15-12.

From the standpoint of higher fitness conferred by heterozygotes, it is common to find fitness superiority expressed under conditions of stress as in extreme environments, high temperature or cold (Langridge, 1962, 1968; Pederson, 1968; Spiess, 1967), desiccation (Parsons, 1973), or where environmental demands are aimed at favoring a phenotype with greater range of tolerance than in less demanding areas. It is often difficult to distinguish operationally between genuine genotype-environment overdominance (heterozygote superiority in the stress condition) and "marginal overdominance" with multiple niche selection and no heterozygote superiority in any niche (Chapter 16). Wills, Phelps, and Ferguson (1975) stated that "marginal overdominance or something very akin to it is the chief source of heterozygote advantage in Drosophila populations." However, that statement is premature in attempting to generalize a very complex concept.

Haldane (1954*a*) proposed a biochemical basis for superiority in Darwinian fitness; he suggested that if homozygotes control only a single protein form (allozyme, globin, etc.), which might be too restrictive or narrowly specialized, the heterozygote by allowing two forms could function in more diverse ways in a wider range of environment conditions. Such increased functional flexibility would be expected to be the equivalent of having superior fitness. A demonstration of that principle was made in the fungus *Neurospora* by Emerson (1952), who produced heterokaryons (haploid nuclei derived from two separate parent plants) in which parent strains differed in ability to synthesize the vitamin p-aminobenzoic acid. One parent strain grew slowly because it synthesized too little of the vitamin, while the second strain synthesized a deleterious excess of the same substance. The balanced amount in the heterokaryon promoted optimum growth. Thus, superior fitness is not synonymous with overdominance. At the molecular level, allozyme heterozygotes may display some property with greater fitness value than either homozygote form, but that property need not be derived from extreme over- or underproduction of a substance. Some notable documentation of molecular heterozygote advantage is presented in the following references:

1. The literature on increased functional diversity of protein variants has been summarized through 1968 by Manwell and Baker (1970). In protein molecules that are dimers or higher-order (multimers) associations of subunits, heteromeric molecules may form in heterozygotes with greater stability over more conditions than homomeric molecules. For example, hybrid hemoglobin in certain hybrid freshwater

sunfish produced a better fit of subunits in the hemoglobin molecule resulting in stronger haeme-haeme interactions and thus improved gas transport than homozygotes. In the snail *Cepaea hortensis*, the enzyme NADP-dependent isocitrate dehydrogenase displayed from four to five times as much activity in heterozygotes as in homozygotes for "fast" migrating alleles. Similarly, in vertebrate lactate dehydrogenases, which exist as tetrameric molecules, genetic hybrids may show hybrid zones in electrophoretic gels with elevated activity in conversion of pyruvate to lactate.

2. Schwartz and Laughner (1969) demonstrated a form of allelic interaction that results in increased enzyme activity in certain heterozygotes, which could be a basis for molecular heterosis. In maize, alcohol dehydrogenase (ADH) is a dimer composed of two subunits; of four allozymes extracted from mature kernels, three homozygous forms have subunits that are unstable at high pH, and the fourth is stable but has low activity on a substrate. Heterozygotes between an unstable form and a stable but inactive form have both stability and activity.
3. Koehn (1969) described allozymes of serum esterase in populations of catastomid fishes in which an allozyme having a southern distribution (warm water) has higher-temperature enzyme activity than the common allozyme from northern (cool) climates. Heterozygotes showed wider range of enzyme activity over temperatures with greater rate of activity at intermediate temperature than would be the case for either homozygote at that temperature.
4. Singh, Hubby, and Lewontin (1974) surveyed allozyme loci in *Drosophila pesudoobscura* for sensitivity to heat (60°C), and they discovered that the octanol dehydrogenase-1 (Odh-1) locus was polymorphic for resistant and sensitive alleles that share the same electrophoretic mobility in acrylamide gel. F_1 progeny from crosses between strains retained more in vitro enzyme activity after being treated for a short time (5 or 10 minutes) than the more heat-resistant parent strain. Thus, a heterosis for enzyme heat stability was demonstrated. Their results suggested that Odh-1 protein was at least a dimer and that the heterozygote probably produced a heteromer more heat resistant than either of the homomers of pure-line parents.

Further summary of molecular overdominance can be found in Johnson (1975). It is important for the student to realize that at the molecular level heterozygotes for enzyme variants have often displayed intermediate functions. Vigue and Johnson (1973), for example, found alcohol dehydrogenase allozymes to show intermediate heat stability in heterozygotes, while in activity (maximum rate of converting substrate into product) heterozygotes displayed dominance of the lower-temperature activity allozyme. Of course, intermediate kinetics or other enzyme parameters may have superior Darwinian fitness, as Haldane pointed out.

It should be obvious that a major technical problem in ascertaining molecular overdominance is to isolate a single locus from closely linked fitness loci that may actually be responsible for the heterotic effect. A technique used to circumvent the problem employs inducing numerous mutations throughout the genome by mutagens to increase heterozygosity at random loci. This technique was used first by Wallace and then by many others including Mukai and his colleagues (Example 19-1).

At first, the experiments of Wallace, in which a small increase in viability accompanied the increase in heterozygosity by exposure to 500 r of X-rays, were contradicted by the results of Falk and Muller (see Falk, 1961; Falk and Ben-Zeev, 1966), who first found deleterious effects in induced heterozygotes but later found no significant lowering of mutant lethal heterozygotes and a very small and nonsignificant increase in viability among nonlethal heterozygotes. Mukai and his colleagues (1965–1966) helped to reconcile the apparent contrast in results of new mutant heterozygotes when they tested for viability on homozygous or on heterozygous genetic backgrounds. If the genic background was homozygous when irradiated, induced mutations improved viability as Wallace had found; while if the background was heterozygous, induced mutations were neutral or deleterious. Therefore, the level of heterozygosity itself may be beneficial to a point, but increasing mutagenic disturbance of that optimal level could be detrimental.

Complementation and favorable epistatic effects on fitness, as well as single-locus overdominance, are similar in their ultimate effect of promoting integration within individual and populational genomes. Genetic interactions, at higher levels than single-locus overdominance, undoubtedly contribute to total genome unification, as we discussed in Chapters 17 and 18.

Developmental Homeostasis

As a fitness trait, the tendency for a steady-state in development that is likely to ensue from a "buffered" ontogenetic sequence of events in development is of fundamental importance to maintenance of a uniform phenotype with cryptic genetic diversity. If the genotypes of wild populations are revealed by inbreeding or special marker techniques (Example 9-5), the phenotypic uniformity, which can merit naming as a species taxonomically, is found to be most often a facade, an end product of development arrived at presumably via diverse genetic pathways. Individuals within a population present to the observer what has been called the "adaptive norm" (Dobzhansky, from Schmalhausen, 1949), which is reflected in taxonomic keys. The adaptive norm is a well-adapted, more or less stable complex of diverse genotypes within the population. As pointed out by Mather (1953) and Thoday (1955), developmental flexibility is the capacity of the individual to develop specifically via internally buffered sequences of processes to counteract the upsetting tendencies of environmental stresses. The implication is not that such developmental buffering brings a rigid stability to development; on the contrary, as stressed by Dobzhansky, "homeostatic changes enable the body to follow its normal developmental paths, but they do not necessarily result in morphological stability. . . . One of the most important kinds of homeostatic reactions is that which makes the organism react to environmental change by switching from one historically determined developmental path to an alternative path" permitting life and development of the organism to proceed unimpeded by outside distrubances.

Example 19-2 illustrates the buffering function of chromosomal heterozygotes over homozygotes as demonstrated in *Drosophila pseudoobscura* under various conditions of food and temperature (macroenvironments) as well as uncontrolled random environmental forces (microenvironments). Random pairs of different wild chromosomes (heterozygotes) gave more uniform survival despite environmental diversity than flies homozygous for wild chromosomes. Dobzhansky and his colleagues proposed as an extension of the coadaptation principle that homeostatic adjustments are more common to heterozygotes than to homozygotes; thus, genotypes in Mendelian populations would be further integrated by the autoregulation of development. Lerner (1954) and Waddington (1957) amassed much evidence in favor of that view. Even when homozygotes from outcrossing species are selected for higher-than-average viability, they are less well buffered to withstand environmental diversity. Coadapted chromosomes "are more often many-sided and versatile in their adaptedness, hence able to live successfully in a broader range of environments" than homozygotes, which tend to be "narrow specialists" if they do well at all in a restricted range of environments (Dobzhansky and Levene, 1955).

Developmental homeostasis in plant heterozygotes has been demonstrated in outcrossing species such as the primrose (*Primula*) by Mather (1950). In contrast, for plants such as highly inbred species of grasses, cotton, and tobacco, outcrossing of homozygous lines does not decrease phenotypic variability (Lerner, 1954; Grant, 1975). As pointed out by Grant (1975), Stebbins (1974), and others, homeostasis does not always appear to be positively correlated with heterozygosity in plants, but instead with the genetic structure of the population based on its breeding system. The plant species that tend to self-pollinate gain no homeostasis from outcrossing. In addition, plants tend to adjust their growth habits and gross morphology under environmental conditions more flexibly than do animals in general (see Bradshaw, 1972). Versatility and buffering during preadult stages of growth is more critical for animals, while plants are more flexible in modifying themselves with a less limiting growth system. Thus, the relationship between the "mere diversity of alleles" (Lerner, 1954), coadaptation, and superior homeostasis of genotypes depends on the reproductive biology of the population and species of organism.

Developmental buffering of gene-environment interactions during growth, or a balance between a flexible ontogenetic process and an inflexible final phenotype to insure maximal fitness in a wide variety of environments, has been described by Waddington (1957) under the term *canalization* as one of the products of stabilizing selection, the selection of individuals near the mean of a population. Such selection may lead to two different responses: (1) elimination of genotypes that "render the developing individual sensitive to the potentially disturbing effects of environmental stresses" (improved buffering of development) and (2) elimination of genotypes leading to abnormalities or extreme phenotypes (normalizing selection). It would seem that selection of parents near the population mean might involve both responses to some extent, but it may only produce the normalizing responses if the environment is unchanging or if no specific selection is performed to reduce phenotypic variance produced by environmental fluctuation or diversity. The general consequences of selection for phenotypic intermediates (Robertson, 1956) with elimination of extreme deviants include fixation of alleles in homozygous balanced state

$$\left(\frac{+ \; - \; + \; - \cdots}{+ \; - \; + \; - \cdots}\right).$$

Such a response amounts to a linear one toward the mean of the phenotypic distribution, and genetic variance will be reduced by selection. On the other hand, if extreme phenotypes have low fitness *because* they are homozygotes and heterozygotes are not only intermediate but also highest in fitness, genetic variability will be preserved in a balanced state. Consequently, normalizing selection outcome depends entirely on the relationship between the particular genotype and phenotype of highest fitness (see A. Robertson, 1955, 1967).

Selection for canalized development has had some success, but at the end of selection the evidence points more toward a homozygous end product genotype than a heterozygous balanced one. The aim of selection for increased canalization has been either (1) to decrease the difference between subdivided cultures raised under dissimilar macroenvironmental conditions (for example, warm or cool) or (2) to decrease the amount of phenotypic variance due to within-culture microenvironmental or unpredictable conditions. To illustrate the first, Waddington and colleagues (1960, 1966) found selection effective for the least difference in number of ommatidia (eye facets) in Bar-eyed *Drosophila melanogaster* raised at warm (25°C) and cool (18°C) temperatures. Cool temperatures tend to increase facet number and size of the eye compared with warm temperatures. Using family selection techniques each generation, subdividing the progeny within lines, raising half at each temperature, and choosing the progenies that gave the least difference when subdivided (greatest canalization), Waddington was able to reduce the difference from about 50 facets to less than 15 on the average over six generations. His anticanalization lines did not respond as well in the opposite direction. However, microenvironmental variation (measured by the coefficient of variation) within selected lines increased in both canalized and anticanalized lines over the foundation stocks. Although the variation in the latter lines was greater, no clear improvement in microenvironmental homeostasis was demonstrated.

To illustrate the selective response to microenvironmental factors, an extensive series of experiments begun in 1959 was described and summarized by Rendel in a book (1967) that outlines much of a model for genetic control of canalized development. By choosing a very constant phenotype (four scutellar bristles) in *D. melanogaster* and a mutant (*scute*^{1}) that reduces the number of bristles to two, these investigators were able not only to change the mean number by selection but also to change the distribution of flies having particular bristle numbers. By selecting for low-variance progenies measured in probit units (proportion of flies with specified frequencies of each bristle number), they achieved canalized wild type with two bristles and *scute*^{1} mutant flies with four bristles. They also had some success at increasing variance in anticanalization lines. Finally, they did find a correlation between sensitivity to macroenvironmental conditions (temperatures from 18°C to 30°C) and the phenotypic variance due to microenvironmental conditions. Thus, developmental buffering was achievable, and the authors pointed out that they had probably produced responses in two different ontogenetic systems—a normalizing system (low variance under small environmental changes) and a canalizing system (least difference between specified environmental conditions).

Genotypes with sensitivity to particular environments (lower thresholds for expression of specific phenotypes) can be revealed by employing environmental stress conditions (changing heat, changing salt content of food, adding other stressful substances, or limiting nutritional requirements) or by incorporating major gene mutations (*Bar eye* or *scute*) that lower the threshold of the developmentally buffered "wild type." Once revealed by such stresses, underlying genetic variation modifying a trait can be selected up to a plateau of response; often when the organism is transferred out of the stress situation (or major gene removed), the selected phenotype may be retained as an "assimilated" genotype (Waddington, 1961). Milkman (1964) points out that the "degree and variety of response of all phenes to genetic variation is dependent on their developmental stability." He was able to "recruit" genotypes to modify posterior crossvein formation in the drosophila wing by using heat treatment on early pupae as an efficient stress to reveal crossvein defects; by selecting those adults with greatest defects after the treatment, Milkman isolated a number of selected lines that had become decanalized for crossvein-making ability. But the trait of making normal crossveins was affected not only by these "recruited genotypes" but also by a considerable number of other modifiers, many of which did not require heat shock to enhance the penetrance of the crossveinless trait. Thus Milkman (1970) was able to show that the wild-type trait (normal crossveins) is controlled and affected in common by a set of diverse functional units. The large number of these units making up the genetic system reinforce each other in making the phenotype canalized.

To sum up stabilizing selection, phenotypic variance may be decreased via at least three avenues: lowered variance to (1) macro- or (2) microenvironmental conditions or simply (3) by an optimum intermediate phenotype. In most selection experiments analyzed genetically, the end product of any of these three avenues is more homozygous than the base population, a result that contrasts with the well-known homeostatic tendencies of heterozygotes in outcrossing populations. It is possible that a technique such as reciprocal recurrent selection for the *combining abilities* of haploid genomes might attain a stabilized and canalized heterozygous goal (Kojima and Kelleher, 1963), although no laboratory experiments have yet achieved what could be described as a well-buffered heterozygote product. What these selection experiments have done is to increase considerably our knowledge of the ubiquity and variety of genetic systems capable of achieving diverse goals that might be expected to have high fitness in nature. Just how stabilizing selection acts to achieve a highly homeostatic heterozygous genotype in natural populations is not yet understood in any detail.

Populational Homeostasis

One of the major cornerstones in the concept of a population's genome as an integrated system is derived from the tendency of genotype and phenotype frequencies to return to a steady state and constant level after an abrupt change in the genetic system has occurred. When artificial selection is superimposed on a well-integrated genetic system, it is commonly observed that the selected lines undergo inbreeding depression. In fact, the most general correlated response to selection is the reduction in fitness following the selection of any metric trait. By relaxing the selection regimen and sacrificing the gains made under selection, an investigator can usually restore normal fitness. As documented by Lerner (1954, 1958), Mather (Example 7-2), Dobzhansky and Spassky (Example 7-4), and A. Robertson (1967), attempts to shift a phenotypic balance by artificial selection are usually resisted by the genotypic system of the base population from which the selection was initiated. Genetic combinations, if predominantly existing in a state of relational balance with potential variability bound in linkages and epistatic combinations, will tend to be restored under natural selection once the artificial regimen or disturbance has been removed. This principle is illustrated in Example 19-3.

Once new balanced genetic combinations can be released by recombination, free genetic variation is available for natural (or artificial) selection to proceed for further change. For the plant or animal breeder, there is a conflict between the necessity to fix useful traits by inbreeding and selection and the need to maintain sufficient variation for fitness (if high fitness in the genetic system is dependent on allelic and nonallelic interactions).

Natural selection probably forces a population to compromise and obtain maximum fitness for the net effects of all its properties and capabilities. Too high an attainment of one fitness property (or trait under selection) may easily bring about negative correlations in other traits (A. Robertson, 1955). A resistance to change is therefore demonstrated when attempting to perturbate a genetic system—a homeostasis at the population level.

Further Stabilizing Systems

Epistatic interaction between portions of a total genome was discussed at some length in Chapter 17. Special combining abilities of alleles from diverse loci can lead to an immense variety of phenotypes with fitness values depending on genic action, behavioral expression, and environmental conditions in which a species exists or is given the opportunity to exist. All such interactions constitute internal mechanisms promoting integration of the population's composite set of genotypes as a unit, expressed by Dobzhansky's "coadapted genic complex."

In Chapter 16, several mechanisms promoting stable genetic equilibrium were explored, including frequency-dependent selection and environmental heterogeneity. Ecological and behavioral relationships between genotypes are likely to be reinforced by any balanced mutualism or facilitation among coexisting forms. Even a simple reduction in competition (lowered interference) between forms may lead to such mutually beneficial interaction that genotypes could evolve to reinforce the lowered competition by their coexistence so that stable polymorphism could result (see Harper, 1968; Beardmore, 1970; Lewontin and Matsuo, 1963; Sokal and Karten, 1964). All these mechanisms would tend to promote establishment of genetic systems for perpetuation of cohesive properties within local populations.

EXAMPLE 19-1

ARE NEWLY INDUCED MUTATIONS OVERDOMINANT IN VIABILITY?

Wallace (1958*a*, 1959) irradiated sperm carrying quasi-normal homozygous second chromosomes obtained from a large experimental population of *Drosophila melanogaster* with 500 r of X-rays. He then tested the heterozygous effects on viability of newly induced mutations in the second chromosome, and he obtained a random array of viability mutations with an average positive increment of 1.5 percent on the viability of heterozygotes compared with unirradiated controls.

On the other hand, Falk (1961) irradiated spermatogonia carrying quasi-normal third chromosomes of the same species with 4000 r of X-rays and examined effects of induced mutations. Contrary to the results of Wallace, reduction of viability was found from 0 to 3 percent in heterozygotes.

Wallace (1963) then extended his tests to include different X-ray dosages, and viability was tested in flies with heterozygous as well as homozygous genetic backgrounds. The experiments with homozygous backgrounds agreed with his previous results, but the viability was slightly depressed when mutations were newly induced into heterozygous backgrounds.

	Homozygous genetic background	Heterozygous genetic background
G_0	$\frac{Cy\ +_A}{Pm\ +_A} \times \frac{+'_A\ Sb'}{+'_A\ Ubx'}$ R	$\frac{Cy\ +_A}{Pm\ +_A} \times \frac{+'_A\ Sb'}{+'_A\ Ubx'}$ R
G_1	$\frac{Cy\ +_A}{Pm\ +_A} \times \frac{Cy\ Sb'}{+'_A\ +_A}$ (1)	$\frac{Cy\ +_A}{Pm\ +_A} \times \frac{Cy\ Sb'}{+'_A\ +_A}$ (1)
G_2	$\frac{+_A\ +_A}{+_A\ +_A} \times \frac{Cy\ +_A}{+'_A\ +_A}$	$\frac{+_B\ +_A}{+_B\ +_A} \times \frac{Cy\ +_A}{+'_A\ +_A}$
G_3	$\frac{Cy}{+_A}$ versus $\frac{+_A}{+'_A}$	$\frac{Cy}{+_B}$ versus $\frac{+'_A}{+_B}$

Figure A. Mating scheme for testing the viability of heterozygotes containing a second chromosome irradiated (marked with prime ') and an unirradiated otherwise isogenic second chromosome in G_3. Symbols: $+_B$ represents a different genetic background from $+_A$. Cy = Curly wing (with inversion), Pm = Plum, on chromosome 2; Sb = Stubble bristles, Ubx = Ultrabithorax, on chromosome 3. From Mukai et al., 1966.

In view of these discrepancies and the implication of genetic backgrounds, Mukai, Yoshikawa, and Sano (1966) carried out an extensive series of tests particularly to control the genetic backgrounds of irradiated flies with respect to the second chromosome of *D. melanogaster*. Flies were first made isogenic for chromosomes 2 and 3; then males were X-rayed with 500 r and mated immediately to obtain sperm stage mutants. Figure A from the Mukai, et al., paper illustrates on the left the mating scheme for producing flies that were wild type but contained one second chromosome that had been irradiated (marked with a prime ') and one unirradiated and identical to its homologue except for induced mutations. The right side of Figure A shows the introduction of a second chromosome from a different stock (B) in the G_3 so that wild-type flies in G_4 will have an irradiated chromosome paired with unirradiated homologue from strain B. Controls consisted of wild-type flies from completely unirradiated G_0 parent flies.

Table A summarizes the main data from Mukai's experiments. Mean viability "on an individual basis" refers to *per fly* within a culture, while "on a chromosome line basis" refers to the sum of fly counts over four cultures pooled. The positive differences for irradiated chromosomes minus the control with homozygous (isogenic) background are both significant (AA versus AA'), while for the heterozygous background (AB versus $A'B$) the difference on an individual basis is not significantly less, although on a chromosome line basis it is significantly less. Thus, induced mutations were slightly beneficial in an otherwise homozygous background directly extracted from a natural population at equilibrium.

Mukai, et al., also utilized a base stock derived as a multihybrid from a large number of unrelated wild-type strains around the world. Induced mutations were either neutral or at least not heterotic in chromosomes derived from the multihybrid. When induced mutations

TABLE A *Heterozygous effects of radiation-induced mutations on viability, ± s.e. A refers to a wild-type strain derived from a natural population. B refers to an unrelated strain extracted from a laboratory stock. Primes (′) designate irradiated second chromosomes. From Mukai, Yoshikawa, and Sano (1966)*

Genotype Derivation	Number of Chromosome Lines Tested	Average Number Flies Emrged per Chromosome Line	Average Viability per Individual
AA (control)	287	766.63	1.0135
AA′	286	789.74	1.0437
Difference (experimental − control)		+23.11 ± 12.89*	+0.0302 ± 0.0103**
AB (control)	317	1211.53	1.1556
A′B	313	1180.35	1.1549
Difference (experimental − control)		−31.18 ± 15.86*	−0.0007 ± 0.0102 (n.s.)

* Significant at the 5 percent level.
** Significant at the 1 percent level.

were incorporated as heterozygotes into any sort of heterologous crosses (intra- or interpopulational), they seemed slightly deterimental to viability.

Thus, it was proposed that overdominance of newly induced mutants is manifested only when the genetic background of a natural population is almost completely homozygous, but manifestation becomes nearly impossible for this trait of viability when a slight increment is made in the number of heterozygous loci over already coadapted heterozygous background.

EXAMPLE 19-2

MICRO- AND MACROENVIRONMENTAL FACTORS IN DEVELOPMENTAL HOMEOSTASIS

Environmental variance of traits (both morphological traits and fitness) has often been demonstrated to be lower in heterozygotes than in homozygotes. Dobzhansky and Wallace (1953) compared viabilities of homozygotes and heterozygotes for certain chromosomes from populations of four species of drosophila, and in all four the mean viability of homozygotes was not only consistently lower but also more variable than that of heterozygotes. The hypothesis of genome coadaptation within gene pools of Mendelian populations emerged from that study.

In 1955, Dobzhansky and Levene described more extensively the ways in which environmental conditions could bring about phenotypic variation (viability) on strains of *Drosophila pseudoobscura* taken from a natural population and made homozygous or heterozygous for the second chromosome. Ten strains were chosen with normal or supervital viability (denoted as *H* for "high") and nine strains with subvital viability (denoted as

L for "low") when grown at 25°C with baker's yeast (*Saccharomyces cerevisiae* (*S*)) supplied to the food medium. For the *macroenvironmental* tests, conditions were varied by supplying two other species of yeast (*Klockeraspora* (*K*) and *Zygosaccharomyces* (*Z*)) as well as baker's yeast and also raising the flies in from two to four temperatures: 27°, 25°, 16°, 4°C. Specifically, the macroenvironments were as follows:

Yeast	Temperature (°C)
S	27, 25, 16, 4
K	27, 25, 16
Z	27, 25

Six replicate cultures of each chromosomal combination were made within each environmental condition (yeast × temperature). Replicate cultures vary because of slightly different amounts of food, degrees of crowding among larvae, amounts of microorganisms, and other unpredictable factors. Thus, variance in viability between replicates within chromosome strain combinations and macroenvironments was a measure of microenvironmental homeostasis. However, sampling error is also a component of variance between replicates. Since viability was measured as a proportion of wild-type flies emerging out of cultures segregating for a dominant marker mutant (as in Example 9-5), the binomial sampling variance can be computed as a fixed component of the observed variance between replicates—$V_O = V_B + V_M$, where V_O is observed variance between six replicates within a chromosomal combination and environment, V_B is binomial sampling variance, and V_M is the "real" component of variance due to unpredictable small differences between replicate cultures.

TABLE A *Mean variance for components of viability variation for all chromosomal combinations*

Microenvironmental variance (V) between replicate cultures raised under nine environmental conditions

	Homozygotes		Heterozygotes		
	L Strains	*H* Strains	$L \times L$	$L \times H$	$H \times H$
Mean V	18.20	13.81	3.53	5.00	5.46
± *s.e.*	±3.88	±1.96	±1.45	±1.59	±1.89

Macroenvironmental variance (V) between series of cultures raised in nine different environments

	Homozygotes		Heterozygotes		
	L Strains	*H* Strains	$L \times L$	$L \times H$	$H \times H$
Mean V	37.40	9.14	4.36	2.10	4.46
V-range for chromosomal strains and combinations	2.1–156.4	0.2–27.9	1.2–8.3	0–6.2	2.6–10.5

The V_M component can be estimated by subtraction after calculating V_B as follows:

If P_i = observed percentage of wild type (out of total including marker flies counted) in the ith culture.

n_i = number of flies in the ith culture counted (usually a constant sample size for each of six replicates) and

$\bar{P} = \sum P_i/6$, as unweighted mean percentage of wild type, then

$$V_B = \left(\sum \frac{\bar{P}(100 - \bar{P})}{n_i}\right) \Big/ 6$$

For the macroenvironmental variance, the combined replicate cultures of a single chromosome combination were treated as a unit within each environment, and differences between environments constituted the main analysis.

Table A summarizes the mean variances for microenvironmental factors (upper) and for macroenvironments (lower). Homozygotes for chromosomal strains selected either for subvitality (L) or normal-supervitality at 25°C (H) were consistently more variable, or less homeostatic, than heterozygotes in both aspects of environmental subdivision.

EXAMPLE 19-3

A HETEROTIC MODEL OF POPULATIONAL HOMEOSTASIS

A model of populational homeostasis is based on that given by Lerner (1954). For simplicity of fitness values, we illustrate a heterotic model, but the student might consider generalization to include any genetic system based on fitnesses leading to a polymorphic stable equilibrium with or without complex genetic interactions of nonallelic sort. Extensions of this illustration to such complex systems as those envisaged by Mather's relational balance concept and Lewontin's (1974) selection of entire genomes should be intuitively obvious.

Assume a quantitative trait (Y) to be determined by additive alleles at two loci (A and B) that are tightly linked. First we assume A-B never recombine. Y could be a trait desirable for improvement under artificial selection by a breeder (for example, high vitamin content in corn or high milk production in cattle), or it could be bristle number in flies, but it is not obviously related to fitness under natural selection. However, we shall further assume that genotypes controlling the Y trait either do have fitness values with heterosis in heterozygotes or that they are completely linked to genes that do have these fitness values. Assume that the fitness values of the A locus are multiplicative when combined with locus B. Let each substitution of A for a add two units to trait Y, so that if $aa = 0$, $Aa = 2$, $AA = 4$; and let each substitution of B for b add one unit to trait Y over minimum genotype $aabb$, so that if $bb = 0$, $Bb = 1$, $BB = 2$. Thus, $AABB = 6$, etc.

Let net fitness (W_i) of each genotype within each locus be as follows:

$$\begin{array}{ll} AA = 1.0 & BB = 1.0 \\ Aa = 1.1 & Bb = 1.2 \\ aa = 0.8 & bb = 0.6 \end{array}$$

We shall define these W's as the amount of reproductive replacement each genotype makes to the following generation (that is, normalized fitness values when dividing each relative fitness by $\bar{W}$ in any generation). Nine genotypes for both loci with their fitnesses (W) as products of the separate loci and their trait values (Y) would be as in the following (W = upper value, Y = lower value):

		BB	Bb	bb	Average $W(A)$
AA	W =	1.00	1.20	0.60	1.0
	Y =	6	5	4	
Aa	W =	1.10	1.32	0.66	1.1
	Y =	4	3	2	
aa	W =	0.80	0.96	0.48	0.8
	Y =	2	1	0	
Average $W(B)$		1.0	1.2	0.6	

If linkage is completely in repulsion (*trans* phase) so that only the three genotypes occur (Ab/Ab, Ab/aB, and aB/aB) under natural conditions (diagonal from upper right to lower left) as a case of relational balance, a stable equilibrium will exist with the following frequencies and Y trait values (formula (14-15) gives $\hat{q}_{aB} = 0.58$):

Genotype	Frequency	Y
Ab/Ab	0.1764	4
Ab/aB	0.4872	3
aB/aB	0.3364	2

so that trait Y mean ($\bar{Y}$) is 2.84 and net reproductive replacement ($\bar{W}$) is 1.02. Thus, the population will have a slight healthy 2 percent increase each generation.

Suppose the breeder wishes to select for an increase in trait Y. If reproductive rates remain the same, selection will bring about depression of fitness because the Ab/Ab genotype has only a fitness of 0.6, and the population size will drop to the point of extinction. The penalty attached to creating an increase in Y lies in the fact that reproductive potential cannot be maintained unless there are a few heterozygotes present. A maximum Y value of 6 could never be attained as long as the linkage remains constant and tight. Therefore, some Ab/aB individuals are essential to keep up the required reproductive level. The student should verify that if only Ab/Ab and Ab/aB genotypes were preserved (maximizing Y for these limitations), the reproductive fitness would be 1.00 (constant N) and an average for the selected trait ($\bar{Y}$) up to 3.45. If those two genotypes were allowed to mate at random, their progeny would have $\bar{Y} = 3.44$, which would be the highest average $\bar{Y}$ attainable. However, what would happen if the artificial selection regimen were to be relaxed? $\bar{Y}$ would quickly fall back to the value of 2.84, which was its value under natural selection.

Now let us remove the limitation on crossing over and assume that some recombination takes place between the A and B loci, producing a few coupling (*cis*) linkages. Since there is no epistasis, the student can easily verify the equilibrium state as $\hat{q}_a = \hat{v}_b = 0.25$, and the new $\bar{Y} = 4.50$ with $\bar{W} = 1.08$ under natural selection. The conversion of potential variation into free has released more genotypes and has allowed not only a natural increase in fitness (reproductive replacement) but also an increase in the selectable trait Y without any directed artificial selection being applied. Finally, of course, the desired maximum value of $Y = 6$ could be attained, although the fitness would fall slightly to $\bar{W} = 1.00$.

CONSEQUENCES OF MIXING GENE POOLS

If integration of a local populational genetic system exists, conceivably a demonstration of its existence could be made by upsetting its integrity. Certainly, it would be unlikely to find two gene pools of separate populations from the same species with identical genetic systems, although they could be mutually compatible. By hybridizing populations or by allowing a small migration of genetic material to flow into one population from another, the balance (coadaptation, if it exists) within a system might be upset sufficiently to recognize as a breakdown of coadaptation if the two systems were not mutually compatible.

Dobzhansky and Pavlovsky demonstrated a loss in chromosomal balanced polymorphism when gene arrangements of *Drosophila pseudoobscura* from widely separate localities were mixed in population cages, as in Example 15-3. An extension of those experiments was devised by Brncic (1954), who described the changes in viability produced after mixing identical chromosomal arrangements from separate gene pools; although first-generation hybrids were luxuriant, the second and third generations following inbreeding were depressed in viability compared with parental viability averages. This breakdown of high within-population fitness was shown to be a result of intrachromosomal recombination between chromosomes of diverse origin, with the lowest viability being found in the individuals with both chromosomes as a pair of crossover products. Whatever internal balance had been selected within the two respective localities for high fitness was thereby upset.

Wallace and Vetukhiv (1955) proposed that the entire gene pool is highly integrated in a hierarchical manner derived from lower orders of integration based on overdominant loci, complementation, homeostasis, and epistatic interactions. Vetukhiv extended Brncic's work to many other fitness properties besides viability in three drosophila species (*pseudoobscura*, *willistoni*, and *paulistorum*), and he showed that all properties tested displayed hybrid breakdown after recombination of gene pools. Wallace's experiments with *D. melanogaster* described in Example 19-4 give an overall view of the expectation from the mixing of integrated gene pools and subsequent recombination between them. The amount of heterozygosity defined as the proportional amount of the genome taken from diverse populations was covaried with the amount of recombination at three levels defined as the amount of "derivation" (whole haploid sets of chromosomes from natural populations and/or crossover chromosomes from interpopulation hybrids). Three important facts emerge from those experiments: (1) interpopulational hybrids in the first generation had higher viabilities than did flies obtained by crossing strains within populations; (2) among the classes that were 100 percent heterozygous were found the highest and the lowest viabilities of the entire experiment, so that clearly the highest viability is not simply conferred by heterozygosity alone; (3) increasing derivation (recombination) between populational genomes tends to decrease viability, as would be expected if balanced complexes were breaking down when recombining.

It is by no means always the case, however, that interpopulational crosses display the features found by Wallace and Vetukhiv; we may learn more of the genetic system and internal balance necessary as conditions for displaying coadaptation and hybrid breakdown from two experiments with contrasting base populations. For example, J. C. King (1955) had lines of *D. melanogaster* selected for resistance to DDT from a certain base population (Syosset, N.Y.) that had not been coadapted or consciously made so in any way. Intercrossing different resistance-selected lines produced no increase in resistance among F_1 progeny but, surprisingly, all his F_2 progeny (from inbreeding F_1's) had lower mean tolerance to DDT and increased variance; these facts were interpreted as evidence that the selected lines had become integrated gene pools for efficient resistance as well as line fertility. On inbreeding, the F_2's subsequent generations (F_3 and F_4) regained the high resistance of the original lines and F_1. Thus, the "breakdown" was of very short duration, but the fact that it had occurred in the first place was evidence that the genetic systems being selected in the different lines must have been geared for an epistatic expression of DDT resistance that broke down after some recombination. Presumably, those flies utilized for propagation subsequently must have preserved more of the selected genetic combinations and not become recombined further. In contrast, Crow (1957) obtained DDT-resistant strains that were wholly additive and not epistatic; crossing of resistant strains and obtaining recombinant progeny did not decrease resistance but simply displayed the expectation from approximately additive systems of polygenes. Crow suggested that his strains, which had been selected from a large genetically heterogeneous population, consisted of genotypes that increase resistance in a large number of diverse background genotypes ("good mixer" resistance factors). King had selected from a homogeneous base population, and his strains' genotypes conferring resistance had been selected for fitness in a much more specific background ("narrow specialists"). King's type of selection is more likely to lead to acquisition of specific epistatic interactions than Crow's. Thus, coadaptation may depend on a diversity of genetic phenomena and the selection history of each population.

A clear distinction must be made between superiority in traits that have been selected and those important to fitness. There seems to be little doubt that outcrossing selected lines or widely separated populations often is luxuriant either for the trait selected or for whatever traits are distinctive for each population. Reproductive and survival abilities are usually luxuriant as well, but fitness as defined by any population's adaptedness is another matter. While some fitness component traits can be increased or improved by outcrossing, the total, or net, fitness for a population has to be a matter of long-term adjustment. The coadapted genetic system probably does not incorporate more superiority of one trait than a compromise of fitness traits will allow for the conditions under which the population exists. The experiments of Wallace in Example 19-4 indicate that a distinction must be made between heterozygosity and genome balance attained by "accidental outcrossing" and that attained by selection, "relational balance," and long-term adjustment. The former may be luxuriant for a number of traits, and the luxuriance may be utilized and retained to some extent in the coadaptation process, especially if a hybrid population is to persist. The outcome of allowing such a hybrid population to survive in any case will probably be indeterminate (as in Example 15-3), because the particular genotypic combinations that emerge and their adaptive functions in the environment where the hybrids find themselves will be enormously complex. A population whose genetic system is coadapted with a selection history should have a determinate trajectory through time if its genotypes are altered or perturbated by artificial means. After an intercrossing of balanced systems, finding lowered fitness recombinants favors the hypothesis that a high level integration must exist within each gene pool. But finding the opposite (highly fit recombinants) does not disprove that the gene pools are integrated.

Finally, we may cite evidence that a coadapted gene pool may be efficient at eliminating specific genetic variants introduced into the population that tend to lower the population's adaptedness. A number of experiments testing whether particular mutants are more likely to be retained at quasi-stable equilibrium or to be eliminated slowly or rapidly from a population indicate that elimination of the mutant is more likely and at faster rates when introduced into polymorphic and naturally coadapted genetic backgrounds than if introduced into monomorphic or unrelated (noncoadapted) genetic backgrounds (Polivanov, 1964; Merrell, 1965; Wallace, 1966*b*; Chung, 1967). It is likely that coadaptation and greater polymorphism for alleles and epistatic combinations express their cohesive quality by eliminating "foreign" genetic material efficiently. Such action would tend to preserve the homeostatic ability of the population as well. It would also be an important factor in counteracting disruptive effects of gene flow between populations in nature.

EXAMPLE 19-4

INTEGRATION OF THE GENOTYPE AND RECOMBINATION BETWEEN GENOMES

Wallace (1955) designed experiments with *Drosophila melanogaster* to measure a component of fitness (preadult viability) under three levels of recombination along with varying amounts of "heterozygosity" (defined as proportional amounts of loci with alleles from diverse populations). If integration of the genome was necessary for high fitness, a component of fitness such as preadult viability would be expected to become reduced increasingly with greater amounts of recombination between portions of the genome taken from two or more unrelated populations. At the same time, the amount of interpopulational diversity ("heterozygosity") was controlled by using parent flies with suitable derivation from particular populations; that is, G_1 progeny from a cross between females from population 1 × males from population 2 would be maximally "heterozygous," while their G_2 progeny would contain intrachromosomal (from crossing over in G_1 females) plus interchromosomal (from assortment of whole chromosomes in G_1 males) recombinants. Thus, Wallace sought to separate the amount of vigor obtained by populational hybridity from that obtained by integration of the genome within a natural population.

We describe here a summary of Wallace's experiment II, in which he designed matings between flies descended from five widely separated natural populations just before the start of the experiments: (1) Blacksburg, Virginia, (2) Riverside, California, (3) Santiago, Chile, (4) Jerusalem, Israel, and (5) Syosset, New York. Parent flies were grown under regular yeasted food medium, but viability under low nutritional stress was measured by transferring first-instar larvae (25 larvae per test vial) into small vials containing 2 ml of molasses-agar food without supplementing the food with yeast, although some yeast accompanied the larvae in transfer. For each cross between or within populations, 12 vials of 25 larvae per vial (300 larvae) were initiated, and 150 crosses total were done.

Each type of mating within or between populations is outlined in Table A. We let each digit represent the populational derivation (from 1 to 5): thus, $\frac{1}{1} \times \frac{2}{2}$ represents any female from population 1 crossed to a male from population 2, or any female from any other population outcrossed to a male from a different population. Their G_1 progeny larvae tested will be $\frac{1}{2}; \frac{1}{2}$ represents populational "heterozygosity" throughout the lengths of both major autosomes (chromosome 2 on the left of the semicolon and chromosome 3 on the right). If a G_1 male from the cross of population 2 $\times$ population 3 (that is, $\frac{2}{3}$) is then crossed to a female from population 1, progeny larvae would all be "heterozygous" but there would be assortment of whole major autosomes so that half of the time, for example, the second and third chromosomes would be derived from different original populations ($\frac{1}{2}; \frac{1}{3}$). Finally, if a heterozygous female is used, progeny will have their matroclinous chromosomes recombined by crossing over between loci from two diverse populations. If the male parent is from a third population, progeny will be on the average with 50 percent recombined chromosomes, 100 percent heterozygous, and haploid sets not reassorted ((1-2)/3; (1-2)/3). In the middle column of Table A, when the male parent is from one of the same populations as the female's immediate ancestor, half of the loci will be heterozygous ((1-2)/1; (1-2)/1). The

TABLE A *Chromosomal origin by population designated by numbers.*

Intrachromosomal Recombined	"Heterozygous" for populational genome: 100% — Parents ♀ × ♂	100% — Tested	50% — Parents	50% — Tested	0%
0% N	$\frac{1}{1} \times \frac{2}{2} \rightarrow$	$\frac{1}{2}; \frac{1}{2}$			$\frac{1a}{1b} \times \frac{1c}{1d}$—all population 1 ⋮
D	$\frac{1}{1} \times \frac{2}{3} \rightarrow$	$\frac{1}{2}; \frac{1}{3}$			$\frac{5a}{5b} \times \frac{5c}{5d}$—all population 5
50% N	$\frac{1}{2} \times \frac{3}{3} \rightarrow$	$\frac{1\text{-}2}{3}; \frac{1\text{-}2}{3}$	$\frac{1}{2} \times \frac{1}{1} \rightarrow$	$\frac{1\text{-}2}{1}; \frac{1\text{-}2}{1}$	
D	$\frac{1}{2} \times \frac{3}{4} \rightarrow$	$\frac{1\text{-}2}{3}; \frac{1\text{-}2}{4}$	$\frac{1}{2} \times \frac{1}{2} \rightarrow$	$\frac{1\text{-}2}{1}; \frac{1\text{-}2}{2}$	
100% N	$\frac{1\text{-}2}{1\text{-}2} \times \frac{3\text{-}4}{3\text{-}4} \rightarrow$ (F_2♂)	$\frac{1\text{-}2}{3\text{-}4}; \frac{1\text{-}2}{3\text{-}4}$			

N = natural chromosomes *not* reassorted.
D = derived interchromosomal reassortment of haploid sets.
(1-2)/3; (1-2)/4 = left of semicolon autosome pair 2; right autosome pair 3.
1-2 = crossed over between populations 1 and 2.

TABLE B *Mean number of flies emerging (out of 300) per cross ± s.e.*

Intrachromosomal Recombined		"Heterozygous" for populational genome		
		100%	50%	0%
0%	*N*	171.80 ± 1.94		156.00 ± 2.95
	D	159.13 ± 1.13		
50%	*N*	165.30 ± 1.17	161.00 ± 1.52	
	D	148.43 ± 1.05	159.00 ± 1.90	
100%	*N*	146.73 ± 1.61		

remaining combinations ought to be self-evident. The maximum of recombination can occur when both parents have chromosomes recombined, but one parent is derived from two populations different from the other parent to make 100 percent heterozygosity in their progeny. At the top right corner of the table, all tested larvae have come from parents within the same population, but 12 substrains (*a*, *b*, . . . , *k*, *l*) had been maintained from each population so that extreme inbreeding would be avoided by always taking parents from different substrains. Thus, all larvae tested would be "homozygous" for populational derivation but not more homozygous genetically than random flies within any population.

Resulting average numbers of flies emerging out of 300 larvae per cross are given in Table B. Interpopulational hybrids (upper left corner) exhibit heterosis, the highest average viability under the stress of low nutrition. This average is significantly greater than that of average individuals within populations (upper right corner). As Wallace pointed out, that may amount to "the price paid by local populations for an integrated gene pool." While 100 percent heterozygous, natural and unassorted (*N*) chromosomes diminish in viability with increasing amounts of recombination (crossing over) between populational genomes. Second, with reassortment of whole haploid sets from G_1 male parents (*D*), there is also immediate significant lowering of viability. Thus, interpopulational hybrids carrying two different haploid sets are more viable than individuals carrying one set from one population but a second set from two different populations assorting at random but without crossing over. Interchromosomal epistasis then must influence viability of hybrids.

In this experiment and another parallel one, Wallace found by far the greatest portion of decrease in viability followed from recombination rather than from "homozygosity" for the intrapopulational genome. (Homozygosity in the usual sense, of course, would have brought about far more loss of viability.) However, in every case but one in each experiment, there was slight lowering in the direction toward "homozygosity." (The exception here is at the 50 percent level for both recombination and for heterozygosity, with *D* haploid sets.)

DIVERGENCE OF GENE POOLS

If a population splits into two subpopulations that become reproductively isolated from each other, their genomes will diverge. Even if their respective genetic systems had any portion of loci with alleles entirely neutral to selection, those systems would diverge purely by accidents of gametic sampling with limited population size. Since it is inconceivable that there would not be some difference in selective factors between two isolated populations, divergence due to the influence of selection as well as random genetic drift is to be expected. Complete stoppage of gene flow is not a necessary ingredient of genetic divergence, however. Population size (N_e) is no doubt restricted most often simply by the distance between likely parents. Random union of gametes over a large area cannot actually be achieved unless parents are able to disperse in high density easily; even if individuals of a population were to be evenly spaced, a random mixing of gametes independently of distance seems a theoretical ideal and not a reality. Restriction of dispersion then brings about a clustering of likely parents in what Wright has termed "neighborhoods" in his analysis of isolation by distance (1943, 1946). Neighborhood size (N_e) depends on number and density of potential parents, their orientation in groups (types of territory occupied—single strips such as shores along rivers or in more circular areas), and the system of mating. Without going into Wright's derivation, we may summarize as follows:

1. If a population has uniform density along a linear range (as along a shoreline) and if offspring from any number (n) of parents become distributed roughly as in a normal distribution, then these potential parents occupy a strip of length 2σ so that the density per unit distance is $d = n/2\sigma$, and the effective number of reproducing individuals becomes $N_e = n\sqrt{\pi} = 2\sigma d\sqrt{\pi} = 3.545\sigma d$.

2. If a population occurs over an area with the number of reproducing individuals occupying a square with 2σ on a side, the density per unit area is $d = n/4\sigma^2$, and the effective number become $N_e = 4\pi(\sigma^2)(d) = 12.566\ \sigma^2 d$, or N_e is the equivalent of the number of parents in a circle of radius 2σ. Such a circle would include 86.5 percent of the parents in the neighborhood.

Levin and Kerster (1974) used Wright's neighborhood size estimation method for a number of plant species by expanding the N_e relation in the second case above to include dissemination of both pollen and seeds. Some of their estimates include higher values of N_e ranging from 75 to 282 in *Phlox pilosa*, pollinated by Lepidoptera and with explosive capsular dissemination of seeds, to low values of N_e at 2.2 to 5.4 for *Lithospermum caroliniense*, pollinated by bees but with two kinds of flowers, outcrossed as well as selfed, and with no special seed-dispersal mechanism. Many of these estimates are pointed out by Levin and Kerster to be overestimates (and thus will underestimate the inbreeding coefficient) because local pollen is ignored and no account is taken of annual fluctuations in proportions of potentially breeding individuals.

Dispersal rate in animals is far more complex than in plants because behavior with respect to sources of food, density of individuals, reproductive cycles, and related factors prevents dispersal from being a random process. Genetic differentiation by "island neighborhoods" is thus a function of dispersal between attractive sites, as illustrated by studies done with drosophila (see Wallace, 1968*b*, 1970; Johnston and Heed, 1975, for summaries). Wright (1968) pointed out that from dispersal observed in *D. pseudoobscura* and the allelism of lethals, neighborhood size in favorable populations might include an area about 250 meters in radius (see also Dobzhansky and Powell, 1974). But the tendency of flies to cluster around food sites makes it apparent their population structure is more like the "stepping stone" than a continuously distributed population. In the opposite extreme, marine animals, such as the blue mussel, *Mytilus edulis* (Koehn, Milkman, and Mitton, 1976), spawn in great profusion into sea water; after fertilization mobile larvae are free to migrate considerable distances for many weeks. Although settling down in favorable sites for adult sedentary habit may be limiting, the population structure must approach very large size and be closer to the panmictic ideal than the cases of land plants and insects so far analyzed.

DIVERGENCE COUNTERACTED BY MIGRATION

Gene flow between populations achieves several effects: (1) introduction of alleles from outside any population is a source of new variation analogous to mutation; (2) equalizing of genetic differences between populations enlarges the effective size of each and reduces the chance of fixation for neutral alleles; (3) accommodation or adjustment on the part of a recipient population to genes external to any integrated genetic system is brought about either by utilizing the extra variability to advantage or shutting it out if it tends to lower fitness. Two simplified models of gene flow, one deterministic and the other stochastic, are discussed below.

A Simple Model of Gene Flow With Large Population Size

If migration of alleles is made possible by incorporation of individuals into a population (X) from outside population (Y), we may treat the simplest situation as follows. Let population X be an "island" group surrounded by a larger population Y. Each generation a fraction m of population X is made up of individuals that originated in population Y but have moved in and then bred in population X. The resident remainder of population X is simply $1 - m$. The outsiders displace that fraction of the recipient population's genes. Any allele with frequency q_0 in population X will have a new frequency (q_1) as follows:

$$
\begin{aligned}
q_1 &= (1 - m)q_0 + m\bar{q} \\
q_1 &= q_0 - m(q_0 - \bar{q}) \qquad (19\text{-}1)
\end{aligned}
$$

where $\bar{q}$ is the average allele frequency in the outside population Y. We may represent the model as in Figure 19-1A. Also, in a similar way, the recipient population (X) could be then considered a hybrid population (H_{X-Y}), because if the migrant individuals mate with resident ones, ensuing offspring will be hybrids, as in Figure 19-1B. Finally, if migration continues at the same rate each generation into population X (or population H_{X-Y}), as in Figure 19-1C, the difference between q in X and $\bar{q}$ in Y will approach zero—$\Delta q = q_1 - q_0$; then, from (19-1), we have

$$
\begin{aligned}
\Delta q &= m(\bar{q} - q_0) \\
&= m\bar{q} - mq_0 \qquad (19\text{-}2)
\end{aligned}
$$

where q_0 is the allele frequency in population X or in the hybrid population H_{X-Y} in any successive generation before more migrants enter. Each successive q_0 will be closer to $\bar{q}$.

Owing to the fact that q_0 in the recipient population is being displaced, we may think of formula (19-2) as analogous to that for change under forward and reverse mutation in Chapter 13 (13-10): $\Delta q = up - vq$. At equilibrium, $\Delta q = 0$, and by setting (19-2) equal to zero, we see that the recipient population will eventually have the allelic frequency of

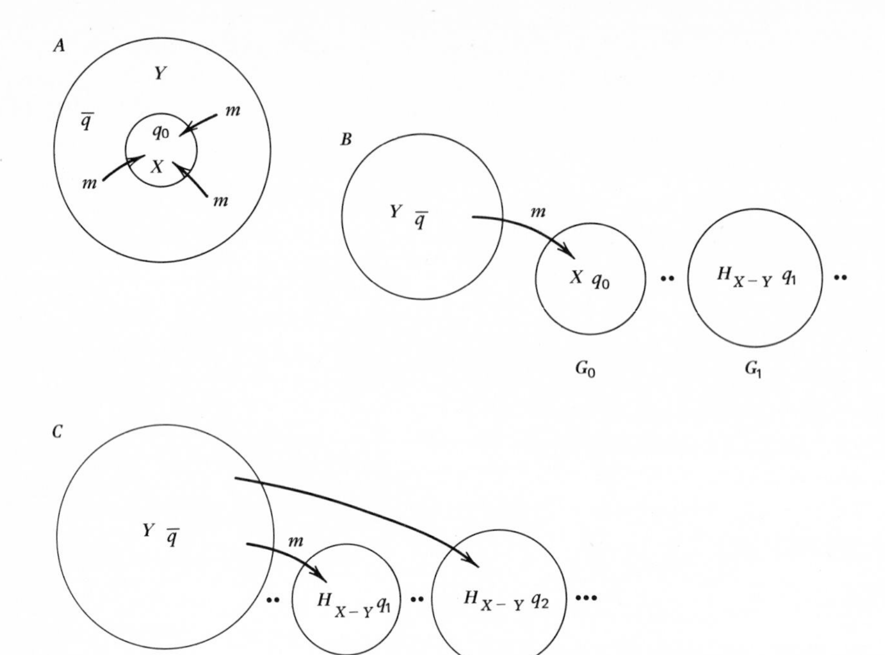

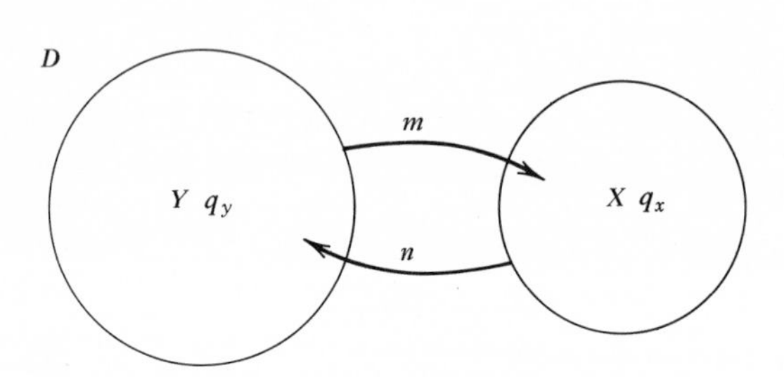

Figure 19–1. Representation of simple models of gene flow.

A. Migrants from a large population (Y) into an island smaller group (X). Allelic frequency $\overline{q}$ in the large population, q_0 initially in the small population, which incorporates the migrants m as a fraction of breeding population X.

B. Once the migrant individuals have hybridized in population X, G_1 individuals as hybrids (H_{X-Y}) will have a new allelic frequency q_1.

C. If migration continues at the same rate each generation into population X, successive generations will have new allelic frequencies q_2.

D. If there is reciprocal exchange between populations X and Y, migration back to Y from X is represented as n. See Exercise 7.

the donor population—$q_0 \rightarrow \bar{q}$. See Exercise 7 for illustration of the case given in Figure 19-1D on reciprocal migration between populations.

In human populations, a few studies have been made to ascertain the average amount of gene flow (m) into one population, or "race," from another over a period of many generations, even though some of the assumptions that have to be made are perhaps overly simple. Nevertheless, we illustrate the method in Example 19-5 with reasoning as follows: if m is reasonably constant over generations, a recurrence of change in q will be, from (19-1) by successive substitution,

$$\begin{aligned} q_2 &= (1 - m)q_1 + m\bar{q} = (1 - m)[q_0 - m(q_0 - \bar{q})] + m\bar{q} \\ &= (1 - m)^2 q_0 + m(1 - m)\bar{q} + m\bar{q} \end{aligned}$$

By factoring the last two terms and adding $(+1 - 1)$ within parentheses, we have (from Glass and Li, 1953):

$$q_2 = (1 - m)^2 q_0 + \bar{q}[1 - (1 - m)^2]$$

or, more generally, if m is constant over t generations,

$$q_t = (1 - m)^t q_0 + \bar{q}[1 - (1 - m)^t] \tag{19-3}$$

and in terms of m

$$(1 - m)^t = \frac{q^t - \bar{q}}{q_0 - \bar{q}} \tag{19-3A}$$

This model of change in gene frequencies due to constant gene flow from one population into another has been invoked in human studies of hybridization between racially distinct groups. Example 19-5 illustrates certain instances and presents a few of the problems involved in assessing accurately the amount of gene flow that has taken place in the past.

Gene Flow Between Populations of Limited Size (Stochastic Considerations)

Analogous with the increase of allelic frequency due to mutation, that due to receiving immigrants with an increased frequency is based on the principle of limited-size N_e. Neutral alleles will tend to be fixed or lost, but migration counteracts the tendency by preserving genetic variability. If we imagine several populations ("islands") all of size N_e forming a network exchanging individuals with adjacent populations at a rate of m, when there are no selective differences between alleles (neutrality), the approximate probability for an individual to be homozygous is given as a modification of the mutation formula (18-14) applied to migration:

$$F \approx \frac{1}{4N_e m + 1} \tag{19-4}$$

probability for homozygosity under migration and drift when m is small. Thus, a small amount of migration will still allow for some fixation over all the populations receiving migrants. Wright (1931) noted that if $m > 1/(2N_e)$, migration will effectively order the gene frequency distribution among populations of neutral genotypes; if the opposite is true, fixation would predominate due to random genetic drift.

In Chapter 18, we noted one of the neoclassic arguments for the selective neutrality of protein variant alleles when their frequencies are nearly equal over broad areas and widely separated populations. Random drift would be expected to produce an array of population gene frequencies with considerable diversity, but fixation at random frequencies in separate populations is easily counteracted by small amounts of migration. In fact, it is expected to take just a little migration to obliterate any differentiation of frequencies that might arise from random fixation of neutral alleles. For example, if two populations of equal size exchange equal proportions of genes each generation (m), the difference (d) in allelic frequency between them is expected to be on the average as follows: if $d = p_1 - p_2$, we use Student's $t = d/s$, as in Table 2-3, where s is the standard error of the difference between means, to describe the expected average difference at the approximate 95 percent confidence level ($t = 2$); thus, $d = 2s$. Taking the average $\overline{p},\overline{q}$ over the two populations, we proceed by using the Wahlund expression of F for the variance between populations (12-7A) and its square root for standard error

$$d = 2\sqrt{\frac{\overline{pq}}{4N_e m + 1}} \tag{19-5}$$

expected difference in q (neutral allele frequency) between two populations after exchanging a fraction m each generation for N_e generations.

If $N = 10^4$, $m = 1 \times 10^{-3}$, and the average allelic frequency in both populations as $\overline{p} = 0.5$, then

$$d = 2\sqrt{\frac{(0.5)(0.5)}{1 + 40}} = 0.156$$

If we refer back to Table A in Example 3-4 to look at the order of magnitude for gene frequency differences observed between natural populations of *D. pseudoobscura* from North America, many polymorphic loci have differences comparable with that expectation: Est-5, allele 1.00; Pt-8, alleles 0.81 and 0.83; XDH, allele 1.00; Amy-1, allele 1.00 (Pt-10 and Pt-12 show greater diversity, however). Consequently, as small a migration rate as one in a thousand could be sufficient to offset differentiation of populations of moderate size if genes are completely neutral. By itself, then, homogenity of gene frequencies between widely separated populations may not be considered proof of consistent adaptive function for such genes in those populations.

Nevertheless, it cannot be assumed that gene flow is so potent as to dominate the forces that tend to bring about differentiation of populations. The relative importance of intra-populational selection, gene flow, and population size must all be considered in ascertaining the amount of differentiation a population may achieve.

EXAMPLE 19-5

GENE FLOW FROM ONE HUMAN POPULATION INTO ANOTHER

Glass and Li (1953) first estimated the generation-by-generation gene flow from one major human population (American Caucasians) into another (American Negroes) by using formulas (19-3) and (19-3A) with estimates of gene frequencies for the Rh blood antigen alleles R^0(cDe) and R^1(CDe). During the period from the early 1700s to the present, following the immigration of African slaves, the larger Caucasian population, by hybridizing with slaves and their descendants, effectively introduced their gene pool into the smaller Negro population. At the time these authors reported their estimates, there was little information on the genetic structure of African populations, and gene frequencies were assumed to be relatively homogeneous throughout equatorial Africa. Such an assumption was premature, and better estimates of gene frequencies in regions of Africa from which slaves originated as well as in American populations were made by Adams and Ward (1973). To illustrate the method of Glass and Li without questioning the homogeneity of populations, we list here estimates of allele frequencies in the groups concerned:

Location Population	R^0(cDe)	R^1(CDe)
Baltimore		
U.S. Negroes	0.446	0.145
U.S. Caucasians	0.028	0.420
East Africa	0.630	0.050

The R^0 allele was reduced and the R^1 allele was increased for American Negroes by hybridization. If we use formula (19-3A) to estimate the amount of gene flow per generation that effectively entered the American Negro population, we assume about 10 generations (t) in about 250 years from the start of the hybridization process, and we would have then

$$(1 - m)^{10} = \frac{0.446 - 0.028}{0.630 - 0.028} = 0.6944$$

Letting q^t = present American Negro allele frequency, $\bar{q}$ = American Caucasian allele frequency, q_0 = African allele frequency, and Glass and Li concluded that about 3.6 percent gene flow had taken place per generation from the Caucasian into the Negro gene pool over this period of history, so that the present American Negro gene pool is approximately 70 percent African and 30 percent European.

Adams and Ward (1973) attempted a careful survey of the regions in Africa from which slaves were imported into the United States and also included 16 different genic loci in their analysis for 5 American Negro populations. The amount of admixture (total gene flow) based on their improved basic data gave weighted mean values over all loci for American Negroes from 4.0 percent for Charleston, South Carolina, to 21.9 percent for Oakland, California—considerably less than Glass and Li's estimate of 30 percent. However, the heterogeneity over all loci was significantly too great to be reliable from two populations in Georgia. Thus, heterogeneity, which pervaded most of their estimates to some extent, was indicative of random genetic drift, although the possibilities for selection or different gene flow in various sections of the United States cannot be ruled out in determining populational frequencies.

This method was used by Workman (1973), who summarized his own work and that of several anthropologists on populational hybrids: (1) in Chile, average Chileans are derived from American Indian-Auracanians × Spanish and Basques, with about 36 percent genes from Indians and 64 percent from Europeans; (2) in the United States, African × Europeans averaged about 10–27 percent admixture; (3) in Brazil, triple hybridity is common between West Africans × Portuguese × American Indians (Tupi-Guarini and Ge linguistics)—about 26 percent African plus 65 percent Caucasian plus 9 percent American Indian estimated.

POPULATIONAL DIFFERENTIATION

In the past it has usually been assumed (Mayr, 1963) that gene flow is a "swamping" influence preventing differences between populations of a species from evolving to a degree sufficient to lead into racial and perhaps ultimately specific levels. On the contrary, in field sampling of wild populations and in laboratory and computer-simulated populations, the evidence for the opposite is found; differentiation of genotypes will occur in the presence of gene flow if the selection regime is sufficiently potent. Also, the opposite may occur; isolated populations may not diverge if selection aims at the same stable point in both. The most notable studies are the following: (1) Ehrlich and Raven (1969) and Brussard and Ehrlich (1970, 1974) on butterfly populations; (2) Thoday and his colleagues (1951–1967) on disruptive selection with gene flow in drosophila (Example 16-4), Thoday and Gibson (1970), and Gibson (1973) on mobility between groups of flies and people; (3) Jain and Bradshaw (1966) and Antonovics, Bradshaw, and Turner (1971) (in Chapter 16) on flowering plant populations growing in soils contaminated by heavy metals; (4) Dobzhansky, Spassky, and others (1967 and later) on combining the effects of selection and migration on behavioral traits (Example 7-4) in *D. pseudoobscura*; (5) Endler (1973) on experimental genetic clines in *D. melanogaster* and computer simulations of selection and migration. The latter experiments are described in Example 19-6. Other studies on cases of microdifferentiation with gene flow between subpopulations include those by Koehn and his colleagues on marine pelecypods (1972, 1973, 1976) and by McKenzie and Parsons (1974, 1975) on drosophila.

Since populational differentiation depends predominantly on selection, we might ask the function of gene flow at all. Obviously, it does serve to introduce extrapopulational genetic material to the coadapted gene pool analogous to the process of mutation; it allows for modulation of gene frequencies between adjacent gene pools and continuity throughout the total species distribution. But under usual circumstances, "swamping" would not necessarily be likely nor would selectively attained intrapopulational equilibrium be much changed even by maximum amounts of gene flow. The effect of gene flow is greatest where there are relatively few subpopulations in a cline, the maximum effect being in the extreme case of two subpopulations, where the problem is essentially the same as that discussed under polymorphism maintained by heterogeneity of environments in Chapter 16. In the case of McKenzie and Parsons' subpopulations inside or outside a wine cellar in Australia, migration of flies into the cellar during vintage time suggests that the most alcohol-tolerant genotypes are the most successful migrants. Consequently, under microenvironmentally distinct circumstances, migration is an immensely vital force in allowing the species to expand and explore diverse niches. The student should also note that two papers on gene flow and subpopulation differentiation (Maynard Smith, 1970; P. T. Spieth, 1974) point out that mere intrapopulational drift can be ruled out as a cause of significant populational differentiation; differing selective pressures are mostly responsible. We should not be ready to assume that local groups are necessarily at equilibrium, because recent shifts in population structure may bring about genotype differentiation without difference in selection regimes between locales.

EXAMPLE 19-6

EFFECTIVENESS OF GENE FLOW AND SELECTION IN A CLINE

Endler (1973) described two series of experiments, one with drosophila and one simulated by a computer, intended to help answer the question: how much does gene flow retard the development of populational differences within a species? He chose a model based on a clinal gradient in genotype frequencies. Graduated changes in selection applied from one end to the other of a linear series of subpopulations ("demes") was expected to determine the cline, but gene flow might be expected to counteract selective forces sufficiently to nullify genetic differences among the demes or to attenuate the cline.

The drosophila X-linked dominant mutant *Bar* was introduced into a standard wild-type ("Kaduna") laboratory population to randomize the genetic background and was then reextracted in specific frequencies according to 15 demes in linear series. Each deme differed from the next adjacent one by an increment of 4 percent in frequency of *Bar* being selected by the observer out of the surviving proportion of that genotype each generation. Emerging adults were scored for phenotype [B/B and $B/+$ ♀♀ with B/Y ♂♂] *Bar* genotypes versus wild type [$+/+$ ♀♀ with $+/Y$ ♂♂] with 50 pairs per deme retained:

Deme Position (X)	Frequency Retained Out of Surviving Proportion	
	Bar Genotypes	Wild Type
1	0.42	0.58
2	0.46	0.54
3	0.50	0.50
4	0.54	0.46
⋮		
15	0.98	0.02

Thus, in deme 1 at G_1, for example, 3 B/B, 18 $B/+$, 29 $+/+$ ♀♀ and 21 B/Y, 29 $+/Y$ ♂♂.

Natural selection against *Bar* is known to be sufficient to eliminate it within a few generations if it occurs in a single large panmictic population, and deme 8 (0.70 *Bar* genotypes: 0.30 wild) was intended to come to equilibrium at about $p_B = 0.50$, by combination of artificial and natural selection.

When gene flow was introduced into the experiment, adjacent demes exchanged 20 percent of their numbers (that is, 40 percent of each deme was migrant); 10 flies of each sex were introduced into the deme on the left and 10 into that on the right so that a given deme received 20 flies combined with resident 30 to make 50 pairs, except for the end demes 1 and 15, where would-be emigrants were returned to the deme from which they came. Following exchange of migrants, flies were allowed to mate and a new progeny generation started. The genotype proportion contributed by migrants then was 40 percent of each

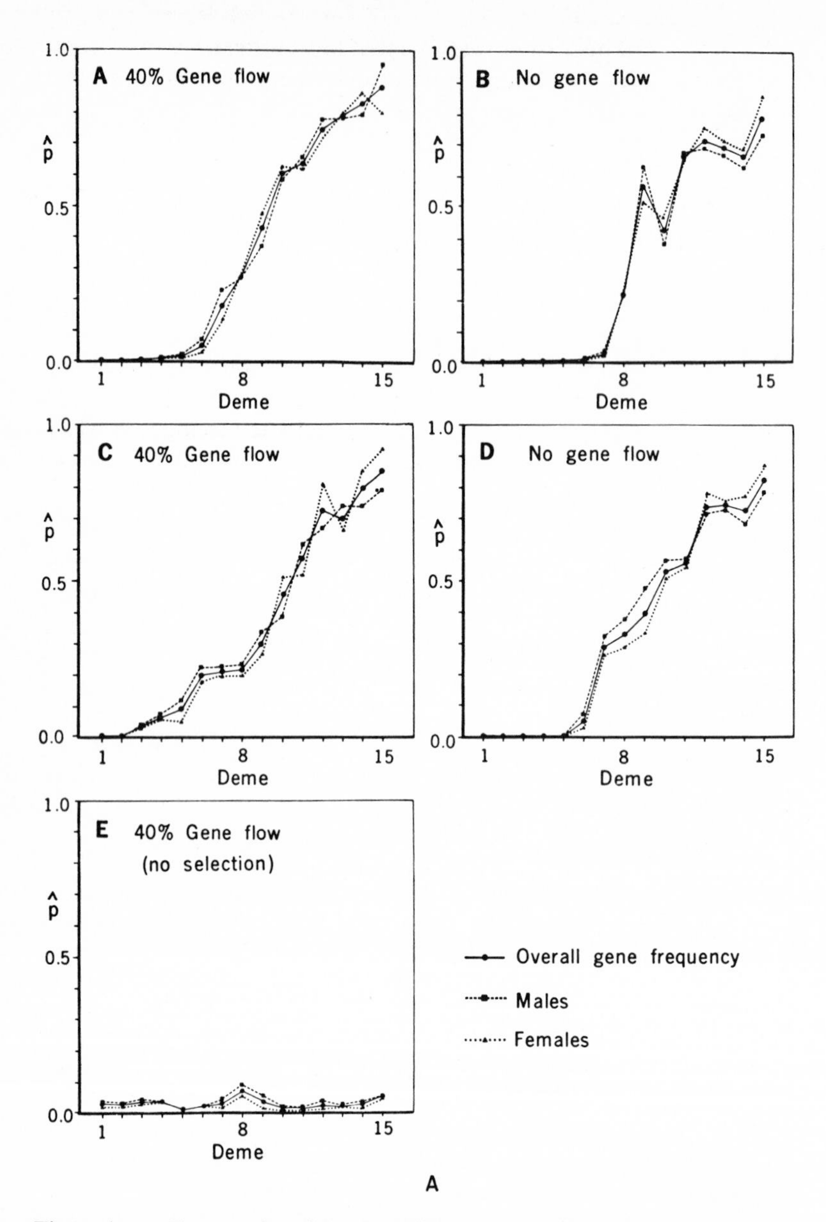

Figure A. Frequencies of *Bar* (p_B) allele at G_{35} in five sets of experimental clines. From Endler, 1973, *Science*, 179:243–250. Copyright 1973 by the American Association for the Advancement of Science.

surviving genotype's proportion. Five sets of experiments were accomplished as follows:

Set	Artifical Selection	Gene Flow	Control
A	Yes	Yes	
B	Yes	No	For gene flow
C	Yes	Yes	
D	Yes	No	For gene flow
E	No	Yes	For artificial selection

At G_{35}, well past the attainment of stable equilibrium among the demes within each series (very little change had taken place from G_{20} on), total enclosing adults were counted and p_B for demes recorded as shown in Figure A. Demes 1 through 6 reached fixation for wild type within the first few generations, but they became polymorphic again subsequently as a result of gene flow in sets *A* and *C* and remained monomorphic in sets *B* and *D* with no gene flow. (In G_{15}, sets *B* and *D* had gene flow by mistake and subsequently remained at a low *Bar* frequency.) Except for a slightly smoother gradient between demes for the gene flow sets *A* and *C* versus no gene flow sets *B* and *D*, these were the only detectable effects of gene flow. Set *E* with no artificial selection favoring *Bar* showed no sign of a cline. Slopes of the clines determined from regressions of p_B on deme position over successive generations are given in Figure B. Slopes of all sets except *E* became significantly greater than zero im-

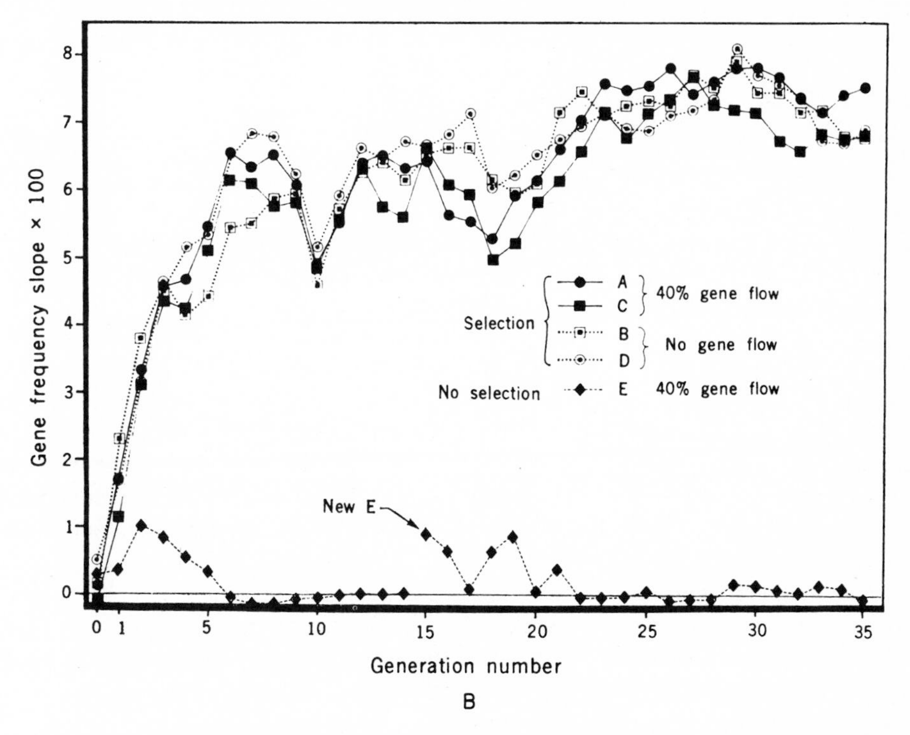

Figure B. Slopes of experimental clines from regressions of p_B on deme position over generations G_0 to G_{35}. From Endler, 1973, *Science*, 179:243–250. Copyright 1973 by the American Association for the Advancement of Science.

mediately and attained approximately stable values from about G_{23} on. Set E had no significant slope except for a few early generations; it was reconstituted in G_{15} since it had reached fixation at wild type in G_{14}. There was no consistent or significant difference between the selective clines with 40 percent gene flow and those without gene flow in slope of the gradient.

The details of the computer simulation models are omitted from this example. Deterministic and stochastic simulations executed on the computer did not differ significantly, so that the results reported were for simulations with the same number of individuals in demes as with the drosophila experiment—$N = 50$ for 50 demes in linear series. Gene flow was varied from 0 to 1 (from zero exchange to complete migration between adjacent demes). Selection (W values) depended on the position of the deme in its series, and four models of selection were used: (1) partial dominance in fitness (heterozygote intermediate), (2) heterozygous advantage, (3) local heterozygous advantage, and (4) frequency dependent. The greatest effect of gene flow was found at levels of weakest selection, but the clinal slope was nevertheless substantial throughout.

Endler pointed out that many possible spatial effects of selection and gene flow could produce a given clinal structure, so that from a given cline in nature we cannot say how much selection or gene flow may be going on merely from the cline itself. Yet it is obvious that a large portion of selective differentiation of genotype frequencies can occur in spite of gene flow. The "swamping" by migration is effective with weak selection, but then it appears to be mostly modulating as a secondary effect rather than a primary one.

CONCEPTS OF POPULATIONAL GENETIC SIMILARITY AND DISTANCE

The problem of measuring the magnitude of genetic identity ("similarity") or its complement ("distance") between populations has occupied several theoretical and experimental investigators, especially since perfection of techniques for measuring allelic differences causing protein variation. If we are to study the process of populational genome differentiation as a result of various combinations of selection, migration, random genetic drift, mutation, and nonrandom mating, we ought to agree on some convenient way to measure the amount of that differentiation. This kind of measure will be particularly essential if we are to learn how levels of evolution are attained to lead to populational units that are reproductively isolated from each other. The units would be (1) partially separated gene pools that may achieve successive levels of ecological niche adaptation (ecotypes, or races), (2) subspecies, (3) semispecies, and (4) species (units that do not exchange genetic material in nature and thus are likely to be reproductively separated and evolving genetically as closed systems).

Conceptually, it would be easy enough to consider one locus at a time. We might use the "coefficient of transmission" (Malecot's "coefficient de parenté," or coefficient of kinship), discussed in Chapter 9, to measure the genetic identity on the average between alleles at one locus—one allele taken from population X, the other from population Y. After several loci have been considered in that way, the average identity over all loci would be an index of genetic similarity. That is the index (I_s) proposed by Sneath (Sokal and Sneath, 1963, p. 157):

$$I_s = (p_x \cdot p_y) + (q_x \cdot q_y) + (r_x \cdot r_y) \tag{19-6}$$

for a single locus or

$$I_s = \frac{(p_{ix} \cdot p_{iy}) + (q_{ix} \cdot q_{iy}) + (r_{ix} \cdot r_{iy})}{L}$$

where L is the total number of loci, each ith locus with particular p,q,r . . . allelic frequencies in populations X and Y, or, more formally,

$$I_s = \frac{1}{L} \sum_{i}^{L} \sum_{j}^{k} p_{ijx} \cdot p_{ijy} \tag{19-6A}$$

where p_{ijx} and p_{ijy} are corresponding frequencies of the jth allele in X and Y populations at the ith locus; k = total alleles at ith locus.

This coefficient would refer to identical genotypes if one gamete is taken from population X and one from population Y. However, the concept of genetic similarity must include the entire set of one populational genome with respect to another, including the similarity of gene frequencies between the two and over all loci since polymorphic as well as fixed loci should be included. Two populations may have identical allelic frequencies and yet not have a maximum similarity I_s value. For example, if both populations X and Y have alleles equal at a single locus ($p = q = 0.5$), then $I_s = [(0.5)(0.5) + (0.5)(0.5)]/1 = 0.50$, which is less than the similarity index for some populations that are clearly different. For example, if X has $p = 0.7$, $q = 0.3$, while Y has $p = 0.9$, $q = 1$, then $I_s = [(0.7)(0.9) + (0.3)(0.1)]/1 = 0.66$. These numerical examples and others are presented in Table 19-1.

TABLE 19-1 *Comparisons of genetic similarity coefficients (from Sneath and from Nei) and the Nei genetic distance estimate for six single loci and for all six together*

Locus	Allele	Allele Frequency		Similarity		Distance
		Population X	Population Y	I_S	I_N	D
1	A	0.7	0.9	0.66	0.957	0.0439
	a	0.3	0.1			
2	B	0.6	0.4	0.48	0.923	0.0800
	b	0.4	0.6			
3	C	1	0	0	0	undefined (very large)
	c	0	1			
4	D	1	1	1	1	0
	d	0	0			
5	E	0.5	0.5	0.5	1	0
	e	0.5	0.5			
6	F	0.2	0.4	0.44	0.924	0.0791
	f	0.1	0.1			
	f'	0.7	0.5			
All 6 loci				0.513	0.733	0.311

This problem of finding an efficient measure of genetic similarity and several others of theoretical importance were summarized by Rogers (1972). Many attributes of other measures of similarity and genetic distance were devised and reviewed by Cavalli-Sforza and Edwards (1967), Edwards (1971), Hedrick (1971, 1975), and Nei (1972, 1975). (See Crow and Denniston, 1974, for a summary of a workshop on genetic distance held at the Fourth International Congress of Human Genetics in 1971.) We shall describe only the similarity and distance measures of Nei because they are more intuitively easy to grasp and have been employed most commonly in the literature (see Exercise 11 on Hedrick's index).

Nei's measure of genetic distance was developed to estimate the number of codon differences per structural genic locus for proteins and the divergence time between closely related species on the supposition of neutrality to selection. Beginning with Nei's measure of similarity, the numerator is the same as that in Sneath's index (19-6) and (19-6A)—namely, the "coefficient of kinship" on the expectation of drawing one gamete from each of two populations. Nei's improvement is to normalize this index by dividing by the geometric mean of identical homozygote frequencies as follows for a single locus:

$$I_N = \frac{\sum(p_{jx} \cdot p_{jy})}{\sqrt{(\sum(p_{jx})^2)(\sum(p_{jy})^2)}} \tag{19-7}$$

(Nei's similarity index; symbols are defined as before for 19-6A).

The student should verify the locus-by-locus results given in Table 19-1. When the two populations have identical frequencies of the same alleles, their similarity index (I_N) is 1, irrespective of whether the locus is polymorphic or monomorphic, while it is zero when there are no alleles in common. Finally, when all genic loci are taken into account, including monomorphic ones, the numerator will be the arithmetic mean coefficient of transmission, or kinship, from (19-6A), and the denominator will be the geometric mean of identical homozygote frequencies averaged over all loci for populations X and Y.

$$I_N \text{ (all loci)} = \frac{I_s}{\sqrt{\left(\sum_i \sum_j p_{ijx}^2\right)\left(\sum_i \sum_j p_{ijy}^2\right) \Big/ L^2}} \tag{19-7A}$$

For example, in Table 19-1, all six loci together in populations X and Y give the similarity index of Nei as follows:

$$I_N = \frac{0.5133}{\sqrt{[(0.7)^2 + (0.3)^2 \cdots (0.1)^2 + (0.7)^2][(0.9)^2 + (0.1)^2 \cdots (0.1)^2 + (0.5)^2]/6^2}}$$

$$= \frac{0.5133}{\sqrt{(0.69)(0.71)}} = 0.7334$$

Genetic distance in the case of Sneath's index would simply be the complement of the similarity ($1 - I_s$). According to Nei, the proportions of different genic loci relative to the amount of diversity between two randomly chosen genomes (populations X and Y) are unequal and their variances are not additive. His measure of "standard genetic distance" then is estimated as the minus natural log of his similarity index:

$$D = -\log_e I_N \tag{19-8}$$

In Table 19-1, it is apparent that when I_N is close to 1, Nei's genetic distance is close to $1 - I_N$, but the genetic distance is greater than the simple complement of the similarity index as I_N decreases. Of course, as I_N approaches zero, the genetic distance measure becomes unreliably high. Nei gave as an estimate of minimum genetic distance (see Exercise 10).

$$D_{\min} = (1 - I_s) - \tfrac{1}{2}[\sum\sum p_x^2 + \sum\sum p_y^2]$$

SOME EXAMPLES OF GENETIC SIMILARITY AND DISTANCE

Nei (1972) applied his method to data of Selander, Hunt, and Yang (1969) on protein variation in two subspecies of house mice, including 36 different proteins in four populations of *Mus musculus musculus* and two populations of *Mus musculus domesticus.* Estimates of I_N and D are given in Table 19-2; below the diagonal are estimates of I and above are D, with values on the diagonal in parentheses of average locus homozygosity $(\sum\sum p_{ijx}^2/L)$. Within each subspecies, identity is high; 98–99 percent of genes are shared with a genetic distance average at 0.0137. However, between subspecies, 82–88 percent of genes are shared with the genetic distance average at 0.1714, about 12 times that within subspecies. Thus, there is a significant genetic differentiation of the subspecies; if electrophoretic detection of protein differences only accounts for one-third of amino acid substitutions, both populational and subspecies differences may be considerably more than this minimum estimate.

Ayala, et al. (1974) made an extensive survey of natural populations from Venezuela of five *Drosophila* species (*D. willistoni, D. tropicalis, D. equinoxialis,* and *D. paulistorum* as sibling species plus *D. nebulosa* as a related but morphologically distinguishable species) by assaying enzymatic variants with starch gel electrophoresis over 30 genic loci. All

TABLE 19-2 *Estimates of I_N (below the diagonal in parentheses), D (above the diagonal), and average homozygosity $\sum p_i^2$ (on the diagonal) for six populations of house mice of two European subspecies. Source data from Selander, Hunt, and Yang (1969). Estimate from Nei (1972)*

	M. m. musculus				*M. m. domesticus*	
Population	1	2	3	4	5	6
1	(0.9319)	0.0178	0.0256	0.0210	0.1941	0.1959
2	0.9823	(0.9031)	0.0147	0.0094	0.1701	0.1713
3	0.9749	0.9854	(0.9155)	0.0057	0.1906	0.1859
4	0.9792	0.9906	0.9943	(0.8899)	0.1337	0.1202
5	0.8236	0.8436	0.8264	0.8748	(0.9319)	0.0018
6	0.8221	0.8425	0.8303	0.8787	0.9982	(0.9301)

species were highly polymorphic (average 69 percent with high heterozygosity per individual (18 percent). Seven local populations were sampled. Calculations of genetic similarity and distance were made according to Nei's methods between several levels of evolutionary divergence: (1) between local populations within the first three species listed above, (2) between the semispecies of *D. paulistorum* (between which reproductive isolation is nearly complete), (3) between the four sibling species, and (4) between *D. nebulosa* and each of the other species. At the first level, pairs of local populations of *D. willistoni*, *D. tropicalis*, and *D. equinoxialis* have I_N at about 1.00 (complete genetic similarity) at 90 percent of their loci; thus, average I_N was 0.968 and average D was 0.033. Between semispecies of *D. paulistorum*, genetic differentiation had taken place so that I_N averaged 0.798 and D was 0.226. Finally, average similarities and distances were as follows:

	$\bar{I}_N$	$\bar{D}$
Between sibling spp.	0.587 ± 0.028	0.538 ± 0.049
Between nonsibling spp.	0.299 ± 0.020	1.214 ± 0.064

Thus, at the level of reproductive isolation (semispecies of *D. paulistorum*), genetic differentiation was about seven times greater than between local populations, while divergence more than doubles up to levels of sibling species and doubles again to the morphologically distinguishable level of species.

In contrast with the *D. willistoni* group of sibling species, *D. pseudoobscura* and *D. persimilis* differ largely only by chromosomal rearrangements while their protein loci have a small genetic distance (Nei gives $D = 0.05$ for these species calculated from data by Prakash, 1969). Also, many other groups of sibling species have low genetic distances; Hubby and Throckmorton (1968) found little divergence between *D. victoria* and *D. lebanonensis* (Nei's $D = 0.18$), for example. Therefore, speciation need not be accompanied by a large number of genic changes.

It is important to realize that increasingly more distant levels of evolutionary divergence in groups of animals such as mice and flies (see data compiled by Nei, 1975) may indicate increasing genetic distance, but the conventionally described human races indicate the opposite trend. Nei and Roychoudhury (1974) surveyed protein variant loci and blood group loci from the literature on populations of Caucasoid, Negroid, and Mongoloid peoples (see also Harris's survey in Example 3-5). The minimum genetic distances ($1 - I_S$) and standard distances between populations within each race and between these races are given in Table 19-3. The interracial net gene locus differences are small (about $\frac{1}{10}$th) compared with the intraracial (between populations), or randomly chosen genomes from any two populations of the same race. Apportionment of total genetic variation between populations within races and between human races indicates that the so-called major races are only slightly genetically divergent if compared with the large amount of variation already evolved within each race between local ethnic or national groups. Such a conclusion was reached by Lewontin (1972) using a different method of apportioning genic diversity. Perhaps this fact indicates that the so-called major races are superficially distinguishable but not genetically separable groups, even though physical anthropologists worked very hard in the days before 1950 to describe their morphological distinctive qualities. Whatever evolutionary forces brought about genetic divergence between local human populations

TABLE 19-3 *Minimum (upper = D_{min}) and standard genetic distance (lower = D) estimates between the major human "races" (above the diagonal) and within these "races" (on the diagonal). Data from Nei and Roychoudhury (1972, 1974) and Nei (1975) include a survey of 62 protein loci for Caucasoids and Negroids and 35 loci for Mongoloids common to the other two "races"*

"Race"	Caucasoid	Negroid	Mongoloid
Caucasoid	0.104 ± 0.023 0.110	0.010 ± 0.003 0.011 ± 0.004	0.010 ± 0.004 0.011 ± 0.005
Negroid		0.092 ± 0.019 0.097	0.017 ± 0.008 0.019 ± 0.009
Mongoloid			0.098 ± 0.027 —

have not operated on the same dimensions between the immensely large three major groupings. These major groupings are perhaps not analogous with groups we refer to as races, ecotypes, or subspecies in other animal species or in plants. If ever in the history of human origin the major "races" had been distinguished by some portion of the genome selected for adaptive traits in particular ecological conditions, then there are at least three possible explanations for the observed facts about the genetic diversity we have just discussed; (1) by our present techniques for assaying protein variation, we have not yet detected those genotypes that were important to the primitive adaptations; (2) the tendencies for gene flow between the races must be greater than tendencies for gene flow between local populations (which seems absurd); (3) the major race classification is quite superficially based on what are fundamentally heterogeneous groups within which there is no phylogenetic uniformity of descent within recent times. This latter possibility seems most likely, although it is conceivable that the first alternative may eventually be shown to be more probable if genuine adaptive traits could be shown to have a causal relation to any functional difference of protein variants that are most contributory to genetic distance between these major "races."

EVOLUTIONARY GENETICS AND POPULATION GENETICS

This book has been designed to emphasize the basic theorems and principles of statics and dynamics of genes in sexually reproducing Mendelian populations. We have been concerned with the forces at work on allelic and genotype frequencies between individuals and groups of individuals that breed together, sharing a common gene pool. In this chapter particularly, we have considered some aspects of gene flow between populations, and we have emphasized that, with restriction of gene flow, divergence of the genome is likely to occur. If gene flow is completely restricted between two subgroups of a species population, those two subgroups may well become permanently diverse. It is our view that any genetically conditioned mechanism that restricts gene flow between distinctive populations of a species constitutes a mechanism that is an essential ingredient for the evolution of that level in organic divergence—what we may term the *species* level. Species have been defined in various ways, often with the definition conditioned by the author's scientific and historical perspective. Consequently, because we are now conditioned by population genetics principles, we may adopt either one or both of two biologically meaningful definitions for species of sexually reproducing organisms: (1) Mayr (1942): "groups of actually or potentially interbreeding natural populations which are reproductively isolated from other such groups"; (2) Dobzhansky (1951): "groups or populations the gene exchange between which is limited or prevented in nature by one, or by a combination of several, reproductive isolating mechanisms. In short, a species is the most inclusive Mendelian population." Maintenence of separation between gene pools can be achieved by numerous biological and physical barrier devices, or isolating mechanisms, consideration of which is beyond the scope of this book but is very thoroughly considered by the two authors mentioned.

As pointed out by Lewontin (1974), "we know virtually nothing about the genetic changes that occur in species formation." Much of what is known about the magnitude of genetic differences between closely related species or races presently evolving into the species level and forces controlling gene flow has been summarized by Mayr (1963) for animals, by Jain and Bradshaw (1966), Grant (1963), and Stebbins (1974) for plants, and by Manning (1965) and Parsons (1973) for drosophila ethological isolation and for other insects. Lewontin (1974), after summarizing much of what is known about protein variation between various levels of speciation in many organisms, concluded that "the speciation process itself resulted in very little differentiation." Of course, genotypes that determine formation of reproductive isolation (control of behavioral preferential mating, for example) are probably not as yet detectable by our present protein separation methods. In short, our view of the total genotypes, with all our biochemical expertise, is simply not yet able to discern and resolve detailed loci in a large enough sample of the genome of a species with a Mendelian life cycle. Logically, it might be quite possible for two populations of a species to differ in just a few genic loci by a few critically important alleles that could determine their isolation from each other and thus start them on their way toward independent acquisition of separate gene pools.

Speciation is one of the most critical processes in nature, but we are a long way from describing the origin of species in the field or with methods of experimental population genetics. The rise of gene pool isolation can be conditional on persistent differences in environment for populations that have diverged at least to the extent of local adaptation. Divergence could continue as the outcome of environmental heterogeneity (disruptive, or diversifying, selection). For a majority of animal species, a primary factor in keeping gene pools delimited must be dependence on mutual attraction between the sexes within the diverging populations. For plants, gametes and gametophytes must be mutually compatible within separately evolving populations. Muller in 1939 proposed that genes controlling behavior and interbreeding are incorporated as by-products of genetic divergence in the course of building up adaptive complexes. Dobzhansky (1951) stressed the role of natural selection in reinforcing divergence already attained between populations within a species; if heterogamic matings are selected against (by eliminating hybrids after mating has occurred, by selecting for homogamic mating propensities, or both, for example) genotypes that favor or ensure intrapopulational breeding would become established. Such response to selection might be called the elimination of genotypes that foster promiscuity. As a corollary to Dobzhansky's coadaptation hypothesis of integrated gene pools, phenomena of lowered fertility in hybrids or the lowered fitness of backcross or inbred (F_2) hybrids compared with original parent populations could serve as important mechanisms in such a process. Then (Dobzhansky, 1950) "the process of speciation must be regarded as an evolutionary adaptation which permits the development of an immense diversity. . . . [It] is a device which enables life to exploit the multiform opportunities offered by the environment. Speciation is accordingly a form of integration of Mendelian populations engendered by natural selection in response to the challenge of the diversity of sympatric environments."

Unfortunately, the criterion of reproductive isolation is not easy to establish. In plants, it is often difficult to decide on boundaries between two gene pools, and isolating mechanisms are often weak between plant populations. Nevertheless, in principle, the permanence of a level in the diversification process must be established if we are to delimit species biologically. Establishment of reproductive isolation both as a by-product of genetic divergence between two or more populations and as a response to selection for homogamy has been demonstrated in laboratory populations (see Spiess review, 1968; Example 10-2).

Carson (1968, 1975) stressed that the classic view of speciation as a gradual microevolutionary (adaptive) process of gene pool divergence is not by any means the only or necessary ingredient of attaining reproductive isolation. In Carson's analysis of Hawaiian drosophila as species formed after invasions of new territory by single or very few founding individuals (one fertile inseminated female perhaps), there is considerable evidence that interisland founders share a common set of chromosomal arrangements and are then "homosequential" for a genic order. Following a "founding" in a new island or territory, chromosomal rearrangement may take place, and a descendent population may become established as a fixed monomorphic, or less often a polymorphic, new species. Thus, about 600 species largely endemic to specific islands and local sites within each island have evolved as a result of genetic reorganization or a series of recombinations following a founding event. (Species of Galapagos finches and Hawaiian honeycreepers are other classic examples of island founder populations, although the details of their genetic divergence have not yet been described.) Carson proposed that the main speciation process for these forms probably is a result of recombination, or "disorganization of a closed system of genetic variability." If the coadapted, or relationally balanced, portion of the species' genome is "closed" to recombination in that it is so integrated that fitness will be lost by recombination within that portion, then ordinarily recombination is "open," or freely available only in the remainder of the genome. Within the species population, the closed system would not be available for variation to any degree since it would be highly stabilized.

Occasionally, populations experience a set of advantageous conditions so that natural stabilizing selection may be temporarily relaxed (K-type selection virtually eliminated during that phase), and the population will experience a flush of numbers—a large, more or less unrestricted increase in N. During that expansion phase, many individuals would survive including recombinants from the "closed" as well as "open" portions of the species' genome. Population flushes are usually followed by a crash in population numbers. The reduction in numbers is likely to be nonselective, but whether individuals migrate to new areas as founders of new populations or remain in situ, the chance of the survivor(s) gaining a recombination in the "closed" portion of the genome is greater than it had been before the flush-crash. If the individual founder had migrated either to a different island during the flush phase or to a different ecological zone within the same island, its reproduction in that new area would produce new allelic frequencies by chance plus a recombined genome. In the area, descendants of the founder would have the opportunity to achieve a newly organized genome with about the same total genetic resources in terms of specific alleles and chromosomal sequencies as its progenitor. New selective forces may then operate within a relatively short time to build a new species genome as a coadapted system. An evolutionary geneticist who compared the total genetic distance between the new species and its progenitor might find little or no evidence of "divergence" in terms of allelic differences or chromosomal sequences. Only new linkages, juxtapositions of genic loci, and organization of the total set of genic loci would have taken place, probably making new epistatic interactions and complementations possible. Thus, the level of *species* need not have included a prodigious alteration of genetic variation or divergence of allelic frequencies, and the composite indices of genetic distance would not have been adequate to disclose the differences between such closely related species. Thus, Carson (1975) pointed out that "organizational shifts due to a flush-founder event may produce genetic changes of a qualitatively different sort from those produced by ordinary geographical subspeciation."

The kinds of events particular to the origin of any species are probably not identical. The gradual divergence by microevolution in the process of ecotype (or subspecies) formation, illustrated best by circular chains of subspecies over large geographic areas (Mayr, 1963; Grant, 1963), bespeaks evidence that species may well originate by a continual process. The incipient isolation between "races" of *Drosophila paulistorum*, described as "a cluster of species *in statu nascendi*" (Dobzhansky and Spassky, 1959; Dobzhansky, 1972), documents a critical point in the process of speciation well enough, although the exact mechanism by which successive subspecies have evolved, whether initially by flush-crash-founder effect or gradually by microevolution, is a matter of conjecture.

EXERCISES

1. Discuss the results of Mukai's experiments in Example 19-1 in the light of the hypothesis from Chapter 18, p. 612 that a plateau of high fitness may exist at a certain optimal level of heterozygosity not far above the mean level for a regularly outcrossing population.
2. (a) How might genic action determine each of the three phenotypic outcomes of stabilizing selection mentioned on p. 638?
   (b) Would you expect homozygotes or heterozygotes to be selected in natural populations for these phenotypic outcomes? Why?
   (c) What is reciprocal recurrent selection, and how might you design a selection regime to accomplish a developmentally well-buffered heterozygous genotype?
3. Kearsey and Barnes (1970) measured the incidence of drosophila bristle number phenotypes raised under low density (near optimal conditions) and under high density in population cages. After correcting bristle numbers for depression by raising progenies

from the same parents under both conditions, they found the corrected bristle numbers for 250 females were as follows:

	$\bar{Y}$	Variance
Population cages	20.47	8.23
Low density	21.91	36.53

(a) What is the effect of high density? Does bristle number seem neutral to selection in population cages?
(b) Comment on the kind of selection likely to be acting on this phenotype in population cages.
(c) Haldane (1954) showed that if selection acted to reduce variance, the intensity of selection would be approximately

$$I = \log_e \left[\frac{\sigma_o}{\sigma_s} \right]$$

where σ_o = standard deviation of phenotype in the original population, and σ_s = standard deviation after selection. With the changes in variance of bristle number given above, what would be the estimated intensity of selection?

4. (a) List all the mechanisms that could promote establishment of genetic systems to help integrate or make more cohesive the genotypes of natural populations.
   (b) How would you account for the results of Wallace in Example 19-4 in terms of integrated genotypes? How might the first-generation heterosis be accounted for following a cross between widely separated populations if there is a threshold plateau of optimal heterozygosity within local adapted populations?
5. (a) In Example 19-3, verify the equilibrium genotype frequencies in the case with only repulsion linkages for the fitness values given.
   (b) Consider what would be the mean Y phenotypic value and genotype frequencies if only coupling linkages occurred in the population.
   (c) How might you account for inbreeding depression and for lowering of fitness when artificial selection is performed on a normally outcrossed population?
6. Assume an island population X is surrounded by five equal-sized populations. Let the gene frequency on the island be $q_X = 0.90$, while the five surrounding populations have $q_i = 0.2, 0.7, 0.9, 0.6, 0.1$.
   (a) If the surrounding populations contribute 0.1 migrants onto the island each generation, which then breed on the island with the resident population, what will be the q_{1x} (gene frequency on the island) after one generation of migration?
   (b) If the island population never sends migrants out, and the outside populations continue to migrate in, how many generations will it take to change the q_X to 0.7? What would be the ultimate q_X on the island?
   (c) Show how the rate of change in q due to migration is similar to that due to mutation (13-10).
7. Suppose migration is going on reciprocally between two populations, as in Figure 19-1D, at different rates: m from population Y into population X and n from X back to Y.

(a) Find the change in $q_X(\Delta q_X)$ from the relationship $q_{1X} = mq_Y + q_X n$, where $q_{1X} =$ gene frequency in X after 1 generation of migration and $q_X, q_Y =$ gene frequencies in each population before migration.

(b) Let $m = 0.2$ and $n = 0.3$, and let $q_X = 0.5$, and $q_y = 0.1$. What will be q_{1X} (after one generation of migration)?

(c) After many generations, if the rates of migration between populations continue at constant amount, what would be the ultimate frequency in the X population relative to the Y population? (Hint: Set $\Delta q_X = 0$.)

8. (a) In Example 19-5, what does $(1 - m)^{10} = 0.6944$ represent? Is the assumption realistic?

   (b) Use the $R'(CDe)$ frequencies for United States Negroes and Caucasians and Africans to estimate the amount of gene flow per generation. Do the rates of migration agree? If not, are they reasonably close to give an approximate magnitude of migration?

9. Nei (1975, p. 182) stated: "Migration retards gene differentiation considerably and even a small amount of migration is sufficient to prevent any appreciable gene differentiation." What qualifications must be added to this statement?

10. Nei showed that for genetic distance between two populations (X and Y), a *minimum* estimate can be simply the equivalent of the complement of Sneath's index minus the average heterozygosity for those populations, as follows:

$$D_{\min} = (1 - I_s) - (1 - p_{jx}^2)(1 - p_{jy}^2)/2 = \left[\frac{p_{jx}^2 + p_{jy}^2}{2}\right] - I_s$$

   (a) Verify that for locus 1 in Table 19-1, $D_{\min} = 0.04$.

   (b) Find the $D_{\min}$ for all six loci where I_s will be the overall similarity index and average homozygosity will be $\sum_i \sum_j p_{ij}^2/L$ in each population.

11. Hedrick (1971, 1975) proposed an index (I_H) of genic similarity based on genotype frequencies rather than allelic frequencies because under various breeding systems, genotypes can be more diagnostic than haploid frequencies:

$$I_H = \frac{\sum p_{jx} \cdot p_{jy}}{\frac{1}{2}[\sum p_{jx}^2 + \sum p_{jy}^2]}$$

   where $p_{jx} =$ genotype frequency in the x population, and $p_{jy} =$ corresponding genotype frequency in the y population.

   (a) Assume random mating and verify that $I_H = 0.8505$ for locus 1 in Table 19-1.

   (b) Assume inbreeding with $F = 0.5$ first just in population x with random mating in y. Then assume $F = 0.5$ in both populations.

   (c) How does the mating system affect the genetic similarity index?

12. Powell, Levene, and Dobzhansky (1972) used the following measure (C) for genetic distance in the case of chromosomal polymorphism in populations of *Drosophila pseudoobscura*:

$$C = \sum_i \frac{|x_i - y_i|}{2}$$

   where x_i and $y_i =$ frequency's of the ith allele in x and y populations respectively—that is, the sum of half of each allelic (chromosomal haploid) difference between populations x and y, measured in absolute value.

(a) Compare this measure with that of Nei, using loci from Table 19-1.
(b) What relationship does this measure C have to the identity of alleles between the two populations?
(c) It has been stated this measure of genetic distance is a "true" distance measure because it obeys the law of triangular inequality: $AB + BC \geq AC$ (see Jacquard, 1973). Explain.

13. Comment on the conditions bringing about genetic differentiation between populations versus the amount of gene flow that counteracts differentiation. How do you visualize that gene flow can assist in the adaptive process of microevolution within a local population?

APPENDIX

These statistical and mathematical notes are included for convenience to assist the reader in becoming acquainted with symbols and concepts found in the main text. For more rigorous proofs, derivation of statistical functions, and their utility, the reader should consult the references recommended in Appendix A-14.

A-1

A Few Primary Ideas

Living material is *variable* owing in part to many uncontrollable, or indeterminate, factors. Experimental treatment or observations of naturally occurring individuals should be designed to ascertain whatever constant factors may exist in spite of indeterminate factors. Sample data *estimate* these main constants or *parameters*. Experimental or observational data are *samples* of a "universe," or population of infinite size, characterized and described by parameter quantities or values. By collecting data in an unbiased manner, we wish to estimate those parameters from samples.

We cannot specify any parameter exactly, but we can rigorously specify the limits within which it lies and the reliability of our estimate—that is, its degree of uncertainty. Tests of significance are designed to measure these limits in probability terms. For example, a coin, if perfectly balanced and unbiased, is expected to fall heads one-half of the time. As tosses of the coin (trials) increase, one may confirm or deny the lack of bias in the coin and probability for lack of bias, because the probable limits within which the parameter lies come closer to a single value, such as the a priori hypothesis of heads one-half of the time.

Relevant parameters may be *central tendencies* (means, mode, median) and *distributions* (how the variable is spread or scattered). Three general types of *variates* are (1) continuous (dimensions, scales with implied limits of accuracy, (2) discontinuous (separate items, integers), and (3) frequencies of variates (proportions, rates, ratios). A *frequency distribution* serves not only to describe the observed scatter of the variate in relative amounts from lower to upper limit but also to estimate the probability of scatter for the "universe" of that variate—that is, to estimate the parameter distribution.

A-2

Basic Probability

For events with some degree of uncertainty, the frequency of occurrence—or proportion of the event's happening out of a total number of trials in which it may or may not happen—is its probability (p).

$$p = \frac{A}{A + B}$$

where A = the number of ways the event happens in a series of trials (empirical) or the number of ways the event can happen (a priori), B = the number of ways the event does not happen (or cannot happen) in a series of trials.

$$q = \frac{B}{A + B} \qquad \text{and} \qquad p + q = 1$$

The probability of an event may be predictable from a priori evidence, as from the Mendelian hypothesis, or by empirical determination of a parameter from a number of trials.

A EVENTS, SAMPLE SPACE, AND PROBABILITIES

A more formalistic definition of probability for an event follows from the concept of *sample space*, or the set of all possible outcomes for a series of trials or experiments. Within the sample space, each outcome is called a *sample point*. An *event* (E) is then defined as the *sum of sample points* with a particular property. The probability for the event, symbolized here as $p\{E_i\}$, is the proportion of the event's sample points of the total sample space. In current texts, probability for an event is often symbolized as $Pr\{E_i\}$, and probability for the event's *not* occurring, or *complementary* event (sample points outside the event), is symbolized as $Pr\{E_i^c\}$, or $1 - Pr\{E_i\}$. We use the symbol p instead of Pr and q instead of $1 - Pr\{E_i\}$. For the complementary event, we use E_i^c.

Other symbols in this section are $p\{E_i \cap E_j\}$ (*intersect*, or sample points common to both events) and $p\{E_i \cup E_j\}$ (*union*, or sum of sample points in the two events).

To clarify these concepts, a deck of 52 playing cards can be a sample space. If we draw a single card, there are 52 sample points, as represented in the following diagram:

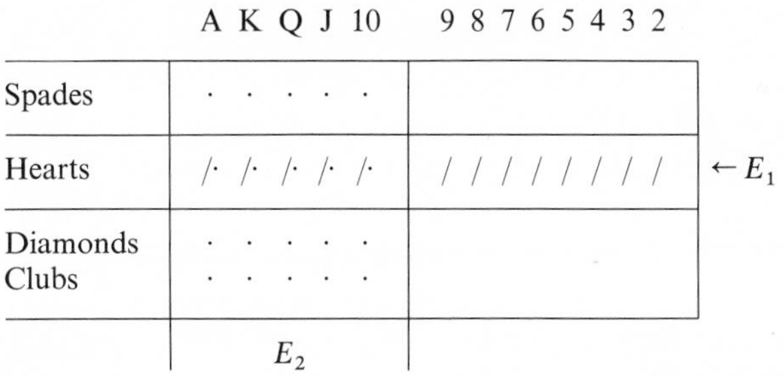

	A K Q J 10	9 8 7 6 5 4 3 2	
Spades	· · · · ·		
Hearts	/· /· /· /· /·	/ / / / / / / /	$\leftarrow E_1$
Diamonds Clubs	· · · · · · · · · ·		
	E_2		

Let E_1 = the event "a heart," and let E_2 = the event "an honor." Then,

$$p\{E_1\} = \tfrac{13}{52} = \tfrac{1}{4} \qquad \text{(crosshatched in diagram)}$$
$$p\{E_2\} = \tfrac{20}{52} = \tfrac{5}{13} \qquad \text{(stippled in diagram)}$$

Overlapping Events. Events that can have sample points in common are said to be overlapping; that is, their intersect exists; $p\{E_1 \cap E_2\} = \frac{5}{52}$ in this example where *heart* and *honor* occur together. Also, it should be noted that this joint probability is the product (horizontal proportion $\times$ vertical proportion) $= (\frac{1}{4})(\frac{5}{13}) = \frac{5}{52}$, since the two events are "independent" (to be defined below).

The sum of overlapping events can be expressed as the probability for *either* one event *or* the other—their union; if we draw a single card, the probability for it to be either a heart or an honor or a heart-honor is $p\{E_1 \cup E_2\} = \frac{1}{4} + \frac{5}{13} - \frac{5}{52} = \frac{28}{52}$. The double event (intersect) cannot be counted twice in adding the two events together, so in general the union of overlapping events is $p\{E_i \cup E_j\} = p\{E_i\} + p\{E_j\} - p\{E_i \cap E_j\}$. For the union of more than two events, we take the sum of single event probabilities *minus* all double (intersects) *plus* triple (intersects), and so on.

Mutually Exclusive Events. Events with no sample points in common are said to be mutually exclusive; that is, their intersect is zero.

$$p\{E_i \cap E_j\} = 0$$

In the deck of cards, the probability for getting a heart and a club in a single draw is impossible.

The union of mutually exclusive events is their sum, since the intersect term is zero: $p\{E_i \cup E_j\} = p\{E_i\} + p\{E_j\}$. The probability of getting either a heart or a club is $\frac{1}{4} + \frac{1}{4} = \frac{1}{2}$.

Conditional Events. If an event's happening (E_1) is conditional on a different event's happening previously (E_2), our attention is confined to that subsample space (E_2) delimited by the previous event. The now-restricted series of sample points constitutes the sample space of interest; probabilities must then be adjusted to sum to 1 in the new space simply by dividing by $p\{E_2\}$. The probability of E_1, given that E_2 has occurred, is symbolized

$$p\{E_1 | E_2\} = \frac{p\{E_1 \cap E_2\}}{p\{E_2\}}$$

For example, if we draw a card and get a peek at the corner to see that it is an honor, what is the probability that it is also a heart? We are then confined to the 20-honors subsample space, and the probability of getting a heart honor is no longer taken out of the total sample space, so that $p\{E_1 | E_2\} = \frac{5}{52} \div \frac{20}{52} = \frac{1}{4}$, which is evident from the diagram. In the condition of the honors' sample space, $\frac{1}{4}$ are hearts. This fact is true whether we consider honors or *not* honors (a further criterion for "independence").

The meaning of conditional probability becomes clearer, however, when illustrated with a reduced deck of cards. Suppose we discard the club honors, leaving 47 cards, 15 of them the remaining honors. Then $p\{E_2\} = \frac{15}{47}$. Now if we peek and see that the card we drew is an honor, the probability that it is also a heart is $\frac{5}{15} = \frac{1}{3}$:

$$p\{E_1 | E_2\} = \frac{p\{E_1 \cap E_2\}}{p\{E_2\}} = \frac{5}{47} \div \frac{15}{47} = \frac{1}{3}$$

However, if we peek and see that the card is *not* an honor, the probability that it is also a heart is $\frac{8}{32} = \frac{1}{4}$:

$$q\{E_1 | E_2^c\} = \frac{p\{E_1 \cap E_2^c\}}{p\{E_2^c\}} = \frac{8}{47} \div \frac{32}{47} = \frac{1}{4}$$

The condition of an honor makes a difference as to the probability for the suit of the card in the reduced deck.

Independent Events. If an event (E_i) has a constant probability whether a second event has already occurred or not (E_j), we say these two events are independent. In the full deck of cards (52), it makes no difference whether we pick an honor or a card that is not an honor; the probability for a heart is the same ($\frac{1}{4}$). However, in the reduced deck, the events "heart" and "honor" are *not* independent, although they are overlapping. The probabilities for their intersects are different whether the second event (E_2) or its complement (E_2^c) has occurred. It can be shown algebraically that when this criterion is met—namely,

$$p\{E_i \mid E_j\} = p\{E_i \mid E_j^c\},$$

the intersect (double event) is the product of the two single events: $p\{E_i \cap E_j\} = p\{E_i\} \times p\{E_j\}$, as is the case for drawing a heart-honor in the full deck of cards. In the reduced deck, it is evident that the probability for drawing a heart-honor is $\frac{5}{47}$, or 0.106 and *not* ($\frac{13}{47}$) × ($\frac{15}{47}$), or 0.088! It is not correct to take the product of two overlapping event probabilities unless the events are independent.

EXAMPLE 1. A population of flies is comprised of 500 females and 500 males with 375 wild type and 125 recessive mutants in each sex. Thus, 750 wild type and 250 mutants total comprise the population. If a single fly is sampled from this population, what is the probability for it to be

1. A wild-type female? Ans. (0.75)(0.50) = 0.375. Why?
2. Either a female or wild type or both wild type and female? Ans. = 0.875
3. Either a wild-type female or a wild-type male? Ans. = 0.75

If 75 wild-type males are discarded from this population, it will then comprise 500 females and 425 males, but 375 wild type: 125 mutants among females and 300 wild type: 125 mutants among males. In this reduced population, what is the probability for any fly to be

1. Wild type? Ans. 675/925 = 0.73
2. A female? Ans. 500/925 = 0.54
3. A female if it is wild type? Ans. 375/675 = 0.556
4. A female if it is a mutant? Ans. 125/250 = 0.50
5. A wild-type female? Ans. 375/925 = 0.405 (compare with question 1 above)
6. Is it correct to take the product of probabilities for being wild type and for being female in this reduced population—that is (0.73)(0.54) = 0.394—to answer question 1 above? Ans. No. Why not?

B PERMUTATIONS AND COMBINATIONS

If one event can happen in *a* different ways and a second event can happen in *b* different ways, a third in *c*, . . . , and their happenings are independent of each other, then the number of ways they may happen jointly is their product: (*a*) (*b*) (*c*) . . . .

Permutations (Arrangements, Orders of Objects). If we are to put objects into linear order, there are n choices for position No. 1, $(n-1)$ for position No. 2, $(n-2)$ for position No. 3, etc. The rth position can be filled with any of the $n-(r-1)$ remaining objects, or $n-r+1$. Thus, if r objects out of n total objects are to be arranged in order, ${}_nP_r = n(n-1)(n-2)\cdots(n-r+1)$, ${}_nP_r$ is permutations of n objects taking r at a time.

Also, because $n! = n(n-1)(n-2)\cdots(n-r+1)(n-r)(n-r-1)\cdots(2)(1)$ and $(n-r)! = (n-r)(n-r-1)\cdots(2)(1)$, then

$$ {}_nP_r = n(n-1)(n-2)\cdots(n-r+1) = \frac{n!}{(n-r)!} \quad \text{(A-2-1)} $$

When $n = r$ (that is, all objects in the array are ordered), then ${}_nP_r = {}_nP_n = n!/0! = n!$, because $0! = 1$.*

EXAMPLE 2. If we put 20 virgin females ($a, b, c, \ldots s, t$) and 5 males into a bottle, how many possible orders of females mating in sequence could occur? Assume each male mates just once and in sequence. (Orders of mating could be, for example, *a-b-f-j-q*, *b-a-f-j-q*, *f-a-b-j-q*, ...) Ans. ${}_{20}P_5 = 20(19)(18)(17)(16) = 1{,}860{,}480$ orders of mating.

Combinations. Combinations are permutations of n objects with r of these indistinguishable. When a subgroup within a larger set contains objects that are identical or indistinguishable, we divide the permutations of all objects in the set by the permutations within the subgroup.

EXAMPLE 3. If we have AAB where the two A's are indistinguishable, $(A_1 = A_2)$, in how many ways can these three letters be arranged in line? Ans. Three—AAB, ABA, BAA. However, if $A_1 \neq A_2$ (distinguishable), there would be six:

$$ \begin{array}{ccc} A_1A_2B & A_1BA_2 & BA_1A_2 \\ A_2A_1B & A_2BA_1 & BA_2A_1 \end{array} $$

When $A_1 = A_2$ (indistinguishable), their permutations within their subgroup are not considered, and we divide the total ${}_nP_n = {}_3P_3 = 3!$ by the subgroup permutations (${}_rP_r = {}_2P_2 = 2!$) since there are $r!$ different subgroup permutations in each group. This is the principle of combinations† (used as the coefficient of binomial probability distribution):

* To demonstrate that $0! = 1$, it can easily be shown that $n! = n(n-1)!$ Then, any integer! = that integer × (integer − 1)! Thus $1! = 1(1-1)!$ and $1! = 1 \cdot 0!$ and $1 = 0!$

† An amusing and informative example of permutations and combinations can be found in R. A. Fisher's "Mathematics of a Lady Tasting Tea" (1960). Fisher suggests an experiment for a lady who says she can detect if a cup of tea made with milk can be distinguished as to whether "the milk or the tea infusion was first added to the cup." The experiment consists of mixing eight cups of tea, four in one way and four in the other. The lady is asked to separate them into two sets of four after being presented with them in random order. If she chooses four in succession, she will do that in $8 \times 7 \times 6 \times 5 = 1680$ ways. But that number includes every possible set of four and all orders within each set (assuming each cup were distinguishable). Within each set of four, the order of cups is not important ($4 \times 3 \times 2 \times 1 = 24$ orders in a row), so the number of possible choices the lady has will be $\frac{1680}{24} = 70$ sets of four. A random choice of two sets of four each would be correctly made then $\frac{1}{70}$ times on the average.

$$_nC_r = \frac{_nP_r}{r!} = \frac{n!}{(n-r)!r!} \tag{A-2-2}$$

also symbolized as $\binom{n}{r}$

EXAMPLE 4. A nucleic acid molecule has eight bases of four types: *C*, *A*, *G*, *U*. If there are 2 *C*'s, 4 *U*'s, 1 *A*, and 1 *G*, how many molecular arrangements are possible (assume a single strand such as mRNA)?

$$\text{Ans. } \frac{_8P_8}{2!4!1!1!} = 840$$

EXAMPLE 5. How many possible combinations of five females out of 20 total from Example 2 above could be mated to the five males, assuming each male mates but once? (Note the three orders given for Example 2 are a single combination: *a-b-f-j-q*.)

$$\text{Ans. } _{20}C_5 = \frac{20!}{15!5!} = 15{,}504 \text{ combinations of five females.}$$

C PROBABILITY FOR COMPOUND EVENTS

When two or more independent events occur together, the proportionality rule follows: probability for their happening jointly is the product of their single event probabilities. For any combination of two or more events involving more than one way in which the combination can occur (that is, the sequences of occurrence—permutations—are indistinguishable), the permutations of the combination may be regarded as mutually exclusive, and the total probability for that combination is the sum of all identical combination probabilities (sum of all sequence probabilities).

EXAMPLE 6. In a Mendelian F_2 generation produced by inbreeding an F_1 *Aa* genotype, the *aa* genotype is expected to occur in $\frac{1}{4}$ of the total progeny. Each individual has the same probability for being *aa*, since each is independently produced. If the F_2 family consists of just four individuals, the probability for all four to be *aa* will be $(\frac{1}{4})^4 = \frac{1}{256}$. Then what is the probability for the combination of $3A{:}1a$ phenotypes in the F_2 family of four? Since $p_A = \frac{3}{4}$ and $q_A = \frac{1}{4}$, the product $(\frac{3}{4})^3(\frac{1}{4})^1$ would account for the probability of $3A{:}1a$ occurring in that sequence (3*A*'s followed by 1*a*); however, the same combination can occur in four ways (the *a* phenotype can occur last, third, second, or first). We then add the four identical probabilities making $4(\frac{3}{4})^3(\frac{1}{4})^1$. The coefficient 4 is $_4C_1$ or $\binom{4}{1}$ from (A-2-2). Expressing all possible combinations for F_2 families of four we obtain:

Four child family ratios	$4A$	$3A{:}1a$	$2A{:}2a$	$1A{:}3a$	$4a$
Single-order probability	p^4	p^3q^1	p^2q^2	p^1q^3	q^4
$_nC_r$ = number of orders	1	4	6	4	1
Total probability P	$1p^4$	$4p^3q^1$	$6p^2q^2$	$4p^1q^3$	$1q^4$
If $p_A = \frac{3}{4}$, $q_a = \frac{1}{4}$, then P =	$\frac{81}{256}$	$\frac{108}{256}$	$\frac{54}{256}$	$\frac{12}{256}$	$\frac{1}{256}$

$$P = {_nC_r}\, p^r q^{n-r} = \frac{n!}{r!(n-r)!} p^r q^{n-r} = \binom{n}{r} p^r q^{n-r} \tag{A-2-3}$$

Note that the coefficient ($_nC_r$) follows the *combinations* rule (A-2-2) because the r genotypes within the n size family are indistinguishable; that is, all dominant phenotypes look alike, and the $n - r$ genotypes (recessive phenotypes) also look alike.

D SAMPLING WITH OR WITHOUT REPLACEMENT

If each trial (or sample) is made with the same set of independent objects (or the same expectation of objects), then each sampling will have the same probability. If *replacement* is made or if the probabilities do not change after sampling in any trial, p and q are the same at each sampling.

EXAMPLE 7. There are four black and six white mice in a cage. If we draw out two mice by taking one at a time and putting the first one back after it is identified, what is the probability of getting two black mice (note: they could be the same mouse)? Ans. $p = (\frac{4}{10})^2 = \frac{4}{25} = 0.16$.

If each trial (or sample) were made *without replacing* the sampled objects, then p would change with each sample.

EXAMPLE 8. What is the probability of drawing out two black mice from the cage in Example 7 if the first mouse is not replaced? Ans. $\frac{2}{15} = 0.133$.*

This result can be expressed in terms of sets of combinations that include all the ways in which the sample can occur with

$$\text{No. ways for two black mice} = {}_4C_2 = \frac{4!}{2!2!} = \frac{(4)(3)}{2}$$

$$\text{No. ways for zero white mice} = {}_6C_0 = \frac{6!}{6!0!} = 1$$

$$\text{Total ways for two mice} = {}_{10}C_2 = \frac{10!}{8!2!} = \frac{(10)(9)}{2}$$

$$\text{Then } p = \frac{\binom{4}{2}\binom{6}{0}}{\binom{10}{2}} = \frac{(4)(3)/2}{(10)(9)/2} = \frac{2}{15} = 0.133$$

* Note that this probability is less than that in Example 7. However, as sample and population sizes increase, the probabilities of sampling with or without replacement become approximately equal; for example, with 10 times the number of mice and the same frequencies of black and white,

$$p\,(\text{with replacement}) = (40/100)^2 = 0.16$$

$$p\,(\text{without replacement}) = \frac{(40)(39)}{100(99)} = 0.158$$

Probability of sampling *without replacement*, then, can be represented as follows:

Let n = sample size
N = size of population from which sample is taken
r = number of certain type (A) in population
$N - r$ = number of other type (B) in population
x = number of type A being considered in the sample

Then

$$P = \frac{\binom{r}{x}\binom{N-r}{n-x}}{\binom{N}{n}} \tag{A-2-4}$$

Note that when we say sampling is "at random," we simply mean the probability of choosing any single object is $\frac{1}{{}_NC_n}$.

In this example we may also express the probability in simpler terms for drawing out two black mice without replacement as the product:

$p\{E_1 \cap E_2\} = p\{E_1\} \cdot p\{E_2|E_1\} = (\frac{4}{10})(\frac{3}{9}) = \frac{2}{15}$

This product may also be expressed as a branch diagram:

First Probability	Second Probability	Product
Black $\frac{4}{10}$	Black $\frac{3}{9}$	$\frac{12}{90} = \frac{2}{15} = 0.133$
	White $\frac{6}{9}$	$\frac{24}{90} = \frac{4}{15} = 0.267$
White $\frac{6}{10}$	Black $\frac{4}{9}$	$\frac{24}{90} = \frac{4}{15} = 0.267$
	White $\frac{5}{9}$	$\frac{30}{90} = \frac{5}{15} = 0.333$

A-3

Central Tendencies

A ARITHMETIC MEAN, OR AVERAGE

The sample arithmetic mean (denoted by $\bar{Y}$) estimates the parameter mean (μ), and is defined as

$$\bar{Y} = \frac{\sum Y_i}{n} \tag{A-3-1}$$

where Y_i = an individual variate value, and n = sample size. In more complete notation,

$$\bar{Y} = \frac{\sum_{i=1}^{n} Y_i}{n}$$

When variate values are grouped by frequencies (f), the mean is defined as

$$\bar{Y} = \frac{\sum f Y_i}{n} \tag{A-3-1A}$$

It is the point in a distribution around which the deviations add up to zero. This fact is fundamental and is demonstrated as follows. If each Y value is represented by the mean value ($\bar{Y}$) of the sample array, then

$$\sum Y = n\bar{Y} \tag{A-3-2}$$

EXAMPLE 1. The sum of 5 numbers equals 5 times the mean of the numbers: $7 + 4 + 4 + 3 + 2 = 4 + 4 + 4 + 4 + 4 = 5(4)$ Then, from (A-3-2), $\sum Y - n\bar{Y} = 0$, since the left side of this identity $= (Y_1 - \bar{Y}) + (Y_2 - \bar{Y}) + \cdots$. Then

$$\sum(Y - \bar{Y}) = 0 \tag{A-3-3}$$

Sum of deviations around the sample mean is zero. Another important relationship between deviations and the mean is demonstrated by taking any arbitrary value for the mean ($\bar{Y}_a$) and finding that the sum of deviations around that value does not equal zero. In the sample in Example 1 above, we assume the mean to be $\bar{Y}_a = 5$, $d = Y - \bar{Y}$, and deviation from the assumed mean is $d_a = Y - \bar{Y}_a$.

	Y	$\bar{Y}$	d	$\bar{Y}_a$	d_a
	7	4	+3	5	2
	4	4	0	5	−1
	4	4	0	5	−1
	3	4	−1	5	−2
	2	4	−2	5	−3
$\sum$	20	20	0	25	−5

It can be demonstrated that the constant difference between true mean and assumed mean equals the *average deviation*, as follows: $d_a = Y_i - \bar{Y}_a$ (deviation from assumed mean). Summed, $\sum d_a = \sum(Y_i - \bar{Y}_a) = \sum Y_i - n\bar{Y}_a = n(\bar{Y} - \bar{Y}_a)$. Then:

$$\frac{\sum d_a}{n} = (\bar{Y} - \bar{Y}_a) \tag{A-3-4}$$

the average deviation from the assumed mean $= \bar{d}_a$. The true mean $=$ the assumed mean $+$ the average deviation from the assumed mean:

$$\bar{Y} = \bar{Y}_a + \bar{d}_a \tag{A-3-4A}$$

Note: If the assumed mean $= 0$, then (A-3-4A) reduces to (A-3-1), since the deviations (d_a) then are the Y_i values.

B OTHER MEANS

Geometric Mean. If observed values form a geometric series (logarithmic or exponential series) it is appropriate to calculate the nth root of the product of values:

$$\text{Geometric mean} = \sqrt[n]{Y_1 Y_2 Y_3 \cdots Y_n} = \sqrt[n]{\prod_{i=1}^{n} Y_i} \tag{A-3-5}$$

In practice, the variables must be transformed into logarithms, the mean log taken, and the antilog is the geometric mean (or value at the midway point of an exponential curve).

EXAMPLE 2. A population of cells increases geometrically with time. If a single cell is doubling each hour, then what is the average cell number (at half the time) for a 5-hour interval?

Time (hours)	Cell No.	$\log_{10}$
1	1	0.00000
2	2	0.30103
3	4	0.60206
4	8	0.90309
5	16	1.20412
	$\sum = 31$	$\sum = 3.01030$
	Average = 6.2	0.60206

The arithmetic mean = 6.2 cells, but the geometric mean = 4.0 (antilog 0.60206), or the cells at half time (2.5 hours). (Note: The geometric mean $<$ the arithmetic mean.)

Harmonic Mean. When data are observed at a certain rate (miles per hour, flies per minute, or some measure per time interval), it might be appropriate to average the time as the variable.

$$\text{harmonic mean} = \text{reciprocal of} \frac{1}{H_Y} \tag{A-3-6}$$

where $1/H_Y = (1/n)\sum(1/Y)$ (average reciprocal).

EXAMPLE 3. In studying the rate of flies' mating, the following observations were made in 1-minute intervals. If the reciprocal of time is taken, we get the fraction of a fly per unit time (third column):

Time in Minutes to Mating Y	No. Flies Mating (f)	Reciprocal Time ($1/Y$)	f/Y
1	4	$\frac{1}{1}$	4
2	3	$\frac{1}{2}$	$\frac{3}{2}$
3	2	$\frac{1}{3}$	$\frac{2}{3}$
4	0	$\frac{1}{4}$	$\frac{0}{4}$
5	1	$\frac{1}{5}$	$\frac{1}{5}$
	$\sum = 10$		$\sum f/Y = \frac{191}{30}$

$1/H_Y = (\frac{1}{10})(\frac{191}{30}) = 0.6367$; therefore, the harmonic mean is $H_Y = 1.57$ flies per minute (while arithmetic mean $= \sum f\,Y/n = 2.10$).

C OTHER CENTRAL TENDENCIES

Median. The middle value that divides an ordered array into two halves is the median. It is the $(n + 1)/2$th item in the array. While extreme values in a sample tend to shift the arithmetic mean (as in skewed distributions), the median is less affected by extremes. It is also useful in collecting data where prolonged observation is inefficient (rate of mortality, LD_{50}, after irradiation, for example). Since the arithmetic mean and median coincide in a normal distribution, it is useful in estimating asymmetry in a distribution.

Mode. The most frequent class, or highest ordinate in a unimodal distribution is the mode. It is the best indicator of heterogeneity in the distribution when multimodality is the case; however, it is most sensitive to sample fluctuations, so it is not reliable in small samples. Because mode, median, and mean lie in that order in the direction of a long tail of a skewed distribution, an approximate value can be calculated as follows: mode = 3 (median) − $2\bar{Y}$.

A-4

Binomial Distribution

A BINOMIAL EXPANSION

The expected probability distribution of discrete objects having two attributes, or qualitites, of interest throughout replicate trials, or sets of independent observations (each trial, or set, having the same number of objects), follows the binomial expansion series, with the general term of (A-2-3).

EXAMPLE 1. Suppose we have replicated a monohybrid testcross ($Aa \times aa$) in drosophila over several vial cultures. We sample the emerging flies by just 10 from each vial. If we expect $\frac{1}{2}Aa:\frac{1}{2}aa$ as the individual genotype probabilities, the distribution of vials having particular lots of 10 will be expected as Table A-4-1.

Note that the expected 5:5 ratio would occur in less than one-fourth of the vials (0.246). As the sample size increases, the expected exact ratio based on a priori probabilities ($p:q$) will have decreasing probability of occurrence.

B SAMPLING FROM A BINOMIAL DISTRIBUTION

Estimating the parameters mean and variance is done by sampling from a binomial distribution. If we take a small number of replicate samples (k) of size n (n = number of independent discrete objects per sample), with r of one type and $n - r$ of a second type, then the arithmetic mean ($\bar{r}$) will be

$$\bar{r} = \frac{\sum r}{k} = np \qquad \text{(A-4-1)}$$

where $p = \sum r/nk$. With variance,*

$$V_{\bar{r}} = npq \qquad \text{(A-4-2)}$$

* Variance is the mean squared deviation. See A-6-E for derivation of this formula.

TABLE A-4-1 *Binomial distribution: testcross progenies from Aa × aa in lots of 10*

$Aa:aa$ =	10:0	9:1	8:2	7:3	6:4	5:5	4:6	...
Expected* $(p+q)^n$ =	p^n	$+ np^{n-1}q^1 +$	$\frac{n(n-1)}{2!}p^{n-2}q^2 +$	$\frac{n(n-1)(n-2)}{3!}p^{n-3}q^3 +$	...	...	...	
Probability fraction P =	$(\frac{1}{2})^{10} = \frac{1}{1024}$	$+ 10(\frac{1}{2})^9(\frac{1}{2})^1 = \frac{10}{1024} +$	$45(\frac{1}{2})^8(\frac{1}{2})^2 = \frac{45}{1024} +$	$120(\frac{1}{2})^7(\frac{1}{2})^3 = \frac{120}{1024}$	$+ 210(\frac{1}{2})^6(\frac{1}{2})^4 = \frac{210}{1024} +$	$252(\frac{1}{2})^5(\frac{1}{2})^5 = \frac{252}{1024} +$	$210(\frac{1}{2})^4(\frac{1}{2})^6 = \frac{210}{1024}$	...
Probability decimal =	0.00098	0.00977	0.04395	0.11719	0.20508	0.24609	0.20508	...

* General term $= \frac{n!}{r!(n-r)!} p^n q^{n-r}$ (see formula (A-2-3)).

in terms of numbers of r type, or variance in terms of *proportion*

$$V_p = \frac{pq}{n} \tag{A-4-3}$$

EXAMPLE 2. If we were sampling testcrosses as in Example 1 in just 5 vials of 10 flies per vial, we might observe the following:

Vial Sample	(r) Aa	$(n-r)$ aa	n	
1	7	3	10	
2	4	6	10	
3	8	2	10	
4	5	5	10	
5	3	7	10	
Sums	27	23	50	(k = 5 samples)

Mean $(\bar{r}) = \frac{27}{5} = 5.4$ (estimates μ, the parameter mean). Sample probability for $Aa(p) = \frac{27}{50} = 0.54$ (estimates the parameter $p = 0.50$). Variance $V = 10(0.54)(0.46) = 2.484$ or in terms of proportion $= (0.54)(0.46)/10 = 0.2484$.

Formulas (A-4-1), (A-4-2), and (A-4-3) can be demonstrated by expanding the binomial (expected frequency, or probability) for each r type beginning with the zero-r end of the distribution as in Table A-4-2. To obtain the *mean* $(\bar{r})$, use (A-3-1A), letting $r = Y_i$ (observed value) and k samples = the denominator. Thus, $\sum fr$ is the sum of the third column in Table A-4-2:

$$\bar{r} = \frac{\sum fr}{k} = \frac{1}{k}[knq^{n-1}p + kn(n-1)q^{n-2}p^2 + \cdots + nkp^n] = np(q+p)^{n-1}.$$

TABLE A-4-2 *Mean of the binomial distribution* (*Right column with* r^2 *is to be used for variance in A-6*)

r	Frequency or Expected No. (f)	fr	fr^2
0	kq^n	0	0
1	$knq^{n-1}p$	$knq^{n-1}p$	$knq^{n-1}p$
2	$k\frac{n(n-1)}{2!}q^{n-2}p^2$	$k(n)(n-1)q^{n-2}p^2$	$2kn(n-1)q^{n-2}p^2$
3	$k\frac{n(n-1)(n-2)}{3!}q^{n-3}p^3$	$k\frac{n(n-1)(n-2)}{2!}q^{n-3}p^3$	$3k\frac{n(n-1)(n-2)}{2!}q^{n-3}p^3$
⋮			
n	kp^n	nkp^n	n^2kp^n

Since $(q + p) = 1$, and k cancels out, then $\bar{r} = np$ (A-4-1). Therefore, a single sample $(k = 1)$ is sufficient to estimate p.

EXAMPLE 3. A certain genetic condition is found to be rather common in a population. All families with exactly 10 children (n) are sampled until 80 such families (k) are checked for the condition. The distribution of families is given:

No. Children with Condition in Family (r)	Frequency of Family (f)	Children (fr)
0	6	0
1	20	20
2	28	56
3	12	36
4	8	32
5	6	30
6	0	0
7	0	0
Total	80 families	174 children

Mean $(\bar{r})$ children per family with condition $= \sum fr/k = \frac{174}{80} = 2.175 = np$. Estimate probability for child to have condition $(p) = 0.2175$. Variance$_{\bar{r}}$ $(V_{\bar{r}}) = 10(0.2175)(0.7825) = 1.7019$. To obtain the expected random distribution of families based on sample size of 10 children per family, one may substitute these p,q values into Table A-4-1. For example, the top three expected frequencies would be as follows:

$80 \times$ exp. P (family with 0 children affected): $q^{10} = (0.7825)^{10} = 6.89$

$80 \times$ exp. P (family with 1 child affected): $10q^9p = 10(0.7825)^9(0.2175) = 19.14$

$80 \times$ exp. P (family with 2 children affected): $45q^8p^2 = 45(0.7825)^8(0.2175)^2 = 23.94$

The student should complete the array of expected frequencies.

C POISSON DISTRIBUTION (SPECIAL CASE OF BINOMIAL DISTRIBUTION)

With discrete objects with two attributes when the probability for a particular attribute is small $(p < 0.01)$ and the number of trials (samples), or "events," is large $(n > 100)$, and if we wish to know whether a particular distribution conforms to the binomial, the computation becomes very cumbersome and inaccurate even with logs. The Poisson distribution should be applied.

If n is large $(n \to \infty)$ and r is small, then for each binomial term,

$$\frac{n!}{r!(n-r)!}p^r q^{n-r}, \qquad q^{n-r} \text{ approximately } = q^n \approx e^{-np}$$

(where e = base of natural logs and $np = \bar{r}$ = mean).* So that the approximate binomial term is the general Poisson term:

$$\frac{n!p^r e^{-np}}{r!(n-r)!} = \frac{(np)^r e^{-np}}{r!} \qquad \text{(A-4-4)}$$

Because $n!/(n-r)! \approx n^r$.

Since e^{-np} is constant for all terms in the distribution, (with np = mean ($\bar{r}$), estimating that parameter μ) and r alone changes for each term, the Poisson distribution is

$$e^{-np}\left[1 + np + \frac{(np)^2}{2!} + \frac{(np)^3}{3!} \cdots\right] \qquad \text{(A-4-5)}$$

for $r = 0, 1, 2, 3 \ldots$ with variance (see formula (A-4-2)).

$$V_{\bar{r}} = np = \text{mean } (\bar{r}) \qquad \text{(A-4-6)}$$

because $q \approx 1$.

EXAMPLE 4. A brief example that fits a Poisson distribution was quoted by R. A. Fisher from data collected by Bortkewitch on the mortality of Prussian cavalrymen in 10 army corps from being kicked by a horse each year for a period of 20 years.

Deaths/Corps/Year	No. Corps Frequency	Probability	Expected No.
0	109	0.5433	108.7
1	65	0.3314	66.3
2	22	0.1011	20.1
3	3	0.0205	4.1
4	1	0.0031	0.6
5	0		
Total Corps	200		199.9

* The base of natural logarithms is

$$e = \left(1 + \frac{1}{n}\right)^n = 1 + \frac{n}{n} + \frac{n(n-1)}{n^2(2!)} + \frac{n(n-1)(n-2)}{n^3(3!)} + \cdots$$

As $n \to \infty$,

$$e = 1 + \frac{1}{1} + \frac{1}{2!} + \frac{1}{3!} + \cdots$$

Also,

$$e^x = 1 + x + \frac{x^2}{2!} + \frac{x^3}{3!} + \cdots$$

The following expression is often found:

$$(1 - x)^n = 1 - nx + \frac{n(n-1)}{2!}x^2 - \frac{n(n-1)(n-2)}{3!}x^3 \ldots$$

which approximately equals

$$e^{-nx} = 1 - nx + \frac{n^2x^2}{2!} - \frac{n^3x^3}{3!} \cdots$$

In the Poisson derivation, if x is p, then $1 - x = q$ so that $q^n \approx e^{-np} = e^{-\bar{r}}$.

Mean deaths/corps/year $= \frac{122}{200} = 0.6100$. Note: $e^{-0.61} = 0.5433$. Variance $= 0.6109$ (note nearly equal to mean).

EXAMPLE 5. In 27 cities of population ranging form 10,000 to 20,000, the total deaths due to typhoid were 19 in a certain year early in this century. Total population for all these cities combined was 373,000 with the average city's population = 13,815 and the average deaths per city = 0.704. Thus, the probability for individual death in those cities was $p = 0.704/13{,}815$. If these deaths were distributed according to a Poisson, then, because $e^{-0.704} = 0.4946$:

Deaths per city	0	1	2	3	...
Expected no. of cities	13.3	9.7	3.3	0.7	...

D MULTINOMIAL DISTRIBUTION

When an object may have more than two attributes, or qualities, throughout replicate trials, the probability distribution of trials follows the multinomial, an extension of (A-2-3):

$$(p_1 + p_2 + p_3 + \cdots p_j) = \frac{n!}{r_1!r_2!r_3!\ldots r_j!}(p_1)^{r_1}(p_2)^{r_2}(p_3)^{r_3}\cdots(p_j)^{r_j} \qquad \text{(A-4-7)}$$

which has $[(n + 1)(n + 2)(n + 3)\cdots(n + j - 1)]/(j - 1)!$ terms.

EXAMPLE 6. In a family with parents of an A blood type father ($I^{A}i$) and a B blood type mother ($I^{B}i$), there are 8 children. What is the probability that the children will occur in the following ratio of blood types: 1A:3B:2AB:2O? Note: each child has equal probability for any of the four blood types.

$$\text{Ans. } P = \frac{8!}{(1!)(3!)(2!)(2!)}\left[\left(\frac{1}{4}\right)^1\left(\frac{1}{4}\right)^3\left(\frac{1}{4}\right)^2\left(\frac{1}{4}\right)^2\right] = \frac{1680}{(4)^8} = 0.256.$$

E NEGATIVE BINOMIAL DISTRIBUTION

Negative binomial distribution (this discussion modified from Li, 1976, pp. 392–393) results if there is a *constant* probability (p) for "success" and a specified number (n) of "successes" as parameters of the distribution, while a variable number (r) of "failures" generates the probability distribution. What we wish to know is the probability of having r events (failures) followed by the nth event (success). If there are $n + r$ trials, the last trial must produce the nth success. In the previous $n + r - 1$ trials, there are only $n - 1$ successes and r failures with probability $P = \binom{n + r - 1}{r} p^{n-1}q^r$, analogous to (A-2-3). The last trial is a success with probability p so that the final probability is

$$P = \binom{n + r - 1}{r} p^n q^r \qquad \text{(A-4-8)}$$

for $r = 0, 1, 2, 3, \ldots$.

The general term for combinations, or coefficient, in (A-4-8) is from (A-2-2):

$$\frac{(n+r-1)!}{r!(n+r-1-r)!} = \frac{(n+r-1)!}{r!(n-1)!}$$

$$= \frac{(n+r-1)(n+r-2)(n+r-3)\cdots(n+1)(n)}{r!} \tag{A-4-9}$$

See (A-2-B) on permutations. Note that terms to the right in the numerator cancel $(n-1)!$ This distribution can also be represented as

$$P = \binom{-n}{r} p^n(-q)^r \tag{A-4-8A}$$

which is algebraically equivalent to (A-4-8). The general combination term for this form is

$$\frac{(-n)!}{r!(-n-r)!} \tag{A-4-9A}$$

which is less cumbersome to write than (A-4-9).

In contrast with the Poisson distribution, the variance is greater than the mean in this negative binomial distribution:

$$\text{mean } \bar{r} = \frac{nq}{p} \tag{A-4-10}$$

$$\text{variance } V = \frac{nq}{p^2} \tag{A-4-11}$$

EXAMPLE 7. For a negative binomial distribution, let $n = 2$, $p = q = \frac{1}{2}$, so that $p^n = \frac{1}{4}$. Then (A-4-8) becomes $P = \binom{r+1}{r}\left(\frac{1}{4}\right)\left(\frac{1}{2}\right)^r$. The combination term (A-4-9) becomes $[(r+1)!]/r!$ so that the distribution is as follows:

Variable r	0	1	2	3	...
Combinations	$\frac{1}{0!}$	$\frac{(2)(1)}{1!}$	$\frac{(3)(2)(1)}{2!}$	$\frac{(4)(3)(2)(1)}{3!}$	...
	$= 1$	$= 2$	$= 3$	$= 4$	
q^r	1	$(\frac{1}{2})^1$	$(\frac{1}{2})^2$	$(\frac{1}{2})^3$	...
p^n (constant)	$\frac{1}{4}$	$\frac{1}{4}$	$\frac{1}{4}$	$\frac{1}{4}$	
Probability distribution	$\frac{1}{4}$	$\frac{2}{8}$	$\frac{3}{16}$	$\frac{4}{32}$	...
$\bar{r} = 2$, $V = 4$					

EXAMPLE 8. Kojima and Kelleher (1962) showed that the distribution of human family size is more closely fitted by a negative binomial than by a Poisson distribution. The parameters they found to give a good fit were a mean $\bar{r} = 3$ children per family and variance $V = 7$. Therefore, the distribution for r family size was approximated by letting $n = 2.25$, $p = \frac{3}{7}$, $q = \frac{4}{7}$. Using (A-4-8A) as follows:

Variable r	0	1	2	3	...
Combinations	+1	−2.25	+3.655	−5.1797	...
$-q^r$	+1	−0.5714	+0.3265	−0.1866	...
p^n (constant)	+0.1486	0.1486	0.1486	0.1486	...
Probability distribution	0.1486	0.1911	0.1774	0.1436	...

A-5

Normal Distribution

Many kinds of data cannot be treated as binomial or multinomial distributions because they cannot easily be accounted for by proportions of discrete objects with two or more specific qualities; that is, they produce a *continuous* distribution within which it is nearly impossible to detect the number of factors acting on the variate and determining the distribution. For example, characteristics such as height, weight, size, number of leaves on a plant, number of hairs on a head, yield of crops, egg production, or intelligence quotient are all determined by so many genetic and environmental conditions that we cannot usually isolate any single factor operating on them. Thus, they tend to be "normally distributed"—that is, according to the *probability density function*, which is symmetrical around the mean, mode, and median that coincide. The normal distribution is the limit to the binomial and multinomial distributions when expanding $(p + q + r \ldots)^n$ as n approaches infinity. While n is small, if $p \neq q$, a binomial is skewed; but if neither p nor q approaches the magnitude of $1/n$ or less, the expansion will approach normality.

A THE NORMAL CURVE

The normal curve distribution formula is as follows:

$$Y = \frac{1}{\sigma\sqrt{2\pi}}(e)^{-(X - \mu)^2/2\sigma^2} \tag{A-5-1}$$

Where Y = height of the ordinate for a given X value, X = the variate value, e = base of natural logarithms, π = constant 3.14159+, μ = the mean, and σ = the standard deviation.

This curve extends from minus to plus infinity above and below the mean symmetrically; it depends on two measures: (1) the mean (μ), or central tendency, and (2) the standard deviation (σ), or dispersion (how wide apart the "tails" are). Since any pair of values for these parameters may describe a unique curve, an infinity of curves is possible. To overcome that difficulty, all curves can be *standardized* by using the ratio $\tau = (X - \mu)/\sigma$, or number of standard deviations in each deviation $(X - \mu)$, the *standardized normal deviate.* Sample estimates of τ are designated t. The scale is in new units, and the σ has the value of unity and the portion of the exponent $(X - \mu)^2/\sigma^2$ becomes τ^2, so that the formula becomes

$$Y = \frac{1}{\sqrt{2\pi}}(e)^{-\tau^2/2} \tag{A-5-2}$$

The area under this curve is equal to unity; for any value of τ, that portion of the area under the curve from any limit (such as the mean, for example) to the observed deviation point on the X-axis can be calculated as the probability for that deviation. Areas under the normal curve are tabulated in most of the statistical references given; because the distribution is symmetrical, only half the probabilities are tabulated. While the exact probability for a particular X value is infinitesimally small, we can determine the probability that X will lie between any two limits.

EXAMPLE 1. In a large human population the intelligence quotient (IQ) mean $= 100$, with a standard deviation of 13. What is the probability that an individual in that population will have an IQ between 80.5 and 106.5? These limits correspond to $t = -1.5$ and $+0.5$. From the mean to the lower limit, the area $= 0.4332$; to the upper limit, area $= 0.1915$, or a total $= 0.6247$, which is the desired probability.

It should be noted that in a normal distribution

$$\pm 1\tau = 2(0.3413) = 0.6826 \text{ of all observations}$$
$$\pm 2\tau = 2(0.4773) = 0.9546 \text{ of all observations}$$
$$\pm 3\tau = 2(0.4987) = 0.9974 \text{ of all observations}$$

or, to express the more well-known "confidence limits" in terms of τ:

$$0.95 \text{ of observations (95 percent confidence)} = 1.960\tau$$
$$0.99 \text{ of observations (99 percent confidence)} = 2.576\tau$$

It is also conventional to express the probability for estimating the limits within which a parameter (the mean, for example) may lie in terms of the error involved in calculating limits within which the parameter may lie. Thus, if we wish to calculate upper and lower limits within which a parameter mean may lie, we may choose to be 95 percent confident in those limits, and so be in error 5 percent of the time if those limits are chosen. With wider limits, we would be in error less often, and so on.

While the parameter (τ) is normally distributed (or nearly so in large populations), the parameter (σ) is seldom known, so that it must be estimated (s) from samples. The distribution of $(X - \bar{X})/s$, known as "Student's t," must be used to estimate limits of confidence for sample means or other sample statistics—a distribution that is narrower than normal (leptokurtic) but approaches normal with increase in sample size above $30 = n$. (See Mather, 1965, p. 43 for the t distribution formula.) A chart of t distribution is provided in Figure A-8-1.

A-6

Standard Deviation, Variance, and Degrees of Freedom

The measure of dispersion in a normal distribution is the point measured along the X-axis from the mean where the acceleration in steepness of the curve becomes zero (where the second derivative equals zero) and is known as the *standard deviation*, symbolized as σ for the parameter population or as s for the sample estimate of σ. It is the square root of the *variance*, or mean squared deviation ("mean square"):

$$\sigma^2 = \frac{\sum(Y - \mu)^2}{n} = \frac{\text{sum of deviations squared}}{\text{population size}} \qquad \text{(A-6-1)}$$

In a sample of the population, $\bar{Y}$ estimates μ and the degrees of freedom $(n - 1)$ is the divisor. The numerator is the "sums of squares" around the sample mean $(\bar{Y})$ so that *sample variance* is

$$s^2 = \frac{\sum(Y_i - \bar{Y})^2}{n - 1} = \frac{\sum d^2}{n - 1} \qquad \text{(A-6-2)}$$

where Y_i = observed value. (We may symbolize variance as V when a distinction between sampling and parameter variance does not need to be made.)

A SUMS OF SQUARES, OR $\sum d^2$

It is cumbersome and inefficient to calculate the sample mean, then the deviations, and then square them. It can be shown in the following identity that it is far better to square each Y_i (observed value) and subtract a correction factor. The deviation from an assumed mean given above formula (A-3-4), omitting subscript i, will be as follows:

Subtracting and adding $\bar{Y}$

$$d_a = Y - \bar{Y}_a = (Y - \bar{Y}) + (\bar{Y} - \bar{Y}_a)$$

Squaring

$$(Y - \bar{Y}_a)^2 = (Y - \bar{Y})^2 + 2(Y - \bar{Y})(Y - \bar{Y}_a) + (\bar{Y} - \bar{Y}_a)^2$$

Summing

$$\sum(Y - \bar{Y}_a)^2 = \sum(Y - \bar{Y})^2 + 2(Y - \bar{Y}_a)\sum(Y - \bar{Y}) + n(\bar{Y} - \bar{Y}_a)^2$$

Because $\sum(Y - \bar{Y}) = 0$, according to (A-3-3), then

$$\sum(Y - \bar{Y}_a)^2 = \sum(Y - \bar{Y})^2 + n(\bar{Y} - \bar{Y}_a)^2 \qquad \text{(A-6-3)}$$

This is the sums of squared deviations from the assumed mean. If $\bar{Y}_a$ is assumed to be zero, then Y_i values become the deviations so that

$$\sum Y^2 = \sum(Y - \bar{Y})^2 + n\bar{Y}^2$$

$$\sum d^2 = \sum(Y_i - \bar{Y})^2 = \sum Y^2 - n\bar{Y}^2 \text{ "sums of squares"*} \qquad \text{(A-6-4)}$$

* The student may prove the identity (A-6-4) by expanding the expression as follows:

$$\begin{aligned}
(Y_1 - \bar{Y})^2 &= Y_1^2 - 2Y_1\bar{Y} + \bar{Y}^2 \\
(Y_2 - \bar{Y})^2 &= Y_2^2 - 2Y_2\bar{Y} + \bar{Y}^2 \\
(Y_3 - \bar{Y})^2 &= Y_3^2 - 2Y_3\bar{Y} + \bar{Y}^2 \\
&\vdots \\
\sum(Y_i - \bar{Y})^2 &= \sum Y_i^2 - 2\bar{Y}\sum Y_i + n\bar{Y}^2 = \sum Y_i^2 - 2\left(\frac{\sum Y_i}{n}\right)(\sum Y_i) + \frac{(\sum Y)^2}{n} \\
&= \sum Y_i^2 - \frac{(\sum Y)^2}{n}
\end{aligned}$$

Alternatively, because $\bar{Y} = \sum Y_i/n$, the sums of squares may be represented as follows:

$$\sum(Y - \bar{Y})^2 = \sum Y^2 - \frac{(\sum Y)^2}{n} \qquad \text{(A-6-4A)}$$

Sum of Y values squared minus the total squared and divided by n. Thus,

$$s^2 = \frac{\sum Y^2 - n\bar{Y}^2}{(n - 1)} \qquad \text{(A-6-2A)}$$

the sample variance, estimate of σ^2.

The "sums of squares" has two components: (a) = sum of squared observed values ($\sum Y_i^2$) plus (c) = the correction factor ($n\bar{Y}^2$), or $(\sum Y)^2/n$, the total squared and divided by sample size. It is always true that $a > c$, so that the sum of squares is always positive (except when it is zero) but never negative.

For example, the following inequality demonstrates that if any two values (x, y) are squared, the sum of their squares (a) is greater than their squared sum divided by $n(c)$ unless $x = y$, in which case the variance is zero.

$$x^2 + y^2 > \frac{(x + y)^2}{2}$$

$$2x^2 + 2y^2 > x^2 + 2xy + y^2$$

$$x^2 + y^2 > 2xy$$

$$(x - y)^2 > 0$$

This inequality is true for any number (n) of items:

$$x^2 + y^2 + z^2 + \cdots > \frac{(x + y + z + \cdots)^2}{n}$$

$$(x - y)^2 + (y - z)^2 + (x - z)^2 \cdots > 0$$

EXAMPLE 1. In Appendix A-3-A the sample table gave Y_i as 7, 4, 4, 3, 2. The sums of squares would then be, using (A-6-4A)

$$49 + 16 + 16 + 9 + 4 - \frac{(20)^2}{5} = 94 - 80 = 14$$

Note the identical result with taking the actual deviations and squaring them.

$$3^2 + 0^2 + 0^2 + (-1)^2 + (-2)^2 = 14$$

The sample variance would then be $s^2 = \frac{14}{4} = 3.5$.

B DEGREES OF FREEDOM (n − 1) FOR SAMPLE VARIANCE DENOMINATOR

At least two values are required to afford an estimate on the dispersion of a distribution, and any two values will supply only a *single* difference. The number of differences between n values that are not "fixed" $= n - 1$. Therefore, n observations contribute $n - 1$ *independent* differences to the estimation of variance.

EXAMPLE 2. For four objects A, B, C, D, we know the following three differences:

$$B - A = +1$$
$$C - A = +3$$
$$D - A = -1$$

What are the differences between $C - B, D - B, C - B$? Ans. $C - B = +2, D - B = -2, D - C = -4$. Therefore, $n - 1$ differences are sufficient and independent for n objects.

A proof that the sample variance denominator is $(n - 1)$ can be obtained from using "expected" algebra (according to the method illustrated by Li (1964). The symbol $E\{\ \}$ refers to "the expected value of" or "the long-range average value of." For example,

1. $E\{Y_i\} = \mu$, expected value of any item is the *mean.*
2. $E\{\sum Y_i\} = n\mu$, expected value of the sum of items is $n \times$ mean.
3. Parameter variance = expected squared deviation, or "mean square."

$$\sigma^2 = E\{Y - \mu\}^2 = E\{Y^2 - 2\mu Y + \mu^2\} \qquad \text{(omitting subscript } i)$$
$$\sigma^2 = E\{Y^2\} - 2\mu E\{Y\} + \mu^2 \qquad \text{(because } \mu \text{ is constant)}$$
$$\sigma^2 = E\{Y^2\} - 2\mu^2 + \mu^2 = E\{Y^2\} - \mu^2 \qquad \text{(because of 1 above)}$$

Therefore, $E\{Y^2\} = \mu^2 + \sigma^2$ and $E\{\sum Y^2\} = n(\mu^2 + \sigma^2)$ = first component of (A-6-4) sum of squares.

4. Expected *second* component of (A-6-4A) is $\left[\frac{(\sum Y)^2}{n}\right]$

Omitting $1/n$ temporarily,

$$E\{(Y_1 + Y_2 + Y_3 \cdots)^2\} = E\{Y_1^2 + Y_2^2 + Y_3^2 \cdots + 2Y_1Y_2 + 2Y_1Y_3 \cdots\}$$

Each square term has $E\{Y_i^2\} = \mu^2 + \sigma^2$ (as above). Each product term $E\{Y_iY_j\} = \mu^2$, because each $E\{Y_i\} = \mu$. So $E\{(\sum Y)^2\} = n(\mu^2 + \sigma^2) + n(n - 1)\mu^2$, because there are $n(n - 1)$ product terms.

Then, using $1/n$,

$$E\{(\textstyle\sum Y)^2/n\} = n\mu^2 + \sigma^2 = E\{n\bar{Y}^2\}$$

Finally, using items 3 and 4, expected value of sample variance $= E\{s^2\} = \sigma^2$. $E\{\sum Y^2\} = n\mu^2 + n\sigma^2$, using 3, and $E\{(\sum Y)^2/n\} = n\mu^2 + \sigma^2$, using 4. Subtracting to obtain (A-6-4A),

$$E\left\{\sum Y^2 - \frac{(\sum Y)^2}{n}\right\} = \sigma^2(n - 1)$$

Therefore,

$$E\left\{\frac{\sum Y^2 - \frac{(\sum Y)^2}{n}}{n-1}\right\} = \sigma^2 \tag{A-6-5}$$

$$E\{s^2\} = \sigma^2 \tag{A-6-5A}$$

Thus, dividing the sample sums of squares by degrees of freedom estimates the parameter variance.

Because the number of degrees of freedom refers to the number of independent differences contributing to an estimate of variance, it should be noted that each degree of freedom can have its own unique contribution and that isolation of degrees of freedom can be of great importance in statistical analysis. Partition of these in the components of variance or of chi-square, then, must be given special attention; however, analysis of variance and partitioning of chi-square are elaborate statistical techniques that go beyond the purposes of this appendix.

With the calculation of totals, averages, or constants of hypothetical parameters in variance analysis, chi-square tests, or other tests of significance, *in general*, the number of degrees of freedom is reduced by one for each such total, constant, or hypothetical parameter calculated from the data. When a variance is estimated, for example, there are $n - 1$ degrees of freedom because the mean has been calculated to obtain sums of squared deviations. Similarly, if a regression coefficient is estimated from observations of two variables on each individual (where Y may be a function of X), then another degree of freedom is lost, making $n - 2$ remaining for the significance of the regression. Also, in chi-square tests, each degree of freedom represents a unique comparison between one class and another; when there are multiple comparisons as in a contingency table with rows (r) and columns (c), the following degrees of freedom would hold in testing significance of differences:

Rows	$r - 1$
Columns	$c - 1$
Interactions	$(r - 1)(c - 1)$
Total	$rc - 1$

C VARIANCE WHEN OBSERVED VALUES (Y_i) MULTIPLIED BY A CONSTANT (c)

If all observed items in an array are multiplied by a constant (c), all deviations squared would be as follows. $(Y_i c - \bar{Y}c)^2 = (Y_i c)^2 - 2Y_i c + (\bar{Y}c)^2$, when summed, gives $\sum(Y_i c - \bar{Y}c)^2 = \sum(Y_i c)^2 - n(\bar{Y}c)^2$ (from formula (A-6-4), which becomes

$$c^2[\sum(Y_i)^2 - n(\bar{Y})^2] \tag{A-6-7}$$

(constant-squared multiplied by sums of squares). Thus, the variance is $c^2 V$.

Adding (or subtracting) a constant to all items in an array does not affect the variance because all deviations will be unchanged.

D VARIANCE OF A SUM $(X + Y)$ OR A DIFFERENCE $(X - Y)$ BETWEEN OBSERVED VALUES

If a sample includes measures X and Y variates (normally distributed) and we are interested in the variance of the sums $(X_i + Y_i)$, we proceed as follows:

$$V_X = \frac{\sum(X - \bar{X})^2}{n - 1} \quad \text{and} \quad V_Y = \frac{\sum(Y - \bar{Y})^2}{n - 1}$$

variances of X and Y variates respectively.

$$V_{(X+Y)} = \frac{\sum[X + Y - (\overline{X + Y})]^2}{n - 1}$$

but

$$(\overline{X + Y}) = \frac{\sum(X + Y)}{n} = \frac{\sum X}{n} + \frac{\sum Y}{n} = \bar{X} + \bar{Y}$$

so that

$$V_{(X+Y)} = \frac{\sum[X + Y - \bar{X} - \bar{Y}]^2}{n - 1} = \frac{\sum[(X - \bar{X}) + (Y - \bar{Y})]^2}{n - 1}$$

$$= \frac{\sum(X - \bar{X})^2 + \sum(Y - \bar{Y})^2 + 2\sum(X - \bar{X})(Y - \bar{Y})}{n - 1} \qquad \text{(A-6-8)}$$

thus,

$$V_{(X+Y)} = V_X + V_Y + 2CoV_{XY}$$

where CoV_{XY} = covariance of X and Y

$$CoV_{XY} = \frac{\sum(X - \bar{X})(Y - \bar{Y})}{n - 1}$$

or the sum of products for deviations in X by deviations in Y.

Thus, the variance of a sum (V_{X+Y}) equals the sum of the variances of X and Y plus twice their covariance. If X and Y are completely independent, they will be uncorrelated and their covariance will be zero. Thus, for independent variables, the variance of the sum $(X + Y)$ simply equals the sum of their respective variances.

EXAMPLE 3. Suppose five observations are made on variables X and Y, and we proceed to find deviations from the respective means $(X_i - \bar{X} = d_X)$ and $(Y_i - \bar{Y} = d_Y)$ as in the following table. First, X and Y are independent:

	X	d_X	d_X^2	Y	d_Y	d_Y^2	$d_X d_Y$	XY	$X + Y$	$d_{(X+Y)}$
	7	+3	9	10	+3	9	9	70	17	+6
	4	0	0	1	−6	36	0	4	5	−6
	4	0	0	3	−4	16	0	12	7	−4
	3	−1	1	12	+5	25	−5	36	15	+4
	2	−2	4	9	+2	4	−4	18	11	0
Sum	20	0	14	35	0	90	0	140	55	0
Mean $\bar{X} = 4$				$\bar{Y} = 7$					$\overline{X + Y} = 11$	

Sum of squares in $Y = \sum d_Y^2 = 90$, $V_Y = \frac{90}{4} = 22.5$

Then $V_{X+Y} = V_X + V_Y$ (since $\sum d_X d_Y = 0$); V_X is identical with the variance in Example 1: $26 = 3.5 + 22.5$.

Now we change the Y column slightly and find that X and Y are no longer independent:

	X	d_X	d_X^2	Y	d_Y	d_Y^2	$d_X d_Y$	XY	$X + Y$	$d_{(X+Y)}$
	7	+3	9	10	+3	9	9	70	17	+6
	4	0	0	4	−3	9	0	16	8	−3
	4	0	0	6	−1	1	0	24	10	−1
	3	−1	1	8	+1	1	−1	24	11	0
	2	−2	4	7	0	0	0	14	9	−2
Sum	20	0	14	35	0	20	8	148	55	0
Mean	$\bar{X} = 4$			$\bar{Y} = 7$					$\overline{X+Y} = 11$	

Sum of squares in $Y = \sum d_Y^2 = 20$, $V_Y = \frac{20}{4} = 5$

Then $V_{X+Y} = V_X + V_Y + 2CoV_{XY}$, with $CoV_{XY} = \sum d_X d_Y/(n-1) = \frac{8}{4} = 2$, becomes $12.5 = 3.5 + 5 + 2(2)$.

Covariance, analogous to the variance (A-6-2A), can be shown to be calculated from the following identity for sum of products (analogous to the sum of squares):

$$CoV = \frac{\sum d_X d_Y}{n-1} = \frac{\sum(XY) - n(\bar{X})(\bar{Y})}{n-1} = \frac{\sum(XY) - \left(\frac{\sum X}{n}\right)\left(\frac{\sum Y}{n}\right)}{n-1} \qquad \text{(A-6-9)}$$

In the two cases given above

$$CoV = \frac{140 - 5(4)(7)}{4} = 0$$

$$CoV = \frac{148 - 5(4)(7)}{4} = 2$$

The student should use the same reasoning and show that the variance of differences (V_{X-Y}) between two variables equals the sum of the separate variances *minus* twice the covariance:

$$V_{X-Y} = V_X + V_Y - 2CoV_{XY} \qquad \text{(A-6-10)}$$

Finally, it may be noted that in section C, multiplying by a constant (c) is identical to summing the variable to itself c times. For example, let $c = 3$, then $V_{X+X+X} = V_X + V_X + V_X + 2V_{XX} + 2V_{XX} + 2V_{XX}$, where V_{XX} is the covariance of a variable with itself, or 1, $= 9V_X = (3)^2 V_X$.

E VARIANCE OF A BINOMIAL

To obtain the variance of the binomial distribution (A-4-2), we use the column of r^2 from Table A-4-2. Letting r be the Y variable for sums of squares (A-6-4), we shall work with the first and second terms of (A-6-4) separately. The number of replicate samples (k) is the denominator of the variance (representing n in formula (A-6-1)), and it cancels out in the first step.

$$\frac{1}{k}[knq^{n-1}p + 2kn(n-1)a^{n-2}p^2 + \cdots + n^2kp^n] - kn^2p^2$$

where the first term is the sum collected from the fourth column of Table A-4-2 ($= \sum fr^2 = \sum Y^2$ from (A-6-4)), and where the second term is the mean-squared times k ($= n\bar{Y}^2$ from (A-6-4)). Thus, we have

$$np[q^{n-1} + 2(n-1)q^{n-2}p + \cdots + np^{n-1}] - n^2p^2$$

Writing the expression in brackets as sum of two terms and grouping,

$$np\{[q^{n-1} + (n-1)q^{n-2}p + \cdots p^{n-1}] + [(n-1)q^{n-2}p + \cdots (n-1)p^{n-1}]\} - n^2p^2 =$$
$$np[(q+p)^{n-1} + (n-1)p(q+p)^{n-2}] - n^2p^2 =$$
$$np[(1 + (n-1)p] - n^2p^2 = np + n^2p^2 - np^2 - n^2p^2 = np(1-p)$$

Therefore, in terms of discrete objects:

$$\text{variance of binomial } (s^2_{\text{bin}}) = npq \qquad \text{(A-6-11)}$$

given in (A-4-2).

To obtain variance of the binomial in terms of proportion in (A-4-3), if each term of Table A-4-2 (r column) is multiplied by $1/n$, each observed frequency will be a proportion of sample size. When all scores of a distribution are multiplied by a constant (c), the variance of the distribution becomes c^2 times as great (from (A-6-7)):

$$s^2 \text{ (in proportion)} = npq\left(\frac{1}{n}\right)^2 = \frac{pq}{n} \qquad \text{(A-6-11A)}$$

F COEFFICIENT OF VARIATION

Because large organisms have large standard deviations and small organisms have small ones, that does not mean that large organisms are actually more variable than small. For their size, their *relative variation* may be the same as that of the small. In order to put all sizes on the same scale, so to speak, the standard deviation is divided by the mean and usually multiplied by 100 to make the ratio a percentage. Thus the coefficient of variation (CV) becomes

$$CV = \frac{\sigma}{\mu} \times 100 \qquad \text{(A-6-12)}$$

estimated by $[s/\bar{Y}]$ (100).

For example, if elephants have $s = 50$ mm and mice have $s = 0.50$ mm, for some dimension such as body length, their CV's would be nearly identical if elephants are 100 times longer than mice.

Lewontin (1966) pointed out that if two variables X and Y are really identical except that Y is k times larger than $X(Y = kX)$, then from formula (A-6-7), we would have the variance k^2 as large: $s_Y^2 = k^2 s_X^2$ and the coefficient of variation for Y would be $CV_Y = k(s_X)/\bar{X}$.

If logarithms are taken of both X and Y, the variance of the logs of the measurements becomes invariant; if $Y = kX$, then $\log Y = \log k + \log X$. Since $\log k$ is a constant (with no variance), then we would not need the CV since

$$s^2_{\log Y} = s^2_{\log X} \quad \text{and} \quad s_{\log Y} = s_{\log X} \qquad \text{(A-6-13)}$$

It may also be noted that if logs are taken to the base e, the variance of the $\log_e(X) = (CV)^2$, approximately.

G COVARIANCE, REGRESSION, AND PRODUCT MOMENT CORRELATION BETWEEN X AND Y

If measurements in X and in Y values are taken in a sample, we may wish to know whether any constant relationship exists between X and Y. For example, in the growth of an individual over time, age (in time) will be the independent variate (X) and the amount of growth (height or weight) will be the dependent (Y). We wish to find a "moving average" with smallest amount of deviation for all pairs of X and Y measures: $(X_1 - \bar{X})(Y_1 - \bar{Y}) + (X_2 - \bar{X})(Y_2 - \bar{Y}) \cdots (X_i - \bar{X})(Y_i - \bar{Y})$ products are to be minimized. This constant relationship can be expressed as follows:

$$(Y_i - \bar{Y}) = b(X_i - \bar{X}) \qquad \text{or} \qquad b = \frac{(Y_i - \bar{Y})}{(X_i - \bar{X})} = \frac{d_Y}{d_X} \tag{A-6-14}$$

where d_Y and d_X represent deviations in Y and X, respectively.

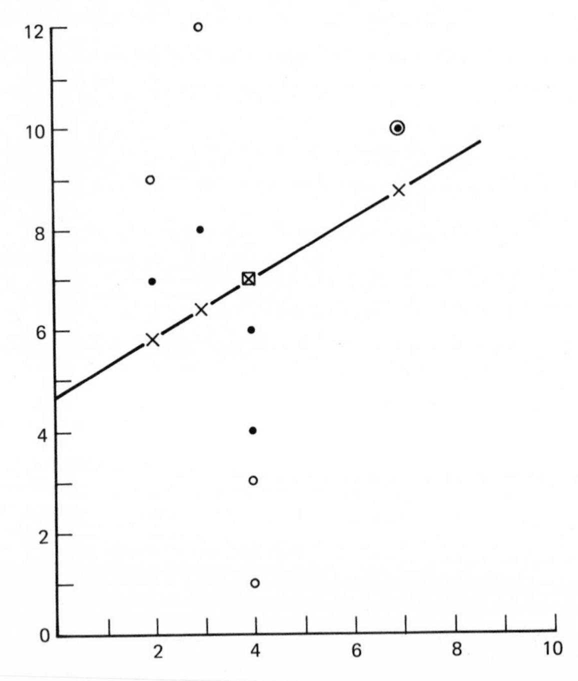

Figure A-6-1. Regression of five sample X, Y values from the second case with correlated Y (solid dots). Linear Y_L values entered on the line as X. Open dots are values from the first case with independent Y. The box ⊠ = $\bar{X}$, $\bar{Y}$. Both independent and correlated cases coincide at 7, 10.

However, if the deviations in X or Y are summed over the sample array, we have zero (from (A-3-3)). Therefore, in (A-6-14), we multiply and divide each sample ratio by the corresponding deviation in the independent variate $(X_i - \bar{X})$ before summing, so that

$$\frac{\sum[(Y_i - \bar{Y})(X_i - \bar{X})]}{(\sum X_i - \bar{X})^2} = \frac{\sum d_Y d_X}{\sum d_X^2} = b \quad \text{(A-6-15)}$$

This constant relationship b is the linear slope of a *regression*, the ratio of (sum of products)/(sum of squares (X)), which is also equal to the *CoV* (A-6-9)/(variance in X (A-6-2A)).

In the second table in Example 3, the data can be represented on a graph as in Figure A-6-1 with solid dots, while the data from the first table in Example 3 are open dots. We use formula (A-6-15) to calculate a regression slope: $b = 2/3.5 = 8/14 = 0.5714$. Then we use (A-6-14) to calculate expected Y values based on b and find $(Y_i - 7) = 0.5714(X_i - 4)$, since $\bar{Y} = 7$, $\bar{X} = 4$. Then $Y_i = +4.7143 + 0.5714X_i$.

When $X = 0$, Y is the point of intersection, usually symbolized as a. Thus, the general linear regression gives an estimated linear Y value (Y_L) in terms of constants $a + b$ for each X value:

$$Y_L \doteq a + bX \quad \text{(A-6-16)}$$

Note that for the point of $\bar{X}, \bar{Y} = 4, 7$, the regression coincides with that point. The solid sample points in Figure A-6-1 cluster more closely to the regression than the points for the sample with open dots where the covariance was zero.

Nevertheless, for this small sample, the regression of Example 3 is actually not significantly greater than zero. We may test confidence limits and the null hypothesis by finding the standard error of the regression coefficient as follows. Using the variance of the regression s_{YX}^2,

$$s_{YX}^2 = \frac{\sum d_{YX}^2}{n - 2} \quad \text{(A-6-17)}$$

where $d_{YX} = Y_i - Y_L$, as in Example 3.

	X	Y_i	Y_L	d_{XY}	d_{YX}^2
	7	10	8.7143	+1.2857	1.6530
	4	4	7.0000	−3.0000	9.0000
	4	6	7.0000	−1.0000	1.0000
	3	8	6.4286	+1.5714	2.4693
	2	7	5.8571	+1.1429	1.3062
Sum	20	35	35.0000	0.0000	15.4285

Note that the last column (sum of squares for "vertical deviations" $(Y_i - Y_L)^2$) may be obtained more easily by the following identity:

$$\sum d_Y^2 - \frac{(\sum d_X d_Y)^2}{\sum d_X^2} = \sum d_{YX}^2 \quad \text{(A-6-18)}$$

from Example 3, $20 - (8)^2/14 = 15.4285.$

Then the standard error of the regression comes from the variance of the regression, using (A-6-17) from Example 3,

$$15.4285/3 = 5.14283$$

The standard error is then

$$s_b = \sqrt{\frac{s_{YX}^2}{\sum d_X^2}} \tag{A-6-19}$$

from Example 3, $s_b = \sqrt{5.14283/14} = 0.60609$.

The null hypothesis for the b coefficient to be zero can then be tested with Student's t (Appendix A-5)—$t = (b - 0)/s_b = 0.57143/0.60609 = 0.943$, with 3 degrees of freedom $(n - 2)$. Thus, the regression coefficient is not significantly greater than zero, since $P = 0.42$ for the null hypothesis.

H CORRELATION (PRODUCT-MOMENT)

Often we may be interested not only in the regression of Y on X, b_{YX}, with Y dependent on X, but also in the regression of X on Y, b_{XY}, letting X be dependent on Y. For example, in parent-offspring traits, while it is usually the case that we are interested in how progeny are determined by parent heredity, we may also want to know the parents' hereditary determination from offspring segregations. If the second regression (b_{XY}) is equally important, then we may obtain the geometric mean of the two regression coefficients, which we call the correlation coefficient:

$$r = \frac{\text{covariance } XY}{\text{geometric mean of } V_X \text{ and } V_Y} = \frac{\sum d_X d_Y}{\sqrt{(\sum d_X^2)(\sum d_Y^2)}} \tag{A-6-20}$$

In the second table in Example 3, we calculate the correlation coefficient

$$r = \frac{8}{\sqrt{(14)(20)}} = 0.4781.$$

Correlation, then, is similar to regression in that the numerator is the sum of products. It can range from -1 through zero to $+1$.

Confidence limits of the r coefficient may be tested for significance from the null hypothesis by using t after obtaining the variance of r as follows:

$$s_r^2 = \frac{(1 - r^2)}{n - 2}, \tag{A-6-21}$$

so that the standard error is

$$s_r = \sqrt{\frac{(1 - r^2)}{n - 2}} \tag{A-6-22}$$

and $t_{(n-2)} = r/s_r$. For the example, $s_r = \sqrt{(0.7714)/3} = 0.50709$ so that $t = 0.4781/0.5071 = 0.9428$, which is not significant for $d.f. = 3$. (See Mather, 1965, Chapter 10, for a discussion of the correlation.)

A-7
Sampling

As pointed out in Appendix A-1, sampling has two main purposes: (1) estimation of parameters by specifying probability confidence limits from a given distribution (normal, binomial, etc.) and (2) estimating the reliability for a hypothesis if differences between observed and expected are due to chance alone. Tests of significance are designed for the latter purpose; for example, the *chi-square test* fixes the variance of the expected distribution, as does that of the *normal deviate* or any other *fixed parameter* distribution test, while *analysis of variance* (or Student's t, which is a special case for a single-contrast degree of freedom) relies usually on the nature of the observed material and its distribution for tests of significance. (See Sokal and Rohlf, 1969, Chapter 7, for a clear discussion of these two purposes: estimation and hypothesis testing.)

A CONFIDENCE LIMITS: THE NORMAL DEVIATE

The utility of the normal deviate (τ) in obtaining the probability for a particular observation within limits defined by the normal distribution was illustrated in Example 1 in Appendix A-5. It is apparent that the area cut off by perpendiculars in the normal curve at deviation $(Y_i - \mu)$ is a function of $(Y_i - \mu)/\sigma$, so that τ is the number of σ's in the observed deviation; that is, the denominator must be a fixed hypothetical value or a known parameter. We let t be an estimate of τ in a sample.

EXAMPLE 1. In using dominant mutants (D) in drosophila that are homozygous lethal, a geneticist may wish to test for the viability of a homozygous chromosome by inbreeding $D/+ \times D/+$ (where the + chromosome was uniparentally derived). The geneticist counts the following progeny: $99D/+ : 36 +/+$, total = 135. What is the probability for obtaining this result if exactly $\frac{2}{3}$ of $D/+ : \frac{1}{3}$ of $+/+$ was expected?

Using the normal deviate, $\tau = (Y - \mu)/\sigma$, estimated by $t = (Y - \bar{Y})/s$, where $s = npq$, from (A-6-11), because npq is fixed by hypothesis, $p = \frac{2}{3}, q = \frac{1}{3}$, we have $t = (99 - 90)/135(\frac{2}{9}) = 1.643$, and because the sample size is large ($n = 135$), $P = 0.10$ for the null hypothesis.

For this normal deviate value, random sampling would have produced that amount of departure from expected in about 10 percent of cases, or a nonsignificant deviation has been observed. Another way of stating the probability would be that if we reject the hypothesis of $\frac{2}{3}:\frac{1}{3}$ distribution for these genotypes, we would be in error 10 percent of the time (Appendix A-7-C). If we wish more reliability—that is, less chance of error on rejection—we may increase the sample size (denominator) or decrease the deviation (numerator) by improving uniformity of conditions. A value of either 101 or 79 for the $D/+$ class could be considered a significant departure from the expected of $\frac{90}{135}$, since the normal deviate would be ± 1.96 (which is the 95 percent level of confidence); consequently, these would be known as the upper and lower limits at that level. (Note: the student should realize that $\tau^2 = d^2/\sigma^2 = \chi^2$, since chi-square is the ratio of observed variance/hypothetical variance; see Appendix A-8 for application to this same problem.)

Confidence levels based on samples smaller than about 120 should be obtained from the t distribution because the normal deviate is a special case of t in which sample size is very large and the distribution is normal (Appendix A-5).

B SAMPLE MEANS AND VARIANCES OF MEANS

Without a hypothesis to provide a fixed mean and distribution, we must rely on the nature of the material observed to provide its own parameters or estimates of parameters. Usually the latter must be obtained from experiments; however, it is important for the student to realize what the data have estimated and what relationships they have to a parameter distribution. An artificial parameter population with normal distribution is provided in many statistical texts; in Sokal and Rohlf (1969), p. 109, there is a population with wing lengths of 100 houseflies; in Snedecor (1956), p. 55, there is an array of gains in weight for 100 swine. For simplicity and ease of arithmetic, the latter has been chosen here for illustrative examples as reproduced in Table A-7-1. We may sample this population, which has $\mu = 30$ lb, $\sigma^2 = 100$, $\sigma = 10$ lb, and range from 3–57 lb. If we take a small sample ($n = 5$) from this population, how do we expect the mean of this sample to vary, or what is the distribution of means of samples?

EXAMPLE 2. Ten samples of five observed values each were made from the population of 100 using a set of random numbers to choose the Y values. The means of the samples ($\bar{Y}$) are as follows:

26.6
29.8
27.0
36.0
36.6
31.6
29.2
30.8
24.0
24.4

Mean of means 29.60

Note: each mean is an unbiased estimate of the parameter mean of 30. The range of means is considerably less than that of the population itself and their standard deviation = 4.34.

Means of samples tend to be normally distributed even when samples are taken from distributions that deviate from normality, although the above 10 means in Example 2 tend to have too many extreme values (24, 24.4, 36, and 36.6) and the grand mean is a bit low at 29.6, because this is a small sample of means. However, increase in number of samples would be expected to estimate the population mean and to produce a normal distribution much more closely. What is the variance of sample means? Two explanations may be offered:

1. If we assume each observed value is subject to the same variance, then variance of each Y totaled would be $n \cdot s^2$, where n = number of Y in the sample. If each

TABLE A-7-1 *A "population" with 100 individuals of approximately normal distribution, grouped by frequency (data from Snedecor, 1956: an array of weight gains in lb from 100 swine)*

Y	Frequency	Y	Frequency	Y	Frequency	Y	Frequency
3	1	20	2	31	4	42	3
7	1	21	3	32	2	43	2
11	1	22	2	33	5	44	1
12	1	23	2	34	3	45	1
13	1	24	3	35	3	46	1
14	1	25	3	36	3	47	1
15	1	26	4	37	2	48	1
16	1	27	3	38	2	49	1
17	2	28	3	39	3	53	1
18	3	29	4	40	2	57	1
19	3	30	10	41	3		

observation is multiplied by a constant $(1/n)$ (A-6-7), then the variance of sample mean is $(1/n)^2 \cdot n \cdot s^2$. The variance of sample mean is

$$s_{\bar{Y}}^2 = \frac{s^2}{n} \tag{A-7-1}$$

The standard error of sample mean is

$$s_{\bar{Y}} = \frac{s}{\sqrt{n}} \tag{A-7-2}$$

Note that $s_{\bar{Y}}^2$ can be calculated by taking a single sample.

2. A derivation of the variance of sample mean (A-7-1) can be obtained by using "expected" algebra, as illustrated by Li (1964). The parameter variance of a sample mean = expected squared deviations of sample means from the parameter mean. Using item 3 from Appendix A-6-B and letting $\bar{Y}_i$ stand for Y_i in the expected for deviations of sample means we would have

$$\sigma_{\bar{Y}}^2 = E\{\bar{Y}_i - \mu\}^2 = E\{\bar{Y}_i^2\} - \mu^2$$

Then item 4 in Appendix A-6-B is n times too great for the E term in this expression that we divide by n to obtain

$$E\{\bar{Y}_i^2\} = E\left\{\left(\frac{\sum Y}{n}\right)^2\right\} = \mu^2 + \frac{\sigma^2}{n}$$

Then, combining both terms,

$$E\{s_{\bar{Y}}^2\} = \sigma_{\bar{Y}}^2 = \mu^2 + \frac{\sigma^2}{n} - \mu^2 = \frac{\sigma^2}{n} \tag{A-7-3}$$

Expected variance of sample means equals parameter variance divided by sample size.

Thus, because sample means estimate σ^2/n, we have formula (A-7-1), and any single sample estimates the variance of sample means without taking more than a single sample. In the parameter "population" given in Table A-7-1, however, we may consider how the estimates of variance of sample means and the standard error of means compares with the expected variance and standard error.

EXAMPLE 3. If we take samples of five items from the parameter population of 100 (Table A-7-1), the expected variance of the sample means would be as given above in (A-7-3): $\sigma_{\bar{Y}}^2 = \frac{100}{5} = 20$. Then the expected standard error of sample means would be the square root, $\sigma_{\bar{Y}} = 4.472$. Looking back at Example 2, we find that for the 10 samples, the standard deviation for the 10 means was 4.34, which is an estimate of the expected standard error, 4.472.

For confidence limits of means, the standard error is frequently used as in the method of t illustrated in Chapter 2, Table 2-3B, where the standard error of a binomial distribution was used. In Example 2, the sample mean of 24.0 is the lowest. We may use the estimate of standard error of 4.34 to test the null hypothesis, that this lowest sample is not significantly different from the parameter mean of 30.00.

$$t = \frac{24.0 - 30.00}{4.34} = 1.38 \text{ and } P = 0.20\ (d.f. = 10 - 1 = 9 \text{ for 10 sample means})$$

Thus, the departure of this sample mean from expected is not significant. We disregard the minus sign in the t value because we call this a two-tailed test; that is, the sample could have been greater or less than the parameter. If we are asking whether the sample mean was significantly *below* the parameter mean, the t test is called one-tailed, and the probability would be half that for the two-tailed test. A chart of t distribution with probabilities for the two-tailed test is given in Figure A-8-1.

EXAMPLE 4. If a single sample of five items is taken randomly from the parameter population of Table A-7-1, an estimate of the variance of the sample mean can be made. The following items were chosen at random using a table of random numbers to choose "without bias": 39, 20, 35, 27, 28. $\bar{X} = 29.8$ and $s^2 = 54.7$ (which is an estimate of the parameter $\sigma^2 = 100$). For this sample, using formulas (A-7-1) and (A-7-2), we have $s_{\bar{Y}}^2 = 54.7/5 = 10.94$ (variance of the sample mean) and its square root $s_{\bar{Y}} = 3.31$ (standard error of the mean), which estimates 4.472, the expected standard error.

C HYPOTHESIS TESTING

If we calculate the probability from a variance analysis, a t test, or a chi-square test, either testing departure from a random distribution or fitting expectancies to observations, we are usually invoking the "null hypothesis"—that is, that no difference exists, or that we have sampled from the same parameter population so that the observed variation between samples is due to chance alone. In testing that hypothesis, we could make either of two types of error as follows: (1) Type I is the error made if one rejects the null hypothesis when the null hypothesis is in fact true, (2) Type II is the error made if one accepts the null hypothesis when the null hypothesis is false.

The conventional familiar significance test provides us with the probability for making a Type I error. For example, in Table A of Example 2-4 using data from Dobzhansky's *ST-CH* populations of *Drosophila pseudoobscura*, the $p = 0.13$, meaning that if we reject the null hypothesis (hypothesis that there is no departure from panmixia), we would be wrong 13 percent of the time. This probability is usually represented as *alpha* (α), and $(1 - \alpha)$ represents the acceptance region of the null hypothesis. This 13 percent error is not sufficiently small to make us confident in rejecting the null hypothesis. Generally, 5 percent error is the usual level of confidence for rejecting the null hypothesis.

The probability of a Type II error is expressed as *beta* (β), or the probability for accepting the null hypothesis when an alternative hypothesis is correct. The alternative to β, $(1 - \beta)$, represents the probability for rejecting the null hypothesis when in fact it is false and is called the *power of the test.*

These two types or errors are related to each other with a fixed hypothesis such as a Mendelian ratio and a fixed number of observations. Then $\alpha = 1 - \beta$, so that when α is decreased, β is increased. However, an alternative hypothesis must be specified in order to estimate the magnitude of β. To decrease both types of error, the number of observations must be increased. (See Sokal and Rohlf, 1969, Chapter 7, for further discussion and utility of these probabilities and types of error.)

A-8

Goodness of Fit and Chi-Square

The chi-square distribution (χ^2) is a continuous probability density function with values ranging from zero to plus infinity. It is complex, and for each number of degrees of freedom the distribution is different; with a single degree of freedom it is highly skewed, but with increased degrees of freedom it approaches a symmetrical and more normal distribution. See the accompanying chart in Figure A-8-1.

The simple test for "goodness of fit" to an expected hypothetical set of proportions is based on discrete frequencies as a set of discontinuities, but its values approach the theoretical probability function sufficiently for its utility when we need to judge whether to reject the null hypothesis. Conventionally, we symbolize this statistic as X^2 to differentiate it from the theoretical function. It is generally calculated as a squared deviation (observed-expected)^2 in proportion to the expected value determined by some hypothesis and summed for all groups:

$$X^2 = \sum \frac{[O - E]^2}{E}$$

where O = observed value and E = expected determined by hypothesis.

A SINGLE DEGREE OF FREEDOM (TWO GROUPS)

If two observed classes have an expected proportional outcome in any experiment or population, then X^2 will be as follows. Let p,q = expected proportions of classes I and II,

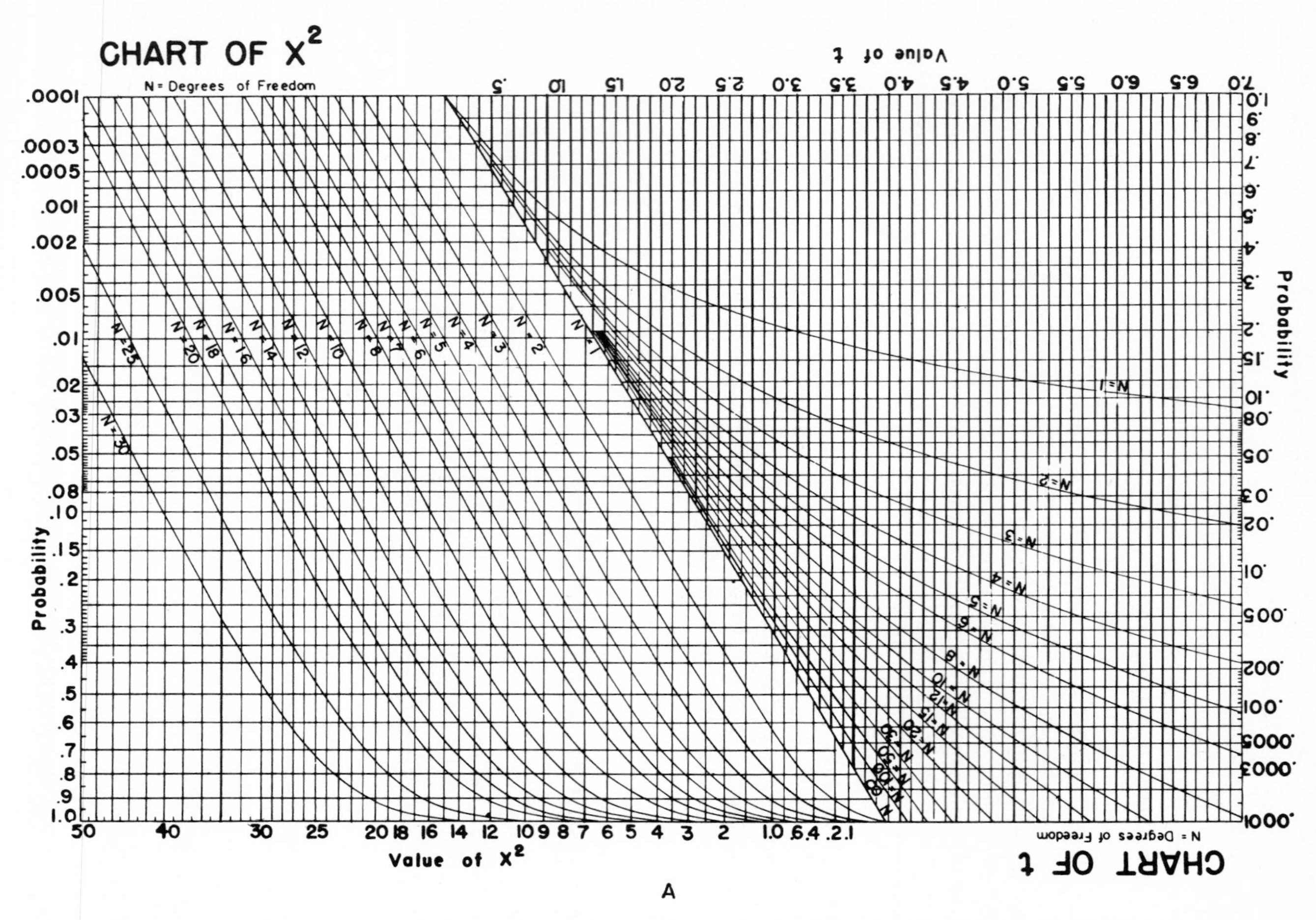

Figure A-8-1. Chart of t and X^2 distributions according to degrees of freedom (N) with α probability. Courtesy of James F. Crow, University of Wisconsin.

respectively: $p + q = 1$; a,b = observed numbers in classes I and II, respectively. Then

	Class I	II	Σ
O	a	b	n
E	pn	qn	n
$O - E$	$a - pn$	$b - qn$	0

$$X^2 = \frac{(a - pn)^2}{pn} + \frac{(b - qn)^2}{qn} = \frac{q(a - pn)^2 + p(b - qn)^2}{pqn} \qquad \text{(A-8-1)}$$

Since $q = 1 - p$ and $b = n - a$, and since $(a - pn)^2 = (-a + pn)^2$, X^2 simplifies to

$$X^2 = \frac{(a - pn)^2}{pqn} \qquad \text{(A-8-2)}$$

which avoids having to calculate deviations.

EXAMPLE 1. In Section A-7-A, Example 1, a geneticist counted 99 $D/+$: 36 $+/+$, total 135 progeny from a cross of $D/+ \times D/+$, where D/D was a lethal. Let the null hypothesis be tested with chi-square as follows:

	$D/+$	$+/+$	Total
O	99	36	135
E	90	45	135
$O - E$	+9	−9	0

Using (A-8-1), $X^2 = \frac{81}{90} + \frac{81}{45} = 0.9 + 1.8 = 2.70$. Using (A-8-2), $X^2 = (99 - 90)^2/(\frac{2}{9})(135) = 2.70$.

From the distribution of chi-square, for 1 degree of freedom, the probability for making a Type I error is $P = 0.10$, which we say is *not significant*—that is, there is no significant departure from the null hypothesis (no difference between observed and expected). We would be wrong to reject the null hypothesis 10 times out of 100, which is not sufficiently small to make us confident in rejecting it.

It should be noted that the denominator (pqn) is the hypothetical variance for a binomial (two classes). In other words, the X^2 here is a ratio of squared deviations (sampling variance) to fixed variance (parameter). Referring back to the section on sampling and the "normal deviate" (Appendix A-7-A), it should be noted that the X^2 for a single degree of freedom is t^2; therefore, the square root of $X^2 = 1.643$, which equals t (the number of standard errors in a difference between means). This t value has the same probability as the X^2 ($P = 0.10$) when it is considered as the normal deviate (population size very large, and $d.f. = 1$).

B CHI-SQUARE AND EXACT PROBABILITY

The X^2 (chi-square) measures the probability for observing a particular ratio (frequency of results) plus the probabilities for observing all greater departures from the expected result. The total probability so measured includes the probabilities for *both* sides of a distribution; that is, if an observed result in a sample of 10 is 4:6, the X^2 calculated will give the probability for 4:6 + 3:7 + 2:8 + 1:9 + 0:10 + 6:4 + 7:3 + 8:2 + 9:1 + 10:0. The two sides of the distribution are measured from the *deviation* away from the expected—that is, ± 1 from 5:5 expected, if $p = q = \frac{1}{2}$. See Example 2 below.

One additional complication. For small samples (where the expected = 5 or less), a correction factor (Yates) should be used by subtracting $\frac{1}{2}$ from +deviations and adding $\frac{1}{2}$ to −deviations—that is, decreasing the deviation in absolute value by 0.50 before squaring. The Yates correction is appropriate only for cases with a single degree of freedom.

EXAMPLE 2

Chi-Square (exp. hypoth. = $\frac{1}{2}:\frac{1}{2}$)			Exact Probability
	Obs.	4:6	$p_{4:6}$ or $p_{6:4}$ or greater departure
if $p = q$	Exp.	5:5	$0.377 + 0.377 = 0.754$
	d	$-1:+1$	
d(Yates correction) = $-0.5:+0.5$			
d^2/exp. = $0.25/5 + 0.25/5 = 0.10$ X^2, $p = 0.73$			
	Obs.	3:7	$p_{3:7}$ or $p_{7:3}$ or greater departure
	Exp.	5:5	$0.172 + 0.172 = 0.344$
	d	$-2:+2$	
d(Yates correction) = $-1.5:+1.5$			
d^2/exp. = $2.25/5 + 2.25/5 = 0.90$ X^2, $p = 0.33$			
	Obs.	2:8	$p_{2:8}$ or $p_{8:2}$ or greater departure
	Exp.	5:5	$0.055 + 0.055 = 0.110$
	d	$-3:+3$	
d(Yates correction) = $-2.5:+2.5$			
d^2/exp. = $6.25/5 + 6.25/5 = 2.50$ X^2, $p = 0.11$			

The student should verify that the 1:9 and 0:10 probabilities also check.

When the expected ratio is different from $\frac{1}{2}:\frac{1}{2}$, the same rule applies; for example, when the expected ratio is $\frac{3}{4}:\frac{1}{4}$, the exact probabilities to be added in order to equal the chi-square probability included the results with the *same deviation* on either side of the expected value.

EXAMPLE 3

Chi-Square (exp. hypoth = $\frac{3}{4}:\frac{1}{4}$)			Exact Probability
	Obs.	4:4	Sum $p_{1:7} + p_{2:6} + p_{3:5} + p_{4:4} = 0.1135$
if $p = \frac{3}{4}$	Exp.	6:2	$p_{8:0} = 0.1001$
$q = \frac{1}{4}$	d	$-2:+2$	Sum = 0.2136
d(Yates correction) = $-1.5:+1.5$			
d^2/exp. = $2.25/6 + 2.25/2 = 1.5$ X^2, $p = 0.21$			

The student should verify that for

$$\text{if } p = \tfrac{3}{4} \quad q = \tfrac{1}{4} \qquad \begin{array}{ll} \text{Obs.} & 3:5 \\ \text{Exp.} & 6:2, \text{the } X^2 = 4.167, p = 0.04 \\ d & -3:+3 \end{array} \qquad \left.\begin{array}{l} p_{3:5} = 0.0230 \\ p_{2:6} = 0.0038 \\ p_{1:7} = 0.0004 \end{array}\right\} = 0.0272$$

C MORE THAN TWO GROUPS

With three or more classes of data observed, the X^2 calculation can be made as follows. Let $p, q, r \ldots$ = expected proportions for classes I, II, III ... with $p + q + r + \cdots = 1$. $a, b, c \ldots$ = observed numbers in those classes. Then

	Class I	II	III	...	Σ
O	a	b	c	...	n
E	pn	qn	rn	...	n
$O - E$	$a - pn$	$b - qn$	$c - rn$	...	0

Then, as with (A-8-1),

$$X^2 = \frac{(a - pn)^2}{pn} + \frac{(b - qn)^2}{qn} + \frac{(c - rn)^2}{rn} \cdots$$

$$= \frac{a^2 - 2apn + p^2n^2}{pn} + \frac{b^2 - 2bqn + q^2n^2}{qn} \cdots$$

Let a_i = any observed value (a, b, *or* c ... as above), and p_i = hypothetical probability of the ith class. Then

$$X^2 = \sum \frac{a_i^2 - 2a_ip_in + p_i^2n^2}{p_in}$$

$$= \sum \frac{a_i^2}{p_in} - 2\sum \frac{a_ip_in}{p_in} + \sum \frac{p_i^2n^2}{p_in}$$

Since $\sum a_i = n$ and $\sum p_in = n$,

$$X^2 = \sum \frac{a_i^2}{p_in} - 2n + n = \sum \frac{a_i^2}{p_in} - n \qquad \text{(A-8-3)}$$

In Example 1, we can of course use (A-8-3), even though there are only two classes: $X^2 = [(99)^2/90 + (36)^2/45] - 135 = 2.70$ (as before). This formula avoids calculation of deviations as in (A-8-1) and is by far the simplest version of X^2.

D GENERAL CONTINGENCY TEST USING CHI-SQUARE

If n samples with k classes (genotypes or phenotypes) within each sample are to be compared in order to determine whether the individual k classes are in the same relative proportions throughout the n samples, calculate the expected on the assumption that from row to row (sample to sample) the column entries are in the same proportions [and alternatively from column to column (class to class) the row entries are in the same proportions]—that is, all columns and rows are homogeneous. This is a convenient statistical test to ascertain probabilities for differences between populations (two or more at a time), as illustrated in Chapter 2. The null hypothesis being tested is that from sample to sample the frequencies within classes are not different, as expressed in the following diagram:

		C_1	C_2	C_3	$\cdots \sum R_i$
R_1	Obs. Exp.	No. 11 $\cdots$ $\frac{(\sum R_1)(\sum C_1)}{T}$	No. 12 $\cdots$ $\frac{(\sum R_1)(\sum C_2)}{T}$	No. 13 $\cdots$ $\frac{(\sum R_1)(\sum C_3)}{T}$	$\cdots \sum R_1$
R_2	Obs. Exp.	No. 21 $\cdots$ etc.	No. 22 $\cdots$ $\frac{(\sum R_2)(\sum C_2)}{T}$	No. 23 $\cdots$ etc.	$\cdots \sum R_2$
R_3	Obs. Exp.	No. 31 etc.	No. 32 etc.	No. 33 etc.	$\cdots \sum R_3$
$\vdots$ $\sum C_i$		$\vdots$ $\sum C_1$	$\vdots$ $\sum C_2$	$\vdots$ $\sum C_3$	$\sum\limits_{CR}\sum = T$

$$\text{Expected value in No. 11} = T\left[\frac{(\sum R_1)}{T} \cdot \frac{(\sum C_1)}{T}\right] = \frac{(\sum R_1)(\sum C_1)}{T}$$

So for each box $(R_i C_i)$ the expected = row sum × column sum divided by the total. Deviations for each box are then squared and divided by the expected in the usual way for X^2:

$$X^2 = \sum \frac{(\text{obs.} - \text{exp.})^2}{\text{exp.}}$$

It can be shown algebraically that if the observed values in the $R_i C_i$ boxes are represented as

	C_1	C_2	C_3	$\cdots$
R_1	a	b	c	$\cdots$
R_2	f	g	h	$\cdots$
R_3	k	l	m	$\cdots$

with total = n, then

$$X^2 = \sum \left[\frac{a_i^2}{p_i q_i n}\right] - n \qquad \text{(A-8-4)}$$

where a_i is each observed value ($a, b, c \ldots$) in a box and $p_i q_i$ is expected proportion for the ith box based on marginal proportions.

For example, in Table 2-2B, the December 1944 ST/ST entry would be

$$\frac{(41)^2}{(0.5)(0.37)(300)} = 30.29$$

and the last entry CH/CH June 1945 would be

$$\frac{(9)^2}{(0.5)(0.1367)(300)} = 3.95$$

All four other entires would be found the same way and summed to obtain the first component of formula (A-8-4): 320.72. Thus, $X^2 = 320.72 - 300 = 20.72$ as given in Table 2-2B.

The two marginal classes (row sums and column sums) have no expected values because the contingency expectations are all based on homogeneity within the rows and columns and the independence of rows and columns. If marginal expectations did exist, they would take this form:

Items	Degrees of Freedom
Rows	$R - 1$
Columns	$C - 1$
Interaction	$(R - 1)(C - 1)$
Total	$(R)(C) - 1$

The first two items cannot be considered because the marginal totals are used to estimate the proportions within themselves, and their contributions to X^2 are zero; only the interaction degrees of freedom remain: $d.f. = (R - 1)(C - 1)$. This contingency test is illustrated in Table 2-2B of the text.

E SIMPLE 2 × 2 CONTINGENCY TEST

When there are two classes (treatments, genotypes, or phenotypes) in two samples (or two groups classified in a second way), we may test for the independence or interaction of the two classifications in a 2 × 2 contingency table (illustrated in Table 2-2A in the text). An algebraic simplification of the general contingency test (Appendix 8-D) can be symbolized and summarized as follows:

	Class 1	Class 2	Sum
Sample 1	a	b	$a + b$
Sample 2	c	d	$c + d$
Sum	$a + c$	$b + d$	n

The X^2 can be shown to be given by the following:

$$X^2 = \frac{(ad - bc)^2(n)}{(a + b)(a + c)(c + d)(b + d)} \qquad \text{(A-8-5)}$$

For Table 2-2A, the form of the X^2 would then be:

$$\frac{[(159)(89) - (141)(211)]^2(600)}{(300)(370)(300)(230)} = 19.06$$

This X^2 has just 1 degree of freedom (for the interaction test).

F HETEROGENEITY CHI-SQUARE

This test is based on asking if the ratio is homogeneous over two or more samples of an experiment (or samples of a population) in which some hypothesis has been tested for confirmation of a particular ratio (such as a Mendelian ratio or some constant p,q value with square law proportions). Each sample X^2 can be considered an independent estimate with its own degrees of freedom. Then the grand totals over all samples may be used to calculate X^2 as if there had been no sampling, assuming a single large homogeneous group or population. The difference between this X^2 on the summed (pooled) data of all samples and the sum of the sample X^2 is called the heterogeneity X^2 with degrees of freedom obtained by subtraction of the *d.f.* for the pooled X^2 from the total *d.f.* of all sample X^2's.

A-9

Maximum-Likelihood Method

A PROBABILITY VERSUS LIKELIHOOD

When a hypothesis is tested by observation, the exact probability (P) for the occurrence of observed combinations of events is given by formula (A-2-3) for a binomial or (A-4-7) for a multinomial, as below:

$$P = \frac{N!}{X!Y!Z!\cdots} p^X q^Y r^Z \cdots \tag{A-9-1}$$

where $p,q,r \ldots$ are probabilities of $X, Y, Z \ldots$ classes fixed by hypothesis.

Suppose we do not know what $p, q, r \ldots$ should be but only observe a set of phenotypes or genotypes such as $X, Y, Z \ldots$, or $D, H, R \ldots$. How do we estimate what the parameters p, q, r may be?

Fisher (1922) pointed out the above expression is a function both of observed numbers numbers ($X, Y, \ldots$) and of the hypothetical frequency p. Regarded as a function of p, q, P is the probability with which $X, Y, \ldots$ will be observed. Regarded as a function of $X, Y, \ldots$, it is not a probability. Fisher called it a *likelihood* function (L), which provided a method for estimating the parameter $p, q, \ldots$ frequencies for any population.

B LIKELIHOOD FUNCTION, THE SQUARE-LAW CONDITION FOR ONE PAIR OF ALLELES

Consider a sample of N individuals in three classes (genotypes or phenotypes) with observed numbers A, B, C (as in formulas (1-3) and (1-3A) in the text). If the proportions expected in each genotype on the basis of random mating are p^2 for A, $2pq$ for B, and q^2

for C, then we can estimate p,q values by maximization of the information in our observation from the likelihood function (L):

$$L = \frac{N!}{A!B!C!}(p^2)^A(2pq)^B(q^2)^C \tag{A-9-2}$$

Estimation of p by the method of maximum likelihood requires that a value be found to make this expression maximum. We should differentiate this expression with respect to p (or to q), equate to zero, and solve for p (or q). But this expression is not easy to differentiate. So we resort to taking first the log of the expression and then the derivative of the log. The expression and its log will have the same maximum at the same value of estimated p. Therefore, finding the value of p to maximize the log L will solve the problem. Taking the log of L, we have

$$\log(L) = \log\frac{N!}{A!B!C!} + A\log(p^2) + B\log(2pq) + C\log(q^2) \tag{A-9-3}$$

Changing q to p, we take the derivative of $\log(L)$ with respect to p and set it equal to zero. The combination term vanishes because it is independent of p. Note: if $y = \log v$, where v is a function of x, the derivative of $\log v$ with respect to x is $d(\log v)/dx = (dv/dx)/v$.

$$\frac{d(\log(L))}{dp} = A\frac{2p}{p^2} + B\frac{2-4p}{2pq} + C\frac{-2+2p}{q^2} \tag{A-9-4}$$

On simplifying and changing q to p in the denominators or back to q for convenience,

$$\frac{d(\log(L))}{dp} = \frac{2A}{p} + B\left[\frac{1}{p} - \frac{1}{q}\right] - \frac{2C}{q} = \frac{2A+B}{p} - \frac{2C+B}{q} \tag{A-9-4A}$$

Then, setting the derivative equal to zero gives

$$\frac{2C}{q} + \frac{B}{q} = \frac{2A}{p} + \frac{B}{p}$$

Thus, changing q to p, $p(2C + B) = 2A + B - p(2A + B)$; solving for p gives

$$p = \frac{2A+B}{2N} \tag{A-9-5}$$

because $N = A + B + C$, and $q = (2C + B)/2N$, which are in agreement with text formulas (1-3) and (1-3A). Thus, $p = D + \frac{1}{2}H$, $q = R + \frac{1}{2}H$.

The variance of this estimate of p is obtained, according to Fisher, by obtaining the second derivative of the likelihood function (A-9-4A) and setting that as the negative reciprocal of the variance; that is,

$$\frac{d^2(\log L)}{dp^2} = -\frac{1}{V_p} \tag{A-9-6}$$

If we take the second derivative of (A-9-4A), remembering that the derivative of an algebraic sum of functions is the sum of the derivatives of the functions, we obtain

$$\frac{d^2(\log L)}{dp^2} = -\left[\frac{2A+B}{p^2} + \frac{2C+B}{(1-p)^2}\right] \tag{A-9-7}$$

Having estimated p and q, we supply into (A-9-7) expected values for A, B, and C: Np^2 for A, $N2pq$ for B, and Nq^2 for C so that

$$\frac{d^2(\log L)}{dp^2} = -\frac{2N}{pq} \tag{A-9-7A}$$

then, from (A-9-6), the variance of the estimated p becomes

$$V_p = \frac{pq}{2N} \tag{A-9-8}$$

in agreement with (A-6-11A) and text formula (2-3).

A-10

Graphical Representation of Population Gene and Genotype Frequencies

A A SINGLE PAIR OF ALLELES

An equilateral triangle (three-coordinate system) has been convenient to represent a population with particular gene and genotype frequencies (Li, 1955*b*, 1976, and references cited therein). One method, attributed to de Finetti (1927) and to Streng (1926) by Li, is to place a point inside the triangle, as shown in Figure A-10-1A at P. Perpendicular distances to the three sides represent the relative frequencies of D, H, and R for a single pair of alleles. Letting H be the perpendicular to the base of the triangle at G, the relative lengths of the segments cut off along the base by the position of G become p,q allelic frequencies. These relationships may be proved by the following reasoning:

1. Height of an equilateral triangle $=$ (side/2) $\sqrt{3}$ (from basic geometry).
2. Letting any point of the triangle be P, with perpendiculars to each of three sides be D, H, and R in relative lengths.
3. We construct parallels to the three sides through P to produce three similar triangles.
4. Let a, b, and c be segments of side BC, $a + b + c = BC = AC = AB$ because the triangle is equilateral.

Since the height of triangle $ABC = (BC/2)\sqrt{3} = (b/2)\sqrt{3} + (c/2)\sqrt{3} + (a/2)\sqrt{3}$, if the height of the triangle $= 1.00$, then sum of perpendiculars $= 1.00$ and $D + H + R = 1.00$.

By inspection, the student may see that $AG:GC = D + \frac{1}{2}H:R + \frac{1}{2}H = p:q$, because all sides of triangles are related to their heights by the common factor $\frac{1}{2}\sqrt{3}$.

The student may use a triangular coordinate graph to plot the points for the "square law" from the root of text formula (1-5): $H = 2\sqrt{DR}$. The parabola thus formed gives the expected projection of genotypes from random mating in any population.

B MULTIPLE ALLELES OF ONE LOCUS

The equilateral triangle coordinate system can be applied to representing three alleles and their genotypes (Li, 1955, 1976), as in Figure A-10-1B. Here the perpendiculars represent allele frequencies p, q, and r, with the sum of perpendiculars equal to 1.00. Construction

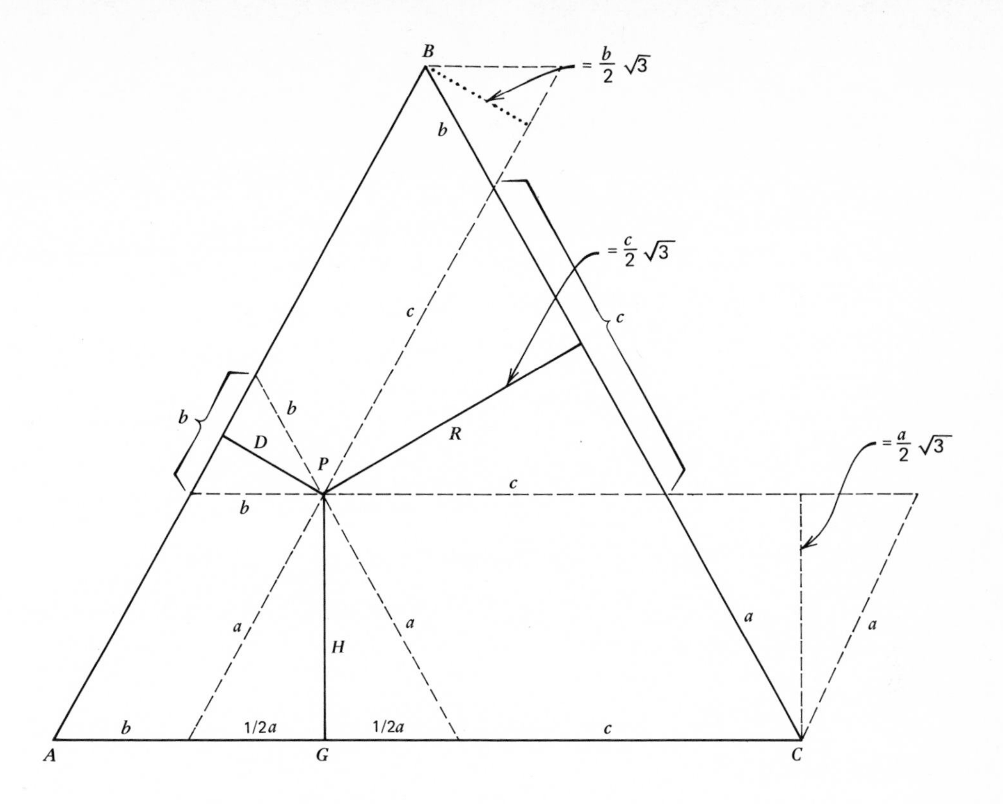

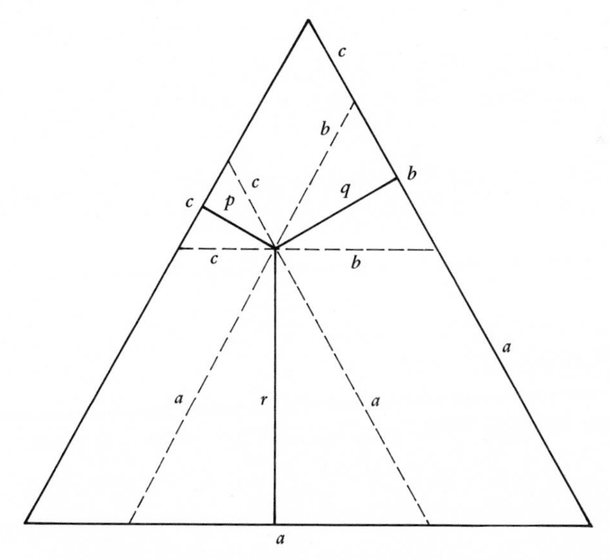

Figure A-10-1A. Equilateral triangle method of graphing the frequencies of dominant (*D*), recessive (*R*), and heterozygote (*H*) frequencies in a population as relative lengths of solid lines to point *P* from each of three sides. Side *BC* is dissected into three segments by parallel dashed lines through point *P*. Segments *a*, *c*, and *b* thus become sides of three similar triangles for geometric proof that at point *G* the $AG{:}GC$ segments are as $p{:}q$. From Li, 1955, 1976.

Figure A-10-1B. Equilateral triangle method of graphing the frequencies of *p*, *q*, and *r* for a population with three alleles. From Li, 1955, 1976. See Appendix text for explanation.

of parallels to the three sides now produces three triangles and three parallelograms whose relative areas represent homozygote and heterozygote proportions, respectively.

Letting segments $a + b + c$ be portions of one side cut off by the three parallels, we first find the areas of the three parallelograms. The area of the ab parallelogram $= a(q)$. Since the height $r = (a/2)\sqrt{3}$, then $a = 2r/\sqrt{3}$. Thus, parallelogram areas will be (relative to each other and to triangles)

$$\square ab = \frac{2rq}{\sqrt{3}} \rightarrow 2rq$$

$$\square ac = \frac{2rp}{\sqrt{3}} \rightarrow 2rp$$

$$\square bc = \frac{2pq}{\sqrt{3}} \rightarrow 2pq$$

The areas of the three triangles will be similarly as follows (relative to each other and to parallelograms) ($\frac{1}{2}$ base $\times$ height):

$$\triangle a = \tfrac{1}{2}a(r) = \frac{r^2}{\sqrt{3}} \rightarrow r^2$$

$$\triangle b = \frac{q^2}{\sqrt{3}} \rightarrow q^2$$

$$\triangle c = \frac{p^2}{\sqrt{3}} \rightarrow p^2$$

The student may use triangular graph paper to plot the population point for $q = \frac{1}{6}$, using formula (3-4) to obtain the case when total homozygosity equals total heterozygosity.

A-11

Scale Transformation

Basically, when comparing groups or populations, we must generally proceed with two assumptions: (1) error variance is the same for all groups and (2) genetic treatments or environmental effects are additive, or linear, in scale. Many variates when obtained by sampling do not conform to these assumptions, but they can be often adjusted, or transformed, in an appropriate way to accomodate these assumptions. We list some of the convenient transformations below.

A ANGULAR TRANSFORMATION

Because the variance (V_p) of a proportion is pq/n (A-4-3) and (A-6-11A), it is dependent on p. The angular transformation overcomes this handicap by making the variance more independent of p. The proportion is transformed into an angle: $p = \sin^2 \theta$, where θ is an angle of a right triangle. For example, if $p = 0.10$, $\theta = 18.4$ degrees. At $p = 1.00$, $\theta = 90$ degrees. Percentage data should usually be transformed in this way before applying statistical tests to them that rely on variance and analysis. (See references for tables of this transformation.)

B SQUARE ROOT TRANSFORMATION

When observations tend to be distributed in a Poisson rather than normally, taking their square root plus $\frac{1}{2}$ is preferred: $\sqrt{Y + \frac{1}{2}}$. This transformation is related to the Poisson distribution as the angular transformation is to the binomial (proportionality).

C LOGARITHMIC TRANSFORMATION

This transformation changes a multiplicative model into a linear one, and it is the most commonly used transformation.

D PROBIT TRANSFORMATION

Percentages, particularly for truncated variables such as percentage of survivors (or not surviving) at a certain point of time, may be converted into the normal deviates (Appendixes A-5 and A-7) on the assumption that an underlying normally distributed function has a threshold beyond which an individual survives or does not survive. Thus, the percentage is converted into the area under the normal curve and measured by the normal deviate. Because the normal deviate has a mean of zero, the arbitrary value of 5 is added to the normal deviate value: probit $= \tau + 5$, thus avoiding the problem of negative numbers. For example, if 20 percent is observed, the normal deviate is -0.84 and the probit $= 5.00 - 0.84 = 4.16$. If 80 percent is observed, the probit would be 5.84.

A-12

Determinants

Cramer's rule follows for solutions of simultaneous equations.

A SECOND-ORDER DETERMINANTS

$$\begin{aligned} a_1x + b_1y &= k_1 \\ a_2x + b_2y &= k_2 \end{aligned} \qquad \text{(A-12-1)}$$

Multiplying upper equation by b_2 and lower equation by b_1,

$$\begin{aligned} a_1b_2x + b_1b_2y &= k_1b_2 \\ a_2b_1x + b_1b_2y &= k_2b_1 \end{aligned} \qquad \text{(A-12-2)}$$

Thus,

$$x = \frac{k_1 b_2 - k_2 b_1}{a_1 b_2 - a_2 b_1}$$

Expressions in numerator and denominator are second-order determinants of the type

$$\begin{vmatrix} a_1 & b_1 \\ a_2 & b_2 \end{vmatrix} = a_1 b_2 - b_1 a_2$$

or $|ab|$ in brief notation. Thus,

$$x = \frac{|kb|}{|ab|}$$

and similarly

$$y = \frac{|ak|}{|ab|}$$

provided that $|ab| \neq 0$.

[Note the upper left to lower right product is positive, while the opposite diagonal product is negative.]

B THIRD-ORDER DETERMINANTS

$$\begin{aligned} a_1 x + b_1 y + c_1 z &= k_1 \\ a_2 x + b_2 y + c_2 z &= k_2 \\ a_3 x + b_3 y + c_3 z &= k_3 \end{aligned} \qquad \text{(A-12-3)}$$

Thus

$$x = \frac{|kbc|}{|abc|}, \qquad y = \frac{|akc|}{|abc|}, \qquad z = \frac{|abk|}{|abc|} \qquad \text{provided that } |abc| \neq 0.$$

Where

$$\begin{vmatrix} a_1 b_1 c_1 \\ a_2 b_2 c_2 \\ a_3 b_3 c_3 \end{vmatrix} = (a_1 b_2 c_3 + a_2 b_3 c_1 + a_3 b_1 c_2 - a_3 b_2 c_1 - a_2 b_1 c_3 - a_1 b_3 c_2)$$

$= |abc|$ in brief notation. [Note all products from upper left to lower right are positive, while opposite diagonal products are negative.]

A-13

Recurrent Series

If we can recognize a series, or sequence of related values, we can prescribe the tth term by means of an expression relating the initial term and explicit constants between members of the sequence. A series may be symbolized as the sequence: $u_0 u_1 u_2 u_3 \ldots u_t$. We wish to find (1) the relationship of one term to the next in sequence and (2) any general term (t = number in sequence) by combining the initial term with a constant or set of constants relating members of the sequence. Finally, we wish to determine whether the terms of the series infinitely increase, decrease, oscillate, or converge on a particular value.

A ARITHMETIC SERIES

An example of an arithmetic series is 1, 3, 5, 7, ... Between successive terms there is a common difference (symbol: λ):

$$u_t = u_{t-1} + \lambda \tag{A-13-1}$$

In this example, $\lambda = 2$. A general term (u_t) equals the initial term (u_0) plus $t \times \lambda$, the common difference:

$$u_t = u_0 + t\lambda \tag{A-13-2}$$

For example, $7 = 1 + 3(2)$ for u_3.

B GEOMETRIC SERIES

An example of a geometric series is 1, 3, 9, 27, ... Between successive terms there is a common ratio: $\lambda = u_t/u_{t-1}$. Here $\lambda = 3$.

$$u_t = \lambda u_{t-1} \tag{A-13-3}$$

For example, $27 = (3)^3(1)$ for u_3.

$$u_t = (\lambda)^t u_0 \tag{A-13-4}$$

C RECURRENT SERIES

When successive terms bear a relationship to more than a single preceding term, it is called "recurrent." An example is 1, 3, 7, 17, 41, ... The given series cannot be called arithmetic or geometric because successive terms have neither a common difference nor a common ratio between them. Instead, each term is dependent on two previous terms:

$$u_t = bu_{t-1} + cu_{t-2} \tag{A-13-5}$$

where b and c are called "scales of relation."

We may find b and c by solving two pairs of sequence terms simultaneously. In the example,

$$\begin{cases} 41 = b17 + c7 \\ 17 = b7 + c3 \end{cases} \qquad \begin{matrix} b = 2 \\ c = 1 \end{matrix}$$

As an example, the student should find the scales of relation for the "Fibonacci sequence" from heterozygotes in sib mating (Chapter 9): $1, \frac{2}{2}, \frac{3}{4}, \frac{5}{8}, \frac{8}{16}, \ldots$. Also, see Exercise 3 in Chapter 4 for sex linkage sequence approach to equilibrium.

To obtain a general expression for any term in a recurrent sequence without determining all preceding terms, we proceed through three axioms:

Axiom I. The sums of corresponding terms of two recurrent series, each with the same b and c scales of relation, also form a recurrent series with the same scales. For example,

X series	1, 3, 7, 17, 41, . . .	$b = 2$
Y series	0, 5, 10, 25, 60, . . .	$c = 1$
$X + Y$	1, 8, 17, 42, 101, . . .	

Conversely, any recurrent series can be assumed to be constitued of components whose sum forms the series.

$$u_t = X_t + Y_t = b(X_{t-1} + Y_{t-1}) + c(X_{t-2} + Y_{t-2}) \qquad \text{(A-13-6)}$$

Axiom II. Any geometric series can be written in the form of a recurrent series with two or more scales of relation. Using formula (A-13-3), adding and subtracting the same term, we have

$$u_t = \lambda u_{t-1} - u_{t-1} + u_{t-1} = (\lambda - 1)u_{t-1} + \lambda u_{t-2} \qquad \text{(A-13-7)}$$

Thus, $b = (\lambda - 1)$ and $c = \lambda$. For example, 2, 10, 50, 250, . . . , $\lambda = 5$, and for u_3, $250 = (4)(50) + (5)(10)$.

Axiom III. Any recurrent series must have terms made up of additive components that form two (or more) geometric series with the same scales of relation and common ratios (λ) for each component.

Component	X	$X\lambda_1$	$X(\lambda_1)^2$	$X(\lambda_1)^3$	$\cdots$
sequences	Y	$Y\lambda_2$	$Y(\lambda_2)^2$	$Y(\lambda_2)^3$	$\cdots$
Recurrent sequence	u_0,	u_1,	u_2,	u_3,	$\cdots$

$$= (X + Y), (X\lambda_1 + Y\lambda_2), [X(\lambda_1)^2 + Y(\lambda_2)^2], [X(\lambda_1)^3 + Y(\lambda_2)^3], \ldots$$

Thus, the general expression for any term with two components is

$$u_t = X(\lambda_1)^t + Y(\lambda_2)^t \qquad \text{(A-13-8)}$$

To solve (A-13-8), we let $u_t = u_0\lambda^t = bu_0\lambda^{t-1} + cu_0\lambda^{t-2}$ by using (A-13-5) and replacing each u of previous terms on the right with its geometric equivalent from (A-13-4). Then u_0 cancels, and we have a simple quadratic if $t = 2$, $\lambda^2 = b\lambda + c$. We obtain the characteristic equation of a recurrent series:

$$\lambda_1 = \frac{b + \sqrt{b^2 + 4c}}{2} \qquad \lambda_2 = \frac{b - \sqrt{b^2 + 4c}}{2} \qquad \text{(A-13-9)}$$

Taking the first terms of any recurrent series we find b and c as above under formula (A-13-5). Solving (A-13-8) for X and Y is done simply by the same method, once λ_1 and λ_2 are obtained.

Finally, we may discover whether the series converges by making t very large ($\infty = t$). For example, from Table 4-2, the successive female gene frequencies were $q = 0.60$ and 0.36. Verify that $b = \frac{1}{2}$, $c = \frac{1}{2}$, $\lambda_1 = 1$ and $\lambda_2 = -\frac{1}{2}$. Then $X = 0.44$ and $Y = 0.16$. Ultimately, q becomes $= 0.44$.

A-14

Useful References in Statistics

Batschelet, E. 1971. *Introduction to Mathematics for Life Scientists.* Springer-Verlag.
Brownlee, K. A. 1965. *Statistical Theory and Methodology.* Wiley, N.Y., 2nd edition.

Elandt-Johnson, R. C. 1967. *Probability Models and Statistical Methods in Genetics.* Wiley, N.Y.
Fisher, R. A. 1950. *Statistical Methods for Research Workers.* Oliver & Boyd, Edinburgh.
Fisher, R. A. 1960. *The Design of Experiments.* Oliver & Boyd, Edinburgh, 7th edition.
Li, C. C. 1964. *Introduction to Experimental Statistics.* McGraw-Hill, N.Y.
Mather, K. 1947. *Statistical Analysis in Biology.* Reprinted 1965 in paperback, Methuen.
Peters, C. C. and VanVoorhis, W. R. 1940. *Statistical Procedures and Their Mathematical Bases.* McGraw-Hill, N.Y.
Simpson, G. G., A. Roe, and R. C. Lewontin. 1960. *Quantitative Zoology.* Harcourt, Brace, N.Y.
Snedecor, G. W. 1962. *Statistical Methods.* Iowa State University, Ames, 6th edition.
Sokal, R. R., and F. J. Rohlf. 1969. *Biometry.* Freeman and Co., 1969.

TABLES

Fisher, R. A., and F. Yates. 1963. *Statistical Tables for Biological, Agricultural and Medical Research.* Oliver & Boyd, London.
Owen, D. B. 1962. *Handbook of Statistical Tables.* Addison-Wesley, Reading, Mass.
Rohlf, F. J., and R. R. Sokal. 1969. *Statistical Tables.* Freeman and Co.

BIBLIOGRAPHY

The following abbreviations for periodical titles are used:

Adv. Gen.	*Advances in Genetics*, Academic Press
Adv. Hum. Gen.	*Advances in Human Genetics*, Plenum Press
Am. J. Hum. Gen.	*American Journal of Human Genetics*
Am. J. Phys. Anth.	*American Journal of Physical Anthropology*
Am. Nat.	*The American Naturalist*
Am. Sci.	*American Scientist* (Society of the Sigma Xi)
Ann. Hum. Gen.	*Annals of Human Genetics*
Ann. Rev. Ecol. Syst.	*Annual Review of Ecology and Systematics*
Ann. Rev. Gen.	*Annual Review of Genetics*
Behav. Gen.	*Behavior Genetics*
C.R.A.S.	*Comptes Rendus de l' Académie des Sciences* (Paris)
C.S.H. Symp.	*Cold Spring Harbor Symposia on Quantitative Biology*
Dros. Info. Serv.	*Drosophila Information Service*
Ecol.	*Ecology*
Evol.	*Evolution*
Evol. Biol.	*Evolutionary Biology*, Appleton-Century-Crofts (Vol. 1-9), Plenum Press
Gen.	*Genetics*
Gen. Res.	*Genetical Research*
Her.	*Heredity*
J. Gen.	*Journal of Genetics*
J. Hered.	*Journal of Heredity*
J. Mol. Biol.	*Journal of Molecular Biology*
Molec. Gen. Gen.	*Molecular and General Genetics* (formerly *Z. f. Vererbungslehre*)
Phil. Trans. Roy. Soc. (B)	*Philosophical Transactions of The Royal Society of London* (Series B)
P.N.A.S.	*Proceedings of the National Academy of Sciences* (U.S.A.)
Proc. Roy. Soc. (B)	*Proceedings of The Royal Society of London* (Series B)
Sci.	*Science* (American Association for the Advancement of Science)
Z.f.A.u.V.	*Zeitschrift für induktive Abstammungs und Vererbungslehre*

Adams, J., and R. H. Ward. 1973. Admixture studies and the detection of selection. *Sci.* 180:1137–1143.

Adelberg, E. A. 1966. *Papers on Bacterial Genetics*. 2nd Ed. Little, Brown, Boston.

Allard, R. W. 1960. *Principles of Plant Breeding*. Wiley, N.Y.

______. 1965. Genetic systems associated with colonizing ability in predominantly self-pollinated species. *The Genetics of Colonizing Species*. H. G. Baker and G. Ledyard Stebbins (eds.). Academic Press, N.Y.

______. 1975. The mating system and microevolution. *Proc. XIIIth International Congress of Genetics*, Part II Genetics 79, Supplement. 115–126.

Allard, R. W., and J. Adams. 1969. Population studies in predominantly self-pollinating species. XIII. Intergenotypic competition and population structure in barley and wheat. *Am. Nat.* 103:621–645.

Allard, R. W., S. K. Jain, and P. L. Workman. 1968. The genetics of inbreeding populations. *Adv. in Gen.* 14:55–131.

Allard, R. W., and A. L. Kahler. 1972. Patterns of molecular variation in plant populations. *Proc. 6th Berkeley Symp. Math. Stat. and Prob.* 5:237–254.

Allen, J. A., and B. Clarke. 1968. Evidence for apostatic selection by wild passerines. *Nature.* 220:501–502.

Allison, A. C. 1954. Protection by the sickle-cell trait against subtertian malarial infection. *Brit. Med. J.* 1:290–294. Reprinted in *Papers on Human Genetics*, 1963, edited by S. H. Boyer, Prentice-Hall, Englewood Cliffs, N.J.

______. 1956*a*. The sickle-cell and haemoglobin C genes in some African populations. *Ann. Hum. Gen.* 21:67–89.

______. 1956*b*. Population genetics of abnormal human haemoglobins. *Acta Gen.* 6:430–434.

______. 1964. Polymorphism and natural selection in human populations. *C.S.H. Symp.* 29:139–149.

Anderson, W. W. 1969*a*. Polymorphism resulting from the mating advantage of rare male genotypes. *P.N.A.S.* 64:190–197.

______. 1969*b*. Genetics of natural populations. XLI. The selection coefficients of heterozygotes for lethal chromosomes in Drosophila on different genetic backgrounds. *Gen.* 62:827–836.

Anderson, W. W., Th. Dobzhansky, and C. D. Kastritsis. 1967. Selection and inversion polymorphism in experimental populations of *Drosophila pseudoobscura* initiated with the chromosomal constitutions of natural populations. *Evol.* 21:664–671.

Anderson, W. W., Th. Dobzhansky, and O. Pavlovsky. 1972. A natural population of *Drosophila* transferred to a laboratory environment. *Her.* 28:101–107.

Anderson, W. W., and C. E. King. 1970. Age-specific selection. *P.N.A.S.* 66:780–786.

Andrewartha, H. G. 1961. *Introduction to the Study of Animal Populations.* Univ. Chicago Press.

Andrewartha, H. G., and L. C. Birch. 1954. *The Distribution and Abundance of Animals.* Univ. Chicago Press.

Antonovics, J. 1971. The effects of a heterogeneous environment on the genetics of natural populations. *Am. Sci.* 59:593–599.

Antonovics, J., A. D. Bradshaw, and R. G. Turner. 1971. Heavy metal tolerance in plants. *Adv. Ecol. Res.* 7:1–85.

Atwood, K. C., L. K. Schneider, and F. J. Ryan. 1951. Selective mechanisms in bacteria. *C.S.H. Symp.* 16:345–354 (also reprinted in Dawson, P. S., and C. E. King. 1971. *Readings in Population Biology*. Prentice-Hall, Englewood Cliffs, N.J.).

Auerbach, C., and B. J. Kilbey. 1971. Mutation in eukaryotes. *Ann. Rev. Gen.* 5:163–218.

Ayala, F. J. 1969. Genetic polymorphism and interspecific competitive ability in *Drosophila*. *Gen. Res.* 14:95–102.

______. 1972. Frequency-dependent mating advantage in *Drosophila*. *Behav. Gen.* 2:85–91.

Ayala, F. J., and C. A. Campbell. 1974. Frequency-dependent selection. *Ann. Rev. Ecol. Syst.* 5:115–138.

Ayala, F. J., M. L. Tracey, L. G. Barr, J. F. McDonald, and S. Perez-Salas. 1974. Genetic variation in natural populations of five *Drosophila* species and the hypothesis of the selective neutrality of protein polymorphisms. *Gen.* 77:343–384.

Babcock, E. B., and R.E. Clausen. 1927. *Genetics in Relation to Agriculture*. 2nd Ed. McGraw-Hill, N.Y.

Baglioni, C. 1967. Molecular evolution in man. *Proc. 3rd Intern. Congr. Hum. Gen.* J. F. Crow and J. V. Neel (eds.). Johns Hopkins Press, Baltimore.

Bajema, C. J. (ed.). 1971. *Natural Selection in Human Populations*. (The Measurement of Ongoing Genetic Evolution in Contemporary Societies.) Wiley, N.Y.

Baker, H. G. 1959. Reproductive methods as factors in speciation in flowering plants. *C.S.H. Symp.* 24:177–190.

Baker, H. G., and G. L. Stebbins. 1965. *The Genetics of Colonizing Species*. Academic Press, N.Y.

Barker, J. S. F., and L. J. Cummins. 1969*a*. Disruptive selection for sternopleural bristle number in *Drosophila melanogaster*. *Gen.* 61:697–712.

______. 1969*b*. The effect of selection for sternopleural bristle number on mating behavior in *Drosophila melanogaster*. *Gen.* 61:713–719.

Barnes, B. W., and M. J. Kearsey. 1970. Variation for metrical characters in *Drosophila* populations. I. Genetic Analysis. *Her.* 25:1–10.

Barrai, I., L. L. Cavalli-Sforza, and A. Moroni. 1964. Testing a model of dominant inheritance for metric traits in man. *Her.* 19:651–668.

Bateman, A. J. 1959. The viability of near-normal irradiated chromosomes. *Int. J. Rad. Biol.* 2:170–180.

Beardmore, J. A. 1963. Mutual facilitation and the fitness of polymorphic populations. *Am. Nat.* 97:69–74.

______. 1960. Ecological factors and the variability of gene-pools in *Drosophila. Essays in Evolution and Genetics in Honor of Th. Dobzhansky*. M. Hecht and W. C. Steere (eds.). Appleton-Century-Crofts, N.Y.

Beardmore, J. A., Th. Dobzhansky, and O. Pavlovsky. 1960. An attempt to compare the fitness of polymorphic and monomorphic experimental populations of *Drosophila pseudoobscura*. *Her.* 14:19–33.

Beckman, L. 1966. *Isozyme Variations in Man*. S. Karger, Basel.

Bennett, J. 1960. A comparison of selective methods and a test of the pre-adaption hypothesis. *Her.* 15:65–77.

Bennett, J. H., and F. E. Binet. 1956. Association between Mendelian factors with mixed selfing and random mating. *Her.* 10:51–56.

Berg, K., and A. G. Bearn. 1968. Human serum protein polymorphisms. *Ann. Rev. Gen.* 2:341–362.

Bernstein, F. 1925. Zusammenfassende Betrachtungen über die erblichen Blutstrukturen des Menschen. *Z.f.A.u.V.* 37:237–270.

______. 1930. Fortgesetzte Untersuchungen aus der Theorie der Blutgruppen. *Z.f.A.u.V.* 56:233–273.

Berry, R. J., and H. N. Southern (eds.). 1970. *Variation in Mammalian Populations*. Symp. Zool. Soc. London, No. 26. Academic Press, London and N.Y.

Birch, L. C. 1955. Selection in *Drosophila pseudoobscura* in relation to crowding. *Evol.* 9:389–399.

Bodmer, W. F., and L. L. Cavalli-Sforza. 1970. Intelligence and race. *Sci. Am.* 223:19–29.

Bodmer, W. F., and J. Felsenstein. 1967. Linkage and selection: Theoretical analysis of the deterministic two locus random mating model. *Gen.* 57:237–265.

Bodmer, W. F., and P. A. Parsons. 1962. Linkage and recombination in evolution. *Adv. in Gen.* 11:1–100.

Bottini, E., P. Lucarelli, R. Agostino, R. Palmarino, L. Businco, and G. Antognoni. 1971. Favism: Association with erythrocyte acid phosphatase phenotype. *Sci.* 171:409–411.

Bowman, J. C., and D. S. Falconer. 1960. Inbreeding depression and heterosis of litter size in mice. *Her.* 1:262–274.

Boyd, W. C. 1950. *Genetics and the Races of Man*. Little, Brown, Boston.

Bradshaw, A. D. 1971. Plant evolution in extreme environments. *Ecological Genetics and Evolution* (*Essays in Honour of E. B. Ford*). Robert Creed (ed.). Blackwell Scientific Publ., Oxford and Edinburgh.

______. 1972. Some of the evolutionary consequences of being a plant. *Evol. Bio.* 5:25–48.

Breese, E. L., and K. Mather. 1957. The organization of polygenic activity within a chromosome in *Drosophila*. I. Hair characters. *Her.* 11:373–395.

______. 1960. The organization of polygenic activity within a chromosome in *Drosophila*. II. Viability. *Her*. 14:375–399.

Brncic, D. 1954. Heterosis and the integration of the genotype in geographic populations of *Drosophila pseudoobscura*. *Gen*. 39:77–88.

______. 1961. Non-random association of inversions in *Drosophila pavani*. *Gen*. 46:401–406.

Brookhaven Symposium in Biology. 1969. *Structure, Function and Evolution in Proteins.* No. 21. Brookhaven National Laboratory. Upton, N.Y.

Brower, J. V. Z. 1960. Experimental studies of mimicry. 4. The reactions of starlings to different proportions of models and mimics. *Am. Nat*. 94:271–282.

Brower, L. P., J. Alcock, and J. V. Z. Brower. 1971. Avian feeding behaviour and the selective advantage of incipient mimicry. *Ecological Genetics and Evolution*. R. Creed (ed.). Blackwell Scientific Publ., Oxford and Edinburgh.

Brower, L. P., L. M. Cook, and H. J. Groze. 1967. Predator responses to artificial Batesian mimics released in a neotropical environment. *Evol*. 21:11–23.

Brown, D. D., P. C. Wensink, and E. Jordan. 1972. A comparison of the rDNA's of *Xenopus laevis* and *X. mulleri*: The evolution of tandem genes. *J. Mol. Biol*. 63:57–73.

Brues, A. M. 1964. The cost of evolution vs. the cost of not evolving. *Evol*. 18:379–383.

Brussard, P. F., and P. Ehrlich. 1970. The population structure of *Erebia epipsodea* (Lepidoptera: Satyrinae). *Ecol*. 51:119–129.

Brussard, P. F., P. R. Ehrlich, and M. C. Singer. 1974. Adult movements and population structure in *Euphydryas editha*. *Evol*. 28:408–415.

Buettner-Janusch, J. 1970. Evolution of serum protein polymorphisms. *Ann. Rev. Gen*. 4:47–68.

Buettner-Janusch, J., R. Reisman, D. Coppenhaver, G. A. Mason, and V. Buettner-Janusch. 1973. Transferrins, haptoglobins, and ceruloplasmins among tribal groups of Madagascar. *Am. J. Phys. Anth*. 38:661–670.

Buri, P. 1956. Gene frequency in small populations of mutant *Drosophila*. *Evol*. 10:367–402.

Burt, C. 1958. The inheritance of mental ability. *Am. Psychol*. 13:1–15.

Buzzati-Traverso, A. A. 1955. Evolutionary changes in components of fitness and other polygenic traits in *Drosophila melanogaster* populations. *Her*. 9:153–186.

Cain, A. J., and J. D. Currey. 1963. Area effects in *Cepaea*. *Phil. Trans. Roy. Soc*. (*B*). 246:1–81.

Cain, A. J., J. M. B. King, and P. M. Sheppard. 1960. New data on the genetics of polymorphism in the snail *Cepaea nemoralis* (L.) *Gen*. 45:393–411.

Cain, A. J., and P. M. Sheppard. 1954. Natural selection in *Cepaea*. *Gen*. 39:89–116.

Cain, A. J., P. M. Sheppard, and J. M. B. King. 1968. Studies on *Cepaea*. I. The genetics of some morphs and varieties of *Cepaea nemoralis* (L.). *Phil. Trans. Roy. Soc.* (*B*). 253: 383–396.

Cannon, G. B. 1963. The effects of natural selection on linkage disequilibrium and relative fitness in experimental populations of *Drosophila melanogaster*. *Gen.* 48:1201–1216.

Carpenter, G. D. H., and E. B. Ford. 1933. *Mimicry*. Methuen, London.

Carson, H. L. 1957. The species as a field for gene recombination. *The Species Problem*, E. Mayr (ed.). Symposium A.A.A.S., Publ. 50, Washington, D.C.

______. 1958. The population genetics of *Drosophila robusta*. *Adv. in Gen.* 9:1–40.

______. 1961. Heterosis and fitness in experimental populations of *Drosophila melanogaster*. *Evol.* 15:496–509.

______. 1967. Inbreeding and gene fixation in natural populations. *Heritage from Mendel*. R. A. Brink (ed.). Univ. of Wisconsin Press., Madison.

______. 1968. The population flush and its genetic consequences. *Population Biology and Evolution*. R. C. Lewontin (ed.). Syracuse Univ. Press.

______. 1975. The genetics of speciation at the diploid level. *Am. Nat.* 109:83–92.

Carson, H. L., D. E. Hardy, H. T. Spieth, and W. S. Stone. 1970. The evolutionary biology of the Hawaiian *Drosophilidae*. *Essays in Evolution and Genetics in Honor of Theodosius Dobzhansky*. M. K. Hecht and W. C. Steere (eds.). Appleton-Century-Crofts, N.Y.

Carson, H. L., and H. D. Stalker. 1968. Polytene chromosome relationships in Hawaiian species of *Drosophila*. I. The *Drosophila grimshawi* group. *Univ. Texas Publ.* 6818:335–353.

Carter, M. A. 1968. Studies on *Cepaea* II. Area effects and visual selection in *Cepaea nemoralis* (L.) and *Cepaea hortensis*. *Phil. Trans. Roy. Soc.* (*B*). 253:397–446.

Castle, W. E. 1903. The laws of Galton and Mendel and some laws governing race improvement by selection. *Proc. Amer. Acad. Arts. and Sci.* 35:233–242.

______. 1921. An improved method of estimating the number of genetic factors concerned in cases of blending inheritance. *Sci.* 54:223.

Cavalli-Sforza, L. L. 1966. Population structure and human evolution. *Proc. Roy. Soc.* (*B*). 164:362–379.

Cavalli-Sforza, L. L., I. Barrai, and A. W. F. Edwards. 1964. Analysis of human evolution under random genetic drift. *C. S. H. Symp.* 29:9–20.

Cavalli-Sforza, L. L., and W. F. Bodmer. 1971. *The Genetics of Human Populations*. Freeman, San Francisco.

Cavalli-Sforza, L. L., and A. W. F. Edwards. 1967. Phylogenetic analysis. Models and estimation procedures. *Am. J. Hum. Gen.* 19:233–257.

Charlesworth, B. 1971. Selection in density-regulated populations. *Ecol.* 52:469–474.

Chung, Y. J. 1967. Persistence of a mutant gene in *Drosophila* populations of different genetic backgrounds. *Gen.* 57:957–967.

Church, R. B., and F. W. Robertson. 1966. Biochemical analysis of genetic differences in the growth of *Drosophila. Gen. Res.* 7:383–407.

Clarke, B. 1962*a*. Balanced polymorphism and the diversity of sympatric species. *Taxonomy and Geography*. D. Nichols (ed.). Systematics Assoc. Publ. No. 4.

Clarke, B. 1962*b*. Natural selection in mixed populations of two polymorphic snails. *Her.* 17:319–345.

Clarke, B. 1964. Frequency-dependent selection for the dominance of rare polymorphic genes. *Evol.* 18:364–369.

______. 1969*a*. Darwinian evolution of proteins. *Sci.* 168:1009–1011.

______. 1969*b*. The evidence for apostatic selection. *Her.* 24:347–352.

______. 1972. Density-dependent selection. *Am. Nat.* 106:1–13.

______. 1975. The contribution of ecological genetics to evolutionary theory: detecting the direct effects of natural selection on particular polymorphic loci. *Gen.* (suppl.). 79:101–113.

Clarke, B., and J. Murray. 1971. Polymorphism in a polynesian land snail *Partula suturalis vexillum. Ecological Genetics and Evolution* (*Essays in Honor of E. B. Ford*). R. Creed (ed.). Blackwell Scientific Publ., Oxford.

Clarke, B., and P. O'Donald. 1964. Frequency-dependent selection. *Her.* 19:201–206.

Clarke, B., and M. Williamson. 1958. Interaction between genetic drift and natural selection. *Evol.* 12:418–419.

Clarke, C. A. 1959. The relative fitness of human mutant genotypes. *Natural Selection in Human Populations.* D. F. Roberts and G. A. Harrison (eds.). Sympos. of the Soc. for the Study of Hum. Bio., Vol. 2. Pergamon Press.

______. 1964. *Genetics for the Clinician.* 2nd Ed. Blackwell Scientific Publ., Oxford.

______. 1970. *Human Genetics and Medicine.* St. Martin's Press. N.Y.

Clayton, G. A., J. A. Morris, and A. Robertson. 1956. An experimental check on quantitative genetical theory. I. Short-term responses to selection. *J. Gen.* 55:131–180.

Clayton, G. A., and A. Robertson. 1956. An experimental check on quantitative genetical theory. II. The long-term effects of selection. *J. Gen.* 55:131–180.

Clayton, G. A., G. R. Knight, J. A. Morris, and A. Robertson. 1956. An experimental check on quantitative genetical theory. III. Correlated responses. *J. Gen.* 55:131–180.

Clegg, M. T., and R. W. Allard. 1972. Patterns of genetic differentiation in the slender wild oat species *Avena barbata. P.N.A.S.* 69:1820–1824.

Cockerham, C. C. 1954. An extension of the concept of partitioning hereditary variance for analysis for covariances among relatives when epistasis is present. *Gen.* 39:859–882.

Cockerham, C. C., P. M. Burrows, S. S. Young and T. Prout. 1972. Frequency-dependent selection in randomly mating populations. *Am. Nat.* 106:493–575.

Connell, J. H., D. B. Mertz, and W. W. Murdoch. 1970. *Readings in Ecology and Ecological Genetics.* Harper Row, N.Y.

Cook, L. M. 1971. *Coefficients of Natural Selection.* Hutchinson, London (Humanities Press, N.Y.).

Cook, S. C. A., C. Lefevre, and T. McNeilly. 1972. Competition between metal tolerant and normal plant populations on normal soil. *Evol.* 26:366–372.

Cotter, W. B. 1974. On the male reproductive behavior and sustained polymorphism of the RT locus in laboratory populations of *Ephestia Künniella. Evol.* 28:109–123.

Cotterman, C. W. 1954. Estimation of gene frequencies in nonexperimental populations. *Statistics and Mathematics in Biology.* O. Kempthorne, et al. (eds.). Iowa State College Press, Ames.

Crawford, M. H., and P. L. Workman. 1973. *Methods and Theories of Anthropological Genetics.* Univ. New Mexico Press, Albuquerque.

Creed, Robert (ed.). 1971. *Ecological Genetics and Evolution* (*Essays in Honour of E. B. Ford*). Blackwell Scientific Publ., Oxford and Edinburgh. Appleton-Century-Crofts, N.Y.

Crossley, Stella. 1974. Changes in mating behavior produced by selection for ethological isolation between ebony and vestigial mutants of *Drosophila melanogaster. Evol.* 28:631–647.

Crow, J. F. 1954. Breeding structure of Populations. II. Effective population number. *Statistics and Mathematics in Biology.* O. Kempthorne, et al. (eds.). Hafner, N.Y. (reprinted 1964).

———. 1957. Genetics of insect resistance to chemicals. *Ann. Rev. Entom.* 2:227–246.

———. 1958. Some possibilities for measuring selection intensities in man. *Hum. Biol.* 30:1–13.

———. 1965. Problems of ascertainment in the analysis of family data. *Genetics and the Epidomiology of Chronic Diseases.* J. V. Neel, M. W. Shaw, and W. J. Schull (eds.). U.S. Dept. H.E.W., P.H.S. Publ. No. 1163.

Crow, J. F., and C. Denniston (eds.). 1974. *Genetic Distance.* Plenum Press, N.Y.

Crow, J. F., and J. Felsenstein. 1968. The effect of assortative mating on the genetic composition of a population. *Eugen. Quart.* 15:85–97.

Crow, J. F., and M. Kimura. 1965. The theory of genetic loads. *Proc. XIth Intern. Congress Genetics*, The Hague, Netherlands. 3:495–505.

———. 1970. *An Introduction to Population Genetics Theory.* Harper & Row, N.Y.

Crow, J. F., and N. E. Morton. 1955. Measurement of gene frequency drift in small populations. *Evol.* 9:202–214.

Crowe, L. K. 1965. The evolution of outbreeding in plants. I. The angiosperms. *Her.* 19: 435–457.

Crumpacker, D. W. 1967. Genetic loads in maize (*Zea mays.* L.) and other cross-fertilized plants and animals. *Evol. Bio.* 1:306–424.

Crumpacker, D. W., and J. S. Williams. 1973. Density, dispersion, and population structure in *Drosophila pseudoobscura. Ecol. Monogr.* 43:499–538.

______. 1974. Rigid and flexible chromosomal polymorphisms in neighboring populations of *Drosophila pseudoobscura. Evol.* 28:57–66.

Dahlberg, G. 1929. Inbreeding in man. *Gen.* 14:421–454.

Davenport, C. B. 1917. Inheritance of stature. *Gen.* 2:313–389.

Dawood, M. M., and M. W. Strickberger. 1969. The effect of larval interaction on viability in *Drosophila melanogaster.* III. Effects of biotic residues. *Gen.* 63:213–220.

DeFries, J. C., and G. E. McClearn. 1972. Behavioral genetics and the fine structure of mouse populations: A study in microevolution. *Evol. Biol.* 5:279–291.

DeGroot, M. H., and C. C. Li. 1960. Simplified method of estimating the MNS gene frequencies. *Ann. Hum. Gen.* 24:109–115.

Dempster, E. R. 1955. Maintenance of genetic heterogeneity. *C.S.H. Symp.* 20:25–32.

Dempster, E. R., and I. M. Lerner. 1950. Heritability of threshold characters. *Gen.* 35: 212–236.

de Souza, H. M. L., A. B. da Cunha, and E. P. dos Santos. 1972. Assortative mating in polymorphic laboratory populations of *Drosophila willistoni. Egypt. J. Gen. Cytol.* 1: 225–230.

Dewey, W. J., I. Barrai, N. E. Morton, and M. P. Mi. 1965. Recessive genes in severe mental defect. *Am. J. Hum. Gen.* 17:237–256.

Dews, D. L. 1970. A model for frequency-dependent mating success. *Dros. Info. Serv.* 45:113–114.

Dobzhansky, Th. 1937, 1941, 1951. *Genetics and the Origin of Species.* 1st, 2nd, and 3rd Eds. Columbia Univ. Press, N.Y.

______. 1947. Genetics of natural populations. XIV. A response of certain gene arrangements in the third chromosome of *Drosophila pseudoobscura* to natural selection. *Gen.* 32:142–160.

______. 1950. Mendelian populations and their evolution. *Am. Nat.* 84:401–418.

______. 1954. Evolution as a creative process. *Caryologia* (Vol. suppl.): 435–449 (Proc. 9th Intern. Congr. Gene.).

______. 1955. A review of some fundamental concepts and problems of population genetics. *C.S.H. Symp.* 20:1–15.

______. 1957. Genetic loads in natural populations. *Sci.* 126:191–194.

______. 1961. On the dynamics of chromosomal polymorphism in *Drosophila. Insect Polymorphism.* J. S. Kennedy (ed.). Roy. Entom. Soc., London.

______. 1962. *Mankind Evolving.* Yale Univ. Press, New Haven.

______. 1962. Rigid vs. flexible chromosomal polymorphism in *Drosophila. Am. Nat.* 96:321–328.

______. 1965. Evolutionary and population genetics. *Genetics Today. Proc. of XIth Intern. Congress Gen.* The Hague, Netherlands, 1963.

______. 1967. Sergei Sergeerich Tshetverikov (1880–1959). *Gen.* 55:1–3.

______. 1968*a*. Adaptedness and fitness. *Population Biology and Evolution.* R. C. Lewontin (ed.). Syracuse University Press.

______. 1968*b*. On some fundamental concepts of Darwinian biology. *Evol. Biol.* 2:1–34.

______. 1970. *Genetics of the Evolutionary Process.* Columbia Univ. Press, N.Y.

______. 1971. Evolutionary oscillations in *Drosophila pseudoobscura. Ecological Genetics and Evolution* (*Essays in Honour of E. B. Ford*). R. Creed, (ed.). Blackwell Scientific Publ., Oxford and Edinburgh.

______. 1972*a*. Natural selection in mankind. *The Structure of Human Populations.* G. A. Harrison and A. J. Boyce (eds.). Oxford University Press, Oxford.

______. 1972*b*. Species of Drosophila. *Sci.* 177:664–669.

Dobzhansky, Th., and C. Epling. 1944. Contributions to the genetics, taxonomy, and ecology of *Drosophila pseudoobscura* and its relatives. *Carnegie Inst. Wash. Publ.* 554: 1–183.

Dobzhansky, Th., R. Felix, J. Guzman, L. Levine, O. Olvera, J. R. Powell, M. E. de la Rosa, and V. M. Salceda. 1975. Population genetics of Mexican *Drosophila.* I. Chromosomal variation in natural populations of *Drosophila pseudoobscura* from Mexico. *J. Hered.* 66:203–206.

Dobzhansky, Th., and H. Levene. 1951. Development of heterosis through natural selection in experimental populations of *Drosophila pseudoobscura. Am. Nat.* 85:247–264.

______. 1955. Genetics of natural populations. XXIV. Developmental homeostasis in natural populations of *Drosophila pseudoobscura. Gen.* 40:797–808.

Dobzhansky, Th., H. Levene, and B. Spassky. 1972. Effects of selection and migration on geotactic and phototactic behaviour of *Drosophila.* III. *Proc. Roy. Soc.* (*B*). 180:21–41.

Dobzhansky, Th., H. Levene, B. Spassky, and N. Spassky. 1959. Release of genetic variability through recombination. III. *Drosophila prosaltans. Gen.* 44:75–92.

Dobzhansky, Th., R. C. Lewontin, and O. Pavlovsky. 1964. The capacity for increase in chromosomally polymorphic and monomorphic populations of *Drosophila pseudoobscura. Her.* 19:597–614.

Dobzhansky, Th., and O. Pavlovsky. 1953. Indeterminate outcome of certain experiments on *Drosophila* populations. *Evol.* 7:198–210.

______. 1957. An experimental study of interaction between genetic drift and natural selection. *Evol.* 11:311–319.

Dobzhansky, Th., and J. R. Powell. 1974. Rates of dispersal of *Drosophila pseudoobscura* and its relatives. *Proc. Roy. Soc.* (*B*). 187:281–298.

Dobzhansky, Th., and B. Spassky. 1947. Evolutionary changes in laboratory cultures of *Drosophila pseudoobscura. Evol.* 1:191–216.

Dobzhansky, Th., and B. Spassky. 1953. Genetics of natural populations. XXI. Concealed variability in two sympatric species of *Drosophila. Gen.* 38:471–484.

______. 1954. Genetics of natural populations. XXII. A comparison of the concealed variability in *Drosophila prosaltans* with that in other species. *Gen.* 39:472–487.

______. 1959. *Drosophila paulistorum*, a cluster of species in *statu nascendi. P.N.A.S.* 45:419–428.

______. 1960. Release of genetic variability through recombination. V. Breakup of synthetic lethals by crossing over in *Drosophila pseudoobscura*. Zool. Jahrb. 88:57–66.

______. 1963. Genetics of natural populations. XXXIV. Adaptive norm, genetic load, and genetic elite in *Drosophila pseudoobscura. Gen.* 48:1467–1485.

______. 1967*a*. Effects of selection and migration on geotactic and phototactic behaviour of *Drosophila*. I. *Proc. Roy. Soc.* (*B*). 168:27–47.

______. 1967*b*. An experiment on migration and simultaneous selection for several traits in *Drosophila pseudoobscura. Gen.* 55:723–734.

______. 1969. Artificial and natural selection for two behavioral traits in *Drosophila pseudoobscura. P.N.A.S.* 62:75–80.

Dobzhansky, Th., B. Spassky, and J. Sved. 1969. Effects of selection and migration on geotactic and phototactic behaviour of *Drosophila*. II. *Proc. Roy. Soc.* (*B*). 173:191–207.

Dobzhansky, Th., B. Spassky, and T. Tidwell. 1963. Genetics of natural populations. XXXII. Inbreeding and the mutational and balanced genetic loads in natural populations of *Drosophila pseudoobscura. Gen.* 48:361–373.

Dobzhansky, Th., and B. Wallace. 1953. The genetics of homeostasis in *Drosophila. P.N.A.S.* 39:162–171.

Dowdeswell, W. H., and E. B. Ford. 1955. Ecological genetics of *Maniola jurtina* in the Isles of Scilly. *Her.* 9:265–272.

Dowdeswell, W. H., E. B. Ford, and K. G. McWhirter. Further studies on the evolution of *Maniola jurtina* in the Isles of Scilly. *Her.* 14:333–364.

Drake, John W. 1970. *The Molecular Basis of Mutation.* Holden-Day, San Francisco.

______ (ed). 1973. *The Genetic Control of Mutation* (Proc. Intern. Workshop sponsored by John E. Fogarty International Center for Advanced Study in the Health Sciences, N.I.H.). *Gen.* 73: Supplement (April).

Dronamraju, K. R. (ed.). 1968. *Haldane and Modern Biology*. Johns Hopkins Press, Baltimore.

Dufour, A. P., R. A. Knight, and H. W. Harris. 1964. Genetics of isoniazid metabolism in Caucasian, Negro, and Japanese populations, *Sci.* 145: 391.

DuMouchel, W. H., and W. W. Anderson. 1968. The analysis of selection in experimental populations. *Gen.* 58:435–449.

Dunn, L. C. 1965. *A Short History of Genetics*. McGraw-Hill, N.Y.

Edelman, G. M. 1970. The structure and function of antibodies. *Sci. Am.* 233:34–42.

Edelman, G. M., and J. A. Gally. 1968. Antibody structure, diversity, and specificity. *Brookhaven Symp. Biol.* 21:328–344.

Edwards, A. W. F. 1971. Distance between populations on the basis of gene frequencies. *Biometrics.* 27:873–881.

Ehrlich, P. R., and P. H. Raven. 1969. Differentiation of populations. *Sci.* 165:1228–1232.

Ehrman, L. 1966. Mating success and genotype frequency in *Drosophila*. *Anim. Behav.* 14:332–339.

______. 1967. Further studies on genotype frequency and mating success in *Drosophila*. *Am. Nat.* 101:415–424.

______. 1968. Frequency dependence of mating success in *Drosophila pseudoobscura*. *Gen. Res.* 11:135–140.

______. 1970. Simulation of the mating advantage of rare *Drosophila* males. *Sci.* 167: 905–906.

______. 1972*a*. Genetics and Sexual Selection. *Sexual Selection and the Descent of Man*. B. Campbell (ed.). Aldine, Chicago.

______. 1972*b*. A factor influencing the rare male mating advantage in *Drosophila*. *Behav. Gen.* 2:69–78.

Ehrman, L., and C. Petit. 1968. Genotype frequency and mating success in the *willistoni* species group of *Drosophila*. *Evol.* 22:649–658.

Ehrman, L., B. Spassky, O. Pavlosky, and Th. Dobzhansky. 1965. Sexual selection, geotaxis, and chromosomal polymorphism in experimental populations of *Drosophila pseudoobscura*. *Evol.* 19:337–346.

Ehrman, L., and E. B. Spiess. 1969. Rare type mating advantage in *Drosophila*. *Am. Nat.* 103:675–680.

Elens, A. A., and J. M. Wattiaux. 1964. Direct observation of sexual isolation. *Dros. Info. Serv.* 39:118–119.

Emerson, S. H. 1952. Biochemical models of heterosis in Neurospora. *Heterosis*. J. W. Gowen (ed.). Iowa State Univ. Press, Ames.

Endler, J. A. 1973. Gene flow and population differentiation. *Sci.* 179:243–250.

Erlenmeyer-Kimling, L., J. Hirsch, and J. M. Weiss. 1962. Studies in experimental behavior genetics. III. Selection and hybridization analyses of individual differences in the sign of geotaxis. *J. Comp. Physiol. Psy.* 55:722–731.

Ewing, A. W. 1964. The influences of wing area on the courtship behavior of *Drosophila melanogaster*. *Anim. Behav.* 12:316–320.

Falconer, D. S. 1960. *Introduction to Quantitative Genetics*. Ronald Press, N.Y.

______. 1965. The inheritance of liability to certain diseases estimated from the incidence among relatives. *Ann. Hum. Gen.* 29:51–76. (Appendix: Table of normal deviate by q; $a = z/q$).

______. 1967. The inheritance of liability to diseases with variable age of onset, with particular reference to diabetes mellitus. *Ann. Hum. Gen.* 31:1–20.

Falk, R. 1961. Are induced mutations in *Drosophila* overdominant? II. Experimental results. *Gen.* 46:737–757.

Falk, R., and N. Ben-Zeev. 1966. Viability of heterozygotes for induced mutations in *Drosophila melanogaster*. II. Mean effects in irradiated autosomes. *Gen.* 53:65–77.

Finney, D. J. 1948*a,b*. The estimation of gene frequencies from family records. I. Factors without dominance. II. Factors with dominance. *Her.* 2:199–218, 369–390.

Fisher, R. A. 1918. The correlation between relatives on the supposition of Mendelian inheritance. *Trans. Roy. Soc. Edinburgh.* 52:399–433.

______. 1930. *The Genetical Theory of Natural Selection*. 1st ed. (completed in 1929). Clarendon Press, Oxford. 2nd ed. 1959, Dover, N.Y.

______. 1941. Average excess and average effect of a gene substitution. *Ann. Eugen.* 11: 53–63.

Fisher, R. A., and E. B. Ford. 1947. The spread of a gene in natural conditions in a colony of the moth *Panaxia dominula* (L.) *Her.* 1:143–174.

Ford, E. B. 1964. *Ecological Genetics*. Wiley, N.Y. 4th ed. 1975.

Fraser, A. S., W. R. Scowcroft, R. Nassar, H. Angeles, and G. Bravo. 1965. Variation of scutellar bristles in *Drosophila*. IV. Effects of selection. *Austral. J. Biol. Sci.* 18:619–641.

Garrison, R. J., V. E. Anderson, and S. C. Reed. 1968. Assortative marriage. *Eugen. Quart.* 15:113–127.

Giblett, E. R. 1969. *Genetic Markers in Human Blood*. Blackwell Scientific Publ., Oxford and Edinburgh.

Gibson, J. B. 1973. Biosocial aspects of life in Britain: Social mobility and the genetic structure of populations. *J. Biosoc. Sci.* 5:251–259.

Gibson, J. B., and J. M. Thoday. 1962. Effects of disruptive selection. VI. A second chromosome polymorphism. *Her.* 17: 1–26.

______. Effects of disruptive selection. VIII. Imposed quasi-random mating. *Her.* 18: 513–524.

______. 1964. Effects of disruptive selection. IX. Low selection intensity. *Her.* 19:125–130.

Gillespie, J. H., and K. Kojima. 1968. The degree of polymorphisms in enzymes involved in energy production compared to that in nonspecific enzymes in two *Drosophila ananassae* populations. *P.N.A.S.* 61:582–585.

Gillespie, J. H., and C. H. Langley. 1974. A general model to account for enzyme variation in natural populations. *Gen.* 76:837–848.

Glass, B. 1954. Genetic changes in populations, especially those due to gene flow and genetic drift. *Adv. Gen.* 6:95–139.

Glass, B., and C. C. Li. 1953. The dynamics of racial intermixture—An analysis of the American Negro. *Am. J. Hum. Gen.* 5:1–20.

Glass, B., M. S. Sacks, E. F. Jahn, C. Hess. 1952. Genetic drift in a religious isolate: An analysis of the causes of variation in blood group and other gene frequencies in a small population. *Am. Nat.* 86:145–159.

Gowen, J. W. (ed.). 1952. *Heterosis.* Iowa State Univ. Press, Ames.

Grant, V. 1963. *The Origin of Adaptations.* Columbia Univ. Press, N.Y.

______. 1975. *Genetics of Flowering Plants.* Columbia Univ. Press., N.Y.

Greenberg, R., and J. F. Crow. 1960. A comparison of the effect of lethal and detrimental chromosomes from *Drosophila* populations. *Gen.* 45:1153–1168.

Greenwalt, T. J. (ed.). 1967. *Advances in Immunogenetics.* Lippincott, Philadelphia.

Grubb, R. 1970. *The Genetic Markers of Human Immunoglobulins.* Springer-Verlag, N.Y., Heidelberg, Berlin.

Hadler, N. M. 1964. Heritability and phototaxis in *Drosophila melanogaster. Gen.* 50: 1269–1277.

Haldane, J. B. S. 1932. *The Causes of Evolution.* 1st Ed. Harper and Bros., London. Cornell Univ. Press, 1966.

______. 1937. The effect of variation on fitness. *Am. Nat.* 71:337–349.

______. 1949a. Disease and evolution. *Ricerca Sci.* 19(Suppl.): 3–10.

______. 1949b. The rate of mutation of human genes. *Proc. Eight Int. Cong. Gen.*: 267–273.

______. 1954*a*. *The Biochemistry of Genetics.* G. Allen and Unwin, London.

______. 1954*b*. The measurement of natural selection. *Proc. 9th Intern. Cong. Gen. Caryologia.* (Vol. Suppl.): 480–487.

______. 1957. The cost of natural selection. *J. Gen.* 55:511–524.

______. 1960. More precise expressions for the cost of natural selection. *J. Gen.* 57:351–360.

______. 1962. Conditions for stable polymorphism at an autosomal locus. *Nature.* 193: 1108.

______. 1964. A defense of beanbag genetics. *Perspect. Bio. and Med.* 7:343–359.

Haldane, J. B. S., and S. D. Jayakar. 1964. Equilibria under natural selection at a sex-linked locus. *J. Gen.* 59:29–36.

______. 1965. The nature of human genetic loads. *J. Gen.* 59:143–149.

Haldane, J. B. S., and P. Moshinsky. 1939. Inbreeding in Mendelian populations with special reference to human cousin marriage. *Ann. Eugen.* 9:321–340.

Hamrick, J. L., and R. W. Allard. 1972. Microgeographical variation in allozyme frequencies in *Avena barbata. P.N.A.S.* 69:2100–2104.

Hamrick, J. L., and R. W. Allard. 1975. Correlations between quantitative characters and enzyme genotypes in *Avena barbata. Evol.* 29:438–442.

Hanna, B. L. 1965. Genetic studies of family units. *Genetics and the Epidemiology of Chronic Diseases* (see Neel, J. V., M. W. Shaw, and W. J. Schull).

Hardy, G. H. 1908. Mendelian proportions in a mixed population. *Sci.* 28:49–50.

Harper, J. L. 1968. The regulation of numbers and mass in plant populations. *Population Biology and Evolution.* R. C. Lewontin (ed.). Syracuse Univ. Press.

Harris, H. 1966. Enzyme polymorphisms in man. *Proc. Roy. Soc.* (*B*) 164:298–310.

______. 1969. Enzyme and protein polymorphism in human populations. *Brit. Med. Bul.* 25:5–13.

______. 1970, 1975. *The Principles of Human Biochemical Genetics.* North-Holland, Amsterdam and London. American Elsevier, N.Y.

Harrison, G. A., and J. J. T. Owen. 1964. Studies on the inheritance of human skin color. *Ann. Hum. Gen.* 28:27–37.

Hatt, D., and P. A. Parsons. 1965. Association between surnames and blood groups in the Australian population. *Acta Gen.* 15:309–318.

Hayes, H. K., F. R. Immer, and D. C. Smith. 1955. *Methods of Plant Breeding.* McGraw-Hill, N.Y.

Hayman, B. I. 1958. The theory and analysis of diallele crosses. II. *Gen.* 43:63–85.

Hedrick, P. W. 1971. A new approach to measuring genic similarity. *Evol.* 25:276–280.

______. 1975. Genetic similarity and distance: Comments and comparisons. *Evol.* 29: 362–366.

Henning, V., and C. Yanofsky. 1963. An electrophoretic study of mutationally altered A proteins of the tryptophan synthetase of *E. coli*. *J. Mol. Biol.* 6:16–21.

Herrnstein, R. 1971. I. Q. *Atlantic.* 228:44–64.

Herzenberg, L. A., H. O. McDewitt, and L. A. Herzenberg. 1968. Genetics of antibodies. *Ann. Rev. Gen.* 2:209–244.

Hexter, W. M. 1955. A population analysis of heterozygote frequencies in *Drosophila*. *Gen.* 40:444–459.

Hirsch, J. 1967. *Behavior-Genetic Analysis.* McGraw-Hill, N.Y.

Hirsch, J., and J. Boudreau. 1958. The heritability of phototaxis in a population of *Drosophila melanogaster*. *J. Comp. Physiol. Psy.* 51:647–651.

Hirsch, J., and L. Erlenmeyer-Kimling. 1962. Studies in experimental behavior genetics. IV. Chromosome analyses for geotaxis. *J. Comp. Physiol. Psy.* 55:732–739.

Hogben, L. 1946. *An Introduction to Mathematical Genetics.* Norton, N.Y.

Holling, C. S. 1965. The functional response of predators to prey density and its role in mimicry and population regulation. *Mem. Ento. Soc. Can.* 45:1–60.

Holt, S. B. 1961. Dermatoglyphic patterns. *Symposia of the Society for the Study of Human Biology*, Vol. IV. Genetical Variation in Human Populations. G. A. Harrison (ed.). Pergamon Press.

______. 1961. Inheritance of dermal ridge patterns. *Recent Advances in Human Genetics.* L. S. Penrose (ed.). Little, Brown, Boston.

______. 1961. Quantitative genetics of fingerprint patterns. *Brit. Med. Bul.* 17:247–250.

______. 1968. *The Genetics of Dermal Ridges.* C. C. Thomas, Springfield, Ill.

Hood, L., and D. W. Talmage. 1970. Mechanism of antibody diversity: Germ line basis for variability. *Sci.* 168:325–334.

Hopkinson, D. A., and H. Harris. 1971. Recent work on isozymes in man. *Ann. Rev. Gen.* 5:5–32.

House, V. L. 1953. The use of the binomial expansion for a classroom demonstration of drift in small populations. *Evol.* 7:84–87.

Huang, S. L., M. Singh, and K. Kojima. 1971. A study of frequency-dependent selection in the esterase-6 locus of *Drosophila melanogaster* using a conditioned media method. *Gen.* 68:97–104.

Hubby, J. L., and **R. C. Lewontin.** 1966. A molecular approach to the study of genic heterozygosity in natural populations. I. The number of alleles at different loci in *Drosophila pseudoobscura*. *Gen.* 54:577–594.

Hubby, J. L., and L. H. Throckmorton. 1968. Protein differences in *Drosophila*. IV. A study of sibling species. *Am. Nat.* 102:193–205.

Huehns, E. R., N. Dance, G. H. Beaven, F. Hecht, and A. G. Motulsky. 1964. Human embryonic hemoglobins. *C.S.H. Symp.* 29:327–331.

Hyde, R. R. 1924. Inbreeding, outbreeding, and selection with *Drosophila melanogaster. J. Exp. Zool.* 40:181–215.

Ingram, V. M. 1963. *The Hemoglobins in Genetics and Evolution.* Columbia Univ. Press, N.Y.

Jacquard, Albert. 1974. *The Genetic Structure of Populations.* Springer-Verlag, Heidelberg, Berlin.

Jain, S. K., and R. W. Allard. 1966. The effects of linkage, epistasis, and inbreeding on population changes under selection. *Gen.* 53:633–659.

Jain, S. K., and A. D. Bradshaw. 1966. Evolutionary divergence among adjacent plant populations. I. Evidence and its theoretical analysis. *Her.* 21:407–442.

Jensen, A. R. 1974. Kinship correlations reported by Sir Cyril Burt. *Behav. Gen.* 4:1–28.

Jinks, J. L., and D. W. Fulker. 1970. A comparison of the biometrical genetical MAVA and classical approaches to the analysis of human behavior. *Psy. Bul.* 73:311–349.

Johnson, G. B. 1973. Enzyme polymorphism and biosystematics: The hypothesis of selective neutrality. *Ann. Rev. Ecol. Syst.* 4:93–116.

______. 1974. Enzyme polymorphism and metabolism. *Sci.* 184:28–37.

______. 1975. Enzyme polymorphism and adaptation. *Stadler Symp.* 7:91–116. Univ. of Missouri, Columbia.

Johnston, J. S., and W. B. Heed. 1975. Dispersal of *Drosophila*: The effect of baiting on the behavior and distribution of natural populations. *Am. Nat.* 109:207–216.

Jukes, T. H. 1966. *Molecules and Evolution.* Columbia Univ. Press, N.Y.

Kabat, D., and R. D. Koler. 1975. The thalassemias: Models for analysis of quantitative gene control. *Adv. Human Gen.* 5:157–222.

Kalmus, H. 1965. *Diagnosis and Genetics of Defective Color Vision.* Pergamon Press, Oxford.

Karlin, S., and F. M. Scudo. 1969. Assortative mating based on phenotype. II. Two autosomal alleles without dominance. *Gen.* 63:499–510.

Kastritsis, C. D., and D. W. Crumpacker. 1966. Gene arrangements in the third chromosome of *Drosophila pseudoobscura.* I. Configurations with tester chromosomes. *J. Hered.* 57:151–158.

______. 1967. Gene arrangements in the third chromosome of *Drosophila pseudoobscura.* II. All possible configurations. *J. Hered.* 58:112–129.

Kearsey, M. J., and B. W. Barnes. 1970. Variation for metrical characters in *Drosophila* populations. II. Natural Selection. *Her.* 25:11–21.

Kearsey, M. J., and K. Kojima. 1967. The genetic architecture of body weight and egg hatchability in *Drosophila melanogaster*. *Gen.* 56:23–37.

Keeler, Clyde. 1968. Some oddities in the delayed appreciation of "Castle's Law." *J. Hered.* 59:110–112.

Kempthorne, O. 1955. The correlations between relatives in random mating populations. *C.S.H. Symp.* 20:60–75.

Kerr, W. E., and S. Wright. 1954. Experimental studies of the distribution of gene frequencies in very small populations of *Drosophila melanogaster*. I. Forked. *Evol.* 8:172–177. II. Bar (Wright and Kerr). *Evol.* 8:225–240. III. Aristapedia and spineless. *Evol.* 8:293–302.

Kessler, S. 1969. The genetics of *Drosophila* mating behavior. II. The genetic architecture of mating speed in *Drosophila pseudoobscura*. *Gen.* 62:421–433.

Kettlewell, H. B. D. 1973. *The Evolution of Mechanism*. Oxford Univ. Press.

Kimura, M. 1955. Solution of a process of random genetic drift with a continuous model. *P.N.A.S.* 41:144–150.

______. 1962. On the probability of fixation of mutant genes in a population. *Gen.* 47:713–719.

______. 1968. Evolutionary rate at the molecular level. *Nature.* 217:624–626.

______. 1970. Stochastic processes in population genetics, with special reference to distribution of gene frequencies and probability of gene fixation. *Mathematical Topics in Population Genetics*. K. Kojima (ed.). Springer-Verlag, Berlin.

Kimura, M., and J. F. Crow. 1964. The number of alleles that can be maintained in a finite population. *Gen.* 49:725–738.

Kimura, M., and T. Ohta. 1971. *Theoretical Aspects of Population Genetics*. Princeton Univ. Press.

______. 1972. Population genetics, molecular biometry, and evolution. *Proc. Sixth Berkeley Symp. on Mathematical Statistics and Probability*. 5:43–68. Univ. California Press, Berkeley.

______. 1973. Mutation and evolution at the molecular level. *Gen.* (Suppl., April) 73: 19–35.

King, H. D. 1918–1921. Studies on inbreeding. I–IV. *J. Exp. Zool.* 26:1–54, 29:71–112.

King, J. C. 1955. Evidence for the integration of the gene pool from studies of DDT resistance in *Drosophila*. *C.S.H. Symp.* 20:311–317.

______. 1965. Genetic implications in the origin of higher levels of organization. *System. Zool.* 14:249–258.

King, J. L. 1967. Continuously distributed factors effecting fitness. *Gen.* 55:483–492.

King, J. L., and T. H. Jukes. 1969. Non-Darwinian evolution: Random fixation of selectively neutral mutations. *Sci.* 164:788–798.

Kirkman, H. N. 1971. Glucose-6-phosphate dehydrogenase. *Adv. Hum. Gen.* 2:1–60.

Kiser, C. V. 1968. Assortative mating by educational attainment in relation to fertility. *Eugen. Quart.* 15:98–112.

Knight, G. R., and A. Robertson. 1957. Fitness as a measurable character in Drosophila. *Gen.* 42:524–530.

Knight, G. R., A. Robertson, and C. H. Waddington. 1956. Selection for sexual isolation within a species. *Evol.* 10:14–22.

Koehn, R. K. 1969. Esterase heterogeneity: Dynamics of a polymorphism. *Sci.* 163:943–944.

Koehn, R. K., R. Milkman, and J. B. Mitton. 1976. Population genetics of marine pelecypods. IV. Selection, migration, and genetic differentiation in the blue mussel *Mytilus edulis*. *Evol.* 30:2–32.

Koehn, R. K., and J. B. Mitton. 1972. Population genetics of marine pelecypods. I. Ecological heterogeneity and adaptive strategy at an enzyme locus. *Am. Nat.* 106:47–56.

Koehn, R. K., F. J. Turano, and J. B. Mitton. 1973. Population genetics of marine pelecypods. II. Genetic differences in microhabitats of *Modiolus demissus*. *Evol.* 27:100–105.

Kojima, K. (ed.). 1970. *Mathematical Topics in Population Genetics*. Biomathematics, Vol. 1. Springer-Verlag, N.Y., Heidelberg, Berlin.

______. 1971. Is there a constant fitness value for a given genotype? No! *Evol.* 25:281–285.

Kojima, K., J. Gillespie, and Y. N. Tobari. 1970. A profile of *Drosophila* spp. enzymes assayed by electrophoresis. I. Number of alleles, heterozygosities, and linkage disequilibrium in glucose-metabolizing zystems and some other systems. *Bioch. Gen.* 4:627–637.

Kojima, K., and S. L. Huang. 1972. Effects of population density on the frequency-dependent selection in the esterase-6 locus of *Drosophila melanogaster*. *Evol.* 26:313–321.

Kojima, K., and T. M. Kelleher. 1961. Changes of mean fitness in random mating populations when epistasis and linkage are present. *Gen.* 46:527–540.

______. 1962. Survival of mutant genes. *Am. Nat.* 96:329–346.

______. 1963. Selection studies of quantitative traits with laboratory animals. *Statistical Genetics and Plant Breeding*. NAS-NRC 982.

Kojima, K., and R. C. Lewontin. 1970. Evolutionary significance of linkage and epistasis. *Mathematical Topics in Population Genetics*. K. Kojima (ed.). Springer-Verlag, N.Y., Heidelberg, Berlin.

Kojima, K., and Y. N. Tobari. 1969*a*. Selective modes associated with karytopes in *Drosophila ananassae*. II. Heterosis and frequency-dependent selection. *Gen.* 63:639–651.

______. 1969*b*. The pattern of viability changes associated with genotype frequency at the alcohol dehydrogenase locus in a population of *Drosophila melanogaster*. *Gen*. 61:201–209.

Kojima, K., and K. M. Yarbrough. 1967. Frequency dependent selection at the esterase-6 locus in *Drosophila melanogaster*. *P.N.A.S*. 57:645–469.

Krimbas, C. B. 1961. Release of genetic variability through recombination. VI. *Drosophila willistoni*. *Gen*. 46:1323–1334.

______. 1964. The genetics of *Drosophila subobscura* populations. II. Inversion polymorphism in a population from Holland. *Zeitschr. Vererb*. 95:125–128.

Lamotte, M. 1959. Polymorphism of natural populations of *Cepea nemoralis*. *C.S.H. Symp*. 24:65–86.

Langridge, J. 1962. A genetic and molecular basis for heterosis in *Arabidopsis* and *Drosophila*. *Am. Nat*. 96:5–27.

______. 1968. Thermal responses of mutant enzymes and temperature limits to growth. *Molec. Gen. Gen*. 103:116–126.

Latter, B. D. H., and A. Robertson. 1962. The effects of inbreeding and artificial selection on reproductive fitness. *Gen. Res*. 3:110–138.

Lee, B. T. U., and P. A. Parsons. 1968. Selection, prediction, and response. *Biol. Rev*. 43:193–174.

Lehmann, H., and R. W. Carrell. 1969. Variations in the structure of human hemoglobin. *Brit. Med. Bul*. 25:14–23.

Lerner, I. M. 1954. *Genetic Homeostasis*. Wiley, N.Y.

______. 1958. *The Genetic Basis of Selection*. Wiley, N.Y.

______. 1972. Polygenic inheritance and human intelligence. *Evol. Biol*. 6:399–414.

Levene, H. 1953. Genetic equilibrium when more than one ecological niche is available. *Am. Nat*. 87:331–333.

Levene, H. 1959. Release of genetic variability through recombination. IV. Statistical theory. *Gen*. 44:93–104.

______. 1963. Inbred genetic boads and the determination of population structure. *P.N.A.S*. 50:587–592.

Levene, H., I. M. Lerner, A. Sokoloff, F. K. Ho, and I. R. Franklin. 1965. Genetic loads in *Tribolium*. *P.N.A.S*. 53:1042–1050.

Levene, H., and R. E. Rosenfield. 1961. ABO incompatibility. *Progress in Medical Genetics*. A. G. Steinberg and A. G. Bearn (eds.). 1:120–157. Grune and Stratton, N.Y.

Levin, D. A. 1972. Low frequency disadvantage in the exploitation of pollinators by corolla variants in *Phlox*. *Am. Nat*. 106:453–460.

Levin, D. A., and H. W. Kerster. 1970. Phenotypic dimorphism and populational fitness in *Phlox*. *Evol*. 24:128–134.

______. 1974. Gene flow in seed plants. *Evol. Biol.* 7:139–220.

Levine, P. 1958. The influence of the ABO system on Rh hemolytic disease. *Hum. Biol.* 20:14–28.

Levins, R. 1965. Genetic consequences of natural selection. *Theoretical and Mathematical Biology*, T. H. Waterman and H. J. Morowitz (eds.). Blaisdell Publ.

______. 1966. The strategy of model building in population biology. *Am. Sci.* 54:421–431.

______. 1968. *Evolution in Changing Environments*. Princeton Univ. Press, Princeton.

Levins, R. and R. MacArthur. 1966. Maintenance of genetic polymorphism in a heterogeneous environment: Variations on a theme by Howard Levene. *Am. Nat.* 100:585–590.

Levitan, M. 1955. Studies of linkage in populations. I. Associations of second chromosome inversions in *Drosophila robusta*. *Evol.* 9:62–74.

______. 1973. Studies of linkage in populations. VII. Temporal variation and *X* chromosomal linkage disequilibrium. *Evol.* 27:476–485.

Levitan, M., and A. Montague. 1971. *Textbook of Human Genetics*. Oxford Univ. Press, N.Y., London, Toronto.

Lewis, D. 1949. Incompatibility in flowering plants. *Biol. Rev.* 24:472–496.

Lewontin, R. C. 1955. The effects of population density and composition on viability in *Drosophila melanogaster*. *Evol.* 9:27–41.

______. 1958. A general method for investigating the equilibrium of a gene frequency in a population. *Gen.* 43:419–434.

______. 1961. Evolution and the theory of games. *J. Theoret. Biol.* 1:382–403.

______. 1964*a*. The interaction of selection and linkage. I. General considerations. Heterotic models. *Gen.* 49:49–67.

______. 1964*b*. The interaction of selection and linkage. II. Optimum models. *Gen.* 50:757–782.

______. 1965. Selection for colonizing ability. *The Genetics of Colonizing Species*. H. G. Baker and G. L. Stebbins (eds.). Academic Press, N.Y.

______. 1966. On the measurement of relative variability. *Syst. Zool.* 15:141–142.

______. 1967. An estimate of average heterozygosity in man. *Am. J. Hum. Gen.* 19:681–685.

______. 1970. The units of selection. *Ann. Rev. Ecol. Syst.* 1:1–18.

______. 1971. The effect of genetic linkage on the mean fitness of a population. *P.N.A.S.* 68:984–986.

______. 1972. The apportionment of human diversity. *Evol. Biol.* 6:381–398.

______. 1973. Population genetics. *Ann. Rev. Gen.* 7:1–17.

______. 1974. *The Genetic Basis of Evolutionary Change*. Columbia Univ. Press, N.Y.

Lewontin, R. C., and C. C. Cockerham. 1959. The goodness-of-fit test for detecting natural selection in random mating populations. *Evol.* 13:561–564.

Lewontin, R. C., and J. L. Hubby. 1966. A molecular approach to the study of genic heterozygosity in natural populations. II. Amount of variation and degree of heterozygosity in natural populations of *Drosophila pseudoobscura*. *Gen.* 54:595–609.

Lewontin, R. C., D. Kirk, and J. Crow. 1968. Selective mating, assortative mating, and inbreeding: Definitions and implications. *Eugen. Quart.* 15:141–143.

Lewontin, R. C., and K. Kojima. 1960. The evolutionary dynamics of complex polymorphisms. *Evol.* 14:458–472.

Lewontin, R. C., and J. Krakauer. 1973. Distribution of gene frequency as a test of the theory of the selective neutrality of polymorphisms. *Gen.* 74:175–195.

Lewontin, R. C., and Y. Matsuo. 1963. Interaction of genotypes determining variability in *Drosophila busckii*. *P.N.A.S.* 49:270–278.

Lewontin, R. C., and M. J. D. White. 1960. Interaction between inversion polymorphisms of two chromosome pairs in the grasshopper, *Moraba scurra*. *Evol.* 14:116–129.

L'Heritier, P., and G. Teissier. 1937. Elimination de formes mutantes dans les populations de Drosophiles. *C. R. Soc. Biol.* Paris. 124:880–882.

Li, C. C. 1953. Some general properties of recessive inheritance. *Am. J. Hum. Gen.* 5:269–279.

______. 1955*a*. The stability of an equilibrium and the average fitness of a population. *Am. Nat.* 89:281–295.

______. 1955*b*. *Population Genetics*. 2d Ed. Univ. of Chicago Press. *First Course in Population Genetics*. 1976. Boxwood Press, Pacific Grove, Cal.

______. 1959. Notes on relative fitness of genotypes that forms a geometric progression. *Evol.* 13:564–567.

______. 1961. *Human Genetics* (*Principles and Methods*). McGraw-Hill, N.Y.

______. 1962. On "reflexive selection." *Sci.* 136:1055–1056.

______. 1963*a*. Decrease of population fitness upon inbreeding. *P.N.A.S.* 49:439–445.

______. 1963*b*. Equilibrium under differential selection in the sexes. *Evol.* 17:493–496.

______. 1963*c*. The way the load ratio works. *Am. J. Hum. Gen.* 15:315–321.

______. 1964. *Introduction to Experimental Statistics*. McGraw-Hill, N.Y.

______. 1967*a*. Castle's early work on selection and equilibrium. *Am. J. Hum. Gen.* 19:70–74.

______. 1967*b*. Fundamental theorem of natural selection. *Nature*. 214:505–506.

______. 1967*c*. Genetic equilibrium under selection. *Biometrics*. 23:397–484.

______. 1969. Population subdivision with respect to multiple alleles. *Ann. Hum. Gen.* 33:23–29.

______. 1970*a*. Human genetic adaptation. *Essays in Evolution and Genetics in Honor of Th. Dobzhansky*. M. K. Hecht and W. C. Steere (eds.). Appleton-Century-Crofts, N.Y.

______. 1970*b*. Table of variance of ABO gene frequency estimates. *Ann. Hum. Gen.* 34:189–194.

______. 1975. *Path Analysis—A Primer*. Boxwood Press, Pacific Grove, Cal.

Li, C. C., and D. G. Horvitz. 1953. Some methods of estimating the inbreeding coefficient. *Am. J. Hum. Gen.* 5:107–117.

Lilienfeld, A. M. 1959. A methodological problem in testing a recessive genetic hypothesis in human disease. *Am. J. Public Health.* 49:199–204.

Livingstone, F. B. 1967. *Abnormal Hemoglobins in Human Populations*. Aldine, Chicago.

______. 1969. The founder effect and deleterious genes. *Am. J. Phys. Anth.* 30:55–60.

Loehlin, J. C., G. Lindzey, and J. N. Spuhler. 1975. *Race Differences in Intelligence*. Freeman, San Francisco.

Lush, J. L. 1945. *Animal Breeding Plans*. 3d Ed. Iowa State College Press, Ames.

Luzzatto, L., E. A. Usanga, and S. Reddy. 1969. G6PD deficient red cells: Resistance to infection by malarial parasites. *Sci.* 164:839.

Lyon, Mary F. 1961. Gene action in the X-chromosome of the mouse (*Mus musculus* L.) *Nature*. 190:372–373.

______. 1968. Chromosomal and subchromosomal inactivation. *Ann. Rev. Gen.* 2:31–52.

MacArthur, R. H., and E. O. Wilson. 1967. *The Theory of Island Biogeography*. Princeton Univ. Press, Princeton.

MacIntyre, R. J., and T. R. F. Wright. 1966. Responses of esterase-6 alleles of *Drosophila melanogaster* and *Drosophila simulans* to selection in experimental populations. *Gen.* 53:371–387.

Magalhães, L. E., A. B. daCunha, J. S. de Toledo, S. P. De Toledo, H. L. Souza, H. J. Targa, V. Setzer, and C. Pavan. 1965. On lethals and their suppressors in experimental populations of *Drosophila willistoni*. *Mutat. Res.* 2:45–54.

Malécot, G. 1948. *Les Mathématiques de l'Hérédité*. Masson et Cié, Paris.

Malogolowkin-Cohen, C., H. Levene, N. P. Dobzhansky, and A. Solima Simmons. 1964. Inbreeding and the mutational and balanced loads in natural populations of *Drosophila willistoni*. *Gen.* 50:1299–1311.

Malogolowkin-Cohen, C., A. S. Simmons, and H. Levene. 1965. A study of sexual isolation between certain strains of *Drosophila paulistorum*. *Evol.* 19:95–103.

Mandel, S. P. H. 1959. The stability of a multiple allelic system. *Her.* 13:289–302.

______. 1971. Owen's model of a genetical system with differential viability between the sexes. *Her.* 26:49–63.

Mange, A. P. 1964. Growth and inbreeding of a human isolate. *Hum. Biol.* 36:104–133.

Manning, A. 1965. *Drosophila* and the evolution of behaviour. *Viewpoints in Biology.* J. D. Carthy and C. L. Duddington (eds.). Butterworths, London.

______. 1967. Control of sexual receptivity in female *Drosophila. Anim. Behav.* 15:239–250.

Manwell, C., and C. M. A. Baker. 1970. *Molecular Biology and the Origin of Species.* Univ. of Washington Press, Seattle.

Marinkovic, D. 1967. Genetic loads affecting fertility in natural populations of *Drosophila pseudoobscura. Gen.* 57:701–709.

Marshall, D. R., and R. W. Allard. 1970. Maintenance of isozyme polymorphisms in natural populations of *Avena barbata. Gen.* 66:393–399.

Martin, J. 1965. Interrelation of inversion systems in the midge *Chironomus intertinctus* (Diptera: Nemotocera). II. A non-random association of linked inversions. *Gen.* 52:371–383.

Mather, K. 1941. Variation and selection of polygenic characters. *J. Gen.* 41:159–193.

______. 1943. Polygenic inheritance and natural selection. *Biol. Revs.* 18:32–64. Cambridge Phil. Soc.

______. 1949. *Biometrical Genetics.* Dover Publications (1971. 2d Ed., with L. Jinks, Cornell Univ. Press, Ithaca, N.Y.)

______. 1950. The genetical architecture of heterostyly in *Primula sinensis. Evol.* 4:340–352.

______. 1951. *The Measurement of Linkage in Heredity.* 2d Ed. Methuen, London; Wiley, N.Y.

______. 1953. The genetical structure of populations. *Symp. Soc. Exp. Biol.* 7:66–95.

______. 1955. Polymorphism as an outcome of disruptive selection. *Evol.* 9:52–61.

______. 1970. The nature and significance of variation in wild populations. *Variation in Mammalian Populations.* R. J. Berry and H. N. Southern (eds.). Symp. Zool. Soc. London, No. 26. Academic Press, N.Y.

Mather, K., and B. J. Harrision. 1949. The manifold effect of selection. *Her.* 3:1–52; 131–162.

Mather, W. B. 1963. Patterns of chromosomal polymorphism in *Drosophila rubida Am. Nat.* 97:59–64.

Maynard Smith, J. 1962. Disruptive selection, polymorphism, and sympatric speciation. *Nature.* 195:60–62.

______. 1966. Sympatric speciation. *Am. Nat.* 100:637–650.

______. 1970. Population size, polymorphism, and the rate of non-Darwinian evolution. *Am. Nat.* 104:231–237.

______. 1972. *On Evolution*. Edinburgh Univ. Press.

Mayr, E. 1942. *Systematics and the Origin of Species*. Columbia Univ. Press, N.Y.

______. 1954. Changes of genetic environment and evolution. *Evolution as a Process*. J. Huxley, A. C. Hard, and E. B. Ford (eds.). Allen and Unwin, London.

______. 1959. Where are we? *C.S.H. Symp*. 24:1–14.

______. 1963. *Animal Species and Evolution*. Harvard Univ. Press. Cambridge.

McDonald, J. F., and F. J. Ayala. 1974. Genetic response to environmental heterogeneity. *Nature*. 250:572–574.

McKenzie, J. A. 1975. Gene flow and selection in a natural population of *Drosophila melanogaster*. *Gen*. 80:349–361.

McKenzie, J. A., and P. A. Parsons. 1974. Microdifferentiation in a natural population of *Drosophila melanogaster* to alcohol in the environment. *Gen*. 77:385–394.

McKusick, V. A., and G. A. Chase. 1973. Human genetics. *Ann. Rev. Gen*. 7:435–473.

McKusick, V. A., J. A. Hostetler, J. A. Egeland, and R. Eldridge. 1964. The distribution of certain genes in the Old Order Amish. *C.S.H. Symp*. 29:99–114.

Merrell, D. J. 1953. Selective mating as a cause of gene frequency changes in laboratory populations of *Drosophila melanogaster*. *Evol*. 7:287–296.

______. 1965. Competition involving dominant mutants in experimental populations of *Drosophila melanogaster*. *Gen*. 52:165–189.

Mertz, D.B. 1970. Notes on methods used in life-history studies. *Readings in Ecology and Ecological Genetics*. J. H. Connell, D. B. Mertz, W. W. Murdach (eds.). Harper & Row, N.Y.

Mettler, L. E., and T. G. Gregg. 1969. *Population Genetics and Evolution*. Prentice-Hall, Englewood Cliffs.

Mettler, L. E., S. E. Moyer, and K. Kojima. 1966. Genetic loads in cage populations of *Drosophila*. *Gen*. 54:887–898.

Milkman, R. D. 1960. The genetic basis of natural variation. I. *Gen*. 45:35–48.

______. 1964. The genetic basis of natural variation. V. Selection for crossveinless polygenes in new wild strains of *Drosophila melanogaster*. *Gen*. 50:625–632.

______. 1967. Heterosis as a major cause of heterozygosity in nature. *Gen*. 55:493–495.

______. 1970. The genetic basis of natural variation in *Drosophila melanogaster*. Adv. Gen. 15:55–114.

Millicent, E., and J. M. Thoday. 1961. Effects of disruptive selection. IV. Gene flow and divergence. *Her*. 15:199–217.

Mitton, J. B., R. K. Koehn, and T. Prout. 1973. Population genetics of marine pelecypods. III. Epistasis between functionally related isoenzymes of *Mytilus edulis*. *Gen*. 73:487–496.

Mohler, J. D. 1965. Preliminary genetic analysis of crossveinless-like strains of *Drosophila melanogaster Gen.* 51:641–651.

______. 1967. Some interactions of crossveinless-like genes in *Drosophila melanogaster*. *Gen.* 57:65–77.

Moody, P. A. 1947. A simple model of "drift" in small populations. *Evol.* 1:217–218.

Moran, P. A. P. 1962. *The Statistical Processes of Evolutionary Theory*. Clarendon Press, Oxford.

______. 1964. On the non-existence of adaptive topographics. *Ann. Hum. Gen.* 27:383–393.

Morton, N. E., J. F. Crow, and J. J. Muller. 1956. An estimate of the mutational damage in man from data on consanguineous marriages. *P.N.A.S.* 42:855–863.

Morton, N. E., and M. Yamamoto. 1973. Blood group and haptoglobins in the Eastern Carolines. *Am. J. Phys. Anthro.* 38:695–698.

Motulsky, A. G. 1964. Hereditary red cell traits and malaria. *Am. J. Trop. Med. Hyg.* 13:147–158.

Motulsky, A. G., J. Vandepitte, and G. R. Fraser. 1966. Population genetic studies in the Congo. I. Glucose-6-phosphate dehydrogenase deficiency, hemoglobin S, and malaria. *Am. J. Hum. Gen.* 18:514–537.

Mourad, A. K. 1964. Lethal and semi-lethal chromosomes in irradiated experimental populations of *Drosophila pseudoobscura*. *Gen.* 50:1279–1287.

Mourant, A. E. 1954. *The Distribution of the Human Blood Groups*. Blackwell Scientific Publ. Oxford.

Mourant, A. E., A. C. Kopec, and K. Domaniewska-Sobezak. 1958. *The ABO Blood Groups. Comprehensive Tables and Maps of World Distribution*. Blackwell Scientific Publ. Oxford.

Mukai, T. 1964. The genetic structure of natural populations of *Drosophila melanogaster*. I. Spontaneous mutation rate of polygenes controlling viability. *Gen.* 50:1–19.

______. 1969. The genetic structure of natural populations of *Drosophila melanogaster*. VII. Synergistic interaction of spontaneous mutant polygenes controlling viability. *Gen.* 61:749–761.

Mukai, T., S. Chigusa, and I. Yoshikawa. 1964. The genetic structure of natural populations of *Drosophila melanogaster*. II. Overdominance of spontaneous mutant polygenes controlling viability in homozygous genetic backgrounds. *Gen.* 50:711–715.

______. 1965. The genetic structure of natural populations of *Drosophila melanogaster*. III. Dominance effect of spontaneous mutant polygenes controlling viability in heterozygous genetic backgrounds. *Gen.* 52:493–501.

Mukai, T., and T. Yamazaki. 1968. The genetic structure of natural populations of *Drosophila melanogaster*. V. Coupling-repulsion effect of spontaneous mutant polygenes controlling viability. *Gen.* 59:513–535.

Mukai, T., I. Yoshikawa, and K. Sano. 1966. The genetic structure of natural populations of *Drosophila melanogaster*. IV. Heterozygous effects of radiation-induced mutations on viability in various genetic backgrounds. *Gen.* 53:513–527.

Muller, H. J. 1939. Reversibility in evolution considered from the standpoint of genetics. *Biol. Reviews*. 14:261–280.

______. 1950. Our load of mutations. *Am. J. Hum. Gen.* 2:111–176.

Murdoch, W. W. 1969. Switching in general predators: Experiments on predator specificity and stability of prey populations. *Ecol. Monog.* 39:335–354.

Murray, J. 1966. *Cepaea nemoralis* in the Isles of Scilly. *Proc. Malac. Soc.* London. 37:167–181.

______. 1972. *Genetic Diversity and Natural Selection*. Hafner, N.Y.

Neel, J. V. 1950. The population genetics of two inherited blood dyscrasias in man. *C.S.H. Symp.* 15:141–158.

______. 1951. The inheritance of the sickling phenomenon with particular reference to sickle cell disease. *Blood.* 6:389–412.

______. 1962. Mutations in the human population. *Methodology in Human Genetics*. W. J. Burdette (ed.). Holden-Day, San Francisco.

______. 1971. The detection of increased mutation rates in human populations. *Persp. Biol. and Med.* 14:522–537.

Neel, J. V., S. S. Fajans, J. W. Conn, and R. T. Davidson. 1965. Diabetes Mellitus. *Genetics and the Epidemiology of Chronic Diseases* (see next listing).

Neel, J. V., M. W. Shaw, and W. J. Schull (eds.). 1965. *Genetics and the Epidemiology of Chronic Diseases*. U.S. Dept. H.E.W., P.H.S. Publ. 1163.

Neel, J. V., and W. J. Schull. 1954. *Human Heredity*. Univ. of Chicago Press.

______. 1968. On some trends in understanding the genetics of man. *Persp. Biol. and Med.* 11:565–601.

Neel, J. V., W. J. Schull, T. Kimura, Y. Tanigawa, M. Yamamoto, and A. Nakajima. 1970. The effects of parental consanguinity and inbreeding in Hirado, Japan. III. Vision and hearing. *Hum. Her.* 20:129–155.

Neel, J. V., W. J. Schull, M. Yamamoto, S. Uchida, T. Yanase, and H. Fujiki. 1970. The effects of parental consanguinity and inbreeding in Hirado, Japan. II. Physical development, tapping rate, blood pressure, intelligence quotient, and school performance. *Am. J. Hum. Gen.* 22:263–286.

Nei, M. 1965. Variation and covariation of gene frequencies in subdivided populations. *Evol.* 19:256–258.

______. 1972. Genetic distance between populations. *Am. Nat.* 106:283–292.

______. 1975. *Molecular Population Genetics and Evolution.* (North Holland Research Monographs, Frontiers of Biology, Vol. 40. A Newberger and E. L. Tatum (eds.). North Holland Publ. Co., Amsterdam-American Elsevier, N.Y. and Oxford.

Nei, M., T. Maruyama, and R. Chakraborty. 1965. The bottleneck effect and genetic variability in populations. *Evol.* 29:1–10.

Nei, M., and A. K. Roychoudhury. 1974. Genic variation within and between the three major races of man, Caucasoids, Negroids, and Monogloids. *Am. J. Hum. Gen.* 26:421–443.

Neyman, J. 1967. R. A. Fisher (1890–1962): An Appreciation. *Sci.* 156:1456–1460.

O'Brien, S. J., and R. J. MacIntyre. 1971. A biochemical genetic map of *Drosophila melanogaster. Dros. Info. Serv.* 46:89–93.

O'Brien, S. J., R. J. MacIntyre, and W. Fine. 1969. An analysis of gene-enzyme variability in natural populations of *Drosophila melanogaster* and *Drosophila simulans. Am. Nat.* 103:97–113.

O'Brien, S. J., R. I. MacIntyre, and W. Fine. 1968. A linkage disequilibrium between two gene-enzyme systems in an experimental population of *Drosophila melanogaster. Gen.* 60:208–209.

O'Donald, P. 1960. Assortative mating in a population in which two alleles are segregating. *Her.* 15:389–396.

Ohba, S. 1967. Chromosomal polymorphism and capacity for increase under near optimal conditions. *Her.* 22:169–189.

Ohta, T., and M. Kimura. 1971. On the constancy of the evolutionary rate of cistrons. *J. Mol. Evol.* 1:18–25.

Parsons, P. A. 1962. The initial increase of a new gene under positive assortative mating. *Her.* 17:267–276.

______. 1965. Assortative mating for a metrical characteristic in *Drosophila. Her.* 20:161–167.

______. 1967. *The Genetic Basis of Behaviour.* Methuen, London.

______. 1971. Extreme-enviroment heterosis and genetic loads. *Her.* 26:479–483.

______. 1973. *Behavioural and Ecological Genetics: A Study in Drosophila.* Oxford University Press.

Pasternak, J. 1964. Chromosomal polymorphism in the blackfly *Simulium vittatum* (Zett.). *Canad. J. Zoo.* 42:135–138.

Pavlovsky, O., and T. Dobzhansky. 1966. Genetics of natural populations XXXVII. The coadapted system of chromosomal variants in a population of *Drosophila pseudoobscura. Gen.* 53:843–854.

Pearson, K., and A. Lee. 1903. Inheritance of physical characters. *Biometrika.* 2:357–462.

Pederson, D. G. 1968. Environmental stress, heterozygote advantage and genotype-environment interaction in *Arabidopsis. Her.* 23:127–138.

Penrose, L. S. 1963, *The Biology of Mental Defect.* 3d Ed. Grune and Stratton N.Y.

______. 1969. Dermetoglyphics. *Sci. Am.* 221:72–84.

Perutz, M. F. 1964. The hemoglobin molecule. *Sci. Am.* 211:64–76.

Petit, C. 1951. Le rôle de l'isolement sexuel dans l'évolution des populations de *Drosophila melanogaster. Bul. Biol. Fr. Belg.* 85:392–418.

______. 1958. Le déterminisme génétique et psycho-physiologique de la compétition sexuelle chez *Drosophila melanogaster. Bul. Biol. Fr. Belg.* 92:248–329.

______. 1972. Qualitative aspects of genetics and environment in the determination of behavior. *Genetics, Environment, and Behavior.* L. Ehrman, G. S. Omenn, E. Caspari (eds.). Academic Press, N.Y.

Petit, C., and L. Ehrman. 1969. Sexual selection in *Drosophila. Evol. Biol.* 3:177–223.

Petras, M. L. 1967*a*. Studies of natural populations of *Mus.* I. Biochemical polymorphisms and their bearing on breeding structure. *Evol.* 21:259–274.

______ 1967*b*. Studies of natural populations of *Mus.* III. Coat color polymorphisms. *Canad. J. Gen. and Cytol.* 9:287–296.

Pianka, E. R. 1972. *r* and *K* selection or *b* and *d* selection? *Am. Nat.* 106:581–588.

Polivanov, S. 1964. Selection in experimental populations of *Drosophila melanogaster* with different genetic backgrounds. *Gen.* 50:81–100.

Post, R. H. 1962. Population differences in red and green color vision deficiency. *Eug. Quart.* 9:131–146. (*Eug. Quart.* 12:28–29 (1965) correction.)

______. 1965. Notes on relaxed selection in man. *Anthrop. Anz.* 29:186–195.

______. 1971. Possible cases of relaxed selection in civilized populations. *Humangenetik.* 13:253–284 (Springer-Verlag).

Powell, J. R. 1971. Genetic polymorphisms in varied environments. *Sci.* 174:1035–1036.

______. 1973. Selection of enzyme alleles in laboratory populations of *Drosophila. Gen.* 75:557–570.

______. 1975*a*. Isozymes and non-Darwinian evolution: A reevaluation. *Isozymes, Vol. IV. Genetics and Evolution.* C. L. Markert (ed.). Academic Press, N.Y.

______. 1975*b*. Protein variation in natural populations of animals. *Evol. Biol.* 8:79–119.

Powell, J. R., H. Levene, and Th. Dobzhansky. 1972. Chromosomal polymorphism in *Drosophila pseudoobscura* used for diagnosis of geographic origin. *Evol.* 26:553–559.

Powell, J. R., and R. C. Richmond. 1974. Founder effects and linkage disequilibria in experimental populations of *Drosophila. P.N.A.S.* 71:1663–1665.

Prakash, S. 1973. Patterns of gene variation in central and marginal populations of *Drosophila robusta. Gen.* 75:347–369.

Prakash, S., and M. Levitan. 1973. Associations of alleles of the esterase-1 locus with gene arrangements of the left arm of the second chromosome in *Drosophila robusta. Gen.* 75:371–379.

______. 1974. Association of alleles of the malic dehydrogenase locus with a pericentric inversion in *Drosophila robusta. Gen.* 77:565–568.

Prakash, S., and R. C. Lewontin. 1968. A molecular approach to the study of genic heterozygosity in natural populations. III. Direct evidence of coadaptation in gene arrangements of *Drosophila. P.N.A.S.* 59:398–405.

______. 1971. A molecular approach to the study of genic heterozygosity in natural populations. V. Further direct evidence of coadaptation in inversions of *Drosophila. Gen.* 69:405–408.

Prakash, S., R. C. Lewontin, and J. L. Hubby. 1969. A molecular approach to the study of genic heterozygosity in natural populations. IV. Patterns of genic variation in central, marginal and isolated populations of *Drosophila pseudoobscura. Gen.* 61:841–858.

Prakash, S., and R. B. Merritt. 1972. Direct evidence of genic differentiation between sex ratio and standard gene arrangements of X chromosome in *Drosophila pseudoobscura. Gen.* 72:169–175.

Price, J. 1967. Human polymorphism. *J. of Med. Gen.* 4:44–67.

Prout, T. 1965. The estimation of fitnesses from genotypic frequencies. *Evol.* 19:546–551.

______. 1969. The estimation of fitnesses from population data. *Gen.* 63:949–967.

______. 1971*a*. The relation between fitness components and population prediction in *Drosophila.* I. The estimation of fitness components. *Gen.* 68:127–149.

______. 1971*b*. The relation between fitness components and population prediction in *Drosophila.* II. Population prediction. *Gen.* 68:151–167.

Provine, W. B. 1971. *The Origins of Theoretical Population Genetics.* Univ. of Chicago Press.

Putnam, F. W. 1968. Structure and evolution of light and heavy chains of immunoglobulins. *Brookhaven Symposia in Biology.* 21:306–327.

Race, R. R., and R. Sanger. 1968, 1975. *Blood Groups in Man.* 5th and 6th Eds. F. A. Davis, Philadelphia.

Rasmuson, M., B. Rasmuson, and L. R. Nilson. 1967. A study of isoenzyme polymorphism in experimental populations of *Drosophila melanogaster. Hereditas.* 57:263–274.

Rasmussen, D. I.1964. Blood group polymorphism and inbreeding in natural populations of the deer mouse, *Peromyscus maniculatus. Evol.* 18:219–229.

______. 1970. Biochemical polymorphisms and genetic structure in populations of *Peromyscus. Symp. Zool. Soc. London.* 26:335–349.

Reed, E. W., and S. C. Reed. 1965. *Mental Retardation: A Family Study.* Saunders, Philadelphia.

Reed, T. E. 1969. Critical tests of hypotheses for race mixture using *Gm* data on American Caucasians and Negroes. *Am. J. Hum. Gen.* 21:71–83.

Reepmaker, J., L. E. Nijenhuis, and J. J. Van Loghem. 1962. The inhibiting effect of ABO incompatibility on Rh immunization in pregnancy. A statistical analysis of 1,742 families. *Am. J. Hum. Gen.* 14:185–198.

Rendel, J. M. 1967. *Canalization and Gene Control.* Academic Press, N.Y. (Logos Press, London).

Richmond. R. C. 1969. Heritability of phototactic and geotactic responses in *Drosophila pseudoobscura. Am Nat.* 103:315–316.

Roberts, R. M., and W. K. Baker. Frequency distribution and linkage disequilibrium of active and null esterase isozymes in natural populations of *Drosophila montana. Am. Nat.* 107:709–726.

Robertson, A. 1955. Selection in animals: Synthesis. *C.S.H. Symp.* 20:225–229.

______. 1956. The effect of selection against extreme deviants based on deviation or on homozygosis. *J. Gen.* 54:236–248.

______. 1967. The nature of quantitative genetic variation. *Heritage from Mendel.* R. A. Brink (ed.). Univ. of Wisconsin Press, Madison.

Robertson, F. W. 1963. The ecological genetics of growth in *Drosophila.* 6. The genetic correlation between the duration of the larval period and body size in relation to larval diet. *Gen. Res.* 4:74–92.

______. 1964. The ecological genetics of growth in *Drosophila.* 7. The role of canalization in the stability of growth relations. *Gen. Res.* 5:107–126.

______. 1966. The ecological genetics of growth in *Drosophila.* 8. Adaptation to a new diet. *Gen Res.* 8:165–179.

Robinson, D. N. (ed.). 1970. *Heredity and Achievement.* Oxford Univ. Press, N.Y.

Rogers, J. S. 1972. Measures of genetic similarity and genetic distance. *Studies in Genetics VII.* Univ. Texas Publ. 7213:145–153.

Rothschild, M. 1971. Speculations about mimicry with Henry Ford. *Ecological Genetics and Evolution.* R. Creed (ed.). Blackwell Scientific Publ., Oxford and Edinburgh.

Roughgarden, J. 1971. Density-dependent natural selection. *Ecol.* 52:453–468.

Rucknagel, D. L., and J. V. Neel. 1961. The hemoglobinopathies. *Prog. Med. Gen.* 1: 158–260. A. G. Steinberg (ed.). Grune and Stratton.

Sang, J. H. 1964. Nutritional requirements of inbred lines and crosses of *Drosophila melanogaster. Gen. Res.* 5:50–67.

Sankaranarayanan, K. 1965*a*. Further data on the genetic loads in irradiated experimental populations of *Drosophila melanogaster. Gen.* 52:153–164.

______. 1965*b*. Inbreeding and genetic loads in irradiated experimental populations of *Drosophila melanogaster*. *Nature*. 207:1216–1217.

______.1966. Some components of the genetic loads in irradiated experimental populations of *Drosophila melanogaster*. *Gen*. 54:121–130.

______. 1967. Influence of selection on the viability of irradiated experimental populations of *Drosophila melanogaster*. *Gen*. 57:687–690.

Scarr-Salapatek, S. 1971*a*. Unknowns in the I.Q. equation (a review of articles by Jensen, Eysenck, and Herrnstein). *Sci*. 174:1223–1228.

______. 1971*b*. Race, social class, and I.Q. *Sci*. 174:1285–1295.

______. 1974. Genetics and the development of intelligence. *Review of Child Development Research*. F. Horowitz, E. M. Hetherington, S. Scarr-Salapatek, and J. Siegel (eds.). Vol. IV. Univ. of Chicago Press.

Schaffer, H. E. 1968. A measure of discrimination in mating. *Evol*. 22:125–129.

______. 1970. Survival of mutant genes as a branching process. *Mathematical Topics in Population Genetics*. K. Kojima (ed.). Springer-Verlag, N.Y.

Scharloo, W., M. Den Boer, and M. S. Hoogmoed. 1967. Disruptive selection on sterno-pleural chaeta number in *Drosophila melanogaster*. *Gen. Res*. 9:115–118.

Scharloo, W., M. S. Hoogmoed, and A. Terkuile. 1967. Stabilizing and disruptive selection on a mutant character in *Drosophila*. I. The phenotypic variance and its components. *Gen*. 56:709–726.

Schull, W. J., T. Furusho, M. Yamamoto, H. Nagano, and I. Komatsu. 1970. The effect of parental consanguinity and inbreeding in Hirado, Japan. IV. Fertility and reproductive compensation. *Humangenetik*. 9:294–315.

Schull, W. J., H. Nagano, M. Yamamoto, and I. Komatsu. 1970. The effects of parental consanguinity and inbreeding in Hirado, Japan. I. Stillbirths and pre-reproductive mortality. *Am. J. Hum. Gen*. 22:239–262.

Schull, W. J., and J. V. Neel. 1965. *The Effects of Inbreeding on Japanese Children*. Harper & Row, N.Y.

Schwartz, D., and W. J. Laughner. 1969. A molecular basis for heterosis. *Sci*. 166:626–627.

Scowcroft, W. R. 1966. Variation of scutellar bristles in *Drosophila*. IX. Chromosomal analysis of scutellar bristle selection lines. *Gen*. 53:389–402.

Scudo, F. M., and S. Karlin. 1969. Assortative mating based on phenotype I. Two alleles with dominance. *Gen*. 63:479–498.

Seiger, M. B. 1967. A computer simulation study of the influence of imprinting on population structure. *Am. Nat*. 101:47–57.

Selander, R. K. 1970. Behavior and genetic variation in natural populations (*Mus musculus*). *Am. Zoologist*. 10:53–66.

Selander, R. K., W. G. Hunt, and S. Y. Yang. 1969. Protein polymorphisms and genic heterozygosity in two European subspecies of the house mouse. *Evol.* 23:379–390.

Selander, R. K., and W. E. Johnson. 1973. Genetic variation among vertebrate species. *Am. Rev. Ecol. and Syst.* 4:75–91.

Selander, R. K., M. H. Smith, S. Y. Yang, W. E. Johnson, and J. B. Gentry. 1971. Biochemical polymorphism and systematics in the Genus *Peromyscus*. I. Variation in the old-field mouse (*P. polionotus*). *Studies in Genetics VI.* Univ. Texas Publ. 7103:49–90.

Selander, R. K., and S. Y. Yang. 1969. Protein polymorphism and genic heterozygosity in a wild population of the house mouse (*Mus musculus*). *Gen.* 63:653–667.

Selander, R. K., S. Y. Yang, and W. G. Hunt. 1969. Polymorphism in esterases and hemoglobin in wild populations of the house mouse (*Mus musculus*). *Studies in Genetics V.* Univ. Texas Publ. 6918:271–338.

Sheba International Symposium. 1974. *Genetic Polymorphism and Diseases in Man.* B. Ramot, A. Adam, B. Bonné, R. M. Goodman, and A. Szeinberg (eds.). Academic Press, N.Y.

Sheppard, P. M. 1951. A quantitative study of two populations of the moth *Panaxia dominula* (L.). *Her.* 5:349–379.

______. 1952. A note on non-random mating in the moth *Panaxia dominula* (L.). *Her.* 6:239–241.

______.1953. Polymorphism and population studies. *Symp. Soc. Exp. Biol.* 7:274–289.

______. 1958, 1967. *Natural Selection and Heredity.* Hutchinson, London.

Sheppard, P. M., and L. M. Cook. 1962. The manifold effects of the *medionigra* gene in the moth *Panaxia dominula* and the maintenance of a polymorphism. *Her.* 17:415–426.

Shreffler, D. C., C. F. Sing, J. V. Neel, H. Gershowitz, and J. A. Napier. 1971. Studies on genetic selection in a completely ascertained Caucasian population. I. Frequencies, age and sex effects, and phenotype associations for 12 blood group systems. *Am. J. Hum. Gen.* 23:150–163.

Shull, G. H. 1948. What is "heterosis"? *Gen.* 33:439–446.

Simpson, G. G. 1953. *The Major Features of Evolution.* Columbia Univ. Press. N.Y.

Sing, C. F., and G. J. Brewer. 1971. Evidence for non-random multiplicity of gene products in 22 plant genera. *Bioch. Gen.* 5:243–251.

Sing, C. F., G. J. Brewer, and B. Thirtle. 1973. Inherited biochemical variation in *Drosophila melanogaster*: noise or signal? I. Single-locus analysis. *Gen.* 75:381–404.

Sing, C. F., D. C. Shreffler, J. V. Neel, and J. A. Napier. 1971. Studies on genetic selection in a completely ascertained Caucasian population. II. Family analyses of 11 blood group systems. *Am. J. Hum. Gen.* 23:164–198.

Singh, R. S., J. L. Hubby, and R. C. Lewontin. 1974. Molecular heterosis for heat-sensitive enzyme alleles. *P.N.A.S.* 71:1808–1810.

Sinnock, P., and C. F. Sing. 1972*a*. Analysis of multilocus genetic systems in Tecumseh, Michigan. I. Definition of the data set and tests for goodness-of-fit to expectations based on gene, gamete, and single-locus phenotype frequencies. *Am. J. Hum. Gen.* 24:381–392.

______. 1972*b*. Analysis of multilocus genetic systems in Tecumseh, Michigan II. Consideration of the correlation between non-alleles in gametes. *Am. J. Hum. Gen.* 24:393–415.

Smith, C. A. B. 1968. Testing segregation ratios. *Haldane and Modern Biology*. K. R. Dronamraju (ed.). Johns Hopkins Press, Baltimore.

Smithies, O. 1955. Zone electrophoresis in starch gels: Group variations in the serum proteins of normal human adults. *Biochem. J.* 61:629–641.

______. 1959. Zone electrophoresis in starch gels and its application to studies of serum proteins. *Adv. in Prot. Chem.* 14:68–113.

______. 1968. Genetics of Variability. *Brookhaven Symp. Biol.* 21:243–258.

Snyder, L. H. 1932. Studies in human inheritance. IX. The inheritance of taste deficiency in man. *Ohio J. Sci.* 32:436–440.

______. 1947. The principles of gene distribution in human populations. *Yale J. Biol. & Med.* 19:817–833.

Sokal, R. R., and I. Huber. 1963. Competition among genotypes in *Tribolium castaneum* at varying densities and gene frequencies (the sooty locus). *Am. Nat.* 97:169–184.

Sokal, R. R., and I. Karten. 1964. Competition among genotypes in *Tribolium casteneum* at varying densities and gene frequencies (the black locus). *Gen.* 49:195–211.

Sokal, R. R., and P. H. A. Sneath. 1963. *Principles of Numerical Taxonomy*. Freeman, San Francisco.

Spassky, B., Th. Dobzhansky, and W. W. Anderson. 1965. Genetics of natural populations. XXXVI. Epistatic interactions of the components of the genetic load in *Drosophila pseudoobscura*. *Gen.* 52:623–664.

Spassky, B., N. Spassky, H. Levene, and Th. Dobzhansky. 1958. Release of genetic variability through recombination. I. *Drosophila pseudoobscura*. *Gen.* 43:844–867.

Spencer, W. P. 1947. Mutations in wild populations of *Drosophila*. *Adv. in Gen.* 1:359–402.

Spickett, S. G., J. G. M. Shire, and J. Stewart. 1967. Genetic variation in adrenal and renal structure and function. *Mem. Soc. Endocrinology*. 15:271–288.

Spikett, S. G., and J. M. Thoday. 1966. Regular responses to selection. 3. Interaction between located polygenes. *Gen. Res.* 7:96–121.

Spiess, E. B. 1957. Relation between frequencies and adaptive values of chromosomal arrangements in *Drosophila persimilis*. *Evol.* 11:84–93.

______. 1958*a*. Chromosomal adaptive polymorphism in *Drosophila persimilis*. II. Effects of population cage conditions of life cycle components. *Evol.* 12:234–245.

______. 1958*b*. Effects of recombination on viability in *Drosophila*. *C.S.H. Symp.* 23: 239–250.

______. 1959. Release of genetic variability through recombination. II. *Drosophila persimilis*. *Gen.* 44:43–58.

______. 1967. Temperature sensitivity and heterosis. *Am. Nat.* 101:93–95.

______. 1968*a*. Experimental population genetics. *Ann. Rev. Gen.* 2:165–208.

______. 1968*b*. Low frequency advantage in mating of *Drosophila pseudoobscura* karyotypes. *Am. Nat.* 102:363–379.

______. 1970. Mating propensity and its genetic basis in *Drosophila*. *Essays in Evolution and Genetics in honor of Th. Dobzhansky*. M. K. Hecht and W. C. Steere (eds.). Appleton-Century-Crofts, N.Y.

Spiess, E. B., and A. C. Allen. 1961. Release of genetic variability through recombination. VII. Second and third chromosomes of *Drosophila melanogaster*. *Gen.* 46:1531–1553.

Spiess, E. B., R. B. Helling, and M. R. Capenos. 1963. Linkage of autosomal lethals from a laboratory population of *Drosophila melanogaster*. *Gen.* 48:1377–1388.

Spiess, E. B., and B. Langer. 1961. Chromosomal adaptive polymorphism in *Drosophila persimilis*. III. Mating propensity of homokaryotypes. *Evol.* 15:535–544.

______. 1964*a*. Mating speed control by gene arrangements in *Drosophila pseudoobscura*. homokaryotypes. *P.N.A.S.* 51:1015–1019.

______. 1964*b*. Mating speed control by gene arrangements in *Drosophila persimilis*. *Evol.* 18:430–444.

Spiess, E. B., B. Langer, and L. D. Spiess. 1966. Mating control by gene arrangements in *Drosophila pseudoobscura*. *Gen.* 54:1139–1149.

Spiess, E. B., and R. J. Schuellein. 1956. Chromosomal adaptive polymorphism in *Drosophila persimilis*. I. Life cycle components under near optimal conditions. *Gen.* 41:501–516.

Spiess, E. B., and L. D. Spiess. 1964. Selection for rate of development and gene arrangement frequencies in *Drosophila permisilis*. *Gen.* 50:863–877.

Spiess, E. B., and L. D. Spiess. 1966. Selection for rate of development and gene arrangement frequencies in *Drosophila persimilis*. II. Fitness properties at equilibrium. *Gen.* 53:695–708.

Spiess, E. B., and A. J. Stankevych. 1973. Mating speed selection and egg chamber correlation in *Drosophila persimilis*. *Egypt. J. Gen. and Cytol.* 2:177–194.

Spiess, L. D., and E. B. Spiess. 1969. Minority advantage in interpopulational matings of *Drosophila persimilis*. *Am. Nat.* 103:155–172.

Spieth, H. T. 1968. Evolutionary implications of sexual behavior in *Drosophila*. *Evol. Biol.* 2:157–193.

Spieth, P. T. 1974. Gene flow and genetic differentiation. *Gen.* 78:961–965.

Sprott, R. L., and J. Staats. 1975. Behavioral studies using genetically defined mice—A bibliography. *Behav. Gen.* 5:27–82.

Spuhler, J. N. 1968. Assortative mating with respect to physical characteristics. *Eugen. Quart.* 15:128–140.

______. 1972. Genetic, linguistic, and geographical distances in native North America. *The Assessment of Population Affinities in Man.* J. S. Weiner and J. Huizinga (eds.). Clarendon Press, Oxford.

Stalker, H. D. 1960. Chromosomal polymorphism in *Drosophila paramelanica* Patterson. *Gen.* 45:95–114.

______. 1964. Chromosomal polymorphism in *Drosophila euronotus. Gen.* 49:669–682.

Stebbins, G. L. 1971. *Chromosomal Evolution in Higher Plants.* Addison-Wesley, Reading, Mass.

______. 1974. *Flowering Plants* (*Evolution Above the Species Level*). Harvard Univ. Press, Cambridge.

Stebbins, G. L., and R. C. Lewontin. 1972. Comparative evolution at the levels of molecules, organisms, and populations. *Proc. of 6th Berkeley Symposium on Mathematical Statistics and Probability.* 5:23–42. Univ. of Calif. Press, Berkeley.

Steinberg, A. G. 1959. The genetics of diabetes: A review. *Ann. of N.Y. Acad. Sci.* 82: 197–207.

______. 1969. Globulin polymorphisms in man. *Ann. Rev. Gen.* 3:25–52.

Steinberg, A. G., H. K. Bleibtreu, T. W. Kurczynski, A. O. Martin, and E. M. Kurczynski. 1967. Genetic studies on an inbred human isolate. *Proc. 3rd Intern. Cong. Hum. Gen.* 267–289.

Steinberg, A. G., and N. E. Morton. 1973. Immunoglobulins in the Eastern Carolines. *Am. J. Phys. Anth.* 38:699–702.

Stent, G. S. 1971. *Molecular Genetics.* Freeman, San Francisco.

Stern, C. 1962. Wilhelm Weinberg (1862–1937) Biography. *Gen.* 47:1–5.

______. 1970. Model estimates of the number of gene pairs involved in pigmentation variability of the Negro-American. *Hum. Hered.* 20:165–168.

______. 1973. *Principles of Human Genetics.* 3d Ed. Freeman, San Francisco.

Stone, W. S., M. R. Wheeler, F. M. Johnson, and K. Kojima. 1968. Genetic variation in natural island populations of members of the *Drosophila nasuta* and *Drosophila ananassae* subgroups. *P.N.A.S.* 59:102–109.

Stone, W. S., F. D. Wilson, and V. L. Gerstenberg. 1963. Genetic studies of natural populations of *Drosophila: Drosophila pseudoobscura*, a large dominant population. *Gen.* 48:1089–1106.

Sturtevant, A. H. 1918. An analysis of the effects of selection. *Carnegie Inst. of Wash. Publ.* 264.

______. 1965. *A History of Genetics.* Harper & Row, N.Y.

Sutton, H. E. 1967. Human genetics. *Am. Rev. Gen.* 1:1–36.

Sved, J. A., T. E. Reed, and W. F. Bodmer. 1967. The number of balanced polymorphisms that can be maintained in a natural population. *Gen.* 55:469–481.

Sweet, E. E., and E. B. Spiess. 1962. Frequency of sterility in a laboratory population of *Drosophila melanogaster*. *Gen.* 47:1519–1534.

Tamarin, R. H., and C. J. Krebs. 1973. Selection at the transferin locus in cropped vole populations. *Her.* 30:53–62.

Taylor, C. E. 1975. Genetic loads in heterogeneous environments. *Gen.* 80:621–635.

Teissier, G. 1943. Apparition et fixation d'un gène mutant dans une population stationnaire de *Drosophiles*. *C.R.A.S.* 216:88–90.

______. 1954*a*. Conditions d'equilibre d'un couple d'alleles et superiorité des heterozygotes. *C.R.A.S.* 238:621–623.

______. 1954*b*. Selection naturelle et fluctuation génétique. C.R.A.S. 238:1929–1931.

Temin, R. G. 1966. Homozygous viability and fertility loads in *Drosophila melanogaster*. *Gen.* 53:27–46.

Temin, R. G., H. U. Meyer, P. S. Dawson, and J. F. Crow. 1969. The influence of epistasis on homozygous viability depression in *Drosophila melanogaster*. *Gen.* 61:497–519.

Thoday, J. M. 1955. Balance, heterozygosity, and developmental stability. *C.S.H. Symp.* 20:318–326.

______. 1959. Effects of disruptive selection. I. Genetic flexibility. *Her.* 13:187–204.

______. 1960. Effects of disruptive selection. III. Coupling and repulsion. *Her.* 14:35–49.

______. 1961. Location of polygenes. *Nature* 191:368–370.

______. 1965. Effects of selection for genetic diversity. *Genetics Today*, Proc. 11th Intern. Congress of Genetics (1963). 3:533–540.

______. 1967. The general importance of disruptive selection. *Gen. Res.* 9:119–120.

______. 1972. Disruptive selection (reviews lecture). *Proc. Roy. Soc.* (*B*). 182:109–143.

Thoday, J. M., and T. B. Boam. 1959. Effects of disruptive selection. II. Polymorphism and divergence without isolation. *Her.* 13:205–218.

______. 1961*a*. Effects of disruptive selection. V. Quasi-random mating. *Her.* 16:219–223.

______. 1961*b*. Regular responses to selection. I. Description of responses. *Gen. Res.* 2:161–176.

Thoday, J. M., and J. B. Gibson. 1962. Isolation by disruptive selection. *Nature.* 193: 1164–1166.

______. 1970. Environmental and genetical contributions to class difference: A model experiment. *Sci.* 167:990.

Thorneycroft, H. B. 1968. A cytogenetic study of the white-throated sparrow, *Zonotrichia albicollis* (Gmellin). Ph.D. thesis, Univ. of Toronto.

Tobari, Y. N., and K. Kojima. 1967. Selective modes associated with inversion karyotypes in *Drosophila ananassae*. I. Frequency-dependent selection. *Gen.* 57:179–188.

Torroja, E. 1964. Genetic loads in irradiated experimental populations of *Drosophila pseudoobscura*. *Gen.* 50:1289–1298.

Tshetverikov, S. S. 1926. On certain aspects of the evolutionary process from the standpoint of modern genetics. (Russian: *Zhurnal Eksperimental'noi. Biologii* A2:3–54.) M. Barker (trans.). I. M. Lerner (ed.). *Proc. Am. Philos. Soc.* 105:167–195.

Turner, J. R. G. 1967. Why does the genome not congeal? *Evol.* 21:645–656.

______. 1972. Selection and stability in the complex polymorphism of *Moraba scurra*. *Evol.* 26:334–343.

Van Valen, L., and G. W. Mellin. 1967. Selection in natural populations. 7. New York babies. *Ann. Hum. Gen.* 31:109–127.

Vigue, C., and F. M. Johnson. 1973. Isozyme variability in *Drosophila*. VI. Frequency-property-environment relationships of allelic alcohol dehydrogenases in *Drosophila melanogaster*. *Biochem. Gen.* 9:213–227.

Vogel, F., and M. R. Chakravartti. 1966. ABO blood groups and smallpox in a rural population of West Bengal and Bihar (India). *Humangenetik*. 3:166–180.

Vogel, F., and R. Rathenberg. 1975. Spontaneous mutation in man. *Adv. Hum. Gen.* 5:223–318.

Waddington, C. H. 1957. *The Strategy of the Genes*. Allen and Unwin, London.

______. 1961. Genetic Assimilation. *Adv. Gen.* 10:257–293.

Wahlund, S. 1928. Zusammensetzung von Population and Korrelationserscheinungen von Standpunkt der Vererbungslehre ans betrachtet. *Hereditas*. 11:65–106.

Wallace, B. 1948. Studies on "sex ratio" in *Drosophila pseudoobscura*. I. Selection and sex ratio. *Evol.* 2:189–217.

______. 1954. Genetic divergence of isolated populations of *Drosophila melanogaster*. *Proc. Ninth Inter. Cong. Genetics*. 1:761–764.

______. 1955. Inter-population hybrids in *Drosophila melanogaster*. *Evol.* 9:302–316.

______. 1956. Studies on irradiated populations of *Drosophila melanogaster*. *J. Gen.* 54:280–293.

______. 1958*a*. The average effect of radiation-induced mutations on viability in *Drosophila melanogaster*. *Evol.* 12:532–552.

______. 1958*b*. The comparison of observed and calculated zygotic distributions. *Evol.* 12:113–115.

______. 1959. The role of heterozygosity in *Drosophila* populations. *Proc. 10th Intern. Cong. Gen.* 1:408–419.

______. 1962. Temporal changes in the roles of lethal and semilethal chromosomes within populations of *Drosophila melanogaster. Am. Nat.* 96:247–256.

______. 1963. Further data on the overdominance of induced mutations. *Gen.* 48:633–651.

______. 1965. The viability effects of spontaneous mutations in *Drosophila melanogaster. Am. Nat.* 99:335–348.

______. 1966*a*. Distance and allelism of lethals in a tropical population of *Drosophila melanogaster. Am. Nat.* 100:565–578.

______. 1966*b*. The fate of *sepia* in small populations of *Drosophila melanogaster. Genetica.* 37:29–36.

______. 1968*a*. Polymorphism, population size, and genetic load. *Population Biology and Evolution.* R. C. Lewontin (ed.). Syracuse Univ. Press.

______. 1968*b*. *Topics in Population Genetics.* Norton, N.Y.

______. 1970. *Genetic Load.* Prentice-Hall, Englewood Cliffs.

______. 1975. Hard and soft selection revisited. Evol. 29:465–473.

Wallace, B., and Th. Dobzhansky. 1962. Experimental proof of balanced genetic loads in *Drosophila. Gen.* 46:1027–1042.

Wallace, B., and J. C. King. 1952. Genetic analysis of the adaptive values of populations. *P.N.A.S.* 38:706–715.

Wallace, B., J. C. King, C. V. Madden, B. Kaufmann, and E. C. McGunnigle. 1953. An analysis of variability arising through recombination. *Gen.* 38:272–307.

Wallace, B., and C. Madden. 1953. The frequencies of sub- and supervitals in experimental populations of *Drosophila melanogaster Gen.* 38:456–470.

Wallace, B., and M. Vetukhiv. 1955. Adaptive organization of the gene pools of *Drosophila* populations. *C.S.H. Symp.* 20:303–310.

Watanabe, T. K., and C. Oshima. 1970. Persistance of lethal genes in Japanese natural populations of *Drosophila melanogaster. Gen.* 64:93–106.

Watt, W. B. 1972. Intragenic recombination as a source of population genetic variability. *Am. Nat.* 106:737–753.

Watterson, G. A. 1959. A new genetic population model and its approach to homozygosity. *Ann. Hum. Gen.* 23:221–232.

Weatherall, D. J. 1969. The genetics of the thalassemias. *Brit. Med. Bul.* 25:24–29.

Weatherall, D. J., and J. B. Clegg. 1969. Disorders of protein synthesis. *Selected Topics in Medical Genetics.* C. A. Clarke (ed.). Oxford Univ. Press, London.

______. 1972. *The Thalassemia Syndromes.* 2d Ed. Blackwell, Oxford.

Wehrhahn, C., and R. W. Allard. 1965. The detection and measurement of the effects of individual genes involved in the inheritance of a quantitative character in wheat. *Gen.* 51:109–119.

Weinberg, W. 1908. Über den Nachweis der Vererbung beim Menschen (On the demonstration of heredity in man). *Jahreshefte des Vereins für Vaterlandische Naturkunde in Württemberg.* Stuttgart 64:368–382. S. H. Boyer (trans.). 1963. *Papers on Human Genetics.* Prentice-Hall, Englewood Cliffs.

Weir, B. S., R. W. Allard, and A. L. Kahler. 1972. Analysis of complex allozymes polymorphism in a barley population. *Gen.* 72:505–523.

Weisbrot, D. R. 1966. Genotypic interactions among competing strains and species of *Drosophila. Gen.* 53:427–435.

Wheeler, L. L., and R. K. Selander. 1972. Genetic variation in populations of the house mouse, *Mus musculus,* in the Hawaiian Islands. *Studies in Genetics VII.* Univ. Texas Publ. 7213:269–296.

White, M. J. D. 1957. Cytogenetics of the grasshopper, *Moraba scurra.* II. Heterotic systems and their interactions. *Australian J. Zool.* 5:305–337.

______. 1958. Restrictions on recombination in grasshopper populations and species. *C.S.H. Symp.* 23:307–317.

______. 1965. J. B. S. Haldane (1892–1964). *Gen.* 52:1–7.

______. 1969. Chromosomal rearrangements and speciation in animals. *Ann. Rev. Gen.* 3:75–98.

______. 1945, 1954, 1973. *Animal Cytology and Evolution.* 1st, 2d, 3d Eds. Cambridge Univ. Press.

White, M. J. D., R. C. Lewontin, and L. E. Andrew. 1963. Cytognetics of the grasshopper *Moraba scurra.* VII. Geographic variation of adaptive properties. *Evol.* 17:147–162.

Whittaker, R. H., S. A. Levin, and R. B. Root. 1973. Niche, habitat, and ecotope. *Am. Nat.* 107:321–338.

Willerman, L., A. F. Naylor, and N. C. Myrianthopoulos. 1970. Intellectual development of children from interracial matings. *Sci.* 170:1329–1331.

Wills, C. 1966. The mutational load in two natural populations of *Drosophila pseudoobscura. Gen.* 52:281–294.

______. 1968. Three kinds of genetic variability in yeast populations. *P.N.A.S.* 61:937–944.

______. 1973. In defense of naive pan-selectionism. *Am. Nat.* 107:23–34.

Willis, C., J. Phelps, and R. Ferguson. 1975. Further evidence for selective differences between isoalleles in *Drosophila. Gen.* 79:127–141.

Wilson, E. O., and W. H. Bossert. 1971. *A Primer of Population Biology.* Sinauer Associates, Stamford, Conn.

Wolstenholme, D. R., and J. M. Thoday. 1963. Effects of disruptive selection. VII. A third chromosome polymorphism. *Her.* 18:413–431.

Woolf, C. M., and K. Church. 1963. Studies on the advantage of heterokaryotypes in the tumorous-head strain of *Drosophila melanogaster*. *Evol.* 17:486–492.

Workman, P. L. 1969. The analysis of simple genetic polymorphisms. *Hum. Biol.* 41: 97–114.

Workman, P. L. 1973. Genetic analysis of hybrid populations. *Methods and Theories of Anthropological Genetics.* M. H. Crawford and P. L. Workman (eds.). Univ. of New Mexico Press, Albuquerque.

Workman, P. L., and R. W. Allard. 1962. Population studies in predominantly self-pollinated species. III. A matrix model for mixed selfing and random outcrossing. *P.N.A.S.* 48:1318–1325.

Wright, S. 1917. Color inheritance in mammals. *J. Hered.* 8:521–527.

______. 1921. Systems of mating. I–V. *Gen.* 6:111–178.

______. 1921. Correlation and causation. *J. Agricul. Res.* 20:557–585.

______. 1931. Evolution in Mendelian populations. *Gen.* 16:97–159.

______. 1934. The results of crosses between inbred strains of guinea pigs differing in number of digits. *Gen.* 19:537–551.

______. 1940. Breeding structure of populations in relation to speciation. *Am. Nat.* 74:232–248.

______. 1943. Isolation by distance. *Gen.* 28:114–138.

______. 1945. The differential equation of the distribution of gene frequencies. *P.N.A.S.* 31:382–389.

______. 1946. Isolation by distance under diverse systems of mating. *Gen.* 31:39–59.

______. 1948. On the roles of directed and random changes in gene frequency in the gametes of populations. *Evol.* 2:279–295.

______. 1949. Adaptation and selection. *Genetics, Paleontology and Evolution.* G. L. Jepsen, E. Mayr, and G. G. Simpson (eds.). Princeton Univ. Press.

______. 1951. The genetical structure of populations. *Ann. Eugen.* 15:323–354.

______. 1952. The genetics of quantitative variability. *Quantitative Inheritance.* E. C. R. Reeve and C. H. Waddington (eds.). H. M. Stationery Office, London.

______. 1955. Classification of the factors of evolution. *C. S. H. Symp.* 20:16–24.

______. 1956. Modes of selection. *Am. Nat.* 90:5–24.

______. 1960. Genetics and 20th century Darwinism. *Am. J. Hum. Gen.* 12:365–372.

______. 1963. William Ernest Castle (1867–1962) Biography. *Gen.* 48:1–5.

______. 1964. Stochastic process in evolution. *Stochastic Models in Medicine and Biology*. J. Gurland (ed.). Univ. of Wisconsin Press, Madison.

______. 1965. The interpretation of population structure by F-statistics with special regard to systems of mating. *Evol.* 19:395–420.

______. 1967. The foundations of population genetics. *Heritage from Mendel*. R. A. Brink (ed.). Univ. of Wisconsin Press, Madison.

______. 1968, 1969, 1976. *Evolution and the Genetics of Populations*, Volumes I, II, and III. Univ. of Chicago Press.

Wright, S., and Th. Dobzhansky. 1946. Genetics of natural populations. XII. Experimental reproduction of some of the changes caused by natural selection in certain populations of *Drosophila pseudoobscura*. *Gen.* 31:125–156.

Wright, S., Th. Dobzhansky, and W. Hovanitz. 1942. Genetics of natural populations. VII. The allelism of lethals in the third chromosome of *Drosophila pseudoobscura*. *Gen.* 27:373–394.

Yamaguchi, O., and D. Moriwaki. 1971. Chromosomal variation in natural populations of *Drosophila bifasciata*. *Japanese J. Gen.* 46:383–391.

Yamamoto, M., and L. Fu. 1973. Red cell isozymes in the Eastern Carolines. *Am. J. Phys. Anthr.* 38:703–708.

Yamazaki, T. 1971. Measurement of fitness at the esterase-5 locus in *Drosophila pseudoobscura*. *Gen.* 67:579–603.

Yanofsky, C. 1967. Structural relationships between gene and protein. *Ann. Rev. Gen.* 1:117–138.

Yanofsky, C., B. C. Carlton, J. R. Guest, D. R. Helinski, and U. Henning. 1964. On the colinearity of gene structure and protein structure. *P.N.A.S.* 51:266–272.

Yarbrough, K., and K. Kojima. 1967. The mode of selection at the polymorphic esterase-6 locus in cage populations of *Drosophila melanogaster*. *Gen.* 57:677–686.

Young, S. S. Y. 1970. Direct and associate effects of body weight and viability in *Drosophila melanogaster*. *Gen.* 66:541–554.

______. 1971. The effects of some physical and biotic environments on heterosis of direct and associate genotypes in *Drosophila melanogaster*. *Gen.* 67:569.

Zouros, E., and C. E. Krimbas. 1973. Evidence for linkage disequilibrium maintained by selection in two natural populations of *Drosophila subobscura*. *Gen* 73:659–674.

Zuckerkandl, E. 1964. Compensatory effects in the synthesis of hemoglobin polypeptide chains. *C. S. H. Symp.* 29:357–374.

______. 1965. The evolution of hemoglobin. *Sci. Am.* 212:110–118.

Zuckerkandl, E., and L. Pauling. 1965. Evolutionary divergence and convergence in proteins. *Evolving Genes and Proteins*. V. Bryson and H. J. Vogel (eds.). Academic Press, N.Y.

INDEX OF AUTHORS

INDEX OF SUBJECTS

Enzyme heat stability, 634

Enzyme variants, *See* Allozymes; Blood enzymes

Epistasis (nonallelic interaction), 34, 37, 139-141, 144-146, 173, 175, 194, 283, 545, 556-559, 563, 569, 573-579, 594, 606, 632, 635, 639, 646

Equilibrium, genetic, 4, 9, 20; migration, 651-653; mutation-drift, 614; mutation-selection, 13, 385, 599-600, 602; negative assortative mating, 309-311; physical example, 30; properties of, 28-30; 86-70, 246-250; random mating, gametic, 125-134, 570; random mating-assortative mating combined, 300-303; random mating-inbreeding combined, 246-250, 267-268; random mating, multiple alleles, 68-70; random mating, one pair alleles, 22-33, 50-51; random mating, sex-linked, 113-117; recurrent-reverse mutation, 398-399; selection, constant fitness, 421-429, 453, 456, 458-462, 547-555, 555-559, 565, 572-586, 596-608; selection, minority advantage, 500-505, 511; selection, niche heterogeneity, 529-536; selection, sex-linked, 483, 486-487

Escherichia coli, see Bacteria

Ethological factors, 305-306, 508

Eugenics, 13, 228

Euthenics, 228

Evolution, 5, 14, 335; rate of, 611, 618-626

Evolutionary theory, 9, 14, 156, 336, 356-364, 592, 665-668